Lecture Notes in Compu  52

Edited by G. Goos, J. Hartmanis,

Springer

Berlin
Heidelberg
New York
Hong Kong
London
Milan
Paris
Tokyo

Hyun-Kook Kahng (Ed.)

Information Networking

Networking Technologies for Enhanced Internet Services

International Conference, ICOIN 2003
Cheju Island, Korea, February 12-14, 2003
Revised Selected Papers

 Springer

Series Editors

Gerhard Goos, Karlsruhe University, Germany
Juris Hartmanis, Cornell University, NY, USA
Jan van Leeuwen, Utrecht University, The Netherlands

Volume Editor

Hyun-Kook Kahng
Korea University
Dept. of Electronics and Information Engineering
208 Suchang-Dong Chochiwon, Chungnam, Korea 339-700
E-mail: kahng@korea.ac.kr

Cataloging-in-Publication Data applied for

A catalog record for this book is available from the Library of Congress

Bibliographic information published by Die Deutsche Bibliothek
Die Deutsche Bibliothek lists this publication in the Deutsche Nationalbibliografie;
detailed bibliographic data is available in the Internet at <http://dnb.ddb.de>.

CR Subject Classification (1998): C.2, H.4, H.3, D.2.12, D.4, H.5

ISSN 0302-9743
ISBN 3-540-40827-4 Springer-Verlag Berlin Heidelberg New York

Springer-Verlag Berlin Heidelberg New York
a member of BertelsmannSpringer Science+Business Media GmbH

http://www.springer.de

© Springer-Verlag Berlin Heidelberg 2003
Printed in Germany

Typesetting: Camera-ready by author, data conversion by DA-TeX Gerd Blumenstein
Printed on acid-free paper   SPIN 10952191   06/3142   5 4 3 2 1 0

Preface

The papers in this book were prepared for and presented at the International Conference on Information Networking 2003 (ICOIN 2003), which was held from February 12 to 14, 2003 at Jeju Island, Korea. It was organized by the KISS (Korean Information Science Society) SIG-IN in Korea, in cooperation with IPSJ (Information Processing Society of Japan) SIG-IN and IEEE Region 10. The papers were selected through two steps: (1) refereeing by TPC members and reviewers; and (2) on-site presentation review by session chairs.

Although we believe that the world is becoming ever smaller through the information networking, the conventional IT and Internet technologies seem to be in their mature stage while the advent of the new technologies required seems to be delayed. To find a new path, ICOIN 2003 posted a call for papers on the technologies needed for end-to-end and networked QOS in wired and wireless Internet, focusing on the enhanced TCP/IP Internet protocol mechanism, its implementation, and the technology required to support wired and wireless Internet. The papers of this book concentrated on IP technologies, which will be the basis of ubiquitous networking.

This book includes the following subjects related to information networking, from the low-layer transmission technologies to higher-layer protocols and services:

- High-Speed Network Technologies: Concerned with switching, routing and integration. Especially optical network technologies are covered.
- QOS on the Internet: Concentrated on the enhancement of the Internet QOS. RSVP/DiffServ-related algorithms and protocols are presented.
- Enhanced Protocols on the Internet: Examines the enhancement or improvement of the conventional Internet protocols/services and network architectures: wireless LAN and PAN; multicast; next-generation architecture; and contents distributions.
- Mobile Internet: Concerned with mobile Internet and ad hoc networks: wireless access technologies, mobile IP, TCP over mobile IP.
- Network Security: Examines the network-related security matters from algorithms to IP security-related network architectures.
- Network Management: Examines the management technologies for networks and services.
- Network Performance: Concerned with performance issues on the Internet. Especially a paper dealing with problems of simulation for performance studies is presented.

The papers are complemented by invited presentations by Krysztof Pawlikowski (University of Canterbury, New Zealand) and Roy Sumit (University of Washington, USA). And all papers in this book, we believe, will prove rewarding for all computer scientists working in the area of information networking.

May 2003   Hyun-Kook Kahng

Organization

Committee Members

General Chair: Yong-Jin Park (Hanyang Univ., Korea)

Steering Committee: Prof. Sunshin An (Korea Univ., Korea)
Prof. Yanghee Choi (Seoul Natl. Univ., Korea)
Dr. Haruhisa Ishida (IRI, Japan)
Prof. Cheeha Kim (POSTECH, Korea)
Prof. Jaiyong Lee (Yonsei Univ., Korea)
Prof. Yong-Jin Park (Hanyang Univ., Korea)
Dr. Jiangpin Wu (CERNET, China)

Organizing Committee

Chair: Ilyoung Chong (Hankuk Univ. of FS, Korea)
Vice-Chairs: Youn Kwan Kim (LG Telecom, Korea)
Kijoon Chae (Ewha Womans Univ., Korea)

Local Arrangements

Co-chairs: Kwangsue Chung (Kwangwoon Univ., Korea)
Kijung Ahn (Cheju Univ., Korea)
Publicity Co-Chairs: Seong-Ho Jeong (Hankuk Univ. of FS, Korea)
Sumit Roy (Univ. of Washington, USA)
Martina Zitterbart (Universität Karlsruhe, Germany)
Richard Lai (La Trobe University, Australia)
Krzysztof Pawlikowski (Univ. of Canterbury,
New Zealand)
Osamu Nakamura (Keio Univ., Japan)
Hideki Sunahara (Nara Sentan dai Univ., Japan)
Suress Ramadass (USM, Malaysia)
Jianping Wu (CERNET, China)
Jae Kim (Boeing Co., USA)
Ray-Shung Chang (Dong Hwa Univ., Taiwan)

Publication Chair: Choong Seon Hong (Kyung Hee Univ., Korea)
Registration Chair: Sanghyun Ahn (Univ. of Seoul, Korea)
Financial Chair: Jongwon Choe (Sookmyung Women's Univ., Korea)
Patron Co-chairs: Hyun Park (LG Electronics, Inc., Korea)
Yongtae Shin (Soongsil Univ., Korea)

Program Committee

Chair: Hyun-Kook Kahng (Korea Univ., Korea)

Vice-Chairs: Chung-Ming Huang (Natl. Cheng Kung Univ., Taiwan)
                  Kyungshik Lim (Kyungpook Natl. Univ., Korea)
                  Krzysztof Pawlikowski (Univ. of Canterbury, New Zealand)
                  Sumit Roy (Univ. of Washington, USA)
                  Changjin Suh (Soongsil Univ., Korea)
                  Martina Zitterbart (Univ. of Karlsruhe, Germany)

Members: Byungjun Ahn (ETRI, Korea)
                  Ray-Shung Chang
                  (Natl. Tung-Hwa Univ., Taiwan)
                  Wen-Tsuen Chen
                  (Natl. Tung-Hwa Univ., Taiwan)
                  Jorge A. Cobb (Univ. of Texas at Dallas, USA)
                  Jon Crowcroft (Cambridge Univ., UK)
                  Sajal K. Das (Univ. of Texas at Arlington, USA)
                  Shigeki Goto (Waseda Univ., Japan)
                  Leduc Guy (Université de Liege, Belgium)
                  Jack Holdsworth (King's College London, UK)
                  Choong Seon Hong (Kyung Hee Univ., Korea)
                  Seong-Soon Joo (ETRI, Korea)
                  Byung G. Kim (Univ. of Massachusetts, USA)
                  Hwa Sung Kim (Kwangwoon Univ., Korea)
                  Kane K.H. Kim (Univ. of California, Irvine, USA)
                  Keecheon Kim (Konkuk Univ., Korea)
                  Richard Lai (La Trobe Univ., Australia)
                  Hyukjoon Lee (Kwangwoon Univ., Korea)
                  Meejeong Lee (Ewha Womans Univ., Korea)
                  Won Jun Lee (Korea Univ., Korea)
                  Bo Li (Univ. of Science and Tech., Hong-Kong China)
                  Ying-Dar Lin (Natl. Chiao-Tung Univ., Taiwan)
                  Sang Won Min (Kwangwoon Univ., Korea)
                  Yuji Oie (Kyushu Institute of Technology, Japan)
                  Koji Okamura (Kyushu Univ., Japan)
                  Suress Ramadass (USM, Malaysia)
                  Shinji Shimojo (Osaka Univ., Japan)
                  Harsha R. Sirisena (Univ. of Canterbury, New Zealand)
                  Tatsuya Suda (Univ. of California, Irvine, USA)
                  Chai-Keong Toh (TRW, USA)
                  Mi Ae Woo (Sejong Univ., Korea)
                  Kenichi Yoshida (Univ. of Tsukuba, Japan)

Table of Contents

I   High-Speed Network Technologies

II   Enhanced Protocols in Internet

III   QOS in Internet

IV   Mobile Internet

V   Network Security

VI   Network Management

VII   Network Performance

Part I

High-Speed Network Technologies

Providing Delay Guarantee in Input Queued Switches: A Comparative Analysis of Scheduling Algorithms*

Sayed Vahid Azhari, Nasser Yazdani, and Ali Mohammad Zareh Bidoki

Router Lab., ECE. Department
Univ. of Tehran
Tehran, Iran
vahidazhari@yahoo.com
yazdani@ut.ac.ir
Zare_b@ece.ut.ac.ir

Abstract. Real-time applications require switching components to deliver packets in bounded delay. Most of such switches are designed as output-queued switches, which suffer from bandwidth limitation. Input queued switches remedy this by introducing buffers at inputs, however due to input/output contention providing delay guarantee is difficult. Based on maximum weighted matching, we present several classes of schedulers, which provide best-effort delay guarantee for various delay classes. Each scheduler uses some combination of Round Robin and Earliest Deadline First policy at various service points. Comparative analysis is performed and some key parameters in the design of efficient schedulers with delay guarantee are provided. These parameters are: average match per time slot, average weight per match, and fairness. Finally, the effect of speedup is studied. Our schedulers can support variable length IP packets as well as multicast packets, the latter being supported by fanout splitting.

1 Introduction

Along with growing in size, Internets growth has also been in the domain of applications and services and their requirements. Real-time applications are one such example requiring bounded deterministic delays. Delay guarantee should be provided per-flow. Each flow, before entering the network, should negotiate with the routing and switching components in its way in order to allocate enough resources to get the guarantee it wants. This process is called Call Admission Control (CAC) [1] [2] [3]. If a flows required delay cannot be guaranteed, CAC should notify that flow of the situation.

Many scheduling algorithms trying to provide delay bounds utilize a scheduling policy called Earliest Deadline First, or EDF for short [4] [5] [6]. EDF and

---

* This work has been partially supported by ITRC (Iranian Telecommunication Research Center).

H.-K. Kahng (Ed.): ICOIN 2003, LNCS 2662, pp. 3–13, 2003.

most other scheduling algorithms that provide bounded delay have been proposed for output queued switches [4] [8]. Output queued switches are promising from a service policy implementation viewpoint, but suffer from a bandwidth limitation problem. Here, each output queue should operate N times (N is the number of input ports) faster than the line rate, which makes it hardly possible to implement regarding todays memory technology (100Gbps) and line rates (10-40Gbps). On the other hand, input queued (IQ) switches, which buffer cells at the line cards, require their memories to run at the line rate. There is typically no buffering performed in the switch fabric which is usually a crossbar. Input queued switches are known to suffer from HoL (Head of Line Blocking) problem [9] that reduces throughput to 58.6%. By using virtual output queuing (VOQ) [10] instead of FIFO queuing at the inputs, this problem can be alleviated and near 100% throughput can be achieved [11]. In virtual output queuing (VOQ), a separate queue is assigned for each output at each of the inputs.

There have also been some solutions for input queued switches, one of which is the SIMP scheduler [12]. There have also been some other approaches based on maximum weighted matching (MWM) and stable matching [13] [7]. Other approaches have also been given which aim on emulating an OQ switch using a Combined Input Output Queued (CIOQ) switch with some speedup [14] [15].

We propose a class of EDF schedulers that are designed for IQ switches. All of the algorithms employ the well-known request-grant-accept-scheduling scheme used in iSLIP [16] with some variations. Each cell has an associated label describing its Maximum Allowed Delay (MAD). A sorter is required in our designs to sort the requests and/or grants with respect to their MADs. Our approaches are shown to perform well comparing to existing non-EDF schedulers like iSLIP as well as other EDF schedulers like SIMP. Our design can also support multicast flows using fanout splitting. Variable IP packet switching either with assembly or without it can also be supported by our schedulers [17].

The rest of the paper is organized as follows: In the next section the delay guarantee problem is discussed. In section 3, we briefly consider some related work. Our schedulers are treated in section 4, and section 5 gives a comparative analysis of our schedulers based on computer simulations as well as considering the effect of speedup. Finally, we conclude in section 6.

2   Problem Statement and Framework

When speaking of delay guarantee, the goal is to put an upper bound on the amount of delay a cell experiences. There are various sorts of guarantees, namely, hard and soft. In hard guarantees, the resource will never break its promise as long as the flow remains faithful to its service contract. However, in soft guarantees, the resource does its best to provide the flow with its negotiated service contract. In this paper, we give soft guarantee on the amount of delay each cell experiences. Each cell has a MAD (Maximum Allowed Delay) field associated with it at the moment of arrival to the switch. This MAD value is used to differentiate between different delay classes. The switch supports at most

256 delay classes. The scheduling algorithm used in this paper is based on the EDF scheduler [4] [5] [6], which schedules cells according to their deadlines with the nearest deadlines having highest priority.

Such a scheduler needs to sort all requests and grant them according to earliest deadline first algorithm. Sorting of requests introduces a major bottleneck in the design. We use a batcher sorter network to sort the requests. Considering IQ switches, there is also a contention problem. Contention can be at inputs between various VOQs or at outputs between different input ports having a cell for the same output. The input/output contention existing in IQ switch architectures makes providing delay guarantees difficult, i.e., cells can't be scheduled independently. So one cell might have to sit in its queue and wait for another cell from the same input to get scheduled to another output, or it might have to wait for a cell to go from another input to the same output.

There are two other main parameters affecting the average delay of the cells belonging to each delay class and the average delay of total cells. The first and the most obvious one is the weight of each match. The match weight is the MAD label associated with each match. It should be noted that MAD values of all cells are decreased during each time slot they reside in the input buffers. This aging mechanism prevents starvation and makes the scheduling fair across all delay classes. The match weight should be minimized for the scheduling algorithm to obey EDF. The other parameter is the average number of matches per scheduling phase. The higher this is, the more cells are scheduled and thus, the lower the total delay for each delay class.

We use policy decomposition technique to achieve delay guarantee. That is, specific service policies are used in all three service points namely, the VOQs, inputs, and outputs. For instance in VOQs, cells are placed such that VOQs remain sorted with respect to MAD value.

3   Related Work

iSLIP [16] is one of existing schedulers with no delay guarantee. This scheduler is based on the request/grant/accept paradigm and utilizes Round Robin (RR) arbiters for issuing grant and accepts. An approach supporting delay guarantee is SIMP [12]. For SIMP, one after the other, outputs select their most critical input and make a match. The first output to select its critical input is determined on a RR basis thus making the approach fair. This approach requires for sorting of all requests at the outputs and issuing grants repeatedly for N iterations, thus it takes much time to find a match using this scheduler.

Another interesting approach presented in [7] uses an $O(N^2)$ variation of the stable matching algorithm to find a matching between inputs and outputs. This approach utilizes a credit-based bandwidth reservation scheme to provide per-flow transmission rate and cell delay guarantee. There it is shown that under certain traffic constraints bounded delay can be achieved. This approach, however, is complex for hardware implementation. Other approaches are mostly applicable to OQ switches. A number of these are proposed in [3] [4].

A different class of approaches focus on emulating output queued (OQ) switch using a combined input output queued (CIOQ) switch [14] [15]. They use a scheduling algorithm that schedules cells based on their departure time. They also assume some amount of speedup for the switch. In [14] and [15] it has been proven that using a proposed scheduling algorithm with a speedup of two, it is possible to exactly emulate an OQ switch using a CIOQ switch. To the best of our knowledge, these approaches might be difficult to implement in hardware.

4 Scheduling Algorithms

Each scheduling algorithm chooses a combination of input and output service policies with a matching algorithm. There are various matching algorithms, e.g., maximum weighted matching, maximal matching, and stable matching. However, we utilize maximal matching algorithms in our schedulers due to ease of implementation.

In a Maximum Weighted Matching Algorithm (MWM) each input/output pair has a weight assigned as its matching weight, the algorithm finds a match between inputs and outputs having maximum sum of weights. In a Stable Matching Algorithm (SM) each input has a preference list of all outputs and vice versa. The algorithm finds a match between inputs and outputs such that never do both partners prefer another option to their current mate.

Unlike the algorithms mentioned above, in Maximal Matching Algorithm, matches made at some iteration would not be removed at further iterations. After a match has been found by a maximal matching algorithm, no more matches can be made. However, the match might not have maximum size. In the following subsections, we describe different scheduling algorithms followed by our schedulers.

4.1 Greedy Maximum Weighted Matching Scheduler (GMWM) and Stable Matching (SM)

Weighted matching algorithms can achieve full throughput under admissible load [11]. The MWM approach results in minimum delay among all IQ schedulers, but is too complex to be implemented in hardware and its execution time takes $O(N^3 LogN)$ serial iterations, thus, SIMP approach presented in [12] as an alternative.

A less complex approach could be a greedy version of MWM called GMWM. In this approach, all the input/output pairs (VOQs) are sorted according to their MAD value in increasing order. Then, we proceed by iteratively selecting the match with least MAD. Although GMWM's time complexity is $O(N^2 LogN)$, the match produced might not have minimum (maximum) weight.

A stable matching algorithm [13] can also be employed as a scheduler. The algorithm takes N time steps in parallel. The SM scheduler is only included for the sake of comparative analysis with our own schedulers.

4.2   Grant/Accept Sorted Schedulers (G/A/S or I/O/S)

A Batcher sorter is used in our methods. A Batcher sorter with N inputs has LogN(LogN+1)/2 stages each consisting of N 2x2 sorting elements. Consequently a Batcher sorter is composed of NlogN(LogN+1)/2 sorting elements which are actually M bit comparators (M is the number of bits used to represent each number). For a 32x32 switch, this would be 15 stages each consisting of 32 2x2 sorter cells. We have synthesized such a sorter with the conventional synthesis tools at hand and have found its delay to be around 32nsec, allowing us to support line rates of up to 10Gbps.

Here we propose our first class of schedulers. They relay on the request-grant-accept-matching scheme employed in a number of well-known schedulers [16]. The main difference between our schedulers and the previous ones is their ability to provide different average delay for each delay class. The scheduler can put together any combination of RR and EDF policies for each service point resulting in four alternative scheduling algorithms. Of these four alternatives, only the most important ones will be considered here. These are RR at inputs (accept) EDF at outputs (grant), and EDF at both inputs and outputs.

There are two schedulers presented in this section. For the first one, each output maintains a grant list of all inputs having a request for that output sorted according to MAD values for each request. Each output issues a grant to the first input in its grant list. The inputs receive grant signals and respond according to RR or EDF policy. When outputs receive accept signals, they stop issuing further grants. The inputs that have been granted will also stop accepting any other grants, but the outputs will not be informed of it, so unmatched outputs will continue issuing grants until they come to the end of the list or receive an accept from some input. Thus, the scheduler would take at most N iterations to find a maximal match. We call this scheduler the OS scheduler.

The second scheduler, in addition to using a sorted list for issuing grants, maintains a sorted accept list of VOQs (outputs) based on their MAD, for each input. Here, each input port upon receiving grants would accept the one that has been issued by the output nearest to the beginning of the accept list. In addition, the sorted grant list is not per output as before. Each grant list contains VOQs with the same priority in the accept list, e.g., the first grant list contains VOQs having highest priority among all inputs, sorted in increasing order and so on. In this way, no list has two VOQs belonging to a common input. We call this scheduler the IOS scheduler.

There is a difference between the scheduling process in the IOS and in the OS method. In the IOS scheduler during each iteration, instead of performing scheduling for VOQs in the same position in all grant lists, it is performed for VOQs in a single grant list starting from the first one. This minor difference results in a major advantage concerning the hardware implementation of the IOS approach, i.e., the sorting process need not sort all grant lists at once. Instead, it can sort them one after another, as they are needed by the scheduler, thus reducing the amount of sorting hardware needed to 1/N. It is also worth noting that considering unicast traffic there would be no need to use a complex

Table 1. Various schedulers belonging to DiSLIP class

		Input Service Policy	
		RR	**EDF**
Output Service Policy	**RR**	**DiSLIP_ERR**	**DiSLIP_EER**
	EDF	**DiSLIP_ERE**	**DiSLIP_EEE**

batcher sorter at each input for IOS. Instead, a simple insertion sorter can be utilized which has to insert at most one VOQ in the accept list each time slot, since at each input at most one VOQ is serviced per time slot.

4.3 DiSLIP Schedulers

Finally, the last and best performance of our scheduler classes is the Delay_iSLIP (DiSLIP) class. These schedulers function based on the iSLIP scheduler [16], but we have modified them to maintain sorted grant and accept lists, thus, enabling them to provide various delay classes with different delay bounds. iSLIP is used as the scheduling engine in some of todays routers, e.g., Ciscos GSR12000, which is a 640Gbps router. It uses the request/grant/accept scheme for scheduling a crossbar fabric with RR service policy at both inputs and outputs.

DiSLIP schedulers can have any combination of RR and EDF service policies at inputs and outputs. Table. 1 shows various combinations of input and output policies. These schedulers are simple to implement like the IOS and OS schedulers, but require LogN rather than N iterations to compute a match. This is the most promising advantage of the DiSLIP schedulers.

Fig. 1 shows the pseudo code for the DiSLIP_ERE. This scheduler uses EDF service policy to issue grants and RR policy to issue accepts. The scheduler executes four stages. In the first stage (R) inputs send their requests to outputs for which they have a cell. This is known as the request phase. In the sorting phase denoted by S, the scheduler sorts requests for each output based on increasing

```
R: for all inputs i do in parallel
       If (VOQ_j has cell) send request to output j;
S: for all outputs j do in parallel
       grantList[j] = Sorted requests according to increasing MAD order;
repeat
       G: for all outputs j do in parallel
              i = first input in grantList[j] which has a request;
              issue grant signal to input i;
       A: for all inputs i do in parallel
              using RR, choose output j among grants received;
              issue accept to output j;
              update RR;
              disable all requests;
until (no more matches can be made)
```

Fig. 1. Pseudo code for DiSLIP_ERE scheduler

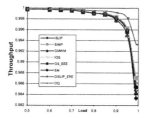

Fig. 2. Throughput under iid traffic

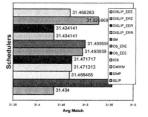

Fig. 3. Average Normalized Match under 99% load and iid traffic

MAD order. Sorting is performed in parallel for all outputs. The first two phases, R and S, are done only once during each scheduling phase. Then, in a loop that continues until no more matches can be found, two other phases are executed, namely, the grant and accept phases denoted by G and A respectively.

During the grant phase, for each output port, the scheduler issues grant signals to the most critical input having a cell for that output. After that, in the accept phase, the scheduler would select a granting output port based on RR policy and issue accept to it for each input. Simulations show that at most LogN iterations should be performed for an NxN switch. Other schedulers of this class can be implemented in the same manner with minor changes in the grant and accept phases according to their service policy.

5 Comparative Analysis

We report results obtained from simulating our schedulers. The results will be compared to each other as well as to an OQ switch. The scheduler runs for 5000 time slots for uniform iid traffic and 100,000 timeslots for 2-state markov bursty traffic. Loads less than 50% will not be considered and the range of MAD values are between 1 and 30 unless stated otherwise. The MAD value for various cells is chosen from this range with equal probability.

Fig. 2 shows throughput for various schedulers under iid traffic for a 32x32 switch. Obviously, throughput of the OQ switch is the highest. Since no cell is stuck in the input buffers. After the OQ scheduler, SIMP and DiSLIP_ERE have the highest throughput. The worst throughput scheduler is iSLIP, which has fixed delay for all delay classes under a certain load.

The higher the average match per time slot the higher is the throughput since more cells get scheduled each time slot. This becomes more apparent for high loads (≥90%) for which the arrival rate is higher. Fig. 3 shows match count normalized to load for each of the schedulers. SIMP scheduler has the most number of matches per time slot after that its DiSLIP_ERE and the scheduler with least match per time slot is iSLIP. This observation is in accordance with the results shown in Fig. 2 concerning throughput.

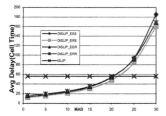

Fig. 4. Average delay for various MAD classes under 99% load, iid traffic

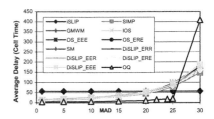

Fig. 5. Average delay for various MAD classes under 99% load, iid traffic

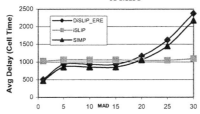

Fig. 6. Average delay for various MAD classes under 99% load, bursty traffic, burst length=32, load=99%

Average match also affects the amount of total and per class average delay. The more the number of matches the less is average delay. Average delay is shown in Fig. 4, Fig. 5 for iid traffic and in Fig. 6 for bursty traffic. As shown by the curves, aside from OQ scheduler, SIMP has the least delay followed by DiSLIP_ERE, which has a delay tightly close to SM. The worst of all is DiS-LIP_EEE. Even the total delay across all MAD classes for DiSLIP is less than that of iSLIP. For example, at 99% load, average delay over all cells for DiSLIP and iSLIP schedulers are 48 and 56 time slots respectively.

Our schedulers provide each class with its required delay as long as no over-loading conditions exist. This is verified in Fig. 4 and Fig. 5, which sketch average delay for each delay class (MAD), under iid traffic and a load of 99%. The results have been compared to a conventional non-delay-guaranteeing scheduler such as iSLIP. It can be seen that for more than 2/3rd of the MAD values, average delay is less than that of iSLIP. For other MAD values, however, the amount of average delay is much higher, since they have required higher delay than others have. The amount of average delay for various delay classes is higher than their MAD value, however, this is due to overload, which means there is more traffic than the EDF scheduler can guarantee delay bounds for. This effect can be remedied by incorporation of a suitable CAC component with some sort of traffic shaping mechanism at inputs [7].

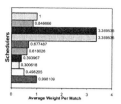
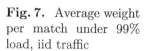

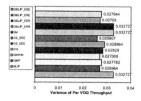

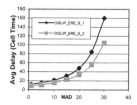

Fig. 7. Average weight per match under 99% load, iid traffic

Fig. 8. Fairness under 99% load, iid traffic

Fig. 9. Average delay with and without speedup for DiS-LIP_ERE

Another important factor in providing each MAD class with its required delay guarantee is the average weight per match discussed in section 2. As explained before, the lower average weight per match (AWPM), the more, most critical (MC) cells get serviced. Fig. 7 shows AWPM for a number of our schedulers under iid traffic for a load of 99%.

As can be observed there, the IOS approach has the best AWPM, and the worst belongs to DiSLIP_EER and DiSLIP_ERR. It seems to be in contrast to the average delay of each scheduler. The reason lies in the average match for each scheduler, i.e., the average number of matches made per scheduling run. It is true that IOS has the least AWPM, but this might be due to its relatively low matches made per time slot. Low average match results in cells to wait longer in the VOQs and, consequently their weight will decrease accordingly.

From the discussion given above, we might conclude that for a scheduler to provide cells with low bound on their delay, it must have a high match count along with a low AWPM, however, it seems that having a high match count is more important, especially as the load increases. This conclusion can be verified by noting that the SIMP has a relatively high AWPM, still its delay is less than any other approach due to its highest match count.

One of the vital requirements for a scheduler is fairness, meaning providing each flow in each VOQ, with somewhat the same service. We measure fairness of our schedulers by calculating the variation of throughput among all VOQs. We name this parameter unfairness factor, since it is the least for the fairest. Fig. 8 shows the unfairness factor for our schedulers. It can be observed that the SM, DiSLIP_ERE, and SIMP are the fairest schedulers among all.

Now we propose DiSLIP_ERE as the best scheduler, which can be implemented easily in hardware. This scheduler uses the DiSLIP algorithm with EDF policy at the VOQs, RR policy for issuing accepts, and again, EDF policy for issuing grants. It should be noted that the DiSLIP_EER version using RR for grants and EDF for accepts is not much good. Since grant signals are the most important ones in a matching problem. That is because the result of the match is completely dependant on the criteria on which grants are issued [13].

It might seem that using EDF policy for issuing accepts should improve the delay bound, however, as shown by the curves, this is not true. The reason is that issuing accept signals according to RR policy increases the average number of matches. We believe the reason is related to the input/output contention existing in IQ switches. Since putting the most-critical-cell-first burden on the decisions made might narrow the solution space, it is likely for contention to occur. Thus, by slightly deviating from this burden, we can achieve more matches and provide cells with less delay bounds.

Finally we consider the effect of speedup. Speedup increases match count and hence decreases average delay. We assume that a switch with a speedup of S can accept S cells for each of its outputs during each time slot. We have considered a small speedup of 2, which turns out to decrease average delay by a large amount. Fig. 9 shows the effect of speedup on DiSLIP_ERE scheduler under iid traffic with 99% load. It should be noted, however, that giving a speedup of 2 to our switch implies using a double speed/width cross bar and some amount of memory inside the switch fabric which complicates implementation.

6   Conclusion

Using policy decomposition and RR and EDF policies, and based on request-grant-accept matching scheme, a number of schedulers providing best effort delay guarantees for input queued switches have been devised, the best of which is DiS-LIP_ERE. Our scheduler along with delay differentiation has its total average delay across all delay classes less than the well-known iSLIP. By simulations, it has been shown that fairness, average match, and weight per match are among the most influencing parameters of a delay guaranteing scheduler. We have also considered the effect of speedup. Currently we are trying to implement our scheduler on a 320Gbps 32x32 switch with 10Gbps lines.

References

[1] Firoiu, V., et al.: Efficient admission control for piecewise linear traffic envelopes at EDF schedulers, IEEE/ACM Trans. Networking, vol. 6, pp. 558-570, Oct. 1998.
[2] Liebeherr, J., et al.: Exact admission control in networks with bounded delay services, IEEE/ACM Trans. Networking, vol. 4, pp. 885-901, Dec. 1996.
[3] Zhang, H.: Service disciplines for packet-switching integrated services networks, Ph.D. dissertation, Univ. California, Berkley, CA, Nov. 1993.
[4] Liebeherr, J., et al.: Priority queue schedulers with approximate sorting in output-buffered switches, IEEE Journal on Selected Area, vol. 17, 1127-1144, June 1999.
[5] Schoenen, R., et al.: Distributed cell scheduling algorithms for virtual-output-queued switches, Proc. IEEE GLOBECOM'99 1211-1215.
[6] Schoenen, R.: An architecture supporting quality-of-service in virtual-output-queued switches, IEICE Trans. Commun., vol. E83-B, No. 2, Feb. 2000.
[7] Kam, A. C., et al.: Linear complexity algorithms for QoS support in input-queued switches with no speedup, IEEE Journal Selected Areas Comm., June 1999 vol. 17, no. 6, 1040-56.
[8] Lee, S. W., et al.: Improved dynamic weighted cell scheduling algorithm based on Earliest Deadline First scheme for various traffics of ATM switch, In Proc. IEEE Globecom '96, pages 1959-1963, London, 1996.

[9] Karol, M., et al.: Input versus output queuing on a space division switch, IEEE Trans. Comm., vol. 6, 1347-1356, 1988.

[10] Tamir, Y., et al.: High performance multiqueue buffers for VLSI communication switches, Proc, of 15th Annual Symposium on Computer Architecture, June 1988.

[11] McKeown, N., et al.: Achieving 100% throughput in an input-queued switch, IEEE Infocom '96.

[12] Schoenen, R., et al.: Weighted arbitration algorithms with priorities for input-queued switches with 100% throughput, Proc. IEEE Broadband Switching Systems, 1999.

[13] Gale, D., et al.: College Admissions and the stability of marriage, American Mathematical Monthly, vol.69, 9-15, 1962.

[14] Stoica, I., et al.: Exact emulation of an output queueing switch by a combined input output queueing switch, 6th IEEE/IFIP IWQoS '98.

[15] Chaung, S. T., et al.: Matching output queuing with a combined input output queued switch, Technical Report, CSL-TR-98-758, Apr. 1998.

[16] McKeown, N.: The iSLIP scheduling algorithm for input-queued switches, IEEE/ACM Trans. Networking, vol. 7, No. 2, Apr. 1999.

[17] McKeown,N. , et al.: Scheduling multicast cells in an input-queued switch. Proc IEEE Infocom 1996.

Performance Analysis of a VC Merger Capable of Supporting Differentiated Services in an MPLS over ATM Switch

Ahmad Rostami[1] and Seyed Mostafa Safavi[2]

[1] Iran Telecommunication Research Center (ITRC)
North Karegar St., P.O.Box 14155-3961, Tehran, Iran
[2] Amir-Kabir University of Technology
Hafez Ave., P.O.Box 15875-4413, Tehran, Iran
arostami@itrc.ac.ir

Abstract. MPLS is a new technology that has been proposed by the IETF community and is one of the most appropriate techniques to provide connectionless IP services over powerful ATM switches. One of the most important issues in designing MPLS switches based on ATM is VC merging. With VC merging all incoming cells belonging to the same destination receive the same outgoing VC values. There are two kinds of VC merging called "full merging" and "partial merging". Full merging does not have the ability to support different QoS classes. However, in partial merging we are able to support multiple QoS classes. In this paper, we analyze the performance of a partial VC merger with the ability to support Differentiated Services and derive some equations for calculating the mean waiting time in the output ports of a VC merger.

1 Introduction

During recent years, Internet has had an exponential growth in terms of increasing number of the users and demands for more bandwidth. This trend has caused an increase in the gap between transmission capacity of media (such as fiber optic) and switching capacity of switch-routers. Therefore, specialists in this field have begun deep researches in order to keep pace of the transmission speed. Introduction of ATM networks in middle of the 1980s attracted a lot of attention to this technology and many specialists assumed that full router-based networks (IP networks) would be replaced by this technology. The reason for this assumption was the good characteristics of the ATM technology including guaranteed QoS and high speed of switching, which are the major bottlenecks of IP-based networks. However, inasmuch as number of the IP protocol users has rapidly increased and the majority of Internet applications are written based on this protocol, as well as complexity of the routing and control protocols of ATM, precluded the formation of data networks fully based on the ATM technology. On the other hand, outstanding aspects of the ATM technology became a motivator to use ATM switching technique in IP-based routers. As a result,

H.-K. Kahng (Ed.): ICOIN 2003, LNCS 2662, pp. 14–23, 2003.

some techniques such as Cell Switching Routers (CSR), IP switching and Tag switching were proposed and finally the Internet Engineering Task Force (IETF) community published drafts of a novel protocol named MPLS (Multi-Protocol Label Switching) in 1997.

Despite the fact that MPLS is independent of any specific layer-2 and layer-3 protocols, it seems that using the IP protocol as the network layer protocol and using ATM as layer-2 protocol for MPLS switches is the best way of integrating the advantages of both IP protocol and ATM switching.

When ATM is used as a switching technology, the label is encapsulated in the VPI and/or VCI fields of the header of the ATM cell and VCs are formed by the MPLS control protocol. An important problem in this case is VC merging technique, which is a critical task to Label Switching Routers (LSRs) in MPLS domain in term of scalability. In VC merging, all incoming cells belonging to the same FEC receive the same outgoing VC values (VPI/VCI).

When we use VC merging, we have to implement some techniques to prevent interleaving of cells which belong to different frames. That is, by implementing VC merging, we have to store all cells belonging to a specific frame in a separate buffer until the last cell of that frame is received. After the last cell of the frame is received we can transfer all cells contiguously without permission to any other cell, belonging to another frame, to be interleaved by cells of the frame. As stated in [1], we may have two kinds of VC merging. If all transit traffic is best effort, we can implement full merging, in which all incoming cells with the same destination network are mapped to the same outgoing VC and so receive the same VC values. On the other hand, if we have different kinds of transit traffic, we may implement partial merging, in which all incoming cells with the same destination network and QoS are mapped to the same outgoing VC.

In [1], all performance issues of full VC merging have been studied. Also, in [2] performance issues of output buffers in case of partial merging have been studied. In [2], it is assumed that we have N different kinds of traffic classes and therefore N different queues in the output. Also, it is assumed that output queues are served with a server that implements N-level strict priority service discipline. However, when we want to implement Differentiated Services in a more realistic situation , we have only one queue that has strict priority over the other N-1 queues and this queue handles the real-time and interactive traffic. Also, we implement Weighted Round-Robin (WRR) service discipline on N-1 queues that have lower priority.

In this paper, we study the performance of partial VC merging that contains N output queues. The Nth queue has strict priority over the rest of the queues and the rest of the queues are served based on the WRR service discipline by the server.

2   Structure of a VC Merger in Partial Merging Mode

Structure of the switch that we analyze is the same as the structure of the switch in [1]. That is, we assume to have an output-buffered switch fabric. However,

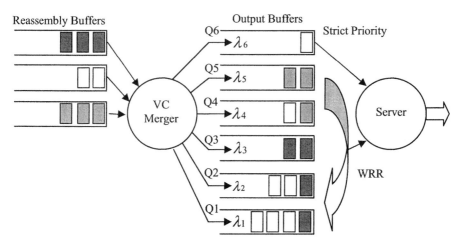

Fig. 1. Structure of an output module capable of supporting partial VC merging

structure of the output modules in our model is different from that of [1]. As depicted in Fig. 1, each output module contains some Reassembly Buffers (RBs), which save the cells belonging to a specific frame until the last cell of the frame is received. Then all cells of a completed frame are contiguously transferred to the merger. If the VC values are unique to the entire switch rather than to the port, then an RB would correspond to an incoming VC [1]. In the VC merger, VC values of the received cells are translated to the outgoing VC values. In this point, as stated above, all cells belonging to the same destination network and QoS class are mapped to the same VC values. Then, after determining the QoS class of the frame, all cells are transferred to the corresponding output buffer contiguously.

In Sect. 3, we propose a model for output buffers and analyze their performance in term of the mean waiting time for each output buffer.

3   Analyzing the Output Buffers

As shown in Fig. 1, output buffers consist of six independent buffers. In this article, we assume that our switch supports Differentiated Services model, so that we have a queue dedicated to Best Effort (BE) traffic, four queues dedicated to Assured Forwarding (AF) traffic (AF1y, AF2y,AF3y and AF4y, where y indicates drop precedence of cells belonging to the same AF class) and a queue dedicated to Expedited Forwarding (EF) traffic. Service disciplines of the output queues are as explained in the following sentences. The Q6 (EF) has strict priority over the other five queues. However, this priority is not preemptive. That is, if the server is serving any of the queues from Q1 to Q5 and a cell is received in Q6, then it must wait until the service phase of the queue which is being served is completed. Whenever the server completes serving any of the queues among Q1 to Q5, it polls Q6 and serves it until Q6 becomes empty. Service disciplines of Q1 to Q5 are based on the WRR scheduling. It means, we dedicate a weight to each of these five queues and these weights are proportional to a certain fraction of the output link capacity. Thus, we have a 5-tuple vector $m = (m1, m2, m3, m4, m5)$ in which mi determines the maximum number of cells that each queue can transmit each time that is visited by the server. Arrival process to each of these six queues is a batch type arrival, which contains multiple ATM cells. Arrival times to each of these queues depend on the completing time of the frames in RBs and thus, depend on input ports speed and transmission rate of the frames from upstream LSRs. Hence, arrivals to each of the output buffers can occur in any arbitrary time slot. Thus, we can assume that the arrival process to each of the six queues is a Poisson process. In our analysis, we will use the PASTA (Poisson Arrivals See Time Average) property of the arrivals [3].

In order to calculate the mean waiting time for each of the queues, we need some definitions as follows. We define λ_i ; (i=1,...,6) (frames/cell-time) as the arrival rate of the class- i frame, $\overline{n}_i$ as the mean number of the cells in a class-i frame, $\rho_i = \lambda_i \overline{n}_i$ as the mean offered load to the server due to class-i traffic and $\rho = \sum_{i=1}^{6} \rho_i$ as the total offered load to the server. The unit of time in all of our calculations is cell time. Also, we assume that the server is work conserving. It means server is not idle if there are cells present in each of the queues.

First, we calculate the mean waiting time for Q6 (EF traffic). Due to the strict priority service discipline for Q6, the mean waiting time for a tagged cell, which enters Q6 in an arbitrary time slot can be calculated as:

$$W_6 = R_6 + L_6 + M_6 \quad \text{(cell times)} . \tag{1}$$

where M_6 equals to the mean number of the cells, which are in front of the tagged cell in a same frame, thus receive service before the tagged cell. M_6 can be easily calculated using:

$$M_i = \frac{\overline{n}_i^2}{2\overline{n}_i} - \frac{1}{2} \quad \text{(cells)} . \tag{2}$$

where $\overline{n}_i$ and $\overline{n}_i^2$ are the first and second moments of number of the cells in a class-i frame, respectively. L_6 in (1) is the average length of Q6 and in steady state conditions can be calculated using Little's formula (here as: $L_i = (\lambda_i \overline{n}_i)W_i$)[3]. Also, R_6 in (1) is the mean residual time of completion of service for the queue which is being served by the server when the tagged cell enters Q6. By using the mean residual time formula [3], we can write:

$$R_6 = \frac{\overline{W}}{2}(1 + C_6^2) \quad \text{(cell times)} . \tag{3}$$

where $\overline{W}$ is the mean life time and C_6^2 is the squared coefficient of variation ($C_6^2 = \frac{\sigma_{W^2}}{\overline{W}^2}$). In our system, the mean life time($\overline{W}$) can be written as:

$$\overline{W} = (\sum_{k=1}^{5} \rho_k m_k) + \rho_6 \quad \text{(cell times)} . \tag{4}$$

Substituting L_6 in (1) by Little's formula and solving it for W_6, we have:

$$W_6 = \frac{R_6 + M_6}{1 - \rho_6} \quad \text{(cell times)} . \tag{5}$$

where R_6 and M_6 can be calculated from (2) through (4).

Now, we derive equations for the mean waiting time of Q1 to Q5. In deriving Wi (i=1,...,5), we are dealing with a cyclic type multi-queue system with the limited service discipline [3], which is interrupted by a high priority queue (Q6). In general, there is not any solution for multi-queue systems under limited service discipline and only some complex algorithms are proposed (such as algorithm which is proposed by Tedijianto in [5]). In this article, in order to calculate the mean waiting time for Q1 to Q5, we use a model called "Vacation model". A vacation model is composed of a single queue and a single server. Before proceeding, we present some definitions about the vacation models. A queue in a vacation model may have two phases. Either it is being served by the server , which we call this phase as the service phase or the server has stopped serving the queue and has gone to the vacation, which we call this phase as the vacation phase. Whenever the server goes to the vacation, it remains in the vacation for duration of V (cell times), which is called the vacation period. Also, when it comes back from the vacation, it serves the queue for the duration of m (cell times), which is called the service period. In addition, we consider summation of the service period and the vacation period $(m + V)$ as total period of the vacation system.

By using the above-mentioned model, we can begin deriving equations for calculating the mean waiting time of Q1 to Q5. We model our system by considering a separated vacation model for each of our five queues. We use elements of vector $(m1, m2, m3, m4, m5)$, as defined previously, as the service period for each of the assumed vacation models. Also, the mean vacation period for each vacation system can be calculated using:

$$\overline{V}_i = \sum_{k=1; k \neq i}^{5} \rho_k m_k + L_6 \quad ; \ i = 1, ..., 5 \ \text{(cell times)} . \tag{6}$$

Indeed, we have five separated vacation models with service periods and mean vacation periods as defined above. Whenever a tagged cell enters Qi (i=1,...,5), it may find the queue either in the vacation phase or in the service phase. We will analyze these two situations separately. The probability that a tagged cell finds the server in a vacation phase when it enters into the queue can be calculated as follows:

$$P_{v,i} = \frac{\sum_{k=1; k \neq i}^{5} \rho_k m_k + L_6}{\sum_{k=1}^{5} \rho_k m_k + L_6} \quad ; \ i = 1, ..., 5 . \tag{7}$$

Clearly, the probability that the tagged cell finds the queue in the service phase when it enters the queue equals to $(1 - P_{v,i})$.

Situation 1. Tagged Cell Arriving during the Vacation Phase: The mean waiting time for the tagged cell in this situation includes three components. First component equals to mean number of the cell times which the tagged cell has to wait until the server comes back from the vacation. This time equals to the mean residual time of the vacation period. Again using the mean residual time, we can write:

$$R_{v,i} = \frac{\bar{V_i}}{2}(1 + C_i^2) \quad ; \quad i=1,...,5 \text{ (cell times)} . \tag{8}$$

where C_i^2 is the squared coefficient of variation.

Second component of the mean waiting time in this situation equals to the mean number of the total periods which the tagged cell should wait before the service period in which it receives service. This mean time can be written as $[\frac{M_i+L_i}{m_i}]$, which is the largest integer contained in [ ] .

The final component of the mean waiting time in this situation equals to the mean number of the cells which are in front of the tagged cell and receive service before the tagged cell but in the same service phase. We can write this time as $(M_i + L_i)mod(m_i)$.

Situation 2. Tagged Cell Arriving during the Service Phase: Similarly, the mean waiting time in this situation is composed of three components. First component equals to the mean number of the cells which receive service in the time interval between the tagged cell entering to the queue and completion of the service phase of the queue. We can write this as $\rho_i \frac{m_i}{2}$.

Second component of the mean waiting time in this situation, includes two parts: first part is equal to a vacation period of the queue, provided that $(L_i + M_i) > \rho_i\frac{m_i}{2}$.(The situation in which the current service phase is finished but the tagged cell is yet in the queue and has to wait for at least another vacation period). Hence, this time can be written as:$Prob((L_i + M_i) > \rho_i\frac{m_i}{2})\bar{V_i}$. Second part is similar to the second component in situation 1 and equals to the number of the total periods which the tagged cell should wait before the service period in which the tagged cell is served. It should be mentioned that this time is considered from the next time that the queue is visited by the server. This time can be written as $[\frac{M_i+L_i-\rho_i\frac{m_i}{2}}{m_i}]$.

The last component of this situation is completely similar to that in situation 1, and can be written as $(M_i + L_i - \rho_i\frac{m_i}{2})mod(m_i)$.

In addition to the waiting time due to the situations 1 and 2, the total mean waiting time for Qi (i=1,...,5), includes another component. The last component is the mean time that the tagged cell has to wait due to the service time of the traffic of Q6 that has strict priority over the other queues. This time is composed of two parts. First part is the mean number of the cells which are present inside Q6 when the tagged cell enters to Qi (i=1,...,5) and can be written as L_6. Second part equals to mean number of the cells which enter Q6 while the tagged cell waits in Qi (i=1,...,5) to receive service. This time can be written as $\rho_6 W_i$. Now, we are ready to derive the mean waiting time for Q1-Q5 as follows.

$$W_i = P_{v,i}\{R_{v,i} + [\frac{M_i + L_i}{m_i}](\bar{V}_i + m_i) + (M_i + L_i)mod(m_i)\} + L_6 + \rho_6 W_i$$

$$+(1 - P_{v,i})\{\rho_i \frac{m_i}{2} + Prob((L_i + M_i) > \rho_i \frac{m_i}{2})\bar{V}_i + [\frac{M_i + L_i - \rho_i \frac{m_i}{2}}{m_i}](\bar{V}_i + m_i)$$

$$+(M_i + L_i - \rho_i \frac{m_i}{2})mod(m_i)\} \; ; \quad i = 1, ..., 5 \text{ (cell times) .} \quad (9)$$

Substituting $P_{v,i}$ by (7), L_i by Little's formula and $R_{v,i}$ by (8) in (9), we can calculate W_i . As it seems from (9), computation of W_i is too complex and needs a sophisticated numerical algorithm. However, by applying some practical considerations we can simplify (9). In practical situations, we are usually concerned with studying the performance of the system under heavy load, which implies that the relation between $(L_i + M_i)$ and m_i is so that we can approximate some of the complex terms in (9) as follows.

$$[\frac{M_i + L_i}{m_i}](\bar{V}_i + m_i) + (M_i + L_i)mod(m_i) \approx \frac{M_i + L_i}{m_i}(\bar{V}_i + m_i) \; . \quad (10)$$

$$[\frac{M_i + L_i - \rho_i \frac{m_i}{2}}{m_i}](\bar{V}_i + m_i) + (M_i + L_i - \rho_i \frac{m_i}{2})mod(m_i) \quad (11)$$

$$\approx \frac{M_i + L_i - \rho_i \frac{m_i}{2}}{m_i}(\bar{V}_i + m_i).$$

$$Prob((L_i + M_i) > \rho_i \frac{m_i}{2}) \approx 1 \; . \quad (12)$$

Substituting (10), (11) and (12) in (9) and solving it for W_i , we can write:

$$W_i \approx \frac{P_{v,i}(R_i + \frac{M_i(\bar{V}_i + m_i)}{m_i}) + (1 - P_{v,i})(\rho_i \frac{m_i}{2} + \frac{M_i - \rho_i \frac{m_i}{2}}{m_i}(\bar{V}_i + m_i) + \bar{V}_i) + L_6}{1 - \frac{\rho_i(\bar{V}_i + m_i)}{m_i} - \rho_6}$$

$$; \; i = 1, ..., 5 \text{ (cell times) .} \quad (13)$$

4   Numerical Results

In this section, we use the derived equations ((5) and (13)) for the mean waiting time of multiple traffic classes, in order to analyze the performance of the system which was defined in Sect. 3. In our computations we have used three different combinations of service disciplines and input traffic. We refer to each of these combinations as a scenario. In analyzing the system under all scenarios, we assume that number of the cells in a frame has geometric distribution with mean 8 (cells/frame). Specifications for all scenarios are listed in Table 1 .According to this table, scenarios differ with each other in terms of weight vector and contribution of each class in the total offered load. Second column in Table 1 represents weight vector (in cells), which is used by WRR scheduler. Also, third

Table 1. Scenarios which have been used to study the performance of the system

Scenario	m_1	m_2	m_3	m_4	m_5	Q1	Q2	Q3	Q4	Q5	Q6
I	6	6	7	8	10	50	11	11	11	11	6
II	9	5	7	9	11	50	11	11	11	11	6
III	6	6	7	8	10	30	15	15	15	15	10

column in Table 1 represents distribution of the total offered load over different traffic classes (in percentages).

Fig. 2-a through Fig. 2-f show the results of using (5) and (13) with the traffic scenarios mentioned above for Q6 through Q1, respectively.

As expected, the mean waiting time for EF class (Fig.2-a) is completely bounded and small compared with that of other classes (Fig.2-b through Fig.2-f). This is because only a small fraction of total offered load belongs to EF class; moreover, EF class has strict priority over the rest of the classes. In addition, the mean waiting time for AF classes (Fig.2-b through Fig.2-e) are less than that of BE class and greater than that of EF class. Furthermore, the mean waiting time increases when we move from AF1 toward AF4 class, due to decrease of their weights. Finally, BE class has the largest mean waiting time among the other classes, because it has access to a small fraction of the output link capacity despite the big fraction of total offered load that belongs to this class.

Now, we study the results which are due to differences among scenarios. Scenarios I and II differ from each other only in terms of the weights of Q1 to Q5. Comparison between the results of applying scenario I and II to Q1 (Fig.2-f) shows that the mean waiting time for scenario II is noticeably less than that in scenario I, because the fraction of the output link capacity which is dedicated to Q1 is increased in scenario II . Also, Fig.2-a shows that the mean waiting time of Q6 for scenario I is less than that in scenario II, because combination of weights in scenario II is changed so that the mean residual time of service phase of other queues ($R6$ in (5)) is increased. The difference between Scenario I and scenario III is only in distribution of total offered load over different classes. Comparison between the results for scenario I and III shows that the mean waiting time of Q2 through Q6 increase and that of Q1 decrease, when we apply scenario III instead of I. Clearly, reason of this behavior is due to increasing in load of Q2 to Q6 and decreasing in load of Q1 in scenario III.

A noticeable point in depicted figures is that the mean waiting time of BE class (Fig.2-f) has a rapid raise when the total offered load becomes close to 1 in scenario I. This behavior can be interpreted as follows. When the total offered load is small, the BE queue has the chance to use the unused fraction of the output link capacity to send its traffic. Indeed, in this situation the BE traffic has access to the output link capacity more than the weight which is dedicated to it by the WRR scheduler. When the server utilization becomes close to 1, despite the fact that the BE traffic constitutes the dominant fraction of the total offered load, it has only access to a limited fraction of the output port capacity and the unused fraction of the link capacity no longer exists.

Finally, we can easily use these results with Little's formula to compute the average length of the queues.

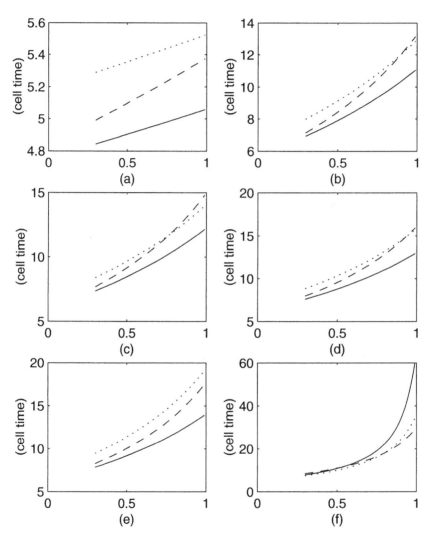

Fig. 2. The mean waiting time for (a)Q6(EF) (b)Q5(AF1) (c)Q4(AF2) (d)Q3(AF3) (e)Q2(AF4) (f)Q1(BE) versus total offered load under scenario I (solid curves), II (dotted curves) and III (dashed curves)

5   Conclusions

In this paper, we analyzed the performance of a VC merger in an ATM-LSR in MPLS domain in terms of the mean waiting time and the average queue length. We assumed that the output port supports Differentiated Services model, so that it has six separated queues for different traffic types. Also, we assumed that mixed service discipline including strict priority scheduling and WRR scheduling are applied to these six queues. Then we derived some equations for the mean waiting time of each of these queues. For the queues which are served by the WRR scheduler we considered some practical points and simplified the equation and finally proposed an approximate equation, which can be easily computed.

By using different sample traffic scenarios, we analyzed the proposed equations to observe the behavior of the system and recognized that the results are completely consistent with a practical system. The derived equations in this article can be used in order to investigate the performance of the similar systems and also can be extended to more complex service models. That is, we can apply the method used in this paper for analyzing queuing systems with more complex service disciplines.

References

[1] Widjaja, I., Elwalid, A. I. : Performance Issues in VC-Merge Capable Switches for Multiprotocol Label Switching. IEEE JSAC., Vol. 17., No.6, (1999)
[2] Dou, C., Lin, C., Wang, S., Leu, K.:Performance Analysis of Packet-Level Scheduling in an IP-over- ATM Network with QoS Control. IEICE Trans. Communications, Vol.E83-B, No.7 , (2000)
[3] Akimaru, H., Kawashima, K.:Teletraffic ,Theory and Applications. Springer-Verlag, (1993)
[4] Davie, B., Rekhter,Y.:MPLS Technology and Applications. MORGAN KAUF-MANN, (2000)
[5] Tedijianto : NonExhaustive Policies in Polling Systems and Vacation Models, Qualitative and Approximate Approaches., Ph.D. thesis, University Of Maryland. (1990)
[6] Davie, B., Lawrence, J., et al.:MPLS using LDP and VC Switching. IETF, RFC 3035, (2001)
[7] Rosen, E., Viswanathan, A., Callon, R.:Multiprotocol Label Switching Architecture. IETF, RFC 3031, (2001)

Design of Load-Adaptive Queue Management for Internet Congestion Control

Seungwan Ryu[1] and Christopher Rump[1]

Department of Industrial Engineering
University at Buffalo
State University of New York
Buffalo, NY14260-2050, USA
{sryu,crump}@eng.buffalo.edu

Abstract. In this paper, we propose the *Proportional-Integral-Derivative (PID)-controller*, which can provide proactive congestion avoidance and control using an adaptive congestion indicator and control function. The goals of PID-controller are to control congestion proactively, to stabilize the queue length around a desired level and to give smooth and low packet loss rates. An extensive simulation study suggests that PID-controller outperforms other active queue management (AQM) algorithms such as Random Early Detection (RED) [1] and Proportional-Integral (PI) controller [2] in terms of the queue length dynamics and the packet loss rates.

1 Introduction

Active queue management (AQM) is a group of FIFO-based queue management mechanisms in a router to support end-to-end congestion control in the Internet. Two main functions are used in AQM: one is the congestion indicator (to detect congestion) and the other is the congestion control function (to avoid and control congestion). Until Random Early Detection (RED) [1] was proposed by the Internet Engineering Task Force (IETF) for deployment [3], FIFO-based tail drop (TD) was the only AQM mechanism used in the network. The TD mechanism uses the instantaneous queue length as a congestion indicator, and controls congestion by dropping packets when the buffer becomes full. Although simple and easy to implement, TD has two well-known drawbacks, the lock-out and full queue phenomena [3].

To overcome the drawbacks of TD, RED enhanced the control function by introducing probabilistic early packet dropping to avoid the full queue phenomena [1]. RED also enhanced the congestion indicator by introducing the exponentially-weighted moving average (EWMA) of the queue length not only to detect incipient congestion but also to smooth the bursty incoming traffic. Following RED, many variants such as Adaptive-RED [4] and REM [5] have been proposed. However, many AQM proposals have shown severe problems with the (average) queue length as a congestion indicator [2, 6, 8]. For example, RED can detect and respond to long-term traffic patterns using the EWMA queue length.

H.-K. Kahng (Ed.): ICOIN 2003, LNCS 2662, pp. 24–34, 2003.

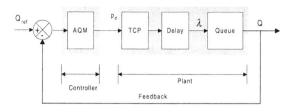

Fig. 1. Feedback control modelling of TCP flow dynamics with an AQM algorithm

However, it is unable to detect incipient congestion caused by short-term traffic load changes. As a result, AQM parameter configuration has been a main design issue since RED was first proposed in 1993.

To address these problems, congestion should be avoided or controlled *proactively* before it becomes a problem. Thus, both the congestion indicator and control function of an AQM should be adaptive to changes in the traffic environment such as the amount of traffic and the fluctuation of traffic load.

This paper is organized as follow. In Section 2, we introduce and analyze feedback control modeling approaches for TCP/AQM dynamics including RED and PI-controller. In Section 3, we propose the proportional-integral-derivative (PID)-controller using PID feedback control to overcome the reactive congestion control of existing AQM algorithms. In Section 4, we compare the performance of PID-controller with other AQM algorithms such as RED and PI-controller under various traffic environments via simulations using the ns-2 simulator [9]. In Section 5, we conclude this study and suggest directions for future study.

2   Control Theoretic Modelling of TCP/AQM

2.1   Feedback Control and TCP Flow Dynamics with AQM

TCP flow dynamics with an AQM algorithm can be modelled as a feedback control system (Figure 1) consisting of: 1) a *plant* which represents a combination of subsystems such as TCP sources, routers and TCP receivers, 2) the queue length at a router as a plant variable denoted by Q, 3) a desired queue length at a router (i.e., a *reference input*) denoted by Q_{ref}, 4) a *feedback signal* which is a sampled queue length used to obtain the error term, $Q_{ref} - Q$, 5) an *AQM controller* which controls the packet arrival rate to the router by generating a packet drop probability, p_d, as a control signal.

In [7], a simplified TCP flow dynamics model was developed. There, the open-loop transfer function (OLTF) of the plant was given by

$$P(s) = P_{TCP}(s) \cdot P_{Queue}(s) = \left(\frac{\frac{R_0 C^2}{2N^2}}{s + \frac{2N}{R_0^2 C}} \right) \cdot \left(\frac{\frac{N}{R_0}}{s + \frac{1}{R_0}} \right), \qquad (1)$$

where N is a load factor (i.e., number of TCP connections), R_0 is the round-trip time, and C is the link capacity.

Example 1: A Sample Network Configuration The TCP flow dynamics $(P(s))$ are shown to be stable when $N \geq N^-$ and $R_0 \leq R^+$ [7]. If we set $N^- = 60$ flows, $R^+ = 246$ ms and $C = 3750$ packets/sec. similar to the network configuration in [2, 7],

$$P(s) = \frac{\frac{C^2}{2N}}{(s + \frac{2N}{R_0^2 C})(s + \frac{1}{R_0})} = \frac{117187.3}{(s + 0.53)(s + 4.05)}.$$

Then, the undamped system frequency (ω_n), the damping ratio (ξ), and time domain performance specifications [10] are

- $\omega_n = 342.3$ rad./sec., $\quad \xi = 4.5793/(2\omega_n) = 0.0067$,
- $e_{ss} =$ the steady-state error $= 1/(1 + \lim_{s \to 0} P(s)) = 1/54985.1 > 0$,
- Maximum overshoot (MOS) (%) $= 100e^{-\pi\xi/\sqrt{1-\xi^2}} = 97.92\%$,
- Rise time, $t_r \simeq 1.8/\omega_n = 5.26$ ms,
- Settling time, $t_s \simeq 4/(\xi\omega_n) = 1.75$ sec.

Because of the long settling time relative to the very short rise with very small damping ratio, the TCP flows show severely oscillating dynamics. Therefore, a well designed AQM controller should be able to compensate the oscillatory TCP dynamics and give satisfactory control performance.

2.2   RED and PI-controller

With RED [1], a link maintains the EWMA queue length, $Q_{avg} = (1 - w_Q) * Q_{avg} + w_Q * Q$, where Q is the current queue length and w_Q is a weight parameter, $0 \leq w_Q \leq 1$. When Q_{avg} is less than the minimum threshold (min_{th}), no packets are dropped. When it exceeds the maximum threshold (max_{th}), all incoming packets are dropped. When it is in between, a packet is dropped with probability p_d that is an increasing function of Q_{avg}, i.e., $p_d = max_p(Q_{avg} - min_{th})/(max_{th} - min_{th})$, where max_p is a maximum value of p_d. RED attempts to eliminate the steady-state error by introducing the EWMA error terms as an integral (I)-control to the TCP flow dynamics. However, since RED introduces a range of reference input, i.e., $[min_{th}, max_{th}]$, rather than an unique reference input value for I-control, the TCP/RED shows oscillatory system dynamics. Moreover, a very small value of $w_Q = 0.002$ (or $1/512$) may bring an effect of large integral time in I-control and may cause a large overshoot [10]. As a result, RED shows oscillatory queue length dynamics and gives poor performance under a wide range of traffic environments.

The Proportional-Integral (PI)-controller has been proposed in [2] to overcome these drawbacks of RED with the introduction of a constant desired queue length (Q_{ref}). PI-controller has been designed based on (1) not only to improve responsiveness of the TCP/AQM dynamics but also to stabilize the router queue length around Q_{ref}. The latter can be achieved by means of I-control, while the former can be achieved by means of proportional (P)-control using the instantaneous queue length rather than using the EWMA queue length.

2.3   Limitations of Currently Proposed AQM Algorithms

Since each TCP source controls its sending rate through window size adjustment, the aggregate input traffic load (the offered load) at time $t \geq 0$, λ_t, is proportional to the total window size of all connections, W, i.e., $W \propto \lambda_t(R + Q_t/C)$, where R is the average propagation delay of all connections, Q_t/C is the queueing delay at a router, C is the output link capacity and Q_t is the current queue length. Because of limited buffer size and output link capacity, the *carried traffic load*, λ'_t, will be a fraction of offered traffic load that is not dropped at a router, i.e., $\lambda'_t = \lambda_t(1 - P_d)$, where, P_d is the packet drop probability.

In a time-slotted model[1], Q_t is a function of λ'_t, i.e, $Q_t = (\lambda'_t - C)\Delta t + Q_{t-1}$, where Δt is the unit length of a time slot. However, the incipient congestion will be a function of the queue length of the next time slot, Q_{t+1}, not a function of Q_t. Therefore, to detect *incipient* congestion *proactively* not *current* congestion *reactively*, P_d should be an increasing function of Q_{t+1} (or λ_{t+1} equivalently). Unfortunately, most AQM algorithms such as RED [1], REM [5], or PI-controller [2] use only the past traffic history such as Q_t (or the average queue length $\overline{Q}$) as a congestion indicator. As a result, these AQM algorithms are unable to detect incipient congestion adaptively to the traffic load variations.

3   Proportional-Integral-Derivative (PID)-controller

The proportional (P), integral (I) and derivative (D) feedback in the PID control is based on the *past (I), current (P)* and *future (D)* control error [11]. Thus, PID control is able to regulate the queue length (Q_t) around the desired level (Q_{ref}) and to provide fast speed of response simultaneously with acceptable stability and damping. A generic PID control equation is

$$u(t) = K_P e(t) + K_I \int e(\tau)d\tau + K_D \frac{d}{dt}e(t) \tag{2}$$

where $u(t)$ is control signal at time $t \geq 0$, K_P is a proportional gain, K_I is a integral gain, and K_D is a derivative time.

3.1   Design of a PID-controller

Consider a PID-controller consisting of a PI control portion connected in serial with a PD control portion. PD control can improve damping and rise time of a control system, but can not eliminate the steady-state error. In contrast, PI control can improve the steady-state error at the expense of an increase in rise time. The PID control equation (2) (in s−domain) is

$$D(s) = K_P + K_D s + \frac{K_I}{s} = D_{PD}(s) \cdot D_{PI}(s) = (K_{P1} + K_{D1}s)\left(K_{P2} + \frac{K_{I2}}{s}\right), \tag{3}$$

---

[1] In this model, time is divided into small time slots. At the end of each time slot, the queue size, Q_t, and total amount of queued input traffic, λ'_t, is calculated.

where, $K_P = K_{P1}K_{P2} + K_{D1}K_{I2}$, $K_D = K_{D1}K_{P2}$ and $K_I = K_{P1}K_{I2}$.

Since the controlled system (1) is a second-order system, time domain performance specifications such as rise time, settling time, maximum overshoot (MOS), time constant, and steady-state error are available analytically [10]. Thus, we will design PID-controller using time-domain design and analysis method. In general, a faster speed of response of a system is accomplished by reduction of the *time constant*, T_C, which was suggested to be bounded by $R_0/2$ in [7] by assuming the existence of only persistent *elephant* FTP flows in equilibrium. However, to accomplish a fast decay of transient response in the presence of short-lived connections in size and life time (so called *mice*) [6], the bound for T_C should be smaller than $R_0/2$. In designing the PD control part, we choose a proper value for T_C that provides satisfactory control dynamics in terms of both speed of response and steady-state error.

PD control. PD control can give fast decay of transient response by maintaining small T_C and MOS. Thus, performance specifications of a PD control are transient performance specifications such as rise time, settling time and MOS. From (1) and (3), the OLTF and closed-loop transfer function (CLTF) of a PD control system, $G(s) = D_{PD}(s)P(s)$ and $G_C(s)$, are

$$G(s) = \frac{\left(\frac{C^2}{2N}\right) K_{D1}\left(s + \frac{K_{P1}}{K_{D1}}\right)}{s^2 + \left(\frac{2N}{R_0^2 C} + \frac{1}{R_0}\right)s + \frac{2N}{R_0^3 C}} \quad \text{and} \quad G_C(s) = \frac{G(s)}{1 + G(s)} \qquad (4)$$

Since (4) is a second-order system and performance specifications of a PD control are functions of ξ and ω_n, PD control parameters, K_{P1} and K_{D1}, can be determined for given acceptable bounds on performance specifications.

Example 2: For a given sample network configuration (Example 1), we set the desired MOS to 5% (i.e., $\xi \cong 0.69$ equivalently) and t_s to the time to reach and stay within $\pm 2\%$ of the steady-state value. As shown in Figure 2, the unit step response of a PD control system with $T_C = R_0/2 = 123$ ms shows better control dynamics than with $T_C = R_0 = 246$ ms in terms of speed of response and the steady-state error. Thus, from the relation between the characteristic equation of a standard second-oreder system, $q(s) = s^2 + 2\xi\omega_n s + \omega_n^2 = 0$, and $G_C(s)$ in (4), PD control parameters for $T_C = R_0/2$ are

$$K_{P1} = 1.166 * 10^{-3} \text{ and } K_{D1} = 9.97 * 10^{-5}. \qquad (5)$$

PI control. From (4), the OLTF of a PI control system, $P_F(s)$, is

$$P_F(s) = G(s)D_{PI}(s) = \frac{\left(\frac{C^2}{2N}\right) K_{D1}K_{P2}\left(s + \frac{K_{P1}}{K_{D1}}\right)\left(s + \frac{K_{I2}}{K_{P2}}\right)}{s\left(s + \frac{2N}{R_0^2 C}\right)\left(s + \frac{1}{R_0}\right)}. \qquad (6)$$

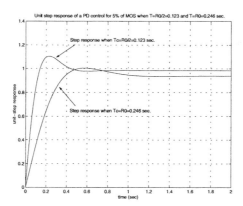

Fig. 2. Unit-step responses of a PD control system when $T_C = R_0$ and $T_C = R_0/2$

where K_{D1} and K_{P1}/K_{D1} are constants (5) obtained in the design of the PD control.

In (6), the dominant pole is $-\frac{2N}{R_0^2 C}$ because $\frac{2N}{R_0^2 C} \ll \frac{1}{R_0}$ in general [7]. Since the TCP flow dynamics show severe oscillatory step response due to the dominant pole, location of the corner frequency, K_{I2}/K_{P2}, should be selected to reduce the transient of the TCP flow dynamics. Thus, the location of K_{I2}/K_{P2} is selected to cancel the dominant pole, $s = -\frac{2N}{R_0^2 C}$, of $P_F(s)$, i.e., $\frac{K_{I2}}{K_{P2}} = \frac{2N}{R_0^2 C}$. Then, the CLTF, $P_C(s) = P_F(s)/(1 + P_F(s))$, becomes a second-order system with

$$P_F(s) = \left(\left(\frac{C^2}{2N} \right) K_{D1} K_{P2} \left(s + \frac{K_{P1}}{K_{D1}} \right) \right) \Big/ \left(s \left(s + \frac{1}{R_0} \right) \right) \tag{7}$$

PID control parameters are obtained from (3) using the relation between characteristic equations of $P_C(s)$ and of a standard second-order system [10], i.e.,

$$s^2 + \left(\frac{1}{R_0} + \left(\frac{C^2}{2N} \right) K_{D1} K_{P2} \right) s + \left(\frac{C^2}{2N} \right) K_{P1} K_{P2} = s^2 + 2\xi \omega_n s + \omega_n^2 = 0. \tag{8}$$

Example 3: Since $\omega_n = 2/(\xi R_0) = 11.78$ rad./sec. for $\xi = 0.69$, for a given network (example 1), from (8), $K_{P2} = 1.02$ and $K_{I2} = 0.5258 * K_{P2} = 0.534$. Finally, PID control parameters for 5% of target MOS are

$$K_P = 1.24 * 10^{-3}, \quad K_D = 1.02 * 10^{-4}, \text{ and } K_I = 6.23 * 10^{-4}. \tag{9}$$

Digitization. PID-controller can be implemented at a router by finding an equivalent discrete controller from a continuous model by *emulation* [12]. Particularly, once the sampling frequency ($f_s = 1/T_s$) is decided analytically or empirically, the digitized PID control equation is obtained using *Tustin's method* for integration and the *backward* rectangle method for differentiation [10]. The packet drop probability at time $k = \lfloor t/T_s \rfloor = 0, 1, \ldots$ is

$$p_k = p_{k-1} + \Delta p_k = p_{k-1} + a_1 e_k - b_1 e_{k-1} + c_1 e_{k-2}, \qquad (10)$$

where $a_1 = \left(K_P + \frac{K_D}{T_s} + \frac{T_s}{2T_I} \right)$, $b_1 = \left(K_P + \frac{2K_D}{T_s} - \frac{T_s}{2T_I} \right)$, $c_1 = \frac{K_D}{T_s}$ and $e_k = Q_k - Q_{ref}$.

3.2 Tuning of PID Control Parameters

Although we can explore the key factor governing the TCP/AQM dynamics through simplified theoretical analysis, this simplification may introduce substantial error [14]. In general, three external signals the reference input (Q_{ref}), load disturbance, and measurement noise, affect a control system [13]. However, the simplified TCP dynamic model (1) does not include load disturbance such as the slow-start and time-out mechanisms. As a result, the analytically designed PID-controller generates aggressive control signal (i.e., aggressive packet drop probability) than necessary, and shows poor control performance under realistic IP traffic environments via ns-2 [9] simulation. Thus, PID control parameters (9) should be tuned to generate proper control signals.

Since simulation is used not only to check the correctness of analytic approaches but also for allowing exploration of complicated network situations that are either difficult or impossible to analyze [14], PID control parameters are tuned and a proper sampling frequency is decided empirically via extensive simulation studies. First, we use simple tuning method for PID control parameters in which these parameters are tuned by multiplying a fractional value (called a tuning constant), κ, to make PID-controller generate a less aggressive control signal. Second, a proper sampling frequency is chosen to provide better control performance with tuned control parameters. We select $\kappa = 0.05$ and $f_s = 10f = 29.5$Hz as a tuning constant and a sampling frequency for digital implementation.

4   Simulation Study

4.1   Simulation Setup

We use a simple network topology shown in Figure 3. All TCP/Reno sources are connected to the routers, nc0 and nc1, with link speeds of 50 Mbps. The propagation delay between TCP source i and nc0 as well as between nc1 and destination i is $5i$ ms, $i = 1, \ldots, 9$. The bottleneck link between nc0 and nc1 is assumed to have a link speed of 30 Mbps and a propagation delay of 10 ms. nc0 maintains an AQM algorithm with buffer size of 800 packets, which is about twice the bandwidth-delay product (BDP). Each packet is assumed to have an average size of 1000 bytes. Two types of flows are considered: 33% are *elephants* (long-lived FTP flows) and 67% are *mice* (short-lived flows) with 3 seconds of average life time.

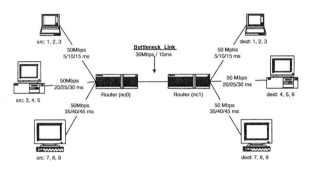

Fig. 3. Simulation network topology

We compare control performance of PID-controller with RED and PI-controller under packet drop mode. In RED and PI-controller, recommended parameter values are used. Q_{ref} for PID-controller and PI-controller is set to 200 packets. Control parameters for each AQM algorithms are

RED: $w_Q = 0.002, max_p = 0.1, max_{th} = 200, min_{th} = 70,$
PI-controller: $a = 1.822 \cdot 10^{-5}, b = 1.816 \cdot 10^{-5}, Q_{ref} = 200, f_s = 160\text{Hz},$
PID-controller: $K_P = 6.2 \cdot 10^{-5}, K_D = 5.1 \cdot 10^{-5}, K_I = 3.12 \cdot 10^{-5}, f_s = 29.5\text{Hz}$

Performance Metrics. Control performance of an AQM algorithm can be measured by two measures: the *transient performance* and the *steady-state error control*. We use the instantaneous queue length as a transient performance metric. For the steady-state control performance, we use the *quadratic average of control deviation (QACD)* [15] defined as follow

$$S_e = \sqrt{\frac{1}{N+1}\sum_{i=0}^{N} e_k^2} = \sqrt{\frac{1}{N+1}\sum_{i=0}^{N}(Q_i - Q_{ref})^2}, \qquad (11)$$

where Q_i is the ith sampled queue length and N is the number of sampling intervals. In addition, to achieve higher throughput, or to accommodate more traffic, maintaining low packet loss rate and high link utilization is important.

4.2   Control Performance of AQM Algorithms

Figure 4 shows the queue length dynamics of PID-controller, PI-controller and RED respectively under 189 flows (upper) and 378 flows (lower). PID-controller shows good control performance under two different traffic load levels in terms of the queue length dynamics staying around $Q_{ref} = 200$. The queue length of PI-controller stays below Q_{ref} most of time under two different traffic load levels. Thus, PI-controller behaves like a TD with buffer size of Q_{ref} with severe fluctuation. RED maintains the (average) queue length between $min_{th} = 70$ and

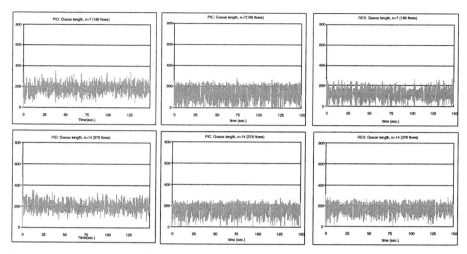

Fig. 4. The queue lengths of PID-controller, PI-controller and RED under 189 (top) and 378 flows (bottom)

Table 1. Mean and variance of QACD under different load levels

number of flows	PID-controller		PI-controller	
	mean	variance	mean	variance
189	43.5	1.89	90.6	4.31
270	42.9	1.14	78.5	5.05
378	42.0	0.56	65.1	2.14

$max_{th} = 200$ effectively with 189 flows. However, RED behaves like a TD with buffer size of Q_{ref} with 378 flows similar to PI-controller.

The steady-state control performance of PID-controller and PI-controller are evaluated in terms of QACD[2] for three different traffic load levels, 189, 270, and 378 flows. As shown in Table 1, PID-controller shows robust steady-state control performance to different traffic load levels in terms of QACD (mean and variance). In contrast, QACD of PI-controller is sensitive to the traffic load.

Figure 5 shows the average packet loss rates and the link utilization of each AQM algorithm under different traffic load levels. PID-controller shows significantly lower average packet loss rate and higher link utilization for all cases of traffic load levels than other AQM algorithms.

---

[2] Since only PID-controller and PI-controller maintain unique desired queue length, Q_{ref}, we use QACD to compare steady-state control performance of these two AQMs.

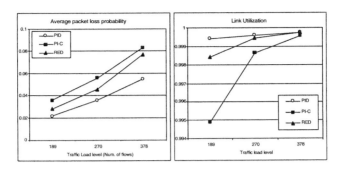

Fig. 5. The average packet loss probabilities and the link utilization of AQM algorithms under several different load levels

5   Conclusions and Further Study Issues

We designed an adaptive and proactive AQM algorithm, called PID-controller, using PID feedback control to overcome the reactive control behaviors of existing AQM proposals. PID-controller showed robust and adaptive congestion control performance and outperformed other AQM algorithms such as RED and PI-controller in terms of queue length dynamics, the packet loss rates, and the link utilization in ns-2 [9] simulation studies.

However, there are issues for further study. Since the traffic process assumed in (1) is different from complicated real IP traffic situations because of the existence of load disturbance and measurement noise, PID control parameters should be tuned to generate proper control signals. Thus, we are trying to find a more precise TCP flow dynamic model than (1) to reduce the effect of load disturbance and find better PID control parameters as a result. We are also working on finding a load adaptive sampling time interval to reduce the measurement noise. Since the plant dynamics (1) vary dynamically under changing traffic situations, we are also examining an adaptive PID-controller such as a *self-tuning* or *model-reference adaptive* PID-controller.

References

[1] Floyd, S., V. Jacobson, "Random Early Detection gateways for congestion avoidance," *IEEE/ACM Trans. Networking*, 1(4), pp. 269–271, 1993.

[2] Hollot, C. V., V. Misra, D. Towsley, and W. Gong, "On designing improved controllers for AQM routers supporting TCP flows," *Proc. of INFOCOM'2001*, 2001.

[3] Braden, B., et al., "Recommendations on Queue Management and Congestion Avoidance in the Internet," *IETF RFC2309*, April 1998.

[4] Feng, W., D. Kandlur, D. Saha, and K. Shin, "A Self-configuring RED gateway," *Proc. of INFOCOM'99*, 1999.

[5] Lapsley, D. E., and S. Low, "Random early marking for Internet congestion control," *Proc. of Globecom'99*, 1999, pp66-74.

[6] Christiansen, M., K. Jaffey, D. Ott, and D. Smith, "Tuning RED for web traffic," *IEEE/ACM Trans. Networking*, 9(3), pp. 249-264, June 2001.

[7] Hollot, C. V., V. Misra, D. Towsley, and W. Gong, "A Control theoretic analysis of RED," *Proceedings of INFOCOM'2001*, 2001.

[8] Ryu, S., C. Rump and C. Qiao, "Advances in Internet Congestion Control," *IEEE Comm. Survey and Tutorial*, accepted, 2002.

[9] McCanne, S., and S. Floyd, "UCB/LBNL/VINT Network Simulator - ns (version 2)," http://www.isi.edu/nsnam/ns/, 1996.

[10] Kuo, B. C., *Automatic Control Systems*, John Wiley & Sons, 1995.

[11] Åstrom, K., and T. Hagglund, "The Future of PID Control," *Control Engineering Pratice*, vol. 9, pp. 1163-1175, 2001.

[12] Franklin, G., J. Powell, and M. Workman, *Digital Control of Dynamic Systems*, Addison-Wesley, Third edition, 1998.

[13] Åstrom, K., and T. Hagglund, *PID Controllers: Theory, Design, and Tuning*, Instrument Society of America, Second edition, 1995.

[14] Floyd, S, and V. Paxson, "Difficulties in simulating the Internet," *IEEE/ACM Trans. Networking*, 9(4), pp. 392-403, August 2001.

[15] Isermann, R., *Digital Control Systems Volumn I: Fundamentals, Deterministic Control*, Springer-Verlag, Second Revised version, 1989.

Performance Analysis of an IP Lookup Algorithm for High Speed Router Systems

Min Young Chung[1], Jaehyung Park[2*], Byung Jun Ahn[3],
Namseok Ko[3], and Jeong Ho Kim[4]

[1] School of Information and Communication Engineering
Sungkyunkwan Univ.
300 Chunchun-dong, Jangan-gu, Suwon, Kyunggi-do, 440-746, Korea
mychung@ece.skku.ac.kr
[2] Dept. of Electronics, Computer & Information Eng.
Chonnam National Univ.
300 Yongbong-dong, Puk-gu, Gwangju, 500-757, Korea
hyeoung@chonnam.ac.kr
[3] Network Laboratory
Electronics and Telecommunications Research Institute
162 Gajeong-dong, Yuseoung-gu, Daejeon, 305-350, Korea
{bjahn,nsko}@etri.re.kr
[4] Dept. of Information Electronics Eng.
Ewha Womans University
11-1 Daehyun-dong, Seodaemun-gu, Seoul, 120-750, Korea
jho@ewha.ac.kr

Abstract. IP lookup is an important issue in designing high speed router systems. The most important function of router is to determine the output port of an incoming IP packet according to its destination address. In this paper, we evaluate the performance of the IP lookup algorithm in IQ2200 Chipset for high speed routers in terms of the memory required for storing lookup information and the number of memory access on constructing the forwarding information and then propose a scheme to enhance the IP lookup performance.

1 Introduction

The increment of hosts and users cause the Internet traffic to be on radical increase. Due to such increment, physical mediums with high bandwidth and network routers with high capacity are required for fast transmitting and processing IP packets. Faster transmission mediums can be achieved by replacing them from copper wires to optical fibers, and this work is on progress rapidly in the Internet backbone. However, the performance improvement of routers has been lagging behind. Consequently, the router becomes the key issue on Internet performance [1].

---

[*] Corresponding author.

H.-K. Kahng (Ed.): ICOIN 2003, LNCS 2662, pp. 35–45, 2003.

A router consists of four major components, which are a routing control processor, input and output interfaces, switching fabrics, and forwarding engines. A forwarding engine determines packets' output ports by referencing forwarding table which maps between packets' destination addresses and output ports. This process is called as IP lookup. An introduction of a CIDR(Classless Inter Domain Routing) [2] makes this lookup be performed as a longest prefix matching which has more computational overhead than an exact match operation [3].

To perform the longest prefix matching faster, IP lookup has been studied into largely two directions. The first one is software schemes on trie data structure which are used as lookup methods in early routers [4, 5, 6]. However, software-based lookup schemes are not appropriate for current high speed routers. The second one is hardware-based schemes [7, 8, 9, 10]. For fast hardware-based IP address lookups, the number of memory access to lookup routes in forwarding table and the size of memory required to store information in forwarding table should be as small as possible. Also it is required to easily update information stored in forwarding table.

In this paper, according to above two performance criteria, we evaluate the performance of the IP lookup algorithm implemented in IQ2200 Chipset [11]. By analyzing the performance of the IP lookup scheme on practical routing entries, the performance of the scheme on IQ2200 Chipset depends on the number of expansion bits. The scheme using 20 bits as expansion bit yields better performance than using the recommended number of expansion bits (18 bits) on IQ2200 design manual. The paper is organized as follows. In Section 2 we describe the IP lookup algorithm. In Section 3 we analyze the performance of the IP lookup algorithm by the number of memory access and the size of required memory. In Section 4 we evaluate the performance by practical routing entries in Internet. Finally we conclude in Section 5.

2   Description of Algorithms

The IP lookup algorithm implemented in IQ2200 Chipset makes use of the two tables, the primary and the secondary tables. In the secondary table, the size of one block associated with an expansion length with L_{exp} bits varies according as the longest prefix length of routes having the same expansion bits.

2.1   Lookup Algorithm

The primary table consists of $2^{L_{exp}}$ entries and each entry contains three pieces of information, i.e., a valid bit, a shift count, and an index. A valid bit is used to indicate whether any information in an entry is valid or not. A shift count is used to select bits of IP address which is being looked up and to determine the position corresponding IP address in the secondary table. An index is used to indicate the base address corresponding the high L_{exp} bits of IP address.

The secondary table consists of blocks with different sizes. One block includes several entries with fixed size. For the given number of expansion bits, the size

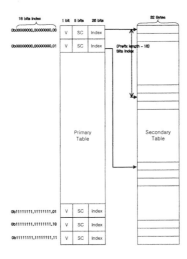

Fig. 1. Architecture of the primary and the secondary tables used in the scheme implemented in IQ2200 Chipset. In case that the length of expansion bits, L_{exp} is 18 and sizes of an entry in the primary and the secondary tables are 4 and 32 [bytes], respectively

of one block depends on the characteristics of route entries in a routing table such as the relationship between these route entries. Each entry in the secondary table be able to contain various information fields according as the considered services, i.e., IP SA and DA filtering, MPLS, DS, VLAN, etc. However, information representing a next hop address on lookup IP address has to be stored within the entry in the secondary table.

In case that the length of expansion bits, L_{exp} is 18 and sizes of an entry in the primary table and the secondary tables are 4 and 32 [bytes], respectively. Architecture of the primary and the secondary tables used in the scheme implemented in IQ2200 Chipset is shown in Figure 1. Arriving an IP packet for being looked up, the high 18 bits of IP address is used to index in the primary table. If the valid bit in the indexed entry is not set, the packet is forwarded to a default route. Otherwise, the high 18 bits of the IP address are masked off and then the remaining bits are shifted for right direction as the value of the shift count. For a given IP address, the sum of the shifting value and the value of the index field in the primary table indicates the location of information, such as next hop address, in the secondary table. The looked-up IP packet is transferred to the next hop as the information in the secondary table. In this scheme two times of table access is always required for looking up an IP packet.

2.2   Table Update Algorithm

Information of route entries in the primary and the secondary table should be updated whenever routing information within a routing table is changed. If the next hop address of a route entry in the routing table is changed, look-up information within the corresponding entry included in the secondary table is simply updated without changing other information within the two tables. However, in case that a route entry is inserted/deleted into/from the routing table. Some entries may be needed to insert/delete into/from the secondary table and information within the primary and the secondary tables should be updated. Update fields within the two tables are different as the relationship between the insertion/deletion route and other routes in the routing table. We classify the relationship into seven cases.

Case 1: The prefix length of an(a) insertion/deletion route, n, is smaller than the number of expansion bits, L_{exp}, and the leftmost n bits of the insertion/deletion route differ from those of any route in the routing table.

Case 2: n is smaller than L_{exp} and the high n bits of the insertion/deletion route are the same as those of other routes in the routing table.

Case 3: n is equal to L_{exp} and the high L_{exp} bits of the insertion/deletion route differ from those of any route in the routing table.

Case 4: n is equal to L_{exp} and the high L_{exp} bits of the insertion/deletion route are the same as those of other routes, whose longest prefix length, h is larger than L_{exp}, in the routing table.

Case 5: n is larger than L_{exp} and is smaller than the largest prefix length of all routes, except the inserted/deleted route, whose high L_{exp} bits are the same as those of the insertion/deletion route.

Case 6: n is larger than L_{exp} and is equal to the largest prefix length of all routes, except the insertion/deletion route, whose high L_{exp} bits are the same as those of the insertion/deletion route.

Case 7: n is larger than both L_{exp} and the largest prefix length of all routes, except the insertion/deletion route, whose high L_{exp} bits are the same as those of the insertion/deletion route.

Suppose that a route insertion is requested. In *Cases 1* and *3*, three pieces of information within $2^{L_{exp}-n}$ entries included in the primary table, indicated by the high n bits of the requested route, are updated. The valid bit and the shift count are set to *valid* and $32 - L_{exp}$, respectively. The index indicates the base address of one block in the secondary table. The block with single entry is newly inserted into the secondary table, and the related information is written into the entry. In *Case 2*, three pieces of information within one of $2^{L_{exp}-n}$ entries indicated by the high n bits of the requested route in the primary table, are updated if the entry is related with a route with the same high n bits of the requested route. The procedures of updating information within the indexed entries in the tables are the same as in *Cases 1* and *3*. However, if the entry is not related with any route with the same high n bits of the requested route, information in the entry is not changed. In *Cases 4, 5*, and *6*, three pieces of

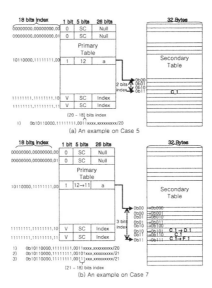

Fig. 2. Examples of inserting route entries into the primary and the secondary tables in the scheme. In case that the length of expansion bits, L_{exp} is 18 and sizes of an entry in the primary table and the secondary tables are 4 and 32 [bytes], respectively. (a) The prefix of the inserted route is 0b10110000,11111111,0011 (prefix length 20 bits) and its next hop address is C_1. (b) After completing (a), two routes are newly inserted. Prefixes of two inserted routes are 0b10110000,11111111,00101 and 0b10110000,11111111,00111 (prefix length 21 bits, respectively, and their next hop addresses are D_1 and F_1, respectively

information within the entry included in the primary table, indicated by the high L_{exp} bits of the requested route, do not changed. The information, such as a next hop address, within the corresponding 2^{h-n} entries in the secondary table is updated. In *Case 7*, the shift count is set to $32 - n$ and the index is newly assigned as the base address of one block with $2^{n-L_{exp}}$ entries in the secondary table. Information within the $2^{n-L_{exp}}$ entries is newly constructed based on the routing table. In *Cases 5* and *7*, examples of updating information within the primary and the secondary tables are shown in Figure 2.

Considering that a route deletion with prefix length n is requested. In *Cases 1* and *3*, information of the valid bit within $2^{L_{exp}-n}$ entries included in the primary table, indicated by the high n bits of the requested route, is changed as *invalid*. In *Case 2*, if information within the entries related with other routes with the same high n bits of the requested route are kept. However, if the entry is not related with any route with the same high n bits of the requested route, information of the valid bit within the entry is changed as *invalid*. In *Cases 4*,

5, and 6, three pieces of information within the entry included in the primary table, indicated by the high L_{exp} bits of the requested route, do not changed. The information related with the deleted route, such as a next hop address is deleted from the corresponding 2^{h-n} entries in the secondary table and then the entries is filled as information supporting longest prefix matching lookups. In *Case 7*, the shift count is set to $32 - h$ and the index is newly assigned as the base address of one block with $2^{h-L_{exp}}$ entries in the secondary table. Information within the $2^{h-L_{exp}}$ entries is newly constructed based on the routing table.

Total number of entries included in the secondary table varies as a route insertion/deletion request occurred for *Cases 1, 2, 3,* and *7*. From this reason, unused segments are wholly distributed in usable memory region as time goes. To solve this problem in the considered scheme, both the primary and the secondary tables are fully reconstructed whenever a route insertion/deletion request occurred for *Cases 1, 2, 3,* and *7*.

3   Mathematical Analysis

To calculate the required memory size and the memory access complexity we make the following notations.

- L_{exp} : Expansion prefix length
- N_n : Number of routes with prefix length n in a routing table
- $N_{n,L_{exp}}$: Number of routes with prefix length n such that high L_{exp} bits of the routes are different from those of other routes
- $N_{n,L_{exp},h}$: Number of routes with prefix length n such that high L_{exp} bits of the routes are the same as those of other routes and the largest prefix length among them is equal to h

Since the total number of entries contained in the primary table depends on L_{exp}, the total size of the primary table [bytes] is simply calculated as

$$M_{total}^{1st} = 2^{L_{exp}} \cdot B_{1st}, \tag{1}$$

where B_{1st} denotes the size of an entry in the primary table.

To obtain the total size of the secondary table, we calculate the number of entries required for storing information on all routes with prefix length n in the secondary table as follows

$$M_n^{2nd} = \begin{cases} N_n, & \text{if } 8 \le n \le L_{exp}, \\ N_{n,L_{exp}} \cdot 2^{n-L_{exp}}, & \text{if } L_{exp} < n \le 32. \end{cases} \tag{2}$$

From Eq. 2, we obtain the total number of entries required for storing information on all routes in the secondary table as the sum of M_n^{2nd} for all n ($8 \le n \le 32$).

The total memory size [bytes] is calculated as

$$M_{total}^{Scheme\ 1} = 2^{L_{exp}} \cdot B_{1st} + \sum_{n=8}^{L_{exp}} N_n \cdot B_{2nd} + \sum_{n=L_{exp}+1}^{32} N_{n,L_{exp}} \cdot 2^{n-L_{exp}} \cdot B_{2nd}, \tag{3}$$

where B_{2nd} denotes the number of bytes of an entry in the secondary table. The first term represents the total size of the primary table and both the second and the third terms represent that of the secondary table.

Information is stored/deleted into/from the primary or the secondary tables whenever information on routes in a routing table is updated. The number of memory access depends on the characteristics of IP route entries, such as the prefix length, the size of expansion bits, and the relationship between prefixes of routes. To analyze the performance of the scheme under the worst condition, we consider that the primary and the secondary tables are newly constructed for a route insertion/deletion request.

Updating a route with prefix length n, the numbers of memory access to the primary and the secondary tables can be obtained as

$$A_n^{1st} = \begin{cases} N_n \cdot 2^{Lexp-n}, & \text{if } 8 \leq n < L_{exp}, \\ \sum_{h=n+1}^{32} N_{n,L_{exp},h}, & \text{if } L_{exp} \leq n \leq 32, \end{cases} \tag{4}$$

and

$$A_n^{2nd} = \begin{cases} N_n, & \text{if } 8 \leq n < L_{exp}, \\ \Delta(n), & \text{if } n = L_{exp}, \\ N_{n,L_{exp}} \cdot 2^{n-Lexp} + \Delta(n), & \text{if } L_{exp} < n \leq 32, \end{cases} \tag{5}$$

where $\Delta(n) = \sum_{h=n+1}^{32} [N_{n,L_{exp},h} \cdot 2^{h-n}]$, respectively.

From Eqs 4 and 5, the total number of memory access for constructing the primary and the secondary tables is calculated as

$$A_{total}^{Scheme\ 1} = \sum_{n=8}^{32} [A_n^{1st} + A_n^{2nd}]. \tag{6}$$

4   Numerical Results

For evaluating the performance of the considered scheme in terms of the required memory size and the number of table access, we use five samples of routing tables on different AS domains in Table 1 [12]. Routing tables on AS domains 1668 and 3277 contain 111,943 and 39,725 routes, respectively.

Table 1. Total number of route entries in the BGP routing tables on five different AS domains

AS number	Total number of routes	Measurement time
1	110,962	Aug. 08, 2002
234	95,232	Aug. 06, 2002
1668	111,943	Sep. 05, 2002
3277	39,725	Sep. 05, 2002
6079	104,831	Sep. 05, 2002

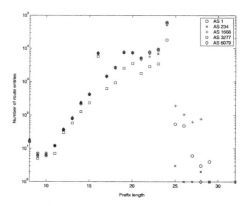

Fig. 3. Prefix length distributions of five different AS domains (*AS 1* on Aug. 08, 2002, *AS 234* on Aug. 06, 2002, *AS 1668, 3277,* and *6079* on Sep. 05, 2002)

Figure 3 shows prefix length distributions on five different AS domains. For AS 1, 234, 1668, 3277, and 6079, 99.89, 99.99, 99.61, 99.99, and 99.99 % of the prefixes are 24 bits or less, respectively. Especially, the portion of routes with prefix length 24 is 55.34, 52.92, 55.42, 45.22, and 54.24 % in AS 1, 234, 1668, 3277, and 6079, respectively.

For AS 1, 1668, and 6079, the numbers of routes with prefix length n whose high L_{exp} bits are different from those of other routes are shown in Figure 4. For AS 1, when L_{exp} is equal to 16, 18, 20, and 22, the number of routes with prefix length 24 is 4,707, 10,594, 20264, and 26,951, respectively. In case that $L_{exp} = 18$, 20, and 22, the number of routes increases about 130, 90, and 30 %, respectively, compared with that in $L_{exp} = 16, 18$, and 20. Increasing portion of the routes decreases as L_{exp} increases. However, in case that prefix length is equal to 25 or more. The number of route is rarely changed as L_{exp} varies. For AS 1668 and 6079, the characteristics of the routes are similar to those for AS 1, as shown in Figure 4. Not shown in Figure 4, routes on AS 234 and 3277 have the same trend as those on other ASs.

Figure 4 shows the required memory size for storing lookup information varying L_{exp}. When L_{exp} is fixed, the required memory size is larger as the ascending order of numbers of routes in routing tables. In case that $L_{exp} = 24$, the required memory size rarely depends on the number of routes in a routing table because the required memory in the primary table is more larger than that in the secondary table. In $L_{exp} = 20$, the required memory sizes on AS 1, 234, 1668, 3277, and 6079 are reduced about 36, 30, 40, 12, and 35 %, respectively, compared with those in $L_{exp} = 18$. They are also reduced about 28, 33, 33, 58, and 29 %, respectively, compared with those in $L_{exp} = 22$. When L_{exp} is equal to 20, the required memory sizes on AS 1, 234, 1668, 3277, and 6079 are about 16, 14, 17, 8, and 15 [Mbytes], respectively. In general, we can efficiently use memory for storing routing information as L_{exp} is set to 20.

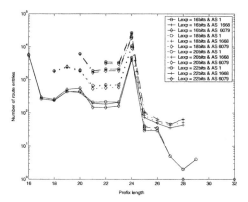

Fig. 4. Number of routes with prefix length n whose high L_{exp} bits are different from those of other routes varying prefix length

Figure 6 shows the number of memory access for writing/deleting lookup information into/from the forwarding information base varying the number of expansion bits. General trend of the number of memory access is similar to that of the required memory size. For $L_{exp} = 20$, numbers of memory access on AS 1, 234, 1668, 3277, and 6079 are about 900,000, 780,000, 940,000, 430,000, and 860,000, respectively. They are reduced about 28, 21, 41, 2, and 25 %, respectively, compared with those in $L_{exp} = 18$. They are also reduced about 41, 49, 49, 58, and 44 %,respectively, compared with those in $L_{exp} = 22$.

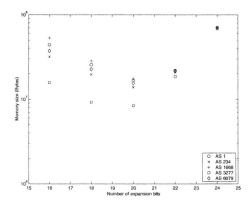

Fig. 5. Memory size for storing lookup information varying *the number of expansion bits* in the scheme

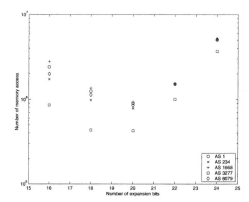

Fig. 6. Number of memory access for writing/deleting lookup information into/from the forwarding information base varying *the number of expansion bits* in the scheme

5 Conclusions

In this work, one of hardware-based schemes, which had implemented in IQ2200 Chipset, was considered and the performance of the scheme was analyzed. The performance of the considered scheme making use of the primary and the secondary tables depended on both the characteristics of IP route prefixes and the length of expansion bits. In the primary table, both the required memory and the number of memory access increase as the length of expansion bits increases. In the secondary table, however, they decrease as the length of expansion bits increases. Since the increasing and the decreasing portions vary according as the length of expansion bits, there exists the length of expansion bits minimizing the required memory size and the number of memory access. From our results, when the length of expansion bits is 20, the required memory size and the number of memory access are minimized. Using 20 bits as expansion bits, the required memory size and the numbers of memory access are reduced about 35% or more and 25% or more, respectively, compared with those using 18 bits.

The scheme can efficiently use memory, but wastes processor resource for fully re-constructing forwarding table due to route updates. For further studies, researches on lookup algorithms efficiently using processor resource are required.

References

[1] Keshave, S., and Rharma, R.: Issues and Trends on Router Design. IEEE Comm. Mag. Vol. 36, No. 5 (1998) 144–151
[2] Fuller, V., Li, T., Yu, J., and Varadhan, K.: Classless Inter-Domain Routing (CIDR): and Address Assignment and Aggregation Strategy. IETF RFC1519 (1993)
[3] Ruiz-Sanchez, M. A., Biersack, E. W., and Dabbous, W.: Survey and Taxonomy of IP Address Lookup Algorithms. IEEE Network Vol. 15 No. 2 (2001) 8–23

[4] Doeringer, W., Karjoth, G., and Nassehi, M.: Routing on Longest Matching Prefixes. IEEE/ACM Trans. Networking Vol. 4 No. 1 (1996) 86–97

[5] Waldvogel, M., Varghese, G., Turner, J., and Plattner, B.: Scalable High Speed IP Routing Lookups. Proc. ACM SIGCOMM (1997) 25–36

[6] Nilsson, S. and Karlsson, G.: IP-Address Lookup using LC-Tries. IEEE JSAC Vol. 17 No. 6 (1999) 1083–1092

[7] Degermark, M., Brodnick, A., Carlsson, S., and Pink, S.: Small Forwarding Tables for Fast Routing Lookups. Proc. ACM SIGCOMM (1997) 3–14

[8] Gupta, P., Lin, S., and Mckewon, N.: Routing Lookups in Hardware at Memory Access Speeds. Proc. IEEE INFOCOM (1998) 1240–1247

[9] Huang, N.-F., Zhao, S.-M.: A Novel IP Routing Lookup Scheme and Hardware Architecture for Multigigabit Switching Routers. IEEE JSAC Vol. 17 No. 6 (1999) 1093–1104

[10] Wang, P.-C., Chan, C.-T., and Chen, Y.-C.: A Fast IP Routing Lookup Scheme. Proc. IEEE ICC (2000) 1140–1144

[11] Design Manual: $IQ2200^{TM}$, Family of Network Processors. Revision 2. Vitesse Semiconductor Corporation (2002)

[12] BGP Reports. http://bgp.potaroo.net

Scalable IP Routing Lookup
in Next Generation Network

Chia-Tai Chan[1], Pi-Chung Wang[1], Shuo-Cheng Hu[2],
Chung-Liang Lee[1], and Rong-Chang Chen[3]

[1] Telecommunication Laboratories, Chunghwa Telecom Co., Ltd.
7F, No. 9 Lane 74 Hsin-Yi
Rd. Sec. 4, Taipei, Taiwan 106, R.O.C.
{ctchan,abu,chlilee}@cht.com.tw
[2] Department of Info. Management
Ming-Hsin University of Science and Technology
1 Hsin-Hsing Rd. Hsin-Fong, Hsinchu, Taiwan 304, R.O.C.
schu@mis.must.edu.tw
[3] Department of Logistics Engineering and Management
National Taichung Institute of Technology
No. 129, Sec. 3, Sanmin Rd., Taichung, Taiwan 404, R.O.C.
rcchens@ntit.edu.tw

Abstract. *Ternary content-addressable memory* has been widely used
to perform fast routing lookups. It is able to accomplish the best matching prefix problem in $O(1)$ time without considering the number of prefixes and its lengths. As compared to the software-based solutions, the
Ternary content-addressable memory can offer sustained throughput and
simple system architecture. It is attractive for IPv6 routing lookup. However, it also comes with several shortcomings, such as the limited number
of entries, expansive cost and power consumption. Accordingly, we propose an efficient algorithm to reduce the required size of Ternary content-
addressable memory. The proposed scheme can eliminate 98 percentage
of Ternary content-addressable memory entries by adding tiny DRAM.
We also address related issues in supporting IPv6 anycasting.

1   Introduction

Due to the advance of the World Wide Web and the promise of future e-
commerce, it has shown that the Internet access continues to grow exponentially.
The Internet has been facing the depletion of IPv4 address. Network administrators must increasingly rely on network address translation (NAT) technologies
to deploy network. However, it complicates the network management and breaks
the end-to-end principle of the Internet. Some applications cannot work across
a NAT device, such as IPsec. Furthermore, the Internet hosts are no longer just
computers, but a whole new range of information appliances, that will require
global IP addresses.

All these issues are the main driving force of IPv6 for its large address space.
For example, the current IPv6 address allocation policy recommendation is to

H.-K. Kahng (Ed.): ICOIN 2003, LNCS 2662, pp. 46–55, 2003.

allocate a 48-bit prefix to every site on the Internet, whether homes, small offices, or large enterprise sites. The 48-bit prefix allows 65,000 subnets within each site, each of which could accommodate a virtually infinite number of hosts. IPv6 also brings such benefits as stateless auto-configuration, more efficient mobility management and integrated IPsec.

The major obstacle for the design of the high-speed router is the relatively slow IP lookup scheme. To forward packets toward their destinations, a router must perform forwarding decision, the next hop for the incoming packet, based on the information gathered by the routing protocols. Since the development of CIDR in 1993 [1], IP routes have been identified by a (route prefix, prefix length) pair, where the prefix length varies from 1 to 32 bits. Due to the variable prefix lengths, the search of best match prefix (BMP) may be time consuming for a backbone router with a large number of table entries. The exponential growth of the Internet hosts has further stressed the routing system. It is difficult for the packet-forwarding rate to keep up with the increased traffic demand. Specifically, the address lookup operation is a major bottleneck in the forwarding performance of today's routers.

Ternary content-addressable memory (TCAM) is one popular hardware device to perform fast IP lookups. As compared to the software-based solutions, the TCAM can offer sustained throughput and simple system architecture, thus makes it attractive for IPv6 routing lookup. However, it also comes with several shortcomings, such as the limited number of entries, power consumption and expansive cost. Specifically, it will need fourfold size of TCAM to process the IPv6 address (128-bit) with identical IPv4 entries. In this article, we propose an efficient algorithm to eliminate 98 percentage of TCAM entries by adding tiny DRAM and also address related issues in supporting IPv6 anycasting.

The rest of the paper is organized as follows. Firstly, the related algorithms are introduced in Section 2. Section 3 presents the proposed algorithm. The experiment results are presented in Section 4. Finally, a summary is given in Section 5.

2   Related Works

There has been an extensive study in constructing the routing tables during the past few years. The proposals include both hardware and software solutions. In [2], Degermark et al. use a trie-like data structure. The main idea of their work is to quantify the prefix lengths to levels of 16, 24 and 32 bits and expand each prefix in the table to the next higher level. It is able to compact a large routing table with 40,000 entries into a table with size 150-160 Kbytes. The minimum and maximum numbers of memory accesses for a lookup are two and nine, respectively in hardware implementation. Gupta et al. presented fast routing-lookup schemes based on a huge DRAM [3]. The scheme accomplishes a routing lookup with the maximum of two memory accesses in the forwarding table of 33 megabytes. By adding an intermediate-length table, the forwarding table can be reduced to 9 megabytes; however, the maximum number of memory accesses for a lookup is increased to three. When implemented in a hardware pipeline, it can achieve one route lookup every memory access. This furnishes 20 million packets per second (MPPS) approximately. Huang et al. further improve it by fitting the forwarding table into SRAM [4].

Regarding software solutions, algorithms based on tree, hash or binary search have been proposed. Srinivasan et al. [5] present a data structure based on binary tree with multiple branches. By using a standard trie representation with arrays of children pointers, insertions and deletions of prefixes are supported. However, to minimize the size of the tree, dynamic programming is needed. In [6], Karlsson et al. solve the BMP problem by LC tries and linear search. Waldvogel et al. propose a lookup scheme based on a binary search mechanism [7]. This scheme scales very well as the size of address and routing tables grows. It requires a worst-case time of $log_2(address\ bits)$ hash lookups. Thus, five hash lookups are needed for IPv4, and seven for IPv6 (128-bit). The software-based schemes can be further improved by using multiway and multicolumn search techniques [8]. Although these approaches feature certain advantages, however, they either use complicated data structures [1, 3, 5, 8] or are not scalable for IPv6 [2, 3, 6].

3   IPv6-aware Router Design

3.1   Router Architecture

Figure 1 schematically depicts the architecture of the IPv6-aware router. With the advent of the high-speed switch capacity and increased traffic volume, distributed routing architecture has became a significant improvement to achieve high capacity in router design. It mainly consists of a network processor module and forwarding engines interconnected to the switching fabric. The network processor executes the routing protocols, such as BGP and OSPF, and maintains a routing table (RT). In each line card, the forwarding engine employs a forwarding table (FT) to make the routing decision for packet forwarding. The forwarding table is derived from the routing table and contains an index of IP prefix associated with an outgoing interface. This separation is used to ensure the routing instability does not impact the performance of packet forwarding engine.

The conceptual configuration of the FE is shown in Figure 2. For an incoming IP packet, the header verification, header update, and route-lookup process (based on the destination IP address) are initiated simultaneously. If the IP header is not correct, the packet is dropped, and the lookup is terminated. Otherwise, the packet header is updated (TTL decrement and checksum update), and the route-lookup process provides the next hop (port number) where the packet should be forwarded. The MAC-address substitution module then substitutes the source-MAC address and the destination-MAC address of the packet before it is forwarded into the interface port. The bottleneck of the FE is the route lookup process, so that our study focuses on the design of a fast and scalable IP lookup scheme.

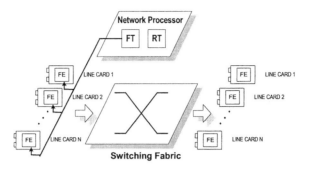

Fig. 1. Architecture of the IPv6-aware router

3.2 TCAM Entry-Reduction Algorithm

With the compelling technical advantages, TCAM based networking devices become a preferred solution for fast, sophisticated IP packet forwarding. The original of the trend is in humble beginnings. The first binary Content Addressable Memories brought to market in the early 1990s and suffered from various limitations. It was too expensive and the software based look-up solutions were adequate to modest traffic-forwarding loads at that time. However, as faster line rates began choking the legacy table lookup solutions, CAM based design became increasingly attractive due to their massively parallel and highly deterministic search characteristics to rapidly perform IP lookup.

The introduction of TCAM opened new possibilities, particularly for the BMP problems. TCAM devices are able to prioritize search results in such a way that multiple search matches, corresponding to different prefix lengths, could be resolved in accordance with BMP requirements. We use an example to explain the TCAM operation, as shown in Figure 3. There are six routing prefixes expressed as the form $< prefix/prefixlength >$. For an incoming packet with destination address "140.113.215.207", the TCAM performs searching within all the prefixes in parallel. Several prefixes may match the destination address (P1

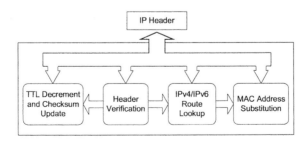

Fig. 2. Function Components of the IPv4/IPv6 forwarding engine

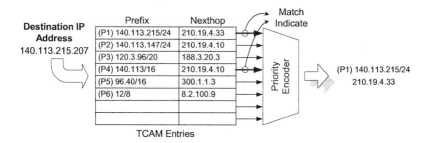

Fig. 3. Solve the BMP problem by sorting the prefixes in decreasing order of lengths

and P4 in this case). A priority encoder then selects the longest matching entry as the result.

Even though the application of TCAM technology is growing gradually, it still comes with some limitations. It is clear that TCAM operates with lower clock rate, much higher power consumption/price and larger package as compared with SRAM. In next generation network, the TCAM will be suffering from a limited number of entries. Table management is another issue of TCAM. As described above, the prefixes in the TCAM are listed in sorted order. However, forwarding tables in routers are dynamic; prefixes can be inserted or deleted due to the changes in network status. These changes can occur at the rate as high as 1,000 prefixes per second [9], and hence it is desirable to obtain quick TCAM updates and keep the incremental update time as small as possible.

Forwarding table updates complicate the keeping of sorted prefixes in the TCAM. This issue is explained with the example in Figure 3. Assume that a new prefix $< 120.3.128/18 >$ is inserted into the forwarding table. It must be stored between prefixes P3 ($< 120.3.96/20 >$) and P4 ($< 140.113/16 >$) to maintain the sorted order. In the worst case, the inserted prefix is compared to all existed prefixes. Furthermore, inserting the prefix in its correct place involve moving other elements. The naive solution is not efficient for a forwarding engine with large amount of entries. Another solution is to keep few locations for following updates. In [9], Gupta et al. proposed a PLO_OPT algorithm. By keeping all the unused entries in the center of the TCAM, each prefix insertion/deletion will cause prefixes swap between different lengths. Hence, the worst-case update time is $W/2$.

The reduction of TCAM entries can improve the TCAM performance in terms of power consumption, price and board usage. Consequently, we introduce a novel TCAM entry-reduction algorithm. The new scheme can reduce the TCAM entries, and also make the current TCAM to support large IPv6 routing table. The basic idea is to merge multiple routing prefixes into one representative prefix. The generated prefixes are inserted into the TCAM and the original multiple prefixes are recorded in the extra memory, such as DRAM. The routing

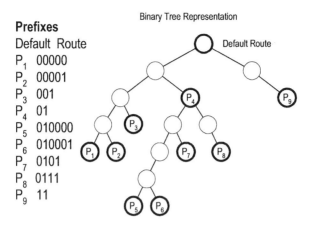

Fig. 4. Representation of the routing prefixes with binary tree

lookup consists of one TCAM and one DRAM access. Although one extra memory access is required, it can use pipeline design to alleviate the performance degradation. However, the system design cost can be reduced dramatically since DRAM is much cheap and power saving.

To begin with the proposed algorithm, we have to construct the binary tree from the routing prefixes. Assume there are 10 prefixes in the routing table, including the default route. The TCAM will need ten entries to record the complete routing information. The binary tree is constructed according to the bit stream of the routing prefixes, as shown in Figure 4.

From the binary tree, we can divide the tree into multiple subtrees. Let the height of the subtrees equal to 2, the binary tree can be divided into four

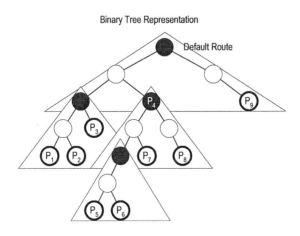

Fig. 5. Four subtrees with a height of two-bit

subtrees, as shown in Figure 5. Each routing prefix will be contained in at least one subtree. Consequently, the four bit streams corresponding to the roots of the subtree are inserted into TCAM, rather than the original prefixes. The detailed algorithm of the subtree construction is listed below. It selects the subtree from the leaves of binary tree. It is because that each leaf (i.e., routing prefix) must be covered. The bottom-up construction can obtain minimum number of subtrees.

Subtree Construction Algorithm:
Input: The root of the constructed binary tree from the routing table
Output: The constructed subtrees and the corresponding bit stream
Constructor *(Parent, Current.Root, Current.Depth);*
Begin
    If (*(the depth of the deepest child in the subtree - Current.Depth) == H)*
       Generate_Subtree (Parent, Current.Root);
    Else
       Constructor(New.Parent, Current.Root— >Left.Child, Current.Depth+1);
       Constructor(New.Parent, Current.Root— >Right.Child, Current.Depth+1);
       If (*(the length of the longest unprocessed prefix - Current.Depth) == H)*
          Generate_Subtree (Parent, Current.Root);
    Endif
End

Routing Lookup: The routing lookup procedure is divided into two parts. The first one is to find out the best match "subtree". The second step is to extract the best match prefix within the subtree. Since the size of the subtree is much smaller, we can expand it completely and put it into the DRAM. In our example, each subtree will be expanded to four entries. The entries of TCAM will indicate their corresponding addresses of subtrees, as shown in Figure 6. Consequently, the follow-up two bits behind the length of matched bits are used to select the coincident entry. For example, we perform the search for address 0101. It will match the P3 in the TCAM whose length is 2 and corresponding subtree is S2. Consequently, the third and fourth bits ('0' and '1' respectively) indicate the second entry in the S2 and the best matching prefix is P2.

Route Update: Since the original routing prefixes have been encapsulated in the subtrees, each route update must enquire the best matching subtree to check whether the updated prefix can be covered. If yes, the related entries in the DRAM will be modified. Otherwise, the routing prefix is inserted into the TCAM directly. After accumulating a certain number of such entries, we can rebuild the subtrees to keep the number of TCAM entries few. The worst-case update time is equal to $max(W/2, 2^{height\ of\ subtree})$.

Routing Lookup for IPv6 Anycasting: Anycasting is a new addressing scheme in IPv6. Each anycast address could indicate a set of servers. Any con-

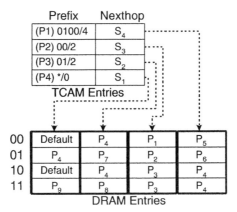

Fig. 6. Reorganize the binary tree into four subtrees with a height of two-bit

nection to the anycast address will be routed to the nearest server. This can help to balance the server load and improve network resiliency. A route to an anycast address can be treated as a host route. Thus the routing lookup for anycasting is exact-matching, rather than best-matching. We could handle the anycast addresses by using an additional CAM (or hash table). The routing lookup for each packet is performed for both TCAM and CAM while the result of CAM has higher priority..

4   Performance Evaluation

Through experiments, we demonstrate that the proposed scheme features much lower TCAM entries with less extra DRAM. Currently, the IPv6 routing tables consists only a few hundred prefixes. To realize the scalability of the proposed scheme, we use the real data available from the IPMA [10] and NLANR [11] projects for comparison, these data provide a daily snapshot of the routing tables used by some major Network Access Points (NAPs). The major performance metrics include the number of generated TCAM entries and the required DRAM storage.

Figure 7 shows the process results for different routing tables. We set the height of the subtree as four and eight. For the largest table with 102271 prefixes (NLANR), it generates 30,260 and 6,010 TCAM entries accompanying with 473 Kbytes and 1.5 Mbytes DRAM, respectively. Intuitively, the number of TCAM entries is reduced with larges subtree, but the required DRAM is also increased as well. Both performance metrics are proportionate to the number of prefixes. By changing the height of the subtree, the TCAM entries and the required DRAM can be adjusted according to the practical environments.

It has been observed that there are a large amount of single-path prefixes in the routing tables [6]. The sparseness caused by the single-path prefixes will make the subtree construction un-efficient. We can adopt the path compression

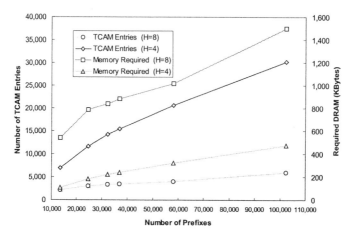

Fig. 7. The performance metrics for different routing tables

to eliminate the single-path prefixes before the subtree construction. As shown in Figure 8, the performance is improved significantly. The TCAM entries can be reduced to as low as 3,536 entries with only 884 Kbytes DRAM (Subtree Height = 8) which shows a dramatically improvement. Note that the lookup result may not be the matched prefixes due to the path compression, thus an extra memory access to its shorter prefix would be necessary.

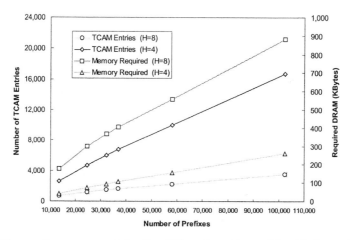

Fig. 8. The performance metrics for different routing tables.(Subtree Height = 4, 8; Path Compression Enabled)

5   Conclusion

This study investigates the related issues in TCAM-based FE design. To make use of the TCAM in IPv6 routing lookup, we need a more efficient approach. The proposed algorithm reduces the number of TCAM entries by merging routing prefixes into subtrees according to the associative positions. The subtrees' roots and their interior information are stored in the TCAM and DRAM, respectively. Each routing lookup procedure consists of one TCAM and one DRAM access which can proceed in pipelining. By adjusting the height of subtrees, we can decide the number of generated TCAM entries and the required DRAM size. If the technique of path compression is applied, both the required TCAM and DRAM can be further reduced. But the scheme will need one extra memory access due to possible "incorrect" match. In the experiments, we illustrate various performance metrics with different settings. In the best results, the proposed algorithm can eliminate 98 percentage of TCAM entries with adding only 2.2 Mbytes DRAM. We also discuss the process for anycast addresses. We believe that this scheme would simplify the design of the IPv6 routers by alleviating the TCAM cost dramatically.

References

[1] Y. Rekhter, T. Li, "An Architecture for IP Address Allocation with CIDR." RFC 1518, Sept. 1993.
[2] M. Degermark, A. Brodnik, S.Carlsson, and S. Pink. "Small Forwarding Tables for Fast Routing Lookups," In Proc. ACM SIGCOMM '97, pages 3-14, Cannes, France, Sept. 1997.
[3] P. Gupta, S. Lin, and N. McKeown, "Routing Lookups in Hardware at Memory Access Speeds," In Proc. IEEE INFOCOM '98, San Francisco, USA, March 1998.
[4] N. F. Huang, S. M. Zhao, and J. Y. Pan, "A Fast IP Routing Lookup Scheme for Gigabit Switch Routers," In Proc. IEEE INFOCOM '99, New York, USA, March 1999.
[5] V. Srinivasan and G. Varghese: Fast IP lookups using controlled prefix expansion. ACM Trans. On Computers, Vol. 17. (1999) 1-40.
[6] S. Nilsson and G. Karlsson, "IP-Address Lookup Using LC-Tries," IEEE JSAC, 17(6):1083-1029, June 1999.
[7] M. Waldvogel, G. vargnese, J. Turner, and B. Plattner, "Scalable High Speed IP Routing Lookups," In Proc. ACM SIGCOMM '97, pages 25-36, Cannes, France, Sept. 1997.
[8] B. Lampson, V. Srinivasan and G. Varghese, "IP Lookups Using Multiway and Multicolumn Search," IEEE/ACM Trans. On Networking, 7(4):324-334, June 1999.
[9] D. Shah and P. Gupta, "Fast updating Algorithms for TCAMs," IEEE Micro Mag., 21(1):36-47, Jan.-Feb. 2001.
[10] Merit Networks, Inc. Internet Performance Measurement and Analysis (IPMA) Statistics and Daily Reports. See $http : //www.merit.edu/ipma/routing_table/$.
[11] NLANR Project. See $http : //moat.nlanr.net/$.

A Constrained Multi-path Finding Mechanism Considering Available Bandwidth and Delay of MPLS LSP*

SangSik Yoon and DeokJai Choi

Department of Computer Science
Chonnam National University
300 YongBong-dong, Buk-gu, GwangJu, South Korea
ssyoon@tyranno.chonnam.ac.kr,dchoi@chonnam.ac.kr
http://tyranno.chonnam.ac.kr/index.html

Abstract. The Multi-Protocol Label Switching(MPLS) technology has been believed to be good for Traffic Engineering(TE) in Internet, which enables effective use of network resources, and maximizes network performance. The Constraint-based Shortest Path First(CSPF), which is the extension of SPF that is used in IP routing protocol, currently finds only one path satisfying the required bandwidth using link's speed as a cost. That method is not proper for the voice traffic and other real-time applications, which require strict guarantee of packet delay. To solve this problem, we propose new path computation method, which considers current delay and available bandwidth simultaneously. If there may exist many paths satisfying the given constraints, the proposed method can find them in the order of low cost of each path. In case of multipath, this method can also distribute traffic load among multiple paths to avoid network congestion. To evaluate the performance of the new proposed method, we compare the result of average available bandwidth rate and average delay of Label Switched Path(LSP)s with that of existing CSPF using our own developed path calculation simulator. Since MPLS TE supports explicit path establishment, by using this new proposed method, Internet Service Provider(ISP)s can optimize network performance and minimize packet delay for severe delay required traffic such as Video on Demand(VoD), Voice over IP(VoIP), etc.

1 Introduction

The increase of Internet user and appearance of various application services result in explosive traffic growth. Among those applications, especially real-time applications(i.e., video conference, VoD, VoIP) require high bandwidth and Quality of Service(QoS) guarantee. MPLS technology was introduced to solve this traffic growth problem by Internet Engineering Task Force(IETF). It integrates layer 3 routing protocol and layer 2 packet switching technology. The

---

* This work is supported by KT Operations Support System Laboratory under grants the project 2002 arts and science services.

H.-K. Kahng (Ed.): ICOIN 2003, LNCS 2662, pp. 56–65, 2003.

ingress Label Switching Router(LSR) may assign a packet to a particular Forwarding Equivalent Class(FEC) just once, as the packet enters the network. The FEC is encoded as a shorted fixed-length value known as a *label*. In the core LSR, the labeled packet is simply forwarded by label swapping[1, 2].

ISPs should solve the rapid growth problem of Internet traffic to survive in the competitive Internet market. One solution would be to increase capacity of link or processing power of router equipment, but this is not best for ISPs since it takes long time and unaffordable expense. Instead of increasing network capacity, it would be more desirable to manage limited network resources and use them efficiently. For example, if a congestion or failure takes place on the specific link, it can be solved by redistributing packets to other links, which do not experience congestion and failure without increasing another network capacity.

From the viewpoint of ISPs, the common goal of TE is to optimize usage of network resources so that they can maximize the network performance[3]. However, it was not so easy to perform TE in Internet, since the existing IP considers only one cost to decide next hop[4]. In order to reroute the traffic from the over-utilized or failed link to other link, the cost of alternative link is adjusted to be the best shortest path. But the adjusting of a specific link cost makes network oscillate and unstable[5].

However, constraint-based routing of MPLS that provides explicit LSP setup can establish alternate path using Constraint-based Routing Label Distribution Protocol(CR-LDP)[6, 7]. Ingress LSR can find a path that satisfies a given constraints using Constraint-based Shortest Path First(CSPF), and if there may exists a path, it establishes explicit LSP using CR-LDP or RSVP-TE[8]. The current CSPF finds explicit path using basically existing SPF method modified by adding constraint parameter[9, 10].

Since the current CSPF computes one path constrained by bandwidth of link only, it is not easy to achieve other TE goal such as minimization of packet delay and maximization of link usage. For real-time traffic that requires strict delay guarantee, it is desirable to consider packet delay constraint in addition to available bandwidth of link.

Another important problem for ISPs is to avoid traffic congestion in a specific link. To do that, it is needed to distribute traffic load among multiple paths. But, since CSPF computes only one path that satisfies required bandwidth, it cannot avoid congestion phenomena by distributing load. Therefore, the current CSPF may need to be modified by finding multi-paths for distributing traffic properly.

The algorithm to find optimal solution under more than two constraints is known to be NP-complete[11]. In this paper, we try to find multi-path(s) heuristic solution, which is close to an optimal solution, satisfying both delay and available bandwidth constraints.

Our new proposed algorithm finds all possible multi-paths that satisfy bandwidth requirement from ingress LSR to egress LSR. After that, it applies delay constraint over the set of found paths one by one. If there is a path that does not satisfy delay constraint, the path should be removed from the selected paths

Table 1. Network parameter definition

Name	Description
R_e	The Reserved bandwidth of link e
A_e	The Available bandwidth of link e
C_e	The Capacity of link e, $R_e + A_e$
ABR_e	The Available Bandwidth Rate of link e, $\frac{A_e}{C_e}$
D_e	The Delay of link e, propagation delay + transmission delay
T_e	The TE cost of link e
$L_{s,t}$	The ordered set of links of LSP from ingress s to egress t
$BL_{s,t}$	The Bandwidth of LSP $L_{s,t}$
$DL_{s,t}$	The Delay of of LSP $L_{s,t}$, $\sum D_e, \forall e \in L_{s,t}$
$ML_{s,t}$	The set of $L_{s,t}$, namely the multi-LSPs from ingress s to egress t
$CML_{s,t}$	The set of Constraints of $ML_{s,t}$

set. For the remained paths, after we apply weights of available bandwidth and delay, finally we decide the distributed bandwidth rate of each path.

MPLS LSP path computation can be done by either on-line or by off-line. In case of on-line computation, it is done by ingress LSR where fast response is required, while the computation should be simple. However, our approach takes long time in large network since it calculates all possible paths. So our method is better suited for off-line path calculation than on-line case. Off-line path computation is used in real world to analyze network status or to provision link which satisfies many constraints[12].

In the next chapter, we define MPLS network model, and propose a new algorithm that computes multi-LSPs considering delay and available bandwidth. In the last, we evaluate the performance of the existing CSPF and new proposed CSPF in terms of average available bandwidth rate and average delay of multi-LSPs.

2   Network Model

MPLS network is represented as a directed graph $G = (V, E)$. V denotes the set of LSRs and E denotes the set of links $e = (u, v)$. In order to describe G in detail, we define some network parameters in Table 1.

The $CML_{s,t}$ contains some constraints of LSP. These constraints are listed in the following;

r_bw :   the required bandwidth of $ML_{s,t}$.
r_color :   the link resource color of $ML_{s,t}$.
r_delay :   the required delay of $ML_{s,t}$.
max_hop :   the maximum hop count of $ML_{s,t}$.
max_path :   the maximum number of multi-path of $ML_{s,t}$.
$metric\_weights$:   the weight of ABR and delay of $ML_{s,t}$.

balance_policy :   the load balancing policy of $ML_{s,t}$.

In order to apply Available Bandwidth Rate(**ABR**) and delay of LSP, we defined *metric_weights* variable. The *metric_weights* of $ML_{s,t}$ is the pair of (α, β) which means each weight of ABR and total delay of $L_{s,t} \in ML_{s,t}$ respectively. The α and β range from 0.0 to 1.0 on condition that $\alpha + \beta = 1$.

For considering both ABR and delay of LSP simultaneously, the normalization of the two values is required. The following (1) and (2) are defined to normalize them between 0 and 1.

$$CABRL_{i,j} = 1 - \frac{(\sum ABRe, \quad \forall e \in L_{i,j})}{sizeof(L_{i,j})} \, . \tag{1}$$

$$MaxDML_{i,j} = Max(DL_{i,j}, \quad \forall L_{i,j} \in ML_{i,j}) \, .$$

$$DRL_{i,j} = 1 - \left(\frac{((MaxDML_{i,j} + 1) - DL_{i,j})}{(MaxDML_{i,j} + 1)} \right) \, . \tag{2}$$

The above $CABRL_{i,j}$ stands for Current Available Bandwidth Rate(**CABR**) of $L_{i,j}$ and $DRL_{i,j}$ means the relative Delay Rate(**DR**) of $L_{i,j}$ within $ML_{i,j}$. Since the ABR and DR is reverse to goodness of $L_{i,j}$, we subtract ABR and delay from 1 respectively. Now we can define the Total Cost(**TC**) function of $L_{i,j}$ which is considered ABR and Delay of LSP simultaneously by using (1) and (2).

$$TCL_{i,j} = \begin{cases} \sum T_e, \quad \forall e \in L_{i,j} & \text{if } (\alpha, \beta) = (0,0), \\ \alpha \cdot CABRL_{i,j} + \beta \cdot DRL_{i,j} & \text{otherwise} \end{cases} \, . \tag{3}$$

If $(\alpha, \beta) = (0.0, 0.0)$, then the path cost is set to summation of all TE-cost of links in $L_{i,j}$. Otherwise, we apply the each weight of (α, β) to $CABRL_{i,j}$ and $DRL_{i,j}$ for total cost of path. By using this $TCL_{i,j}$, we are able to select the best path(s) which cost is small.

The *balance_policy* is used to decide the amount of bandwidth of each $L_{i,j}$ within $ML_{i,j}$. It may have a value either of { No Balance(**N_B**), Cost Equal Balance(**C_E_B**), Bandwidth Equal Balance(**BW_E_B**), or Bandwidth Priority Balance(**BW_P_B**)}. The value of N_B means that the only one single path is selected in any case. In case of the second C_E_B value, for all equal cost $L_{i,j}$ in $ML_{i,j}$, the total bandwidth of $ML_{i,j}$ is distributed into each $L_{i,j}$ with the same rate. For the third BW_E_B value, the bandwidth is distributed with the same rate regardless of cost of $L_{i,j}$, and the last BW_P_B value indicates that the bandwidth of $L_{i,j}$ must be allocated in proportion to the cost of that. For example, if we suppose the number of multi-path(s) is n, and the bandwidth distribution rate of maximum cost path is γ, then the that of second largest cost path is $2 \cdot \gamma$, and the next is $3 \cdot \gamma$, and so on. Finally, the rate of the minimum cost path is $n \cdot \gamma$.

3   A CMPF Mechanism Considering Available Bandwidth and Delay of MPLS LSP

The first key idea of our new proposed Constrained Multi-path Finding(**CMPF**) mechanism is to find all possible multi-paths from ingress to egress, and then distributes load with equal rate or proportional to the cost of each path. All possible paths are simply calculated by the following algorithm.

The pseudo program code of function that finds non-equal cost all multi-LSPs from ingress s to egress t

```
MLs,t find_nec_all_ml( s, t )
{
  Q = {};
  MLs,t = {};
  do {
    if( Q is empty ) /* Is the first iteration? */
      j = s;
    else /* Get first element(path) from Q */
      Ls,j = delete from Q;

    for all k, such that link[e = (j, k)] in E
    {
      Ls,k = add link[e=(j, k)] into Ls,j at the end;
      if( Ls,k is cyclic )
        continue;
      else if( k == t )
        /* a~path is found, add it into MLs,t. */
        add Ls,t into MLs,t;
      else
        add Ls,k into Q;
    }
  } while( Q is not empty );

  return MLs,t;
}
```

The above find_nec_all_ml() function finds all possible paths regardless of path cost using Queue data structure if it does not occur cycle of path.

With the given MPLS network $G = (V, E)$, the proposed CMPF mechanism now can be defined that it finds the set $ML_{s,t}$ from network G, with the input set $CML_{s,t}$ which is constraints of $ML_{s,t}$. The CMPF algorithm is described as the following procedures.

(**Step 1.**) Remove all links e which bandwidth is smaller than $CML_{s,t}(r\_bw)$ or color is not $CML_{s,t}(r\_color)$.

(**Step 2.**) Find the set $P_{s,t}$ of all possible multi-paths from ingress s to egress t using above find_nec_all_ml() function.

(**Step 3.**) Remove $L_{s,t} \in P_{s,t}$ which hop count is greater than $CML_{s,t}(max\_hop)$ or $DL_{s,t}$ is larger than $CML_{s,t}(r\_delay)$.

(**Step 4.**) For all $L_{s,t} \in P_{s,t}$, calculate the total cost $TCL_{s,t}$ with applying (3). And sort the $P_{s,t}$ in ascending order using $TCL_{s,t}$ as key value.

(**Step 5.**) Decide the number of $ML_{s,t}$ and distributed bandwidth of each $L_{s,t} \in ML_{s,t}$ according to $CML_{s,t}(balance\_policy)$. We let $PL_{s,t}$ be the result of (step 4.) and n be the number of final multi-path(s) of $ML_{s,t}$. For each value of $CML_{s,t}(balance\_policy)$, we follow the below procedure.

$$\mathbf{N\_B} : \begin{cases} n = 1; \\ ML_{s,t} = \{PL^0_{s,t}\}; \\ BML^0_{s,t} = CML_{s,t}(r\_bw); \end{cases}$$

$$\mathbf{C\_E\_B} : \begin{cases} n = min(\text{number\_cost\_equal}(PL_{s,t}), CML_{s,t}(max\_path)); \\ ML_{s,t} = \{PL^0_{s,t}, \cdots, PL^{n-1}_{s,t}\}; \\ BML^k_{s,t} = \frac{CML_{s,t}(r\_bw)}{n}, \text{ for } 0 \le k < n; \end{cases}$$

$$\mathbf{BW\_E\_B} : \begin{cases} n = min(\text{sizeof}(PL_{s,t}), CML_{s,t}(max\_path)); \\ ML_{s,t} = \{PL^0_{s,t}, \cdots, PL^{n-1}_{s,t}\}; \\ BML^k_{s,t} = \frac{CML_{s,t}(r\_bw)}{n}, \text{ for } 0 \le k < n; \end{cases}$$

$$\mathbf{BW\_P\_B} : \begin{cases} n = min(\text{sizeof}(PL_{s,t}), CML_{s,t}(max\_path)); \\ ML_{s,t} = \{PL^0_{s,t}, \cdots, PL^{n-1}_{s,t}\}; \\ BML^k_{s,t} = \frac{(n-k) \cdot CML_{s,t}(r\_bw)}{\sum_{m=1}^{n} m}, \text{ for } 0 \le k < n; \end{cases}$$

In the above algorithm, $PL^k_{s,t}$ means the k-th $L_{s,t}$ in $PL_{s,t}$, namely the k-th best path. The $BML^k_{s,t}$ represents the bandwidth of $L_{s,t}$, which is the k-th path in $ML_{s,t}$. The function of number_cost_equal() counts the number of path(s) which cost of path is equal.

In case of BW_P_B value, the bandwidth of each path is set in proportion to its relative priority. However, in special case, if there are two or more equal cost paths in $ML_{s,t}$, the bandwidth of them are allocated to relatively equal value. For example, we assume $ML_{s,t}$ is $\{p_0, p_1, p_2\}$ and path cost are 0.5, 0.5, 0.7 respectively, and required bandwidth of $ML_{s,t}$ is 10Mb. Then the bandwidth of path p_2 becomes $10 \times (1 \div (1 + 2 + 3)) \approx 1.6$Mb, and that of p_0 and p_1 are set to $10 \times (2.5 \div (1 + 2 + 3)) = 4.2$Mb equally, where 2.5 is $(2 + 3) \div 2$.

Let the average number of out-degree for all routers be $AvgLink(G)$. Then the time complexity of our new proposed algorithms is $O(AvgLink(G)^{|V|})$. So our proposed algorithm must be suitable for finding path in off-line TE system.

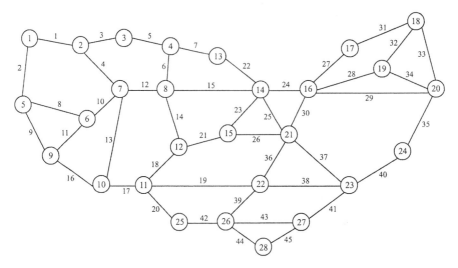

Fig. 1. The USA long haul network topology for simulation

A network operator can figure out their problem that both of link usage and delay are considered simultaneously with our method.

4   Performance Evaluation

To evaluate the our proposed CMPF, we use the two performance parameters. The one is Average Delay(**AvgD**) of $ML_{s,t}$ and the other is Average Available Bandwidth Rate(**AvgABR**). The $ABRL_{s,t}$ is the rate of available bandwidth after establishing $ML_{s,t}$ into the current MPLS network.

$$AvgD(ML_{s,t}) = \frac{\sum(DL_{s,t}, \quad \forall L_{s,t} \in ML_{s,t})}{sizeof(ML_{s,t})} . \tag{4}$$

$$ABRL_{i,j} = \frac{[\sum(1 - \frac{R_e + BL_{i,j}}{C_e}), \quad \forall e \in L_{i,j}]}{sizeof(L_{i,j})} . \tag{5}$$

$$AvgABR(ML_{s,t}) = \frac{\sum(ABRL_{s,t}, \quad \forall L_{s,t} \in ML_{s,t})}{sizeof(ML_{s,t})} . \tag{6}$$

For simulation, we have developed our own multi-path finding simulator in C language. For simulation network, we have used USA long haul network topology shown in Fig. 1. The number of LSRs in simulation network is 28, and the average number of link for all LSRs is about 3.2. And we suppose that all links are capable of TE, and delay of link is only considered propagation delay.

The simulation network is configured with the following Table 2. We basically configured all links according to the value of each link number %(modulo) 3 = {0, 1, 2}. The reserved bandwidth and delay of all links are set randomly.

Table 2. The configuration of simulation network

Link i%3	TE-Cost	BW(Mb)	Rersv_BW(Mb)	Color	Delay(us)
0	1	10	rand(1, 2, 3)	1	rand(10, 20, 30)
1	2	8	rand(1, 2, 3)	1	rand(10, 20, 30)
2	3	6	rand(1, 2, 3)	1	rand(10, 20, 30)

In the USA long haul simulation network, we have simulated for four LSPs($LSP_{1,14}$, $LSP_{5,20}$, $LSP_{9,16}$, and $LSP_{10,18}$). The required bandwidth of all LSPs is 4Mb, the delay is $200\mu s$, the maximum hop count is 30, and the maximum number of multi-path is 4. For each LSP, we calculated multi-path(s) with our proposed CMPF algorithm in case of (α, β)={(0, 0), (1, 0), (0.9, 0.1) $\cdots$ (0.1, 0.9), (0, 1)} and all *balance_policy* values. Table 3 shows the result of AvgABR and AvgD of $LSP_{9,16}$ after simulation.

Figure 2 is showing the average value of AvgABR after the simulation of four LSPs. In case of $(\alpha, \beta) = (0.0, 0.0)$, our proposed method results in better AvgABR than that of N_B and C_E_B. For AvgABR, our method has good performance with the sequence of N_B, C_E_B, BW_E_B and BW_P_B. Especially, if (α, β) is (1.0, 0.0), our new proposed method results in the best performance.

Figure 3 shows average value of AvgD after simulation of four LSPs. The CMPF method has longer AvgD as you can see in Fig. 3. The reason why AvgD was long is that our CMPF selected multi-paths which delay might be more longer than that of N_B and C_E_B for load balancing. Although AvgD was a little long, the overall result of simulation is that we have got more higher ABR than the existing CSPF method. If we consider only ABR without delay, the delay value is very high.

5   Conclusion

In this paper, we proposed a new Constrained Multi-path Finding(CMPF) method, which can consider available bandwidth and delay of MPLS LSP simultaneously. With proposed CMPF methd, network operator can find the best multiple LSPs, which optimize network performance and minimize packet delay. Also the

Table 3. The simulation result of $LSP_{9,16}$

	N_B		C_E_B			B_E_B		B_P_B	
(α, β)	ABR	Delay	ABR	Delay	# of Path	ABR	Delay	ABR	Delay
(0, 0)	30.00	90	36.25	95	2	43.73	100	44.16	100
(1, 0)	38.06	180	38.06	180	1	53.17	165	54.45	165
(0.5, 0.5)	30.00	90	30.00	90	1	46.42	95	46.66	95
(0, 1)	30.00	90	41.93	90	2	46.40	95	46.67	95

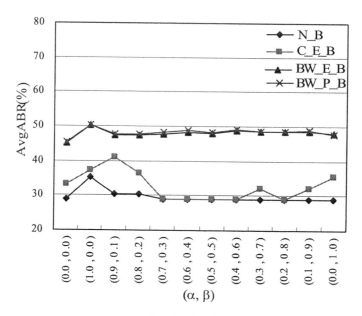

Fig. 2. The average value of AvgABR of $LSP_{1,14}$, $LSP_{5,20}$, $LSP_{9,16}$, and $LSP_{10,18}$

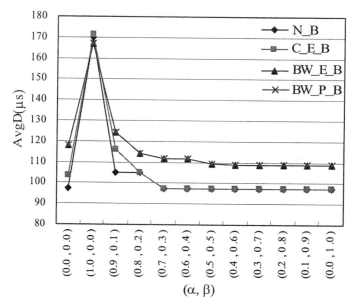

Fig. 3. The average value of AvgD of $LSP_{1,14}$, $LSP_{5,20}$, $LSP_{9,16}$, and $LSP_{10,18}$

proposed method has a function that distributes traffic load among multiple paths in three ways(C_E_B, BW_E_B, BW_P_B) to avoid network congestion.

In order to simulate our CMPF method, we have developed CMPF simulator with C language. After simulation with our simulator, the result of CMPF method is better than that of existing CSPF. Especially, the proposed CMPF method has a good advantage of considering multiple constraints that are available bandwidth and delay of LSP.

Since we have used to find all possible multi-paths, the time complexity of the current CMPF algorithm becomes $O(AvgLink(G)^{|V|})$. However, when the size of network is small(i.e., $| V |< 30, AvgLink(g) < 4$), we expect that our CMPF method may be used to find explicit multi-paths for network optimization in off-line TE system.

References

[1] Eric C. Resen, Arun Viswanathan, Ross Callon: A Framework for Multiprotocol Label Switching. RFC 3031, Jan. 2001
[2] Vivek Alwayn: Advanced MPLS Design and Implementation. CCIE #2995, Sept. 2001
[3] Daniel O. Awduche et al: Requirements for Traffic Engineering Over MPLS. RFC 2702, Sept. 1999
[4] J. Moy: OSPF Version 2. RFC 2328, Apr. 1999
[5] Daniel O. Awduche et al: Overview and Principles of Internet Traffic Engineering. RFC 3272, May 2002
[6] Bilel Jamoussi et al: Constraint-Based LSP Setup using LDP. RFC 3212, Jan. 2002
[7] Dave Katz, Derek Yeung: Traffic Engineering Extensions to OSPF. Inernet draft, Sept. 2000
[8] Daniel O. Awduche et al: RSVP-TE: Extensions to RSPF for LSP Tunnels. RFC 3209, Dec. 2001
[9] Cisco: MPLS Traffic Engineering. http://www.cisco.com/univercd/cc/td/doc/product/software/ios120/120newft/120t/120t7/te120_7t.pdf
[10] Chuck Semeria, Marketing Engineer: Traffic Engineering for the New Public Network. http://www.juniper.net/techcenter/techpapers/200004.html, Juniper Networks, Inc.
[11] Don Fedyk, et al: Multiple Metrics for Traffic Engineering with IS-IS and OSPF. Internet draft, Nov. 2000
[12] Xepeng Xiao et al: Traffic Engineering with MPLS in the Internet. IEEE Network Mar. 2000

A High Speed ATM/IP Switch Fabric Using Distributed Scheduler*

Ali Mohammad Zareh Bidoki, Nasser Yazdani,
Sayed Vahid Azhari, and Siavash Samadian-Barzoki

Router Lab., ECE. Dept., Faculty of Engineering
Univ. of Tehran
Tehran, Iran
Zare_b@ece.ut.ac.ir
yazdani@ut.ac.ir
vahidazhari@yahoo.com
s.samadian@ece.ut.ac.ir

Abstract. An increasing number of high performance IP routers, LANs and Asynchronous Transfer Mode (ATM) switches use crossbar switches in their backplanes. Most of these systems use input queuing to store packets. Obviously, we will encounter Head of Line (HoL) blocking in input queues which decrease throughput to 56.8% under uniform traffic. In this paper, we develop a switch fabric with 16-input/output ports using crossbar architecture with internal partitioned memory in the feedback path. Our design also uses virtual output queuing (VOQ) with adaptive intelligent scheduling in each input port. The system distributes the scheduling algorithm to reach higher aggregated throughput. The switch fabric is internally non-blocking and avoids HoL by using VOQs. It is shown in the paper that its throughput and delay are very well compared to the current high performance switch fabrics. The architecture supports full Multicasting. By using bit-slicing the total capacity of the switch can be extended to 640 Gb/s. Distributed scheduling well maximizes the capacity and processing speed of the switch. Simulation results show that it is possible to reach 100% throughput with our design.

1 Introduction

We have experienced profound transition in the field of data communication in the last decay. Birth of World Wide Web has created an interesting media for business, communication, entertainment, research, etc. It has caused data explosion in Internet such that the amount of data doubling in every six months. On the other hand, there has been exponential growth in the communication link rate. We have come long way in a short time from the initial LAN technology of 10Mbps to the current rate of 10Gbps. All these dictate that the previously used techniques are inadequate to process and switch data in the global network. With

---

* This work has been partially supported by Iran Telecommunication Research Center(ITRC).

H.-K. Kahng (Ed.): ICOIN 2003, LNCS 2662, pp. 66–75, 2003.
© Springer-Verlag Berlin Heidelberg 2003

this trend, we need high capacity switch fabrics to handle the current and future volume of data in communication infrastructure. With development of Asynchronous Transfer Mode (ATM) technology, there have been many attempts to design and develop cell switches with high capacity [19]. For a while, everybody in the data communication and computer network realm thought that ATM will set fire to the world. It sounds this is not happening. Instead, IP technology is getting more ground every day. In an attempt to take advantage of cell-switching capacity, ATM switches and IP protocol (IP) routers have been recently merged by using the ATM switching power in hardware and the flexible functionality of IP as network layer protocol . The idea is already being carried out with cell switches forming the backplanes of high speed IP routers [11], [14]. With explosive growth of Internet itself, these imply that we need more capable switches to keep pace with current data communication trend.

There are two main alternatives for queuing in ATM switches, input queuing and output queuing. In output queuing approach, a shared memory is employed for storing all packets arriving to the switch. A shared memory switch [2], [6], can easily support IP packet switching by performing segmentation and reassembly. In addition, in this architecture multicast packets can be handled efficiently. The main drawback of shared memory architecture is its bandwidth requirement due to sharing memory among all inputs. To remedy this, input queuing (IQ) architectures have been evolved.

Input queued switches buffer packets as they enter input port processors. Therefore, the memories need only to operate at the line rate. Most IQ switches utilize a crossbar as their switching element. A crossbar switch establishes a one to one matching between each input and output ports. Unfortunately only one input (output) can send (receive) packets to (from) an output (input). This is called the input/output contention. IQ switches, which usually use FIFOs at their inputs, suffer from a Head of Line blocking (HoL) problem. This problem can be solved by using Virtual Output Queuing (VOQ), in which there is a separate queue in each input port for each output [16].

Input queued switches need an arbitration algorithm to schedule packets. Such a scheduler should receive requests from all VOQs and find a match between inputs and outputs. This algorithm is mostly implemented in a central manner, i.e., all the scheduling decisions are made by one scheduler. There are a number of central scheduling algorithms presented in literature like iSLIP [9], PIM [1]. Clearly, the arbitration process can be a bottleneck, which reduces the switching speed. In an attempt to remove this bottleneck, a distributed arbitration algorithm can be used.

In this paper, we propose a new switch architecture called UTS (University of Tehran Switch) that uses distributed crossbar for internal arbitration. To overcome output contention, the architecture uses a shared memory consisting of N separated memories where N is the number of input ports. Any cell unabling to exit switch due to contention recirculates in shared memory and feedbacks to the crossbar. The shared memory is partitioned into N separated memories to prevent any port with a heavy load to lockout others. To avoid HoL, we use

Virtual Output Queuing (VOQ) [16] in input buffers since input buffer switches have higher capacity. We also use N pieces of memory at outputs inside the switch. Our simulation results show that by having a memory in each output port which works two times the speed of line, we can achieve close to 100% throughput and zero loss rate. To reduce the cell loss rate, both shared memory and memories in outputs report their statuses to input ports. Our switch supports multicast packets without any overhead. The switch supports variable length IP packets and its throughput is quite independent of burst length. Our design combines good features of both IQ and shared memory output queued switches in order to achieve the lowest possible delay and memory bandwidth requirement.

The rest of the paper is organized as follows: In Sect. 2, first, we review the related work in distributed scheduling. In Sect. 3, our architecture is presented. Sect. 4 contains simulations results for various traffic types. Multicasting will be treated in Sect. 5. Sect. 6 briefly considers the implementation of the switch. Finally, we conclude in Sect. 7.

2   Background and Related Work

In central scheduling such as iSLIP [9] and PIM [1], each input port sends its requests to a central scheduler. The scheduler has to arbitrate between these requests and choose a set of non-conflicting matches. Generally, in this scheme, the scheduler is a bottleneck reducing switching speed and capacity. Clearly, a distributed arbitration algorithm can remove this bottleneck. There are various ways to distribute the arbitration process including distributing the decision between all cross points in a crossbar architecture.

ISLIP is a centralize scheduler which uses modulo N round-robin arbiters, one for each input and one for each output. Each arbiter maintains a pointer indicating the element currently having the highest priority.

We review different proposed solutions to a distributed Scheduler available in the literature [13], [3], [4] in the following. We also discuss the advantages and disadvantages of these architectures in order to shed more light on our design and novelty of its architecture.

At first, we explain the shared memory switch fabric with VOQs in input ports. In this architecture, a shared memory resides in the switch fabric [13]. The memory is shared between all outputs and has been logically partitioned into N pieces. Each partition belongs to one output and includes N separate queues, one corresponds to each input. It does this in order to reassemble IP packets coming from input ports. When a cell arrives at input port i destined for output port j, it is placed into the jth VOQ (VOQ_i^j). Each input contains a separate scheduler making a decision independently. To avoid packet loss inside the switch, the switch must inform inputs when shared buffers are close to full so that the inputs stop sending new cells. To do this, the shared memory grants the permissions to the inputs for sending data with a memory grant signal. When the current size of memory is below a predefined threshold this signal is set.

The proposed solution is distributed and has no HoL problem. Furthermore, it does not need reassembly in outputs so it is appropriate for IP packets. However, this architecture has problem such as not being scalable due to the shared memory nature. It has also problem with multicasting.

The second solution for distributed scheduler is using buffered crossbar switch fabric with VOQ. In the architectures proposed in [3], [4] the switch fabric uses N^2 separate memories one for each cross point, each capable of holding an Ethernet packet with size of 1500 bytes. Each input sends its packets independently to the switch fabric. Nevertheless, the architecture needs a large number of memory partitions and the number of ports can hardly be scaled. It has also difficulty with multicast packets. Finally, its memory utilization is low.

3   University of Tehran Switch (UTS)

Input buffered switches have usually higher capacities. Then, we use them to reach the fourth goal in the above list. Some shared memory switch fabrics such as Washington switch fabric [17] have been fabricated at Tb/s (Terabit per second) using Multistage Interconnection Network (MIN). However, generally, these switches are complex, non-scalable, and costly.

In order to have a distributed scheduler, inputs must be able to send their data independently. Therefore, to prevent packet loss, we need to put some amount of memory inside the switch in contrast to central schedulers like iSLIP and PIM. This memory creates a feed back path for cells unable to go out in the first round. A feed back based switch with N inputs and N outputs constitutes of $N^2(R+1)$ cross points (CX). (R+1)*N is the number of crossbar rows needed to feedback packets in the shared memory and route them again, until they exit the switch fabric. The dotted rectangle in Fig. 1 shows a general view of the architecture.

Upon arriving packet, it is initially sent to the corresponding output port if the port is free. If the port has been reserved by another input port, it is sent to the shared memory in order to be feedbacked in the next cell times. Finally, if there was no room in the shared memory, the packet is dropped.

Unfortunately, this switch has two main problems. First, its loss rate is high. We can fix this problem by increasing the value of R (2.5 at minimum) [7] or by using MIN. Unfortunately, both solutions have drawbacks. Increasing R enlarges the crossbar and increases the total system delay. Meanwhile, due to the limitation in the amount of on-chip memory, it is impossible to reach zero loss rate. We have simulated the loss rate of this architecture with respect to different values of R. The loss rate decreases with increasing R and the loss rate for R=3 is at least two times less than the loss rate for R=2. Using MIN, while increasing delay, requires fast and efficient algorithms for configuring switch and setting paths at the high line speed.

We propose using a combination of shared memories and input buffers to reduce the loss rates. The reader can refer to Fig. 1 for a general view of the design.

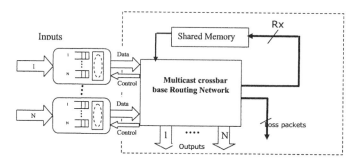

Fig. 1. Combining a feed backed crossbar switch with VOQ buffers

Unfortunately, this does not eliminate cell loss completely. The reason lies in the distributed nature of the scheduler. For instance, when each input sends a packet to output number j during the interval [t, t+3] where t is t-th cell time, some packets will be dropped at time t+4 since the shared memory capacity can be at most R*N (R=3 in this case). Putting buffer in input reduces cell loss, but on the other hand, it also reduces throughput. We can deal with this problem with VOQ discussed later in this section. We can improve cell loss rate problem by placing a memory at each output port inside the switch that works twice the input line speed. Our simulation results show we can achieve even zero loss, and still, maintain distributed management by controlling the size of the memory partitions. This requires each output informs its status to all inputs. One of the remaining problems is the shared memory, which is located in the feedback path. This problem has been solved by using R*N separate. In the scheduling mechanism inside the crossbar, arbitration is done in a RR fashion. Therefore, if all inputs want to send packet to a specific output, only two of them can go and the others must wait. Since each cross point has room only for two cells, the other cells are routed to the shared memory and to the feedback path.

Taking R=3 increases delay. Fortunately, we can reach a good delay and throughput using R=1 as the simulation results show. We discuss this in the next section. To be fair when the size of the memories in the feedback path is equal, the same RR method is used as well.

Simulation results show that the maximum depth of memory partitions in the feedback path would be better to be 5 and for the output memories to be 32. As we discussed previously, a feedback switch leaves to inputs to decide which cell to forward independent of each other. Another important thing is buffer management in inputs. The simplest method is FIFO. However, as we explained before, HoL prevents the throughput to get higher than 60% [5]. Then, the method used here is Virtual Output Queuing. We explain different VOQ selection algorithms in the next subsection.

In the proposed switch architecture, totally, $2N+1$ signals arrive at each input, N signals from the corresponding CX row, N signals from output memory partitions and 1 from the feedback memory partition. Totally, N^2+N+1 signals go out from the switch to the input ports. Each input independently selects (using an O(N) arbiter) the VOQ which may forward a packet.

To ensure fairness across VOQs, we consider four alternative policies: round robin, least recently used (LRU), longest queue first (LQF), and oldest queue first (OQF). Currently, we only consider using RR because its implementation is easier.

In Round-robin (RR) selection method, the outputs are sorted according to increasing output number. Each input i, maintains a pointer to the VOQ_j that was served most recently, say VOQ_i^j starting from $VOQ_i^{(j+1)modN}$. Arbiter selects the next VOQ based on the following criteria:

1. At least one packet is waiting in the queue.
2. Output-queue grant has been received.
3. Feedback-queue grant has been activated.
4. It doesnt have a failure (could not send out a packet in this queue) in the last cycle.

Using a special programmable priority encoder (PPE), the scheduler selects next cell to send in a constant time (about 1.46ns for 16x16 switches) [8], [16].

Our proposed architecture is distributed and eliminates the HoL problem by using VOQ. It is also very simple for implementation in hardware. There is no need to perform packet reassembly in outputs. It supports IP packets and full multicasting as explained later. It scales well to higher speed (up to 100 Gb/s).

4   Simulation Results

The proposed architecture is fairly complex and analyzing its performance even under simple traffic patterns is very difficult. Simulations using a software model of the architecture have been carried out in order to study different system parameters like throughput and delay. The simulation programs work based on discrete event. All programs have been written in C and implemented on Visual C++ compiler. All traffic types are identically distributed across all inputs, and for bursty traffic the destinations are uniformly distributed across all outputs unless stated otherwise. In the following, we study the impact of different system parameters, both in isolation and combination with each other on our architecture.

First we compare the proposed architecture to the conventional FIFO input queues. A system with FIFO performs poorly under all conditions due to HoL blocking. If FIFOs are replaced by VOQ input queues, the delaythroughput characteristic will greatly improve. This can be explained by realizing the fact that using VOQ prevents HoL blocking. Fig. 2 compares the effect of using VOQs to FIFO at input ports under independently and identically distributed (i.i.d.) and bursty traffic.

Using VOQ instead of FIFO at input makes switch more stable and less sensitive to the burst length, i.e., for different burst lengths the throughput of the switch will remain quite the same. Fig. 3 verifies this statement. As can be

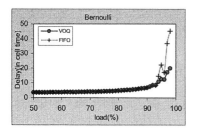
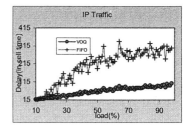

Fig. 2. The effect of FIFO versus VOQ in input ports

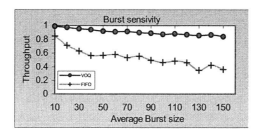

Fig. 3. The effect of size of burst on throughput

seen there, when using FIFOs at inputs, throughput decreases to a low value of 35%. However, using VOQs increases throughput to the minimum 90% in the worst case. The reason behind this is that with FIFOs, the HoL problem will affect more the switch behavior as the burst length increases, whereas with VOQs the HoL problem is eliminated altogether.

In Fig. 4, we have sketched average cell delay versus load for various switch sizes, the number of inputs or outputs. It can be observed that for i.i.d. traffic, average cell delay is independent of switch size. On the other hand, for bursty traffic, the average delay will increase by increasing the switch size. This is due to the fact that more contention happens between various input/output ports.

Fig. 5 compares the performance of the proposed switch with an ideal output-queue switch with infinite output memory for three different switch sizes (N=16,

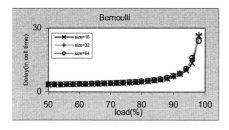
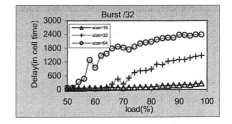

Fig. 4. The effect of switch size on delay in switch by Bernoulli and burst traffic

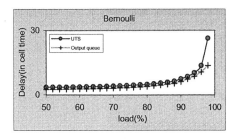
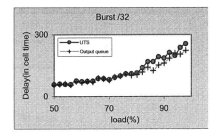

Fig. 5. Comparing the UTS with ideal output queue switch

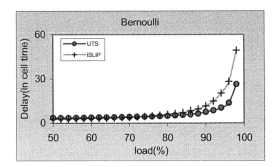

Fig. 6. comparing the UTS with input queue switches like iSLIP

32 and 64) and various traffic types, IP Bursty/32. It is apparent from the curves that UTS architecture performs quite close to the ideal output queued. This is very promising result.

As pointed out above, our approach is essentially Virtual Output Queuing with distributed scheduling. Therefore, we would like to compare its performance to approaches that use VOQ with centralized scheduling (Fig. 6). As a representative of this class, we selected the iSLIP scheduler with log_2N iterations, which achieves a very good tradeoff between complexity and performance [9]. As the Fig. 6 shows, in Bernoulli traffic, our system acts better than iSLIP. In burst, it is like iSLIP.

5   Supporting Multicast

One of the main achievements of our architecture is its efficient and simple support for multicasting. This feature is due to the distributed scheduling nature of the architecture that either forwards a multicast cell to the free outputs, or recirculates the packet back to the feedback path in order to wait for another chance. Thus; there is no need to perform any provisioning in the input ports for multicast packets. The whole process is handled independently and efficiently by the cross-points and the feedback mechanism. However, it is very difficult to support multicast in iSLIP scheduler unless some preprocessing is done. One

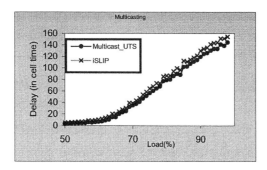

Fig. 7. Multicasting in UTS and iSLIP

approach used in iSLIP is fanout splitting [10], which splits one multicast packet to different queues according to its destinations among corresponding VOQs at the input. Needless to say, fanout splitting introduces some complexity in the input port design.

In Fig. 7, we have compared our architecture to iSLIP under multicast traffic. In our simulation, multicast packets represent 50% of the total packets and have 2 destinations. It can be observed there that the delay of our approach is less than iSLIP.

6   Scalability

The UTS switch is implemented as bit-sliced fashion. Each fabric is 16x16 with 10Gbps inputs line rate and is operating at 300MHz. Using a high-speed serial communication channel [15], the number of pins to interconnect port processors to the fabric can be highly reduced. The width of the interface of each line card to the fabric is 3 bits wide. Up to 4 bit-slices can be stacked on top of each other resulting in 40Gbps line rate and 640Gbps total capacity. The aggregate bandwidth is 1.2Tbps considering input/output. The number of inputs can also be doubled by interconnecting 4 UTS in square like fashion.

7   Conclusion

We proposed UTS switch by combining crossbar and shared memory architectures with a feedback memory mechanism, which increases throughput and decreases packet loss. UTS architecture takes advantages of a distributed scheduler, which decreases packet delay compared to existing IQ switches. Using VOQs at input ports reduces sensitivity to the burst length. The switch supports multicast packets quite efficiently without any provisioning in the input ports. The switch is also considered to be able to operate at 1.2Tbps with 40Gbps line cards using a bit-sliced architecture. Currently, we are trying to implement this switch on ASIC and also on FPGA (Virtex-II pro [18]).

References

[1] Anderson, T., et al. : High speed switch scheduling for local area networks, ACM-Trans. on Computer Systems.(1993) 319-352

[2] Andersson, P., et al.: A VLSI Architecture for an 80 Gb/s ATM Switch Core, in Proc. 8th Annual IEEE Intl Conf. Innovative System in Silicon, Austin, TX (1996) 9-11

[3] Yoshigoe,K., et al.: A parallel- polled Virtual Output Queued Switch with a Buffered Crossbar.(2001)

[4] Del Re, E. , et al.: Performance Evaluation of Input and Output Queueing Techniques in ATM Switching Systems, IEEE Trans. Commun.,vol. 41, no. 10 (1993)

[5] Karol, M., et al.: Input versus output queueing on a space division switch, IEEE Trans.Communications, vol. 35, no.12, (1988) 1347-1356

[6] Katevenis, M., et al. : ATLAS I: A General-Purpose, Single-Chip ATM Switch with Credit-Based Flow Control, in Proc. IEEEHot Interconnects IV Symposium, Stanford, CA, (1996) 15-17

[7] Kim,K., et al.: MASCON:A single IC solution to ATM Multi-Channel switching with Embedded Multicating,IEEE/ACM Trans,networking (1997)

[8] Mckeown, N., Gupta, P.: Design and Implementation of a Fast Crossbar Scheduler (2000)

[9] Mckeown, N. W., Scheduling Algorithms for Input-Queued Switches, Ph.D. Thesis, University of California at Berkeley (1995)

[10] Mckeown, N. and B. Prabhakar : Scheduling Multicast Cells in an Input-Queued Switch, in Proc. INFOCOM 96, San Francisco, CA, (1996) 271-278

[11] Mckeown, N., et al. : The Tiny Tera: A Packet Switch Core, IEEE Micro Magazine, Jan.-Feb.1997, . Packet Switch Core Hot Interconnects V, Stanford University, (1996) 26-33

[12] Mckeown, N. :The iSLIP Scheduling Algorithm for Input-Queued Switches, IEEE/ACM Trans. Networking, vol. 7, no. 2, (1999) 188-201

[13] Minkenberg, C. J. A.: An Integrated Method for Unicast and Multicast Scheduling in a Combined Input- and Output-Buffered Packet Switching System, pending patent application CH (1999) 8-1999-0098

[14] Partridge, C., et al. : A fifty gigabit per second IP router, To appear in IEEE/ACM Transactions on Networking (1998)

[15] Sidiropoulos, S. , et al. : Current Integrating Receivers for High Speed System Interconnects, IEEE Custom Integrated Circuits Conference, (1995)

[16] Tamir, Y. and G. L. Frazier, Dynamically Allocated Multi-Queue Buffers for VLSI Communication Switches, IEEE Trans. Computers, vol. 41, no. 6, (1992) 725-737

[17] Turner, J. S. :Terabit Burst Switching, Washington University Technical Report, (1998) WUCS-98-17

[18] Virtex-II Pro data sheet, htpp://www.xilinx.com

[19] Tobagi, F. : Fast Packet Switch Architectures for Braodband Integrated Services Digital Networks, in Proc. IEEE, vol. 78, no. 1, (1990) 133-167

Scalable Packet Classification for IPv6 by Using Limited TCAMs

Chia-Tai Chan[1], Pi-Chung Wang[1], Shuo-Cheng Hu[2],
Chung-Liang Lee[1], and Rong-Chang Chen[3]

[1] Telecommunication Laboratories
Chunghwa Telecom Co., Ltd.
7F, No. 9 Lane 74 Hsin-Yi Rd. Sec. 4, Taipei, Taiwan 106, R.O.C.
{ctchan,abu,chlilee}@cht.com.tw
[2] Department of Info. Management
Ming-Hsin University of Science and Technology
1 Hsin-Hsing Rd. Hsin-Fong, Hsinchu, Taiwan 304, R.O.C.
schu@mis.must.edu.tw
[3] Department of Logistics Engineering and Management
National Taichung Institute of Technology
No. 129, Sec. 3, Sanmin Rd., Taichung, Taiwan 404, R.O.C.
rcchens@ntit.edu.tw

Abstract. It has been demonstrated that performing packet classification on a potentially large number of filters on key header fields is difficult and has poor worst-case performance. To achieve fast packet classification, hardware support is unavoidable. *Ternary content-addressable memory* (**TCAM**) has been widely used to perform fast packet classification due to its ability to solve the problem in O(1) time without considering the number of entries, mask continuity and their lengths. As compared to the software-based solutions, the TCAM can offer sustained throughput and simple system architecture. It is attractive for packet classification, especially for the ultimate IPv6-based networks. However, it also comes with several shortcomings, such as the limited number of entries, expansive cost and power consumption. Accordingly, we propose an efficient algorithm to reduce the required TCAM by encoding the address portion of the searchable entries. The new scheme could encrypt the 128-bit prefixes of the real-world IPv6 routing tables into 11 bits and still keeps the property of CIDR.

1 Introduction

The major obstacle for the high-speed router ties to the relatively slow Internet lookup, including routing lookup and packet classification. For an incoming packet, a router must perform routing lookup to forward packets toward their destinations based on the information gathered by the routing protocols. In next generation networks, the new services, such as firewall processing, RSVP style resource reservation policies, QoS Routing, and normal unicast and multicast forwarding, require more discriminating forwarding called packet classification.

H.-K. Kahng (Ed.): ICOIN 2003, LNCS 2662, pp. 76–85, 2003.

It allows service differentiation because the router can distinguish traffic based on source/destination address, TCP/UDP port numbers, and protocol flags. Consequently, each packet is distinguished according to the policies (or filters). The forwarding database of a router consists of a potentially large number of policies. Each policy has a given cost. The header of the incoming packet might match multiple policies. The policy with least cost will be used to forward the packet.

To perform packet classification on a potentially large number of policies on key header fields is difficult and has poor worst-case performance. Unlike the routing prefixes, the policy could be un-continuously masked and the length of policy is much longer than that of routing prefix. The routing lookup problem is just a special case of packet classification. As a result, the search of least cost policy (LCP) may be time consuming for a backbone router with a large number of table entries. The exponential growth of the Internet hosts has further stressed the routing system. It is difficult for the packet-forwarding rate to keep up with the increased traffic demand.

1.1   Problem Statement

Essentially, packet classification is a problem of multi-dimensional range match. To describe the problem formally, we have to define the classifier and the policy. A classifier maintains a set of policies to divide an incoming packet stream into multiple classes. A policy $F = (f[1], f[2], \ldots, f[k])$ is called k-dimension if the policy consists of k fields, where each $f[i]$ is either a variable length prefix bit string, a range or a explicit value of the packet header. A policy can be any combination of fields of the packet header, the most common fields are the IP source address, the IP destination address, the protocol type, port numbers of source/destination applications and protocol flags. A packet P is said to match a particular policy F if for all i, the i_{th} field of the header satisfies the $f[i]$. Each policy has an associative action. For example, the policy $F = (140.113.*, *, UDP, 1090, *)$ specifies a rule for flows which address to the subnet 140.113 use the progressive networks audio (PNA) and the action of the rule may assign the packets belonged to these flows with higher queueing priority. Besides the action, the policy is usually given a cost value to define the priority in the database. The action of the least-cost matched policy will be used to process the arriving packets, thus the packet classification problem is a least-cost problem.

1.2   Existing Approaches

Recently, several packet classification algorithms have been proposed in the literature [2, 3, 4, 5, 6]. It can be categorized into following classes: linear search/caching, hardware-based, grid of tries/cross-product, recursive-flow classification, and hash-based solutions. In the following, we briefly described the main properties of these algorithms. Assume that N is the number of the policies, D is the number of classified fields and W is the length of IP address.

Linear Search/Caching: The simplest approach for packet classification is to perform a linear search through all the policies. The space and time complexity is $O(N)$. Caching is a technique often employed at either hardware or software level to improve performance of linear search. However, performance of caching is critically dependent on having large number of packets in each flow. Also, if the number of simultaneous flows becomes larger than cache size, the performance degrades severely.

Bit-Parallelism: Another scheme that relies on very wide memory bus is presented by Lakshamn *et al.* [5]. The algorithm reads Nk bits from memory, corresponding to the BMPs in each field and takes their intersection to find the set of matching policies. Memory requirement for this scheme is $O(N^2)$. This scheme relies on heavy parallelism, and requires significant hardware cost, not to mention that flexibility and scalability of hardware solutions is very limited.

Grid of Tries/Cross-Product: Specifically for the case of 2-field policies, Srinivasan *et al.* [2] presented a trie-based algorithm. This algorithm has memory requirement $O(NW)$ and requires $2W - 1$ memory accesses per policy lookup. Also presented in [2] is a general mechanism called cross-product which involves performing the BMP lookups on individual fields, and using a pre-computed table for combining results of individual prefix lookups. However, this scheme suffers from a $O(N^k)$ memory blowup for k-field policies, including k = 2 field policies.

Recursive-Flow Classification: Gupta *et al.* presented an algorithm, which can be considered as a generalization of cross-product [3]. After BMP lookup has been performed, recursive flow classification algorithm performs cross-product in a hierarchical manner. Thus k BMP lookups and $k - 1$ additional memory accesses are required per policy lookup. It is expected to provide significant improvement on an average, but it requires $O(N^k)$ memory in the worst case. Also, for the case of 2-field policies, this scheme is identical to the cross-producting and hence has memory requirement of $O(N^2)$.

Hash-Based Solution: The basic idea is motivated by the observation that while policy databases contain many different prefixes or ranges, the number of distinct prefix lengths tends to be small [1]. For instance, backbone routers have about 60K destination address prefixes, but there are only 32 distinct prefix lengths. Thus it can divide all the prefixes into 32 groups, one for each length (W). Since all prefixes in a group have the same length, it can use the prefix bit string as a hash key. That leads to a simple IP lookup scheme, which requires $O(W)$ hash lookups, independent of the number of prefixes. The algorithm of Waldvogel [1] performs a binary search over the W length groups, and achieves $O(\log W)$ worst-case time complexity. The tuple space idea generalizes the aforementioned approach to multi-dimension policies [4]. A tuple is a set of policies with specific prefix lengths, and the resulting set of tuples is called as "tuple space". Since each tuple has a specific bit-length for each field, by concatenating these fields in order to create a hash key, which can be used to perform the tuple lookup. Thus, the matched policy can be found by probing each tuple alternately, and keep track of the least cost policy. As an example, the two-dimension policies $F = (10*, 110*)$ and $G = (11*, 001*)$ will both belong to the tuple $T_{2,3}$. When searching $T_{2,3}$, a hash key is constructed by concatenating 2 bits of the source field with 3 bits of the destination field. Since the number of tuples is generally much smaller than the number of policies, even a linear search of the tuple space results, in a significant improvement over linear search of the policies.

Ternary CAM: Ternary content-addressable memory (TCAM) is one popular hardware device to perform fast packet classification. As compared to the software-based solutions, the TCAM can offer sustained throughput and simple system architecture, thus makes it attractive. However, it also comes with several shortcomings, such as the limited size, power consumption and expansive cost. For example, a 9 Mbits TCAM chip (US \$200, 40mm×40mm) running at 100 MHz dissipates about 8.5W. In comparison, a 9 Mbits SRAM (US \$20, 14mm×22mm) running at 250 MHz dissipates only 0.75W. Specifically, the policy length could be as long as 296 bits with IPv6. With the state-of-the-art 9-Mbit TCAM, it could support 16K such entries. In next generation networks, the TCAM will be suffering from a limited number of entries.

In this article, we propose an efficient algorithm to reduce the required TCAM by encoding the address portion of the searchable entries. The new scheme could reduce the length of TCAM entries from W to $(\log N + \beta)$ and still keeps the property of CIDR, where N is the number of the policies, β is the maximum number of levels and W is the length of IP address. In our experiments, it could encrypt the 128-bit prefixes of the real-world IPv6 routing tables into 11 bits.

Table 1. Complexity comparisons

Schemes	Speed	Storage	Scalability
Linear Search	$O(N)$	$O(NW)$	-
Bit Parallelism [5]	$O(DW + N/B)$	$O(DN^2)$	-
Grid of Tries [2]	$O(W^{D-1})$	$O(NDW^2)$	$\surd$
Cross-producting [2]	$O(DW)$	$O(N^D)$	-
RFC [3]	$O(D)$	$O(N^D)$	-
Tuple Space Search [4]	$O(N)$	$O(NW)$	$\surd$
Ternary CAM	$O(1)$	$O(NW)$	-
Proposed scheme	$O(1)$	$O(N \times (\log W + \beta))$	$\surd$

N:the number of prefixes, W:the maximum prefix length
D:the number of dimensions, B: the memory bus width
β:the number of levels.

The rest of the paper is organized as follows. Section 2 presents the proposed algorithm. The experiment results are presented in Section 3. Finally, a summary is given in Section 4.

2   TCAM Entry Encryption Algorithm

From the proposed system architecture, the packet classification could inherit the search result from routing lookup. It motivates us to encode the routing prefix as a much shorter one by replacing the original source/destination addresses in the original policy with the generated keys, so that the required TCAM is also reduced. To achieve the purpose, we have to realize the nature of IP routing prefixes firstly.

The adoption of classless inter-domain routing (CIDR) [10] allows the network administrator to specify a smaller network within an existing network. For example, an ISP network is specified by prefix $\langle 206.95.*\rangle$ whose next hop is A. It might exists a enterprise network, which is specified by prefix $\langle 206.95.130.*\rangle$ and its next hop is B. The encoding scheme must be able to reflect the hierarchical nature of the routing prefixes. Namely, the generated key for the prefix $\langle 206.95.*\rangle$ must be a shorter prefix of that for the prefix $\langle 206.95.130.*\rangle$.

Basic Scheme. To encode the address portion of the searchable entries, a straightforward scheme is to divide the prefixes into several sub-prefixes according to the length of their shorter prefixes, as shown in Figure 1. The prefix P_4 $\langle 010000\rangle$ has two shorter prefixes: P_1 $\langle 0\rangle$ and P_3 $\langle 0100\rangle$. Thus it is divided as three sub-prefixes $\langle 0\rangle$, $\langle 100\rangle$ and $\langle 00\rangle$ that are inserted to bit-stream group "Level I", "Level II" and "Level III", respectively. Clearly, it may derive the same sub-prefixes from different prefixes; such as the "Level II" bit-stream $\langle 01\rangle$ of prefix P_{13} is identical to that of P_2. In each group, the duplicate bit-stream must be eliminated and each bit-stream is assigned a unique IDs. By concatenating the relevant IDs, the encrypted key for each prefix can be generated. In this example, there are three different bit-stream in "Level I", five in "Level II" and four in "Level III". Thus maximum 7 (=2+3+2) bits are required to represent the original prefixes.

Exclusive Scheme. Though the basic scheme is simple, the number of the encrypted bit-streams may be over-estimated. The encoding results are inefficient for the reason that it is without considering the associated relation between prefixes. For example, the bit-streams $\langle 100\rangle$ and $\langle 1010\rangle$ (i.e., corresponding to P_3 and P_5 respectively) in "Level II" only concatenate to $\langle 0\rangle$ (i.e., P_1). Thus we only have to count the number of bit-streams attached to a specific "shorter" bit-stream. In our example, there are three bit-streams connected to $\langle 0\rangle$ (P_1) and two to both $\langle 101\rangle$ (P_6) and $\langle 11\rangle$ (P_{12}). Therefore, the number of bits for "Level II" could be reduced to two and the total length is reduced to six. In Figure 2, we list the ID for each bit-stream. The dotted line is used to separate the bit-streams based on their attached bit-streams.

Adaptive Scheme. The exclusive scheme can be further improved with ingenious encoding of the bit-streams. According to the successive bit-stream length, the exclusive scheme can adjust the length of encoded ID dynamically, as shown in Figure 3. It encodes the bit-stream with bottom-up manner and uses "Huffman Encoding" to reduce the maximum length of concatenated IDs. Thus the bit-streams in "Level III" are encoded at first. The minimum required length for each "Level III" bit-stream is recorded in its preceding prefix. The "Level II" bit-streams are sorted according to the length of their successive bit-streams. Then the longest one, for example, the "Level II" bit-streams $\langle 1 \rangle$ corresponding to P_8, is assigned the shortest ID "0". Another "Level II" bit-stream $\langle 0 \rangle$ attached to P_6 is thus assigned the ID "01". The IDs for each bit-stream is listed in the left part of Figure 3.

In Table 2, we show the encrypted prefixes for different schemes. With the basic scheme, the maximum required length is 7. It can be improved to 6 and 4 by adopting the exclusive and adaptive schemes respectively. Generally speaking, the length of the required bits for the adaptive scheme is quite close to the optimal value ($\log_2 N$). Consequently, we use the prefixes of the real-world routing tables to demonstrate the performance of the proposed scheme.

Usage. In each classifier, the referred routing prefixes are extracted from the policies. Then we construct the prefix tree and execute the encoding algorithm to

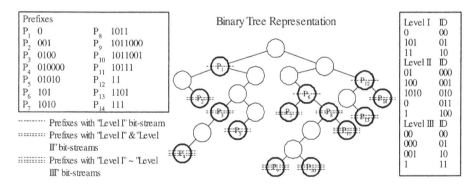

Fig. 1. Encoding the bit-streams according to what they attach to

Table 2. The encrypted prefixes for different schemes

Prefix	Basic	Exclusive	Adaptive	Prefix	Basic	Exclusive	Adaptive
P_1	00	00	10	P_8	01100	0101	10
P_2	00000	0000	1010	P_9	0101	010100	0000
P_3	0001	0001	100	P_{10}	0110	010101	0001
P_4	0000100	000100	1000	P_{11}	0111	010110	0010
P_5	00010	0010	1011	P_{12}	10	10	11
P_6	01	01	0	P_{13}	10000	1000	110
P_7	01011	0100	010	P_{14}	10100	1001	111

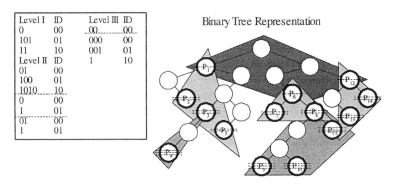

Fig. 2. Encoding the bit-streams according to what they attach to

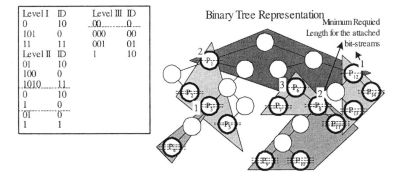

Fig. 3. Encoding bit-streams according to the length of their successive bit-streams

encrypt the prefixes. These results are attached to the routing prefixes as a part of lookup results. For those prefixes which are not referred in the policies, the encrypted results of their referred sub-prefixes are recorded. Also, the address portion of the policy is replaced by the encoded prefixes and inserted into the TCAM.

The routing lookup for each incoming packet will decide the next hop and also the encoded prefixes. Since the routing lookup is only performed for the destination address, an extra lookup for the source address is required. Consequently, the addresses in the packet header are replaced by the encoded prefixes and forwarded to the classifier. Then the classifier performs packet classification to derive the service priority.

3   Performance Evaluation

Through experiments, we demonstrate that the proposed scheme features much less TCAM bits. Currently, the IPv6 routing tables consist of only few hundreds

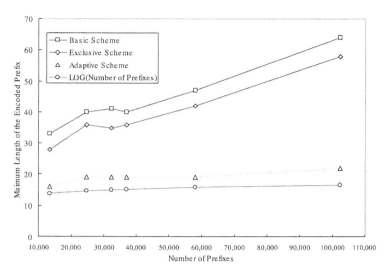

Fig. 4. The maximum length of encoded prefixes for different IPv4 routing tables

prefixes which are download from 6bone. To further realize the scalability of the proposed scheme, we also use the real data available from the IPMA [8] and NLANR [9] projects for comparison, these data provide a daily snapshot of the routing tables used by some major Network Access Points (NAPs). We illustrate the maximum length of the encoded prefixes for different routing tables and different schemes.

Figure 5 shows the encoding results for different routing tables. For the IPv4 routing tables, the basic scheme and exclusive scheme might incur longer encoded-bits than the original prefixes (32 bits). It is because these schemes concatenate maximum bits for each level to generate the encrypted prefixes. Contrarily, the adaptive scheme could complement the longest ID with shortest ID to eliminate the total length. In the experimental results, the adaptive scheme could encrypt the 32-bit prefix to a 22-bit one, which shows a bit-reduction of 70%. Moreover, the proposed scheme has achieved near optimal encoding as compared with the value $\log_2(number\ of\ prefixes)$. The difference between two values is incurred by the round-off error in each level. For example, in the routing table of NLANR, there are 102,271 prefixes and 6 levels. The maximum length for NLANR table is 22-bit which is nearly equal to $\log_2(102,271) + 6=22.6$.

For the IPv6 routing tables, the adaptive scheme still outperforms the rest schemes, as shown in Figure 5. But with fewer prefixes, the number of levels is reduced as well. Thus even with the simplest scheme, it could achieve fairly good results. While the number of prefixes increases, the effect of the adaptive scheme is emerged. In the results, the adaptive scheme could encrypt the 128-bit prefix to an 11-bit one, which shows a bit-reduction of 9%.

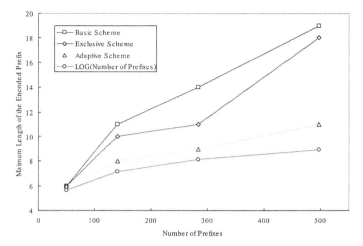

Fig. 5. The maximum length of encoded prefixes for different IPv6 routing tables

According to the experiments, we believe that the required bits are constrained from the increasing prefix length and prefix count. As the routing table contains 1M entries and 12 levels, it would require about 32 bits for encryption, not to mention the route aggregation in IPv6 will largely reduce the number of levels.

4   Conclusions

This study investigates the major issues in TCAM-based router design, including its price, power and size. To make use of the TCAM in IPv6-based packet classification, we propose an efficient approach to utilize the limited bits of TCAM. The scheme is motivated by the necessity of routing lookup for each packet. By encoding the prefixes into a much shorter one, the required bits for the TCAM entry could be significantly reduced. The basic idea is to divide the prefixes according to the length of their shorter prefixes and allocate enough bits for each level. It could be further improved by adopting the concept of exclusion to eliminate the combination in each level. Accordingly, we address how to joint a long ID with a short one by using Huffman Encoding. The length of the generated prefix is dramatically reduced. A typical IPv4 routing prefix needs 22 bits and 11 bits for IPv6 routing prefix in the worst case. The resulted prefix length is nearly equal to $O(\log_2 N + \beta)$, where N is the number of prefixes and β is the maximum number of levels. We also demonstrate this in our experiments. With the route aggregation in IPv6, the value of b tends to be small. Also, only the referred routing prefixes have to be encoded. Thus the required length for an IPv6 prefix is likely less than 32 bits, which save more than 75% storage.

References

[1] M. Waldvogel, G. vargnese, J. Turner, and B. Plattner, "Scalable High Speed IP Routing Lookups," In Proc. ACM SIGCOMM '97, pages 25-36, Cannes, France, Sept. 1997.

[2] V. Srinivasan, G. Varghese and S. Suri, "Packet Classification using Tuple Space Search," in ACM SIGCOMM, September 1999, pp. 135–146.

[3] Pankaj Gupta and Nick McKeown, "Packet Classification on Multiple Fields," in ACM SIGCOMM, September 1999, pp. 147–160.

[4] V. Srinivasan, G. Varghese, S. Suri and M. Waldvogel, "Fast Scalable Level Four Switching," in ACM SIGCOMM, September 1998, pp. 191–202.

[5] T. V. Lakshman and D. Stidialis, "High Speed Policy-based Packet Forwarding Using Efficient Multi-dimensional Range Matching," in ACM SIGCOMM, September 1998, pp. 203–214.

[6] Anja Feldmann and S. Muthukrishnan, "Tradeoffs for Packet Classification," in IEEE INFOCOM, March 2000, pp. 1193–1202.

[7] D. Shah and P. Gupta, "Fast updating Algorithms for TCAMs," IEEE Micro Mag., 21(1):36-47, Jan.-Feb. 2001.

[8] Merit Networks, Inc. Internet Performance Measurement and Analysis (IPMA) Statistics and Daily Reports. See *http://www.merit.edu/ipma/routing_table/*.

[9] NLANR Project. See *http://moat.nlanr.net/*.

[10] Y. Rekhter, T. Li, "An Architecture for IP Address Allocation with CIDR." RFC 1518, Sept. 1993.

Delayed Just-Enough-Time Scheduling: An Approach to Improve Efficiency of WDM Optical Burst Switching Networks[*]

Dooil Hwang, Seongho Cho, and Chongkwon Kim

School of Electrical Engineering and Computer Science
Seoul National University
Seoul, Republic of Korea
{dhwang,shcho,ckim}@popeye.snu.ac.kr

Abstract. Optical Burst Switching (OBS) is a prominent switching paradigm for WDM (Wavelength Division Multiplexing) optical networks. It is known that the JET (Just-Enough-Time) scheme reduces the reservation delay and utilizes network bandwidth efficiently in WDM OBS networks. However, since JET processes the network bandwidth reservation request in FCFS (First Come First Serve) order, it has the potential problem of not considering efficient network resource allocation. Hence we propose a new scheduling algorithm, DJET (Delayed Just-Enough-Time), using an extra offset to improve the performance of the JET scheme. Simulation results show that the DJET scheme can improve network resource utilization by about 40% over the JET scheme without significant delay degradation.

1 Introduction

The rapid and explosive growth in IP (Internet Protocol) traffic has triggered several research activities in the development of new high-speed transmission and switching technologies. The most prominent among these technologies is WDM (Wavelength Division Multiplexing). WDM enables high-speed transmission by multiplexing many wavelengths into a single fiber. Also, an extension form of WDM called DWDM (Dense Wavelength Division Multiplexing) has been developed. DWDM provides a transmission rate of about 400 Gbps by multiplexing 80 - 160 wavelengths into a single fiber [9]. As a result, DWDM has the potential to support today's explosive increasing bandwidth requirements. However, the supportable bandwidth difference between DWDM networks and today's IP router-based networks is still so large that existing IP router-based networks cannot sufficiently meet the bandwidth requirement of DWDM networks. To overcome this limitation, a new network architecture called an IP-over-DWDM network has been proposed. To support IP-over-DWDM networks,

---

[*] This work was supported by the Brain Korea 21 Project of the Korean Ministry of Education, and the University IT Research Supporting Program under the Ministry of Information & Communication of Korea.

several switching paradigms have been suggested. One approach is the optical circuit switching method which sets up an optical lightpath between then optical ingress and egress router using the wavelength routing capability of the optical layer, then transmits packets through that lightpath. However, the optical circuit switching method can be applied only when there is a large volume of traffic between optical the ingress and egress router. Hence optical circuit switching can potentially suffer from poor network utilization. An alternative to optical circuit switching is the optical packet switching method. Optical packet switching can provide the most flexible and efficient use of bandwidth at the optical layer. However, current technologies that support optical packet switching are not mature. Also, optical packet switching has to overcome several technological constraints such as limited optical buffering (optical memory technology) and synchronization [3].

To overcome these limitations of optical circuit and packet switching, the optical burst switching method (OBS) has been proposed by [1]. OBS allows the switching of data channels entirely in the optical domain by allocating network resources in the electrical domain. OBS consists of the control packet, the data burst, and the offset time. The operation of OBS is as follows. First, if the data burst is generated, the corresponding control packet is transmitted to the destination. The control packet visits intermediate nodes and reserves the network resources that will be used by the corresponding data burst. After the offset time, the data burst is transmitted immediately from the source to the destination. Among the several OBS methods, the JET (Just-Enough-Time) scheme is know to be the most efficient mechanism [7]. By [2], JET improves network utilization using an offset and delayed reservation mechanism. The JET scheme processes the network resource reservation request for each incoming control packet in FCFS (First Come First Serve) order. Therefore, there is a potential problem of JET not efficiently allocating network resources. For example, assume there are two control packets 1, 2, and the control packet 1 arrives earlier than 2 but the time difference is small. Also, the length of data burst 1 is much smaller than 2, it makes a collision. Instead of processing control packet 1 first, processing control packet 2 first improves network resource utilization. Thus, when the difference of the control packets' arrival time is small enough and the difference of the data burst length is large enough, FCFS runs into a network efficiency problem.

To solve the problem of JET scheme, we propose a new scheduling method, the DJET (Delayed Just-Enough-Time) scheduling method. DJET uses an extra offset time to schedule each incoming control packet and considers a combination of total hop counts, visited hop counts, and data burst length for the evaluation metric. In the extra offset time interval, DJET selects the control packet which has the greatest evaluation metric value and reserves bandwidth for it. Then, DJET processes the control packet that will not collide with the selected control packet's corresponding data burst's arrival and end time. The simulation results show that the DJET scheduling method outperforms the JET scheme with regards to network resource utilization.

Fig. 1. The basic components of the DJET scheduling algorithm

The rest of the paper is organized as follows. In Section 2, we propose the DJET scheduling method to improve network resource utilization of the JET scheme. In Section 3, we will show the analysis for the length of the extra offset of the DJET scheduling method. In Section 4, we will provide simulation results by comparing JET with DJET using the total data burst drop percentage as a metric. Finally, we conclude the paper in Section 5.

2   DJET: Scheduling Algorithm to Improve Network Resource Utilization

2.1   The Basic Components of DJET

Before presenting the DJET scheduling method, let us first examine the basic components of DJET. The basic components of DJET are presented in Figure 1.

As shown in Figure 1, DJET consists of a control packet, data burst, base offset, priority offset, and an extra offset (which will be called the interval I offset). The data burst, base offset (the offset which is proportional to the total hop counts of each control packet), and priority offset (to achieve class isolation) [2] components are the same components as in the JET scheme. The interval I offset is an extra offset used to schedule each incoming control packet. The value of the interval I offset is static. During the interval I offset, DJET receives each incoming control packet and calculates the evaluation metric value of each incoming control packet using the control packet's information. The control packet of DJET has extended fields for the total hop count and the visited hop count. The visited hop count is incremented into the control packet processing time at the intermediate nodes. This field will be used to calculate the evaluation metric value that will be explained in Subsection 2.2.

2.2   Evaluation Metric of DJET

During the interval I offset, the DJET scheduler calculates an evaluation metric value for each incoming control packet. The evaluation metric is a scheduling policy that indicates how the incoming control packet should be scheduled. Therefore, if the value of the evaluation metric is larger, then the probability of scheduling that control packet will be greater.

We devised the evaluation metric from a combination of total hop counts, visited hop counts of each incoming control packet at the intermediate node, and a data burst length which is included in each incoming control packet. The main factor is the total hop count of the control packet. The control packet which has a greater total hop count will use more network resources than one with a smaller count. The visited hop count at the intermediate node should also be considered as an important factor. The visited hop count indicates how much of the network resources are reserved by the control packet at the intermediate node. If a control packet which has a large visited hop count is discarded, the retransmission overhead of that control packet will be too high. When several incoming control packets have the same total hop count, processing the control packet with the greatest visited hop count first will improve network resource utilization. In addition, the data burst length is also an important factor to consider. If several control packets have the same total hop count and visited hop count value, processing the control packet with the greatest data burst length will increase network resource utilization. Considering these factors, the evaluation metric has to be proportional to the total hop count, visited hop count, and data burst length. The evaluation metric of the ith control packet can be expressed as follows.

$$Evaluation\ Metric\ of\ i^{th}\ control\ packet = \frac{H_i V_i L_i}{H_{max}^2 L_{max}}$$

In the evaluation metric, the terms of H_i, V_i, L_i represent the total hop count, visited hop count and data burst length value, respectively, which are included in each incoming control packet. H_{max} and L_{max} represent the maximum total hop count and maximum data burst length. These two factors are introduced to normalize the evaluation metric value.

2.3   DJET Scheduling Algorithm

In this subsection, we describe the DJET scheduling algorithm. As previously explained, DJET uses an extra offset, i.e., interval I offset, and an evaluation metric that is composed of total hop counts, visited hop counts of each incoming control packet, and a data durst length included in the control packet to schedule each network resource request. The DJET scheduling algorithm operates as follows.

Each intermediate node receives the incoming packets during the interval I offset. Initially, the interval I offset starts when the node receives the first incoming control packet. The visited hop count field of each incoming control packet is incremented by one, and the node calculates the evaluation metric value of the control packet. The resulting evaluation metric value and corresponding control packet ID are stored in some part of the memory using simple data structure X. If the evaluation metric value of a later control packet is greater than the value stored in the data structure X, then data structure X is replaced by the new value. Therefore, among incoming control packets, data structure X will always have the greatest evaluation metric value and the corresponding control packet ID. Meanwhile, each node stores the arrival time and the end time of each data burst which corresponds to the control packet using data structure Y. Data structure Y has the arrival time, end time and the ID of each incoming control packet. Where the interval I offset ends, the ID stored in data structure X becomes the selected control packet's ID. The selected control packet first reserves network resources. Then, the process of checking for the collision between the selected control packet's reservation range and others that were received during the interval I offset period begins. If there are control packets which do not collide with the selected control packet's reservation range, those control packets successfully reserve network resources. The control packets which collide with the selected one are either discarded at that intermediate node or retransmitted. Data structure X and Y are flushed and reinitialized after the interval I offset ends. After the interval offset I ends and the new control packet is received, the interval I offset starts again and DJET operates again as previously explained. However, if the control packet successfully reserves network resources in the previous node, but fails to reserve network resources in the current node, the network resources reserved by that control packet has to be removed. In this case, the cancel control packet which is generated at the current node is transmitted to the destination to cancel network resources reserved by that control packet. This enables canceled network resources to be reused by another control packet that visits the intermediate node.

3   Analysis - Interval I Offset Length

In this section, we analyze the appropriate length of the interval I offset in the DJET scheduling algorithm. If the interval I offset length is too large, the sum of the interval I offset length becomes greater than the propagation delay. Therefore, the performance of DJET would be more degraded than JET. On the other hand, if the interval I offset length is too small, the performance between JET and DJET will be similar, having little effect on the DJET scheduling algorithm. Therefore, analyzing the proper length of the interval I offset is an important issue of the DJET scheduling algorithm. We analyzed the length of the interval I offset for the worst-case in this section.

3.1   The Length of the Interval I Offset

For simplicity, we consider two control packets, the earliest arrived control packet and the last arrived packet in the range of the interval I offset. We assume that the maximum data burst length and the maximum total hop count are L_{max} and H_{max}, respectively. Also, we assume that the data bursts that correspond to the earliest and latest arrived control packets in the interval I offset always make a collision. We consider two cases presented in Figure 2. The first case shows the earlier arrived control packet with the earlier arrived data burst and the second case shows the earlier arrived control packet with the later arrived data burst.

In Figure 2, $Arr(i)$ and $Arr(j)$ are the arrival times of control packets i and j. O_i and O_j represent the offset length included in control packet i and j,

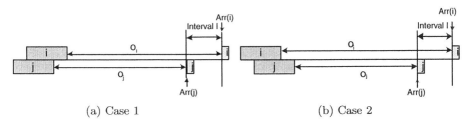

Fig. 2. Analysis model for interval I offset length

respectively. Also, L_i and L_j represent the data burst length included in control packet i and j. For the worst-case analysis, we assume that the control packet arrival time difference for i and j is I. Let's first examine case 1 in Figure 2. If a collision between data burst i and j occurs, the following inequality is satisfied.

$$(I + O_j + L_j) - (O_i + L_i) \le L_j \tag{1}$$

Therefore, the following result can be obtained from inequality (1).

$$I \le L_i + (O_i - O_j) \tag{2}$$

For case 2, we can derive the following inequality when data burst i and j make a collision.

$$(O_i + L_i) - (I + O_j + L_j) < L_i \tag{3}$$

Therefore, the following result can be obtained from (3).

$$I > (O_i - O_j) - L_j \tag{4}$$

In case 2, the offset length of i is greater than that of j. So, $(O_i - O_j)$ has a positive value. Therefore, from inequality (4), if the value of $(O_i - O_j)$ is greater than L_j, I is greater than $\alpha(\alpha = (O_i - O_j) - L_j)$. On the contrary, if the value of $(O_i - O_j)$ is equal to L_j, I is greater than 0. However, the results obtained from (3) and (4) need not be considered since we just consider the worst-case. A valid result for the worst-case interval I offset length can be obtained from (1) and (2). From (2), the upper bound of the interval I offset length is $L_i + (O_i - O_j)$. If we substitute L_i for L_{max} and $(O_i - O_j)$ for $(H_{max} - 1)\Delta$ where Δ is the control packet processing delay, the following value can be the worst-case interval I offset length.

$$I = L_{max} + (H_{max} - 1)\Delta \tag{5}$$

The worst-case length of the interval I offset is represented by the sum of the maximum data burst length and the product of $H_{max} - 1$ and the control packet processing delay in the network. In equation (5), the interval I offset length also increases as the network diameter increases (i.e., the value of H_{max} increases).

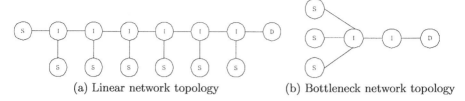

(a) Linear network topology   (b) Bottleneck network topology

Fig. 3. Simulation topology

4   Simulation Results

In this section, we present the simulation results of the DJET scheduling algorithm. The performance metric used in this simulation is the total packet drop percentage. In this simulation, the packet means the control packet. We assumed that the control packets and the data bursts arrive at the node by the Poisson process, and the offsets are exponentially distributed. Also, it is assumed that the control packet header processing delay and the transmission speed are fixed in the network. We compared the DJET and JET scheduling algorithms using several parameters. The parameters used in the simulation are described in the table 1. The duration of the simulation is 10000 time units.

4.1   Simulation Topology

The simulation topology is shown in Figure 3 (a) (linear network topology) and 3 (b) (bottleneck network topology). The nodes which are marked S are source nodes, I are intermediate nodes, and D is a destination node. Control packets and data bursts are generated at random at the source nodes. Each intermediate node performs the DJET scheduling algorithm (does not generate control packets or data bursts) and forwards or discards control packets and data bursts. The destination node just receives the incoming control packets and data bursts.

4.2   Simulation Results

1) Total packet drop percent and interval I offset length
First, we simulated the effect of the interval I offset time to total packet drop

Table 1. Parameters used in the simulation

Average data burst length	40 μs
Maximum data burst length	80 μs
Interval I offset length	Variable
Transmission speed	2.5 Gbps
Average control packet interarrival time	Variable
Link propagation delay	500 μs
Control packet processing delay	10 μs

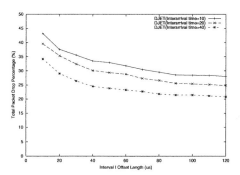

Fig. 4. Simulation results for total packet drop percentage and interval I length

percentage. We simulated this effect by varying the interval I offset length from 10 to 120μs when the interarrival time of the control packets are 10, 20, and 40 μs respectively. The topology used in this simulation is a linear network topology. The simulation results which compare the relationship between the total packet drop percentage and the interval I length are presented in Figure 4.

In Figure 4, the total packet drop percentage decreases as the interval I offset length increases in each case. Therefore, if we increase the interval I offset length, network resource utilization increases. However, if the interval I offset length is greater than 100 μs, the total packet drop percentage becomes nearly constant in each case. Using the results in Section 3, the worst-case interval I offset length is 140 μs. This means there exists an optimal value for the interval I offset length that can achieve efficient network resource utilization. In this simulation, the optimal value of the interval I offset exists in the range of 95-100 μs for each case.

2) Comparing performance between DJET and JET

Next, we compared the DJET scheduling algorithm to the JET scheme. This was simulated by the varying the interarrival time of the control packets. A range for the interarrival time 10 μs to 100 μs was used in this simulation. We simulated the DJET scheduling algorithm with three cases by varying the interval I offset time to 20, 40 and 80 μs. We ran this simulation for both network topologies (linear and bottleneck network topologies). The simulation results which compare the performance between DJET and JET are described in Figure 5 (a) (linear network topology) and 5 (b) (bottleneck network topology).

In Figure 5 (a) (linear network topology case), JET has a much poorer performance compared to any other cases of DJET. In the case of JET, when the network is heavily congested, for example, when the interarrival time of the control packets is 10 μs, the total packet drop percentage is almost 70%. However, the DJET scheduling algorithm results in a 29% (interval I offset length = 80 μs), 33% (interval I offset length = 40 μs), and a 42% (interval I offset length = 20 μs) total packet drop. Hence, when the network is heavily congested, the

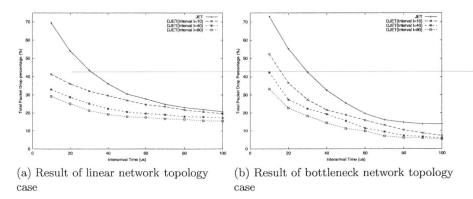

(a) Result of linear network topology case

(b) Result of bottleneck network topology case

Fig. 5. Performance comparison between JET and DJET

DJET scheduling algorithm improves the total packet drop percentage by approximately 28% - 41%. The difference is less dramatic when the network is not congested (interval I offset length is 80 μs. In this case, DJET shows a 7% - 23% improvement in the total packet drop percentage. The degree of performance improvement becomes smaller as the network becomes less congested or in other words, as the interarrival time of the control packet becomes larger. In Figure 5 (b) (bottleneck network topology case), the total packet drop percentage of JET is about 73% when the interarrival time is 10 μs, and it decreases as the interarrival time increases. Generally, when the control packet interarrival time is small, the total packet drop percentage of JET in the case of bottleneck network topology is higher than that of the case of linear network topology. In the case of bottleneck network topology, the total packet drop probability is lower for DJET than JET, just as in the linear topology case. However, in the case of bottleneck network topology, the degree of improvement is smaller than in the linear topology case. The improvement of DJET over JET is about 21% - 40% when the network is congested (the interarrival time of the control packet is 10 μs), and 10% - 22% when the network is not congested.

5   Conclusion

In this paper, we proposed a new scheduling algorithm, DJET, to improve network resource utilization in the WDM optical network. For DJET, we introduced an additional offset time called interval I offset time. Processing control packets with time delay and using an evaluation metric, DJET allocates precious network resources more efficiently than JET. We simulated the effect of interval I offset time on the total packet drop percentage for the DJET scheduling algorithm and the JET scheme to compare their performance. The simulation results show that the DJET scheduling algorithm improves performance by up to 40% over the JET scheme in regards to the total packet drop percentage. This means that the DJET mechanism achieves more efficient network resource utilization compared than the JET scheme. The optimal value for the interval I offset has yet to be solved. This will be explored further at another time.

References

[1] M. Yoo, C. Qiao, "Optical Burst Switching (OBS) - A new paradigm for an optical Internet," in a special issue of *Journal of High Speed Networks(JHSN) on WDM Networks*, Vol. 8, No. 1, pp. 69-84, 1999.

[2] M. Yoo, C. Qiao, and S. Dixit, "Optical Burst Switching for service differentiation in the next - generation optical Internet," *IEEE Communications Magazine*, Vol. 39, Issue 2, pp. 98-104, Feb. 2001.

[3] S. Verma, H. Chaskar, and R. Ravikanth, "Optical Burst Switching : A viable solution for terabit IP backbone," *IEEE Network*, Vol. 14, Issue 6, pp. 48-53, Nov-Dec, 2000.

[4] J. Y. Wei, R. I. McFarland Jr., "Just-In-Time Signaling for WDM Optical Burst Switching Networks," *Journal of Lightwave Technology*, Vol. 18, no. 12, 2000.

[5] M. Yoo, C. Qiao, "Supporting multiple classes of services in IP over WDM networks," Global Telecommunications Conference, 1999. *GLOBECOM '99*, Vol. 1b, 1999.

[6] M. Yoo, C. Qiao, "QoS performance of Optical Burst Switching in IP-over-WDM networks," *IEEE JSAC in Communications*, Vol. 18, Oct. 2000.

[7] M. Yoo, C. Qiao, "A novel switching paradigm for buffer-less WDM networks," *Optical Fiber Communication Conference*, 1999, and the *International Conference on Integrated Optics and Optical Fiber Communication OFC/IOOC* 1999.

[8] Y. Xiong, M. Vandenhoute, H. C. Cankaya, "Control Architecture in Optical Burst-Switched WDM networks," *IEEE JSAC in Communications*, Vol. 18, Oct. 2000.

[9] P. Lin, R. Tench, "The exciting frontier of lightwave technology," *IEEE Communications Magazine*, Mar 1999.

[10] M. Yoo, C. Qiao, "Just-Enough-Time (JET): A high speed protocol for bursty traffic in optical networks," *IEEE/LEOS Conference on Technologies for a Global Information Infrastructure*, Aug 1997.

[11] L. Liu, P. Wan, and O. Freider, "Optical burst switching: The next IT revolution worth multiple billions dollars?," *MILCOM 2000. 21st Century Military Communications Conference Proceedings* , Vol. 2.

[12] J. Turner, "Terabit Burst Switching," *Journal of High Speed Networks*, Vol. 8, No. 1, 1999.

[13] P. Green, "Progress in optical networking," *IEEE Communications Magazine* Jan. 2001.

[14] H. Chaskar, S. Verma, and R. Ravikanth, "A Framework to Support IP over WDM Using Optical Burst Switching," *Nokia Research Center report* Jan. 2000.

Double EIM and Scalar BPM Analyses of Birefringence and Wavelength Shift for TE and TM Polarized Fields in Bent Planar Lightwave Circuits

Won Jay Song[1], Won Hee Kim[2], Byung Ha Ahn[3],
Bo Gwan Kim[2], and Munkee Choi[1]

[1] Optical Internet Research Center and Grid Middleware Research Center
Information and Communications University
305-732, Republic of Korea
songwonjay@ieee.org
mkchoi@icu.ac.kr
[2] VLSI and CAD Labs, Department of Electronics Engineering
Chungnam National University
305-764, Republic of Korea
kimwonhee@ieee.org
bgkim@cnu.ac.kr
[3] Systems Control and Management Labs, Department of Mechatronics
KwangJu Institute of Science and Technology
500-712, Republic of Korea
bayhay@kjist.ac.kr

Abstract. We have applied the effective index method to reduce the two-dimensional refractive index profile into the one-dimensional refractive index structure and modified the wave equations to obtain the paraxial wave equations. Then, TE and TM polarized fields in the curved single-mode planar waveguides are analyzed by using the scalar beam-propagation method employing the finite-difference method with a slab structure. The birefringence for TE and TM polarized fields in bent waveguides is calculated from the phase difference of the optical fields. The wavelength shift due to the birefringence of TE and TM polarized fields in bent waveguides is also calculated.

1 Introduction

Planar lightwave circuits (PLCs) are the technology of optical devices, created by the combination of optical fiber fabrication and semiconductor microfabrication technologies. The technology of PLCs is one of the key technologies for fiber-optic-based high-speed telecommunication systems. Recently, research and development efforts in Japan and the United States are focused on large PLCs, such as planar erbium-doped amplifiers and wavelength division multiplexing (WDM) systems. Therefore, in order to raise the integration density, PLCs including several kinds of the waveguide are needed bent waveguides to connect

H.-K. Kahng (Ed.): ICOIN 2003, LNCS 2662, pp. 96–107, 2003.
© Springer-Verlag Berlin Heidelberg 2003

two straight waveguides. It is well known that optical power is lost at the entrance and exit of the bent waveguides and by bending effect [1]. The former is called the transition loss and the latter the pure bending loss. The bending loss contains both losses. A number of theoretical treatments have been done to determine and reduce the radiation losses in curved waveguides.

In order to analyze the lightwave propagating phenomena in bent waveguides with the two-dimensional (2-D) refractive index profile, the 2-D structure of index profile is converted into the 1-D structure using the EIM for TE and TM polarized fields respectively, we adopt the 2-D FD-BPM, based on a scalar wave equation, to obtain the behavior of the optical field. From these results, in this paper, in the process of the FD-BPM calculation to obtain the propagating field from the initial field, we can obtain the propagation constants through calculating the phase difference between the fields of the two consecutive steps in the propagation direction for each of TE and TM modes. Dividing calculated propagation constants by the free-space propagation constant, the effective guide indices can be obtained. The index difference between the two orthogonal modes is used to calculate the wavelength shift.

2   Bent Waveguides and Numerical Analysis

2.1   TE and TM Modes in Bent Waveguides

An electromagnetic field in bent waveguides is described by two related vector fields: the electric field $\boldsymbol{E}(\mathbf{r})$ and the magnetic field $\boldsymbol{H}(\mathbf{r})$. The modes of the bent slab waveguide can be classified into TE and TM modes. We consider TE and TM modes separately. The generalized Helmholtz equation for $\boldsymbol{E}$ may be derived from Maxwell's equations. In this paper, we used the approximation of a bent dielectric slab waveguide where there is no variation in $\hat{z}$-direction. In the TE configuration, where the electric field is horizontally polarized (the $\hat{r}$-direction), the vector wave equation for $\boldsymbol{E}$ reduces to the scalar equation in terms of the single component E_{TE} in cylindrical coordinates as follows [2];

$$\left[\frac{\partial^2}{\partial r^2} + \frac{1}{r}\frac{\partial}{\partial r} + \frac{1}{r^2}\frac{\partial^2}{\partial \theta^2} + k_0^2 n^2(r) \right] E_{\mathrm{TE}} = 0, \tag{1}$$

where $\epsilon(r)$ is assumed to be a scalar function of the spatial coordinate r. Thus the equation for E_{TE} does not contain the derivative of $n(r)$ and therefore E_{TE} will be continuous and smooth. The generalized Helmholtz equation for $\boldsymbol{H}$ may also be derived from Maxwell's equations. In the TM configuration, where the electric field is vertically polarized (the $\hat{z}$-direction), the vector wave equation for $\boldsymbol{H}$ reduces to the scalar equation in terms of the single component H_{TM} in cylindrical coordinates as follows;

$$\left[n^2(r)\frac{\partial}{\partial r}\frac{1}{n^2(r)}\frac{\partial}{\partial r} + \frac{1}{r}\frac{\partial}{\partial r} + \frac{1}{r^2}\frac{\partial^2}{\partial \theta^2} + k_0^2 n^2(r) \right] H_{\mathrm{TM}} = 0. \tag{2}$$

Thus H_{TM} will be continuous, but may not be smooth across the interfaces where $n(r)$ changes abruptly.

2.2   Beam-Propagation Method

Accurately numerical analysis of guided-wave structures is essential for the development of photonic integrated circuits. The beam-propagation method (BPM) is a powerful technique for obtaining numerical solutions to problems of wave propagation in structures of great complexity [3] [4]. The BPM has been successfully used to analyze a wide spectrum of guided-wave structures [5] [6]. The numerical scheme to solve the paraxial wave equation is to use a finite-difference approximation. This method will be referred to as FD-BPM. The FD-BPM has also been successfully applied to the analysis of nonlinear propagation in a radially symmetric structure [7].

2.3   Transparent Boundary Condition

The BPM is notoriously deficient in modeling structures that scattering radiation, since that radiation tends to reflect from boundaries of the problem back into the solution region. The reflection causes unwanted interference. The way of preventing boundary reflection has been the insertion of artificial absorption regions adjacent to the boundaries [8]. On the other hand, the transparent boundary condition (TBC) algorithm assumes that there is no boundaries. Radiation is allowed to escape freely without any reflection, where there is no radiation flux back into the problem region.

2.4   Application of the Effective Index Method into the BPM

We introduce the effective index method (EIM) and the scalar FD-BPM in cylindrical coordinates for the bent waveguide. In order to find solution of TE modes with a two-dimensional square index profile in bent waveguides which have horizontal polarization, we must calculate effective index by using the TM Helmholtz equation (TM-EIM) and then run the BPM for TE modes (TE-BPM). Similarly, in order to find solution of TM modes with a two-dimensional square index profile in bent waveguides which have vertical polarization, we must calculate effective index by using the TE Helmholtz equation (TE-EIM) and then run the BPM for TM modes (TM-BPM).

3   Characteristics of Bent Waveguides

3.1   Simulation Methods and Waveguide Structures

In order to investigate the bending loss and phase difference between TE and TM waves, we examine the performance of bent waveguides using the beam-propagation method employing the finite-difference method (FD-BPM) combined with the transparent boundary condition (TBC). The three-dimensional (3-D) FD-BPM analysis of a typical device usually requires a large amount of computational power. Therefore, we adopt the 2-D FD-BPM, which is also called the scalar FD-BPM, combined with the effective index method (EIM).

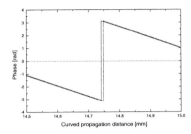

Fig. 1. Phase diagram obtained from the FD-BPM simulations for the TE wave with solid line and for the TM wave with dashed line. The phase difference between the waves is 0.05726333 rad for $\Delta = 0.3$ %, $L = 15$ mm, and $R = 10$ mm

This practice greatly reduces the required computation time, typically by a factor of hundreds.

In the numerical simulation of bent waveguides, the waveguide structure has a step index profile, i.e., the refractive indices are uniform in the channel and substrate regions with different values. Such an index profile is typical in doped-silica or semiconductor slab waveguides in one- and two- dimensions. The waveguides are single-mode and the effective indices are 1.4530690978933 and 1.4530647731523 for TE and TM polarized fields, respectively, at 1.55 μm wavelength. These values are obtained from the well-known eigenvalue equation for the three-layer slab waveguide in the EIM.

3.2 Wavelength Shift

The technology of planar lightwave circuits (PLCs) with bent waveguides will be indispensable for the super-high-speed and large-capacity optical communication networks of the future. However, the PLC devices exhibit polarization dependence because of waveguide birefringence. The birefringence which should be controlled. To solve this problem for the network, we need to calculate wavelength shift precisely.

Birefringence and Wavelength Shift. In straight waveguides with the square cross section made of isotropic dielectric material, the propagation constant for TE wave is exactly equal to that of TM wave. However, in a bent waveguide with the square cross section, the two propagation constants of TE and TM waves are different because of the bending effect. Therefore, the phase velocities for the two orthogonal waves are different and the phase difference between the waves is 0.05726333 rad for $\Delta = 0.3$ %, $L = 15$ mm, and $R = 10$ mm as shown in Fig. 1. The phase velocity of TM wave is faster than that of TE wave, as the effective guide index for TM wave in the channel waveguide is smaller than that of TE wave.

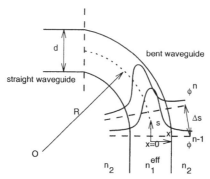

Fig. 2. Schematic diagram to calculate phase difference between the previous and next fields in order to find the propagation constant and the effective guide index

The effective guide index n_g^{eff} is defined by β/k_0 where the propagation constant β has been calculated from the FD-BPM and $k_0 \equiv 2\pi/\lambda$. In Fig. 2, we schematically show the method to calculate the phase difference $\Delta\phi$ between the previous and next fields, which are obtained from the FD-BPM, and β can be found as follows;

$$\Delta\phi = \Delta\beta\Delta s,$$

$$\phi_{\text{max}}^n - \phi_{\text{max}}^{n-1} = (\beta - \beta_{\text{ref}})\Delta s,$$

$$\beta = \beta_{\text{ref}} + \frac{\phi_{\text{max}}^n - \phi_{\text{max}}^{n-1}}{\Delta s}, \qquad (3)$$

where ϕ_{max}^n is the phase of the wave at the position of the maximum optical field intensity in the bent waveguide and β_{ref} is chosen to be $k_0 n_2$. Therefore we can obtain n_g^{eff} from Eq. (3), as

$$n_g^{\text{eff}} = n_2 + \frac{\phi_{\text{max}}^n - \phi_{\text{max}}^{n-1}}{k_0 \Delta s}. \qquad (4)$$

Starting from the reference point, n_g^{eff} approaches to a constant value when the propagation distance becomes larger as shown in Fig. 3. It is interesting to note that n_g^{eff} is smaller than the refractive index of channel waveguide n_1^{eff}, which means n_1^{TE} and n_1^{TM}, obtained from the EIM. The difference of the effective guide index between TE and TM waves, Δn_g^{eff}, is then

$$\Delta n_g^{\text{eff}} \equiv n_g^{\text{TM}} - n_g^{\text{TE}}, \qquad (5)$$

where n_g^{TM} and n_g^{TE} are the effective guide indices for the waves polarized along the fast and slow axes of the channel waveguide. This is also called birefringence and it determines the wavelength shift due to the phase velocity difference of optical waves depending on the direction of polarization of its electric field

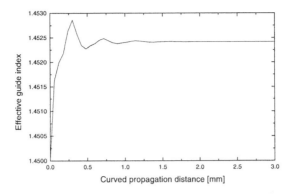

Fig. 3. Calculation result of the effective guide indices for the TE wave with solid line and the TM wave with dashed line at Δ of 0.3 %, L of 3 mm, and R of 4 mm. These indices are almost same

vector [9]. In general the birefringence exists in an anisotropic dielectric crystal without any wave-guided structure. However in the bent waveguides made of isotropic dielectric material, the birefringence can also exist due to the difference of the effective guide index between the two polarized waves.

As a consequence, the spectral response for the two polarizations will be different. This shift can be calculated by comparing the propagation constant β^{TE} and β^{TM} in the waveguide as follows: When TE mode with λ and TM mode with $\lambda + \Delta\lambda$ are propagating in the bent waveguide, their internal propagation constants may become identical for a certain value of $\Delta\lambda$. Therefore,

$$\beta^{TE}(\lambda) = \beta^{TM}(\lambda + \Delta\lambda),$$
$$\left(\frac{2\pi}{\lambda}\right) n_g^{TE} = \left(\frac{2\pi}{\lambda + \Delta\lambda}\right) n_g^{TM}. \tag{6}$$

After some manipulations, $\Delta\lambda$ can be obtained as

$$\Delta\lambda = \lambda \left[\left(\frac{n_g^{TM}}{n_g^{TE}}\right) - 1 \right], \tag{7}$$

where $\Delta\lambda$ is called the wavelength TE/TM shift. Therefore we can calculate the wavelength TE/TM shift from the difference of the effective guide index n_g^{eff} between two orthogonal polarizations.

Correction for Calculation of Wavelength Shift. The refractive indices n_1^{eff} of TE and TM waves for square channel waveguide become different because the different eigenvalue equations for even TE modes and even TM modes are used in the EIM. This contradicts the fact that these values should be same for square

Table 1. Comparison with the effective index difference of TE and TM waves calculated from the single EIM and the double EIM (Eight-digit decimal chopping arithmetic)

Channel materials	Single EIM	Double EIM
Low doped silica	4.3247410×10^{-6}	9.1255095×10^{-7}
High doped silica	1.9070978×10^{-5}	3.3736049×10^{-6}
Compound semiconductor	4.2762337×10^{-3}	8.0234431×10^{-4}

waveguides. Calculation of the effective guide index n_g^{eff} by using the FD-BPM in the bent waveguides with the different refractive index profiles, generates the difference between the guide indices by bending effect as well as by the unwanted refractive index difference for the waveguide core by the EIM. Thus the correction to remove the effective index difference of TE and TM waves is needed to calculate birefringence or wavelength TE/TM shift.

In order to obtain the same refractive indices for TE and TM waves in the waveguide core region, we use the double EIM. The double EIM means that the result of calculation of the first EIM for TE modes replaces n_1 which works an initial value to calculate n_1^{eff} of TM modes in the second EIM and vice versa. In the double EIM, the difference of refractive index becomes about 10 times smaller than that in the single EIM as shown in Table 1. Corrected effective guide index of TM wave is obtained by adding the difference of refractive index in the double EIM to the original effective guide index n_g^{TM}. The effective guide index of TE wave remains unchanged. Finally, the wavelength TE/TM shift is obtained from using the corrected effective guide index of TM wave and unchanged effective guide index of TE wave. The result becomes smaller than uncorrected one.

Low Doped Silica Bent Waveguides. In the case of the doped silica bent waveguides with the refractive index for substrate n_2 of 1.45 and the lower Δ of 0.3 %, the difference of the effective guide index n_g^{eff} between the two orthogonal polarizations is order of 10^{-6}. Corresponding wavelength TE/TM shift is order of

Table 2. Effective guide index obtained from the FD-BPM simulation for TE and TM waves with $\Delta = 0.3$ %. Wavelength TE/TM shift is calculated for doped silica bent waveguides at the propagation distance L of 3 mm (Eight-digit decimal chopping arithmetic)

Radius [mm]	2	6	10
Index for TE wave	1.4530612	1.4522026	1.4520512
Index for TM wave	1.4530566	1.4522014	1.4520502
Index difference $[\times 10^{-7}]$	46.035458	11.692888	9.6997140
Uncorrected shift [pm]	4.9106642	1.2480335	1.0354012
Corrected shift [pm]	3.9372338	0.27402763	0.061293763

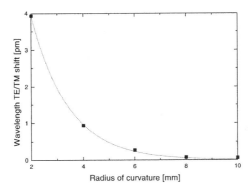

Fig. 4. Wavelength TE/TM shift of the doped silica bent waveguides with $\Delta = 0.3$ % as a function of the radius of curvature R

picometers for light of 1.55 μm as shown in Table 2. This wavelength TE/TM shift is negligible for larger radius of curvature R. For smaller radii R, the wavelength TE/TM shift rises exponentially as is shown in Fig. 4.

High Doped Silica Bent Waveguides. The waveguide birefringence is due to the different field-continuity conditions at the interfaces for the two orthogonal polarizations [10]. Thus, it can increase by increasing the refractive index difference between the waveguide core and the cladding region or by increasing the degree of bending. In case of the doped bent waveguide with Δ of 0.7 % in the same substrate, the difference of effective guide index between the two orthogonal polarizations is larger than the case of the doped bent waveguides with Δ of 0.3 % as shown in Table 3. The uncorrected wavelength TE/TM shift is therefore greater than the case of the low doped bent waveguides. In Fig. 5, $\Delta\lambda$ also exponentially increases as R is reduced.

Table 3. Effective guide index obtained from the FD-BPM simulations for TE and TM waves with $\Delta = 0.7$ %. Wavelength TE/TM shift is calculated for doped silica bent waveguides at the propagation distance L of 3 mm (Eight-digit decimal chopping arithmetic)

Radius [mm]	2	6	10
Index for TE wave	1.4558373	1.4554365	1.4554060
Index for TM wave	1.4558316	1.4554327	1.4554024
Index difference [$\times 10^{-6}$]	5.6962180	3.7912021	3.5369608
Uncorrected shift [pm]	6.0646458	4.0375262	3.7668452
Corrected shift [pm]	2.4728382	0.44472950	0.17397314

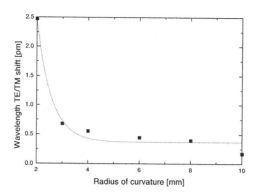

Fig. 5. Wavelength TE/TM shift of the doped silica bent waveguides with $\Delta = 0.7\ \%$ as a function of the radius of curvature R

Compound Semiconductor Bent Waveguides. In general, the cross section of ridge waveguides made of compound semiconductor material is generally rectangular, different from doped silica waveguides. Even for straight waveguides, the difference of effective guide index between the orthogonal polarizations exists in the rectangular cross section where this is non-existent in the square cross section. This difference results in a wavelength shift between TE and TM waves. In addition, above the waveguide is bent, the difference between the two orthogonal polarizations of compound waveguide becomes larger due to the bending effect. In order to obtain the guide-index difference for the bending effect only, we simulate a square waveguide made of InGaAsP of the bandgap index of 3.382 at 1.3 μm embedded in a homogeneous medium InP of constant refractive index of 3.167, though this square compound waveguide is very difficult to manufacture. The index difference Δ is 6.3571851 %. The dimension of cross section for the above single-mode waveguide is assumed to be 0.8 μm $\times$ 0.8 μm. The calculation

Table 4. Effective guide index obtained from the FD-BPM simulation for TE and TM waves with the bandgap index of 3.382 for InGaAsP at 1.3 μm. Wavelength TE/TM shift is calculated for InGaAsP/InP bent waveguides at the propagation distance L of 3 mm (Eight-digit decimal chopping arithmetic)

Radius [mm]	2	6	10
Index for TE wave	3.2705165	3.2704770	3.2704714
Index for TM wave	3.2695253	3.2695010	3.2694964
Index difference [$\times 10^{-4}$]	9.9122627	9.7607304	9.7503116
Uncorrected shift [nm]	0.46977310	0.46259710	0.46210411
Corrected shift [pm]	89.517064	82.336471	81.842825

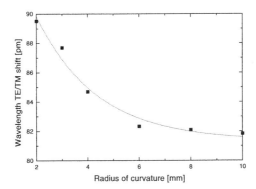

Fig. 6. Wavelength TE/TM shift of the InGaAsP/InP bent waveguides with the bandgap index of 3.382 for InGaAsP at 1.3 μm as a function of the radius of curvature R

results for the wavelength TE/TM shift is as shown in Table 4 and Fig. 6. It can be seen Δn_g^{eff} and $\Delta\lambda$ are much larger than for doped silica waveguides.

Stress Birefringence. Since an isotropic dielectric material is subject to mechanical stresses such as film, thermal, and intrinsic stress, it may become optically anisotropic. Due to this stress effect, the index ellipsoid changes. This phenomenon is known as stress birefringence or the photoelastic effect [11]. Doped silica (SiO_2) waveguides are generally made on crystal silicon (Si) substrates. There is a large difference in the thermal expansion coefficients between doped SiO_2 (0.5 to 1.0×10^{-6} /$^\circ$C) and Si ($\approx 2.5 \times 10^{-6}$ /$^\circ$C) [12]. The film stress in the silica waveguide is compressive, because the thermal expansion coefficient of silicon is larger than that of doped-silica, and a stress-induced birefringence is about 4×10^{-4} [13]. Waveguide birefringence can be reduced by replacing Si substrates with pure SiO_2 substrates, since pure SiO_2 substrates have a thermal expansion coefficient of approximately 0.35×10^{-6} /$^\circ$C. On the other hand, the birefringence can be increased to about 11×10^{-4} by using sapphire (Al_2O_3) substrates which have a larger thermal expansion coefficient of 5.4×10^{-6} /$^\circ$C [13]. A wavelength TE/TM shift by stress effect is not calculated when using the FD-BPM in bent waveguides with the doped silica. Actually, the wavelength TE/TM shift by stress effect is much larger than that by bending effect.

4   Conclusions

In bent waveguides, essential in complex planar lightwave circuits, the scalar TE and TM wave equations in cylindrical coordinates have been converted into the paraxial wave equations on the condition of slowly varying envelope. The effective index method is used to convert a two-dimensional refractive index profile into a one-dimensional refractive index profile for TE and TM polarized fields. In bent slab waveguides with one-dimensional refractive index profile, optical fields with horizontal and vertical polarizations are found from the initial fields using the scalar beam-propagation method employing the finite-difference method to solve the paraxial wave equations of TM and TE modes. A relatively small computational window has been used by adopting the transparent boundary condition.

In bent waveguides with the square cross section, the waveguide birefringence due to the difference of effective guide index between TE and TM polarized fields is calculated. The waveguide birefringence increases with increasing the refractive index difference between the waveguide and the cladding region as well as with increasing the degree of bending. The wavelength shift due to the birefringence of TE and TM polarized fields in bent waveguides is also calculated. The wavelength TE/TM shift for low and high doped silica and InGaAsP/InP bent waveguides are 0.06 pm, 0.44 pm, and 89.52 pm at the radius of 10 mm, 6 mm, and 2 mm respectively. In doped silica bent waveguides, there is the waveguide birefringence by stress effect, which can be larger than that by bending effect. However, this can not be calculated with the scalar beam-propagation method.

Acknowledgements

This work has been supported in part by the Korea Science and Engineering Foundation (KOSEF), the Korea IT Industry Promotion Agency (KIPA), Ministry of Science and Technology (MOST), and Ministry of Information and Communication (MIC) in Republic of Korea.

References

[1] K.Petermann, "Microbending loss in monomode fibers," *Electronics Letters*, vol. 12, pp. 107–109, 1976.

[2] M.Rivera, "A finite difference BPM analysis of bent dielectric waveguides," *IEEE Journal of Lightwave Technology*, vol. 13, no. 2, pp. 233–238, 1995.

[3] M. D.Feit and J. A.Fleck, "Light propagation in graded-index optical fibers," *Applied Optics*, vol. 17, no. 24, pp. 3990–3998, 1978.

[4] M. D.Feit and J. A.Fleck, "Calculation of dispersion in graded-index multimode fibers by a propagating beam method," *Applied Optics*, vol. 18, no. 16, pp. 2843–2851, 1978.

[5] L.Thylen, "Theory and application of the beam propagation method," *Numerical simulation and analysis of guided-wave optics and optoelectronics*, vol. 3, pp. 20–23, 1989.

[6] T. B.Koch, J. B.Davies, and D.Wickramasinghe, "Finite element/finite difference propagation algorithm for integrated optical device," *Electronics Letters*, vol. 25, pp. 514–516, 1989.

[7] S. T.Hendow and S. A.Shakir, "Recursive numerical solution for nonlinear wave propagation in fibers and cylindrically symmetric systems," *Applied Optics*, vol. 25, no. 11, pp. 1759–1764, 1986.

[8] J.Saijonmaa and D.Yevick, "Beam propagation analysis of loss in bent optical waveguides and fibers," *Journal of Optical Society of America*, vol. 73, no. 12, pp. 1785–1791, 1983.

[9] A.Yariv, *Optical Electronics*, Philadelphia: Saunders College Publishing, 1991.

[10] M. R.Amersfoort, *Phased-Array Wavelength Demultiplexers and Their Integration with Photodetector*, PhD thesis, Delft University of Technology, 1994.

[11] M.Born and E.Wolf, *Principles of Optics*, Oxford: Pergamon Press, fifth ed., 1975.

[12] M.Kawachi, "Silica waveguides on silicon and their application to integrated-optic components," *Optical and Quantum Electronics*, vol. 22, pp. 391–416, 1990.

[13] S.Martellucci, A. N.Chester, and M.Bertolotti, *Advances in Integrated Optics*, New York: Plenum Press, 1994.

Performance Analysis of Degree Four Topologies for the Optical Core of IP-over-WDM Networks

Rui M. F. Coelho[1], Joel J. P. C. Rodrigues[2], and Mário M. Freire[2]

[1] Superior School of Technology, Polytechnic Institute of Castelo Branco
Avenida do Empresário 6000-000 Castelo Branco, Portugal
{rmfcoelho}@netvisao.pt
[2] Department of Computer Science, University of Beira Interior
Rua Marquês d'Ávila e Bolama, 6201-001 Covilhã, Portugal
{joel,mario}@di.ubi.pt

Abstract. In this paper, we present a performance analysis of degree four topologies for the optical core of IP-over-WDM networks. For comparison purposes, degree three chordal ring topologies are also considered and the performance analysis is focused on chord lengths that lead to smallest diameters (best performance). It is shown that the increase of the nodal degree from 3 (degree three topology with smallest diameter) to 4 (degree four topology with smallest diameter) improves the network performance if a larger number of wavelengths per link is available. It is also shown that the performance of networks with degree four topologies with diameters slightly higher than the smallest diameter is very close to the performance of networks with smallest diameter. Our results show that there are several degree four topologies, with a performance very close to the optimum, that may be implemented in the optical core of IP-over-WDM networks. These results clearly increase the implementation flexibility of degree four topologies in those environments.

1 Introduction

IP-over-WDM (IP: Internet Protocol; WDM: Wavelength Division Multiplexing) networks are expected to be an infrastructure for next generation Internet, by directly carrying IP packets on WDM-based networks [1]-[2]. Optical networks are already in use to provide WDM point-to-point connections for a multi-layer architecture to transport IP traffic. Although this approach increases the link bandwidth by using WDM, it does not solve the problem of network bottleneck due to the exponential traffic growth driven by Internet-based services, since this solution only shifts the bottleneck problem from the link to the electronic router. A solution to this problem that also leads to lower management costs and lower complexity consists in the use of a two-layer architecture, in which IP traffic is transported directly over optical networks. In this new approach, some of the switching and routing functions, which have been performed by electronics, are incorporated into the optical domain. Therefore, next generation backbone networks should include both IP routers with IP-packet switching capabilities

H.-K. Kahng (Ed.): ICOIN 2003, LNCS 2662, pp. 108–117, 2003.

and optical cross-connects with wavelength-path switching capabilities to reduce the burden of heavy IP-packet switching loads.

Recent technology developments, such as the advent of optical add/drop multiplexers and optical cross-connects, are enabling the evolution from point-to-point WDM links to wavelength routing networks. Optical networks with wavelength routing in a mesh topology are now being under intense research. In [3], it is presented a study of the influence of nodal degree on the fibre length, capacity utilisation, and average and maximum path lengths of wavelength routed mesh networks. It is shown that average nodal degrees varying between 3 and 4.5 are of particular interest. In this paper, we consider WDM networks with degree four topologies.

Recently, Freire and da Silva [4] have investigated wavelength routed optical networks with chordal ring topology. Chordal rings are a well-known family of regular degree three topologies proposed by Arden and Lee in early eighties for interconnection of multi-computer systems [5]. In [4], Freire and da Silva have shown that the best network performance is obtained for the chord length that leads to the smallest network diameter. In [6], the same authors have shown that the performance of a chordal ring network (which has a nodal degree of 3), with a chord length that leads to the smallest diameter, is similar to the performance of a mesh-torus network (which has a nodal degree of 4). Since a (bi-directional) chordal ring network with N nodes has $3N$ (unidirectional) links and a (bi-directional) mesh-torus network with N nodes has 4N (unidirectional) links, the choice of a chordal ring with minimum diameter, instead of a mesh-torus, reduces network links by 25/100. Moreover, since chordal rings have lower nodal degree, they require in each switch, a smaller number of node-to-node interfacing (NNI) ports.

However, there are some restrictions that limit the practical implementation of chordal rings with the smallest network diameter (as well as mesh-torus), when compared with other chord lengths. In fact, the smallest network diameter in a chordal ring is obtained for a square value of N ($N = m^2$) and $N \leq 64$ [5].

In [7], we presented an assessment of the traffic performance in wavelength routing networks with random topologies of average nodal degrees of 2 and 3. It was shown that the performance of a network, with a random topology and an average nodal degree of 2, is better than the performance of rings and some chordal rings. It was also shown that the performance of a network, with a random topology and an average nodal degree of 3, is better than the performance of a network with a chordal ring topology with smallest diameter. This fact led us to the question of the existence of degree three topologies that outperform chordal rings.

In order to try to find degree three topologies that may outperform chordal rings with smallest diameter, in [8], we introduced a general regular degree three family, of which the chordal ring family is a particular case. We analysed all topologies of that general regular degree three family and we showed that there are several regular degree three topologies with the same smallest diameter of the chordal ring with smallest diameter and with exactly the same path blocking performance, but we have not found any degree three topology outperforming chordal rings with smallest diameter.

Since we have not found any regular degree three topology outperforming chordal rings with smallest diameter, we looked for irregular degree three topologies. In [9], we have obtained a lower bound of the traffic performance in networks with irregular degree three topologies and we have shown that this lower bound is very close to the performance of a network with a random topology and an average nodal degree of three, while the performance of a network with a chordal ring topology (regular topology) with smallest diameter is worst than the lower bound of the traffic performance in networks with irregular degree three topologies.

In this paper, we consider wavelength routing networks with degree four topologies. We introduce a general regular topology family for any nodal degree, but we concentrate on the performance analysis of networks with degree four topologies.

The remainder of this paper is organised as follows. Section 2 introduces regular topology families for any nodal degree. Section 3 briefly describes the model used to evaluate the path blocking performance in the optical core of IP-over-WDM networks with shortest path routing over degree four topologies. A performance analysis of WDM networks with degree four topologies is presented in section 4. Main conclusions are presented in Section 5.

2   Degree n Topologies

As described in [5], a chordal ring is basically a ring network, in which each node has an additional link, called a chord. The number of nodes in a chordal ring is assumed to be even, and nodes are indexed as 0, 1, 2, ..., N-1 around the N-node ring. It is also assumed that each odd-numbered node i (i=1, 3, ..., N-1) is connected to a node $(i+w)mod\ N$, where w is the chord length, which is assumed to be positive odd. For a given number of nodes there is an optimal chord length that leads to the smallest network diameter. The network diameter is the largest among all of the shortest path lengths between all pairs of nodes, being the length of a path determined by the number of hopes.

In [8], we introduced a general family of degree three topologies, of which the chordal ring family is a particular case. In each node of a chordal ring, we have a link to the previous node, a link to the next node and a chord. In [8], we assumed that the links to the previous and to the next nodes are replaced by chords. Thus, each node has three chords, instead of one. Let $w1$, $w2$, and $w3$ be the corresponding chord lengths, and N the number of nodes. We represented a general degree three topology by DTT($w1$, $w2$, $w3$). We assumed that each odd-numbered node i (i=1, 3, ..., N-1) is connected to the nodes $(i+w1)mod\ N$, $(i+w2)mod\ N$, and $(i+w3)mod\ N$, where the chord lengths, $w1$, $w2$, and $w3$ are assumed to be positive odd, with $w1 \leq N-1, w2 \leq N-1$, and $w3 \leq N-1$, and $w_i \neq w_j, \forall i \neq j$ and $1 \leq i,j \leq 3$.

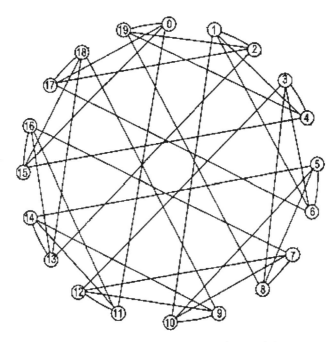

Fig. 1. Schematic representation of a D4T(1,3,5,9) for $N{=}20$ nodes

In this notation, a chordal ring with chord length w is simply represented by DTT(1,N-1,w).

Here, we introduce a general topology for a given nodal degree. We assume that instead of a topology with nodal degree of 3, we have a topology with a nodal degree of n, where n is a positive integer, and instead of having 3 chords we have n chords. We also assume that each odd-numbered node i ($i{=}1,3,...,N$-1) is connected to the nodes $(i+w)mod\ N$, $(i+w2)mod\ N$, ..., $(i+wn)mod\ N$, where the chord lengths, $w1$, $w2$, ... wn are assumed to be positive odd, with $w1 \leq N-1, w2 \leq N-1, ..., wn \leq N-1$, and $w_i \neq w_j, \forall i \neq j$ and $1 \leq i,j \leq n$. Now, we introduce a new notation: a general degree n topology is represented by DnT($w1$, $w2$,...,wn). In this new notation, a chordal ring family with chord length w is represented by D3T(1,N-1,w). As an example, figure 1 shows three topologies for networks with $N{=}20$ nodes: D3T(1,19,5) (degree three chordal ring with a chord length of 5), D3T(1,9,5), and D4T(1,19,9,5).

3   Evaluation of Path Blocking Probability

One of the key performance metrics used in optical networks with wavelength routing is the path blocking probability, i. e., the probability of a connection request being denied due to unavailable optical paths. To compute the path blocking probability in optical networks with wavelength conversion, we have used the model presented in [10], since it applies to topologies with low connectivity, has a moderate computational complexity, and takes into account dynamic traffic and the correlation between the wavelengths used on successive links of a multi-link path.

In [10], it is assumed that, given the loads on links 1, 2, ..., i-1, the load on link i of a path depends only on the load on link i-1 (Markovian correlation model). The analysis presented in [10] also assumes that the hop-length distribution is known, as well as the arrival rates of sessions at a link. The arrival rates at links have been estimated from the arrival rates of sessions to nodes, as in [10].

The hop-length distribution is a function of the network topology and the routing algorithm, and is easily determined for most regular topologies with the shortest-path algorithm. In this work, instead of defining the hop-length distribution through an analytical equation used by the analytical framework to compute the path blocking probability, we developed an algorithm that provides the hop-length distribution for a given DnT($w1,w2,...,wn$) topology.

4   Assessment of Path Blocking Performance

In this section, we present an assessment of the blocking performance in wavelength routing networks with a topology of the type D4T($w1,w2,w3,w4$). The performance analysis is focused on networks with 100 nodes.

We investigated minimum and maximum diameters for all D4T topologies with $w3$ and $w4$ free (both ranging from 1 and 99, with $w1 \neq w2 \neq w3 \neq w4$), for each fixed value of $w1$ and $w2$. Fig. 2 shows minimum (Min) and maximum (Max) diameters for the families D4T(1, 99, $w3$, $w4$) and D4T(1, 3, $w3$, $w4$), as a function of $w3$. The smallest diameter we found for each of these D4T families was 6, for N=100 nodes (see figure 2, for the case of families D4T(1, 99, $w3$, $w4$) and D4T(1, 3, $w3$, $w4$)). We found that each family of the type D4T(1, 99, $w3$, $w4$) and D4T(1, 3, $w3$, $w4$), ..., D4T(97, 99, $w3$, $w4$) has four minima (for a given value of $w3$), equal to 7, which are not the smallest diameter. We found that these minima occur in each D4T($w1$, $w2$, $w3$, $w4$) family, for which $w2=(w1+2)mod\ N$ and $w3 = (w1 + 4)mod\ N$, or $w2 = (w1 - 2)mod\ N$ and $w3 = (w1 + 2)mod\ N$, or $w2 = (w1 - 4)mod\ N$ and $w3 = (w1 - 2)mod\ N$. As may be seen in Fig. 3, each family of this kind, i.e. D4T($w1$, $(w1+2)mod\ N$, $(w1+4)mod\ N$, $w4$), or D4T($w1$, $(w1-2)mod\ N$, $(w1+2)mod\ N$, $w4$), or D4T($w1$, $(w1-4)mod\ N$, $(w1-2)mod\ N$, $w4$), with $1 \leq w1 \leq 99$ and with $1 \leq w4 \leq 99$ and $w1 \neq w2 \neq w3 \neq w4$ has a diameter which is a shifted version (with respect to $w4$) of the diameter of the other two families.

As may be seen in Fig. 3, each of these three families with shifted diameters has 12 minimum diameters (which is 6, for N=100), as a function of $w4$. Moreover, we have found that for i) a family of the type D4T($w1$, $(w1+2)mod\ N$, $(w1+4)mod\ N$, $w4$), those four minimum diameters occur at $w4=(17+w1-1)mod\ N$, $w4=(19+w1-1)mod\ N$, $w4=(23+w1-1)mod\ N$, and $w4=(29+w1-1)mod\ N$, $w4=(41+w1-1)mod\ N$, $w4=(43+w1-1)mod\ N$, $w4=(59+w1-1)mod\ N$, and $w4=(61+w1-1)mod\ N$, $w4=(73+w1-1)mod\ N$, $w4=(79+w1-1)mod\ N$, $w4=(83$

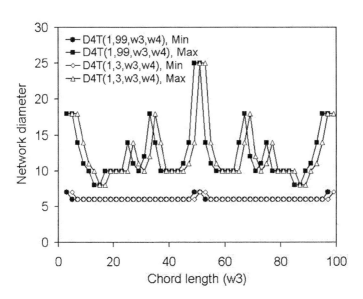

Fig. 2.   Minimum (Min) and maximum (Max) network diameters for D4T(1,99,$w3$,$w4$) and D4T(1,3,$w3$,$w4$), with $3 \leq w3 \leq 99$, and $3 \leq w4 \leq 99$, and $w1 \neq w2 \neq w3 \neq w4$

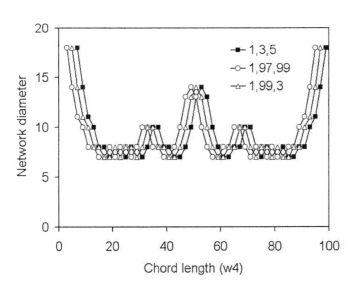

Fig. 3.   Network diameter for D4T($w1$,($w1 + 2$)mod N,($w1 + 4$)mod N,$w4$), D4T($w1$,($w1 - 2$)mod N,($w1 + 2$)mod N,$w4$), and D4T($w1$,($w1 - 4$)mod N,($w1 - 2$)mod N,$w4$), with $w1 = 1, 1 \leq w4 \leq 99$, and $w1 \neq w2 \neq w3 \neq w4$

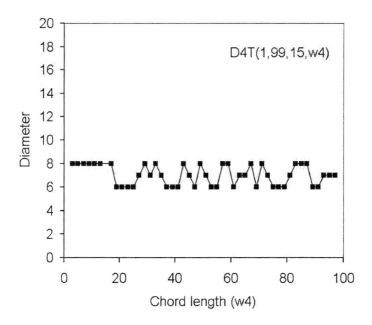

Fig. 4. Network diameters for D4T(1, 99, 15, $w4$), with $1 \leq w4 \leq 99$, and $w1 \neq w2 \neq w3 \neq w4$

$+w1$-1)mod N, and $w4=(85+w1$-1)mod N; *ii*) a family of the type D4T($w1$, $(w1$-2)mod N, $(w1+2)$mod$ N, $w4$), those four minimum diameters occur at $w4=(15+w1$-1)mod N, $w4=(17+w1$-1)mod N, $w4=(21+w1$-1)mod N, and $w4=(27 + w1$-1)mod N, $w4=(39+w1$-1)mod N, $w4=(41+w1$-1)mod N, $w4=(57+w1$-1)mod N, and $w4=(59+w1$-1)mod N, $w4=(71+w1$-1)mod N, $w4 = (77 + w1 - 1)$mod$ N, $w4 = (81+w1-1)$mod$ N, and $w4 = (83+w1-1)$mod$ N; *iii*) a family of the type D4T($w1$, $(w1$-4)mod N, $(w1$-2)mod N, $w4$), those four minimum diameters occur at $w4=(13+w1$-1)mod N, $w4=(15+w1$-1)mod N, $w4=(17+w1$-1)mod N, and $w4=(25 + w1$-1) mod N, $w4=(37+w1$-1) mod N, $w4=(39+w1$-1) mod N, $w4=(55+w1$-1) mod N, and $w4=(57+w1$-1) mod N, $w4=(69+w1$-1) mod N, $w4=(75+w1$-1) mod N, $w4=(77+w1$-1) mod N, and $w4=(81+w1$-1) mod N.

We also observed that each topology of the type D4T($w1$, $w2$, $w3$, $w4$), with a smallest diameter of 6 (for N=100), has the same diameter and the same hop-length distribution of each topology of the type D4T($(w1+k)$mod$ N, $(w2+k)$mod$ N, $(w3+k)$mod$ N, $(w4+k)$mod$ N), where k=2i, being i an odd positive integer. For instance, D4T(1,99,5,23) and D4T(3,1,7,25) topologies have exactly the same diameter (6) and the same hop-length distribution.

In the following, we concentrate the performance analysis on D4T(1,99,3,19), which is a degree-four chordal ring with smallest diameter, and D4T(1,99,5,29). For comparison purposes D3T(1,99,13), which is a degree three chordal ring with smallest diameter, is also considered. The smallest diameter of the D4T(1, 99,

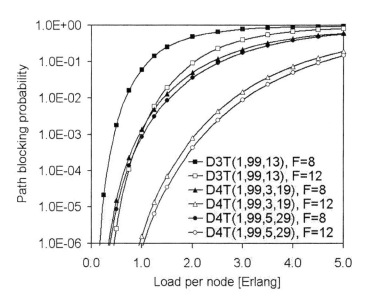

Fig. 5. Path blocking probability for D3T(1,99,13), (chordal ring with minimum diameter) and for D4T(1,99,3,19) and D4T(1,99,5,29), without wavelength interchange. $N=100$; F: number of wavelengths per link

3, $w4$), which is equal to 7, occur at $w4=17$, $w4=19$, $w4=23$, $w4=29$, $w4=41$, $w4=43$, $w4=59$, $w4=61$, $w4=73$, $w4=79$, $w4=83$, and $w4=85$. The smallest diameter of the D4T(1,99,5,$w4$), which is equal to 6, occur at $w4=23$, $w4=29$, $w4=31$, $w4=39$, $w4=41$, $w4=43$, $w4=61$, $w4=63$, $w4=73$, $w4=75$, $w4=79$, and $w4=81$. We observed a small change in the hop-length distribution of each of these two degree four topology families for a given smallest diameter. However, the network performance for both topology families with smallest diameter is very close. Fig. 4 shows the path blocking probability for D3T(1,99,13), D4T(1,99,3,19) and D4T(1,99,5,29), without wavelength interchange. As may be seen, there is an almost insignificant difference between the blocking probabilities of D4T(1,99,3,19) and D4T(1,99,5,29), which is mainly due to the reduction of the network diameter from 7 (diameter of D4T(1, 99, 3, 19)) to 6 (diameter of D4T(1, 99, 3, 19)). However, the reduction of the diameter (of D3T(1, 99, 13)) from 9 to 6 (D4T(1, 99, 3, 19)) or 7 (of D4T(1, 99, 3, 19)) lead to a larger difference on the blocking probability, namely, for a larger number of wavelengths per link. These results are confirmed in Fig. 5, which shows the blocking probability versus converter density for networks with the following topologies: D3T(1, 99, 13), D4T(1, 99, 3, 19) and D4T(1, 99, 5, 29). These results clearly show that there are several degree four topologies that lead to similar performances, which increase the implementation flexibility of degree four topologies in the optical core of IP-over-WDM networks. However, the decrease of the nodal degree from 4 to 3, clearly lead to a performance degradation.

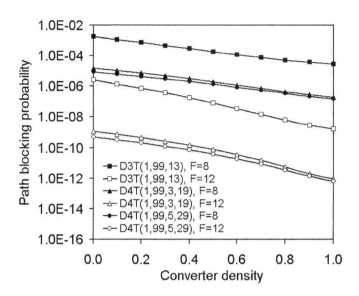

Fig. 6. Path blocking probability versus converter density for networks with D3T(1,99,13) and D4T(1,99,3,19) and D4T(1,99,5,29). $N{=}100$; Load per node: 0.5 Erlang; F: number of wavelengths per link

Another interesting result that we found is concerned with the influence of each D3T topology with smallest diameter (which is 9, for $N{=}100$) on the characteristics of a D4T topology. We found that if we add a fourth chord to each D3T topology with smallest diameter, i.e., D3T($w1$, $(w1\text{-}2)mod\ N$, $w3$), with $w3{=}(13{+}w1\text{-}1)mod\ N$, $w3{=}(15{+}w1\text{-}1)mod\ N$, $w3{=}(85{+}w1\text{-}1)mod\ N$, or $w3{=}(87{+}w1\text{-}1)mod\ N$, and D3T($w1$, $(w1{+}2)mod\ N$, $w3$), with $w3{=}(15{+}w1\text{-}1)mod\ N$, $w3{=}(17{+}w1\text{-}1)mod\ N$, $w3{=}(87{+}w1\text{-}1)mod\ N$, and $w3{=}(89{+}w1\text{-}1)mod\ N$, we obtain a D4T topology with a diameter between 6 and 8, as may be seen in Fig. 6.

5 Conclusions

We presented a performance analysis of degree four topologies to be used in the optical core of IP-over-WDM networks, assuming that the shortest path routing algorithm is used. The performance analysis was focused on networks with smallest diameters or diameters close to the smallest. It was shown that the increase of the nodal degree from 3 to 4 improves the network performance if a larger number of wavelengths per link is available. It was also shown that, for degree four topologies, if the diameter is slightly higher than the smallest (e.g. 7), the corresponding network performance is still very close to the performance obtained with the smallest diameter (6). We found several degree four topologies, with a performance similar to the optimum. The existence of these high performance topologies increase the implementation flexibility of degree four topologies in the optical core of IP-over-WDM networks.

Acknowledgements

Part of this work has been supported by Portuguese Foundation of Science and Technology (Fundação para a Ciência e Tecnologia, Portugal), in the framework of project PIONEER, at IT-Coimbra.

References

[1] Qiao, C., Datta, D., Ellinas, G., Gladisch, A., and Molina, E. (eds): WDM-Based Network Architctures. IEEE Journal of Selected Areas in Communications, **20**/1 (2002).

[2] Dixit, S. S., and Lin, P. J. (eds.): Optical Networking: Signs of Maturity. IEEE Communications Magazine, **40**/2, (2002) 64–167.

[3] Hjelme, D. R.: Importance of Meshing Degree on Hardware Requirements and Capacity Utilization in Wavelength Routed Optical Networks. In: M. Gagnaire and H. R. van As (eds.): Proceedings of the 3rd IFIP Working Conference on Optical Network Design and Modeling (ONDM'99), Paris, France, (1999), 417–424.

[4] Freire, M. M., da Silva, H. J. A.: Influence of Chord Length on the Blocking Performance of Wavelength Routed Chordal Ring Networks. In Jukan, A. (ed.): Towards an Optical Internet - New Visions in Optical Network Design and Modelling, Kluwer Academic Publishers, Boston (2002) 79–88.

[5] Arden, B. W., Lee, H.: Analysis of Chordal Ring Network. IEEE Transactions on Computers **C-30** (1981) 291–295.

[6] Freire, M. M., da Silva, H. J. A.: Performance Comparison of Wavelength Routing Optical Networks with Chordal Ring and Mesh-Torus Topologies. In Lorenz, P. (Ed.): Networking - ICN 2001, Lecture Notes in Computer Science, Vol. 2093. Springer-Verlag, Berlin Heidelberg (2001) 358–367.

[7] Coelho, R. M. F. and M. M. Freire: Optical Backbones with Low Connectivity for IP-over-WDM Networks. In I. Chong (Ed.): Information Netwoking- Wired Communications and Management, Lecture Notes in Computer Science, Vol. 2343, Springer-Verlag, Heidelberg (2002) 327–336.

[8] Coelho, R. M. F., Rodrigues, J. J. P. C., and Freire, M. M.: Performance Assessment of Wavelength Routed Optical Networks with Shortest Path Routing over Degree Three Topologies. In Proceedings of IEEE International Conference on Networks (ICON'2002), Singapore (2002), 3–8.

[9] Coelho, R. M. F., Rodrigues J. J. P. C., and Freire, M. M.: Performance Assessment of Wavelength Routing Optical Networks with Irregular Degree-Three Topologies. In Proceedings of IEEE International Conference on High Speed Networks and Multimedia Communications (HSNMC'02), Jeju, Korea (2002), 392-396.

[10] Subramaniam, S., Azizoglu, M., Somani, A. K.: All-Optical Networks with Sparce Wavelength Conversion. IEEE/ACM Transactions on Networking 4 (1996) 544-557.

Analysis of End-to-End Recovery Algorithms with Preemptive Priority in GMPLS Networks

Hyun Cheol Kim[1], Jun Kyun Choi[2], Seongjin Ahn[1], and Jin Wook Chung[1]

[1] SungKyunKwan University, 440-746 Suwon, Korea
{hckim,sjahn,jwchung}@songgang.skku.ac.kr
[2] Information and Communications University, P.O.Box 77 Daejon, Korea
jkchoi@icu.ac.kr

Abstract. To provide high resilience against failures, optical networks must have an ability to maintain an acceptable level of service during network failures. This paper proposes a new enhanced end-to-end recovery algorithm that guarantees fast and dynamic utilization of network resources in Wavelength Division Multiplexing (WDM) based Generalized Multi-protocol Label Switching (GMPLS) networks. The proposed algorithm guarantees the maximum availability with less burden in finding as many disjointed paths. This paper also has developed an analytical model and performed analysis for the proposed algorithm in terms of two performance factors: mean system time and blocking probability.

1 Introduction

With the advent of WDM technology and the integration of various communication technologies, today's communication networks can provide integrated high quality services. Although WDM technology has greatly increased the transmission capacity and has provided larger bandwidth, it also has some potential problems. The most painful problem is the survivability of networks. Therefore, a key expectation of the WDM based GMPLS network in the aspect of survivability is that it will offer fast recovery, comparable speed to Synchronous Optical Network (SONET), and versatile survivable functions, such as priority based recovery schemes in Asynchronous Transfer Mode (ATM) network [1], [2].

The techniques that have been proposed for survivability can be classified into two general categories: protection and restoration. Protection is a predetermined failure recovery and one or more dedicated protection Label Switched Paths (LSPs) is/are fully established to protect one or more working LSPs. While restoration denotes the paradigm whereby some resources may be pre-computed and signaled a priori, but not cross-connected to restore a working LSP [3]. Although protection mechanisms allow service providers to offer hard guarantees on recovery time, most data services may not require such hard recovery time. Moreover, it is difficult to find numbers of fully disjoint recovery LSP with the working LSP. Restoration mechanisms utilize bandwidth more efficiently than protection and it can naturally handle simultaneous multiple fiber failures [4].

H.-K. Kahng (Ed.): ICOIN 2003, LNCS 2662, pp. 118–127, 2003.
© Springer-Verlag Berlin Heidelberg 2003

This paper proposes a new enhanced recovery algorithm in GMPLS networks that uses both protection and restoration schemes. The proposed algorithm supports protection LSPs that can transform dynamically. The algorithm also provides efficient recovery function for different types of service by supporting dynamic protection LSPs and maximum restoration LSPs. The object of the proposed algorithm is to use the Shared Risk Link Group (SRLG) concept with the expanded method that is proposed in this paper to provide as many backup paths as possible. Moreover, the proposed algorithm does not need an extension of GMPLS signaling to support this mechanism. This paper also proposes a generalized queueing model to represent the enhanced recovery algorithm and a method to apply the queueing model using the extended SRLG concept.

The rest of this paper is organized as follows. Various recovery mechanisms and proposed recovery algorithms are described in section 2 and 3, respectively. The queueing analysis of the priority based restoration model and the recovery timing analysis are also presented. Finally the paper concludes in Section 4.

2   The Existing Recovery Algorithm

Recovery schemes in GMPLS networks can be classified under its coverage: link based and end-to-end based. Link based recovery reroute the disrupted traffic between nodes of the failed link and offer fast failure detection and tries to recover at the expense of efficiency. While in end-to-end recovery, the source and destination nodes of affected connections switch to a backup path upon failure and only the end nodes are responsible for initiating the establishment of backup paths. The end-to-end recovery scheme provides great flexibility to achieve survivability, since it enables the separation between the logical aspects of the paths and physical topologies.

For analysis, we get restoration setup time equations from operation procedures. In our analysis reported here, we assume that the inter-arrival time, service holding time, and duration of the fiber faults are exponentially distributed. Notations deployed in our analysis are as followings:

- $r_{1+1}, r_{m:n}, r_{path}$: Restoration time of $1+1, m : n, path$ recovery, respectively.
- T: Processing time including queueing delay at each node.
- C_{1+1}: The sum of failure detection time and switching over time.
- $C_{m:n}$: The sum of failure detection time and lightpath table lookup time, backup path selection time, and SRLG check time.
- C_{path}: The sum of failure detection time and path search time.
- $S_{m:n}$: Time to configure and setup a Optical Cross Connect (OXC).
- S_{path}: Time to configure and setup a OXC with wavelength reservation.
- $n_{1+1}, n_s, n_{m:n}$ and n_{path}: The number of nodes from a destination node of a failed link to a destination node in a working path, from a source node of a failed link to a source node in a working path, from a source node to a destination node in a backup path, and from a source node to a destination node in a backup path, respectively.
- D, G: Service hold time distribution and revert time distribution.

2.1   Restoration Setup Time Analysis

In $1 + 1$ protection, a destination node of the failed link notifies the destination node of the connection, which immediately switches over a recovery path. Thus, restoration time is (1), where $\mu_{1+1} = \overline{r_{1+1}} + \min(G, D)$.

$$r_{1+1} = (P + T)n_{1+1} + C_{1+1} \qquad (1)$$

In $m : n$ recovery, a set of m specific recovery paths protects a set of up to n working paths. Obviously, if several working LSPs are concurrently affected by some failure, only m of these failed LSPs may be recovered. A source node searches for a recovery path not belonging to the same SRLG and sends a setup message to configure the OXC. The restoration time of $m : n$ recovery is :

$$r_{m:n} = P(n_s - 1) + Tn_s + S_{m:n}n_{m:n} +$$
$$2P(n_{m:n} - 1) + 2Tn_{m:n} + C_{m:n}$$
$$\mu_{m:n} = \overline{r_{m:n}} + \min(G, D)$$

In path restoration, a source node of the connection looks for the backup path that satisfies the bandwidth and SRLG requirements. The destination node, upon receiving the setup message, sends a confirm message back to the source node. Finally, restoration time is the following, where $\mu_{path} = \overline{r_{path}} + \min(G, D)$.

$$r_{path} = P(n_s - 1) + Tn_s + S_{path}n_{path} +$$
$$2P(n_{path} - 1) + 2Tn_{path} + C_{path}$$

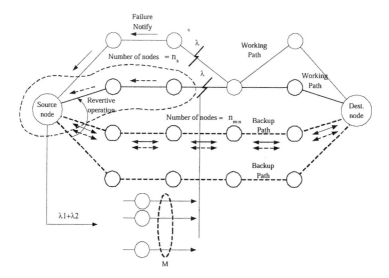

Fig. 1. $m : n$ recovery queueing model

2.2 Blocking Probability Analysis

In $m : n$ recovery, a LSP failure input with parameter λ feeds m identical servers that operate in parallel as shown in Fig. 1. Each server has an independent, identically distributed exponential service holding distribution with parameter $\mu_{m:n}$. However, there is no facility in the system to wait. If an arriving request finds all servers busy, it leaves the composite queue without waiting for service.

$$p_k = p_0 \left(\frac{\lambda}{\mu_{m:n}} \right)^k \frac{1}{k!} \qquad k = 1, 2, ..., m$$

Solving for p_0, the normalization condition $\sum_{k=0}^{n} p_k = 1$

$$p_0 = \left[\sum_{k=0}^{m} \left(\frac{\lambda}{\mu_{m:n}} \right)^k \frac{1}{k!} \right]^{-1}$$

The probability that a failure arrival will find all servers are busy and therefore be lost is (2). In $m : n$ recovery, the blocking probability is mainly affected by service hold time distribution as shown in Fig. 2.

$$p_m = \frac{(\lambda/\mu_{m:n})^m / m!}{\sum_{k=0}^{m} (\lambda/\mu_{m:n})^k / k!} \qquad (2)$$

In $1 : n$ recovery, only one backup path is available for recovery. Thus, $1 : n$ recovery can be analyzed by the $M/M/1/K$ $(K = 1)$ queueing model. The

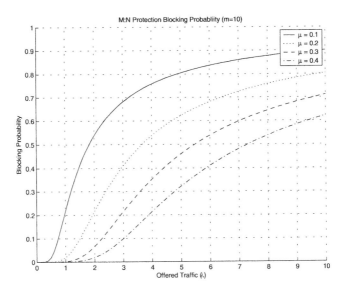

Fig. 2. Blocking probability of m:n recovery

blocking probability of $1 : n$ recovery increases, as failure arrival rate increases as shown in Fig. 3. The blocking probability of $1 : n$ recovery almost linearly increases as failure arrival rate increases regardless of the number of nodes.

$$p_m = \frac{(\lambda/\mu_{1:n})^m/m!}{\sum_{k=0}^{m}(\lambda/\mu_{1:n})^k/k!}$$

3   The Proposed Recovery Algorithm

To minimize the failure interference, there are recent efforts on research of sharing links between recovery paths or between working paths and recovery paths. A set of links may constitute a shared risk link group, if they share a resource whose failure may affect all links in the set. Several algorithms have been presented to find the maximum disjoint paths between source and destination nodes. However, when a link sharing many risks with other links is used as a disjoint route, a problem of not obtaining a sufficient number of disjoint paths will occur [4].

As shown in Fig. 4, if SRLG 1 exists, conventional recovery algorithm try to set up new disjoint path that is not included in the SRLG. So, all the paths from node 1, node 2, and node 3 to node 4 are excluded. But, for example as in failure 2, when an error occurs in transponder of OXC, the paths to the node 2 and node 3 are still available. This denotes that the number of recovery LSP will be very limited to smaller range when applying $m : n$ type recovery scheme.

Fig. 4 shows a extension of SRLG constraints of the enhanced recovery algorithm. The object of SRLG extension is to use the SRLG with the expanded

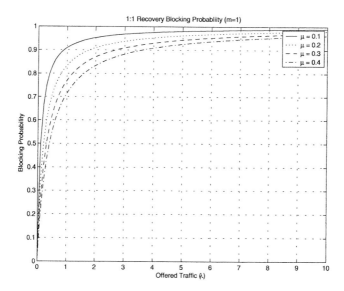

Fig. 3. Blocking probability of 1+1 recovery

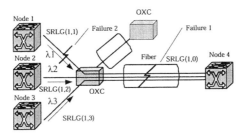

Fig. 4. Example of extended SRLG network configuration

method to satisfy as many demands as possible for a fixed amount of network resources. The object of SRLG extension is to distinguish different recovery LSP in the SRLG group and when there is a SRLG setup request. It continues to requests for LSP setup for each LSP in SRLG group until the LSP are setup. When at least one of the LSP in SRLG group has succeeded in recovery LSP establishment, the other requests in the queues will be deleted. The great advantage of the proposed recovery algorithm is that it provides much more recovery path compared to the conventional $m : n$ type recovery method.

Fig. 5 shows the strategy used in the proposed algorithm to allocate a recovery path. In order to provide fast recovery for the high priority LSP, the minimum number of protection LSP will be setup. However, the number of protection LSP excluding SRLG restoration path area can be adjusted by the network manager. In addition, the protection LSP has higher priority in the preemptive area than restoration LSP, thus the SRLG restoration LSP can be preempted by the protection LSP. SRLG restoration path area is an area in which the restoration LSP can be installed using extended SRLG mechanism. The SRLG restoration LSP can be extended to preemptive area. Since it has lower priority compared to the protection LSP in the preemptive area, when a collision occurs, the restora-

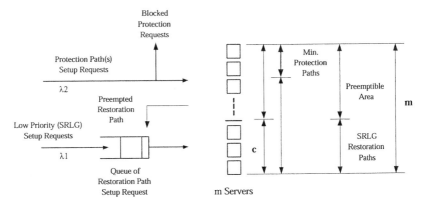

Fig. 5. Path allocation strategy of the proposed algorithm

tion LSP will be discarded and a protection LSP will be setup. The number of minimum protection path and SRLG restoration path is determined by the management policy decision and can be modified dynamically.

3.1 Blocking Probability of Proposed Recovery Algorithm

This paper proposes an improved algorithm that supports the maximum number of recovery paths that can be used by SRLG restoration LSP without preemption. The proposed algorithm gives a preemptive priority to protection path over SRLG path only in the region above the minimum number of SRLG paths. Arriving SRLG restoration path setup request that cannot find idle path are queued for service, but protection path setup requests being denied of services are blocked. The maximum number of paths that SRLG restoration can use without preemption is called the cut-off value c. By adjusting c, the proposed algorithm can have an effective control to adapt to the varying recovery requests.

To use matrix geometric solution, we define the steady-state by $s(n_1, n_2)$, where n_1 is the number of SRLG restoration paths in the system and n_2 is the number of protection paths being served. The state space can be represented by the set $\{s(n_1, n_2) \mid 0 \le n_1, 0 \le n_2 \le m\}$. Also, let the steady-state probability that the system is in state $s(n_1, n_2)$ be $p(n_1, n_2)$. The steady state probability vector is then partitioned as $p = (p_0, p_1, ...)$, where n_2th component of p_{n1} is $p(n_1, n_2)$ [5]. The vector p is the solution of the equations $pQ = 0, pe = 1$.

Q is transition rate matrix of the Markov process. The matrix Q is given by

$$Q = \begin{vmatrix} A_0 & D & & & & 0 \\ B_1 & A_1 & D & & & \\ & \cdots & \cdots & \cdots & & \\ & & \cdots & \cdots & \cdots & \\ & & & B_m & A_m & D \\ & & & & B_m & A_m & D \\ 0 & & & & & \cdots & \cdots & \cdots \end{vmatrix} \tag{3}$$

In equation (3) submatrices are defined for $i, j, l = 0, 1, 2, ..., m$ by (4) where $a_l(i)$ is the value that makes the sum of the row elements of Q equal to zero.

$$A_l(i, j) = \begin{cases} \lambda_2 & if\ i = j - 1\ \ and\ j \le m - k(l) \\ (j+1)\mu_2 & if\ i = j + 1 \\ a_l(i) & if\ i = j \\ 0 & otherwise \end{cases}$$

$$B_l(i, j) = \begin{cases} \min(l,\ m - j)\mu_1 & if\ i = j - 1 \\ 0 & otherwise \end{cases} \tag{4}$$

$$D(i, j) = \begin{cases} \lambda_1 & if\ i = j \\ 0 & otherwise \end{cases}$$

$$k(l) = \begin{cases} l & if\ 0 \le l \le c \\ c & if\ c < l \le m \end{cases}$$

To solve p with transition rate matrix Q, Neut's two-step process can be applied to obtain the minimal nonnegative matrix R of the matrix equation $R^2 B_m +$

$RA_m + D = 0$ [6]. As shown in (5), we can determine the minimal nonnegative matrix of the matrix equation by iteration. Iterations can be made directly until the (6) is satisfied, where $R(n)$ is the nth iteration and ε is the degree of accuracy required. The matrix R gives the relationship (7) [7].

$$R^{(0)} = 0$$
$$R^{(1)} = -DA - 1$$
$$R^{(2)} = -(R^{(1)})^2 BA^{-1} - DA^{-1} \tag{5}$$
$$\cdots$$
$$R^{(m+1)} = -(R^{(m)})^2 B_m A_m^{-1} - DA_m^{-1}$$

$$\max_{i,j} [R_{ij}(m) - R_{ij}(m-1)] < \varepsilon \tag{6}$$

$$p_k = p_{m-1} R^{k-m+1}, \quad k \geq m \tag{7}$$

Once the matrix R and the boundary probability vector $\widetilde{p} = (p_0, p_1, ..., p_m)$ have been computed, as shown in Fig. 6, the mean number of SRLG setup request messages in the system is [8]:

$$N_1 = \sum_{n_2=0}^{m} \sum_{n_1=0}^{\infty} n_1 p(n_1, n_2)$$
$$= p_{m-1} R^2 (I - R)^{-2} e + m p_{m-1} R(I - R)^{-1} e + \sum_{n_2=0}^{m} \sum_{n_1=0}^{m-1} n_1 p(n_1, n_2)$$

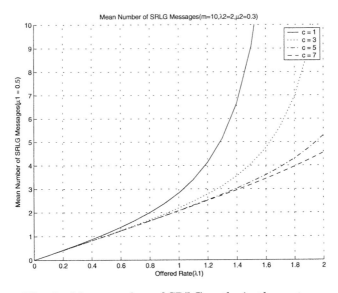

Fig. 6. Mean number of SRLG paths in the system

Then, the mean system time (i.e., the sum of queueing time and service time) for SRLG restoration path setup request messages is $W_1 = N_1/\lambda_1$. The mean system time W_1 for SRLG restoration path request message has only the component of service time, i.e., $W_1 = N_1/\mu_1^{-1}$ [9]. Finally, the blocking probability for protection path setup request message is :

$$P_{N2} = \sum_{n2=(m-c)}^{m} \sum_{n1=\max(0,\,(m-c))}^{\infty} p(n_1, n_2)$$

$$P_{N2} = \sum_{n2=(m-c)}^{m} \sum_{n1=\max(0,\,(m-c))}^{m-1} p(n_1, n_2) + \sum_{n2=(m-c)}^{m} [p_m R(I - R)^{-1}]_{n2}$$

Fig. 7 shows the blocking probability of proposed $m : n$ recovery and also shows the effect of the cut off value. From these figures, we can see that there is a trade-off between protection recovery LSP blocking probability and SRLG restoration LSP setup time. The proposed algorithm can give satisfactory recovery service to each type of recovery requests by choosing appropriately a cut-off value. The cut-off value can be changed as the recovery requests and network loads are changed. Thus, by simply changing the cut-off value, the proposed algorithm can adapt to the varying optical recovery requests more easily.

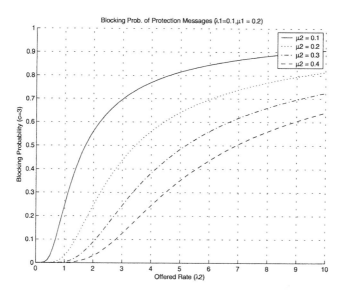

Fig. 7. Blocking probability of proposed recovery

4   Conclusions

As optical networks carry more and more information, even a break in a fiber link or the interruption of service for short periods of time can cause severe service loss. Thus, the prevention of service interruption and the reduction of service loss to a minimum are two of the major critical issues.

Although conventional protection scheme does provide quick recovery time, it has disadvantage of using up too much bandwidth and lack of ability to find sufficient disjoint paths. This paper proposes a new enhanced path recovery algorithm that overcomes these problems of conventional recovery schemes. The great advantage of the proposed recovery algorithm is that it provides much more recovery path compared to the conventional $m : n$ type recovery method.

References

[1] Yinghua Ye, Sudhir Dixit, Mohamed Ali: On Joint Protection/Restoration in IP-Centric DWDM-Based Optical Transport Networks, IEEE Communications Magazine, Vol. 6. (2000) 174–183
[2] S. Ramamurthy, Biswanath Mukherjee: Survivable WDM Mesh Networks, Part I-Protection, INFOCOM '99, Vol. 2. (1999) 744–751
[3] S. Ramamurthy, Biswanath Mukherjee: Survivable WDM Mesh Networks, Part II-Restoration, ICC '99, Vol. 3. (1999) 2023–2030
[4] Eiji Oki, Nobuaki Matsuura: A Disjoint Path Selection Schemes With Shared Risk Link Groups in GMPLS Networks, IEEE Communications Letters, Vol. 6, (2002) 406–408
[5] You Ze Cho, Chong Kwan Un: Analysis of the M/G/1 Queue under a Combined Preemp-tive/Nonpreemptive Priority Discipline, IEEE Transactions on Communications, Vol. 41. (1993) 132–141
[6] Marcel F. Neuts: Matrix-Geometric Solutions in Stochastic Models, (1983) 81–100
[7] Kraimeche. B, Schwartz. M: Bandwidth Allocation Strategies in Wide-Band Integrated Networks, IEEE J. Sel. Areas in Communications, (1986) 1293–1301
[8] B. Ngo, H. Lee: Queueing Analysis of Traffic Access Control Strategies with Preemptive and Nonpreemptive Discipline in Wideband Integrated Networks, IEEE J. on Selected Areas in Communications, Vol. 9. (1991) 1093–1109
[9] YoungHan Kim, ChongKwan Un: Analysis of Bandwidth Allocation Strategies with Access Restrictions in Broadband ISDN, IEEE Transactions On Communications, Vol. 41. (1993) 771–781

An Efficient Restoration Scheme
Using Protection Domain
for Dynamic Traffic Demands in WDM Networks

Chen-Shie Ho*, Ing-Yi Chen **, and Sy-Yen Kuo

Department of Electrical Engineering
National Taiwan University
Taipei, Taiwan
sykuo@cc.ee.ntu.edu.tw

Abstract. Network survivability is a key issue in reliable WDM optical
network design. Rather than conventional connection link or path based
protection switching scheme, in this paper we propose protection domain-
based approach to fast traffic recovery from single/multiple link/node
failures. The simulation results show the locality resource utilization and
restoration time are efficient under heavy load in the dynamic traffic
environment over backbone networks.

1   Background

Wavelength-division multiplexing (WDM) technology offers the capability of
building large wide-area networks with Tb/s order of throughput, and is a viable
solution to meet the dramatic bandwidth demand arising from several emerging real-time broadband service applications. A WDM optical network consists
of a set of reconfigurable wavelength cross-connects (OXCs) interconnected by
bi-directional point-to-point fiber links in an arbitrary topology. This paper considers the wide area wavelength routed mesh-based backbone environment. The
OXCs may possess wavelength conversion capability to convert data on an incoming wavelength into data on an outgoing wavelength in the optical domain,
without optical-electronic conversion. When the optical layer receives a connection or session request, it establishes a lightpath for the session during its service
period. A lightpath is defined as a set of links through intermediate routers from
source to destination, with a wavelength assigned on each link of the path. For
networks without wavelength conversion, wavelength continuity constraint will
be required which implies that the same wavelength has to be used on all links of
the lightpath. In order to establish a lightpath, the network needs to decide on
the route and the wavelengths for the lightpath. Given a set of connections, the

---

* Chen-Shie Ho is also with the Department of Electronics Engineering, Van Nung
  Institute of Technology, Chung-Li, Tao-Yuan, Taiwan.
** Ing-Yi Chen is with the Department of Electronics Engineering, Chung Yuan Chris-
  tian University, Chung-Li, Tao-Yuan, Taiwan.

H.-K. Kahng (Ed.): ICOIN 2003, LNCS 2662, pp. 128–137, 2003.
© Springer-Verlag Berlin Heidelberg 2003

problem of setting up lightpaths by routing and assigning wavelength to each connection is called the Routing and Wavelength Assignment (RWA) problem [1]. The traffic demand assumed by a RWA algorithm can be either static or dynamic. In the static traffic condition, the entire set of connections demands is known ahead of time. The objective is to assign lightpaths to all of them so as to minimize the resources (wavelengths or fibers) required. In the dynamic traffic condition, the connection requests arrive based on some stochastic process. Once a request is provisioned, the connection is held for a random finite time before being terminated. The objective here is to decrease the blocking probability (or to increase the acceptance ratio) of the connections. It has the natural control complexity because of the frequent reconfiguration of the network element in response to changing traffic patterns and variety of service demands. This paper will mainly consider the dynamic condition. A good RWA algorithm is capable of improving the wavelength channel utilization in WDM networks satisfying wavelength continuity constraint. WDM networks are also prone to component failures. A fiber cut causes a link failure. A node failure may be caused due to the failure of the associated OXC. A fiber may fail due to the failure of its connected end components. The nodes adjacent to the failed link can detect the failure by monitoring the power levels of signals on the link in physical layer [2]. Since WDM networks carry high volumes of traffic loads, failures may result in severe loss especially in business transaction. Therefore, it is imperative that these networks have fault tolerance capability. Fault tolerance refers to the ability of the network to reconfigure and re-establish original communication upon failures. The process of reestablishing communication through a lightpath between the end nodes of a failed lightpath is known as lightpath restoration. A lightpath that carries traffic during the normal operation is known as the working lightpath. When a working lightpath fails, the traffic is rerouted over a new lightpath known as the spare or protection lightpath. A lightpath can be protected against failures by pre-assigning resources to its spare lightpath statically. This approach is referred to as static restoration or protection switching [3]. Alternatively, resources can be dynamically searched to establish a protection lightpath after the node or span failure. This approach is referred to as dynamic restoration [2,3].

This work is concerned with semi-static lightpath restoration in WDM networks with dynamic traffic demands. We assume that the networks do not have wavelength conversion capability. We consider both the single link/node failure and the multiple failure models in our work. These models assume that at any instance of time, only one link or node can fail, or there exists several failures of any type simultaneously. In other words, failures may accumulate without repairing and the occurrences of the order of failures are random. We call a connection with fault tolerance requirements a reliable connection. Our algorithms use a dynamic proactive approach wherein a reliable connection is identified with the establishment of a working and a protection lightpath at the time of provisioning the request, and the spare capacity will update according to the changing network status. To assure restorability of the node and multiple failures, the algorithms use node-disjoint protection path with or without multiplexing techniques to share wavelength channels among several backup lightpaths in the same or merged protection domain to improve the channel utilization with comparison to link or path protection. To achieve 100% restoration guarantee upon failures, the ternary spare path information will be exchanged among the participating neighboring nodes. The objective is to maximize the acceptance ratio of reliable connections and resource utilization efficiency with reasonable restoration time. We develop an efficient and computationally simple heuristic to estimate the cost function used by the protection domain selecting decision. The proposed algorithms are flexible to choose a suitable extent to recover from failures by the dynamic available network resource.

The rest of this paper is organized as follows. Section 2 presents the concepts of the protection/restoration solutions in WDM networks. Section 3 describes the proposed algorithms. The results of the simulation experiments are discussed in Section 4. Section 5 summarizes the paper.

2   Protection Domain Concept

The restoration methods can be classified into reactive and proactive methods [5]. In contrast to the proactive method that uses the protection-based approach, the reactive method restores disrupted traffics without preserved resources. The reactive method has the simple and low overhead feature of recovering from failures but cannot guarantee successful recovery due to the resource shortage during restoring. In a proactive method, protection lightpaths are identified and resources are reserved along the protection lightpaths at the time of establishing the working lightpath. A proactive or reactive restoration method is either link-based or path-based [3,4,5]. A link-based method reroutes traffic around the failed component. A new path is selected between the end nodes of the failed link. This path along with the active segment of the working path will be used as the protection path. In a path-based restoration method, a protection lightpath is selected between the end nodes of the entire failed lightpath, which has better resource utilization than the link-based restoration methods. However, it requires excessive signaling information and results in longer restoration time. The primary-backup multiplexing proactive method can be employed to further improve resource utilization at the expense of gaining less than fully restorable guarantee if two primary lightpaths do not fail simultaneously. These concepts are illustrated in Fig. 1.

In Fig. 1, there is a connection request between $(s, d) = (A,D)$. After the setup process the working path is determined to be A-F-E-D. If one chooses path-based proactive strategy to survive the failure event occurred in link E-F or any link located on the working path, the protection or spare path A-B-C-D will be replaced to continue the traffic after short reconfiguration time. The end-nodes have the responsibility to calculate the capacity needed and reserve them for path A-F-E-D during all successive connection demands. One can use K-shortest paths for path selection decision to prevent over-reserved network capacity and avoid wasteful link resource. By link-based proactive scheme, the spare path remains

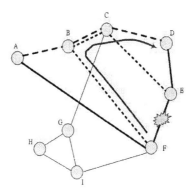

Fig. 1. Illustration of link, path and domain-based protection mechanism

as working path except the failure portion E-F, the end-nodes on both sides of failure link will find an alternate (shortest) path to reconnect them and complete the connection. The rerouting complexity of link-based scheme is lower than path-based approach but it also lacks the flexibility the path-based scheme has. If we can relax the selection extent to a more flexible environment chosen by some resource-oriented criteria, a better efficiency and survivability may be able to be achieved. Using the same example in Fig.1, if one chooses the protection domain A-B-C-D-E-F as the resource pool for any failure restoration located within this area, A-F-B-C-D will be the spare path if there are more available network capacities in C-D than C-E or E-D. Our approach uses this concept with link state protocol that is used to distribute topology information about the network to form a new protection scenario. Given the topology, we first determine the working path, which is not necessary a shortest path from source to destination, and then assign the wavelength. In the mean time we decide a protection path by a weighting criteria that indicates the network capacity metrics, consisting of the hop count of the path, available wavelength on the links, node connectivity and distance or setup cost. The objective of the protection finding algorithm is to determine the suitable protection domain compassed by the working path and protection path based on these weights. This protection domain will provide necessary backup resource for successive reliable connection request. Every time a new connection arrives the protection area size and component may be affected, and it changes dynamically depending on the network capacity states. The motivation behind our work is that we can use the weight functions to bound the connection blocking probability to improve the network performance. The frequent exchange of control signals will be an overhead to each node, and this overhead can be reduced if there exists a domain manager that is responsible to perform the task of collecting information sent from other domain manager by out-of-band control channel. The domain manager then distributes new routing information to members in the domain group, and any new status in this domain will also reflect to the manager for further broadcast-

ing to other domain manager. The election of domain manager is depending on
the loading level and the lightest loading one is the best choose. The protection domain configuration setup is fully dynamic. There are two kinds of main
methods in WDM network protection [6,7,8,9,10]. One is ring-based protection
approach, which apply multiple ring covers to make self-healing activation under failure. This way suffers inefficiency in network capacity resource utilization.
On the other hand, mesh-based protection scheme can achieve optimal resource
usage but complicate the reconfiguration process, which makes it suitable for
static traffic demands. The proposed protection-based approach does not need
the pre-configuring process at the expense of complex distributed information
exchange between network nodes by control channel and expensive but efficient
resource calculation. The formation of the domain is determined by the current
network resource status and considerations included in the weight function. It
can recover single link/node failure within the same domain and multiple failures
in different domains by exploiting resources of all domains efficiently.

3   Proposed Algorithm

This section presents the scheme we propose. When a new reliable connection
request arrives, we find the working path and protection path simultaneously by
two node-disjoint paths. The distributed control signaling mechanism will exchange provision status between adjacent nodes following the forward packets.
If there is any intermediate node found lacking of capacity, it returns a NACK
message to its predecessor. This node may find it solely has the protection link
available, and it will return the same message to its predecessor, too. Following
this procedure the node-disjoint working-protection path pair will be found eventually. When performing protection path finding, we use the weighting function
as follows:

$$
\begin{aligned}
weight = {} & \alpha \times hop\ count + \beta \times available\ wavelength\ in\ a\ link \\
& + \gamma \times \sum_i traffic\ load\ in\ each\ output\ port + \eta \times setup\ cost \\
& + \delta \times vulnerability
\end{aligned}
$$

In this formula, the hop count is the number of links traversed up to this intermediate node; the available wavelength in a link is recorded in the resource table
in each node for wavelength switching; the traffic loads will be summed because
each node may have higher degree but with unbalanced load distribution. The
setup cost depends on the type of switching equipment and may have different
values. The vulnerability is also considered for the weak link, segment or area.
The scaling factor is different for various managing and control strategy. The
node-disjoint path finding process is summarized as follows:

Step1. Assume the connection request (s, d) arrives. Issue two connection request messages from two different output directions which have enough capacity or sufficient resource by Eq.(1) to start exploring all possible con-

nection paths. Mark the attributes of the request in the resource table and return result message (ACK or NACK) to the predecessor node.

Step2. Find node disjoint-path from (s, n_1), (s, n_2),..., (s, n_{d-1}), where d-1 is the predecessor of the destination d. Connect the least cost node from these disjoint paths as the first node-disjoint backup path. For example, $s - s_1 - n_{11} - n_{12} - \cdots - d$ in Fig. 2 is a protection path candidate.

Step3. Along the shortest path from s to d, perform step 2 with transitive intermediate node as the source node each time.

Step4. Choose one (for single link failure recovery) or two (double node failure recovery) paths from these node-disjoint backup paths to form the protection domain protecting all connection requests within it.

In this scenario, the connection demands will utilize all the capacity within the protection domain to achieve reliable working-protection pair. If there occurs a single link or node failure, the protection path corresponding to the failed working path will be replaced along the alternating path in the protection domain. These cases are demonstrated in Fig. 3. As the figure shows all connection requests will be processed in this way. For the multiple failure case, we consider 3 possible conditions. First, if failures occur in the working lightpath, the protection switching process will be executed automatically. Secondly, if all the failures occur on the protection paths, the rerouting mechanism will take action in the background, which will attempt to find a new backup route for the underlying active paths. The third case is more complex that if one failure exists on the working lightpath and the other on the protection path. The path recovery process will restart and the re-computation will result in slow restoration. The ternary protection path finding is an efficient method for this problem because the independent and sharable property of this path to all other protection paths of connections. All active working paths share this backup path that is allowed to convey low priority traffic in normal condition, and the capacity will be reserved for the fast restoration. In fact, the ternary protection path is not necessary in all the cases depending on the location of the failures, which is illustrated in Fig. 4. Although two failures occur simultaneously in the same domain, there still exist protection paths to recover the failed traffic.

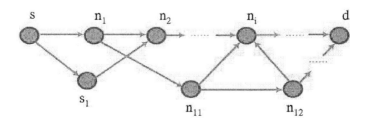

Fig. 2. Procedure to find node-disjoint paths forming the protection domain

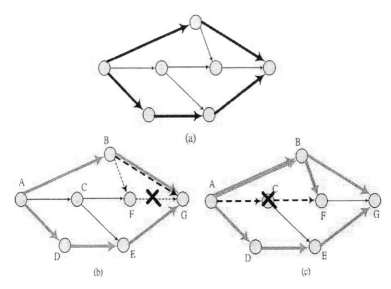

Fig. 3. Protection switching example. (a) protection domain inclusive by 2 node-disjoint paths (b) single link failure restoration. (working: B-F-G, protection: B-G) (c) single link failure restoration. (working: A-C-F, protection: A-B-F)

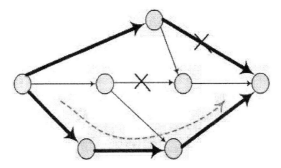

Fig. 4. Illustration to show unnecessary ternary restoration under multiple failure

4 Simulation Results

We evaluate the effectiveness of the proposed algorithms by performing extensive simulation. The network considered here is the EON (European Optical Network System) network with 34 duplex links and 18 nodes. Other representative network benchmark will be simulated extensively in the future. A duplex link is comprised of two simplex links in opposite directions, and each simplex link is assumed to have 32 wavelengths and therefore, a duplex channel consists of 64 wavelength channels. Every node has different connectivity in the target network topology. The connection requests arrive at a node following the Poisson

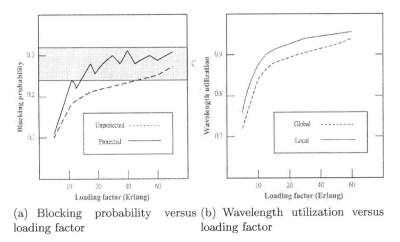

(a) Blocking probability versus loading factor

(b) Wavelength utilization versus loading factor

Fig. 5. Plots of simulation result about various parameters

process with exponentially distributed holding time with unit mean. Every node has the same probability to be a source or a destination node for a connection request. We use 3 metrics, *connection blocking probability*, *resource utilization* and *restoration time* to measure the performance of the proposed algorithms. We compare the performance of our algorithms with that of an algorithm that does not use any spare lightpath. The blocking probability is evaluated by the success ratio that the spare path could be designated during working path establishment phase. The attribute of the traffic will be verified first to classify its priority, spare path assignment procedure will continue if it is a premium-class connection request. If the process failed to assign a proper backup route then block this request. But this block will trigger the resource aggregating process of the protection domain to collect and update the existing allocation to prepare for successive request. The resource utilization is examined within a protection domain and full exploitation can be achieved. The restoration time is evaluated by summing several factors including failed point to end node notification time, path switching time and spare node setup time. Since the destination node will be notified from control channel before the recovered data arrive, the penalty of it will be ignored. We assign some test values to different factors to approach practical values in the actual network.

The blocking probability versus loading factor (in Erlang unit) is plotted in Fig. 5(a). We remark this figure as follows: (i) The probability value is limited on an oscillation period C, which can be adjusted dynamically according to the selection strategy of the protection domain. (ii) The probability is higher when the protection domain size reduces due to the heavy contention in the same domain area. (iii) The protection domain size grows when the loading is higher, so efficient management scheme and exploitation process of the available resource is very important on the final performance. (iv) The value of the blocking probability is the result of dynamic equilibrium of resource contention. When the traffic is congested, some connection demands are rejected and others are accepted. The congestion level also results in reconstructing the protection domain and re-balancing the capacity insufficiency to accommodate more traffic requests. (v) When the traffic loading increases, the volume of control message flow exchanged in the network is also increasing. So the reliable control channel protection is becoming more important to accomplish the control signaling and network status passing. The wavelength utilization ratio versus loading factor (in Erlang unit) is plotted in Fig. 5(b). We also remark this figure as follows. We divide the ratio into 2 types: local and global. The local utilization which means the ratio of utilizing wavelengths and fiber ports within the protection domain is higher than the global ratio which represents the degree of network resource accessing because of the local resource accessing. We can see in the figure the wavelength resource is exhausted soon in each protection domain, which makes the domain re-construction process is activated. This ratio is not only relative to the distribution of the traffic demand but also to the degree (node connectivity) of each node. The influence of node degree to network survivability will be verified in future study. The restoration time versus loading factor is also plotted in Fig. 6. The setup time of the reliable connection is longer in average as the node-disjoint paths finding process. As the prediction when load is increasing, the exploitation of network resource tends to global accessing which then makes the result approaching to path-based protection strategy. The curve becomes smooth when network loading is high since the blocking ratio increases at that time. Also note that the nodes in this protection mechanism have more tasks to complete making them busy to perform control information processing, which then postpone the response to every event. The request/reply messages or network status will have higher priority to handle than the traffic request messages to assure reliable service guarantee.

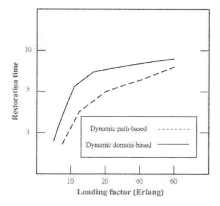

Fig. 6. Simulation result of restoration time versus loading factor

5   Summary

This paper studied the dynamic protection domain-based restoration heuristics in optical WDM networks. Rather than the traditional link/path level restoration schemes, the adaptive protection domain specified by available network capacity is more flexible and properly reflects practical condition due to the variant traffic demands. We incorporate working path with node-disjoint spare path to form new resource pool for the restoring activity. Based on this model we can handle single/multiple link/node failures in the networks. The selection of the spare path is by the criteria formed by hop count, available bandwidth/wavelength, and node with congestion degree as the weight function. Different weighting can be offered by scaling different parameter metric. Results demonstrate that using a combination of the adaptive domain-based strategy with appropriate chosen cost function based on consideration of factors about network capacity will result in better performance for resource utilization on failure restoration.

References

[1] H. Zang, J. P. Jue, and B. Mukherjee: A review of routing and wavelength assignment approaches for wavelength-routed optical WDM networks , Optical Networks Magazine, Vol. 1, No. 1 (2000) 47–60

[2] S.Ramamurthy and B. Mukherjee: Survivable WDM mesh networks, Part II–Restoration, Proc. ICC, Vol. 3 (1999) 2023–2030

[3] S. Ramamurthy and B. Mukherjee: Survivable WDM mesh networks, Part I–protection, Proc. IEEE INFOCOM, (1999) 744–751

[4] B. T. Doshi, S. Dravida, P. Harshavardhana, O. Hauser, and Y. Wang: Optical network design and restoration, Bell Labs. Tech. J., Jan (1999) 58–84

[5] G. Mohan, Arun K. Somani: Routing Dependable Connections with Specified Failure Restoration Guarantees in WDM Networks, Proc. IEEE INFOCOM, Vol. 3 (2000) 1761–1770

[6] M. Medard, S. G. Finn and R. A. Barry: WDM loop back recovery in mesh networks, IEEE INFOCOM, (1999) 744–751

[7] S. Lumetta, M. Medard, and Y.-C. Tseng: Capacity versus robustness: A tradeoff for link restoration in mesh networks, J. Lightwave Technology, Vol. 18 (1999) 1765–1775

[8] O. Gerstel, R. Ramaswami, and G. Sasaki: Cost-effective traffic grooming in WDM ring, IEEE Trans. Networking, Vol. 8, No. 5 (2000) 618–630

[9] W. D. Grover and D. Stamatelakis: Cycle-oriented distributed preconfiguration: ring-like speed with mesh-like capacity for self-planning network restoration, IEEE ICC, (1998) 537–543

[10] W. D. Grover: Distributed restoration of the transport networor, Telecommunications Network Management into the 21st Century, Piscataway, NJ: IEEE Press,(1994) 337–417

Part II

Enhanced Protocols
in Internet

Periodic Communication Support in Multiple Access Networks Exploiting Token with Timer

Young-yeol Choo[1] and Cheeha Kim[2]

[1] Department of Computer Engineering
Tongmyong University of Information Technology, Busan, Korea
yychoo@tit.ac.kr
[2] Department of Computer Science and Engineering
Pohang University of Science and Technology, Pohang, Korea
chkim@postech.ac.kr

Abstract. Timed token medium access protocols are widely adopted in real-time multiple access networks because of bounded access time. However, timed token protocols inadequately provide periodic communication service, although this is crucial for hard real-time systems. We propose an algorithm to guarantee periodic communication service on a timed token protocol network. In this algorithm, we allocate bandwidth to each node so that the summation of each bandwidth allocation is the Target Token Rotation Time (TTRT). Upon receiving the token, a node consumes, at most, pre-allocated time for message transmission, and then passes it to the next node. If the node cannot consume the allocated time, the residual time is allocated to other nodes for non-periodic service using a timer which contains the unused time value and is appended to the token. This algorithm can always guarantee transmission of messages before their deadlines when network utilization is less than 50%.

1 Introduction

In distributed real-time systems, there are many delay elements caused by the stochastic nature of computer systems, communication networks, and errors (see Fig. 1). The communication network is mainly responsible for delays. Besides delay, delay variation can also result in a negative effect on real-time systems. On designing a discrete-time system, the system is considered to operate with an exact period or constant time delay, which makes it relatively easy to analyze the system and compensate for the effects caused by the delay. It is known that time-varying delays can lower control performance and can cause instability [1]. In Fig. 1, time-varying delays occur at the feedback path and the data sampling period. Generally, time-varying delays are very difficult to treat theoretically and decrease the predictability of real-time systems behavior. Therefore, periodic communication service is necessary to facilitate the design and analysis and improve performance. Timed Token Protocol (TTP) is one of the most widely used medium access control (MAC) protocols in real-time multiple access networks because of the bounded medium access time. The examples of timed token

H.-K. Kahng (Ed.): ICOIN 2003, LNCS 2662, pp. 141–150, 2003.

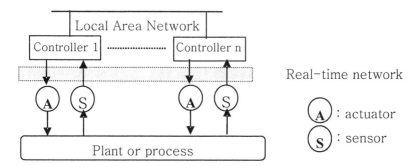

Fig. 1. A configuration of simple distributed computer control systems

protocol include the Fiber Distributed Data Interface (FDDI), IEEE 802.4 Token Passing Bus, IEC Fieldbus, SAFENET, Profibus, and so on. For these protocols, the token visits each node in pre-defined order. However, these TTPs cannot provide a periodic communication service since they cannot guarantee periodic token arrival at each node.

Let t be the time when a node i receives a token. If the node i receives the token exactly at time $t + n$TTRT for all n ($n = 1, 2, ...$), then periodic communication is guaranteed to node i. If the transmission queue of a node is empty, the node passes the token to the next node immediately, although it has not yet consumed the time allowed for. So the minimum inter-arrival time of the token is τ, which occurs when no message is waiting for transmission at each node. The upper bound between two consecutive token arrival at node i is stated as

$$TTRT - \tau - \sum_{j=1,\cdots,n, j \neq i} H_j \geq 2 \cdot TTRT, \qquad (1)$$

where H_j is the synchronous bandwidth allocation to node j [2, 3]. Equation (1) was proved formally in [4]. [2] presented the evaluation of maximum token rotation time of IEEE 802.4 Token Passing Bus protocol considering asynchronous TTRT, which could be equated similarly with equation (1). That is, two consecutive token arrival time varies from token walk time τ to 2·TTRT. The IEC 61158 protocol provides periodic service using a pre-established scheduling table and the concept of an implicit token [5]. This protocol cannot provide periodic service except the first node because the token arrival time at other nodes may be changed depending on the load of previous nodes. A token in TTPs may arrive at a node early or late depending on the token rotation timer value and the message load of the previous nodes. Therefore, a periodic communication service cannot be guaranteed using TTP.

In this paper, we propose an approach to guaranteeing periodic communication service on a timed token protocol network. The remainder of the paper is organized as follows. In section 2, we present the new algorithm guaranteeing periodic service on multiple access networks using a token with a timer. The bandwidth allocation scheme guaranteeing real-time message transmission in this algorithm is proposed and its real-time performance is analyzed in section 3. In section 4, the working of the new algorithm is exemplified using a traffic example. Section 5 contains concluding remarks.

2   New Approach: TwT (Token with Timer) Algorithm

Real-time messages may be classified as either periodic or aperiodic messages. Periodic and aperiodic messages have deadlines while non-real-time messages do not. Assume that a message is transmitted in a frame with fixed size and the transmission time of a frame, c_{fr}, is considerably short compared to TTRT because the frame length in real-time systems is relatively short. A real-time message at node i, M_i, is commonly characterized as (C_i, D_i, P_i), where C_i is message size, D_i is the deadline which is defined as a time interval from the message creation to the completion of its transmission, and P_i is the inter-creation time of messages. If there are m message streams in node i, we divide this node into m sub-nodes, where the token passing time between sub-nodes equals zero. By performing this transformation, a network with multiple message streams in a node can be treated equivalently as one with one message stream per node [6]. We assign the highest priority to aperiodic messages. For real-time messages, we assume that $D_i = P_i$ and $P_{min} = \min\{P_1, P_2, \cdots, P_n\}$. Utilization factor U is given as $\sum_{i=1}^{n} U_i = \sum_{i=1}^{n} C_i/P_i$, where U_i is utilization factor of M_i and n is the number of nodes. As a performance metric in real-time communication, a *worst case achievable utilization* is commonly used [3, 7]. If a bandwidth allocation scheme can meet all the deadlines for a message set whose utilization is less than or equal to U', we say that U' is the achievable utilization of the bandwidth allocation scheme. The *worst case achievable utilization U^** of a scheme is the least upper bound of its achievable utilization U'.

In order to meet all the deadlines for a message set, the total bandwidth allocated to real-time messages must be less than the available network bandwidth (*protocol constraint*) and the bandwidth available for node i should not be less than the amount of transmission time (*deadline constraint*) [7]. In this approach, the bandwidth for node i, B_i, is allocated as follows.

$$\sum_{i=1}^{n} B_i = TTRT - \tau \tag{2}$$

B_i is defined as

$$B_i = H_i + h_i \tag{3}$$

where H_i and h_i are the bandwidths for real-time and non-real-time messages at node i, respectively. h_i is given by the following equation.

$$h_i = (TTRT - \tau - \sum_{i=1}^{n} H_i)/n \tag{4}$$

The equation of H_i will be derived shortly. Equation (2) guarantees periodic communication, if each node consumes all of its bandwidth allocation whenever it receives the token. However, this is not the case. So, in order to have a periodic token visit, we propose a new token control algorithm, as follows. Instead of a token holding timer THT_i and a token rotation timer TRT_i, which are used to control the amount of available bandwidth for message transmission at TTP [3, 6], we define transmission timer TX_i to control how long node i may transmit messages. The TX_i timer always counts down. Hereafter, we assume that the token visits node i following the ascending order of i; hence, first node 1, then node 2, 3, $\cdots$ until node n, then again node 1. The token, which plays the same role as it does in TTP, is named as the *normal token*. Upon receiving the *normal token*, node i sets the transmission timer TX_i to B_i and transmit messages until TX_i expires. If node i does not consume all the bandwidth allocation, instead of holding the token for the residual time r_i, r_i is given to other nodes to prevent a waste of the bandwidth. After complete consumption of r_i, node i passes the token to the next node. This guarantees the token arrival time to a node to be fixed at each period. To give the residual time r_i to others, we use the token carrying a timer named TOT. We call this type of token a *bonus token*. Thus, we divide the token cycle time available for message transmission into two parts: a *normal cycle* and a *bonus cycle* . Using the *normal token*, node i operates in a *normal cycle* as follows.

Op 1) Upon receiving *normal token*, node i sets TX_i to a given constant B_i.

Op 2) Node i may transmit messages until TX_i expires. TX_i continues to count down during message transmission.

Op 3) If $TX_i > 0$ when there is nothing to transmit, node i initiates a *bonus cycle* by creating *bonus token*. Note that at this moment, current *normal cycle* is suspended. Otherwise, release *normal token* to the next node.

The *bonus cycle* is an interval during which other nodes use the surplus bandwidth that a node has not consumed. Bonus cycle begins if $TX_i > 0$ and no messages are waiting for transmission. When node i generates the *bonus token*, it sets the timer value in the token to the current value of $TX_i - \tau$ and passes the *bonus token* to the next node. Upon receiving the *bonus token*, node i can transmit aperiodic or non-real-time messages. In a *bonus cycle* , node j operates as follows.

Op' 1) Upon receiving *bonus token*, node j passes it to the next node immediately if $TOT = 0$. Otherwise, it sets TX_j to TOT.

Op' 2) If $TX_j < c_{fr}$, where c_{fr} is time to transmit a frame, time is killed. Otherwise, node j may transmit aperiodic messages and non-real-time messages. TX_j continues to count down during message transmission.

Op' 3) If $TX_j = 0$ or node j runs out of messages to transmit, it sets TOT to TX_j and passes *bonus token* to the next node.

When the *bonus token* returns to node i which has generated it, node i must examine the TOT in it. If TOT is not completely consumed ($TOT > 0$), another *bonus cycle* may begin with a new TOT which is TOT minus the token walk time τ. Note that if the new TOT is not large enough to transmit one frame, then it must be consumed at node i. If TOT is completely consumed ($TOT = 0$), node i resumes the normal cycle. In this way, each node gets *normal token* periodically every $TTRT$ as long as the token travels over nodes on the ring/logical ring according to a fixed order. Below we present a pseudo-code description of our proposed approach.

```
/* At initialization procedure */
Negotiate TTRT and fix real-time bandwidth allocation of each node;
Node 1 begins normal cycle;
/* Run-time procedure */
At node i, at the token arrival, DO :
IF normal token THEN
   {DO normal cycle}
ELSE
   {DO bonus cycle }
Normal cycle:
   TX_i ← B_i ;
   Start TX_i ; // count-down timer
   WHILE real-time packets in queue DO:
      {Transmit a real-time packet ;}
   WHILE TX_i > c_fr AND pending non-real-time packets DO:
      {Transmit a non-real-time packet ;}
   IF TX_i − τ ≥ c_fr THEN
         {TOT ← TX_i − τ ; // unused time at TOT
      Pass bonus token to the next node (i+1) (modulo n);}
   ELSE
      {Kill the residual time;
      Pass normal token to the next node (i+1) (modulo n);}

Bonus cycle:
   IF TOT = 0 THEN {
      IF initiator of bonus cycle THEN
         Pass normal token to the next node; // resume normal cycle
      ELSE
         Pass bonus token to the next node; // continue bonus cycle
   }
   ELSE {
      TX_i ← TOT ;
      Start TX_i;
      WHILE TX_i < c_fr AND pending real-time asynch.
            or non-real-time packets DO :
         Transmit one of them;
      IF initiator of bonus cycle THEN {
         IF TX_i − τ < c_fr THEN // killing time operation
            {Kill the time;        // until TX_i = 0
            Pass normal token to the next node and return;}
```

> ELSE
>> $TX_i = TX_i - \tau$; // for additional *bonus cycle*
> }
> ELSE // not initiator of *bonus cycle*
>> IF $TX_i < c_{fr}$ THEN {Kill the time;} // until $TX_i = 0$
>
> $TOT \leftarrow TX_i$;
> Pass the token to the next node ;
> }

3 Bandwidth Allocation and Worst Case Achievable Utilization

Bandwidth allocation scheme is crucial for guaranteeing a message set and closely related to the real-time performance. To derive bandwidth allocation H_i, we need to consider the minimum number of times that node i can have token until the deadline of M_i.

Theorem 1. Let $t_i(k)$ ($k = 1, 2, \cdots$) denote the k-th arrival time of *normal token* at node i ($i = 1, 2, \cdots, n$). Then, in TwT, $t_i(k+1) - t_i(k) = TTRT$ for any i and k, provided that the bandwidth of node i is allocated according to equation (2) through (4).

Proof. During the interval from $t_i(k)$ to $t_i(k) + TTRT$, if every node j consumes its bandwidth allocation B_j, obviously node i can receive *normal token* again at time $t_i(k+1) = t_i(k) + TTRT$ by equation (2). This arises when each node is fully loaded with real-time and non-real-time messages. On the other hand, if node i consumes only ε ($0 \le \varepsilon < B_i$), then node i sends *bonus token* with the residual time $r_i = B_i - \varepsilon - \tau$.

i. If $r_i < c_{fr}$, it is killed at this node.
ii. If $r_i > c_{fr}$, the residual time r_i may be consumed by other nodes. If r_i is not consumed completely until *bonus token* returns to node i, the above sequence is repeated.

After all, the residual time is consumed and *normal cycle* will be resumed at the next node immediately after r_i. Therefore, $t_{(i+1) \bmod n}(k) = t_i(k) + B_i$, where $i \bmod n$ is modular operation. By the same operation, $t_{(i+2) \bmod n}(k) = t_{(i+2) \bmod n}(k) + B_{(i+1) \bmod n} = t_i(k) + B_i + B_{(i+1) \bmod n}$. As a result,

$$
\begin{aligned}
t_i(k+1) &= t_i(k) + \tau + \sum_{i=i}^{n} B_i + \sum_{i=1}^{i-1} B_i \\
&= t_i(k) + \tau + \sum_{i=1}^{n} B_i \\
&= t_i(k) + TTRT
\end{aligned}
$$

From Theorem 1, we have $TTRT \le P_{min}$. Theorem 1 can be used to derive Corollary 1.

Corollary 1. In any interval of D_i, the *normal token* will visit node i at least V_i times, where $V_i = \lfloor D_i/TTRT \rfloor$.

That is, node i can use the bandwidth for real-time messages as much as $\lfloor D_i/TTRT \rfloor H_i$ at least during interval D_i. Using the result in [7], we have the minimum amount of transmission time during interval t $(t \geq TTRT)$ on node i.

$$X_i(t) = \lfloor t/TTRT \rfloor H_i + a(t) \tag{5}$$

where $a(t) = \max(0, \min(H_i, t - \lfloor t/TTRT \rfloor TTRT - \sum_{j=1, j \neq i}^{n} B_j - \tau))$.
At the worst case, $a(t)$ gives the amount of available bandwidth during time interval $(\lfloor t/TTRT \rfloor TTRT, t]$. In equation (5), $\lfloor D_i/TTRT \rfloor H_i$ is the minimum amount of transmission time for real-time message at node i until its deadline. It should meet to the message length to be transmitted. $U_i D_i$ is the message length generated during the same interval. Because $D_i = P_i$, $\lfloor D_i/TTRT \rfloor H_i = U_i D_i$ and we have:

$$H_i = \frac{U_i D_i}{\lfloor D_i/TTRT \rfloor} = \frac{U_i P_i}{\lfloor P_i/TTRT \rfloor} \tag{6}$$

In [8], it was shown that the deadline constraint is satisfied if each node allocates its H_i as in equation (6). Therefore, for a message set $M = \{M_1, M_2, \cdots, M_n\}$, if H_i $(i = 1, 2, \cdots, n)$ is allocated as in equation (6), every real-time message can be transmitted always within its deadline. It is also shown in [8] that if H_i $(i = 1, 2, \cdots, n)$ is allocated as in equation (6) and its utilization satisfies the inequality in equation (7),

$$U_i \leq \frac{\lfloor P_{min}/TTRT \rfloor}{\lfloor P_{min}/TTRT \rfloor + 1}(1 - \frac{\tau}{TTRT}) \tag{7}$$

then *protocol constraint* is satisfied. This implies that equation (7) gives the least upper bound of the utilization of TwT protocol.
Consequently, if the bandwidth of a message set M is allocated as in equation (2), (3) and (6), the worst case achievable utilization U^* is (for details, see [8])

$$U^* = \frac{\lfloor P_{min}/TTRT \rfloor}{\lfloor P_{min}/TTRT \rfloor + 1}(1 - \frac{\tau}{TTRT}).$$

For example, when $P_{min} = TTRT$, $U^* = 1/2(1 - \tau/TTRT)$. Because τ is much smaller than $TTRT$, U^* approaches 50%, which is the highest worst case achievable utilization having been reported so far.

4   Illustrative Example

In this section, TwT algorithm will be compared with the FDDI protocol for periodic communication service using an example. In this example, the network parameters are assumed as follows:

Table 1. Message parameters and bandwidth allocation

	Message (packet) model		H_i (time unit)		
	C_i	P_i	FDDI (TTRT = 10)	TwT (TTRT = 20)	B_i
M_p^1	1	20	1	1	5.7
M_p^2	6	40	2	3	7.7
M_p^3	6	60	1.2	2	6.6

- $c_{fr} = 0.2$ time unit ($> 2\tau$)
- τ is considered as 0.
- Order of token rotation: 1, 2, 3, 1, $\cdots$.
- M_P^i and M_N^i are periodic and non-real-time packets to be transmitted by node i, respectively. Each packet arrives at the beginning of its period.

It is assumed that a network consists of three nodes that contain one periodic message M_P^i, respectively. M_P^i is characterized as $M_P^i = (C_i, D_i, P_i)$, where $D_i = P_i$ ($i = 1$, 2, and 3). As was mentioned previously, messages are transmitted by packet having fixed length. The characteristics of periodic messages are presented in Table 1. Non-real-time message M_N^i is assumed that it is generated at each node exhaustively. For the simplicity of illustration, the token transmission delay between adjacent nodes is neglected. Message overrun is not considered in this example.

As defined in section 2, the total utilization U for the periodic message set is

$$U = \sum_{i=i}^{n} U_i = \sum (C_i/P_i)$$
$$= 1/20 + 6/40 + 6/60 = 0.3$$

Note that the total utilization is lower than 33 %, which is the worst case achievable utilization of FDDI [6].

• Calculation of TTRT

In TTP, to satisfy the deadline of a message M_i, it is necessary for node i to have at least one opportunity to send the message before expiration of its deadline. By *Johnson and Sevcik's theorem*, the time interval between two consecutive token visits at a node can be as much as two times TTRT. Since a message with the shortest period, P_{min}, must have at least one opportunity within P_{min}, the TTRT in TTP should not be larger than $P_{min}/2$. On the other hand, TTRT = P_{min} in TwT. Therefore, the TTRT of FDDI and TwT can be chosen as $TTRT_{FDDI} = 10$ and $TTRT_{TwT} = 20$, respectively.

• Bandwidth allocations

The local allocation scheme of FDDI shows nearly the same performance as the global allocation scheme [7]. The local bandwidth allocation H_i in FDDI network is given as

$$H_i = \frac{C_i}{\lceil D_i/TTRT \rceil - 1}.$$

In the TwT algorithm, bandwidth allocation for periodic messages, H_i, is computed from equation (8). That is,

$$H_i = \frac{U_i P_i}{\lfloor P_i / TTRT \rfloor} = \frac{C_i}{\lfloor P_i / TTRT \rfloor}.$$

Then, B_i is obtained through equation (4) through (6). The computation results of both protocols are shown in Table 1. Note that, for FDDI protocol, M_p^2 and M_p^3 request three and five times of token visits within their deadlines, respectively.

5   Concluding Remarks

Although the timed token protocol guarantees bounded medium access delay, the periodic communication service cannot be provided. In real-time applications, collaborating tasks have distributed implementation and are often invoked periodically. Hence, the periodic communication service is important in real-time communication. In this paper, we proposed a new algorithm to guarantee periodic communication service in a multiple access network, based on a timed token protocol. Although the TwT algorithm is based on a timed token protocol, it can be implemented on other multiple access networks such as the Ethernet by initializing a logical ring among nodes. Not as TTP, a fixed period is available to every node without wasting bandwidth. To grant unused bandwidth to other nodes, the token carries a timer. Notably, the worst case achievable utilization of TwT is not less than 50%, while a conventional TTP is 33% [6, 7] and there are no optimal local allocation schemes [9]. In addition, each node is required to maintain only one timer compared to two timers in TTP. We understand it is not trivial to run a timer, so implementation complexity can be reduced significantly.

References

[1] Torngren, M. (1998). Fundamentals of implementing real-time control applications in distributed computer systems, Real-Time Systems, 14, pp 219-250.
[2] Montuschi, P., Ciminiera, L., and Valenzano, A. (1992). Time characteristics of IEEE 802.4 token bus protocol, IEE Proceedings-E, vol. 139, no. 1.
[3] Tovar, E. and Vasques, F. (1999). Cycle time properties of the PROFIBUS timed-token protocol. Computer Communication 22, pp. 1206-1216.
[4] Sevcik, K. and Johnson, M. (1987). Cycle time properties of the FDDI token ring protocol. IEEE Trans. Software Eng. vol. SE-13, no. 3, pp. 376-385.
[5] IEC/SC65C/224/FDIS, (1999). Digital data communications for measurement and control - Fieldbus for use in industrial control systems - Part 4: Data Link protocol specification.
[6] Agrawal, G., Chen, B., Zhao, W., and Davari, S. (1994). Guaranteeing synchronous message deadlines with the timed token medium access control protocol, IEEE Tr. On Computer. vol. 43, no. 3, pp 327-339.
[7] Malcolm, N., Kamat S. and Zhao W. (1996). Real-time communication in FDDI networks, Real Time Systems, 10, pp. 75-107.

[8] Xiong, H., Luo, Z., and Zhang, Q. (1997). Bandwidth allocation for real-time communication with LTPB protocol, Proc. of IEEE WFCS, pp 920-925.

[9] Han, C. Shin, K. G. and Hou C. (2001). Synchronous bandwidth allocation for real-time communications with the timed-token MAC protocol, IEEE Trans. Computers, vol. 50, no. 5, pp.414-431.

An Adaptive Contention Period Control in HFC Networks

Chih-Cheng Lo, Hung-Chang Lai, and Wen-Shyen E. Chen

Institute of Computer Science
National Chung-Hsing University
Taichung, Taiwan
{loremi,hclai,echen}@cs.nchu.edu.tw

Abstract. As the deployment of the cable TV (CATV) network becomes ubiquitous, the CATV networks have emerged as one of primary
technologies to provide broadband access to the home. Due to the natural of CATV network - the radical asymmetric bandwidth, the upstream
channel is critical to allocating and scheduling. The objective of this paper is to propose a new adaptive contention period control algorithm
in HFC networks to improve the utilization and throughput of the upstream bandwidth. Through simulation, we show that in any cases, our
adaptive method performs better than MCNS DOCSIS.

1 Introduction

In this new millennium, multimedia and broadband services are provided widely
over the Internet. Several emerging wireline and wireless access network technologies to provide broadband access to the home, such as HFC, xDSL, FTTx,
and LMDS/MMDS access networks [1]. As ubiquitous deployment of the cable
TV (CATV) network, the CATV network has emerged as one of primary technologies to provide broadband access to the home. There are many organizations
proposed the MAC layer protocols as the standard of modern HFC networks.
In the paper, we propose an adaptive contention slots allocating scheme of HFC
networks to improve the throughput and request delay of DOCSIS HFC networks.

A number of organizations have worked many years in order to define open
standards for CATV network systems [2-3]. The major associations working
on HFC networks are the Multimedia Cable Network System (MCNS) Partners Ltd., the IEEE working group 802.14 [4], the European Cable Communication Association (ECCA), the Digital Audio Video Council (DAVIC) and
the Digital Video Broadcasting (DVB). DVB and DAVIC work closely together
are tightly connected to the European Telecommunication Standards Institute
(ETSI). While both DOCSIS and IEEE 802.14a were developed to facilitate the
interoperability between stations and headends designed by different vendors.
The MCNS, a consortium consisting of predominantly North American cable
operators and media companies, was developed to create a quick and uniform

H.-K. Kahng (Ed.): ICOIN 2003, LNCS 2662, pp. 151–160, 2003.
© Springer-Verlag Berlin Heidelberg 2003

interface specifications for standard, interoperable, for transmission of data over cable networks. Those documents are commonly referred to as the Data Over Cable Service Interface Specification (DOCSIS) [5] and DOCSIS v1.0 was approved as a standard by the ITU on March 19, 1998.

The main goal of this paper is to propose a new adaptive contention period control algorithm in HFC networks to improve the utilization and throughput of the upstream channels. The remainder of this paper is organized as follows. Section 2 presents the MAC protocol adopted in MCNS DOCSIS and IEEE 802.14a will be discussed. Our research, idea and methods, including modeling analysis and algorithms are presented in Section 3. In Section 4, we present our simulation results in two different aspects. Finally, we make a brief conclusion in Section 5.

2   MAC Protocol: MCNS DOCSIS vs. IEEE 802.14a

Since the development of DOCSIS v1.0 followed closely the IEEE 802.14a specifications, the MAC protocols described in the MCNS and IEEE 802.14a specifications are fundamentally similar: including virtual queue, minislot, downstream MPEG-II format, security module, piggybacking, synchronization procedure, and modulation schemes [2]. However, there are two major differences that may have a direct impact on performance, namely the mapping of higher layer traffic and the upstream contention resolution algorithm [10].

The framing structure of the MCNS is significantly different from the one adopted in the IEEE 802.14a. The MCNS DOCSIS proposed a more suitable IP environment. 6 bytes of MAC header are added to every packet regardless of weather it is an ATM cell or an LLC packet. Nevertheless, since IEEE 802.14 specification intends to provide a complete support of ATM environment and in order to minimize the MAC layer overhead, one byte is added to each ATM cell to form a MAC data PDU, where the ATM layer VPI field is used as part of the 14-bit local station ID. Furthermore, every station must be capable of AAL5 segmentation and reassembly in order to carry IP/LLC traffic.

As for the contention resolution algorithms, DOCSIS adopts Binary Exponential Backoff algorithm to resolve collisions in the request minislot contention process. However, it allows flexibility in selecting the Data backoff start (DBS) and Data backoff end (DBE) window sizes to indicate the initial and maximum backoff window size used in the algorithm. The headend controls the initial access to the contention slot by setting Data backoff start and Date backoff end. When a station has data to send, it sets its internal backoff windows according to the data backoff range indicate in the allocation MAP. The station then randomly selects a number within the backoff range and sends out its request. Since station cannot detect whether it is collision or not, it should wait the headend sends feedback either a Data Grant or an Acknowledgement (Ack) in a subsequent allocation MAP. If station does not receive either Data Grant or Ack in the subsequent allocation MAP, it means a collision has occurred. In this case, the station must then increase its backoff windows by a factor of two as long as it is less than the Data backoff end value set in the allocation MAP. Once again, the station randomly selects a number within its new window range and repeats the contention process depicted above. After 16 unsuccessful retries, the station discards the MAC PDU.

The collision resolution algorithm in the IEEE 802.14a consists of two parts. The first part is the first transmission rules that adopt priority plus FIFO algorithm designed for newcomers, while the second part is the retransmission rule that adopts n-ary tree plus p-persistent algorithm designed for collide requests. The headend controls the initial access to the contention slots and manages the Contention Resolution Protocol (CRP) by assigning a Request Queue (RQ) number to each contention slot. When a data packet is received, the station generates a Request Minislot Data Unit (RMDU). The headend controls the stations entry by sending an Admission Time Boundary (ATB) periodically. Only stations that generated RMDU time before ATB are eligible to enter the contention process.

3   Adaptive Contention Period Control

In this section, we propose an adaptive contention period control scheme to predict the number of contention slots in order to better cope with the request contention and to provide a better system performance. Some papers have studied this topic [11-15]. The idea of our research is how to dynamically adjust the number of contention slots to meet the number of requests arrived at the system, and to achieve the maximum bandwidth throughput and minimum request delay time. We first analyze the probability of contention, and then derive the bandwidth utilization and predict the reserved number of contention minislots. Finally, we propose a new adaptive method to dynamically adjust the number of contention slots to realize our ideas.

3.1   Investigation Probability of Contention

For each contention minislot in the upstream direction, there have only three states, *Idle*, *Success*, or *Collision*. It is intuitive that we should expect more Success slots but less Idle and Collision slots, since the latter two cases will waste the bandwidth. The following shows how we derive the probability of those three cases. Suppose we have m minislots and n requests, where m and n denote the number of contention slots and the number of requests, respectively. Then we can derive the total combinations of the permutation as:

$$\binom{n+m-1}{n} \tag{1}$$

And the combinations of the permutation if there is only one empty minislot and exactly one monislot has been selected by one request were shown in equations (2) and (3), respectively.

$$\binom{n+(m-1)-1}{n} \tag{2}$$

$$\binom{(n-1)+(m-1)-1}{n-1} \tag{3}$$

Based on the aforementioned formulas, we can derive the probability of *Idle slots*, *Success slots*, and *Collision slots* as shown in equations (4), (5), and (6), respectively.

$$\frac{(2)}{(1)} = \frac{\binom{n+m-2}{n}}{\binom{n+m-1}{n}} = \frac{\left(\frac{(n+m-2)!}{n!(m-2)!}\right)}{\left(\frac{(n+m-1)!}{n!(m-1)!}\right)} = \frac{m-1}{n+m-1}, \tag{4}$$

$$\frac{(3)}{(1)} = \frac{\binom{n+m-3}{n-1}}{\binom{n+m-1}{n}} = \frac{\left(\frac{(n+m-3)!}{(n-1)!(m-2)!}\right)}{\left(\frac{(n+m-1)!}{n!(m-1)!}\right)} = \frac{n(m-1)}{(n+m-1)(n+m-2)}, \tag{5}$$

and $1 - (4) - (5) =$

$$\begin{aligned} &1 - \frac{m-1}{n+m-1} - \frac{n(m-1)}{(n+m-1)(n+m-2)} \\ &= \frac{n(n-1)}{(n+m-1)(n+m-2)} \end{aligned} \tag{6}$$

The objective of our proposal is to reserve the appropriate number of contention slots that meet the maximum success probability when n number of requests had arrived, such that, we can take differentiation the equation (5) with respect to m, and find out the maximum value of m as shown in equation (7). From (7), we obtain that m approximates n, i.e., if there are number of n requests arrived, the system should reserve m contention slots to achieve the maximum success probability.

$$m = \sqrt{n(n-1)} + 1 \tag{7}$$

3.2   Dynamic Adjustment Contention Slots Method

After analyzing the value of maximized success probability, we will discuss how to dynamically adjust the contention slots based on the previous probability of idle/success/collision. The idea is to make the appropriate number of the contention slots predictable. To illustrate this point, we first define the following notations to depict our algorithms:

D_{map}: time between the CMTS starts transmitting a MAP and when the MAP goes into effect

N_p: the request had been received by CMTS successfully but have not allocate data slot yet

N_{cm}: total number of CMs in this HFC networks

P_{idle}: the probability of slot as idle state in the previously contention slot

$P_{success}$: the probability of slot as success state in the previously contention slot

$P_{collision}$: the probability of slot as collision state in the previously contention slot

T_{idle}: the threshold of idle state's probability

$T_{success}$: the threshold of success state's probability

$T_{collision}$: the threshold of collision state's probability

N_{pc}: the number of contention slots in previously competition

N_c: the number of contention slots in this cycle's competition

N_{max}: the maximum number of contention slots

N_{min}: the minimum number of contention slots

Weight: the weighting value

We can derive from above discussion that the range of contention slots number must be between the following two numbers:

$$N_{max} = N_{cm} - N_p$$
$$N_{min} = D_{map}$$

Since our prediction concept is to observe the previous probability of success/collision and whether that reserved minislots are suitable or not as the basis. Nevertheless, not only previous basis, we also have to include a set of thresholds as the adjusting foundation. The set of values (idle, success, collision) in our simulation is (45%, 40%, 35%), respectively. While the object of weighting function is to accelerate the adjustment of the number of collision slots when the probability of collision is greater than the successes. Based on these concepts, we can derive our algorithms for different conditions as follows:

if ($P_{collision}$ > $P_{success}$)
 	
    weight++;
 	
else
 	
    weight = 0;
	If ($P_{collision}$ > $T_{collision}$)
	{
	    if (weight > 2)
	        $N_c = N_{cm} - N_p$; // set to maximum value
	    else
	        $N_c = N_{pc} + \max(D_{map}, N_{pc})$*weight;
	};

If (P_{idle} > T_{idle})
    $N_c = N_{pc} -$
    $(N_{pc}*P_{idle} - N_{pc}*P_{success})$;
	If ($P_{success}$ > T_{succes})
	    $N_c = N_{pc}$;

4 Simulation Result

In this Section, we will compare our method with DOCSIS specification and point out the difference. For practicality, we measure the throughput, request delay, and drop numbers of the simulated system, in which throughput is defined as how much data (in Mbps) can be transmitted in a unit time period. The request delay is the time it receives a transmission acknowledgement by the headend from the time the request arrives at the station. We assume that data arrives at the MAC layer of a station is small enough to fit into a single data slot. The request delay includes the waiting time for a newcomer slot, delays due to collision resolution, scheduling delay of a grant message at the headend, and transmission delay of the data slot. The drop number are the amount of packets dropped by the queue of a cable modem. In general, the condition of packet dropped is caused by the buffers of CM is full or that retry number due to collision is greater then the upper bound, i.e., 16.

As most of the subscribers attached at the leave nodes of the HFC networks, we assume that all of them have the same distance with the headend, all have the Poisson arrival rate and with the simulation parameters listed in Table 1. The following simulation studies separate the performance of the proposed mechanism in two aspects. The first one fixes the number of CMs, but varies with the offered load, throughput, and request delay. The other group fixes the offered load, but varies with the number of CMs, throughput, request delay, and drop number. Figures 1 (a) to (c) show the results of the simulation for the first aspect discussed above.

As observed in Figure 1, when the offered load is less than 1.28 Mbps, there is no difference between our method and DOCSIS. But when the offered load is greater then 1.28 Mbps, the difference becomes significant gradually. Another checkpoint is when the offered load reaches 1.44 Mbps, the throughput does not change after that value. As a result, that point can be viewed as the watershed to saturation state. Before that point, our adaptive scheme obtains a better throughput than DOCSIS for about 7%.

The other comparisons are the relationship between request delay, drop number and offered load. As shown in Figure 2, the number of CMs is fixed at 50, 300, and 600 respectively, but request delay, offered load, and drop number are variable.

As shown in Figure 2, the solid line represents the request delay, while the dotted line represents the drop packet numbers. In the request delay item, the curve tendency of our method is similar with DOCSIS. When offered load below then 1.28 Mbps, the increase is smoothly. While if offered load over that point, the request delay will take off sharply until offered load equal 1.44 Mbps, and the increasing will about 9 times of original value. After 1.44 Mbps, the curve retain horizontally. Let's observe the other parameter, the drop number almost stays zero when offered load is less then1.28 Mbps, but it increases obviously when offered load becomes greater than that point.

There is an interesting phenomenon, i.e., when the difference between our method and DOCSIS in request delay is small, then the difference in drop number

Table 1. Simulation Parameters : We assume the number of contention slots in DOCSIS is fixed at 40 slots

Parameter	Value
Upstream channel capacity (QPSK)	2.56 Mbps
Downstream channel capacity	26.97 Mbps
Minislot	16 bytes/minislot, 50 usec/minislot
Number of contention slots in a MAP	40 minislot → 40 * 16 = 640 bytes → 40 * 50 = 2 msec
MAP size	50 minislot (100%) ∼ 2048 minislot (2%) → 800 ∼ 32768 bytes → 2.5 ∼ 102.4 msec
Maximum number of IEs in a MAP	240
Number of CMs	50 ∼ 600
One way delay	0.5 msec
DMAP time	2 msec
Data size	64 bytes
Backoff limit (DOCSIS only)	6 ∼ 10
Maximum retry	16
Number of queue for data in CMs	30

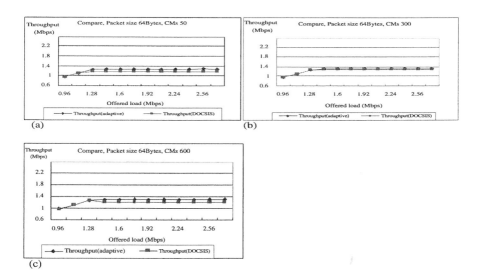

Fig. 1. The relationship between throughput and offered load. (a)CMs = 50 (b)CMs = 300 (c)CMs = 600

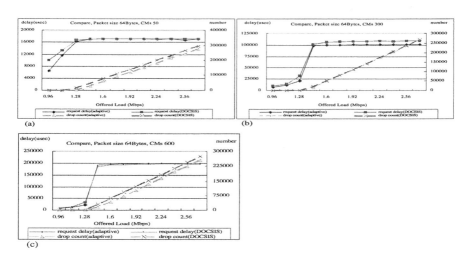

Fig. 2. The relationship between request delay, drop number and offered load. (a)CMs = 50 (b)CMs = 300 (c)CMs = 600

will get larger. But, when the difference between our method and DOCSIS in request delay getting larger, then the difference in drop number gets smaller.

We discuss another issue of our simulation now, to fix the offered load. We depict the simulation result in Figure 3. As we have seen in these figures, when offered load is less then 1.44 Mbps, there are almost no difference between our methods and DOCSIS. But when offered load getting greater than 1.44 Mbps, there is about 7% difference between them.

In our last experiment, it is observed the relationship between request delay, drop number, and the number of CMs while fixed the offered load. Again, the solid line represents the request delay, while the dotted line represents the drop packet numbers.

As shown in Figure 4, when offered load is less than 1.28 Mbps, in the item request delay, our method is about 20% to 50% better than DOCSIS. But the difference diminishes when offered load is greater than 1.44 Mbps. Nevertheless, when offered load is light, the drop numbers of both methods are almost zero. Whereas, when the offered load gradually increases and exceeds 1.44 Mbps, the drop packet number increases dramatically, especially in DOCSIS case.

5   Conclusion

In this paper, we have presented the architecture of the HFC networks and pointed out that the upstream channel is one of the major factors affecting the network performance and the channel access allocation mechanism is very important in improving the overall throughput. We propose an adaptive method to predict the suitable number of contention slots to meet the request numbers arriving at the system; i.e., if there are n arriving requests, then the system should

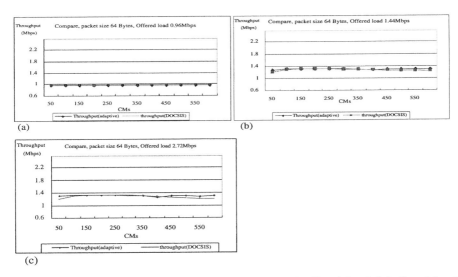

Fig. 3. The relationship between throughput and offered load (a)offered load = 0.96 (b)offered load = 1.44 (c)offered load = 2.72

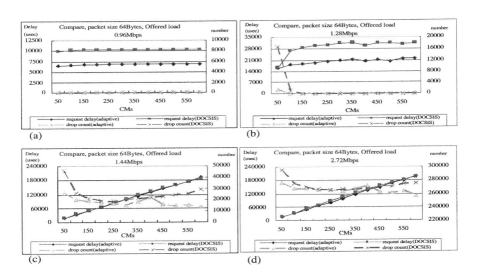

Fig. 4. The relationship between request delay, drop number and offered load. (a)offered load = 0.96 (b)offered load = 1.28 (c)offered load = 1.44 (d)offered load = 2.72

reserve m contention slots to achieve the maximum success probability. We also evaluate the performance of the system we proposed in two different aspects by experiments. From the simulation result, we conclude that our adaptive method has a better performance than that of MCNS DOCSIS in the case studied.

References

[1] Gorge Abe, "Residential Broadband, Second Edition," Cisco Press, 2000.

[2] Y. D. Lin, W. M. Yin, and C. Y Huang, "An Investigation into HFC MAC Protocols: Mechanisms, Implementation, and Research Issues," *IEEE Communications Surveys*, Third Quarter 2000, PP. 2-13.

[3] V. Sdralia, C. Smythe, P. Tzerefos, and S. Cvetkovic, "Performance Characterisation of the MCNS DOCSIS 1.0 CATV Protocol with Prioritised First Come First Served Scheduling," *IEEE Trans. Broadcasting*, vol. 45, no. 2, Jun. 1999, pp. 196-205.

[4] Institute of Electrical And Electronics Engineers, Inc., 345 East 47th Street, New York, NY 10017, USA, IEEE project 802.14/a Draft 3 Revision 1, Apr. 1998.

[5] Cable Television Laboratories, Inc., Data-Over-Cable Service Interface Specifications-Radio Frequency Interface Specification, Version 2.0, Nov. 2001.

[6] M. Ahedand G. Roeck, "IP Over Cable Data Network (IPCDN) Service," IETF Draft, 1996.

[7] Residential Broadband Architecture Framework, ATM Forum Technical Committee, AF-RBB-0099.000, Jul 1998.

[8] ITU-T, Recommendation J.83 Series J: Transmission of Television, Sound Programme and Other Multimedia Signals, ITU-T, April 1997.

[9] ITU-T, Recommendation J.112: Transmission Systems for Interactive Cable Television Services, ITU-T Pre-published, March 1999.

[10] N. Glomie, F. Mouveaux, and D. Su, "A Comparison of MAC Protocols for Hybrid Fiber/Coax Networks: IEEE 802.14 vs. MCNS," *Proc. ICC 1999*, pp. 266-272.

[11] D. Sala, J. O. Limb, "A Protocol for Efficient Transfer of Data over Fiber/Cable Systems," IEEE Trans. on Networking, vol. 5, no. 6, Dec. 1997.

[12] N. Glomie, Y. Saintillan, and D. Su, "A Review of Contention Resolution Algorithms for IEEE 802.14 Networks," *IEEE Communications Surveys*, First Quarter 1999, PP. 2-12.

[13] Y. D. Lin, C. Y Huang, and W. M. Yin, "Allocation and Scheduling Algorithms for IEEE 802.14 and MCNS in Hybrid Fiber Coaxial Networks," *IEEE Trans. on Broadcasting*, vol. 44, Dec. 1998, pp. 427-435.

[14] W. M. Yin, and Y. D. Lin, "Statistically Optimized Minislot Allocation for Initial and Collision Resolution in Hybrid Fiber Coaxial Networks," IEEE Journal on Select Area in Commun., vol. 18, no. 19, Sep. 2000, pp. 1764-1773.

[15] M. D. Corner, J. Libeherr, N. Golmie, C. Bisdikian, and D. H. Su, "A Priority Scheme for the IEEE 802.14 MAC Protocol for Hybrid Fiber-Coax Networks," IEEE/ACM Trans. On Networking, vol. 8, no. 2, Apr. 2000, pp. 200-211.

An Extension of Scalable Global IP Anycasting for Load Balancing in the Internet

Siddhartha Saha, Kamalika Chaudhuri, and Dheeraj Sanghi

Department of Computer Science and Engineering
Indian Institute of Technology
Kanpur, UP - 208016, India
dheeraj@cse.iitk.ac.in

Abstract. IP anycasting was considered to be non-scalable, until Katabi *et al* [5] proposed a scheme called GIA. This scheme makes two important assumptions. One, the anycast addresses have a specific format distinguisable from unicast address. Two, it is only necessary to reach the nearest anycast member only if there is a lot of traffic to the anycast group.

In this paper, we propose an extension of GIA to achieve load balancing in the Internet. Our scheme builds upon GIA by adding load information exchange features. This enables the client to choose a member of the anycast group which is not just routing-wise nearest to it, but the one which actually improves its response time, by a proper balance of routing proximity and load. An advantage of our scheme is that it does on-demand routing information exchange, resulting in less overhead. We evaluate the performance of our design through simulation, and show that it achieves better load balancing, compared to GIA, which sends packets to the nearest server, irrespective of the load.

1 Introduction

Anycasting is a network service in which a data packet addressed to an anycast address can be sent to any member of a group of designated hosts, preferably the host that is nearest to the sender. Anycasting was first introduced by Partridge et al [6]. They define anycasting as a "stateless best-effort delivery of an anycast datagram to at least one host, and preferably only one host, which serves the anycast address." It is preferred that this host be the "nearest" from the sender, compared to all other members of the anycast group. Network can define "nearest" according to any routing metric - hop count, delays, etc. Anycast has been adopted by IPv6 as well [4, 1].

An important use of anycasting will be in providing a mirrored network service in a manner transparent to the user. The network will automatically select a server which is nearest to the client.

Another application can be a sort of auto-configuration [6]. Instead of configuring addresses of different servers for each machine, one may keep all the servers providing the same service in a well-known anycast group.

H.-K. Kahng (Ed.): ICOIN 2003, LNCS 2662, pp. 161–170, 2003.

Traditional approaches to anycasting either impose the restriction that all members of an anycast group should be confined within a single domain, or attempt to distribute information about all the members of each anycast group, to every network in the world. While the first one severely limits the utility of anycasting, the second one does not scale at all.

An interesting design for implementing scalable IP anycasting in the entire Internet was proposed by Katabi et al [5]. They have made two assumptions for this design. One, IP addresses used for anycast groups are of a particular format, which is distinguishable from unicast addresses. Two, it is not necessary to route each packet to the nearest member of the group. It suffices to route the packet to any one server in the group. Their design, Global IP Anycasting (GIA), provides default routes to all anycast groups (to its home domain). The address of home domain can be derived from the anycast address.

However, a plain implementation of GIA does not guarantee load balancing. GIA always tries to find the anycast host which is physically (or routing-wise) nearest to the client. This may not always be the best host in terms of response time for the client. This host may be overloaded, and it might be a better idea to contact the next-nearest server for a particular request. Our scheme extends GIA and provides load balancing features to the same.

Apart from GIA, there have been other schemes to do anycasting in the Internet [8, 2]. All these have examined issues related to the implementation of anycasting. However, none of them have a scalable solution for global anycast groups.

The remaining part of the paper is organized as follows. In Section 2, we describe the design and implementation of our scheme. In Section 4, we discuss simulation experiments, and provide performance results. Finally in Section 5, we conclude.

2   Extension for Load Balancing

The main principle behind our design is to provide an efficient scheme for the exchange of highly dynamic load information between the client and the anycast group members and using this load information decide on the optimum anycast server in such a way to minimize the overhead of load information exchange.

2.1   Design Principles

GIA works well with relatively static or slowly changing metrics such as hop count or routing delays. However, a somewhat different framework is required if we want to do routing based on highly dynamic metrics such as load information. What we need in this case is an efficient mechanism of load information exchange between the server and the client; this load information propagation should be scalable, should involve as little overhead as possible, and should preferably be propagated on-demand.

Our scheme has a fast load information exchange framework to take the load variations of the anycast servers into account. But any framework which adapts itself to a highly dynamic scenario, like change in the load information, has the potential of being unstable and can flood the network with routing data packets. We have avoided this by minimizing the additional routing information that needs to be exchanged between the routers.

Whenever a router conducts a search, it estimates the current load on its reachable servers. After the search is finished, the border router switches its anycast server only if it finds that its current anycast server is much more heavily loaded than most of the other anycast servers in the network. This approach helps reduce the unnecessary switches of anycast server and provides the required stability that we seek.

2.2   Framework

In our model, we assume that each server of the anycast domain resides in an autonomous network, which is connected to other autonomous networks, or, the Internet, through a border router. Each anycast group member estimates its own load periodically, and indicates its load information to its Border Router at regular intervals of time, or, when its load status changes by some threshold amount. The Border Router would in turn provide this load information to the clients who are getting service from that group member.

We call the Border Router associated with the anycast member as the Server Side Border Router and the Border Router associated with the edge domain as the Client Side Border Router.

Client side border routers initiate a search for a nearby member of the anycast group with low load when both the following two conditions are satisfied: One, when the border router has determined that this group is popular (that is, there are several packets going to this group). Two, when the latest load information about the anycast server that it has been in touch, is substantially higher than what was assumed earlier. Search mechanism is described in more detail later in Section 2.4.

The client-side border router would store the anycast routes found during the searches. When the border router has to send a packet to an anycast address which is popular for that domain, it tunnels the packet directly to the Border Router of the domain containing the chosen member of that anycast group. Each client domain also stores the load information of the chosen anycast member.

2.3   Load Information Exchange

As mentioned earlier, the server side will maintain its load information. This information will be sent out only when it is substantially different from the earlier information, or the previous information was sent a long time ago.

The information is not broadcast by the server-side routers, but instead, it is sent as unicast packets only to those client-side routers which need it. Server-side router needs to identify such client-side routers who have old information.

For this purpose, we maintain a sequence number at the server-side router. The number is incremented either when the load changes substantially, or when a timer expires. Whenever server-side router sends its load information, it includes this number in the packet.

We propose adding an extra field to the tunnel packet - the sequence number of the load information packet that it last received from the server. So, whenever a client-side border router tunnels a packet to the server-side border router, the latter knows the sequence number of the latest load information available to the client. If this sequence number is different from the current sequence number, only then the server-side border router sends back a load information packet to the client.

2.4   Search

The mechanism of search is essentially the same as that in GIA. The search is done by the border routers. The border router which initiates the search for a particular anycast address, sends a path query packet to its neighboring border routers, which in turn propagate the query to their neighbors, upto a predefined hop limit. If any of the border routers which received the query packet, knows a path to the anycast address in question, it sends a reply to the initiator router.

But we have made changes in the BGP [7] packets proposed in that scheme. In this case we need to propagate the load information also along with the search reply packet. Thus the BGP search reply packet is modified to accommodate the load information along with the sequence number.

One of the important consideration in designing the system is stability. If the load of one member of an anycast group increases, then a naive approach to load balancing would cause most of the client domains which were associated with that particular anycast member to initiate a search and possibly associate with another anycast member. This kind of behavior can easily lead to large oscillations in the load profile of the anycast members and instability in the network. By instability we mean that the client domains change their anycast servers frequently, thus leading to generation of more and more search packets and consequently, load information packets. In order to curb this oscillating behavior, we have introduced some dampening measures.

When the client-side border router receives a new load information packet from the server-side border router, it compares the current load information of the server with the average load of the anycast group. The average load of the anycast group is calculated as the average of the load values of all the servers for which the client domain received estimates during the previous search. This is admittedly older information, but it gives some idea about what kind of load is there on the group as a whole. And one would switch servers only if the server is more highly loaded than the rest of the group; otherwise we would do better by sticking to our current server by avoiding the overhead of searching.

After a client-side border router has initiated a search, it waits for a fixed time interval T for replies. Since some of the neighboring border routers may have a stale estimate of the load information of a server, we need each border router to transmit the sequence number along with the load information value. If different estimates are received for the same server from two border routers, the more recent one is considered. After time T, the border router looks at all the responses received, and calculates the average load of the network. If the average load is still less than the current server load by a threshold, it has to choose a different server. To do so, it chooses k least-loaded servers out of all the responses received; out of these, it chooses the nearest. Our experiments show that with the current Internet topology, and 0.5 percent of the domains randomly containing an anycast server, a good value for k would be about 3. A larger value would result in simply choosing the nearest server disregarding any consideration for the load and smaller values would ignore any neighborhood information and would result in choosing simply the least loaded server.

If the current load information is higher than the average load value by a threshold (assumed to be 20% in our experiments), the client-side border router should try to switch servers. However, if all clients switch servers whenever the load at a server increases, the load balancing would become unstable. Here, we adopt a probabilistic approach in deciding whether a router should switch server when the load in the server is has increased by a substantial amount. Ideally, only those many routers should switch servers such that the load in the anycast server drops to an amount where its load becomes roughly equal to the average load in the network. We define the probability p_{switch} of initiating search for switching the server when the increase in load of the anycast server is over the threshold value as:

$$p_{switch} = \frac{load_{server} - load_{network}}{load_{server}}.$$ (1)

where $load_{server}$ is the current load of the anycast server and $load_{network}$ is the estimated average load of the network.

Note that, we keep the information of the average load of the network, hence, we do not need any extra information for the calculation of this probability measure. But there may be cases when these searches would not be sufficient to reduce the load in the anycast server, especially when there is a huge change in the load of the network. To protect against this, the routers which do not fire search immediately, waits for an interval of time T_{max} before scheduling the search. T_{max} is estimated to be roughly the time period needed to complete the search. In this period, some clients may have switched their servers, and the load of the overloaded server may have decreased as a result. But if after the time T_{max}, the load of the server still remains greater than the average load of the group by a threshold, the same algorithm is followed; otherwise no action is taken.

2.5   Load Metric

Our method allows an anycast group to use a load metric that is relevant to the application; the only constraint is that the same metric should be used for all members belonging to an anycast group. In our experiments, we had assigned each server a capacity value, which is the maximum number of packets that it can server in an unit time. The load was defined as the ratio of number of packets served in an unit time to this capacity.

3   Experiment

To evaluate the performance of our scheme, and compare it with GIA, we have conducted simulation experiments. We describe below the simulation setup, parameters, and the results obtained.

3.1   Simulator

We implemented a custom simulator to study the performance of our scheme. For the simulation topology, we use a set of snapshots of the Internet inter-domain topology generated by NLANR based on the BGP routing tables [3].

In our experiments, we simulated only one anycast group. This is because, in our scheme, load balancing for each anycast group is independent of the other anycast groups in the Internet. By having only one representative anycast group, we could increase the number of domains, number of members of the anycast group, etc., and run simulations for longer periods of time.

Given a graph representing the Internet domain topology, we randomly assign n domains to have members of the anycast group. Out of the remaining domains, a fraction p of the domains consider that anycast group as a popular group. For the rest of the nodes, the anycast group may be accessed only rarely. One of the domains having anycast group member is randomly marked as the home domain.

As the simulation starts, a series of requests are generated by each domain, and these requests are routed to the appropriate anycast group member, chosen according to the scheme we have proposed. The desired quantities, namely load profiles and processing delays are then measured at each anycast group member. Each node starts sending requests after a random time interval, to avoid synchronization of requests.

3.2   Parameters

To measure the performance of our scheme for various load conditions, we define several parameters. We define C, the capacity of an anycast server, as the number of packets it can process in unit time. We used a simplistic queue management algorithm, in which if an host receives more that C packets in an unit time interval, the extra packets are dropped, and the originator node of that packet is notified. We also keep track of the number of packets dropped in this manner during the simulation period.

In order to compare the performance of our extensions, with the original GIA, we use three parameters. We compute average queuing delays at the server side. When a packet arrives at the server, the number of packets in the queue multiplied by the service time of a single packet, gives the queuing delay.

Our second parameter is the path length (hop count) from the client domain to the chosen anycast server. Finally we measure the load profiles at the anycast hosts. To get a measure of the average load condition of the network, we define another parameter L. L is given as

$$L = \frac{\frac{1}{2}(n_p p + n_u(1 - p)) * N}{n * C} \ . \tag{2}$$

Here p is the fraction of domains which have the anycast group as a popular group, n_p is the maximum number of packets that a client can send to a popular group in unit time, and n_u is the maximum number of packets that a client can send to other groups in unit time. The actual number of requests generated by a source (client) is a uniformly distributed random number between 0 and n_p or n_u (as the case may be). N denotes the total number of nodes in the graph, and n is the number of anycast servers. Number of domains in the Internet (N) is taken as 6474 [3]. The numerator indicates the expected number of packets generated in the network per unit time and the denominator is the total number of packets that can be processed by all the servers in unit time. The ratio thus gives a measure of the overall load in the network.

The other parameters that we use in our simulations is n, the number of domains containing an anycast host and p, the fraction of domains for which the anycast group is popular. Simulations were performed under two different load conditions (medium load, $L = 0.25$ and high load, $L = 0.55$) for different values of n and p.

3.3   Results

We wanted to measure the performance of this load balancing scheme under various scenarios, by varying different parameters of the network model. We performed two sets of experiments. In the first set, we studied the performance of our scheme with increasing values of n, the number of members in the anycast group. n is varied between 20 to 140. In these experiments, it is assumed that 60% of the domains consider the group as popular ($p = 0.6$).

In the second set of experiments, we study the performance of our scheme as p, the fraction of domains for which the anycast group is popular, increases. The value of p is varied between 0.1 to 0.9. It is assumed that the anycast group has members in 60 domains ($n = 60$).

For both sets, we perform simulations for both medium and high load conditions,viz. $L = 0.25$ and $L = 0.55$, for GIA and our scheme. The simulation results are described below.

Load Balancing Figure 1 shows the variance of the loads at the anycast servers for GIA and our scheme for two different load conditions in the network, as a function of number of members in the anycast group.

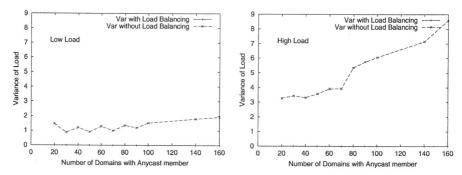

Fig. 1. Variance of the load (with and without load-balancing) among the anycast group members, under low-load and high-load situations, as a function of the number of anycast servers in the network

We notice that when load balancing is implemented, the variance in loads of different servers is always close to 0, irrespective of load as well as the number of members in the anycast group.

This shows that our scheme is actually able to do balance the load between the servers rather well.

Our experiments with varying percentage of the domain which have this anycast group as popular also show same kind of result. i.e., close to zero variation with the load balancing scheme enabled.

Queueing Delays In the case where load-balancing is not done, we find that some servers don't receive any requests, while some other servers have huge queues built up, and the average queue-size is quite high. In case of load-balancing, all servers get roughly equal number of requests, and hardly any queues are built anywhere in the network.

Figure 2 shows the average queueing delays at the anycast servers for GIA and our scheme for two different load conditions in the network, as a function of the number of anycast servers.

As is expected, proposed extension is able to reduce average queueing delays to almost negligible values even under high load.

Figure 3 shows the average queueing delay at the anycast servers, with and without load balancing extension, for two different load conditions in the network, as a function of percentage of domains which consider this group as popular. Again the queueing delays are almost non-existent with load-balancing. Smaller queueing delays are obviously preferable, since it implies smaller response time for the client. Thus, load-balancing results in faster response to the client.

Path Length Since our scheme does not choose the nearest server, the average path length with the load balancing enabled may be more than the pure imple-

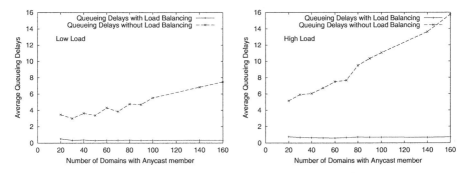

Fig. 2. Average queue size (with and without load-balancing) at the anycast servers, under low-load and high-load situations, as a function of the number of anycast servers in the network

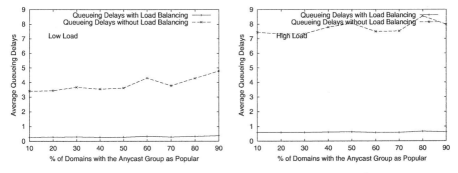

Fig. 3. Average queue size (with and without load-balancing) at the anycast servers, under low-load and high-load situations, as a function of the percentage of domains that consider this group as popular

mentation of GIA. To check this out, we looked at the average path length in case of both GIA and load-balancing extensions.

Figure 4 shows the average path length (number of hops) for GIA and with load-balancing extensions for two different load conditions in the network, as a function of number of members in the anycast group. The average is taken over all clients which consider this anycast group as a popular group.

From the figure, we notice that there is a little increase in path length (one extra hop on average) with the load balancing schemes, as expected. The increase in round-trip time due to this extra hop will be more than compensated by smaller queueing delays and the much less number of dropped packets.

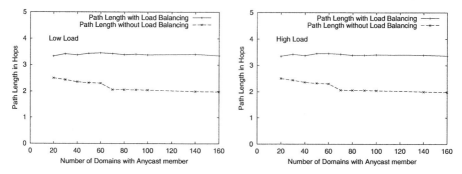

Fig. 4. Average path length (with and without load-balancing), under low-load and high-load situations, as a function of the number of anycast servers in the network

4 Conclusion

Our scheme builds upon GIA [5] by adding features which enable the exchange of dynamic load information. The information exchange happens only on demand, thereby not imposing large overhead on the network. Furthermore, we make sure that a border router does not switch servers until it is absolutely necessary, i.e., when the server load goes significantly higher than other anycast members, thus providing stability in the network. The simulation results show that the scheme is able to distribute the load on the servers evenly, under different load conditions. As a consequence of this, the queuing delays at the servers are much less for our scheme than for GIA. Finally, the average path length to reach an anycast server when using our scheme, is only higher by one hop on an average as compared to GIA.

References

[1] S. Deering and R. Hinden. Internet protocol version 6 (IPv6) specifications. RFC 2460, Dec. 1998.
[2] R. Engel, V. Peris, D. Saha, E. Basturk, and R. Haas. Using ip anycast for load distribution and server location. In *Proc. of Third Global Internet Mini-conference*, Nov. 1998.
[3] T. N. L. for Applied Network Research (NLANR). http://www/moat.nlanr.net/AS/.
[4] R. Hinden and S. Deering. IP version 6 addressing architecture. RFC 2373, July 1998.
[5] D. Katabi and J. Wroclawski. A framework for scalable global ip-anycast (GIA). In *Proc. of ACM SIGCOMM 2000*, pages 3–15, Stockholm, Sweden, Aug. 2000.
[6] C. Partridge, T. Mendez, and W. Milliken. Host anycasting service. RFC 1546, Nov. 1993.
[7] Y. Rekhtar and T. Li. A Border Gateway Protocol 4 (BGP - 4). RFC 1771, Mar. 1995.
[8] D. Xuan, W. Jia, W. Zhao, and H. Zhu. A routing protocol for anycast messages. *IEEE Transactions on Parallel and Distributed Systems*, 11(6), 2000.

ATFRC: Adaptive TCP Friendly Rate Control Protocol[*]

Seongho Cho[1], Heekyoung Woo[2], and Jong-won Lee[3]

[1] School of Electrical Engineering and Computer Science
Seoul National University, Seoul, Republic of Korea
shcho@popeye.snu.ac.kr
http://jack.snu.ac.kr/~shcho/research.html
[2] Division of Information Technology Engineering
Soonchunhyang University, Asan, Republic of Korea
woohk@sch.ac.kr
[3] School of Computer Science and Electronic Engineering
Handong University, Pohang, Republic of Korea
ljw@han.ac.kr

Abstract. TFRC (TCP Friendly Rate Control) is a *rate-based* congestion control protocol for non-TCP flows. TFRC controls the sending rate
by using the TCP throughput equation considering the triple duplicate
ACKs and the timeouts. However, this equation provides the conservative throughput bound of TCP. This conservativeness causes lower
throughput and slower response time for TFRC than those for TCP.
In this paper, we propose the Adaptive TCP Friendly Rate Control (AT-
FRC) protocol. Reflecting the transient behavior of TCP, we divide the
throughput calculation periods into two. For each period, we adapt the
sending rate based on different TCP throughput bounds. Simulation results show that the proposed algorithm gets closer to the long-term fair
share of TCP than that of TFRC. In addition, ATFRC has faster convergence time than TFRC when bandwidth is available.

1 Introduction

Transmission Control Protocol (TCP) has been widely used as a transport layer
protocol with the explosive growth of the Internet. The Internet has operated
stably with the TCP congestion control mechanism which avoids congestion collapse [1] and achieves a reasonable throughput level. In the next generation Internet, new services such as multimedia transmissions and multicast applications
will require different service criteria: a steady bandwidth guarantee, low delay,
and a stable delay jitter. For new services, a transport layer protocol should
guarantee the smooth bandwidth and a low delay jitter. However, TCP is not

---

[*] This work was supported by the Brain Korea 21 Project of the Korean Ministry of
Education, and the University IT Research Supporting Program under the Ministry
of Information & Communication of Korea.

H.-K. Kahng (Ed.): ICOIN 2003, LNCS 2662, pp. 171–180, 2003.

proper for new types of services because the throughput of the AIMD (Additive Increase Multiplicative Decrease)-based TCP congestion control algorithm oscillates rapidly even for slight losses. In addition, TCP has a retransmission mechanism for reliable transmission, and the TCP congestion window allows for the transmission of data packets in bursts. However, these TCP mechanisms increase throughput oscillation and delay jitter. Therefore, new services use UDP (User Datagram Protocol) as a transport layer protocol. But, since UDP has no congestion control mechanism, the increase of the UDP traffic for new services can cause network instability.

TCP friendly protocols have been developed to overcome these defects of the current transport layer protocols for new services. TCP friendly congestion control protocols are designed to satisfy the following design issues [2]. First, friendliness with the existing TCP should be provided. Since the Internet has been operated stably with the TCP congestion control mechanism to prevent congestion collapse, introducing a new protocol should not cause instability of the network. In addition, a TCP friendly protocol should not harm the conformant TCP throughput. Second, smoothness of the transmission rate should be satisfied. Rapid changes in the transmission rate deteriorate the service quality by increasing delay and delay jitter. Also, because the multimedia codec has the difficulty in changing data rate abruptly, the transmission method of TCP is not proper for multimedia communication. Third, responsiveness to a varying network state is required. Traffic multiplexing, router queue length, and routing update changes constantly the network state. A TCP friendly protocol should be able to adapt to various network states. Fourth, convergence to the fair share should be offered. When bandwidth is released, the protocol should converge to the fair share point as fast as possible. These properties conflict in some sense. Especially, smoothness can make the protocol response slow for network congestion and introduce unfairness with other flows. Also, slow convergence makes the protocol experience low long-term throughput and slow responsiveness. Therefore, TCP Friendly protocols should be designed considering these trade-offs.

Several TCP friendly protocols [3, 4, 5, 6, 7, 8, 9, 10, 11, 12] are proposed to improve friendliness, smoothness and responsiveness. These mechanisms are categorized into two classes. The first one is the *window-based* congestion control protocol. The sender manages the sending window size by the feedback information from the receiver, and the sending rate is implicitly determined by the current window size. Many mechanisms are based on Chiu/Jain's convergence model [13], but they modify converging functions to provide better fairness and efficiency point. For example, Binomial [3], GAIMD [4], and SIMD [6] can be classified. These protocols proposed their own convergence pattern from the transient state to the optimal point by adapting the window increasing/decreasing methods. *Window-based* congestion control methods respond rapidly to the current loss state, and therefore they have less smoothness. Also, *window-based* congestion control methods transmit packets in bursts.

The other category is the *rate-based* congestion control protocol. These methods estimate the proper sending rate of TCP from the TCP throughput analysis or TCP emulation. *Rate-based* TCP friendly protocols are RAP [7], TFRC [8] and TEAR [12]. TFRC is based on the long-term throughput of TCP considering triple duplicate ACKs and timeouts. TFRC guarantees a smooth sending rate, however it has a low long-term throughput and slow responsiveness. That's because the TCP throughput bound of TFRC is too conservative. Also, there is a *rate-based* congestion control protocol based on AIMD, called RAP (Rate Adaptation Protocol). RAP responses as fast as TCP, but doesn't provide smoothness. However, RAP gets higher long-term throughput than TCP [7]. Compared to *window-based* congestion control protocols, *rate-based* congestion control protocols have smoother but less responsive properties.

We propose an Adaptive TCP Friendly Rate Control (ATFRC) protocol. The proposed protocol is based on the *rate-based* congestion control method, TFRC. Considering the strength and weakness of RAP and TFRC, we can improve TFRC with an adaptive sending rate control. We divide throughput calculation periods into two: triple duplicate ACK periods without timeout and periods after timeout occurrence. ATFRC calculates the sending rate more accurately for each period. The receiver keeps the two loss event history for triple duplicate ACKs and timeouts. The receiver feedbacks the two loss event rates at each RTT. The sender then adapts the sending rate according to different equations for the triple duplicate ACK and timeout event. The proposed method can achieve a higher throughput than TFRC using the transient throughput bound of TCP. Also, if available bandwidth is provided, the convergence time to a new fair share is decreased.

The rest of the paper is organized as follows. In Section 2, we examine the basic concept of TFRC. It is shown that TFRC is conservative because it is based on Padhye's TCP throughput analysis. We also describe Padhye's TCP throughput analysis based on triple duplicate ACKs and timeouts. In Section 3, we propose an adaptive rate control mechanism by dividing loss indications. From the AIMD throughput analysis and Padhye's analysis, we determine the sending rate using a τ parameter. Also, the sender and receiver functionalities are described. In Section 4, we simulate and compare the proposed algorithm with TCP, RAP and TFRC. From the results, we show that the proposed algorithm satisfies the smoothness with faster responsiveness and less conservativeness than TFRC. In Section 5, we conclude our results and present some future research issues.

2   TFRC: TCP Friendly Rate Control

TFRC is mainly based on Padhye's TCP throughput analysis [14]. In this analysis, TCP throughput is determined by the static loss rate p and Round Trip Time (RTT). The probability of the triple duplicate ACK and timeout occurrence, which are loss indications, can be estimated from the static loss rate p. Padhye's TCP throughput equation is based on the average number of packets sent during the estimated period of the Triple Duplicate ACK Period (TDP) and the Time Out Period (TOP). The throughput of TCP with triple duplicate

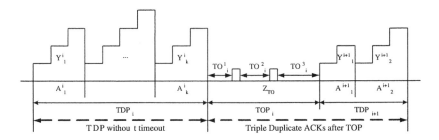

Fig. 1. Window Size of TCP with Triple Duplicate ACKs and Timeouts

ACKs and timeouts is given as

$$B_{TO}(RTT, p) = \frac{1}{RTT\sqrt{\frac{2bp}{3}} + T_0 min(1, 3\sqrt{\frac{3bp}{8}})p(1 + 32p^2)} \tag{1}$$

where b is the number of ACKed packets when TCP uses a delayed acknowledgement option, p is the static loss rate, RTT is the measured round trip time,
and T_o is the Retransmission Timeout (RTO) length which is approximated as
4 x RTT.

Figure 1 shows the TCP window size changing pattern in the triple duplicate
ACK and timeout periods. Because the average number of packets is calculated
over TDP and TOP, the TCP long-term throughput can be underestimated in
high loss rate state. In TOP, there are few packets sent although TOP is a long
period of time. Moreover, the probability of timeout occurrence is increased as
the loss rate p gets higher. Therefore, Padhye's TCP throughput analysis can
be too conservative in a short-term period or in a varying loss rate environment.
This feature makes the sending rate of TFRC close to zero with a high loss
probability [5].

3   ATFRC: Adaptive TCP Friendly Rate Control

We propose a new TCP friendly congestion control protocol, ATFRC (Adaptive
TCP Friendly Rate Control). We use different theoretical throughput bounds:
Padhye's analysis and [15, 16]'s analysis according to the network state. Padhye's
TCP throughput bound calculates the long-term TCP throughput considering
TDPs and TOPs, but underestimates short-term throughput. Meanwhile, there
is previous research [15, 16] on TCP throughput considering triple duplicate
ACKs only. In an ideal case, the TCP throughput oscillates between $\frac{W_{max,i-1}}{2}$
and $W_{max,i}$, half of the previous $(i\text{-}1)$th period's maximum window size and
current ith period's maximum window size, respectively. The throughput of
TCP with only triple duplicate ACKs is derived as follows

$$B_{TD}(RTT, p) = \frac{1}{RTT}\sqrt{\frac{3}{2bp}} \tag{2}$$

where b is the number of ACKed packets when TCP uses a delayed acknowledgement option, p is the static loss rate, and RTT is the round trip time. ATFRC protocol functions are described as follows.

3.1   Sender Functionality

The proposed algorithm divides the loss indication into triple duplicate ACKs and timeouts to reflect the transient behavior of TCP. Each indication can imply a different network status. The triple duplicate ACK event can indicate that the sending rate exceeds fair share, or that there is slight congestion at the network router. Meanwhile, the timeout event signifies that there is significant congestion on the route. We introduce a new variable τ, which can parameterize the timeout effect. The sender determines the proper sending rate with two loss notifications from formula (3).

$$R=\begin{cases} \frac{s}{RTT}\sqrt{\frac{3}{2bp}} & \text{TDP} \qquad\qquad\qquad\qquad\qquad (a) \\ \frac{s}{RTT\sqrt{\frac{2bp}{3}}+T_0 min(1,3\sqrt{\frac{3bp}{8}})L_k} & \tau \text{ Triple Duplicate ACKs after TOP (b)} \end{cases} \quad (3)$$

where s is the average packet size and L_k is the timeout period after k time-outs, as follows

$$L_k = \begin{cases} (2^k - 1) & \text{for } k \leq 6 \\ (63 + 64(k - 6)) & \text{for } k > 6 \end{cases}. \quad (4)$$

As in Figure 1, during the triple duplicate ACK only period, the sending rate is calculated based on equation (2). However, when a timeout occurs, the current sending rate is reduced to the half of the original sending rate, and the next sending rate calculation is based on (1). Until the τ number of triple duplicate ACK events have occurred, the sending rate is calculated with the TCP throughput equation (1). After the effect of the timeout disappears, the sending rate of TCP can be restored to equation (2). Consider the case that τ is 5. Before the timeout occurs, the sending rate calculation is based on equation (3 a). If the l times of timeouts happen, the sending rate is reduced by as much as half. And the sending rate calculation equation is changed to equation (3 b) considered with L_l until the 5th triple duplicate ACK event occurs. Therefore, the sending rate adaptation period is divided into two dotted lines as shown in Figure 1. As described above, our method adaptively calculates the sending rate for each error event.

3.2   Receiver Functionality

The receiver keeps the two loss interval of the triple duplicate ACKs and the timeouts. Like TFRC, we also use the Average Loss Interval (ALI) method for each TDP and TOP. The receiver keeps the received packet sequence number. If a higher sequence packet is received, we regard it as a packet loss. If the receiver

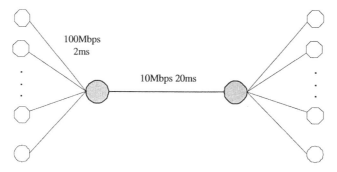

Fig. 2. Simulation Topology

timer expires without receiving a packet, then we consider it a timeout. We obtain the kth loss event rate from the following ALI method

$$S(k) = \frac{\sum_{i=1}^{n} W_i S(k-i)}{\sum_{i=1}^{n} W_i} \qquad W_i = \begin{cases} 1 & 1 \leq i \leq n/2 \\ 1 - \frac{i-n/2}{n/2+1} & n/2 < i \leq n \end{cases} \qquad (5)$$

where n is the number of events kept, $S(k)$ is the kth loss event rate, and W_i is the weight for each loss history.

The receiver should send at least one or more feedback packets to the sender every RTT to prevent the deterioration of the sender throughput accuracy [7]. Each feedback packet contains the current timestamp, the feedback packet processing delay, the receiver-estimated rate, and the two loss event rates of triple duplicate ACKs and timeouts.

4   Simulation Results

4.1   Simulation Environment

In this section, we present the performance of ATFRC compared to TFRC and TCP. We conducted our simulation based on the ns 2 network simulator [17]. All of our experiments use a single bottleneck topology shown in Figure 2, and the bottleneck queue is managed with the RED (Random Early Detection) mechanism. We evaluate the throughput smoothness, long-term throughput, and conservativeness of TCP, RAP, TFRC, and ATFRC in Section 4.2 and compare the convergence characteristics of TCP, TFRC and ATFRC in Section 4.3.

The notations, TCP($1/\beta$), RAP($1/\beta$), TFRC(n), and ATFRC(n, τ) are used. The parameter β of TCP($1/\beta$) and RAP($1/\beta$) is a decrease factor for congestion notification. In TFRC(n) and ATFRC(n, τ), the parameter n denotes the number of loss event histories with the ALI calculation for receiver functionality, and the parameter τ of ATFRC(n, τ) is the number of Triple duplicate ACKs after timeout periods for the sender to reflect the timeout effect.

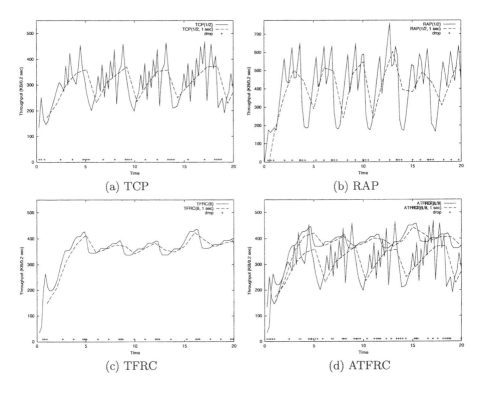

Fig. 3. Throughput Variation in Fixed Loss Environment

4.2 Throughput Analysis

To verify the TCP-friendliness and transient behavior of ATFRC, we inspect the throughput smoothness, long-term throughput pattern, and the conservativeness. We set the explicit drop sequence and simulate over 20 seconds. Each throughput is calculated every 0.2 seconds. To compare the relative smoothness of throughput, we also calculate the throughput over 1 second and compare the long-term pattern of throughput. As in Figure 3 (a), even though we average the throughput per 0.2 seconds, the throughput of TCP(1/2) fluctuates with a saw-tooth shape of bandwidth with light packet drops. Also, RAP(1/2) shows the throughput of the saw-tooth shape in Figure 3 (b). In particular, RAP(1/2) gets higher bandwidth than TCP(1/2). However, in Figure 3 (c) and (d), TFRC(8) and ATFRC(8, 5) both show a smooth bandwidth adaptation for loss indications. This result shows that ATFRC(8, 5) flows satisfy the smoothness like TFRC(8) flows. However, ATFRC(8, 5) gets closer bandwidth allocation to the fair share of TCP(1/2). In the high packet drops simulation, we can find similar results of ATFRC(8, 5) throughput alteration. Comparing with the long term bandwidth shape of TCP(1/2) obtained by averaging over 1 second, ATFRC(8, 5) also follows the fair share of bandwidth of TCP(1/2).

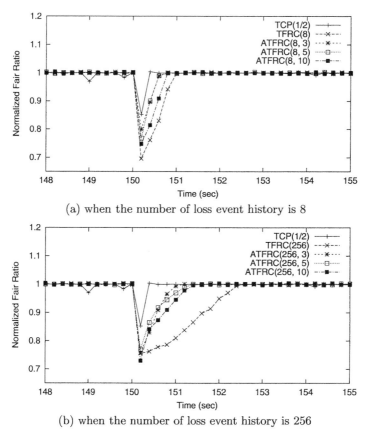

(a) when the number of loss event history is 8

(b) when the number of loss event history is 256

Fig. 4. Convergence Time to the Expected Throughput

4.3   Convergence Time Analysis

To compare the convergence time to the fair share, we introduce the normalized fair ratio that is defined as the ratio of current throughput to expected through-put, where the expected throughput is calculated from the fair share for current bandwidth. Figure 4 shows the normalized fair ratio of each protocol. At the 150 second mark, the CBR traffic stops and available bandwidth is offered. TCP flows response immediately in 1 RTT, but TFRC flows response slowly. In Figure 4 (a), the slow responsiveness of TFRC makes the convergence time as much as 1.4 seconds (about 4 RTTs). However, ATFRC flows response faster than TFRC flows. For τ is 3 or 5, the convergence time is reduced to 0.8 seconds (2 RTTs). Compared with the convergence time of TCP(1/2) which is 0.4 seconds (1 RTT), ATFRC responds more slowly, but this result shows that ATFRC(8, 3 or 5) converges faster than TFRC(8). As the periods after timeouts get longer, the responsiveness of ATFRC flows become closer to that of TFRC flows. On the other hand, as the periods after timeouts are shortened, the response time of ATFRC flows gets closer to that of TCP flows.

Figure 4 (b) shows the convergence time when the number of loss events used to calculate the average loss interval is 256. The convergence time of TFRC(256) is 2.6 seconds (about 8 RTTs) which is two times of TFRC(8). That is, the convergence time of TFRC gets longer as the number of loss events is larger. However, the convergence time of ATFRC(256, 3) is just 1.2 seconds (3 RTTs). Even in case of ATFRC(256, 3) or ATFRC(256, 5), the convergence time is not degraded even though the number of loss events is increased as much as 256. These results show that ATFRC improves convergence time.

5   Conclusion

Next generation Internet will not merely support data services but also multimedia services. TCP is not sufficient for providing multimedia and multicast services. To support new services, several TCP-friendly protocols have been proposed. However, TFRC have been too conservative in utilizing available bandwidth effectively. And unlike TCP protocol, rate-based TCP-friendly protocols response slowly to network congestion notifications. So we developed a new protocol, called ATFRC, to provide faster responsiveness and convergence time with less conservativeness. Since ATFRC reflects the transient status of loss notification, two aspects are improved over TFRC. First, the proposed protocol is less conservative in utilizing bandwidth than TFRC. TFRC bounds the sending rate with the average throughput derived for the whole lifetime of the TCP flow. In some temporary periods, the throughput of TFRC can be too low, especially in a high loss environment. However, ATFRC determines the rate according to the current state of TCP. Therefore, the proposed protocol gets closer to the fair share even in a high loss environment. Second, the proposed protocol provides a faster convergence time than TFRC. ATFRC and TFRC use the loss history to calculate loss event rates. To provide smooth bandwidth allocation, a longer loss events history is needed. TFRC is sensitive to this loss events history. As the loss history gets longer, the convergence time of TFRC takes longer. However, because the proposed method reflects more accurate loss states, ATFRC converges to the fair share faster than TFRC.

For the future work, we will analytically evaluate the performance of the proposed protocol, TFRC, and other TCP-friendly protocols, and we will simulate under a dynamic environment. We also plan to develop a new kind of TCP friendly protocol suitable for a wireless environment.

References

[1] J. Nagle, "Congestion Control in IP/TCP," *IETF RFC 896*, Jan. 1984.
[2] J. WidMer, R. Denda, and M. Mauve, "A Survey on TCP-Friendly Congestion Control," *IEEE Network*, May/June 2001.
[3] D. Bansal, and H. Balakrishnan, "Binomial Congestion Control Algorithms," *In Proceedings of IEEE INFOCOM 2001*, Anchorage, Alaska, Apr. 2001.

[4] Y. Yang and S. Lam, "General AIMD Congestion Control," *In Proceedings of ICNP 2000*, Osaka, Japan, Nov. 2000.

[5] Y. Yang, M. Kim, and S. Lam, "Transient behaviors of TCP-friendly congestion control protocols," *In Proceedings of IEEE INFOCOM 2001*, Anchorage, Alaska, Apr. 2001.

[6] S. Jin, L. Guo, I. Matta, and A. Bestavros, "TCP-friendly SIMD Congestion Control and Its Convergence Behavior," *In Proceedings of 9th IEEE ICNP 2001*, Riverside, CA, November, May 2001.

[7] R. Rejaie, M. Handly, and D. Estrin, "RAP: An End-to-end Rate-Based Congestion Control Mechanism for Realtime Streams in the Internet," *In Proceedings of IEEE INFOCOM 1999*, Mar. 1999.

[8] M. Handley, J. Padhye, and S. Floyd, "TCP Friendly Rate Control (TFRC): Protocol Specification," *IETF Internet Draft*, Jul. 2001.

[9] D. Bansal, H. Balakrishnan, S. Floyd, and S. Shenker, "Dynamic Behavior of Slowly-Responsive Congestion Control Algorithms," *In Proceedings of SIGCOMM 2001*, San Diego, CA, Aug. 2001.

[10] X. Zhang, and K. Shin, "Second-Order Rate-Control Based Transport Protocols," *In Proceedings of 9th IEEE ICNP 2001*, Riverside, CA, November, May 2001.

[11] D. Sisalem, A. Wolisz, "LDA+ TCP-Friendly Adaptaion: A Measure-ment and comparison Study," *In Proceedings of International Workshop Network and Operation System Support for Digital Audio and Video*, Jun. 2000.

[12] I. Rhee, V. Ozdemir, and Y. Yi, "TEAR: TCP emulation at receivers - flow control for Multimedia streaming," Tech. report, Dept. of Comp. Sci., NCSU, Apr. 2000.

[13] D. Chiu, and R. Jain, "Analysis of the increase and decrease algorithms for congestion avoidance in computer networks," *Computer Networks and ISDN Systems*, vol. 17, 1989.

[14] J. Padhye, V. Firoiu, D. Towsley, and J. Kurose, "Modeling TCP Throughput: A simple Model and its Empirical Validation," *In Proceedings of ACM SIGCOMM 1998*, Vancouver, Canada, Sep. 1998.

[15] M. Mathis, J. Semske, J. Mahdavi, and T. Ott, "The Macroscopic Behavior of the TCP Congestion Avoidance Algorithm," *Computer Communication Review*, vol. 27, no. 3, Jul. 1997.

[16] T. Ott, J. Kemperman, and M. Mathis, "The Stationary Behavior of Ideal TCP Congestion Avoidance," Manuscript, Aug. 1996, *ftp://ftp.bellcore.com/pub/tjo/TCPwindow.ps*.

[17] ns-2 Network Simulator, *http://www.isi.edu/nsnam/ns*, 2002.

Energy-Efficient Communication Protocols for Wireless Networks

Amitava Datta and Subbiah Soundaralakshmi

School of Computer Science & Software Engineering
University of Western Australia
Perth, WA 6009, Australia
{datta,laxmi}@csse.uwa.edu.au

Abstract. A wireless network (WN) is a distributed system where each node is a small hand-held commodity device called a *station*. Wireless sensor networks have received increasing interest in recent years due to their usage in monitoring and data collection in a wide variety of environments like remote geographic locations, industrial plants, toxic locations or even office buildings. One of the most important issues related to a WN is its energy constraints. A station is usually powered by a battery which cannot be recharged while on a mission. Hence, any protocol run by a WN should be energy-efficient. We design energy-efficient protocols for one-to-one communication on a single channel and k-channel wireless network. Our protocols can be extended for multiple multicasting as well.

1 Introduction

Wireless and mobile communication technologies have grown explosively in recent years. New demands for enhanced capabilities for these technologies will continue to grow in future. The communication in most cellular systems is based on robust infrastructures. However, wireless networks should be rapidly deployable and self-organizing. Wireless networks are useful for disaster relief, search-and-rescue, collaborative computing and interactive mission planning [8]. The first wireless network was a packet radio network(PRNET) developed in the 1970s [2].

A wireless network (WN) is a distributed system with no central arbiter, consisting of p radio transceivers called *stations*. The stations are usually small hand-held devices running on batteries and the batteries cannot be recharged while on a mission [6]. Hence, it is important that any protocol designed for a WN is power efficient, i.e., the stations spend as little power as possible. We assume that each of the p stations in the WN has a unique integer ID in the range $[1, p]$. Assigning unique IDs to the stations in a WN is a separate problem and has been solved by Nakano and Olariu [9]. They have shown that even if the stations do not have ID numbers initially, it is possible to devise a protocol which assigns a unique ID to each station and terminates in $O(\frac{p}{k})$ time slots with high probability, where k is the number of available channels.

H.-K. Kahng (Ed.): ICOIN 2003, LNCS 2662, pp. 181–191, 2003.

Designing routing protocols for single-hop WNs is important, since routing in a multi-hop WN is often done by decomposing it into multiple single-hop WNs [2, 8]. While the computational power of small hand-held devices is increasing at a rapid rate, the lifetime of batteries is not expected to improve significantly in the near future [11]. Further, recharging batteries may not be possible while on a mission. It is known that a significant amount of energy is spent by a station while transmitting or receiving packets. A station spends power when it receives a packet that is not destined for it [6, 11]. If we want to make a protocol energy-efficient, a station should not receive packets that are not meant for it. However, for a single-hop WN, this is a nontrivial problem to solve as all stations are within the transmission range of each other.

Conflict resolution is one of the main issues in designing protocols for WNs. Carrier Sense Multiple Access (CSMA) is a simple and robust random access method for media access and is suitable for WNs [1]. However, a fraction of the available bandwidth is wasted for resolving random conflicts of messages [2]. Several conflict-free multiple access schemes have been proposed recently for radio networks. The most popular of these schemes is the Demand Assignment Multiple Access (DAMA) schemes proposed for transmission networks [3, 7]. In the DAMA scheme, all the stations that want to transmit a message on a given channel are ordered in a logical ring according to which they are given transmission access to the channel [7]. From the viewpoint of designing energy-efficient protocols, the DAMA scheme is better since transmission contention resolution in the CSMA scheme results in high energy consumption [13]. In a DAMA or reservation based protocol, collisions are avoided by reserving channels. Hence, there is no need for retransmission of packets lost due to collision. We are interested in designing a reservation based DAMA protocol for one-to-one routing on a wireless network.

In this paper, we are interested in low-mobility, single-hop WNs where every station is within the transmission range of every other station and the mobility of the stations is low or, at least much smaller compared to the time taken by a protocol to complete. An example of such a network is a collection of small sensor devices scattered over a terrain for collecting data. We evaluate the performance of a protocol designed for a WN based on two criteria : **(C1.)** the protocol should terminate as quickly as possible, and **(C2.)** the energy consumption of each station should be as low as possible.

Nakano *et al.* [10] have recently published a protocol for energy-efficient permutation routing in a WN. Their protocol routes n packets in a k-channel, p-station WN in at most $(2d + 2b + 1)\frac{n}{k} + k$ time slots with no station being awake for more than $(4d + 7b - 1)\frac{n}{p}$ time slots, where $d = \left\lceil \frac{\log \frac{p}{k}}{\log \frac{n}{p}} \right\rceil$, $b = \left\lceil \frac{\log k}{\log \frac{n}{p}} \right\rceil$, and $k \leq \sqrt{\frac{p}{2}}$. Nakano *et al.* [10] argue that for most practical systems, the number of channels is much less than the number of stations and in turn, the number of stations is much less than the number of packets, i.e., $k \ll p \ll n$. One of the crucial assumption in [10] is that each station sends and receives exactly $\frac{n}{p}$ packets which is not true in many real applications.

Datta and Zomaya [4] have improved upon the protocol by Nakano *et al.* [10]. The protocol in [4] is more efficient both in terms of *total number of slots* and the *number of slots each station is awake* compared to the protocol by Nakano *et al.* [10] when $k \ll p \ll n$. Recently, Datta and Soundaralakshmi [5] have given an alternative protocol for permutation routing that works well when the destination addresses of packets are uniformly and randomly distributed.

We concentrate on one of the most fundamental communication problems in a distributed system. This is the problem of one-to-one communication. In this problem, each of the p stations holds a packet for another station. The task is to route all the p packets correctly to their destination stations. Nakano *et al.* [10] have presented a simple and optimal protocol for a special case of this problem when each destination address is unique. Their protocol does not work when a station is the destination for more than one packet. However, a station may be the destination of more than one packet in most practical situations. All the permutation routing protocols in [10, 4, 5] make a crucial assumption that each station sends and receives an equal number of packets. This is a restrictive assumption in real situations. It is not clear whether the protocols in [10, 4, 5] can be modified to do one-to-one routing when the destination addresses are not unique.

In this paper, we design energy-efficient protocols for the general one-to-one routing problem, i.e., when a station may receive more than one packet. Our protocol works well when the destination addresses are chosen randomly. In this case, a station may receive more than one packet, however, each station receives only a small number of packets. When there is only one communication channel, our protocol solves the general one-to-one routing problem in a p-station WN in $2p \log p + 2 \log p$ slots and each station remains awake for at most $5 \log p$ slots. When $k > 1$ channels are available and $k \leq \sqrt{p}$, our protocol solves this problem in $\frac{2p}{k} \log(\frac{p}{k}) + p + \frac{p}{k} + 2 \log(\frac{p}{k}) + k^2$ slots and each station remains awake for at most $5 \log(\frac{p}{k}) + k^2 + 3k + 1$ slots. Both the completion time and the maximum awake time are quite good for our protocol.

The rest of the paper is organized as follows. In Section 2, we present some preliminaries. We present our protocol for the general one-to-one routing problem using a single channel in Section 3. We present a protocol for one-to-one routing on a k-channel WN in Section 4. Finally, we conclude with some comments in Section 5.

2   Preliminaries

We consider a radio network $WN(p, k)$ with p stations and k channels. The i-th station is denoted by $S(i)$, $1 \leq i \leq p$. As mentioned before, we assume that the stations have been initialized by running an initializing protocol as in [9]. Hence, each station has a unique ID in the range $[1, p]$. Each station holds 1 packet to be routed. Each packet has a unique destination address which is the address of one of the p stations. Station $S(i)$ knows the destination of the packet that it holds. However, $S(i)$ does not know from which stations it will receive packets if it receives any packet at all. As in [9, 10], we assume that each station has a local clock which keeps synchronous time by interfacing with a Global Positioning System (GPS). Time is divided into slots and all packet transmissions take place at slot boundaries.

There are k transmission channels in the $WN(p,k)$. These k channels are denoted by $C(1), C(2), \ldots, C(k)$. One data packet can be transmitted in a time slot. In each time slot, a station can tune to one of the k channels and/or transmit a packet on one of the channels. These two channels may or may not be distinct. As in the paper by Nakano $et~al.$ [10], we assume that $k \leq \sqrt{p}$. This is a reasonable assumption, since in most real life situations, the number of channels is much less than the number of stations, i.e., $k \ll p$. If we want to design a protocol without collision, we need to ensure that only one station transmits a packet over a given channel in a time slot.

The basic protocol in [10] runs in p rounds. In the j-th round, all the packets destined for station $S(j)$ are routed to $S(j)$. Further, in the j-th round, first a reservation protocol is run, when each station reserves the slots it requires for transmitting its packet destined for $S(j)$. Once the reservation protocol is complete, each station $S(i)$ knows exactly when it should wake up to send the packet destined for $S(j)$. Hence, during the packet routing stage, each station needs to be awake for 1 time slot for sending its packet. Nakano $et~al.$ [10] have presented several efficient variations of this basic protocol for the permutation routing problem and we will adopt a simple variant of their protocol for our purpose.

We explain the basic reservation protocol in [10] assuming that each station has multiple packets as in the permutation routing problem. The method discussed below holds for the case when each station has only one packet. The basic reservation protocol in [10] is the following. Consider for the time being that we have an $WN(p,1)$, i.e., a single channel WN populated by p stations. Station $S(i)$ has n_i items to transmit to station $S(j)$ for $1 \leq i \leq p$. The reservation protocol takes $p-1$ slots. In the first slot $S(1)$ sends n_1 to $S(2)$, in the second slot $S(2)$ sends $n_1 + n_2$ to $S(3)$ and in general, in the i-th time slot, $1 \leq i \leq p-1$, station $S(i)$ sends $n_1 + n_2 + \ldots + n_i$ to station $S(i+1)$. Note that each station is awake at most for two time slots, one slot for receiving a packet from the station just before it and one slot for sending a packet to the next station. At the end of the protocol, each station $S(i)$ knows $n_1 + n_2 + \ldots + n_{i-1}$. Hence, during the packet routing stage, station $S(i)$ will wake up at time slot $n_1 + n_2 + \ldots + n_{i-1} + 1$ and start transmitting its n_i packets to station $S(j)$. After $n_1 + n_2 + \ldots + n_i$ time slots, station $S(i)$ will go to sleep again. Note that this is a prefix sum computation over the integers n_i.

The one-to-$one~communication$ problem is defined as follows. There are p stations with each station holding one packet. Each packet has a destination address and the aim of the protocol is to deliver the packets to their correct destinations. Consider station $S(i)$ with a packet destined for station $S(j)$, $1 \leq i, j \leq p$. The communication problem is asymmetric in the sense that though $S(i)$ knows the destination of its packet, $S(j)$ does not know that it should expect a packet from $S(i)$. We do not impose any restriction on the number of packets a station can receive. Though each station is the sender of one packet, it can receive more than one packets. Note that this is a more general communication scheme compared to the permutation routing protocols studied in [10, 4, 5], where each station sends as well as receives an equal number of packets.

If energy-efficiency is not an issue, we can design a simple protocol. Assume that the stations have access to only one channel. Each station transmits its packet according to its serial ID. For example, station $S(i), 1 \leq i \leq p$, will transmit its packet in time slot i. All other stations will remain awake for all the p time slots during this transmission and receive all the p packets. If station $S(j), 1 \leq j \leq p$, finds that a packet is destined for it, it will accept the packet, otherwise it will reject the packet. This protocol completes in p time slots and each station remains awake for p slots. Hence, each station needs to spend a large amount of energy by remaining awake for all the p slots.

In the *multicasting* problem, a station broadcasts a packet to a group of stations. In the *multiple multicasting* problem, multiple stations perform the multicasting operation. There may be overlaps among the recipient groups, i.e., a particular station may receive packets from more than one station performing multicasting.

3   A Single-Channel Protocol for One-to-One Communication

We present a simple divide-and-conquer protocol for one-to-one communication in a single-channel WN. We divide the p stations in the WN into 2 groups with $\frac{p}{2}$ stations in each group. The 2 groups are denoted by $G(1)$ and $G(2)$. The basic idea behind the protocol is the following. We first route all the packets with destination addresses in $G(1)$ to the stations in $G(1)$. The packets are not necessarily sent to their correct destination stations at this stage. Similarly, we route the packets with destination addresses in $G(2)$ to the stations in $G(2)$. Once all the packets with destination addresses in $G(1)$ (resp. $G(2)$) are in the stations in $G(1)$ (resp. $G(2)$), we have two independent routing problems. The first (resp. second) problem is to route all the packets within $G(1)$ (resp. $G(2)$). Hence, we can now recursively solve these two routing problems within $G(1)$ and $G(2)$.

Phase 1. This phase is divided into 2 rounds. In the first round, we consider the packets in the stations in $G(1)$. All packets in $G(1)$ that have their destination addresses in stations in $G(2)$ are sent to the stations in $G(2)$. This is done by executing the two steps below. In the first step, the stations in $G(1)$ calculate the exact slot numbers when they should wake up to transmit packets. Also, the stations in $G(2)$ calculate the exact slots when they should wake up to receive the packets from stations in $G(1)$. The actual packet transmission is done in the second step.

Step 1. Since we are designing a DAMA protocol, our first task is to count the number of packets that are in $G(1)$ and have to be sent to the stations in $G(2)$. Once this count is known, we can reserve the slots for packet transmission so that the senders and receivers will remain awake during these slots.

The packets in $G(1)$ with destination addresses in $G(2)$ are counted using the reservation protocol by Nakano *et al.* [10]. Starting from the first station $S(1)$, each station passes a count to the next station in each time slot. Hence, $S(j)$ receives the count in time slot $j, 1 \leq j \leq \frac{p}{2}$. We denote this count as $count(j)$ to indicate that this is the intermediate count in $S(j)$. Note that $count(j)$ is the total number of packets destined for stations in $G(2)$ in the stations $S(1), S(2), \ldots, S(j-1)$. If $S(j)$ has a packet with destination address in $G(2)$, $S(j)$ increments count by 1 and passes it to $S(j+1)$. After $\frac{p}{2}$ slots, the last station $S(\frac{p}{2})$ in $G(1)$ receives the total count of packets that have destination addresses in $G(2)$. We denote this count by *total_count*. Moreover, each station $S(j), 1 \leq j \leq \frac{p}{2}$, if it holds a packet for $G(2)$, knows the exact number of packets destined for $G(2)$ in the stations $S(1), \ldots, S(j-1)$. This reservation protocol takes $\frac{p}{2}$ slots to complete and each station remains awake for 2 slots. The first and the last stations remain awake for 1 slot each.

$S(\frac{p}{2})$ now broadcasts *total_count* to all the stations in $G(2)$. Our aim is to distribute the packets destined for stations in $G(2)$ equally among the stations. Recall that, the packets are sent to the stations in $G(2)$ and not necessarily to their correct destinations in $G(2)$. Since each station in $G(2)$ knows the number of stations $\frac{p}{2}$ in $G(2)$ as well as *total_count*, each station can decide that it will receive $\lceil total\_count/\frac{p}{2} \rceil$ packets during the packet transmission in the next step.

Moreover, station $S(j), 1 \leq j \leq \frac{p}{2}$, in $G(1)$ knows that it should transmit its packets for $G(2)$ starting at slot $count(j) + 1$. Consider the s-th station $G(2, s), 1 \leq s \leq \frac{p}{2}$ in $G(2)$. Since each station in $G(2)$ receives $\lceil total\_count/\frac{p}{2} \rceil$ packets, station $G(2, s)$ should wake up after all the stations $G(2, 1), G(2, 2), \ldots, G(2, s-1)$ have received their packets. In other words, $G(2, s)$ should wake up at time slot $\lceil total\_count/\frac{p}{2} \rceil \times (s-1) + 1$ to receive its packets.

This step terminates after station $S(\frac{p}{2})$ broadcasts *total_count* to all the stations of $G(2)$, i.e., after $\frac{p}{2}+1$ slots, $\frac{p}{2}$ slots for running the reservation protocol and 1 slot for the broadcast. Each station in $G(1)$ remains awake for at most 2 slots for running the reservation protocol. The last station $S(\frac{p}{2})$ in $G(1)$ remains awake for 1 slot during the reservation protocol and 1 slot for the broadcast.

Step 2. The actual packet transmission takes place in this step. At the end of Step 1, each station $S(j), 1 \leq j \leq \frac{p}{2}$, in $G(1)$ knows exactly the slot when to transmit its packets destined for stations in $G(2)$. Also, each station in $G(2)$ knows exactly the slot when to start listening for packets.

Since each station $S(j)$ in $G(1)$ holds exactly one packet, $S(j)$ keeps awake for one slot to transmit its packet. Each station in $G(2)$ remains awake for $\lceil total\_count/\frac{p}{2} \rceil$ slots. Since we do not have any restriction on the destination addresses of the packets, in the worst case, all the $\frac{p}{2}$ packets in the $\frac{p}{2}$ stations of $G(1)$ will have destination addresses in the stations in $G(2)$. In other words,

$total\_count = \frac{p}{2}$ in the worst case and each station in $G(2)$ has to be awake for $\lceil \frac{p}{2}/\frac{p}{2} \rceil = 1$ slot to receive their share of packets. This step takes p slots to complete since each of the p stations transmit their packets one after another. At the end of step 2, all the packets with destination addresses in $G(2)$ are in the stations in $G(2)$. However, the packets may not reach their correct destination stations.

We now estimate the total slot requirement for the completion of the first phase and the maximum number of slots each station remains awake. Steps 1 and 2 together takes $2 \times \frac{p}{2} + 1 = p + 1$ slots to complete, $\frac{p}{2}$ slots for the reservation protocol, $\frac{p}{2}$ slots for actual packet transmission and 1 slot for the broadcasting by station $S(\frac{p}{2})$. In Step 1, each station in $G(1)$ remains awake for at most 2 slots for running the reservation protocol and 1 slot for transmitting its packet i.e., 3 slots in total. Similarly, each station in $G(2)$ remains awake for 2 slots, 1 slot for receiving the broadcast from $S(\frac{p}{2})$ and 1 slot for receiving its packet from a station in $G(1)$.

This concludes the first round in phase 1. The routing in the second round is identical to the two steps above. All the packets in $G(2)$ that have destination addresses in stations in $G(1)$ are sent to the stations in $G(1)$ in this round. The total slot requirement to complete Phase 1 is the sum of the slots required for the two rounds. Hence, the total slot requirement is $2p + 2$. Each station remains awake for 5 slots from the analysis in the previous paragraph.
End of Phase 1.

Phase 2. We now have two independent routing problems, all the packets in the stations in $G(1)$ are to be routed to the stations in $G(1)$ and similarly, all the packets in the stations in $G(2)$ are to be routed to the stations in $G(2)$. We solve these two routing problems recursively by applying the method in Phase 1. Note that after Phase 1, we now have two independent routing subproblems with $\frac{p}{2}$ stations in each problem. If the total number of slots required for the overall routing is $T(p)$, it follows the recurrence :

$$T(p) = 2T(\frac{p}{2}) + (2p + 2)$$

which gives $T(p) = 2p \log p + 2 \log p$.

The number of subproblems doubles at every phase of the recursion. However, each station takes part only in one subproblem as each station is part of only one group. Hence, the number of slots each station is awake remains the same for all the $\log p$ phases of the recursion. This results in a total awake time of $5 \log p$ for each station.

Lemma 1. *The one-to-one routing problem on a single channel wireless network with p stations can be completed in $2p \log p + 2 \log p$ slots and each station remains awake for at most $5 \log p$ slots.*

3.1   Some Details of the Protocol

The protocol works well if the destination addresses are chosen randomly from a uniform distribution. Since there are p stations and p packets, the probability that a packet has a particular destination address is $\frac{1}{p}$. The probability that $c > 1$ packets have the same destination address is $\frac{1}{p^c}$. Hence, it is extremely unlikely that a particular station will receive a large number of packets if the destination addresses are chosen randomly. We have done extensive experiments with upto $100,000$ stations when each station holds a single packet. When the destination addresses are chosen randomly, the maximum number of packets with the same destination address is less than 10 almost always.

Our protocol works well even in a worst case scenario. Consider the case when all the p stations hold packets for a single station, say station $S(1)$. In every phase of the recursion the packets will be concentrated in half of the stations compared to the previous phase. For example, all the packets will be in the first $\frac{p}{2}$ stations after the first phase, in the first $\frac{p}{4}$ stations after the second phase and so on. Hence, in every phase some of the stations have to receive double the number of packets compared to the previous phase and other stations will be idle. In this situation, station $S(1)$ has to be awake for $\geq p$ slots to receive all its packets, however, a station like $S(2)$ will be active until the last phase of the recursion and receive number of packets which follows a geometric progression $1, 2, 4, 8, \ldots, \log p$ phases, i.e., in total $O(p)$ packets. However, other stations will use less and less energy depending on when they drop out of the recursion.

Since a wireless network is a distributed system without a central arbiter, we need to specify how the stations know when a particular phase ends and the next phase begins. Note that the maximum number of packets that are transferred in each phase is p. In our description of Phase 1, we have two groups. However, in general the number of groups in a particular recursion level will be a power of 2. Still the maximum number of packets that will be transferred in each phase will be p. Hence, the maximum completion time of each phase will be $2p + 2$ as we have shown for Phase 1. Hence, phase $m, 1 \leq m \leq p$, starts at time slot $m \times (2p + 2)$ and each station can compute this slot number by counting the slots from the start of the protocol. We have assumed that p is a power of 2. It is easy to modify our protocol when p is not a power of 2.

In a particular phase, we do the routing for the subproblems one after another. For example, in Phase 1, we first route all the packets that are in $G(1)$ and destined for $G(2)$ to the stations in $G(2)$. Then we route all the packets that are in $G(2)$ and destined for $G(1)$ to the stations in $G(1)$. We follow the same scheme in successive phases. Since each station knows its ID and the phase number, it can determine the group to which it belongs in a particular phase and accordingly wake up to participate in the routing.

4   k-Channel Routing

We now extend our protocol for one-to-one routing in a k-channel wireless network. First, in Phase 1, we divide the routing problem into k independent routing subproblems with $\frac{p}{k}$ stations in each subproblem. Then we use our recursive strategy to route the packets for each subproblem in parallel using a separate channel for each subproblem.

Phase 1.
Step 1. We divide the p stations into k groups $G(i), 1 \leq i \leq k$, with $\frac{p}{k}$ stations in each group. We denote the k channels by $C(i), 1 \leq i \leq k$, and dedicate one channel for each group. This step is done in parallel in each of the groups. We discuss the computation in group $G(1)$.

The reservation protocol of Nakano *et al.* [10] is run k times in the stations in $G(1)$. The purpose of the i-th run of the reservation protocol is to identify the packets that have destination addresses in $G(i)$. At the end of these k runs, the last station in $G(1)$ knows the k sums $total\_count_1(i), 1 \leq i \leq k$, for the total number of packets destined for the k groups, where the subscript 1 denotes that these are counts for $G(1)$. This is also true for the k last processors in all the groups $G(1), G(2), \ldots, G(k)$.

Since each run of the reservation protocol takes $\frac{p}{k}$ slots and it is run k times in each group in parallel, this step takes $\frac{p}{k} \times k = p$ slots to complete and each station remains awake for at most $2k$ slots.

Step 2. We transfer the packets with destination addresses in $G(i)$ to the stations in $G(i)$ in this step. We proceed in the following way. We assign distinct stations in $G(i)$ to accept the packets from a unique group. Recall that there are $\frac{p}{k}$ stations in $G(i)$, we use subgroups of $\frac{p}{k^2}$ stations within $G(i)$ to receive packets from a distinct group. For example, the first $\frac{p}{k^2}$ stations will receive packets from $G(1)$, the next $\frac{p}{k^2}$ stations will receive packets from $G(2)$ etc. This is possible if $\frac{p}{k} \geq k$, i.e., $k \leq \sqrt{p}$. This is a reasonable assumption since the number of stations is usually much larger than the available channels.

The first task in this step is to broadcast the integers $total\_count_i(k), 1 \leq i \leq k$, from the last stations of each group. All the p stations listen to these broadcasts. Recall that there are k such integers $total\_count_i(k)$ in each of the k last stations and hence, there are k^2 such sums. So, this broadcast takes k^2 slots overall and each station remains awake for all of these k^2 slots.

After this broadcast, consider the stations in $G(i)$. The $\frac{p}{k^2}$ stations in the j-th subgroup, $1 \leq j \leq k$, of $G(i)$ have received $total\_count_j(i)$ from the last station in $G(j)$ during the broadcast discussed above. Each station in the j-th subgroup can now decide to receive $\lceil total\_count_j(i)/\frac{p}{k^2} \rceil$ packets from the stations in $G(j)$. All other stations in the other subgroups of $G(i)$ as well as in the other $k - 1$ groups similarly can decide how many packets they will receive and from which group.

Consider again a group $G(m)$ as a representative group. All the stations in $G(m)$ use channel $C(m), 1 \leq m \leq k$, to transmit their packets. The stations send the packets to $G(1), G(2), \ldots, G(k)$ serially one after another. Consider the n-th station $G(m, n)$ in $G(m)$. Suppose, $G(m, n)$ has a packet for $G(i)$. We have to resolve two issues, **(i)** when $G(m, n)$ should send its packet to a station in $G(i)$, and **(ii)** when a station in $G(i)$ should expect a packet from $G(m,n)$. This is done in the following way.

First we resolve when $G(m,n)$ should send its packet. Since all packets in $G(m)$ are sent serially using channel $C(m)$, $G(m,n)$ should send its packet after, **(a)** all the packets in $G(m)$ with destinations in groups $G(1), G(2), \ldots, G(i-1)$ have been sent, and **(b)** all the packets in stations $G(m,1), G(m,2), \ldots, G(m,n-1)$ and with destinations in $G(i)$ have been sent. Note that $G(m,n)$ can determine the integer in **(a)** by computing the sum $\sum_{r=1}^{i-1} total\_count_m(r)$. Recall that all these integers are available to all the stations. $G(m,n)$ already knows the integer in **(b)** due to the running of the reservation protocol in Step 1. Hence, $G(m,n)$ can decide the exact slot when it should transmit its packet.

The subgroup of $\frac{p}{k^2}$ stations in $G(i)$ that are responsible for receiving packets from $G(m)$ can also compute the sum $\sum_{r=1}^{i-1} total\_count_m(r)$. Hence, the first packet from $G(m)$ to one of these $\frac{p}{k^2}$ stations will arrive at slot $\sum_{r=1}^{i-1} total\_count_m(r) + 1$. Since each of these $\frac{p}{k^2}$ stations receive $(total\_count_m(i)/\frac{p}{k^2})$ packets it is easy for each station to decide at which slot it should expect its packets to arrive over channel $C(m)$.

The actual packet transmission starts after this. Since each group has $\frac{p}{k}$ packets, and all groups transmit their packets in parallel, the packet transmission takes $\frac{p}{k}$ slots in total. The maximum value of any $total\_count$ is $\frac{p}{k}$ and hence each station remains awake for at most $\frac{p}{k}/\frac{p}{k^2} = k$ slots to receive its share of packets. Each station remains awake for 1 slot to transmit its packet and k^2 slots to receive all broadcasts of the $total\_count$ values.

Hence, the completion time for this step is $\frac{p}{k} + k^2$ and each station remains awake for at most $k^2 + k + 1$ slots. The total completion time for steps 1 and 2 is $p + \frac{p}{k} + k^2$ and each station remains awake for at most $k^2 + 3k + 1$ slots.

Phase 2. In this phase we solve the routing problem in all the k groups in parallel using a separate channel. Under a random distribution of destination addresses, each group will have roughly $\frac{p}{k}$ packets. Hence, we can apply the method in Section 3 to do this routing. From Lemma 1, this routing takes $\frac{2p}{k} \log(\frac{p}{k}) + 2 \log(\frac{p}{k})$ slots and each station remains awake for at most $5 \log(\frac{p}{k})$ slots. Combining the complexity of Phases 1 and 2 we get the following lemma.

Lemma 2. *The one-to-one routing problem on a k-channel wireless network with p stations can be completed in $\frac{2p}{k} \log(\frac{p}{k}) + p + \frac{p}{k} + 2 \log(\frac{p}{k}) + k^2$ slots and each station remains awake for at most $5 \log(\frac{p}{k}) + k^2 + 3k + 1$ slots.*

5   Conclusion

We have presented simple and energy efficient protocols for the one-to-one communication problem in a wireless network. This is one of the most fundamental communication problems in any distributed system. Our protocols in Sections 3 and 4 can be modified easily for doing multiple multicasting in a wireless network with a single or k channels. Assume that each station $S(i), 1 \leq i \leq p$, needs to send a packet to a group of d stations. In this case, we can assume that $S(i)$ has d packets with different destinations and apply our protocol. We are confident that the protocol will work well for random distribution of destination groups and when a particular destination station is in only a small number of groups for a multicasting operation.

Acknowledgements

The first author's research is partially supported by Western Australian Interactive Virtual Environments Centre (IVEC) and Australian Partnership in Advanced Computing (APAC).

References

[1] N. Abramson, "Multiple access in wireless digital networks", *Proc. IEEE*, Vol. 82, pp. 1360-1370, 1994.

[2] D. Bertzekas and R. Gallager, *Data Networks*, 2nd Edition, Prentice Hall, 1992.

[3] I. Chlamtac and A. Farago, "An optimal channel access protocol with multiple reception capacity", *IEEE Trans. Computers*, Vol. 43, pp. 480-484, 1994.

[4] A. Datta and A. Y. Zomaya, "New energy-efficient permutation routing protocol for single-hop radio networks", *Proc. 8th International Computing and Combinatorics Conference (COCOON '02)*, LNCS Vol. 2387, pp. 249-258, 2002.

[5] A. Datta and S. Soundaralakshmi, "A simple and energy-efficient routing protocol for radio networks", *Proc. 18th ACM Symposium on Applied Computing (SAC '03)*, Melbourne, Florida, March 2003, to appear.

[6] W. C. Fifer and F. J. Bruno, "Low cost packet radio", *Proc. IEEE*, Vol. 75, pp. 33-42, 1987.

[7] M. Fine and F. A. Tobagi, "Demand assignment multiple access schemes in broadcast bus local area networks", *IEEE Trans. Computers*, Vol. 33, pp. 1130-1159, 1984.

[8] C. R. Lin and M. Gerla, "Adaptive clustering in mobile wireless networks", *IEEE J. Selected Areas in Comm.*, Vol. 16, pp. 1265-1275, 1997.

[9] K. Nakano and S. Olariu, "Randomized initialization protocols for radio networks", *IEEE Trans. Parallel and Distributed Systems*, Vol. 11, pp. 749-759, 2000.

[10] K. Nakano, S. Olariu and A. Y. Zomaya, "Energy-efficient permutation routing in radio networks", *IEEE Trans. Parallel and Distributed Systems*, vol. 12, No. 6, pp. 544-557, 2001.

[11] R. A. Powers, "Batteries for low-power electronics", *Proc. IEEE*, Vol. 83, pp. 687-693, 1995.

[12] A. K. Salkintzis and C. Chamzas, "An in-band power-saving protocol for mobile data networks", *IEEE Trans. Communications*, Vol. 46, pp. 1194-1205, 1998.

[13] K. Sivalingam, M. B. Srivastava and P. Agarwal, "Low power link and access protocols for wireless multimedia networks", *Proc. IEEE Vehicular Technology Conference (VTC '97)*, 1997.

Adaptation of IPv6
and Service Location Protocol
to Automatic Home Networking Service

Kwangmo Jung[1], Sookyoung Lee[1], and Sangwon Min[2]

[1] Korea Electronics Technology Institute
Seohyng-Dong, Pundang-Gu, Sungnam-Si, Kyunggi-Do, 463-771, Korea
[2] Department of Telecommunications Engineering, Kwangwoon University
447-1, Wolgye-Dong, Nowon-Gu, Seoul, 139-701 Korea
(Tel: +82-2-940-5552, Fax: +82-2-942-5552)
min@daisy.kw.ac.kr

Abstract. As there has been the growth of many post-PCs and appliances that are capable of networking themselves and providing service via the Internet, home networking needs an effective and user-easily scenario for the less sophisticated users. Some features of the IP version 6 (IPv6) including auto-configuration and Service Location Protocol version 2 (SLPv2) will be especially helpful for this goal. This paper proposes a scheme for providing IP services automatically by those technologies.

1 Introduction

With the rapid growth of the potential of home networking in the technology and the market, many related issues have increasingly been focused and debated all over the fields including not only electric utilities but also diverse network services and protocols. The popularity of the home networking was driven from the continual growth of the number of households with multiple PCs and the introduction and the phenomenal growth of the home-based business and telecommuting. Thus, most services in home are expected to be connected to the Internet beyond home as well as networked with one another inside home[1].

Although lots of studies about home networking have been performed, the following questions are continuously asked: Which new services home users will require near future and How those services will be able to be provided more easily, more friendly, and more effectively. The latter mentions the requirements of home networking, which are low cast points, data type versatility, extensibility, automatic security, and ease of installation[2][3]. The most important factor to gain high position in the home networking market is the ease of installation, which means that simple and easy configuration and furthermore zero configuration.

There have been many studies for this goal in different parts of home networking, especially, home appliances and middleware[4]-[6]. However, their disadvantage is lack of interoperability among different middleware or appliances and difficult to provide the IP services. Therefore, the IPv6 protocol became

H.-K. Kahng (Ed.): ICOIN 2003, LNCS 2662, pp. 192–201, 2003.

a good candidate in home networking with useful functions like the neighbor discovery (ND) and stateless auto-configuration. In addition, as the end-to-end IP services which need more IP addresses are coming to be introduced to home, IPv6 will play an important role in home networking.

In this paper, we propose how the IP services in home can be provided automatically without a well-trained operator using IPv6 and SLPv2. Each device obtains an IP address by auto-configuration scheme of the IPv6 and any user in the Internet can find the location of the IP services and access them using SLP adapted in home networking. Ultimately, devices in home are connected to the Internet automatically and provide their services. This will provide benefits to ordinary customers of home networking.

2   SLPv2 in Home Networking

SLP is a protocol that allows networking applications or users to discover *Existence, Location,* and *Configuration* of the networked services in the enterprise networks[7]. Many vendors such as Apple or Sun Microsystems have adopted SLP for their products because it is not tied to a specific OS and users can obtain service's location information by characteristics as well as service type[8]. Also recently, the SLP has been modified for IPv6[9].

This protocol consists of the following three agents:

- *User Agent (UA)* requests and obtains the location information of a networking service on behalf of client software.
- *Service Agent (SA)* registers the location and attributes to all reachable DAs in advance and responds to each inquiry of UA on behalf of service software.
- *Directory Agent (DA)* aggregates service information from SAs into a stateless repository and there can be more than one DA under the same administration for robustness.

The existence and location of three agents are designed case by case. Figure 1 shows all types of the agents are adopted for home networking. While a SA is embedded in each appliance, a DA and an UA exist in the same node respectively, the representative home server directly connected to the gateway. And, either IPv4 or IPv6 access terminal uses the UA to find an IP service in home.

Also, we adopts *When a DA is present* for home networking between two modes of SLP's operation in which UAs unicast a request to the DA despite UAs directly multicast a request to SAs. And, as a DA discovery method the active discovery is selected in which an UA or a SA sends a request to find where DA is without waiting for DA's unsolicited advertisement. As a result, three agents are behaved as shown in figure 2.

Also, the scope field of SLP plays an important role. Because SAs and an UA are in the different links in figure 1, the IPv6 address scope of the service Uniform Resource Locator (URL) must be at least site-local. This means that the user of Home A can reach services at Home B in the point of IPv6 address's view and there can be a security problem among distinct homes. The scope field of SLPv2

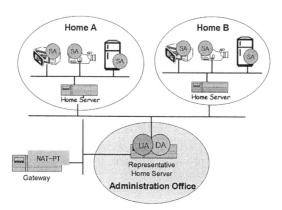

Fig. 1. Configuration of UA, SA and DA in the Home Networking. This shows how SLP agents are configured in real home networking environment. SA agents are located at each appliance. Both UA and DA are placed at a home server in administration office together

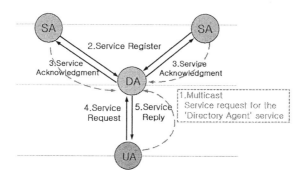

Fig. 2. Behaviors among UA, SA and DA. This shows the simple behaviors among three SLP agents. First SA and UA find out the existence and location of DA using multicast. After the processing, SA agents register their service information to DA periodically by unicast communication and UA requests to and receive response from DA about service information also by the unicast operation

solves this problem. SAs in each home have the unique scope value in string and it is registered to DA together with the information of appliances. Therefore, IP addresses of appliances collected in DA are protected by scope value. Table 1 shows the scope of SLPv2 of our consideration and figure 3 explains the DA existed operation mode and the active DA discovery mechanism.

Table 1. Scope of SLPv2 for This Scenario. This table shows options selected for home networking. DA exists and SA and UA look for it by the active discovery mechanism without DHCP. And the number of DA depends on the size of home network. SLP's scope field is used to make a difference among houses

Function	Scope selected
Operation mode	When a DA is present
DA discovery	Active discovery
DHCP option	Not needed
Duplication of DA	Depending on the scale of the home network
SLP Scope field	Used (UA, DA: all homes, SA: each home only)

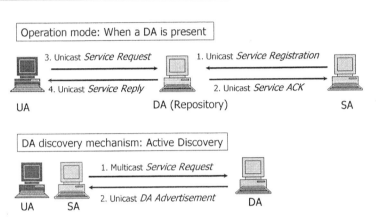

Fig. 3. SLP Operation Mode and DA Discovery Mechanism. There is a DA as a repository for searching the service information quickly and the number of DAs in a network can be adjusted according to its scale. Also, both UAs and SAs find a DA during initialization by themselves

3   IPv6 Auto-configuration in Home Networking

IPv6 supports multiple IP address scopes for unicast and multicast as follows: Link-local, Site-local and Global[10]. Each home appliance has first a link-local address and then a site-local unicast address and two well-known site-local multicast addresses (FF05::123, FF05::116). However, it does not need to be equipped with global address because services in each home are not necessary to directly communicate with nodes beyond home. Likewise, a home server, a set-top box or a residential gateway in each home has the same addresses as appliances. On the other hand, the representative home server between IPv4 and IPv6 must be assigned an IPv6 global unicast address to communicate nodes in the global Internet outside home. Table 2 presents IPv6 address scopes for the scenario.

Also, between two mechanisms of the auto-configuration of IPv6, we adopt the stateless auto-configuration mechanism in which an IPv6 address is configured without any intervention between IPv6 routers and hosts. In figure 4, Home

Table 2. IPv6 Address Scopes in This Scenario. This table shows IPv6 address's type and scope that each home networking node takes. Each appliance in home has unicast and multicast addresses with site-local scope. A home server, a set-top box or a residential gateway in each home also has same address type and scope in inside and outside interfaces. A representative home server in the administration office should be assigned with global IPv6 address toward outside because IPv4 users try to access the web server

Nodes	Address scope
Appliances in home	Site-local unicast, Site-local multicast
A home server, a set-top box or a residential gateway in each home	Both interfaces: Site-local unicast, Site-local multicast
A representative home server in the administration office	Inside home: Site-local unicast, Site-local multicast Outside home: Global unicast

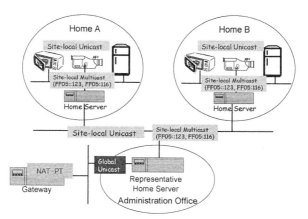

Fig. 4. Assignment of the IPv6 Addresses at Each Interface. This shows configuration of the IPv6 addresses in the interface as the explanation of Table 2

A, Home B and Administration office are combined to one site. Thus, all appliances in each home have the same site-local prefix which a home server in each home advertises in a Router Advertisement (RA) message. This prefix value is already built-in the home server before it operates at home.

In addition, the ND protocol of IPv6 also plays an important role in auto-configuration and corresponds to a combination of ARP, ICMP router discovery, and ICMP redirect of the IPv4 protocols. In this scenario, although a simple IPv6 router in each home is not demanded to perform all functions of the legacy IPv6 router, checking the duplicate address detection (DAD) and finding and keeping neighboring routers should be necessarily required besides the forwarding function.

4   Overall Operation

In this chapter, we explain the behaviors of each device in home networking with functions of the IPv6 and SLP in detail.

4.1   Overview

We assumed that the home infrastructure is constructed with the TCP/IPv6 protocol stack. The trial to minimize the IPv6 stack size for low cost network appliances has been discussed at the 52nd IETF meeting. This draft includes core functions of IPv6 needed in home networking including auto-configuration[11]. Figure 5 is the abstract illustration of the proposed scenario.

For one thing, home appliances such as a refrigerator, a surveillance camera, and a microwave are connected to IPv6 infrastructure. The access terminal to these things can be a PC, a smart phone, a PDA, and so on and they can be located or moving in the both IPv4 and IPv6 network. If they are in IPv6 home network, they take the control of home applications using the IPv6 network. But, if they are outside the home in the Internet especially IPv4, they need translation protocol between IPv4 and IPv6.

4.2   Transition Issue between the IPv4 Internet
and the IPv6 Home Network

For the transition between the IPv4 Internet and the IPv6 home network, the Network Address Translation – Protocol Translation (NAT-PT) mechanism is equipped at the boundary of IPv6 network in our consideration because of its advantage to require no changes at each IPv4 or IPv6 terminal[12]. As shown in figure 6, it is connected to the representative home server in the administration

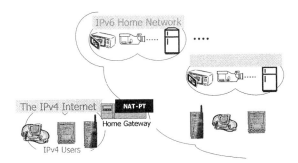

Fig. 5. Abstract Illustration of the Proposed Scenario. This figure shows overall abstract home networking considered in this paper. There are large IPv4 Internet and IPv6 island for home networking infrastructure. Between the two, there is home gateway equipped with NAT-PT for translation. IPv6 appliances can be accessed by IPv4 and IPv6 users both

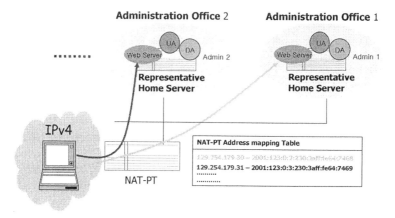

Fig. 6. Transition Issue between IPv4 and IPv6 in the Home Networking. The number of inbound connections or IPv4 users who can access different administration office simultaneously is restricted by the capability of the IPv4 address pool in NAT-PT

office and IPv4 users in the Internet find the IP addresses of appliances in home networking and access them using a web server located in the office. Therefore, traffics through the NAT-PT gateway are not SLP messages but http packets. This removes the need of the additional Application Level Gateway (ALG) for SLP in the NAT-PT gateway as well as reduces the number of competitors over the global IPv4 address pool of NAT-PT.

4.3   Overall Scenario Operation

Figure 7 shows the overall scenario in the real environment. For instance, an IPv4 user in the IPv4 Internet tries to contact the remote control system of home appliances with its URL. IPv4 users can use this system with web browsing because all commands and results are performed through web interface.

When a representative home server in the administration office starts to work, it behaves as follow:

1. IPv6 link-local unicase address and site-local unicast address are configured automatically.
2. An UA and a DA module of SLP start to work.
3. Well-known IPv6 multicast addresses (FF05::123, FF05::116) are configured automatically.
4. The UA finds and keeps the location of the DA in the same machine.
5. The DA maintains the cache of service information registered from SAs.

Next, when a home server in each home starts to work, the IPv6 routing module starts to work and advertises a RA message into devices in home. And when a new home appliance plugs in a home, these things happen as follow:

1. A SA's scope is set as the unique address string manually.
2. IPv6 link-local unicast address is configured and in turn, site-local unicast address is generated automatically by RA of the home server.
3. The site-local unicast address is stored in the URL of service template which a SA will register to the DA.
4. A SA starts to work with service template already built in the device before release.
5. The SA finds the location of the DA and registers attributes of its service including URL.
6. The SA keeps on sending its service information to the DA per lifetime interval.

Lastly, an IPv4 user tries to contact a web server in the representative home server.

1. DNS-ALG in NAT-PT works at this time.
2. The UA in the representative home server retrieves the information of services in the user's house with SLP's scope value, from the DA's repository in the same machine.
3. The service URL is sent to the user in the IPv4 Internet in http messages through NAT-PT gateway and displayed.

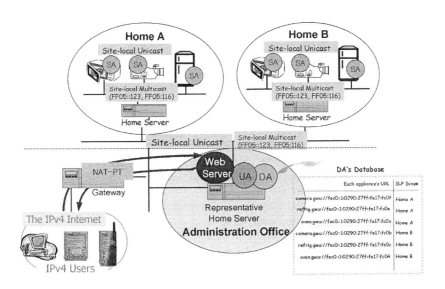

Fig. 7. Overall Scenario. This shows real home networking environment including IPv6 and SLP configuration. An IPv4 user gets an access to the web server in the administration office by NAT-PT with DNS application level gateway. All SLP messages and control messages for handling device remotely are not passed through gateway. Red line represents traffic flows between IPv4 and IPv6l

4. Next, the user in the IPv4 Internet can command for control of the appliance. And the control software in the representative home server starts to work.
5. The control software performs the command using the service's IP address and displays the result to the IPv4 user.

5 Conclusion

Our consideration is a challenge for enabling a zero administration for home networking service using IPv6 and SLPv2. Through the scenario, with the terminals equipped only with a web browser, users can take the control of appliances in their homes anywhere. The features of the scenario are removing the legacy middleware and adopting TCP/IP instead in home appliances and trying to adapt SLPv2 to home networking. The advantage of this scenario is zero-configuration in-home networking for unsophisticated users, and this will be an important factor in the market.

Finally, we should mention the need for the service template definition and registration to IANA for each home appliance and more robust security guaranteed among distinct homes. Also, the work to reduce the size and function of IPv6 protocol stack should be enhanced for home appliances which are relatively cheap and have a small memory.

Acknowledgement

The present research has been conducted by the Research Grant of Kwangwoon University and E-0580 Project in 2003.

References

[1] S. Sharan, "Home networks - getting there," in Proc. IEEE 4th International Workshop, pp. 169 –178, 2002.
[2] D. Marples, and S. Moyer, "Guest Editorial: In-Home Networking," IEEE Communications Magazine, April 2002.
[3] U. Saif, D. Gordon and D. Greaves, "Internet Access to a Home Area Netework," IEEE Internet Computing, January 2001.
[4] K. S. Lee, S. Lee, K. T. Oh, and S. M. Baek, "Network configuration technique for home appliances," Consumer Electronics, 2002. ICCE. 2002 Digest of Technical Papers. International Conference, pp. 180 –181. 2002.
[5] F. Meijs, M. Nikolova, and P. Voorwinden, "Remote mobile control of home appliances," Consumer Electronics, 2002. ICCE. 2002 Digest of Technical Papers. International Conference, pp. 100 –101. 2002.
[6] P. Dobrev, D. Famolari, C. Kurzke, and B. A. Miller, "Device and service discovery in home networks with OSGi," IEEE Communications Magazine, vol. 40, pp. 86-92, 2002.
[7] E. Guttman, C. perkins, J. Veizades, and M. Day, "Service Location Protocol, Version 2," IETF RFC 2608, June 1999.

[8] E. Guttman, "Service Location Protocol: Automatic Discovery of IP Network Services," IEEE Internet Computing, July 1999.

[9] E. Guttman, "Service Location Protocol Modifications for IPv6," IETF RFC 3111, May 2001.

[10] R. Hinden, and S. Deering, "IP Version 6 Addressing Architecture," IETF RFC 2373, July 1998.

[11] A. Inoue, "Minimum Requirements of IPv6 for Low Cost Network Appliances," IETF internet draft, March 2002.

[12] G. Tsirtsis, and P. Srisuresh, "Network Address Translation – Protocol Translation (NAT-PT)," IETF RFC 2766, February 2000.

Router-Assisted TCP-Friendly Traffic Control for Layered Multicast

J. Y. Son[1], K. R. Kang[2], D. Lee[2], S. H. Kang[2], Y. H. Lee[2], and D. W. Han[1]

[1] Electronics and Telecommunications Research Institute
Gajeong-dong, Yusong, Daejon, 305-350, Korea
{jyson,dwhan}@etri.re.kr
[2] School of Engineering, Information and Communications University
58-4 Hwaam-dong, Yusong, Daejon, 305-732, Korea
{korykang,dlee,kang,yhlee}@icu.ac.kr

Abstract. In this paper, we propose an efficient TCP-friendly traffic control scheme for layered multicast with router assistance. The proposed scheme is based on Network-based Layered Multicast (NLM), which dynamically adjusts the traffic on each link at a router to ensure the high quality data reception at the receivers as much as possible the network allows. The proposed scheme enhances TCP-friendliness of NLM by improving the traffic control granularity. The performance results show that the proposed scheme yields better performance compared with the original NLM and RLC, an end-to-end traffic control scheme for layered multicast.

1 Introduction

With the advancement of computer and network technology, multi-party multimedia applications such as video conferencing and video on demand have become of great interest. However, still the network bandwidth and the computing capability of the receivers are various and the efficient adaptation of a given network condition is inevitable to support heterogeneous receivers. The *layered multicast* approaches, such as Receiver-driven Layered Multicast (RLM)[1] and Layered Video Multicast with Retransmission (LVMR)[3], have been widely recognized as an efficient mechanism to handle the receiver heterogeneity.

In the layered multicast, video data are encoded into multiple layers: *base layer* and *enhancement layers*. The sender transmits the encoded data over separate multicast groups, and each receiver determines how many layers to subscribe depending on its capability or desired level of quality of video. However, RLM and LVMR shows poor inter-session fairness [2] and they can harm TCP traffic which is the dominant one of the Internet. The requirement that a rate control mechanism should work similar to TCP is called *TCP-friendliness*.

To enhance TCP-friendliness in layered multicast, RLC[7] and FLID-DL[6] have been proposed. However, they are an end to end approach and require at least round trip time between the sender and the farthest receiver. To reduce the time delay in rate adaptation, network-based layered multicast approaches have

H.-K. Kahng (Ed.): ICOIN 2003, LNCS 2662, pp. 202–211, 2003.
© Springer-Verlag Berlin Heidelberg 2003

been introduced [5, 8, 9]. The network based approaches allow fast adaptation to network traffic changes over time since they determine the number of layers at the router or exploit the information provided by the router. However the granularity of traffic change is much coarser than TCP and TCP traffic may suffer instability.

In this paper, we propose an efficient TCP-friendly traffic control scheme for layered multicast with router assistance. The proposed mechanism exploits the previous work, *Network-based layered multicast (NLM)* [5] and enhances the traffic control granularity to improve the TCP-friendliness. Like NLM, Time-To-Live (TTL) threshold is used in determining whether or not to forward packets and Type-of-Service (TOS) bits of the IP header is used to distinguish the traffic under our control scheme. To enhance the traffic control granularity, we consider both the total number of outgoing sessions and the queue occupation ratio of video traffic at the same time. We simulated and evaluated the proposed scheme using ns-2. The performance evaluation results show that the proposed scheme yields better TCP-friendliness compared with the original NLM and RLC[7].

The remainder of the paper is organized as follows: Section 2 describes the existing TCP-friendliness schemes for layered multicast. In Section 3, we describe design considerations and the details of the proposed scheme and then, in Section 4, we present the simulation results and analysis. Finally, Section 5 offers the conclusion.

2   Related Works

To support TCP-friendliness in a layered multicast, Xue Li [2] proposed layer-based congestion sensitivity rate control. It showed that the basic rate control scheme in the layered approach do not handle inter-session fairness well when there are multiple video sessions competing for bandwidth and that the basic layer adaptation scheme can bring unfairness to the competing TCP traffic as TCP is more sensitive to congestion. To achieve better inter-session fairness, they used equation-based TCP-friendly rate control to provide the fairness with TCP traffic. This scheme guarantees bounded fairness with respect to TCP using equations. However, it is hard to measure the accurate round trip time in a real network environment.

With receiver-driven layered control (RLC), Vicisano et al. [7] developed a scheme in which the receivers join or leave a layer based on their measured loss rates. Using specially flagged packets, the sender indicates synchronization points at which receivers might join or leave a specific layer. With RLC, the sender divides its data into layers and sends them on different multicast sessions. To test resource availability, the sender periodically generates a short burst of packets followed by an equally long relaxation period in which no packets are sent. The data rate of the flow is doubled during the burst. After receiving a packet burst, the receivers can join a higher layer if the burst is lossless; otherwise they remain at their current subscription level. The receivers might leave a layer at any time if losses are measured.

The Fair Layered Increase/Decrease with Dynamic Layering (FLID-DL) proposed in [6] enhances the RLC by using dynamic layering. Dynamic layering reduces the leave latency when dropping a layer, since a receiver has to periodically join additional layers to maintain a non-decreasing rate.

Some network based adaptation algorithms have also been proposed. Bhattacharyya et al.[8] introduced a useful reduction technique at the router using dependencies between video frames, which they refer to as the Group-Of-Picture (GOP)-level discard technique. They also demonstrated that network-based adaptation could yield significant performance gains for multicast video distribution. As another approach, Gopalakrishnan [9] proposed a hybrid scheme of layered and network driven adaptations, which is called Receiver-driven Layered Multicast with Priorities (RMLP). It is based on RLM [1] and a two-priority dropping scheme at the router. They demonstrated that their scheme improves the stability and intra-session fairness over those of RLM. However, as presented in [5], it shows low TCP-friendliness.

3 Proposed Scheme

3.1 Design Considerations

Our design goal is to develop a TCP-friendly layered multicast scheme allowing fine-tunable traffic control dynamically adapting to transient network traffic changes. It is desirable to allow moderate traffic control granularity since too large traffic control granularity may incur highly frequent traffic changes. To the contrary, too small granularity may result in poor TCP-friendliness because TCP sessions adapt their traffic too aggressively. For fine-tunable traffic control, we consider following two factors.

Number of Video Sessions Sharing a Bottleneck Link. Network based layered multicast approaches consider video sessions as a whole and not separately at a given link. As the number of sessions increases, does the number of video sessions that could be affected by the control scheme. This incurs coarse traffic granularity of the traffic control and, as a result, the reception quality of the receivers is destabilized. To avoid this problem, the traffic control scheme should moderate the sensitivity of layer add/drop criteria according to the number of sessions.

Bandwidth Occupation Ratio of Video Traffic. In the layered encoding[1], as the level of layer increases, the data rate does exponentially. Thus, the number of video sessions is not enough for fine grained traffic control. With the same number of sessions, the larger number of layers is allowed for a session, the more video traffic in a link. Dropping a higher layer will incur more traffic change than a lower layer. Therefore, the larger is the bandwidth occupation ratio of video traffic, the more conservatively should the control algorithm work.

3.2   NLM

NLM [5] is designed to deliver layered video to a heterogeneous set of receivers using the network-wide traffic information. Like other layered multicast schemes, it assumes that the source encodes the video signal into multiple discrete layers and each layer is transmitted on a separate multicast group. In NLM, the source also assigns a proper Time-to-Live (TTL) value and Type-of-Service (TOS) bits to each packet. The TTL value is used as a mark to show which layer the flow belongs to and the TOS bits are used to differentiate the packets under NLM scheme. A receiver subscribes to as many layers as its link bandwidth permits and the initial membership may last until the end of the flow. The receiver only reproduces the original data using the received data, and the quality of data is determined by the number of data layers received.

NLM employs a traffic controller including a *filter* and a *measurer* in a router. The *filter* is located in front of the queue and controls the amount of output packets of the link. It checks if the packet is qualified to forward. The *measurer* measures the average queue length, which is used as traffic metric. Based on a queue length, NLM categorizes the level of current link traffic status into three levels: *unloaded, loaded* and *congested*. Each status level has a corresponding tuple of highest and lowest threshold, and change direction. The direction indicates whether the threshold is to be increased or decreased. The value of the direction is -1 for a decrement and +1 for an increment. The set of these tuples is called *Guide*. It exists per output link interface and the direction of the changes of the router is determined by taking a minimum value among the directions of each outgoing interface. Using this traffic state information and the Guide, a traffic controller requests the neighboring controller for a given link to change the TTL threshold. The neighbor controller modifies the corresponding TTL as requested. By changing the value of the TTL threshold to reflect the traffic state, the traffic controller achieves traffic control. It can moderate the traffic by dropping the data of less significant layers, resulting in a change in the number of data layers transmitted through the network interface. These steps are repeated periodically.

3.3   Session-Based Traffic Adaptation Algorithm

As mentioned in Section 3.1, we aim to design a fast adaptive and TCP-friendly traffic control scheme for layered multicast. To support fast adaptation to the network traffic change, the proposed scheme assumes that the source and the router act as specified in NLM[5]: the source encodes the video stream into multiple layers and transmits data packets, assigning a proper TTL value according to its layer; The value of TOS bits is the same as NLM; The router drops the multicast packets whose TTL exceeds the TTL threshold. For improving the TCP-friendliness of NLM, we propose a session-based traffic adaptation algorithm which considers the number of the sessions in a link and bandwidth occupation ratio of each type of traffic.

The algorithm assumes that the traffic in a link consists of NLM traffic and TCP traffic. NLM traffic is identified by the TOS bits of the IP header.

Table 1. TTL threshold adaptation parameters

Parameters	Description
$ttl\_threshold_i(x)$	TTL threshold value for outgoing interface i at time interval x
max_r	maximum change ratio of the TTL threshold
$delta_i(x)$	threshold change ratio for outgoing interface i at time interval x
$q_t(x)$	average queue size of total traffic at time interval x
$q_v(x)$	average queue size of NLM traffic at time interval x
$n(x)$	the number of NLM sessions on a link at time interval x
$direction_i(x)$	threshold change direction for outgoing interface i
	at time interval x

Periodically, the router measures the bandwidth occupation ratio of NLM traffic over the total traffic and the number of NLM sessions in a out-going link. We assume a NLM session as a one-to-many data distribution, and a set of flows of the same source address and different destination addresses is identified as a single session. Using the bandwidth occupation ratio and the number of NLM sessions, the algorithm determines the amount of the TTL threshold change as described in equation (1).

Table 1 presents the parameters used for describing our algorithm. To update the change unit of the TTL threshold adaptively with an upper bound and lower bound, the TTL threshold value for an outgoing interface i at time interval x, $ttl\_threshold_i(x)$ is dynamically changed by the ratio $delta_i(x)$.

$$
\begin{aligned}
ttl\_threshold_i(x) &= ttl\_threshold_i(x-1) + delta_i(x), \\
delta_i(x) &= direction_i(x) \times \tfrac{max\_r}{n(x)} \times \tfrac{q_t(x)}{q_v(x)}
\end{aligned} \tag{1}
$$

The value of the $direction_i(x)$ is one of -1, 0, and $+1$. The condition to determine $direction_i(x)$ is the same as that of NLM. The *Guide* is configured by the network administrator and, $direction_i(x)$ is determined according to the traffic status to which the expected average queue length of time interval x corresponds. Parameter max_r is set to 50, which is the value at which one layer can be dropped or added. The max_r is divided by the number of NLM sessions to moderate the amount of traffic change incurred by the TTL threshold change. The factor $q_t(x)/q_v(x)$ has the same effect, in other words, the larger amount of NLM traffic results in the smaller threshold change. As a result, the total amount of traffic change can be controlled in a fine-grained manner.

Fig. 1 describes the session-based traffic adaptation algorithm applying the above equation in a router. *check_status()* function is to determine the network traffic status based on the expected queue length and returns one of *unloaded, loaded* and *congested*. The *convert_into_direction()* function returns one of -1, 0, and 1, according to the *expected_average_queue_length*.

```
At Router: If it a~receives threshold change request from i-th
r-router, it saves as direction_req(i)

for every network interface i at time interval x {
        it calculates the expected_average_queue_length and
        number of sessions(n), and VBR traffic ratio(qv/qt)
        in the queue;

        traffic_state = checks_status(expected_average_queue_length);
        direction = convert_into_direction(traffic_state);
        for each outgoing interface i {
                if (direction(i) < direction_req(i)
                    direction(i) = direction_req(i);
                delta(i) = direction(i) * (max_r / n) * (qv / qt);
                ttl_threshold(i) = old_ttl_threshold(i) + delta(i);
                old_ttl_threshold{i}=ttl_threshold{i};
        }
        threshold_change_request =
                        the minimum values of direction(i);
        It notifies threshold_change_request to upward routers;
}
```

Fig. 1. Traffic adaptation algorithm in a router

4   Performance Evaluation

We simulated the performance of the proposed scheme using ns-2 [11]. We also compared the proposed scheme with the original NLM and RLC. The simulation topology used in our evaluation is shown in Fig. 2.

In Fig. 2., the white-colored circles represent routers including the traffic controller. Gray-colored circles represent the source, and box-shaped nodes represent receivers. To simulate the receiver heterogeneity, the link bandwidth is selected within the range of 512 kbps and 8 Mbps.

4.1   Simulation Scenarios

We used the Variable Bit Rate (VBR) source model for NLM traffic as described in [4]. This model generates traffic over one second intervals for the base layer. In each interval, n packets are transmitted, where $n = 1$ with probability $1 - 1/P$ and $n = P \times A + 1 - P$ with probability $1/P$. A is the average number of packets per interval, and it is chosen to be four 4 KB packets. n packets are transmitted in a single burst, starting at uniformly distributed random time within the interval P, which represents the burst size of the traffic source, is set to 3 for modeling VBR sources. For each layer, the interval is broken into two subintervals. Each source encodes the data into four layers. The base layer is transmitted at the rate of 32 kbps, with the rate doubling for each subsequent

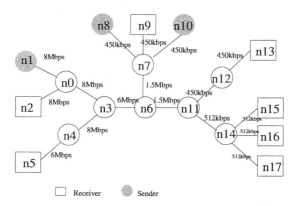

Fig. 2. Simulation network topology

Table 2. Simulation scenarios

Scenario id	Session configuration	Session description
Scenario 1	two VBR sessions (S1,S2) and two ftp sessions (T1,T2)	S1(n1):2-610, S2(n10):50-610, T1(n1,n15):30-400, T2(n2,n15):80-500
Scenario 2	three VBR sessions (S1,S2,S3) and one ftp sessions (T1)	S1(n0):2-610, S2(n10):50-610, S3(n10):120-610, T1(n1,n15):300-600
Scenario 3	one VBR session (S1) and three ftp sessions (T1,T2,T3)	S1(n0):2-610, T1(n1,n15):150-400, T2(n2,n15):250-500, T3(n10,n15):200-600

layer. FTP is used for a TCP session. At each router, the maximum queue size is set to 50.

We extend the idea of [10] to describe the TCP-friendliness in a quantitative manner by adding a time factor. TCP-friendliness at a given time interval x, $F(x)$, is defined as the ratio of the average throughput of their protocol proposed to the average throughput of TCP as follows:

$$F(x) = \frac{T_v(x)}{T_T(x)}, T_v(x) = \frac{\sum_{j=1}^{k_v(x)} T_j^v(x)}{k_v(x)}, T_T(x) = \frac{\sum_{j=1}^{k_T(x)} T_j^T(x)}{k_T(x)} \qquad (2)$$

$k_v(x)$ is the total number of the NLM sessions and $k_T(x)$ is the total number of TCP sessions at the time interval x. A TCP session is identified by the source address and the destination address. $T_1^v(x), T_2^v(x), ..., T_{k_v}^v(x)$ is the throughput of each NLM session and $T_1^T(x), T_2^T(x), ..., T_{k_T}^T(x)$ is that of each TCP session, respectively. The scheme can be said TCP-friendly if $F(x)$ is close to 1.

4.2   Simulation Results

This section examines the TCP-friendliness of the proposed scheme, comparing with the original NLM and RLC. The simulation results are obtained from the receiver node n15.

Fig. 3 presents the TCP-friendliness, $F(x)$, observed in Scenario 1. The proposed scheme shows far better TCP-friendliness compared with the original NLM. The peak at the starting point of the sessions results from the fact that a VBR session starts with higher rate than FTP session. However, as the sessions go on, the proposed scheme shows better TCP-friendliness because the proposed scheme controls the number of the layers. RLC shows larger $F(x)$ than the proposed scheme, which implies the NLM traffic takes smaller portion of the bandwidth. The fluctuation of $F(x)$ is severe and it may cause the reception quality instability at the receivers.

Fig. 4 shows the TCP-friendliness, $F(x)$, of Scenario 2. The proposed scheme shows larger $F(x)$ than the original NLM while maintaining the data rate of VBR sessions doubled. As shown in Fig. 4, the variation of $F(x)$ of RLC is higher than that of Scenario 1.

In Scenario 3, the original NLM and the proposed scheme show the similar result as shown in Fig. 5. RLC still shows frequent fluctuation.

5   Conclusions

Layered multicast has been considered to be an effective mechanism for multimedia data delivery to heterogenous receivers. To incorporate TCP-friendliness and prompt reaction to network traffic changes, the approaches with router assistance have been proposed. However, they control the traffic in a coarsely grained manner and result in instability of coexisting TCP sessions. To allow more efficient traffic control and TCP-friendliness support, in this paper, we propose a router-assisted TCP-friendly layered multicast scheme. The proposed scheme shows prompt reaction to network traffic changes by exploiting the previous

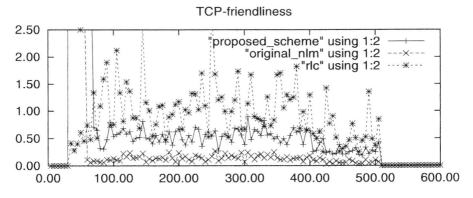

Fig. 3. Scenario 1: two FTP and two VBR sessions

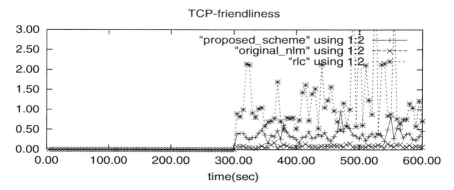

Fig. 4. Scenario 2: one FTP and three VBR sessions

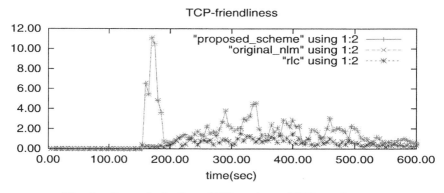

Fig. 5. Scenario 3: three FTP and one VBR sessions

work, NLM, and supports TCP-friendliness by allowing fine traffic control granularity. The performance results show that the proposed scheme provides better TCP-friendliness than the original NLM and RLC.

Acknowledgements

This work was supported in part by the National Research Laboratory Program funded by Ministry of Science and Technology, The Republic of Korea under Grant 2EG1900.

References

[1] S. McCanne, V. Jacobson, and M. Vetterli: Receiver-driven Layered Multicast. Proc. of ACM SIGCOMM96, August(1996) 117-130
[2] X. Li, S. Paul, and M. Ammar: Layered Video Multicast with Retransmissions (LVMR): Evaluation of Hierarchicalrate Control. Proc. of IEEE Infocom'98, March/April (1998) 1062-1072

[3] X. Li, S. Paul, and M. H. Ammar: Multi-Session Rate Control for Layered Video Multicast. Proc. of SPIE/ACM MMCN99. (1999)

[4] S. Bajaj, L. Breslau, and S.Shenker: Uniform versus priority dropping for layered video. Proc. of SIGCOMM98. (1998) 131-143

[5] Kyungran Kang, Dongman Lee, Hee Young Youn, Kilnam Chon: NLM: Network-based Multicast for Traffic Control of Heterogeneous Network. Computer Communications, Vol 24. Elsevier, Netherland (2001) 525-538

[6] J. Byers, M. Frumin, G. Horn, M. Luby, M. Mitzenmacher, A. Roetter, and W. Shaver.: FLID-DL: Congestion Control for Layered Multicast. Proc. of NGC 2000. (2000) 71-81

[7] Lorenzo Vicisano, Jon Crowcroft: TCP-like Congestion Control for Layered Multicast Data Transfer. Proc. of IEEE Infocom'98. (1998) 996-1003

[8] S. Bhattacharjee, K. L. Calvert, E. W. Zegura: Network Support for Multicast Video Distribution. Technical Document of Geogia Tech. GIT-CC-98/16. Atlanta (1998)

[9] R. Gopalakrishnan, J. Griffioen, G. Hjalmtysson, C. Sreenan, S. Wen: A Simple Loss Differentiation Approach to Layered Multicast. Proc. of IEEE Infocom 2000. (2000) 461-469

[10] J. Padhye, J. Kurose, D.Towseley, and R.Koodli,: A Model Based TCP-Friendly Rate Control Protocol. Proc. of NOSSDAV'99. (1999)

[11] The Mash Research Team. The ns network simulator. http://www-mash.cs.berkeley.edu/ns.

The Stability Problem of Multicast Trees in Layered Multicast

Feng Shi[1], Jiangping Wu[1], and Ke Xu[1]

Department of Computer Science and Technology, Tsinghua University
100084 Beijing, China
{shf,jianping,xuke}@csnet1.cs.tsinghua.edu.cn

Abstract. In this paper, we study the stability problem of an one-to-many multicast tree in the context of cumulative layered multicast system. Especially, "How does the number of links change as the number of users in a group changes when congestion occurs?" First, we introduce a stability index to evaluate and quantify the stability of the tree. Then, we develop a simple statistical model. With this model, we analyze the stability problem for various tree topologies. We observe that a class of trees commonly found in IP multicast has the similar stability when congestion occurs. . . .

1   Introduction

Recently, the technique of layered multicast, which employs multiple groups to transmit content at different rates, has been employed as a strategy capable of accommodating diverse client populations. A novel working examples of this approach is receiver-driven layered multicast [1]. This approach uses cumulative layering, which imposes an ordering on the multicast layers and requires clients to subscribe and unsubscribe to layers in sequential order. The desire to keep the number of layers manageable motivates the following widely-used allocation [2]: the multicast group associated with the base layer transmits at a rate B_0, while all other layers i transmit at rate $B_0 * 2^{i-1}$. In such an allocation, subscribing to an additional layer doubles a receiver's effective reception rate; similarly, unsubscribing to a layer halves the reception rate. Therefore congestion control in the context of a cumulative layered organization is possible.

For a layered multicast system using routing mechanism to achieve congestion control, it could result in a major burden for the underlying multicast routing protocol as join and leave decisions occur much more frequently. In this paper, we consider the stability problem of multicast trees in the context of layered multicast congestion control schemes.

Two modeling assumptions are made here. We assume that (a) packets are delivered along the shortest path tree; and (b) if congestion occurs somewhere, all downstream receivers will unsubscribe to the highest layer they have subscribed to. These assumptions allow us to design a simple model without losing the properties we want to observe. Based on these assumptions, we have chosen to

H.-K. Kahng (Ed.): ICOIN 2003, LNCS 2662, pp. 212–221, 2003.

model a layered multicast session as follows: a single source sends packets on different cumulative layers, and all layers follow the same multicast routing that builds a shortest path tree where the unique sender is located at the root and receivers are located at leaves. The transmission is controlled by a receiver-driven congestion control scheme where receivers leave their current groups immediately once they experience congestion.

For analyzing congestion in a multicast tree, we develop a simple statistical model based on the dependent-degree model [3]. The statistical model also builds on random-marking schemes (e.g., REM [4] or RED [5]). The model, which considers the dependence of the marking probabilities between different links, can cover more general cases.

The paper is structured as follows. Sect. 2 describes the definition of the stability index for multicast trees. In Sect. 3, we propose a simple congestion statistical model, used to derive the expressions of multicast bottleneck probability distributions and the stability index in Sect. 4. Sect. 5 describes the numerical evaluation results. Finally, we conclude in Sect. 6.

2   Stability of Multicast Tree

In most layered multicast congestion control schemes, as congestion occurs, all receivers behind the common bottleneck should leave their current highest layer. For evaluating the stability of a multicast tree in such cases, the following definition is introduced.

Definition 1. *If a shortest path tree T with one sender and n receivers corresponds to a layer in a layered multicast session, letting the number of links in the tree T and the number of links changing when congestion occurs be denoted respectively by $L(T)$ and $\Delta(T)$, then, the average number of changes is equal to $E(\Delta(T))$. The stability of the tree may be measured by **a real-valued stability index** $S(T)$, which is determined by $S(T) = 1 - E(\Delta(T))/S(T)$, where $S(T) \in [0, 1]$ and $n \geq 2$.*

Clearly, the larger $S(T)$, the more stable the tree T is when congestion occurs.

3   The Statistical Model for Modeling Congestion in Multicast Trees

3.1   Dependent Random-Marking Multicast Tree Model

In random-marking schemes like RED and REM, the marking/congestion state of a link is a function of its queue length. However, the queue lengths of different links carrying the same flows are generally not independent of each other. For instance, if a large (small) queue is built up at a congested upstream link in a multicast tree, the downstream links carrying the same flows are more likely to have large (small) queues. For layered multicast congestion control with dependent marking probabilities, we develop a simple statistical model for analyzing

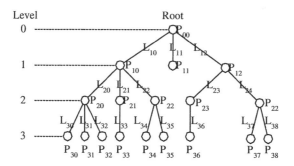

Fig. 1. Dependent random-marking multicast tree model with tree height $m = 3$

the probability distributions of the congestion bottlenecks in a tree. The following definition is introduced.

Definition 2. *A dependent random-marking tree T of height m consists of a set, L, of links that satisfy the following four conditions:*

C1. *All links in L are labeled as shown in Fig. 1 such that $L = \{L_{10}, \ldots, L_{1,j_1}, L_{20}, \ldots, L_{2,j_2}, \ldots, L_{m0}, \ldots, L_{m,j_m}\}$, which connect all nodes in the tree, $\{P_{00}, P_{10}, \ldots, P_{1,j_1}, P_{20}, \ldots, P_{2,j_2}, \ldots, P_{m0}, \ldots, P_{m,j_m}\}$, as shown in Fig. 1. P_{ij} represents the node of level i and L_{ij} represents the link that connect P_{ij} with its neighboring upstream node $P_{i+1,l}$. All receivers are located at leaf nodes.*
C2. *To know whether two different links are neighboring with each other in T, we define three comprehensive operations:*
   - *$NU(T, L_{ij}, L_{i+1,l})$, where L_{ij} and $L_{i+1,l}$ are two links of level i and $i+1$ in the tree, judges if L_{ij} is the neighboring upstream link of $L_{i+1,l}$;*
   - *$NUL(T, P_{ij})$, where P_{ij} is a node of T, get a set of links, L', so that $L' = \{L_{1,j_1}, L_{2,j_2}, \ldots, L_{i-1,j_{i-1}}, L_{i,j_i}\}$ forming a path from the root node P_{00} to P_{ij};*
   - *$NUP(T, P_{ij})$ obtains a set, P', of nodes, where $P' = \{P_{00}, P_{1,j_1}, P_{2,j_2}, \ldots, P_{i-1,j_{i-1}}, P_{i,j_i}\}$, in which all nodes are located in the path of L' where $L' = NUL(T, P_{ij})$.*
   Then the following expressions hold:

$$NU(T, L_{ij}, L_{i+1,l}) = \begin{cases} 1, & \text{if } L_{ij} \text{ is the neighboring upstream link of } L_{i+1,l}; \\ 0, & \text{otherwize} \end{cases};$$

$$NUL(T, P_{ij}) = \{L_{1,j_1}, L_{2,j_2}, \ldots, L_{i-1,j_{i-1}}, L_{i,j_i}\},$$

where $\sum_{l=1}^{i-1} NU(T, L_{l,j_l}, L_{l+1,j_{l+1}}) = i - 1$;

$$NUP(T, P_{ij}) = \{P_{00}, P_{1,j_1}, P_{2,j_2}, \ldots, P_{i-1,j_{i-1}}, P_{i,j}\}$$

Specially, we have $NUP(T, P_{00}) = \{P_{00}\}$.

C3. *The marking state of link L_{ij}, where $1 \leq i \leq m$, is represented by a random variable X_{ij} taking value in $\{0, 1\}$ such that $P_r\{X_{ij} = x_{ij}\} = \begin{cases} p_{ij} & , \quad x_{ij} = 1; \\ 1 - p_{ij}, & x_{ij} = 0; \end{cases}$, where p_{ij} is the marking probability for L_{ij} if a random-marking scheme is used and satisfies $0 < p_{ij} < 1$.*

C4. *The congestion marking states of all links at different multicast-tree levels are dependent and satisfy the following properties:*

$$P_r\{X_{ij} = x_{ij} \mid X_{i-1,0} = x_{i-1,0}, \cdots, X_{i-1,j_{i-1}} = x_{i-1,j_{i-1}},$$
$$X_{i-2,0} = x_{i-2,0}, \cdots, X_{i-2,j_{i-2}} = x_{i-2,j_{i-2}},$$
$$\cdots, X_{10} = x_{10}, \cdots, X_{1,j_1} = x_{1,j_1}\}$$
$$= P_r\{X_{ij} = x_{ij} | X_{i-1,l} = x_{i-1,l}\},$$

where $NU(T, L_{i-1,l}, L_{ij}) = 1$.

C4 is reasonable because one link's congestion state depends most on its immediate upstream link's congestion state. The upstream's influence on a downstream link's congestion state propagates through its immediate upstream link which carries the same flows, and thus, as long as the immediate upstream congestion state is given, the probability distribution at the downstream link is independent of the congestion state at links located above the immediate upstream link as indicated by C4.

3.2   Congestion Bottleneck Statistical Model

Among all existing bottleneck nodes in the tree, the node nearest to the source dominates the number of links leaving from the tree. To explicitly model this feature, the following definition is introduced.

Definition 3. *In a dependent-marking multicast tree of height m as defined in Definition 2, a dominating congestion bottleneck is any node which has experienced congestion but whose upstream nodes have not.*

Based on Definition 2 and 3, the following theorem derives the probability distributions of the dominating congestion bottlenecks.

Theorem 1. *In a dependent-marking multicast tree T of height m as defined in Definition 2, there may exist more than one dominating congestion bottleneck, and the probability distribution, $\Psi_d(T, P_{ij})$, where node P_{ij} becomes the dominating congestion bottleneck, is given by*

$$\Psi_d(T, P_{ij}) = \begin{cases} P_r\{X_{1,j_1} = 1\} & ,if\ i = 1 \quad ; \\ P_r\{X_{1,j_1} = 0\} \times P_r\{X_{i,j_i} = 1 | X_{i-1,j_{i-1}} = 0\} & \\ \times \prod_{l=1}^{i-2} P_r\{X_{l+1,j_{l+1}} = 0 | X_{l,j_l} = 0\} & ,if\ 2 \leq i \leq m; \end{cases}$$

$$(1)$$

where the random variables $X_{1,j_1}, X_{2,j_2}, \ldots, X_{i-1,j_{i-1}}, X_{i,j_i}$ *correspond to the marking states of links* $L_{1,j_1}, L_{2,j_2}, \ldots, L_{i-1,j_{i-1}}, L_{i,j_i}$ *given by* $NUL(T, P_{ij})$.

Proof. The proof is detailed in the [8].

We observe by Eq. (1), the larger i, the smaller $\Psi_d(T, P_{ij})$. This is expected, since a downstream bottleneck is always dominated by a co-existing upstream bottleneck.

To use Eq. (1), we need to derive explicit expressions for $P_r\{X_{i+1,j_{i+1}} = x_{i+1,j_{i+1}} | X_{i,j_i} = x_{i,j_i}\}$ used in Eq. (1). However it is difficult to know the accurate dependency between two random variables. To solve this problem, we introduce **a real-valued dependency-degree factor** α ($\alpha \in [0,1]$) from [3] to quantify all possible degrees of dependency between two random variables. The following Definition and Theorem are introduced.

Definition 4. *Two dependent link marking states* X_{ij} *and* $X_{i+1,l}$ *are said to be positively (negatively) dependent if* $P_r\{X_{i+1,l} = x | X_{i,j} = x\} > P_r\{X_{i+1,l} = \bar{x} | X_{i,j} = x\}$ *(*$P_r\{X_{i+1,l} = x | X_{i,j} = x\} < P_r\{X_{i+1,l} = \bar{x} | X_{i,j} = x\}$*), where* $x \in \{0,1\}$

Based on Definition 4, the following theorem models the degree of dependency between two random variables.

Theorem 2. *Consider a couple of links,* L_{ij} *and* $L_{i+1,l}$*, in a dependent-marking multicast tree* T *of height* m *as defined in Definition 2 where* $NU(T, L_{ij}, L_{i+1,l}) = 1$*, and their marking states are represented respectively by random variables* X_{ij} *and* $X_{i+1,l}$*. If* X_{ij} *and* $X_{i+1,l}$ *is positively dependent, and the link marking probability is equal to* $P_r\{X_{ij}\} = p_{ij}$*, then the following claims hold:*

C1. $\exists \alpha_{il} \in [0,1]$ *such that all possible dependency-degree between* X_{ij} *and* $X_{i+1,l}$ *can be measured by a real-valued dependency-degree factor* α_{il}*, and*

$$\begin{cases} \alpha_{il} = 0, & \textit{iff } X_{ij} \textit{ and } X_{i+1,l} \textit{ are independent} \quad ; \\ \alpha_{il} = 1, & \textit{iff } X_{ij} \textit{ and } X_{i+1,l} \textit{ are perfectly dependent;} \end{cases}$$

C2. *The conditional distribution* $P_r\{X_{i+1,l} = x_{i+1,l} | X_{i,j} = 0\}$*, with* $x_{i+1,l} \in \{0,1\}$*, are determined by*

$$P_r \left\{ X_{i+1,l} = 0 | X_{i,j} = 0 \right\} \tag{2}$$
$$= \begin{cases} 1 - (1 - \alpha_{il})p_{i+1,l} & , \quad \textit{if } p_{ij} \geq p_{i+1,l}; \\ (1 - \alpha_{il})(1 - p_{i+1,l}) + \alpha_{ij}\left(\dfrac{1 - p_{i+1,l}}{1 - p_{ij}}\right), & \textit{if } p_{ij} < p_{i+1,l}; \end{cases}$$

$$P_r \left\{ X_{i+1,l} = 1 | X_{i,j} = 0 \right\} \tag{3}$$
$$= \begin{cases} (1 - \alpha_{il})p_{i+1,l} & , \quad \textit{if } p_{ij} \geq p_{i+1,l}; \\ (1 - \alpha_{il})p_{i+1,l} + \alpha_{ij}\left(\dfrac{1 - p_{i+1,l}}{1 - p_{ij}}\right), & \textit{if } p_{ij} < p_{i+1,l}; \end{cases}$$

Proof. The proof is detailed in the [8].

Applying Eq. (2) and Eq. (3) to Theorem 1, we can obtain the general-case expressions for calculating the dominanting congestion bottlenecks's probability distributions.

4   Statistical Properties of Stability of Multicast Trees under Dependent Markings

The following definition lays a foundation for deriving the stability index of various multicast tree.

Definition 5. *Consider a dependent random-marking tree T of height m as defined in Definition 2. Since T can be represented as a set of links L which equals to $\{L_{10}, \ldots, L_{1,j_1}, L_{20}, \ldots, L_{2,j_2}, \ldots, L_{m0}, \ldots, L_{m,j_m}\}$, we define a series of operations on the set operation, i.e. plus operation and minus operation. T is represented as $T = L = \{L_{10}\} + \cdots + \{L_{1,j_1}\} + \{L_{20}\} + \cdots + \{L_{2,j_2}\} + \cdots + \{L_{m0}\} + \cdots + \{L_{m,j_m}\}$. Similarly, the minus operation is used to prune a sub-tree from T. We design the following comprehensive operations:*

C1. $UPN(T, P_{ij})$ get a node $P_{i'j'}$, which has no branch and belongs to the node set given by $NUP(T, P_{ij}) - \{P_{00}\}$. For example, $UPN(T, P_{ij}) = P_{ij}$ if some branches appear at P_{ij}'s parent node.

C2. $SubT(T, P_{ij})$ obtains a subtree of T rooted at P_{ij}. We have $SubT(T, P_{00}) = T$.

Based on Definition 5, the following theorem derives the means of links changes in a multicast tree.

Theorem 3. *If a dependent-marking multicast tree T of height $m(m \geq 2)$ as defined in Definition 2 is controlled by a receiver-driven layered multicast congestion control scheme, then the following claims hold:*

C1. Let $\Delta(T)$ represent the number of changed links in T when congestion occurs and $\Delta(T, P_{ij})$ represent the number of changed links when the node P_{ij} becomes one of the dominating congestion bottlenecks in T. We have:

$$\Delta(T, P_{ij}) = L(SubT(T, P_{i'j'})) + 1 \tag{4}$$

$$+ \sum_{l=1}^{i} \Delta(SubT(T, P_{i'-1,j'-1}) - SubT(T, P_{l,j_l}) - \{L_{l,j_l}\}),$$

where $P_{i'j'} = NBU(T, P_{ij}), P_{l-1,j_{l-1}} \in NUP(T, P_{i'j'})$

C2. Considering a receiver, R_{ij}, located at the leaf node P_{ij}, the means of the number of changed links is determined by:

$$E(\Delta(T)) = \sum_{l=1}^{i} \Psi_d(T, P_{lj}) \cdot \Delta(T, P_{lj}) \tag{5}$$

$$+ (1 - \sum_{l=1}^{i} \Psi_d(T, P_{lj}))$$

$$* \sum_{l=1}^{i} \Delta\left(SubT(T, P_{l-1,j_{l-1}}) - SubT(T, P_{l,j_l}) - \{L_{l,j_l}\}\right),$$

where $P_{l,j_l} \in NUP(T, P_{ij})$.

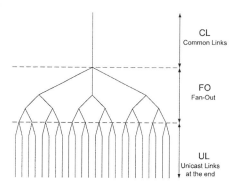

Fig. 2. Trees generic topology

Proof. The proof is detailed in the [8].

According to Definition 1 and Theorems 1 to 3, we can derive the stability index for any tree topology if model parameters (e.g. link-marking probabilities and dependency-degree factors) are given.

5   Numerical Evaluations

5.1   Description of the Evaluations

The stability index can be calculated for any tree based on the previous analytical results. We start the evaluations by studying various stability properties in a binary tree. Then, the evaluations are extended to other topologies. To observe the influence of the model parameters on trees's stability clearly, we chose to uniform these parameters. In the following evaluations, we let $0 \leq \alpha_{ij} = \alpha \leq 1$, $0 < p_{ij} < p < 1$ $(\forall i, j)$.

Modeling the Topology To study the stability of various multicast tree types, we introduce a tree topology model from [6]. These topologies consist of three parts as represented in Fig. 2:

- A first set of CL links, which is common to all receivers.
- A Fan-Out whose total depth is FO. The first step of this fan-out is k-ary (degree k for the first node of the fan out), all the other fan-outs are binary.
- A unicast transmission of depth UL (there is no duplication of the packet in this part of the tree, and no link shared by different receivers).

We study the stability of three basic types of trees represented in Fig. 3. In addition to complete binary trees, we consider:

- Umbrella Trees: these trees end with a long unicast transmission after a short fan-out (large value of UL). It is characteristic of a multicast tree where the receivers only share few links. This kind of topology is identified in [7] as being often found in Mbone sessions.

Fig. 3. Fundamental types of tree topologies

– Reverse Umbrella Trees: packets are forwarded first along a long common path, and then a short fan out ends the transmission (large value of CL).

5.2   Analysis of the Stability for a Complete Binary Tree

We study a complete binary tree with $CL = 0$, $FO > 0$ and $UL = 0$. Evaluation results are shown in Fig. 4(a) and Fig. 4(b).

Fig. 4(a) plots the stability index $S(T)$ calculated according to Definition 1 and Eq. (5) against tree height FO (from 1 to 8) for various dependency-degree factors. This figure is obtained by setting $p = 0.1$ and varying α from 0.1 to 1. We observe that the link-marking dependency, α, has the direct impact on the stability of the tree. The smaller α, the smaller $S(T)$, which implies that the tree may be influenced by congestion more easily. We also observe that the larger the tree height, the less stable it is. Specially, Fig. 4(a) shows that even for smaller link-marking probability, when α is smaller, the multicast tree could change dramatically while it experiences congestion, over 40% links in the tree could leave.

Using Eq. (5), Fig. 4(b) plots the stability index $S(T)$ against FO (from 1 to 8) while varying p from 0.1 to 0.9. We note that $S(T)$ decreases quickly as tree height FO becomes larger with higher link-marking probability.

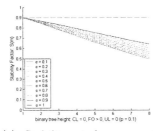

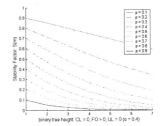

(a) Stability index vs. tree height for various dependency-degree factors

(b) Stability index vs. tree height for various link-marking probabilities

Fig. 4. Stability vs. tree height in the case of complete binary trees

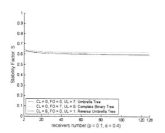

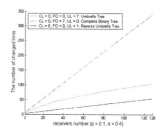

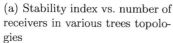

(a) Stability index vs. number of receivers in various trees topologies

(b) Number of changed links vs. number of receivers in various trees topologies

Fig. 5. Stability vs. various tree topologies

5.3 The Influence of Various Tree Topologies on the Stability Index of Trees

We now evaluate the stability of different tree topologies described in Sect. 5.1. The total depth of the tree always equals 7. We only vary the value of CL, FO and UL with the sum of 7. Fig.5(a) and Fig. 5(b) plot the stability index and the number of changed links of three fundamental trees (umbrella tree, binary tree and reverse umbrella tree) respectively while varying the number of receivers in the tree. In particular, for a binary tree, we first fill the tree with enough receivers so that the tree becomes a complete binary tree (e.g. in the case of tree depth equal to 7, there are 128 receivers located at the leaf of the tree), then we choose one receiver randomly and let it leave. This behavior is repeated until the number of receivers decrease to 1.

We found that the topology of three tree types has little influence on the stability indexes of trees. In Fig. 5(a), three lines are very close to each other. However, the stability index of umbrella tree is a little larger than that of other two tree types. The second observation is that the number of changed links in the umbrella tree is much more than that in other two tree types as shown in Fig. 5(b).

6   Conclusion

In this paper, we study the stability problem of a multicast tree in the presence of a layered multicast congestion control scheme. Using a simple statistical model, we accurately capture congestion in a tree when the congestion markings at different links are dependent. Then, we derive general expressions for the probability distributions of individual nodes to be the dominating congestion bottlenecks. We also introduce a dependency-degree factor to quantify and evaluate the dependency between link congestion markings at different links in the tree. With this model, we obtain equations for all one-step transition probabilities as functions of the marginal link-marking probabilities and the dependency-degree factors. Using these analytical results, we derive the general expression for the stability index of the multicast tree in layered multicast congestion control where receivers make leave decisions as they experience congestion. We also introduce a tree topology model to study the stability of various tree types. We found that three fundamental tree types has the similar stability index in our model.

Acknowledgments

We would like to thank the National Natural Science Foundation of China under Grant No. 90104002 and 60203025, and the National High Technology Development 863 Program of China under Grant No. 2001AA121013.

References

[1] McCanne, S., Jacobson, V., Vetterli, M.: Receiver-Driven Layered Multicast. Proc. of ACM SIGCOMM'96, Stanford, CA (1996), pp.117-130.
[2] Vicisano, L., Rizzo, L., Crowcroft., J.: TCP-like Congestion Control for Layered Multicast Data Transfer. Proc. of IEEE INFOCOM'98, San Francisco, CA (1998), pp.996-1003.
[3] Zhang, X., Shin, K. G.: Statistical Analysis of Feedback-Synchronization Signaling Delay for Multicast Flow Control. Proc. of IEEE INFOCOM'2001. Anchorage, Alaska, USA (2001), pp.1133-1142.
[4] Lapsley, D., Low, S.: Random Early Marking: for Internet Congestion Control. Proc. of IEEE GLOBECOM'99. USA(1999), pp. 1747-1752.
[5] Floyd, S., Jacobson, V.: Random Early Detection Gateways for Congestion Avoidance. IEEE/ACM Trans. on Networking, vol. 1, no. 4, pp. 397-413, August 1993.
[6] Chaintreau, A., Baccelli, F., Diot, C. Impact of Network Delay Variation on Multicast Sessions with TCP-like Congestion Control. Proceedings of IEEE INFOCOM'2001, Vol. 2. Anchorage, Alaska (2001), pp.1133-1142.
[7] Stoica, I., Eugene Ng, T. S., Zhang, H.: A Recursive Unicast Approach to Multicast. Proc. of IEEE INFOCOM'2000. Tel-Aviv (2000).
[8] Shi., F., Wu, J. P., Xu, K.: The Stability Problem of Multicast Trees in Layered Multicast. Technical Report, Network Labs., CS Dept., The University of Tsinghua, August 2002.

RAMRP Protocol for Reliable Multicasting in Wireless Ad-Hoc Network Environments

Sang Yun Park[1], Ki-seon Ryu[2], and Young Ik Eom[1]

[1] School of Information and Communications Engineering, Sungkyunkwan Univ.
Chunchun-dong 300, Jangan-gu, Suwon, Kyounggi-do, Korea
{bronson,yieom}@ece.skku.ac.kr
[2] Technology Research Center, Handan BroadInfoComm. Co., Ltd.
Shinsung Plaza 697-11, Yeogsam-dong, Kangnam-gu, Seoul, Korea
ksryu@handan.co.kr

Abstract. In wireless ad-hoc network environments, packet retransmissions and route reconstructions caused by link failures can be more frequent than in the wired network environments. Therefore, new multicasting scheme reducing transmission delays and packet losses due to link changes of a multicast tree is required. In this paper, we propose a RAMRP (Reliable Ad-hoc Multicast Routing Protocol) protocol. It supports a reliable multicasting suitable for wireless ad-hoc networks by reducing the number of route reconstructions and packet retransmissions.

1 Introduction

Multicasting in wireless ad-hoc networks potentially has a trade-off between stability and efficiency. Since many other multicasting schemes in wireless ad-hoc networks depend on group member states managed by the mobile hosts for route construction and packet transmission, those schemes cause heavy control traffic and also require a lot of time for routing:[1]. There are well-known multicast routing protocols such as AMRIS:[2], ODMRP:[3], and AmRoute:[4]. And now, an enhanced multicasting scheme improving both stability and efficiency is needed.

In this paper, we propose a RAMRP protocol that uses a link soundness based route construction method and an agent based ACK strategy. It can provide a reliable multicasting environment in wireless ad-hoc networks. The next section describes a system model and algorithms for our protocol. In section 3, we present the performance analysis with simulations. Finally, the last section discusses our conclusions and areas of future works.

2 RAMRP Protocol

2.1 System Model

Each node in wireless ad-hoc networks is classified into a member node or a non-member node. Member nodes construct a multicast tree with a sender as a center

H.-K. Kahng (Ed.): ICOIN 2003, LNCS 2662, pp. 222–231, 2003.

of the tree. Among non-member nodes, some nodes do not join the multicasting, but they perform passive forwarding of data. We assume that link priority of a node having multi-links is based on stability, and all nodes in a multicast tree should support functions for multicast routing since non-member nodes can be included in the multicast tree for routing.

2.2 Algorithms

Link Soundness Based Route Construction Algorithm The existing schemes mostly made use of hop count and node identifier as criteria for routing. These criteria may have led to fast and simple routing. But, when nodes move frequently and connections among nodes become weak, frequent route reconstructions raised by the reasons can cause packet delays more and more. In order to resolve these problems, RAMRP protocol reduces the number of route reconstructions and packet retransmissions by using link soundness based route construction method. Link soundness can be defined as a degree of link stability. There are representative criteria for link soundness such as signal strength, hop count from a sender, and CMR(Call to Mobility Ratio) based on migration of nodes.

Fig. 1-1 shows procedures of route construction and tree construction in RAMRP protocol. A sender node broadcasts a JOIN_REQ packet as a request for join to member nodes in order to start data transmissions. A receiver node records information about current hop count and signal strength in the packet and broadcasts it again. And, the receiver node selects a route having the highest degree of link soundness among the routes in which JOIN_REQ packets have

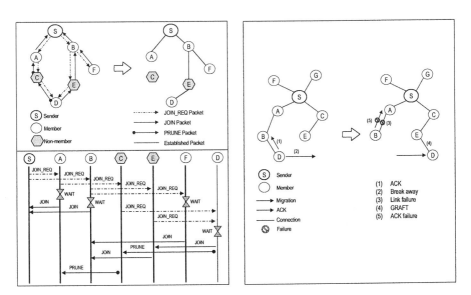

Fig. 1. 1. Route construction procedures, 2. Scenario of ACK packet loss

Table 1. Algorithms for processing RAMRP packets

```
A.1 Algorithm for processing JOIN_REQ packet
    if (first_seen JOIN_REQ.SEQ) { start JOIN_TIMER; }
    store JOIN_REQ.INFO;
    if (HOP_CNT < TTL) { flood JOIN_REQ; }
    while (Tjoin-req-timer)  ;  PREF(JOIN_REQ[].INFO);
    if (NOT_SELECTED_LINK) { send PRUNE(JOIN_NAK=true) to up-link nodes; }
    else { send JOIN to up-link node; }

A.2 Algorithm for processing JOIN packet
    if (this==FORWARDING_NODE) { send JOIN to up-link node along the existing route;
        set ROUTING_TBL with NEW_CHILD; }
    else {
        set FORWARDING_NODE to this node;  set ROUTING_TBL with NEW_CHILD;
        send JOIN to PARENT node; }

A.3 Algorithm for data transmission and ACK processing in a~sender node
    while (packets to send remain) {
        if (event : Tsnd) {
            if (data_cache is not full) { flood next DATA packet; }
        if (event : ACK) {
            if (already seen ACK:diff ACK_NUM) { discard ACK(DATA_NUM);  continue; }
            if (#ACK(DATA.NUM)++ == #MEMBER_TBL_ENTRY) { purge DATA(DATA.NUM) from data_cache;
                flood PURGE(DATA.NUM) packet; }
        }
    }

A.4 Algorithm for processing an ACK packet
    if (same ACK is in ACK_TBL)  { discard ACK packet; }
    else {
        add ACK.INFO to ACK_TBL;  send RE_ACK packet to CHILD(ACK_SRC) node;
        send ACK_TBL packet to PARENT node;  }

A.5 Algorithm for processing a~GRAFT_REQ packet
    GRAFT_RADIUS = 1;  flood GRAFT_REQ(GRAFT_RADIUS) packet;
    while (event : GRAFT packet) {
        while (Tgraft)  ;  GRAFT_RADIUS++;  send GRAFT_REQ(GRAFT_RADIUS) packet;  }
    set SOURCE(GRAFT) node to PARENT node;  send GRAFT_ACK packet to SOURCE(GRAFT) node;

A.6 Algorithm for processing a~GRAFT_ACK packet
    If (Current_Node != DEST(GRAFT_ACK)) {
        set Current_Node to FOWARDING node;  forward GRAFT_ACK to DEST(GRAFT_ACK) node;
        set ROUTING_TBL with NEW_CHILD;
    } else  add GRAFT_ACK.INFO to ROUTING_TBL;
```

been received for a period of time and sends a JOIN packet along the route. The receiver node sends PRUNE packets to the nodes on the other routes except the selected route to inform that other route was selected and to stop the nodes from transmitting packets.

Algorithm A.1 in Table 1 shows an algorithm for processing JOIN_REQ packet in a node receiving a JOIN_REQ packet. When a node receives a JOIN_REQ packet for the first time, the node receives the same JOIN_REQ packets from other routes for a period of time by using a timer. After that, the node stores link soundness information, packet identifier, and parent node information included in the JOIN_REQ packet in its cache. If TTL (Time To Live) value of the packet is lower than hop count limit of the packet, the node increases TTL value in the packet and broadcasts the packet. After a period of time, the node selects a route having the highest degree of link soundness among routes stored in its cache and sends a JOIN packet along the route. And, the node sends PRUNE packets to other nodes to stop the nodes from transmitting packets.

Algorithm A.2 in Table 1 shows an algorithm for processing JOIN packet in a node receiving a JOIN packet from a child node. If the node has performed multicast routing, the node sends the JOIN packet to parent node along the existing route and adds the child node to its routing table. If the node has not performed multicast routing, the node adds itself as a forwarding node to the routing table and forwards the JOIN packet to parent node stored in its cache. A node receiving a PRUNE packet does not forward packets anymore.

Algorithm A.3 in Table 1 shows an algorithm for data transmission and ACK processing in a sender node. Each node on multicast tree manages a buffer with fixed available space in order to deliver data to new grafting nodes as soon as possible. If a node receives the same packet as before, the node ignores the next data. Data stored in a buffer can be removed only after receiving ACKs from all receivers. And then, the node broadcasts a PURGE packet to have other nodes remove the data confirmed by the received ACKs from their buffer.

ACK Processing Algorithm Links in wireless ad-hoc network environments have a high probability of failure. Also, link failures can occur among direct-connected nodes as well as indirect-connected nodes. Link failures among indirect-connected nodes can raise serious problems. Fig. 1-2 shows a scenario of ACK packet loss caused by link failure between indirect-connected nodes. After sending an ACK to parent node B, node D migrates and then becomes a child node of node E. If the link between node A and node B is lost for an interval of the above procedure, node B cannot forward the ACK of node D to node A. Hence, sender S retransmits the data for node D. This situation can occur frequently in wireless ad-hoc network environments with host mobility.

In this paper, we propose an agent based ACK strategy to reduce the number of retransmissions caused by ACK packet losses. Parent node receiving an ACK packet from a child node has a role of forwarding the ACK packet to a sender as an agent. The parent node sends a RE_ACK packet to the child node as a confirmation for forwarding of the ACK packet. If the child node cannot receive the RE_ACK packet, the child node should send an ACK packet to parent node after time-out or in every migration time. Algorithm A.4 in Table 1 shows an algorithm for processing an ACK packet.

When a parent node receives an ACK packet, if the packet is duplicated, then the packet is discarded. Otherwise, information about a packet identifier and an ACK sequence number in the packet is recorded in an ACK packet table. After that, the parent node sends a RE_ACK packet to the child node. The ACK sequence number refers to the number of ACK packets that the child node

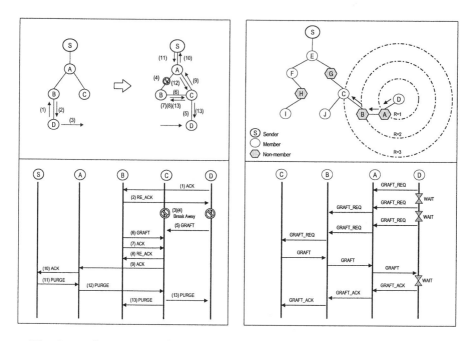

Fig. 2. 1. Agent based ACK procedure, 2. Expand ring search procedure

retransmits because it has not received the RE_ACK packet. Fig. 2-1 shows an agent based ACK procedure. Node D sends an ACK packet to node B, and node B sends a RE_ACK packet to node D. After that, node D migrates, and the connection between node A and node B is cut off. Node B becomes a child node of new parent node C and forwards the ACK packet of node D to node C as an agent. Finally, Node C forwards the ACK packet of node D to sender S and the procedure finishes.

Route Maintenance Algorithm In wireless ad-hoc network environments, a network can be partitioned into several ones due to migration of nodes. Although frequent reconstructions of multicast tree can support route maintenance of the tree, frequent reconstructions can cause considerable construction delays and wastes of bandwidth raised by heavy control traffic. In this paper, in order to reduce those overheads, we use expand ring search scheme that supports re-graft of multicast tree when a network is partitioned.

Fig. 2-2 shows an expand ring search procedure. Node D separated from a multicast tree broadcasts GRAFT_REQ packets to nodes within 1 hop distance (R=1) for re-graft of the tree. Though node A is residing in one hop distance, node D broadcasts again GRAFT_REQ packets to nodes within 2 hops distance (R=2) because node A is not a member of the multicast session. Since B is not a member node also, node D repeats the above process until it meets a member node C in 3 hops distance. Node C sends a GRAFT packet as a reply of the GRAFT_REQ packet to node D through B and A. Node D selects a route of the first received packet and sends a GRAFT_ACK packet as a confirmation for the GRAFT packet along the route. Though node A and B are not member nodes, they update their routing table to support passive forwarding of multicast data. If the expand ring search procedure is repeated, length of the route can be increased more and more. So, stable route construction by periodic global tree construction is required.

Algorithm A.5 in Table 1 shows an algorithm for processing a GRAFT_REQ packet. A separated node repeats transmission of GRAFT_REQ packets from one hop distance (GRAFT_RADIUS=1) until it meets a member node. And, the member node sends a GRAFT_ACK packet to the separated node.

Algorithm A.6 in Table 1 shows an algorithm for processing a GRAFT_ACK packet in the node receiving a GRAFT_ACK packet. If the final destination of the GRAFT_ACK packet is not current node, the current node updates itself as a forwarding node to the routing table and forwards the GRAFT_ACK packet to the final destination. When the GRAFT_ACK packet arrives at the final destination, the destination node adds graft ACK information (GRAFT_ACK.INFO) to its routing table.

3 Performance Evaluation

The simulations are classified into the comparison for link metric and the comparison for ACK strategy and transmission method.

3.1 Comparison for Link Metric

We assume that a sender node transmits data at intervals of every 10ms. The transmission delay from a sender node to a member node is assumed to have exponential distribution with mean having the value of 15ms. We choose link soundness and link delay as components of the link metric. And, we measure packet retransmission count and transmission delay for comparison between a link soundness based route construction scheme and a link delay based route construction scheme.

Fig. 3-1 shows changes of packet retransmission count according to link metric. Transmission failure ratio of 0.2 means that the probability of link failure on transmission is 20%. As transmission failure ratio grows, packet retransmission count of the link soundness based scheme is much smaller than that of the link delay based scheme.

Fig. 3-2 shows changes of delay in the case of finishing transmitting 100 packets and receiving all ACK packets. As transmission failure ratio increases, total transmission delays of both two schemes are increasing almost linearly. In the case where transmission failure ratio is higher than 0.07, this figure shows that the link soundness based scheme can reduce packet retransmission time and can provide faster data transmission than the link delay based scheme.

Table 2. Definitions of state variables for simulations

Class	Notation	Description
Prob.	P_{suc}	Probability of data/ACK transmission without link failure
	P_{fail}	Probability of data/ACK transmission with link failure
	P_{af}	Probability of ACK transmission with link failure
	P_{df}	Probability of data transmission with link failure
Time	$T_{timeout}$	Time intervals of time-out
	T_{data}	Data packet transmission delay
	T_{ack}	ACK packet transmission delay
	T_{graft}	Re-graft delay of a separated node
Etc.	N_{hf-a}	The number of hops with ACK packet transmission failure
	N_{hf-d}	The number of hops with data packet transmission failure

3.2 Comparison for ACK Strategy and Transmission Method

If each node does not support buffering, the node should send next data after receiving all ACK packets for the previous data in order to guarantee reliability. However, in order to optimize transmission delay for new graft node, each node manages a buffer with fixed available space and can send next data before receiving all ACK packets within the buffer limit.

When each node does not use both a buffer and ACK strategy, the node should retransmit data after time-out. In this case, the time-out value is assumed to be 80ms. When each node manages a buffer, a sender node generates data at intervals of every 20ms, and transmission delay of data packet and ACK packet is assumed to have exponential distribution with mean having the value of 20ms. Re-graft delay of a separated node is assumed to have exponential distribution with mean having the value of 40ms which is two times as long as a regular transmission delay. Definitions of the state variables (Table 2) and expressions of formulas for simulations are as follows.

1. One packet transmission delay without agent based ACK and data buffering

   In the case that the first transmission of both data and ACK packets is successful

   $$T_{m-1} = (T_{data} + T_{ack}) \times P_{suc} \tag{1}$$

   In the case that the $N + 1$th transmission of both data and ACK packets is successful after the Nth failures

   $$T_{m-3} = \sum_{k=1}^{n} \{(k \times T_{timeout} + T_{data} + T_{ack}) \times P_{fail}^{k} \times P_{suc}\} \tag{2}$$

   $$T_{m-3} = \left[\left(\frac{P_{fail} \times (1 - P_{fail}^{n})}{(1 - P_{fail})^2} - \frac{n \times P_{fail}^{n+1}}{1 - P_{fail}} \right) \times T_{timeout} \right.$$
   $$\left. + \frac{(T_{data} + T_{ack}) \times P_{fail} \times (1 - P_{fail}^{n})}{1 - P_{fail}} \right] \times P_{suc} \tag{3}$$

2. One packet transmission delay with agent based ACK and without data buffering

In the case that the first transmission of both data and ACK packets is successful

$$T_{m-1} = (T_{data} + T_{ack}) \times P_{suc} \tag{4}$$

In the case that the first data transmission is successful, but the first ACK transmission fails

$$T_{m-2} = (T_{data} + N_{hf-a} \times T_{graft} + T_{ack}) \times P_{fail} \times P_{af} \tag{5}$$

In the case that the $N+1$th transmission of both data and ACK packets is successful after the Nth failures

$$T_{m-3} = \left[\left(\frac{(P_{fail} \times P_{df}) \times (1 - (P_{fail} \times P_{df})^n)}{(1 - P_{fail} \times P_{df})^2} - \frac{n \times (P_{fail} \times P_{df})^{n+1}}{1 - P_{fail} \times P_{df}} \right) \times T_{timeout} \right.$$
$$\left. + \left(\frac{(P_{fail} \times P_{df}) \times (1 - (P_{fail} \times P_{df})^n) \times (T_{data} + T_{ack})}{1 - P_{fail} \times P_{df}} \right) \right] \times P_{suc} \tag{6}$$

In the case that the $N+1$th transmission of both data and ACK packets also fails after the Nth failures

$$T_{m-4} = \sum_{k=1}^{n} \{ (k \times T_{timeout} + T_{data} + N_{hf-a} \times T_{graft} + T_{ack})$$
$$\times (P_{fail} \times P_{df})^k \times P_{fail} \times P_{af} \} \tag{7}$$

$$T_{m-4} = \left[\left(\frac{(P_{fail} \times P_{df}) \times (1 - (P_{fail} \times P_{df})^n)}{(1 - P_{fail} \times P_{df})^2} - \frac{n \times (P_{fail} \times P_{df})^{n+1}}{1 - P_{fail} \times P_{df}} \right) \times T_{timeout} \right.$$
$$\left. + \left(\frac{(P_{fail} \times P_{df}) \times (1 - (P_{fail} \times P_{df})^n) \times (T_{data} + N_{hf-a} \times T_{graft} + T_{ack})}{1 - P_{fail} \times P_{df}} \right) \right] \tag{8}$$
$$\times P_{fail} \times P_{af}$$

3. One packet transmission delay with both agent based ACK and data buffering

In the case that the first transmission of both data and ACK packets is successful

$$T_{m-1} = (T_{data} + T_{ack}) \times P_{suc} \tag{9}$$

In the case that the transmission of data packet and/or ACK packet fails

$$T_{m-2} = (T_{data} + N_{hf-d} \times T_{graft} + T_{ack} + N_{hf-a} \times T_{graft}) \times P_{fail} \tag{10}$$

Fig. 3-3 shows changes of transmission delays according to retransmission methods. Retransmission methods are classified into the three cases. The first is the case of retransmission after time-out, and the second is the case of agent

1. Packet retransmission count vs. link metric 2. Total transmission delay vs. link metric

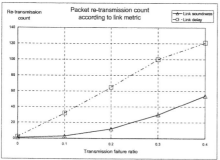 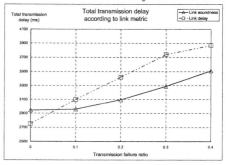

3. Total transmission delays vs. re-trans. methods 4. Transmission delays vs. buffer sizes

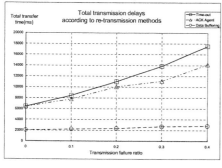 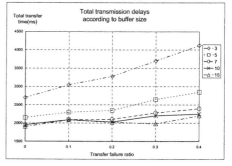

5. Average packet delays vs. buffer sizes

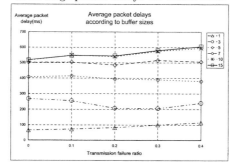

Fig. 3. Simulation results

based ACK forwarding by other node. The last is the case of forwarding from a buffer at next re-graft time. Buffer size is assumed to be 5. We measure delays in the case of finishing transmitting 100 packets and receiving all ACK packets for the packets. The buffering case is little affected by transmission failure ratio and provides efficient transmissions. The time-out case and the agent based ACK case are much affected by transmission failure ratio. However, as transmission failure ratio increases, the agent based ACK case has smaller delay than the time-out case.

Fig. 3-4 shows changes of transmission delays according to buffer sizes. When the buffer size is high, the transmission delay is small. But, when the buffer size approaches to 15, the transmission delay does not be reduced anymore. In other words, an excessive buffer allocation can cause only waste of memory without improvement of performance. Fig. 3-5 shows changes of average packet delays according to buffer sizes. As the buffer size is higher, buffer staying time of an individual packet becomes longer and average packet delay is higher.

4 Conclusions

Interests about multicasting in wireless ad-hoc network environments are rising today. In this paper, we proposed a RAMRP protocol to support an efficient multicasting in wireless ad-hoc network environments. RAMRP protocol can provide more stable route construction, and can reduce the numbers of route reconstruction and packet retransmission using the link soundness based route construction method and the agent based ACK strategy. Also, our protocol can transmit data to new graft node as soon as possible through the route maintenance technology with buffer management. For future works in wireless ad-hoc network environments, development of an enhanced scheme for reducing packet losses and providing robust services against network state changes is required.

References

[1] K. Obraczka, G. Tsudik, *"Multicast Routing Issues in Ad Hoc Networks,"* IEEE 1998 International Conference on Universal Personal Communications, ICUPC'98, 1998.

[2] C. W. Wu, Y. C. Tay and C. Toh, *"Ad-hoc Multicast Routing Protocol utilizing Increasing id-numberS (AMRIS) Functional Specification,"* draft-ietf-manet-amris-spec-00.txt, Internet-Draft, IETF, 1998.

[3] M. Gerla, G. Pei, S. J. Lee, and C. C. Chiang, *"On-Demand Multicast Routing Protocol (ODMRP) for Ad-hoc Networks,"* draft-ietf-manet-odmrp-00.txt, Internet-Draft, IETF, 1998.

[4] M. Liu, R. R. Talpade, and A. McAuley, *"AmRoute : Ad-hoc Multicast Routing Protocol,"* Technical Research Report CSHCN T. R. 99-1, Center for Satellite and Hybrid Communication Networks, 1999.

An MPLS Broadcast Mechanism and Its Extension for Dense-Mode Multicast Support*

Siavash Samadian-Barzoki, Mozafar Bag Mohammadi, and Nasser Yazdani

Router Lab., Dept. of EE & Computer Eng., Univ. of Tehran, Tehran, Iran
{s.samadian,mozafarb}@ece.ut.ac.ir, yazdani@ut.ac.ir

Abstract. Due to the increasing multicast applications and high desire for its deployment, it seems that any new technology should support multicast. However; MPLS technology which enhances IP packet forwarding capability by using layer 2 switching still does not offer any special solution for multicast support. The main difficulty in supporting dense-mode multicast protocols like DVMRP and PIM-DM is lack of a broadcast mechanism in MPLS. We propose a genuine broadcast mechanism for MPLS in this paper. First, the proposed mechanism is based on a central broadcast label assigner called BLAC (Broadcast Label Assignment Center). Then, we propose a new dense-mode multicast protocol for MPLS as an extension to our broadcast mechanism. Our method consumes fewer labels compared to the existing proposals supporting dense-mode multicast groups and has smaller forwarding tables.

1 Introduction

Multicast capability is a very important feature in data networks and supporting it considered as a highly valued services being offered by some Internet Service Providers (ISPs) [3]. In many circumstances, an application must send the same information to many destinations. Using multicast transmission instead of unicast in these cases reduces the transmission overhead on the sender, on the network and saves the time taken for all destinations to receive the data [1].

MPLS technology which enhances IP packet forwarding capability by using layer 2 switching is being standardized in IETF [10] and manifests performance benefits in the backbone of Internet [11]. MPLS tries to tackle the problem of how to integrate the best attributes of the traditional layer 2 and layer 3 technologies. There are many unsolved problems for supporting IP multicast in an MPLS domain [6][7]. Multicast support is not defined in MPLS architecture and it is left for further study [10]. Any efficient solution for supporting multicast in MPLS will accelerate MPLS deployment in the future network.

We introduce a solution to the broadcast problem in MPLS network for the first time in this paper. Our solution is based on the RPF checking mechanism and multicasting in IP [1][15]. It assigns a label to a source of broadcast packet

---

* This work is partially supported by ITRC (Iranian Telecommunication Research Center).

H.-K. Kahng (Ed.): ICOIN 2003, LNCS 2662, pp. 232–242, 2003.

upon its request. The label is then, the identifier of the sources broadcast packets and each LSR (Label Switch Router) uses this label to determine correct incoming interface based on RPF check. If the packet was from right interface, it is copied to the other interfaces of LSR. Broadcast label is assigned and released by a central node named BLAC (Broadcast Label Assignment Center). We propose to define a specific value in layer 2 headers (such as ethertype value in Ethernet and IEEE 802.3 or Protocols field in PPP) to indicate the broadcast label space in the MPLS header. If not possible, our second solution is a globally specific reserved label value to sit at top of the label stack while the source identifying label sits in the next level. In this way, the scheme consumes only identifying labels for broadcasting while unicast label space is remained intact by separating the broadcast and unicast label spaces. Using the broadcasting mechanism, we devise a new dense-mode multicast routing protocol in MPLS environment which is like DVMRP [8]. This protocol consumes fewer labels; only one label from the multicast label space and no label from unicast label space, for a source tree than the existing proposals and has smaller forwarding tables.

The rest of paper is organized as follows. In the next section, the BLAC broadcast mechanism is discussed. We explain our proposal for MPLS dense-mode multicast routing protocol in section 3. Section 4 discusses related work. Finally, we conclude in section 5.

2   Broadcast Mechanism

For broadcasting data in MPLS, we need a means to identify broadcast packets. This can be done by defining a specific value in layer 2 headers (such as ethertype value in Ethernet and IEEE 802.3 or Protocols field in PPP) to indicate whether the label space in MPLS header is broadcast. If this solution was not possible, a globally specific reserved label value is needed among MPLS labels to sit at top of the packet's label stack. The latter solution, of course, consumes more network bandwidth with 4 bytes extra label. Although we strongly recommend the first solution, the rest of the paper is based on the second solution to achieve an independent method and when the first solution is possible for a special layer 2 technology, the needed conversions should be done appropriately.

In this way, each a LSR can examine the packet label and if it was the broadcast label, it starts broadcasting packet and forwarding it to all interfaces except the incoming one. Of course, this is a simple solution. Unfortunately, this way broadcasting is inefficient and broadcast traffic will congest network rapidly. Better methods are presented for IP broadcast, namely RPF, extended RPF and RPB (Reverse Path Broadcasting) [1][15]. These methods use packets' source addresses to filter extra packets and determine output interfaces for a right packet. We consider a packet as right packet if it has come on an interface used to forward the packet to its source (reverse path) based on the routing table. In MPLS, one can't guess source address of a packet by merely examining MPLS header. Then, the LSR must examine the IP header. To avoid this time consuming process, a one to one binding must be defined between source addresses and assigned label and all LSRs must be aware of this correspondence. Then, source of a packet can be identified from the corresponding assigned label in the header.

In our method, when a source S wants to broadcast some packets into the network, first it claims a broadcast label from BLAC (Broadcast Label Assignment Center). BLAC is an LSR which is responsible for assigning and releasing broadcast labels. Therefore, the source sends a *Broadcast_Label_Request* message toward BLAC. When BLAC node receives these messages it acts as follows:

Algorithm *BLAC-BLA*
(∗ Broadcast_Label_Request message processing in BLAC ∗)
1. **if** $\exists$ $Label_B \in Label\_Range_B$
2.   **then** Send a *Broadcast_Label_Assignment* message(X, $Label_B$);
3.   destined to S to all interfaces;
4. $Label\_Range_B \leftarrow Label\_Range_B - \{Label_B\}$;
5.   Insert (IP_S, $Label_B$) in $Table_{Assigned\_Labels}$;
6.   **else** Discard the *Broadcast_Label_Request* message;

If a broadcast label ($Label_B$) is available in the broadcast label space ($Label\_Range_B$), BLAC broadcasts the following message in the MPLS domain:

	Shim Header		Source Address	Destination Address	Payload
L2 header	X	Label$_B$	IP$_{BLAC}$	IP$_S$	Broadcast_Label_Assignment

Consequently, all LSRs in MPLS domain know that $Label_B$ is assigned to S. S can use this label to broadcast its data. Label X at top of the stack identifies BLAC broadcasted messages. When this message reaches S, it knows that $Label_B$ can be used for its data broadcasting. Therefore, it first pushes $Label_B$ in label stack of each broadcast packet followed by another label Y. Label Y has special meaning, and is used by all LSRs for broadcasting conventional broadcast packets. Finally, S broadcasts the resulting packets to all of its interfaces. Important fields of these broadcast packets are as follows:

	Shim Header		Source Address	Destination Address	Payload
L2 header	Y	Label$_B$	IP$_S$	IP$_{Broadcast}$	Broadcast Data

Notice that label Y is needed when we have no identifier in layer 2 header for broadcasting to prevent partitioning the label space in to two parts, one for unicast packets and another for broadcast ones. By using one more level in label stack, we have separated broadcast label space from unicast label space. Therefore, both can use label space independently at the expense of using more bandwidth. This larger packet size (4 bytes) has little effect on the network performance since broadcast event is less frequent. When an LSR receives a packet with label Y, it knows that the next label in the stack identifies the source of packet. Therefore, it acts as the algorithm *LSR-Broadcast-Data* which is shown in the next page.

If the first label in the stack is Y, then the second label ($Label_B$) is looked up in the BFT (Broadcast Forwarding Table). BFT maintains broadcast information in corresponding LSR. It is constructed by means of BLAC broadcast

Algorithm *LSR-Broadcast-Data*
(∗ Broadcast data processing in LSRs ∗)
```
1.   if top label is Y
2.      then Pop the top label;
3.          if Label_B exists in BFT
4.              then iif_RPF ← BFT(Label_B).input_interface;
5.                  if iif ≠ iif_RPF
6.                      then Discard the packet;
7.                      else Push label Y at top of stack in the packet;
8.                          Forward the packet to AllLinks − {iif};
9.              else Push label Y at top of stack in the packet;
10.                 Forward the packet to AllLinks − {iif};
```

messages. Each entry of this table contains assigned label by BLAC and its corresponding source address. It also contains one of LSR interfaces which is used by that LSR to forward received packet toward the source. As in the basic reverse path forwarding algorithm [1][15], a LSR forwards a broadcast packet originating at source S if and only if it arrives via the shortest path from the LSR back to S (i.e., the "reverse path"). We named this interface as iif_{RPF} in the above pseudo code. Required steps for constructing BFT table are explained later in the text. If $Label_B$ is found in the BFT, then we perform RPF check. If the RPF check is succeeded, data will be copied to all interfaces except incoming one ($AllLinks - \{iif\}$). Otherwise, data will be silently discarded.

If incoming label doesn't exist in BFT, data will be copied to all interfaces except incoming one. This occurs when *Broadcast_Label_Assignment* message is received by S but it isn't reached this LSR yet. Therefore, the LSR will be possibly informed about it in future. There are other options to treat with unknown labels. One can simply drop the corresponding packets. Unfortunately, this reduces reliability of the broadcast mechanism. Another option that we selected is to broadcast the packet anyway and don't bother RPF check. This method wastes network bandwidth but increases reliability of broadcast algorithm. In this case, the LSR will shortly be informed about label assignment, and returns to normal efficient broadcast mechanism. The final option is to use the information in the IP header and IP lookup to check RPF which increases the processing time of broadcast packet.

When S no more needs the broadcast assigned label, it must inform BLAC to release that label. Therefore it sends a *Broadcast_Label_Release* message toward BLAC. BLAC deletes the corresponding label from the used label set. It must also inform all other LSRs in MPLS domain to delete it from their BFTs. Therefore, BLAC broadcasts a *Broadcast_Label_Release* message in the domain. This is the same as *Broadcast_Label_Assignment* message.

Since the message is broadcasted, requesting node S is informed about success of the label releasing procedure as well. Careful attention must be paid to the success of S requests because *Broadcast_Label_Request* and *Broadcast_Label_Release* messages may be lost in the network. *Broadcast_Label_Request* may also be dropped by *BLAC* in case of broadcast label shortage. Therefore S activates a timer initially set to $\alpha * RTT_{BLAC}$ when it sends a request toward BLAC. The timer is turned off after receiving the corresponding reply from BLAC. When the timer is expired, it resends the request in an exponential back-off manner if it has not received the reply message yet. Values of α and RTT_{BLAC} are determined base on simulation experiments. RTT_{BLAC} is the round trip time between S and BLAC.

Till now, we considered the BLAC address to be known by LSRs. The question is how a LSR distinguishes the BLAC address? In our solution which is like the monitor election in rings or root election in spanning tree for Ethernet [19], each LSR broadcasts a *BLAC_Address* message when it boots. When a LSR receives such a message it first performs RPF check. If the test succeeded, it compares address contained in the message with a currently stored BLAC address. The stored address is initially set to the LSR's own address. LSR saves the new address as BLAC address and broadcasts *BLAC_Address* message when the new address is less than the old one. Otherwise, the packet is discarded. At last, this procedure converges and all LSRs will select the LSR with smallest IP address in the domain. The selected node is afterwards responsible to broadcast *BLAC_Address* messages periodically after $T_{Broadcast\_Address}$ to refresh the corresponding BFT entry in LSRs. Each LSR has a timer that is reset to $\beta * T_{Broadcast\_Address}$ when it receives *BLAC_Address* message from BLAC node. If the timer is expired, the LSR will broadcast a *BLAC_Address* message to candidate himself as BLAC. Values of β and $T_{Broadcast\_Address}$ are determined from simulation experiments. They are affected by broadcast reliability and fault tolerance. The *BLAC_Address* message format is as follows:

	Shim Header		Source Address	Destination Address	Payload
L2 header	X	S=1	IP$_{BLAC}$	IP$_{Broadcast}$	BLAC_Address

When a LSR receives this message, it finds that it is a BLAC address advertisement message (S=1 denotes the last label in label stack [16]). Therefore, it extracts BLAC address from it and adds the following entry to its BFT:

Label-in	Input interface	IP Address
X	Routing[IP$_{BLAC}$]	IP$_{BLAC}$

Routing[IP_{BLAC}] returns the output interface which is used to reach BLAC node from this LSR using IP lookup. As you see, there is no output interface field in the BFT. Output interfaces can be calculated as *All Links − Input_interface*. Output label for a packet is the same as input label and so is not included in the table as well. We present the pseudo-code for a LSR when it receives a broadcast packet from BLAC in algorithm *LSR-BLAC-Data* in the next page.

On reception of a new *Broadcast_Label_Assignment* which satisfies the code, the following entry will be added to BFT:

Label-in	Input interface	IP Address
Label$_B$	Routing[IP$_S$]	IP$_S$

We finish this section by an example of the broadcast mechanism as shown in figure 1. In figure 1a, S_1 and S_2 want to broadcast some packets into the

Algorithm *LSR-BLAC-Data*
(* BLAC broadcast data processing in LSRs *)
1. **if** *top label is X*
2.   **then if** $BFT(X)$ does not exist and $S = 1$
3.   **then** $iif_{RPF} \leftarrow Routing[IP_{BLAC}]$;
4.   **else if** $BFT(X)$ does not exist and $S = 0$
5.   **then** Discard the packet; Return;
6.   **else** $iif_{RPF} \leftarrow BFT(X).input\_interface$;
7.   **if** $iif \neq iif_{RPF}$
8.   **then** Discard the packet;
9.   **else if** $S = 0$
10.   **then** Pop the top label;
11.   **if** payload is *Broadcast_Label_Assignment*
12.   **then** Insert $(Label_B, Routing[IP_S], IP_S)$ in BFT;
13.   **else if** payload is *Broadcast_Label_Release*
14.   **then** Delete $(Label_B, Routing[IP_S], IP_S)$
                                from BFT;
15.   **else** Discard the packet; Return;
16.   Push label X at top of stack in the packet;
17.   Forward the packet to $AllLinks - \{iif\}$;
18.   **else if** payload is *BLAC_Address*
19.   **then** Insert $(X, iif_{RPF}, IP_{BLAC})$ in BFT;
20.   Forward the packet to $AllLinks - \{iif\}$;
21.   **else** Discard the packet;

network. Both nodes must first request a broadcast label from BLAC by sending a *Broadcast_Label_Request* message. BLAC assigns L_1 and L_2 to S_1 and S_2 respectively and broadcasts two *Broadcast_Label_Assignment* messages as depicted in figure 1b. Now, consider LSR_9. It only accepts broadcast messages from its direct link to BLAC and rejects broadcast messages from LSR_3 due to RPF mechanism. Since the shown network is considered symmetric, the distances between two nodes in two different directions are the same. This is not the case in real networks and most paths are asymmetric [18]. Therefore the RPF interface must be determined in reverse direction in their graphs. Bidirectional arrows in the figure mean that broadcast message is sent in both directions of the link. Upon receiving *Broadcast_Label_Assignment* by S_1 and S_2, they will broadcast their data using labels L_1 and L_2 accordingly. They will issue *Broadcast_Label_Release* messages once they are done with broadcasting (the same

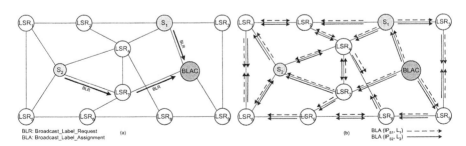

Fig. 1. Two steps of the broadcast protocol (a) broadcast label request by sources (b) label assignment by BLAC

as fig. 1a). *BLAC* will broadcast *Broadcast_Label_Release* messages to other nodes in response (the same as fig. 1b).

BLAC can save network bandwidth and processing power by accommodating *Broadcast_Label_Assignment* and *Broadcast_Label_Release* messages in a single message when broadcasting instead of sending separate reply messages for separate requests. It waits for $T_{Delayed\_Response}$ before responding. If other requests are received, it combines the reply messages.

3   Dense-Mode Multicast Protocol

We use our broadcast mechanism to implement a dense-mode multicast protocol in MPLS. BLAC must assign a broadcast label to each (S, G) pair. Hence, all messages must be changed such that to contain both source and multicast group addresses. When a source wants to send data to group G, it first, sends a modified *Broadcast_Label_Request* message toward BLAC. BLAC responds by a modified *Broadcast_Label_Assignment* message. Upon receiving the message, each LSR adds the following entry to its MFT (Multicast Forwarding Table):

Source Address	Group Address	Label-in	Input interface	Pruned entry
IP_S	IP_G	$Label_B$	$Routing[IP_S]$	NULL

In MFT, *Source Address* and *Group Address* fields together specify a multicast tree. We can consider them as a FEC identifying a multicast tree. Meaning of *Label-in* and *Input interface* fields is as before. *Pruned entry* field is a pointer to PIT (Pruned Interfaces Table) where the first entry of pruned interfaces for the tree is located. Interface index of pruned interfaces and status of their timers are maintained in this table. In this way, we avoid variable length entry in MFT. Therefore, MFT can be efficiently implemented in proper hardware like CAM. PIT can be implemented in RAM. A simple linked list structure can be used for it. An example of MFT and PIT relationship is presented in figure 2.

In dense mode multicast groups, number of pruned interfaces of each LSR is zero or a small fraction of total interfaces of a LSR. Therefore, maintaining state of pruned interfaces is more cost-effective than keeping state of tree links. Hence, the main advantage of PIT is its small memory requirements.

Next, we focus on steps needed for constructing PIT. Once source S acquires a broadcast label from BLAC, it adds it to its MFT. Rather than using reserved label Y for broadcasting, it uses another reserved label Z for broadcasting its multicast data. By this means, all other LSRs know that this packet is a multicast data packet. Also the multicast group address must be placed in *Destination Address* field in data packet. All other steps for processing these broadcast packets are like previously described procedures in broadcast algorithms except that this time we use MFT and PIT instead of BFT. The outgoing link set for a data packet is *AllLinks − {iif} − AllPrunedLinks* in this case.

To avoid the extra label Z, we can do the same thing as the broadcast mechanism. We use the value in layer 2 headers when available to differentiate a multicast MPLS packet from unicast and broadcast. This value already exists in PPP

(Protocol field type 0283 hex for MPLS multicast), Ethernet and IEEE 802.3 (ethertype value 8848 hex for MPLS multicast) [16][20]. We propose to define specific values for MPLS multicast in other layer 2 technologies as well where not available.

When a leaf LSR receives multicast packet of source S and it has no member for that group, it sends a *Prune* message toward S. Each LSR that receives this message from one of its interfaces, it examines PIT for that interface. If there exists such interface, resets associated timer. Otherwise, it is added to the PIT. If all interfaces of a LSR are in prune state, it forwards received *Prune* message toward S. In some occasions, a LSR sends a *Graft* message toward S, to restore pruned link. When a LSR receives such message from one of its interfaces, it deletes that interface from PIT.

There are some other issues regarding our solutions in this paper. We propose label values 4-6 for the three reserved labels X, Y and Z in our mechanisms when there is not possible to have layer 2 supports [16]. We have separated unicast, multicast and broadcast label spaces from each other. Hence, if a label is consumed in one of them, it has no impact on the others. This means that we can support $2^{20} - 16$ sources of broadcast data and (S, G) multicast trees in an MPLS domain.

If BLAC crashes, although existing multicast groups remain unaffected, no new broadcast label can be assigned or released. Therefore, BLAC is a single point of failure and it must be fail-proof. It also can crash when it is overloaded by label requests. We envisage heavily loaded BLACs to be implemented in the form of a computing cluster connected by a fault-tolerant and load-balancing middleware infrastructure. Other protocols like ones in [12] and [17] use the same approach.

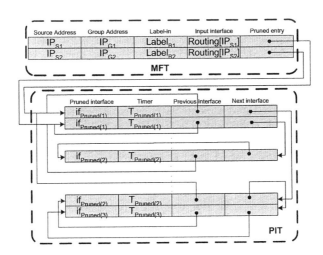

Fig. 2. Linked list structure for PIT

4   Related Works

The first operational IP multicast shortcut over ATM prototype for label switching IP multicast consisted of a Unix workstation and an ATM switch [21]. The proposal needs mixed L2/L3 forwarding. It suggests to set up LSPs based on changes to the Multicast Forwarding Cache (MFC), which can be categorized as a traffic driven trigger [6]. Ooms et. al. in [6] and RFC3353 [7] present detailed framework for multicast support in MPLS.

Protocols such as PIM-SM [2] and CBT [9] have explicit *Join* messages which could carry the label mappings. This approach, called piggy-backing method, is described in [4]. Protocol messages must be changed properly in favor of MPLS. Implementation of their approach in case of dense-mode protocols like PIM-DM and DVMRP is inefficient since these protocols use no explicit messages for piggy-backing labels on them. The pros and cons of piggy-backing labels on multicast routing messages are described in [6][7].

[5] suggests that labels be assigned on a per-flow (source, group) basis in a traffic-driven fashion. The proposal has some disadvantages compared to ours. First, our proposal outperforms their work regarding MFT size, since they store (interface, label) pairs for each multicast tree branch at the LSR. The majority of interfaces in most LSRs are tree branches in dense-mode groups and this leads to larger MFT size in their approach. Second, they use variable length entries in their MFT. This makes hardware implementations more difficult. Third, label binding and distribution is done at each LSR which introduces extra delay in tree construction. The label binding in our method is done at BLAC and labels are distributed to all LSRs. When the binding reaches the source, it broadcasts its traffic according to new label binding, while other LSRs may still be waiting for this binding. Forth, we consume fewer labels when the label pool is common between interfaces in a LSR.

A traffic-driven label distribution method is introduced in [14] and a dense-mode multicast routing protocol is proposed. They have the same weaknesses as [5] compared to our proposal. The MFT size in their method is even larger than [5] since they replicate common information in their table entries to achieve fixed length entries. Our proposal consumes more bandwidth compared to them. This can be avoided by using layer 2 support, delayed responses, expending more processing power in LSRs and using RPB as explained previously.

To make multicast traffic suitable for aggregation, the approach in [13] converts p2mp LSP setup to multiple p2p LSP problems. The protocol assumes multicast members are present only at edge routers which is not valid in general case. Their method results in a bad usage of network resources when the groups are dense. The scheme also prevents end-to-end label switching of data and disturbs the unicast traffic due to layer 3 operations needed at LERs.

5   Conclusion and Future Works

We propose a protocol for MPLS broadcast for the first time. This mechanism can be used in all applications and protocols requiring broadcast. A central node called BLAC is responsible for broadcast label assignment and release. Proposed mechanism has no overhead on unicast label space. We have shown its use in dense mode multicast support in MPLS. Our multicast protocol has many nice properties in comparison with other proposals available for dense mode groups. It consumes only one label for each multicast tree from multicast label space and no label from unicast label space. Also it has smaller multicast forwarding tables by storing an entry for each non-tree link instead of each tree link. Therefore, it consumes LSR memory conservatively. It poses no delay on multicast data distribution except for the label assignment phase in contrast with mechanisms existing so far. Proposed multicast protocol is suitable for intra-domain use, and its extension beyond a MPLS domain through hierarchy of domains is subject to further study. Another thing that needs further study is the impact of link failure on the operation of proposed mechanisms and the way we treat that.

References

[1] Deering, S. E., Cheriton, D. R.: Multicast Routing in Datagram Internetworks and Extended LANs, ACM Transactions on Computer Systems, Vol.8, No.2 (1990)
[2] Deering, S., et al.: The PIM Architecture for Wide-Area Multicast Routing, IEEE/ACM Transactions on Networking, Vol.4, No.2 (1996)
[3] Almeroth, K. C.: The Evolution of Mrulticast: From the MBone to Inter-Domain Multicast to Internet2 Deployment", IEEE Network (2000)
[4] Farinacci, D.,et al.: Using PIM to Distribute MPLS Labels for Multicast Routes, IETF Draft, draft-faranacci-mpls-multicast-03.txt, Nov. 2000
[5] Acharya, A., et al.: IP Multicast Support in MPLS, IEEE Proc. on ATM Workshop (1999)
[6] Ooms, D., Livens, W.: IP Multicast in MPLS Networks, Proc. of the IEEE Conf. on High Performance Switching and Routing (2000)
[7] Ooms, D., et al.: Overview of IP Multicast in a Multi-Protocol Label Switching (MPLS) Environment, RFC 3353 (2002)
[8] Waitzman, D., et al.: Distance Vector Multicast Routing Protocol, RFC 1075 (1988)
[9] Ballardie, A.: Core Based Trees (CBT Version 2) Multicast Routing - Protocol Specification, RFC 2189 (1997)
[10] Rosen, E., et al.: Multiprotocol Label Switching Architecture, RFC 3031 (2001)
[11] Liu, G., Lin, X.: MPLS Performance Evaluation in Backbone Network, IEEE ICC (2002), Vol. 2, 1179-1183
[12] Boudani, A., Cousin, B.: A New Approach to Construct Multicast Trees in MPLS Networks, ISCC (2002) 913-919
[13] Yang, B.,Mohapatra, P.: Edge Router Multicasting with MPLS Traffic Engineering, IEEE ICON (2002)
[14] Zhang, Z., et al.: The New Mechanism for MPLS Supporting IP Multicast, APC-CAS (2000).
[15] Dalal, Y. K., Metcalfe, R. M.: Reverse Path Forwarding of Broadcast Packets, Communications of the ACM, Vol. 21, No. 12 (1978)
[16] Rosen, E., et al.: MPLS Label Stack Encoding, RFC 3032 (2001)
[17] Keshav, S., Paul, S.: Centralized Multicast, Proc. of IEEE ICNP (1999)
[18] Paxson, V.: End-to-End Routing Behavior in the Internet, SIGCOMM'96

[19] Peterson, L. L., Davie, B. S.: Computer Networks: A Systems Approach, Morgan Kaufmann Pub., 2nd Ed. (2000)

[20] Protocol Numbers and Assignment Services, `http://www.iana.org/numbers.html`

[21] Dumortier, P., et al.: IP Multicast Shortcut over ATM: A Winner Combination", IEEE Globecom'98

Error Recovery Algorithm
for Multimedia Stream in the Tree-Based
Multicast Environments

Kiyoung Kim, Miyoun Yoon, and Yongtae Shin

Department of Computer Science, Soongsil University
{ganet89,myyoon,shin}@cherry.ssu.ac.kr

Abstract. This paper describes Tree-based Multicast System which leverages Preceding Error Recovery(PER) algorithm and Media Controller(MC). This mechanism solves jitter and delay problem by reducing control packets over tree-based reliable multicast. Our proposed PER results in suppression of control packets over the network and fast recovery of a buffer space from underflow status by considering constraints of dependency between the frames and the data being played within proper time. Proposed algorithm is operated at improved TMTP(Tree-based Multicast Transport Protocol) and located in each receiver. We also evaluate and simulate the performance of PER. The results are shown that PER much superior to previous works in the point of buffer efficiency and the number of occurred control packet.

1   Introduction

Most of reliable multicast studies has not been considered multimedia stream, so these do not satisfy QoS(Quality of Service) although reliability and QoS should be satisfied in multimedia data communication. Furthermore, they are needed control packets as many as a number of receivers for guaranteeing reliability and QoS. In other words, a number of control packets increase as receivers increase, so that they are required to provide scalability with reliability in multicast environment.

RMT[11] has been researched into sender-initiated, receiver-initiated and local recovery mechanism. In receive-initiated, receiver sends NACK to sender for only lost packets. If this method is based on three structure, it supports not only reliability but also scalability[4, 5, 7]. Local recovery mechanism configures a group to physically close members. If most of members in the group dose not receive a packet, the corresponding same control packets are suppressed through a timer[3].

These technologies have limits not to consider time, spatial and timing condition for multimedia compression algorithm such as H.261[8] and H.263[9]. If the multimedia data is delivered continuously, it causes control packets' explosion and request of data without satisfaction of timing property, so suitable QoS can not be obtained. When we request retransmission of a lost packet, we should

H.-K. Kahng (Ed.): ICOIN 2003, LNCS 2662, pp. 243–252, 2003.

make account of feature of multimedia data; also we have necessity for additive different structure to recognize characteristics of one in routing.

We propose PER algorithm which recoveries average error by considering peculiarity of multimedia and MC. PER is located at end node, and is run over tree-based reliable multicast structure including MC. Retransmission request is delivered to MC depending on PER in multicast tree. The MC is responsible for determining error recovery for its members so that explosion of control packets is prevented, and minimum QoS is guaranteed; furthermore, we can recover quickly network resources.

In section 2, we briefly discuss multicast related works. In section 3, we propose MC and its buffer architecture and PER located in end node. In section 4, we evaluate performance of the proposed the algorithm of PER and compare to existing related studies, and conclusion and future works are illustrated in section 5.

2   Related Works

2.1   RMT (Reliable Multicast Transport)

RMT has been researched into sender-initiated and receiver-initiated. Sender-initiated protocol has each of ACK-list for a receiver. If a packet loss is occurred, the receiver recovers an error relying on selective repeat. If not, the receiver transports ACK to a sender[5]; then, a sender updates the receiver's ACK-list. The sender uses a timer for detection of ACK loss. When the sender has not possessed all ACK until expired time, it regards as packet loss and resets the timer again, and it retransmits the lost packet. This mechanism can bring about ACK explosion for large group.

In receiver-initiated protocol, a receiver is responsible for packet transportation reliability. Source node simply delivers data until it receives NACK from a receiver. When source node receives NACK, it retransmits the requested data. The source makes use of a timer to retransmit lost packets. A receiver detects loss of packet by verifying a sequence number of received packet. If a sequence number of received packet is larger than expected one, receiver can packet detect packet loss easily. Sender should forward its own status information periodically to prevent originating NACK caused timer expiration of receivers if source does not have data to send. This mechanism reduces a number of control packets, but NACK might be exploded to sender if most of receivers lost data.

One solution for sender-initiated scheme is RMTP(Reliable Multicast Transfer Protocol)[11], but many receivers causes ACK centralized to sender. SRM is receiver-initiated protocol; it has difficulty in setting timer considered dynamic network environment.

2.2   Scalable Multicast

TMTP(Tree-based Multicast Transport Protocol)[7] has DM(Domain Manager) which is selected from members. Sender multicasts data, and DM is delegated retransmission for NACK locally. DM guarantees reliability using ACK from its members. When DM can not retransmit data, it delivers NACK to its parent's DM hierarchically, so this protocol is able to prevent NACK explosion.

Local group mechanism supports scalability together with reliability. One group composed of its neighbors physically delivers one NACK to sender. When a member loses a packet in the group, it does not forward to NACK but monitors its network interface before delivery of NACK. If the same NACK existed, the NACK is cancelled; thus, this solution suppresses NACK messages and does not affect group size but NACK messages increments in proportion as a number of groups, so that network capacity becomes higher. Also DM accomplishes retransmission of packet that is requested using NACK.

Fig 1 shows local group error recovery and control packet's flow. When a receive node loses a packet in a group A, it forward NACK to its delegation node. If the delegation has the packet, it forwards the packet to its member. When receive nodes detect error in each group A and B, each node requests retransmission to each delegation. The delegation A, B requests retransmission to their parent P. If the delegation P could not retransmit data, it sends retransmission request to a sender lastly; therefore, a number of control packets are decreased, so NACK explosion can be prevented at sender.

When this mechanism is applied to multimedia data transmission, it might support scalability. However, when the data is transmitted continuously and burst error is occurred, it makes unnecessary control packets without timing characteristics of the data. I.e. the control packet is not meaningless in real-time communication. Our proposed mechanism solves these problems relying on MC and PER.

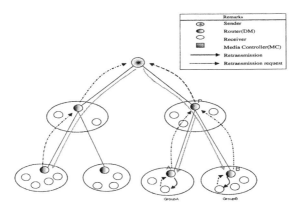

Fig. 1. Error Recovery using Local Group Mechanism and Control Packet's Flow

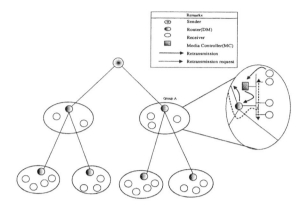

Fig. 2. Proposed MC Architecture in TMTP

3   Media Control System

We propose PER and MC(Media Controller) to support QoS and scalability together with reliability. MC has responsibility for determination of PER's execution by checking of NACK in TMTP. MC is paired off a DM and is managed by DM. Members belongs to an MC are in broadcast network environment like Fig 2.

When MC receives NACK from its member, it checks its own buffer whether it has the corresponding packet. MC performs retransmission of the packets relying on local recovery algorithm if it has the packet. If not, MC determines whether it request to sender after identifying the receiver's buffer status. Thus, MC transmits control packet to DM, only if packet that does not exist in its buffer is requested.

3.1   Error Recovery Algorithm for MC

If MC receives a NACK message, it simply retransmits the data to receivers in its own buffer. This mechanism is similar to existing local recovery algorithm. If MC does not have the packet in the buffer, it performs Primary algorithm 1. If the flag in NACK message indicates argent, it means that frames to play are less than threshold value of a buffer for retransmission time. Then MC executes forward error recovery which request next GOP(Group Of Picture) using its buffer's information. The threshold value is minimum data size. Voice or video data is very sensitive to time; therefore, revocation of an expired packet is one of ways to guarantee QoS.

Error Recovery Algorithm for MC

```
01: While(not fault){
02:  wait until receive NACK from mi;
03:   if(true==seek queue by sequence number)
04:    retransmit packeti correspond to NACK;
05:  else
06:  {
07:    retrieve and check the information on frame type in NACK;
08:    if(Flag==PER)
09:    {
10:      if(I-frame)
11:        request next First of GOP;
12:    }
13:    else
14:      request retranmsmit to sender;
15:  }
16: }
```

When MC received NACK and a requesting data is existed in its buffer, it retransmits the data(4). If MC did not possess the data in its buffer, it means that all members in a group do not receive the data. In this case, MC checks out the flag in NACK information. If the flag is set urgent(8-11), MC discards NACK and requests next GOP to sender additionally. The sender performs retransmission of the data correspondent lost packets to the group using multicast. Imminent request(=setting flag urgent) is determined by each receiver's buffer status independently.

3.2   Buffer Structure for MC

If MC receives NACK message, it does not rely to a sender but retransmits the data(frame) in its buffer. When the frame does not exist in the buffer, the MC applies PER mechanism like a line 8. If flag is not PER, the MC requests retransmission of data to its parent MC hierarchically. It is similar to role of DM at original TMTP. In worst case, the sender receives retransmission request.

Therefore, MC stores each GOP to prepare for retransmission request from a member in order to support hierarchical error recovery like Fig 3. Because GOP is the smallest unit to consist a picture in video, this scheme is a good method for H.263.

Stored GOP is deleted from a buffer to reallocate buffer when a timer is expired. The variable of timer may be estimated by calculating RTT between child MCs by the parent's MC. This scheme improve local recovery for multimedia data in both non-PER and PER mechanism, further it reduces retransmission time of I-frame and a number of control messages.

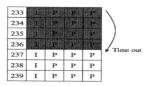

Fig. 3. Buffer Architecture of MC

3.3   Preceding Error Recovery Algorithm for Group Member

Receiver in a group plays media data using streaming method. Although this method is commonly used for voice or video communication, it does not guarantee QoS if burst packet loss is occurred. Also, if we lost I-frame, even though we received P-frame correctly, we might meet drop of QoS. This is commonly referred to error propagation.

A key feature of the PER is a use of error propagation characteristics. Buffer can be quickly recovered from underflow caused by a burst packet loss. Each receipt node saves locating information of a playing frame to B. When receipt node detects a packet loss, it determines whether PER is performed through checking B information.

The receipt node with packet loss checks out B variable. If the value is less than threshold, a flag of NACK sets urgent and retransmission request is delivered to MC. Fig 4 illustrates a formalized architecture resulted from execution of PER.

Let $P(i)$ be a error probability of a receive node, D_T be a threshold value and δ be maximum buffer playing time. We can determine B at time t like (1). If B does not satisfy (2) at time t, receipt node performs Forward Error Recovery. δ is a variable, so we estimate threshold at initial communication.

$$B = \{buff_{size} + \sum_{i=1}^{t} \lfloor \varepsilon(1 - P(i)) - \tau \rfloor\} \tag{1}$$

$$D_t \leq B \times \frac{\delta}{buff_{size}} \tag{2}$$

Proposed system results in suppression of control packets, and fast recovery of a buffer with underflow status as increasing value of B, so it is enable to guarantee QoS. A receipt node delivers a NACK message contained a flag, which is set by a result from executing PER, at MC. The MC carries out retransmission of the packet if there are request packet GOP_n existed. If not, it requests first I-frame of GOP_{n+1} to its parent MC; thus, the MC is responsible for determining and executing final action of PER. Arriving the first packet of the GOP_{n+1}, recipient processes it correctly, because it sets the expected sequence number when PER is executed.

PER Algorithm

```
While(on Session){
  wait until receive error;
  if(decision threshold <= current buffer pointer)
    do usual error recovery;
    continue;
  else
    do PER;
    next sequence number:=sequence of first frame in the next
    GOP;
    continue;
}
```

4 Performance Evaluation of PER

4.1 Environment

We evaluate and analyze scalability of PER and bandwidth applicability depending on occurrence frequency of control packets. Because network error and delay brings out control packets, low control packets is suitable in real-time communication. We assume experimental network as follows:

MC_G defines MCs set - $MC_G = \{MC_1, MC_2, \ldots\ldots, MC_i\}$ for $0 < i < n$, and n is a number of MCs. MCs are composed of complete tree with level k. Source node is root of the tree. We can obtain MC_i's depth as formula (3), and a number of MC also is defined as formula (4) with root MC_i. Also, MR_G is a MCs' set composed to same distance with sender $-MR_G = \{2^{Level(i)} - 1, 2^{Level(i)}, \ldots, 2^{Level(i)} - 1\}$

$$Level(i) = \lfloor \log_2 i \rfloor + 1 \qquad (3)$$

$$Sub\_MC(i) = 2^{k-Level(MC_i)} \qquad (4)$$

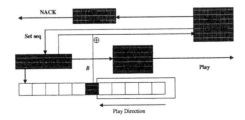

Fig. 4. Error Recovery of Receive Node

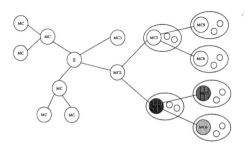

Fig. 5. Experimental Topology

We assume topology like Fig 5 and analyze control packet's size. Here, control packet is made of *MC3*. We estimate packet size relying on depth's change of *MC4*'s sub-tree. k is tree's depth.

4.2   Performance Analysis

The average control packet size T_s is (4) at MC_i.

$$T_s = Sub\_MC(i) \sum_{e=1}^{m} \{ \binom{m}{e} (1 - P_1)^{m-e} P_1^e E(i) \} \tag{5}$$

k is depth of subtree, and m is member's size in subtree. P_1 is an error occurrence probability.

$E(i)$ is an expectation of control packet size when PER is applied. F is size of GOP, and i is a number of I-frame composing GOP. When packet loss is occurred at receive node, P_2 is an I-frame loss probability of the GOP.

$$E(i) = \sum_{i=1}^{F} \binom{F}{i} (1 - p_2)^{F-i} p_2^i P_{size} \tag{6}$$

LGC is total average control packet's size at local group mechanism.

$$LGC = (2^k - 1) \sum_{e=1}^{m} \binom{m}{e} (1 - p_1)^{m-e} p_1^e P_{size} \tag{7}$$

Fig 6 and Fig 7 illustrate comparison proposed scheme with existing local error mechanism.

When P_1 is 0.01, P_2=0.6, and GOP's size is 100, the average control packet size is reduced by comparison with existing studies. The proposed mechanism tries error recovery depending on not retransmission request but considering frame format and buffer's status. Although packet loss is higher control packet's size is larger rapidly, the proposed mechanism can be used for a large number of groups because a number of control packets is maintained if a number of receive node is more than 16, so that network bandwidth can be used efficiently.

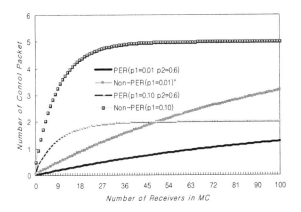

Fig. 6. Comparison of Control packet size by each error rate

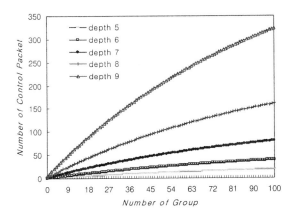

Fig. 7. Comparison of Control packet size by MC level

When P_1 is 0.01, the average control packet size is much lower as increasing receive node's size in a group.

Fig 7 shows packet loss rate by physical distance between a receive node and source node. A depth of tree composed of MC is deeper, control packet size is larger. The performance is the best at $k \leq 8$. When $k \leq 8$ and receiver node's size is 30, a receive node can process control packet size 0.1; thus, we are able to support scalability. We verify that GOP size is larger, control packet size is also increased because I, P-frame is determined by compression rate. GOP's composition is configured as considering group size and level in initial communication at sender, so that communication utilization becomes much higher.

5   Conclusion and Future Works

We presented a PER mechanism, which reduces control packets by considering characteristic of data and showed detailed algorithm and its procedure at receive node. As reducing a number of control packets, PER is more scalable and reliability. Furthermore, in contrast with the existing mechanism, it improves QoS through reducing a number of unnecessary control packets.

The performance showed that the number of control packets increases gradually in proportion to rise in the number of group. Thus this solution can be applied to real-time broadcast, videoconference and so on. To improve recovery of data requested by PER, we need research of a buffer's structure to periodically refresh data in buffer of MC, and study for managing session with considering receiver's buffer status.

References

[1] Maufer, T. A.: eploying IP Multicast in the Enterprise, 1st Ed., Prentice-Hall Inc., pp. 102-144, New Jersey, (1998)

[2] Ballardie, T., Francis, P., and Crowcroft, J.: "Core Based Trees(CBT)," SIG-COMM '93, pp. 85-95, September (1993)

[3] S. Deering, "Host Extensons for IP Multicasting," *Internet RFC 112*, Aug. (1989)

[4] S. Paul, K. K. Sabnani, J. C. Lin, S. Bhattacharyya, "Reliable Multicast Transport Protocol (RMTP),". IEEE Journal on Selected Areas in Communications, Vol. 15 No. 3, April (1997) 407-421

[5] B. Levine and J. J. Garcia-Luna-Aceves, "A comparison of known classes of reliable multicast protocols,," Proc. Conference on Network Protocols (ICNP-96), Columbus, Ohio, 14 IEEE TRANSACTIONS ON NETWORKING Oct.(1996)

[6] M. Hofmann, "A Generic Concept for Large-Scale Multicast", B. Plattner, Ed., Proc. International Zuerich Seminar, volume 1044 of LNCS, pp. 95–106, Springer Verlag, February (1996)

[7] R. Yavatkar, J. Griffioen, and M. Sudan, "A reliable dissemination protocol for interactive collaborative applications," in Proc. ACM Multimedia (1995) 333–44

[8] ITU-T Recommendation H.261, Video codec for audiovisual services at p× 64kbits. December 1990, March (1993)(revised)

[9] ITU-T Recommendation H.263, Video coding for low bitrate communication. (1995)

[10] B. Whetten, T. Montgomery, S. Kaplan: A High Performance Totally Ordered Multicast Protocol. Submitted to INFOCOM'95, April (1995)

[11] J. C. Lin and S. Paul, "RMTP: A reliable multicast transport protocol," in Proc. IEEE Infocom, pp. 1414–1425, March (1996)

[12] Jacobson, V., et al. A Reliable Multicast Framework for Lightweight Sessions and Application-Level Framing. in ACM SIGComm 95. (1995)

Study on Merge of Overlapped TCP Traffic Using Reliable Multicast Transport

Yoshikazu Watanabe[1], Koji Okamura[2], and Keijiro Araki[1]

[1] Graduate School of Information Science and Communication Engineering
Kyushu University
[2] Computing and Communications Center, Kyushu University

Abstract. The amount of data transmitted on the Internet has been increasing every year as the use of the Internet spreads, and situations that multiple streams simultaneously carry the same data often happen. For example, when multiple users access to a web site at once, their streams contain the same data. If it is possible to merge such multiple overlapped streams carry the same data into a single stream, we can use the Internet more efficiently. In this paper, we propose architecture for merging the same data streams by using reliable multicast transport. Then we applied this architecture to FTP traffic and implemented FTP proxy which transfers data by reliable multicast transport between multiple points. Furthermore, we experimented in FTP proxy and showed that FTP proxy are able to save network bandwidth when overlapped traffic occur.

1 Introduction

As the Internet has been evolved, various services are provided over the Internet such as WWW, FTP, E-mail and so on. Especially WWW and FTP services are widely used as methods of information provision and data distribution. In recent years, Internet access has been improved dramatically by spread of new technologies such as ADSL, and users of the Internet have became to be able to retrieve large amount of information or data through the Internet more easily. It follows that traffic of the services on the Internet continues to increase, and overlapped traffic occur in higher probability. If multiple receivers attempt to retrieve the same data on a server at the same time, the server duplicates the data as many times as the number of receivers and sends the data to each receivers individually. If it is possible to eliminate such overlapped traffic, we can use the Internet more efficiently.

Overlapped traffic described above come from the fact that the services use TCP, i.e. unicast, which is one-to-one communication model. On the other hand, IP also supports multicast(IP multicast), which is one-to-many communication model. IP multicast packets sent from a sender to multiple receivers are duplicated at routers in the Internet only if duplications are needed to delivery the packets to all receivers. In this communication model, redundant traffic are eliminated, and it is possible to save bandwidth of networks when data are sent

H.-K. Kahng (Ed.): ICOIN 2003, LNCS 2662, pp. 253–262, 2003.

to multiple destinations. If WWW or FTP services could communicate over IP multicast, it would become possible to eliminate overlapped traffic and redundant traffic in the Internet. However, WWW and FTP are based on TCP and they are already in widespread use. Even if a new facility for IP multicast is added to them, it is impractical to modify existing many systems to use new protocols.

In this paper, we propose architecture for merging overlapped TCP traffic by replacing TCP data stream with IP multicast in the Internet. And we design FTP proxy as a instantiation of applying our architecture to FTP traffic. Moreover we mention the implementation of FTP proxy and an experiment we performed.

The remainder of this paper is organized as follows. Chapter 2 describes the goal of our research. Chapter 3 presents architecture for merge of TCP traffic. Chapter 4 shows FTP proxy as a instantiation of the proposed architecture. Chapter 5 describes implementation and experiment of FTP proxy and chapter 6 concludes this paper.

2   Goal of Our Research

We have two major goals in designing the architecture we propose in this paper. One is a transparency to users and the other is scalability. If some modifications to user applications are required when we deploy the architecture, it becomes difficult to apply the architecture to existing systems. Therefore, the architecture must be transparent to users and all processing must be done in networks. Scalability of the architecture is also important. The more widely the architecture is deployed in the Internet, the more overlapped traffic are able to be merged, and the architecture is able to make the Internet more efficient.

3   Architecture for Merge of Overlapped TCP Traffic

This chapter describes important components of architecture for merge of overlapped TCP traffic. These components are protocol translation server, session and reliable multicast protocol.

Protocol translation servers are the main component of our architecture. They transfer the data of overlapped TCP traffic using IP multicast between multiple points on networks. Protocol translation servers create a multicast session when they merge overlapped TCP traffic. Information of a created session is announced to all protocol translation servers in order to allow other protocol translation servers to join the session. When protocol translation servers send data of overlapped TCP traffic using IP multicast, they use reliable multicast protocol. Reliable multicast protocol reliability can guarantee reliable communications over IP multicast, even if a communication over IP multicast is unreliable. Each component is explained in following sections.

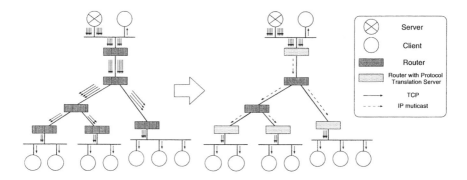

Fig. 1. Overview of replacing TCP streams with IP multicast by protocol translation servers

3.1 Protocol Translation Server

In our architecture, we locate multiple servers at various points on networks. They receive data by TCP and send the data by IP multicast and vice versa in order to realize merge of overlapped TCP traffic transparently for end-hosts. In this paper, we call such a server as protocol translation server. Protocol translation servers keep watch on TCP streams on networks. When they detect overlapped TCP streams such as multiple transfers of the same file, they start protocol translation. Protocol translation server communicates with end-hosts using TCP and transmits the data received from a end-host to other protocol translation servers using IP multicast. The judgment way of whether multiple streams carry the same data depends on the protocol of the higher layer.

Figure 1 shows overview of replacing TCP streams with IP multicast by protocol translation servers.

3.2 Session

In this section, we explain about a multicast session in our architecture.

Multicast Session When a protocol translation server merges overlapped TCP streams, it creates a multicast session to send data using IP multicast. Each multicast session uses a different transport address.

Session Description We need a common format to describe multicast sessions in order to share session information among protocol translation servers. Then we use SDP(Session Description Protocol)[1] for that purpose.

Text format for general real-time multimedia session description purposes is defined in SDP. Though SDP is originally aimed to describe multimedia sessions, which distribute audio or video streams, SDP provides a method of describing application specific attributes as well. We decided to use SDP because it has enough capability for describing multicast sessions we create.

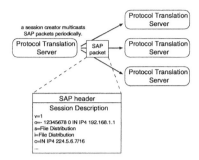

Fig. 2. Relationship between SDP and SAP

Session Announcement A protocol translation server which created a session has to announce the information of the session to other protocol translation servers. We use SAP(Session Announcement Protocol)[2] in order to distribute the session information described using SDP.

Figure 2 shows the relationship between SDP and SAP. A protocol translation server which created a session multicasts SAP packets periodically. A SAP packet consists of a SAP header and a payload. The payload is a session description described using SDP. An example of session descriptions is explained in section 4.5.

3.3 Reliable Multicast Protocol

IP multicast does not guarantee reliability in data delivery. It means that lost packets are not recovered, or receivers may receive duplicated packets, or the order packets arrive may be different from the order packets were sent. On the other hand, TCP provides reliability with communications over IP. Therefore, communications between protocol translation servers using IP multicast must be reliable in order to guarantee reliability of communications between end-hosts using TCP.

As a method of providing IP multicast with reliability, many reliable multicast protocols have been proposed such as AFDP[3], SRM[4] and RMTP[5]. These protocols provide reliability by implementing facilities of detection and recovery of lost packets and sequencing packets. However, methods of implementing such facilities are different per reliable multicast protocols, and they respectively have different characteristics of scalability and throughput. It is needed that the reliable multicast protocol used in our architecture has good scalability because the more extensively protocol translation servers are deployed, the more efficiently we can merge streams. Though there are many reliable multicast protocols as mentioned above, there is no single reliable multicast protocol which always achieves good throughput for various receiver sets.

Therefore, we use dynamic protocol selection architecture for reliable multicast[6, 7], which our laboratory are researching. In this architecture, the system dynamically selects an optimum reliable multicast protocol using information about the type of the application and the number of receivers in the multicast group. Then it sends data received from the application to multicast group members using the selected protocol. Applications can efficiently multicast with reliability under various circumstances by this architecture.

4 FTP Proxy

In this chapter, we propose FTP proxy as a instantiation of applying our architecture to FTP traffic.

4.1 Target Traffic of FTP Proxy

We suppose to overlapped merge download traffic of a file from an anonymous FTP server by FTP proxy. Upload traffic is outside of target of FTP proxy because we consider that there are few opportunity to merge upload traffic. In this paper, we assume that anonymous FTP servers allows users only to download files and upload of files to the servers is prohibited.

4.2 System Configuration

A system which merges download traffic consists of multiple FTP proxies. Each FTP proxy is located at various point on networks. A FTP proxy has the role of protocol translation server described in section 3.1. Though we described that merging of traffic must be completely transparent to users, we let protocol translation server pretend to be FTP server in this system to simplify the implementation. Namely, FTP proxy has two roles. The first is as a protocol translation server, and the second is as a FTP server. This is why we call protocol translation server as FTP proxy here.

FTP clients login to FTP proxy instead of a real FTP server and request files or file lists. FTP proxy itself does not have any files to provide FTP clients with in initial state. FTP proxy retrieves the file requested by a client from a real FTP server according to need, and at that time, it merges traffic using reliable multicast protocol if possible. When FTP proxy retrieves files or file lists from a real FTP server, it saves retrieved data in a local disk as a cache. The behavior of FTP proxy is described in detail later.

4.3 Components of FTP Proxy

A FTP proxy consists of five components. They are FTP server, file list retrieval server, file retrieval server, session management server and resend server. FTP server accepts login from FTP clients and provides them with FTP services. List retrieval server retrieves file lists from a real FTP server and manages retrieved file lists. File retrieval server retrieves files from a real FTP server and manages retrieved files. Session management server creates multicast sessions, announces them, and collects session information using SAP. Resend server accepts resend requests from other FTP proxies.

Figure 3 shows the system configuration and the relationship between each components of FTP proxy.

4.4   Behavior of FTP Proxy

In this section, we explain the behavior of FTP proxy in detail.

Basic Behavior FTP server listens on port 21 and provides clients with FTP services. When a client requests a file list by a LIST or STAT command, FTP server asks file list retrieval server for the requested file list. When a client requests a file by a RETR command, FTP server asks file retrieval server for the requested file. The behaviors of file list retrieval server and file retrieval server are described later in this section.

Session management server maintains two session lists. One is an active session list and the other is a completed session list. Session management server joins the well-known multicast address defined by SAP and collects session information which other FTP proxies are announcing. The collected session information are added to the active session list. When session management server receives a session deletion packet, it moves the ended session from the active session list to the completed session list.

Process for File List Request When a client requests a file list, FTP server asks file list retrieval server for the requested file list. The file list retrieval server checks through a local cache. If the requested file list exists in the cache, the file list retrieval server passes it to the FTP server. Otherwise, the file list retrieval server logins to a real FTP server and retrieves the requested file list. In this

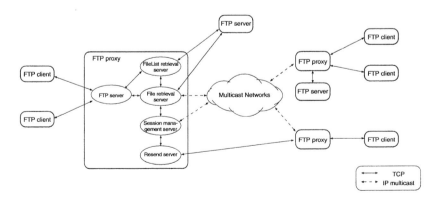

Fig. 3. Illustration of the system configuration and components of FTP proxy system

case, the file list retrieval server uses TCP to connect to the real FTP server. It means that traffic of the file list retrieval is not merged by IP multicast. In general, the size of a file list is not so large, and we consider that it is difficult to merge this kind of traffic. The file list retrieve server passes the retrieved file list to the FTP server and saves it in a local disk as a cache. The FTP server sends the received file list to the client.

Process for File Request When a client requests a file, FTP server asks file retrieval server for the requested file. The file retrieval server checks through a local cache. If the requested file exists in the cache, the file retrieval server passes it to the FTP server. Otherwise, the file retrieval server sends inquiries to session management server about whether there is an active session for the requested file. If there is a session for the file, the file retrieval server joins the session and receives the file using reliable multicast protocol. At that time, the beginning part of the file has been already sent, and the file retrieval server is not able to obtain them from the session. Therefore, the file retrieve server asks resend server running on the sender of the session to resend the missing data. The resend server sends the requested data using unicast or IP multicast. The resend server selects protocol in consideration of the size of the requested data. If the size is smaller than a threshold, the resend server chooses unicast, and otherwise it chooses IP multicast. We are going to investigate appropriate value for the threshold in process of employment of this system. The file retrieval server passes the file received from the session and the resend server to the FTP server.

If there is no active session for the requested file, the session management server looks up in the completed session list. If there was a session for the requested file, the file retrieval server asks resend server running on the sender of the completed session to send the file. The resend server asks session management server running on the same host to create a new session. Then, the resend server sends the requested file to the created session. The file retrieval server joins the created session and passes the received file to the FTP server.

If there is no session for the requested file even in past times, the file retrieval server logins to a real FTP server and retrieves the file. In parallel, the file retrieval server asks the session management server to create a new session in order to inform other FTP proxies that it is possible to merge traffic. The file retrieval server passes the received file to the FTP server.

The FTP server send the received file from the file retrieval server to the client. The File retrieval server saves the file in a local disk as a cache.

4.5   Session Information

As described in section 3.3, when a new session is created, session management server announces the session information described by SDP using SAP. The session description used in this system contains the following information.

- multicast address, port number and TTL used in the session
- the name of the file transfered in the session
- the size of the file transfered in the session

5   Implementation and Experiment

In this chapter, we describe implementation and experiment of FTP proxy.

5.1   Implementation

We implemented FTP proxy servers on Linux. Each server communicates with other servers using UNIX domain socket. As for implementation of FTP server, we utilized wu-ftpd 2.6.2, which is the FTP server widely used on the Internet to provide anonymous ftp service. Main modification applied to wu-ftpd is that codes to read file lists or files are replaced by ones to communicate with file list retrieval server or file retrieval server.

Though we described that our architecture use dynamic protocol selection architecture to multicast with reliability, the implementation of the architecture has not completed yet. Therefore, currently we have employed libsrm, which is an implementation of SRM by MASH project. Libsrm is provided in the form of C++ library and very easy to build into application softwares.

5.2   Experiment

Figure 4 shows the network for this experiment. We set up four FTP proxies in the experimental network. Each Linux PC is equipped with Pentium III 1GHz processor, 256 MB memory and several 100Base-T network interface cards. The original FTP server is located at outside of the experimental network.

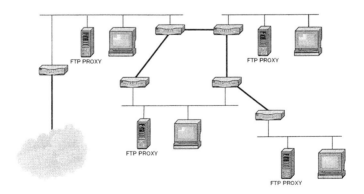

Fig. 4. Illustration of the experimental network

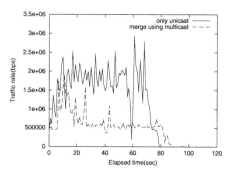 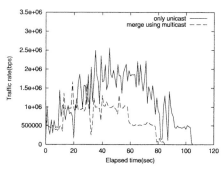

Fig. 5. Traffic rate of the backbone link with the heaviest traffic(interval : 3 seconds)

Fig. 6. Traffic rate of the backbone link with the heaviest traffic(interval : 10 seconds)

We measured traffic rate of backbone links in the experimental network when four clients download the same file from the respectively different FTP proxy. Backbone links are drawn thickly in figure 4. Whether overlapped traffic occur depends on how clients access. In this experiment, each client starts downloading one after another with three or ten seconds interval. Though clients access FTP proxies from outside of the experimental network, packets between clients and FTP proxies are not counted into the result. We also measured traffic rate of backbone links when FTP proxies don't merge overlapped traffic using multicast for comparison. At this moment, we are not able to multicast with libsrm at faster than 500 Kbps due to a problem on implementation. We limited the transfer speed between FTP proxies and clients to 500 Kbps to perform measurements under the same condition.

Figure 5 and figure 6 show traffic rate of the backbone link with the heaviest traffic when clients start downloading with three and ten interval respectively. In figure 5, the traffic rate is reduced to one fourth by merging overlapped traffic in many portion. However, the effect of merging overlapped traffic is not clear until twenty seconds elapsed. This is because FTP proxies resend missing part of files by unicast immediately after the second or later clients start downloading. On the other hand, in figure 6, the traffic rate decreases in fewer degree than in figure 5 when FTP proxies merge overlapped traffic. In this case, the time which clients' accesses overlap is shorter and FTP proxies have to resend more by unicast, then FTP proxies communicate using both multicast and unicast while transferring the file. This makes traffic rate higher. However, when FTP proxies merge overlapped traffic, data transfers finish more quickly than as in only unicast and they eventually save total network bandwidth.

Currently, FTP proxies always resend data using unicast. However, resend traffic may be possible to merge under some situations. For example, if two clients start downloading the same file at once when other client is already retrieving the file, FTP proxy must resend twice and the two resend traffic exactly overlap.

We are now considering to use multicast when FTP proxies resend relatively large amount of data.

6   Conclusion

Our research goal is the realization of efficient use of the Internet by merging overlapped TCP traffic into a single stream using reliable multicast transport. In this paper, we proposed architecture in which multiple protocol translation servers are located at various points on the Internet. When they detect overlapped TCP traffic, they create a multicast session and transfer the data through the created multicast session. We use SDP and SAP to share the information of created multicast sessions among all protocol translation servers. Moreover, they use reliable multicast protocol in order to guarantee reliability of the communication over IP multicast.

We also designed FTP proxy as an instantiation of our architecture for FTP traffic. FTP proxies accept users' login and retrieve files from a real FTP server on behalf of users. When multiple FTP proxies are requested the same file at the same time, they transfer the file using IP multicast between FTP proxies. We experimented in FTP proxy and we succeeded to save bandwidth of backbone links when overlapped traffic occur using FTP proxy.

The result of our experiment in this paper was measured under the condition that overlapped traffic certainly occur. We are currently considering to examine how much FTP proxy can save bandwidth when real ftp users access them.

References

[1] M. Handley and V. Jacobson, "SDP: Session Description Protocol",*IETF RFC2327*, April 1998.
[2] M. Handley and C. Perkins and E. Whelan, "Session Announcement Protocol", *IETF RFC2974*, October 2000.
[3] Jeremy R. Cooperstock and Steve Kotsopoulos,"Why Use a Fishing Line When You Have a Net? An Adaptive Multicast Data Distribution Protocol", *Usenix '96*, 1996.
[4] Sally Floyd and Van Jacobson and Ching-Gung Lin and Steven McCanne and Lixia Zhang, "Sessions and Application Level Framing", *IEEE/ACM Transactions on Networking*, November 1996.
[5] John C. Lin and Sanjoy Paul, "RMTP: A Reliable Multicast Transport Protocol", *IEEE INFOCOM'96*, pp. 1414-1424, March 1996.
[6] Kensuke Shibata and Koji Okamura and Keijiro Araki, "Evaluation of Reliable Multicast Transport with Protocol Selection Architectur", *IPSJ Journal Vol.42 No.12*, pp.3102-3111, Desember 2001.
[7] Kensuke Shibata and Koji Okamura and Keijiro Araki, "Design and Evaluation of Dynamic Protocol Selection Architecture for Reliable Multicast", *Proc. of Symposium on Applications and the Internet (SAINT) 2002*, pp. 262-269, January 2002.
[8] WU-FTPD Development Group, http://www.wu-ftpd.org/
[9] MASH Project, http://www.openmash.org/mash/

Multicast Datagram Delivery Using Xcast in IPv6 Networks

Myung-Ki Shin[1] and Sang-Ha Kim[2]

[1] ETRI,
161 Kajong-Dong, Yusong-Gu, Taejon 305-350, Korea
mkshin@pec.etri.re.kr
http://www.etri.re.kr
[2] ChungNam National Univ.,
220 Gung-Dong, Yusong-Gu, Taejon, 305-764, Korea
shkim@cclab.cnu.ac.kr
http://www.cnu.ac.kr

Abstract. This paper presents an alternative scheme, called Xcast6+, which is an extension of Explicit Multicast (Xcast) for an efficient delivery of multicast datagrams over IPv6 networks. The mechanism incorporates MLDv2 and a new control plane into existing Xcast6, and not only does it provide the transparency of traditional multicast schemes to sources and recipients, but it also enhances the routing efficiency in networks. Since intermediate routers do not have to maintain any multicast states, it results in a more efficient and scalable mechanism to deliver traditional multicast datagrams. Furthermore, the seamless integration of Xcast6+ in Mobile IPv6 can support multicast efficiently for mobile nodes (both sources and recipients) over IPv6 networks by avoiding tunnel avalanches and tunnel convergence. Our simulation results show distinct performance improvements of our approach. Our approach can reduce network resources in many "medium size groups" multicast, particularly as the number of recipients in a subnet increases (i.e., "subnet-dense group"). It also has more performance gain on mobile multicast environments.

1 Introduction

IP multicast, the ability to efficiently send datagrams to a group of destinations, is becoming increasingly important for applications such as IP telephony and video-conferencing. However, while traditional multicast schemes [1] are scalable in the sense that they can support very large groups, there are scalability issues when a network needs to support a very large number of distinct groups [2]. As alternative schemes for multicast, many researches, such as Explicit Multicast(Xcast) [3], Overlay Multicast [4, 5] etc. have been studied for a long time. However, these kinds of approaches are not replaced as standard multicast schemes due to the lack of a consensus. Furthermore, many people believe that IPv6 multicast could be the one of solutions, but this is an illusion

H.-K. Kahng (Ed.): ICOIN 2003, LNCS 2662, pp. 263–272, 2003.

and (in fact) there is no difference between IPv4 and IPv6 to solve the existing multicast scalability issues. This paper presents an alternative scheme for medium size and subnet-dense groups, called Xcast6+, which is an extension of Explicit Multicast (Xcast) for an efficient delivery of traditional multicast datagrams over IPv6 networks. The mechanism incorporates MLDv2(Multicast Listener Discovery version 2) and a new control plane into existing Xcast6(Xcast for IPv6), and not only does it provide the transparency of traditional multicast schemes to sources and recipients, but it also enhances the routing efficiency in networks. Since intermediate routers do not have to maintain any multicast states, it results in a more efficient and scalable mechanism to deliver multicast datagrams. Furthermore, the seamless integration of Xcast6+ in Mobile IPv6 can support multicast efficiently for mobile nodes (both sources and recipients) over IPv6 networks, by avoiding tunnel avalanches tunnel and convergence. Xcast6+ is not intended to replace the existing traditional multicast schemes. Moreover, it complements the existing multicast schemes.

2   Related Works

2.1   Multicast Schemes (ASM and SSM)

Traditional multicast was designed to offer scalable multipoint-to-multipoint delivery necessary for using group communication applications on the Internet. The ASM(Any-Source Multicast)[1] can minimize bandwidth consumption for many-to-many communication. However, while ASM is scalable in the sense that they can support very large groups, there are scalability issues, such as group management, routing complexity (i.e., multicast state management), multicast address allocation, etc. when a network needs to support a very large number of distinct groups. Therefore, the ASM service has been slow commercial deployment by ISPs and Carriers. The SSM(Source-Specific Multicast)[6] which is newly designed to alleviate some of the deployment problems (e.g., multicast address allocation) can minimize bandwidth consumption for one-to-many communication. But, even the SSM cannot alleviate all of inherent scalable issues of traditional multicast. That is, the SSM still creates state and signaling per multicast channel in each on-tree node. In the traditional multicast schemes, the scalability issues of IPv4 are similarly set in IPv6 too.

2.2   Xcast6 (Explicit Multicast for IPv6)

Explicit Multicast techniques had been researched in many contexts: SGM[7], CLM[8], and MDO6[9]. Xcast[3] has been newly proposed, as the common functionalities(both IPv4 and IPv6) had been agreed from these proposals above. It supports a very large number of small groups. The goal of Xcast is to insist on stateless multicast for small groups. In order to achieve this, the datagrams are sent containing a list of destination in Xcast header. Xcast6, which is Xcast scheme for IPv6 uses explicit encoding of a destination list in the IPv6 Routing

Extension header, instead of using multicast routing protocols. Xcast6 scheme has a number of advantages: it can save bandwidth between routers similar to traditional multicast schemes, even if the routers do not need to maintain multicast states. Furthermore, since there is no need to define an additional Xcast header in Xcast6 scheme (Xcast6 uses the existing Routing Extension header (type=Xcast) as Xcast6 header), Xcast6 has more benefits than Xcast4. However, it may suffer from a scalability problem as the number of recipients in a subnet increases, since it is applicable only to small groups. As well, the Xcast6 assumes all of destination list is already known to sources statically before sources start to send datagrams, so dynamic recipients/sources are not compatible with Xcast6 scheme.

2.3 Overlay Multicast

Currently, as another alternative approach, Overlay Multicast or self-configuring host-based distribution scheme has been proposed. End System Multicast[4] is to repartition all multicast related functionality including membership management and datagram replication between hosts(end systems) and routers. By shifting multicast support from routers to end systems, this alternative architecture has the potential to address most problems associated with traditional multicast schemes. Yoid[5], self-configuring host-based distribution scheme is a set of protocols for host-based content distribution. It allows a group of hosts that are receiving/sending content to dynamically self-organize themselves into a distribution topology tunneled over unicast and, where available, multicast. So the hosts can efficiently distribute content synchronously or asynchronously. However, while they are feasible to use these overlay approaches in the case of small and sparse groups, the performance penalty would be high as the number of recipients in a subnet increases.

2.4 Mobile Multicast

There is an increasing need to support user mobility in Today's computing environments. The incorporation of multicast routing support in a mobile computing environment presents several challenges. The Mobile IP proposes two approaches to support mobile multicast, i.e., home agent (HA) based routing using bidirectional tunneling and remote subscription[10]. In the HA-based routing approach, the mobile node receives multicast datagram by way of its HA using the unicast Mobile IP tunnels. While this approach handles both source mobility and recipient mobility by hiding node mobility from all other members of the group, the drawback is the routing path for multicast delivery which can be far from optimal. In addition, it causes the tunnel avalanches and tunnel convergence problems[11]. In remote subscription, (like new recipient) each mobile node can resubscribe to its desired multicast group when it enters a visit network. The main advantage of this approach is that the multicast datagrams are always delivered on the optimal path. However, this approach cannot handle source mobility. As well, the overhead is the cost of reconstructing the multicast tree. The default approach for Mobile IPv6 might be to use remote subscription.

3   Proposed Scheme : Xcast6+

3.1   Xcast6+Basics

Xcast6+ provides an enhanced scheme supporting the strengths of traditional multicast schemes in basic Xcast6. This is achieved by adding MLDv2 subscriptions at recipient's side. In Xcast6+, a recipient initiates MLDv2(S,G) subscription, where S is a source address and G is a multicast group address. When a designated router(DR) at recipient's side receives the request, it sends new Xcast+ controls containing S, G, and own address, DR, to the source using unicast (instead of using multicast routing protocols). These procedures imply the addition of new control plane for Xcast6. Thus, when the DR at the source's side receives the controls, it can keep track of the addresses of all DRs at the recipient's side involved in the multicast session (S,G) in its Xcast6+ cache table. When the source sends multicast datagrams, a DR at the source's side which receives the multicast datagrams, explicitly encodes the addresses of the DRs at the recipient's side in Xcast6 header, and sends the multicast datagram as Xcast6 datagram on unicast path. Each router along the way parses the header, partitions the destinations based on each destination's next hop, and forwards a datagram with an appropriate Xcast6 header to each of the next hops. These procedures comply with the data plane for existing Xcast6, except for encoding addresses of the DRs at the recipient's side in the datagram, instead of addresses of recipients. When the DRs at the recipient's side receive the Xcast6+ datagram, they send the datagrams as traditional multicast datagrams to recipients. For example, suppose that A, B, C, D, E, and F are trying to receive multicast datagrams distributed from S in Figure 1 below: This is accomplished as follows: A, B, C, D, E, and F initiate MLDv2(S,G) subscription. When R4, R8 and R9 receive the request, they send Xcast6+ RReq messages to the source. Meanwhile, when a R1 receives the message, it sends Xcast6+ RRep (it means the acknowledgement of the safe delivery of the RReq message) and does not forward this message to the source. All of these new controls messages will be described in Section 3.3. Therefore, R1 identifies a set of all DRs at the recipient's side <R4, R8 and R9> dynamically. Thus, when R1 receives multicast datagrams (S,G) from A, R1 sends Xcast6+ datagrams with the list of <R4, R8 and R9> in its datagrams to the next hop router, R2.

3.2   M2X (Multicast to Xcast) and X2M (Xcast to Multicast)

When Xcast6+ routers receive traditional multicast datagrams or Xcast6 datagrams, they needs to properly process the datagrams. The processing that routers do on receiving one of these datagrams is exactly the same to basic Xcast6 except for the following processing :

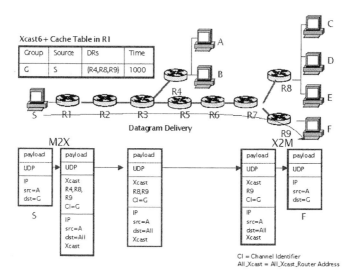

Fig. 1. Xcast6+ Datagram Delivery

X2M: When a router receives an Xcast6+ datagram, if the address of destination is equal to own address (i.e., the router is a DR at the recipient's side), the router sends the datagram as a traditional multicast datagram to the recipients. The channel identifier in Xcast6 header[3] is encoded as the multicast address in IPv6 basic header.

So, in the example above, R7 will send one copy of datagram to destination R8 with Xcast6+ list of <R8> and one copy of datagram to destination R9 with Xcast6+ list of <R9>. When R9 receives the datagram, it will, by the X2M, send the datagram as a traditional multicast datagram to the recipients <F>. In addition, a router receiving a traditional multicast datagram needs the following processing additionally :

M2X: When a DR receives traditional multicast datagrams, if there is an entry for the multicast session in the Xcast6+ cache table, it sends the datagrams as Xcast6+ datagrams including the addresses of DRs at the recipient's side to the next hops. The multicast address in IPv6 basic header is encoded as the channel identifier in Xcast6 header[3].

So, in the example above, when R1 received a multicast data from A, R1 will, by M2X, send the datagram as an Xcast6+ datagram to the R2.

3.3   Xcast6+ Controls

All the Xcast6+ control messages are implemented by four messages: Registration Request (RReq), Registration Reply (RRep), Withdrawal Request (WReq), and Withdrawal Reply (WRep). All of these messages are designed by the Router Alert option within the Hop-by-Hop option header. All these messages have the same format, and they are classified by the Option type fields. The Xcast6+ module in Xcast6+ routers (at least DRs) should identify these messages by checking the value of Router Alert Option, which must be assigned by IANA in the future.

4   Integration of Proposed Scheme in MIPv6

In this section, we discuss the seamless integration of Xcast6+ in Mobile IPv6 (X+MIPv6) which enables efficiently the delivery of multicast datagram on Mobile IPv6 networks. The main problems of MIPv6 multicast are the high cost of reconstructing multicast tree (i.e., datagram loss[10]) and source mobility support (i.e., delivery black-hole), where the default multicast technique in MIPv6 might be remote subscription. As well, if HA-based routing using bi-directional tunneling is supplementarily used in MIPv6 multicast, inherent tunnel avalanches and tunnel convergence problems should be avoided. Actually, multicast in MIPv6 has trade-off between the optimal delivery path and the frequency of the multicast tree reconfiguration. The goal of X+MIPv6 is to clear the way of obstacles when MIPv6 multicast is applied. In order to achieve this, we integrate Xcast6+ in MIPv6(X+MIPv6). X+MIPv6 aims are to maintain the optimal delivery path without multicast tree reconfiguration using stateless multicast between mobile routers. Furthermore, the source mobility is to be provided. The X+MIPv6 uses remote subscription scheme with minimum HA-based routing using Xcast6+. In X+MIPv6, a message called Registration Update(RU), is newly defined, like Binding Update(BU) in MIPv6. The goal of RU is to update Xcast6+ cache table in HA. Consider the example shown in Figure 2 (in case of the recipient mobility support). When mobile nodes(MNs) enter a visit network, RU is sent from the DR(R1) in visit networks to the HA. Therefore, when a HA receives Xcast6+ datagram which should be forwarded to its nodes, the HA reencodes the Xcast6+ datagram according to the updated Xcast6+ cache table and forward it to its MNs which are roaming in the visit network. At the same time, the MNs try to subscribe to its remote multicast routers(R1), so (the same as the proposed Xcast6+) the cache table in the source's side DR(R2) is updated as the RReq messages are sent to the source(CN). After the procedure above, optimal path for X+MIPv6 is created. Let T_i be the time of RU arrival at the HA and T_j be the time of RReq arrival at the source's side DR respectively. Thus,

- If $T_i < T_j$, then the HA services HA-based routing using Xcast6+ during the time of $(T_j - T_i)$. That is, the HA with multiple mobile nodes away from home must forward the datagrams using Xcast6+ to each of visit routers, according to its cache table updated.
- If $T_i \geq T_j$, then X+MIPv6 is the same as remote subscription. That is, the datagrams must be received only by way of optimal path.

Therefore, X+MIPv6 can provide the delivery of multicast datagrams supporting the avoidance of tunnel avalanches and tunnel convergence as well as the optimal path without multicast tree reconstruction.

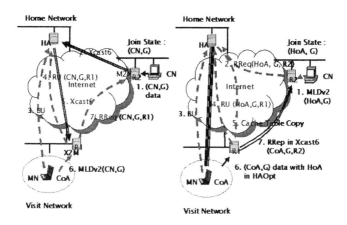

Fig. 2. X+MIPv6 : Recipient Mobility Support

Fig. 3. Source Mobility Support

In addition, the source mobility can be easily achieved. Consider the example shown in Figure 3 (in case of the source mobility support). When a mobile source(MN) enters a visit network, RU is sent to its HA. So, the HA's cache table can be copied to the DR(R1) at source's side. Instead of HA, the DR(R1) at source's side in the visit network sends RReq messages using Xcast6+ (in order to save network resources).

5   Performance Evaluation

In this section, we evaluate the properties and overheads associated with Xcast6+ and X+MIPv6. We are interested in evaluating the effectiveness of our approach for many medium size and subnet-dense groups by comparing with native unicast, multicast and Xcast approaches. We introduce the notion of network resource usage. We define it as $\sum_{i=1}^{L} d_i * s_i + \sum_{j=1}^{R} h_j$, where, L is the number of links active in datagram delivery, d_i is the delay of link i, s_i is the datagram delivery stress of link i, R is the number of routers active in datagram delivery, and h_j is the header processing stress of router j. That is, s_i is to indicate the number of identical copies of a datagram carried by link i (i.e., Multicast, Xcast6 and Xcast6+ have an s of 1 and Unicast has an s of the number of recipients.). Also, h is defined as $f*(0.8*n+0.2)$, where f is the experimental constant number (the ratio of the header processing stress relative to datagram delivery stress) and n is the destination list in header. (we assume that the ratio of overhead of routing table lookups and datagram forwarding are estimated respectively as 0.8*n and 0.2) The network resource usage is a fundamental metric of all of the network resources consumed in the process of datagram delivery to all recipients. To facilitate our comparison, we consider the following advanced performance and overhead metrics :

- Normalized Resource Usage (LRU), defined as the ratio of the network resource usage of Unicast, Xcast6 and Xcast6+ relative to Multicast. NRU is a measure of the additional network resources consumed by Unicast, Xcast6, Xcast6+ compared to traditional Multicast.
- Relative Header Processing (RHP), define as the ratio of the header processing stress of Xcast6 and Xcast6+ relative to native Unicast and Multicast. RHP is a measure of the additional header processing overhead consumed by Xcast6, Xcast6+ compared to native Unicast and Multicast.
- Relative Worst-case Stress (RWS), defined as $max_{i=1..L^s}$, RWS is a measure of the effectiveness of Xcast6+.

We used the real multicast topology consisting of 255 nodes and 266 links which was generated from 1996-MBone trace by ns-2. The topology is reorganized into 45 subnets (i.e, the number of DR is 45). All experiments we report are conducted in the following manner. One source is chosen at random to send datagrams at constant rate. New recipients subscribe to one group at random every 1 second. We allow the simulation to run for 1000 seconds. In addition, we tried to evaluate the benefits of X+MIPv6. We did not experiment all of mobility factors. Among them, we measure the overheads imposed in bi-directional tunneling and X+MIPv6 respectively before the DR at source's side receives RReq from the DR at a visit network (where $T_i < T_j$, in Section 4.) We think after an optimal path is re-routed, the performance improvement of X+MIPv6 is the same as Xcast6+ using remote subscription only. Figure 4 plots the normalized resource usage (NRU) against group size for f = 0.1 and 0.001, respectively. All curves indicate that an increase of group size results in an increase of network resources. Xcast6 and Xcast6+ consume less network resources than native unicast. We observe that Xcast6+ and Xcast6 curves are close to each other for small group size (about < 70) but seem to diverge for larger group size. The diverging point means that the group would be subnetdense (at least more than two recipients in most of a subnet) after the time. For examples, while for a group size of 30, the NRU for Xcast6+ is about 3 - 5, and 5.5 - 6 for Xcast6, for a group size of 164, the NRU for Xcast6+ is about 3.7 - 8 and 22.5 - 27 for Xcast6. These results imply a nearly 55% savings of network resource after the group is subnet-dense (if additional control overheads are ignored). We believe the savings could be more significant, if recipients are clustered in each subnet (subnet-dense). Figure 5 and Figure 6 depict the relative header processing (RHP) overheads and relative worst-case stress (RWS) respectively against group size. The lower and upper curves correspond to Xcast6+ and Xcast6 respectively. First, we observe that whereas the RHP overhead of Xcast6 increases (this is consistent for all group size), the RHP overhead of Xcast6+ does not increase more after group size is about 70 (it is time to be subnet-dense). These results also show that Xcast6+ has the performance gain when the group is subnet-dense. Second, the increase of group size results in an increase of the RWS of Xcast6 ultimately, but the RWS of Xcast6+ is fixed as 1. The meaning above is that if we don't consider the header processing overheads in Xcast6+, the network resource usages of Xcast6+is the same as traditional multicast. Figure 7 plots the network resource (the datagram delivery stress) comparison between X+MIPv6 and bi-directional tunneling during minimum HA-based routing $(T_i < T_j,)$. We observe that HA-based routing using Xcast6+ lowers resources by at least 65% compared to HA-based bi-directional tunneling.

6   Summary and Conclusion

Due to the cost of multicast address allocation, multicast routing state management, control overhead, and scalability issues of traditional IP multicast scheme, ASM leads to a search for other multicast schemes. SSM avoids a multicast address allocation. However, SSM still creates state and signaling per multicast channel in each on-tree node. Both ASM and SSM become expensive for its members if the groups are small. Xcast6+ combines the advantages (stateless multicast) of Xcast6 with the strengths of traditional multicast schemes. Xcast6+

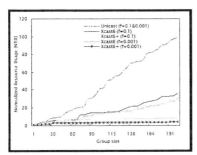

(a) **Fig. 4.** Effect of group size on NRU

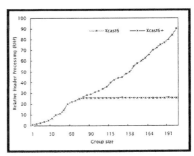

(b) **Fig. 5.** Effect of group size on RHP

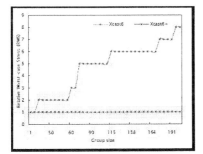

(c) **Fig. 6.** RWS for Xcast6+ and Xcast6

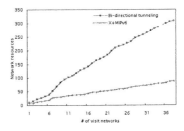

(d) **Fig. 7.** X+MIPv6 vs. bi-directional tunneling

cannot fundamentally perform for many large groups distributed widely as effectively as traditional multicast schemes. However, we believe that we made two contributions in this paper. First, as an alternative scheme, it is feasible to use Xcast6+ to efficiently and effectively support multicast for medium size and subnet-dense groups. We have shown, with simulation results, that the net-work resources are relatively saved as the number of recipients in a subnet increases. Second, we have shown the many challenges in MIPv6 multicast could be alleviated by integration of Xcast6+ in MIPv6. The remote subscription scheme with minimum HA-based routing using Xcast6+ reduces network resources and minimizes the data-gram loss, as well as the support of the source mobility.

References

[1] S. Deering, Multicast Routing in a Datagram Internetwork, Stanford University, Ph.D. thesis, Dec. 1991.
[2] C. Diot et al., Deployment Issues for the IP Multicast Service and Architecture, In IEEE Networks Magazine's Special Issue on Multi-cast, 2000.
[3] R. Boivie et al., Explicit Multicast (Xcast) Basic Specification, IETF Internet-Draft, draft-ooms-xcast-basic-spec-0x.txt, 2002.
[4] Y. Chu et al., A Case for End System Multicast, IEEE Journal on Selected Areas in Communication (JSAC), Special Issue on Networking Support for Multicast, 2002.
[5] P. Francis, Yoid: Extending the Internet Multicast Architecture, Unrefereed report, ACIRI, 2000.
[6] H. Holbrook et al., Source-Specific Multicast for IP, IETF Internet-Draft, draft-ietf-ssm-arch-0x.txt, 2002.
[7] R. Boivie, Small Group Multicast, IETF Internet-Draft, draft-boivie-sgm-0x.txt, 2000.
[8] D. Ooms, Connectionless Multicast, IETF Internet-Draft, draft-ooms-cl-multicast-0x.txt, 2000.
[9] Y. Imai, Multiple Destination Option on IPv6 (MDO6), IETF Internet-Draft, draft-imai-mdo6-0x.txt, 2000.
[10] C. R. Lin et al., Mobile Multicast Support in IP Networks, IEEE INFOCOM 2000, May 2000.
[11] T. G. Harrison et al., Mobile Multicast(MOM) Protocol: Multicast Support for Mobile Hosts, The Third Annual ACM/IEEE International Conference on Mobile Computing and Networking, 1997.
[12] M. Shin et al., Explicit Multicast Extension (Xcast+) for Efficient Multicast Packet Delivery, ETRI Journal, Vol. 23, No. 4, 2001.

Efficient Multicast Supporting in Multi-rate Wireless Local Area Networks*

Yongho Seok and Yanghee Choi

School of Computer Science and Engineering, Seoul National University
{yhseok,yhchoi}@mmlab.snu.ac.kr

Abstract. Wireless local area networks(LAN) have become popular due to their capability of supporting the high data rates required in ubiquitous computing. These high data rates are possible through the use of new modulation schemes that are optimized for different channel conditions. In particular, IEEE 802.11b supports a variety of different data rates, i.e. 11Mbps, 5.5Mbps, 2Mbps and 1Mbps, depending on the channel conditions between the access point(AP) and the mobile station. All mobile stations share the same wireless channel or frequency, so the bottleneck point of wireless LAN system is the traffic load of the lowest data rate connections. Since wireless LAN supports higher data rates, it is more attractive for ubiquitous multimedia services. Multimedia services generally require high bandwidth and most of them utilize the multicast system. Since most multimedia services are provided by real-time communications, reduced end-to-end packet delay and delay jitter are more important than reliable packet transmission. However the data rate of the multicast communication used in most commercial products is generally set to be low by the administrator, who only considers the coverage of the AP, and its value is generally fixed. In this environment, data rate which is selected for the transmission of multicast packets is a very important factor in determining the system performance of the wireless LAN, since multimedia services require a large bandwidth. Thus, a new algorithm is necessary to dynamically select the data rate of the multicast according to the traffic load of the wireless LAN. In this paper, we present two naive data rate selection algorithms which allow the data rate to be dynamically adjusted, and a more novel data rate selection algorithm based on the two primitive algorithms. We show that our suggested algorithm achieves high performance in wireless LAN and efficiently supports high quality multimedia services, through the use of the $NS - 2$ simulation in a wireless LAN environment based on IEEE 802.11b.

1 Introduction

The IEEE 802.11a and IEEE 802.11b media access control(MAC) protocols provide a physical-layer multi-rate capability [1]. The original IEEE 802.11 protocol

---

* This work was supported in part by the Brain Korea 21 project of Ministry of Education and in part by the National Research Laboratory project of Ministry of Science and Technology, 2002, Korea.

H.-K. Kahng (Ed.): ICOIN 2003, LNCS 2662, pp. 273–283, 2003.

supports a single base rate, typically 2Mbps. With the multi-rate enhancement, the data transmission can take place at various rates according to the channel conditions. Higher data rates than the base rate are possible when the signal-to-noise ratio(SNR) is sufficiently high. With IEEE 802.11a the set of possible data rates includes 6, 9, 12, 18, 24, 36, 48 and 54 Mbps whereas for IEEE 802.11b the set of possible data rates includes 1, 2, 5.5 and 11 Mbps. As the multi-rate enhancements are physical layer protocols, MAC mechanisms are required to exploit this capability. The auto rate fallback(ARF) protocol is the commercial implementation of a MAC that utilizes this feature [2]. With ARF, senders attempt to use higher transmission rates after consecutive transmission successes and revert to lower rates after failures. Under most channel conditions, this multi-rate enhancement using ARF provides a performance gain over the pure IEEE 802.11 single base rate.

Since the ARF protocol selects the data rate according to the channel conditions between the AP and mobile station, it can only be used in point-to-point communications, such as in the case of unicast. In the case of the point-to-multipoint communications(e.g., multicast and broadcast), it is difficult to determine the data rate, because the link characteristics between the AP and each mobile station can vary. In current commercial products, the administrator selects the data rate of the point-to-multipoint connection, which is used to provide network connectivity to all mobile stations covered by the AP. Because the coverage of the AP is inversely proportional to the transmission data rate, the administrator selects the proper data rate according to the distance between the different APs. As the distance between the APs is increased, the data rate of the point-to-multipoint connection has to be reduced in order to compensate for the increased range that the AP has to cover. This simple approach dose not efficiently support point-to-multipoint connections, due to the characteristics of wireless LAN explained below.

All mobile stations in the basic service set(BSS) share the same wireless channel or frequency, by using the Carrier Sense Multiple Access-Collision Avoidance MAC protocol. In this environment, the bottleneck point of the wireless LAN system is the traffic load of the lowest data rate connections, which may involve either point-to-point connections or point-to-multipoint connections. When the number of packets transmitted to mobile stations using 1Mbps data-rate is increased, the resulting deterioration in throughput, delay and packet loss will always be greater than that in the case of an 11Mbps data-rate, even if the total amount of transmitted traffic is the same in both cases. Thus, to maximize throughput, we need to consider the performance degradation of wireless LAN resulting from the traffic load of the lowest data rate connections between the AP and mobile stations. In the case of point-to-point communications, the appropriate data rate can be set individually for each mobile station, after taking the current channel conditions, such as the SNR, into consideration. However, in the case of point-to-multipoint communication, the data rate is almost always set to a fixed, low value by the administrator, who only takes the coverage area of the AP into consideration.

We propose an algorithm which can be used to dynamically select data rate of the multicast, which determines the overall performance of the wireless LAN system and the quality of service(QoS) of the multimedia application. The remainder of this paper is organized as follows. We start in Section 2 by providing some background on some data rate selection algorithms used for unicast packet transmission in multi-rate wireless LAN. The proposed data-rate selection algorithms for multicast packet transmission are described in Section 3. Section 4 presents the results of the simulation experiments. Finally, we summarize and conclude this paper in Section 5.

2   Related Work

Most commercial implementations that exploit the multi-rate capability of IEEE 802.11b and IEEE 802.11a are termed ARF [2]. The ARF scheme proposed in [2] is based on keeping track of the timing function and missed acknowledgment(ACK) messages. Operation at the maximum data rate is considered to be the default. When an ACK is missed for the first time following earlier successful transmissions, the first retry transmission is still performed at the same rate. When the ACK is missed again, the second retry and subsequent transmissions are performed at the fallback rate, which is half of the previous date rate. In the [3], a rate adaptive MAC protocol, called the receiver-based auto rate(RBAR) protocol, is presented. The novelty of RBAR is that its rate adaptation mechanism is situated in the receiver instead of in the sender. This is in contrast to the ARF scheme in which the sender decides the data rate. RBAR is based on the RTS/CTS mechanism and it is better because it results in a more efficient channel quality estimation which is then reflected in a higher overall throughput.

3   Data-Rate Selection Algorithm
of Multicast Packet Transmission

The main issue in transmitting multicast packets is selecting the appropriate data rate. Currently, in most commercial APs that support the IEEE 802.11b protocol, the administrator selects a fixed data rate. The drawback of this approach is that selecting a fixed data rate is not an effective solution. In particular,

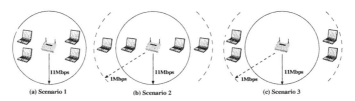

Fig. 1. Multicast issues in multi-rate wireless LAN

it is very inefficient to set the data rate to a certain low value so that all mobile stations can receive the multicast packets from the AP. In most multicast applications, perfect reliability is not needed. In addition, such reliability is not guaranteed even with a low data rate, due to the inherent limitations associated with wireless communication. Since multicast applications generally require high bandwidth, transmitting at a low data rate leads to a loss of throughput. In Figure 1 (a), if the data rate of the multicast is set to 11Mbps, the four mobile stations in the multicast group can receive the multicast packets. However, in Figure 1 (c), the four mobile stations have difficulty in receiving the multicast packets, since they are beyond the coverage of the 11Mbps data rate. In Figure 1 (b), two of these mobile stations are located at a short enough distance from the AP to receive the data, while the other two mobile stations are not able to receive the data. If the data rate is reduced to 1Mbps so that all four stations can receive data, the wireless LAN suffers from throughput degradation problems resulting from the low data rate.

3.1   Fixed Data-Rate

Most currently used wireless AP devices supporting IEEE 802.11b, transmit packets at a fixed data rate. The data rate is determined by the AP administrator according to the distance between the APs. Wider coverage for an AP is achieved by lowering the data rate, if the distance between the APs is great. The shortcoming of this simple approach is to mistakenly use a low data rate, whereas a high data rate could be used even when the all of the mobile stations in the group are located close to the AP. In order to solve this problem, a technique that applies a different data rate to each different multicast group is considered in next subsection.

3.2   Maximum Data-Rate Covering the All Mobile Stations of Group

In this subsection, we consider a technique that maintains the lists of mobile stations for each multicast group. The transmission of multicast packets is carried out at the lowest data rate among data rates of all mobile stations in the list of the multicast group. This technique partially solves the problem resulting from the use of a fixed data rate, however, it also has the disadvantage of having to maintain a list of mobile stations in each group and the lowest data rate among them. Furthermore, the lowest data rate will always be selected, in order to be able to transmit data to all of the mobile stations in the group, even if most multicast applications require higher bandwidth. For example, let us consider the case of a real-time multicast requiring a 1.5Mbps bandwidth. If one of the mobile stations in the multicast group has a 1Mbps link and the others all have 11Mbps links, then the problem of which data rate to select becomes apparent. The 1Mbps data rate would enable every mobile station to receive the data, however selecting this option would result in low throughput due to high resource utilization($\frac{Traffic}{Capacity} \simeq \frac{1.5Mbps}{1Mbps}$). To improve the throughput, selecting the 11Mbps data rate would be better, although some mobile stations would have difficulty in receiving the data. However, a mobile station situated far from the AP is not entirely prevented from receiving the data, even at the 11Mbps data rate. The only problem is that the bit error rate will be high. Consequently, we need to provide flexibility in the selection of the data rate, since most multicast applications do not require high reliability.

3.3 Adaptive Data-Rate Based on the Average Transmission Time of Packets

We propose an algorithm which provides a means of selecting the data rate of a multicast packet according to the traffic load. If setting the data rate such that all mobile stations in a group can receive data results in a degradation of the wireless LAN's performance, we prefer to increase the data rate in order to improve the overall performance of the wireless LAN. However, in this case, some mobile stations in a group may not receive data packets, since the IEEE 802.11 MAC protocol provides reliability using ACK packets in the case of unicast packets, but not in the case of multicast or broadcast packets [5]. This is not a serious drawback since, in multicast applications, reducing delay and delay-jitter is more important than providing reliability, and at application level, UDP is used, rather than TCP, for this reason.

For our proposed technique, we used a new algorithm for maintaining queues, instead of the existing *DropTail* buffer management algorithm. We choose the packet with the longest transmission time for dropping when the queue is full. Hence, by this algrorithm, we can maximize the number of available entries and decrease the probability of dropping packets accordingly. When the network layer has a packet to transmit, it calls the enqueue function of our queue management system. In the enqueue function, if the queue is full, we choose the packet which has the longest transmission time based on the channel state monitor information, such as the data-rate of each mobile station and then drop it. When the MAC protocol wants to transmit a packet, it calls the dequeue function of our queue management system. In this case, the dequeue operation of the unicast packet is simply to transmit the packet with current data rate between the AP and the packet destination. However, the dequeue operation of a multicast packet

Table 1. Notations for algorithm description

Notation	Description
$Q.Length()$	return the queue length (*packets*)
$Q.Insert(P)$	insert the *packet* P into the queue
$Q.Lookup(i)$	return the $i'th$ *packet* in the queue
$Q.Remove(P)$	delete the *packet* P in the queue
$P.Size()$	return the size of *packet* P
$P.NextHop()$	return the next hop address of *packet* P
$Bit\_Rate(k)$	the host k's bit rate

Algorithm 1 $Enqueue(P)$ P : newly arriving packet

1: $Q.Insert(P)$;
2: **if** $Q.Length() \geq QueueLimit$ **then**
3:    $i \leftarrow 1$;
4:    $Max\_TxTime \leftarrow 0$;
5:    **while** $i \leq Q.Length()$ **do**
6:      $P_{tmp} \leftarrow Q.Lookup(i)$;
7:      $TxTime \leftarrow$
                    $P_{tmp}.Size()/Bit\_Rate(\ P_{tmp}.NextHop()\ )$;
8:      **if** $TxTime > Max\_TxTime$ **then**
9:        $Max\_TxTime = TxTime$;
10:       $P_{victim} \leftarrow P_{tmp}$;
11:      **end if**
12:     $i \leftarrow i + 1$;
13:    **end while**
14:    $Avg\_TxLimit \leftarrow w \cdot Max\_TxTime + (1 - w) \cdot Avg\_TxLimit$;
15:    $Q.Remove(P_{victim})$;
16:    $Drop(P_{victim})$;
17: **end if**

---

requires an additional mechanism in order to decide the appropriate data rate. The data rate decision mechanism will be shown in Algorithm 2.

In Algorithm 1, we present the proposed queue management algorithm using the notation shown in Table 1. Algorithm 1 operates when a new packet (P) arrives at its queue (Q). The queue management module inserts the arriving packet (P) into the queue (Q). If the queue is full, we determine which packet to be dropped according to the following rules. The overflow of the queue implies that the number of incoming data packets for the wireless system is greater than the number of outgoing packets. Therefore, we choose the packet which would have the longest transmission time to be dropped, so that we can reduce the packet drop probability. Since the upper bound of the queue length commonly uses the number of packets instead of the number of bytes, reducing the average packet transmission time results in an increase in the available queue space. Lines 5-13 in Algorithm 1 shows how to look for the packet which has the longest transmission time $(P.Size()/Bit\_Rate(k))$ in the queue Q. For multicast packets, the data rate is set to the highest value among the data rates of all the mobile stations in the group. In the case of unicast, there is only one receiver, but in the case of multicast there are several. When comparing the dropping priority between a unicast packet and a multicast packet, the transmission time of the multicast packet is based on the highest data rate, so that just one receiver is able to receive the packet. Since the multicast connection is considered to be a point-to-point connection, as in the case of unicast, this mechanism can provide fairness of throughput between unicast and multicast. At Line 14, the average transmission time of dropped packets is evaluated using

---

Algorithm 2 *Data − rate Selection of Multicast Packet P*

1:   *Multicast_rate* ← 1*Mbps*;
2: **if** *Q.Length*() ≥ *Threshold* **then**
3:    **if** *P.Size*()/1*Mbps* ≤ *Avg_TxLimit* **then**
4:      *Multicast_rate* ← 1*Mbps*;
5:    **else if** *P.Size*()/2*Mbps* ≤ *Avg_TxLimit* **then**
6:      *Multicast_rate* ← 2*Mbps*;
7:    **else if** *P.Size*()/5.5*Mbps* ≤ *Avg_TxLimit* **then**
8:      *Multicast_rate* ← 5.5*Mbps*;
9:    **else if** *P.Size*()/11*Mbps* ≤ *Avg_TxLimit* **then**
10:      *Multicast_rate* ← 11*Mbps*;
11:    **end if**
12: **end if**

---

the exponentially weighted moving average(EWMA). Namely, the most recently dropped packet has the weight of w and the cumulated average has $1 - w$.

The transmission data rate of a multicast packet is determined by this average transmission time as in the Algorithm 2. $Avg\_TxLimit$ presents the upper bound of transmission time that packets in the queue must not exceed. Once the packet (P_{victim}) to be dropped has been chosen, we remove this packet from the queue (Q). This priority queue management algorithm, based on the transmission time of the packet, gives the lower priority to mobile stations with lower-data rates, when the queue is full. However, small sized packets, such as those used in voice or interactive applications, get high priority for transmission, even though their data rate is low, since little time is required for transmitting small-sized packets, even at a low data rate.

In Algorithm 2, the transmission data rate of a multicast packet is determined as follows. As in the case of the queue management algorithm described in Algorithm 1, which drops the packet with the longest transmission time when the queue is full, a similar method needs to be applied here. The transmission time of a multicast packet should be shorter than the average transmission time of all dropped packets ($Avg\_TxLimit$), in order to agree to giving higher priority of entering the queue to the packet with shorter transmission time. In this way, resources are allocated fairly to both multicast and unicast packets. Thus, when the queue length exceeds a certain threshold($Threshold$), the data rate of a multicast packet is set, such that the transmission time of the multicast packet is less than the average transmission time of all dropped packets(Lines 3-11 in Algorithm 2). If there are many such data rates, the lowest one is selected so that the largest number of mobile stations are served.

4   Simulation Experiments

The simulation results in this paper were generated using the $ns - 2$ network simulator, with extensions from NOAH, a wireless routing protocol that supports

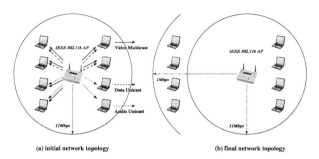

Fig. 2. Simulation scenario

direct communication between APs and mobile stations. In this simulation, the available data rates are based on IEEE 802.11b and are set to 1 Mbps, 2 Mbps, 5.5 Mbps, and 11 Mbps. The values for the data rates were chosen based on the distance ranges specified in the $Orinoco^{TM}$ 802.11b card data sheet [6]. Figure 2 (a) represents the initial network topology. It is composed of one access point(AP) and 8 mobile stations. In Figure 2 (a), all mobile stations are close enough to the AP so that they can all be served by the AP using the 11 Mbps data rate. When the simulation starts, half of the 8 mobile stations begin to move away from the AP with a velocity of 0.5 m/s. Consequently, their data rates are successively reduced firstly from 11 Mbps to 5.5 Mbps, then to 2 Mbps and finally down to 1 Mbps. Figure 2 (b) shows 4 the situation in which the 4 mobile stations move, while the other 4 mobiles stations stay in place, and these stations are served using the 1 Mbps and 11 Mbps data rates, respectively.

We performed our simulation using 3 different applications(Video, Audio, and Data traffic) to show the various internet traffic characteristics. In this simulation, the video application uses multicast routing while the other applications use unicast routing. The mobile stations joining the multicast group for the video application comprise 4 moving mobile stations and 2 stationary mobile stations. Each of the data and audio applications has 4 connections, 2 of them involve moving mobile stations, and the other 2 involve stationary mobile stations. The video traffic is CBR generating 1.5 Mbps with a packet size of 1400 bytes. The audio traffic is exponential On/Off, which can be configured to behave as a Possion process, generating 256 Kbps with a packet size of 128 bytes. The data traffic is Pareto On/Off generating 512 Kbps with a packet size of 512 bytes. All traffic sources in the wired network are 2 hops away from the AP.

4.1   Simulation Result

Figure 3 shows the total throughput of the AP, i.e. the total number of data bytes including Video, Audio and data traffic transmitted from the AP to mobile stations per second. The x-axis represents time ranging from 0 to 1400 seconds and the y-axis represents the throughput(Mbps). In this graph, we show 3 data rate selection algorithms for multicast packets, *Fixed Data-rate* (Section 3.1)

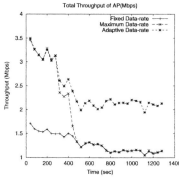

Fig. 3. Total throughput of AP

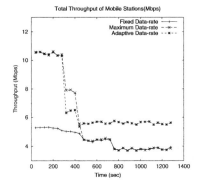

Fig. 4. Total throughput of mobile stations

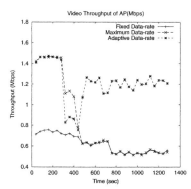

Fig. 5. Video throughput of AP

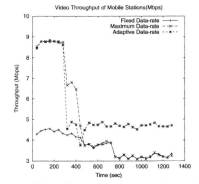

Fig. 6. Video throughput of mobile stations

which implements the multicast packet transmission with the 1Mbps fixed data-rate, *Maximum Data-rate* (Section 3.2) which implements the multicast packet transmission with the maximum data-rate covering all mobile stations in each group and *Adaptive Data-rate* (Section 3.3) which implements the multicast packet transmission with the adaptive data-rate based on the average transmission time of dropped packets.

First, when *Fixed Data-rate* is used, throughput hardly exceeds 1.5Mbps. This is due to the packet's fixed data rate of 1Mbps. Using *Maximum Data-rate* which maintains a list of the multicast packets' data rates for each multicast group, a relatively high throughput is obtained, particularly in the case where all of the mobile stations in a group are close to the AP. However, as the distance from the AP to the mobile stations increases, the packets' data rate decreases and, after 500(sec), it shows a similar throughput to that of *Fixed Data-rate*. Consequently, maintaining separate data rates per multicast group is not the most effective alternative. *Adaptive Data-rate* shows a throughput of over 2Mbps, even after 500s. It is shown that selecting the appropriate data rate for multicast, according to the traffic load of the wireless LAN, can lead to an increase in the throughput. In the proposed *Adaptive Data-rate* algorithms, the number of dropped packets can be reduced by using the packet transmission time for queue management, since in this way it is possible to reduce the length of the queue rapidly. Moreover, it is possible to dynamically adjust to the traffic conditions of the wireless LAN, by setting the data rate of the multicast packets after considering the average transmission time of the dropped packets.

The graph in Figure 4 shows the throughput in terms of the total amount data received by all mobile stations. The y-axis shows the total quantity of packets successfully received by all mobile stations for each packet transmitted by the AP. The graph shows similar results to those in Figure 3. If the data rate of the multicast packets is fixed (*Fixed Data-rate*), the receiving throughput of all mobile stations is maintained at 4 to 5Mbps. On the other hand, it suffers penalty in throughput by over 2Mbps after 500s compared with *Adaptive Data-rate*. Comparing *Maximum Data-rate* with *Adaptive Data-rate*, their throughput up until 500s is similar, as shown in the graphs in Figure 3. This result shows that determining the multicast data rate by means of the data rates of the mobile stations in the multicast group is quite effective, as long as the mobile stations are located close to the AP. However, when using this approach, no more gain in throughput was obtained after 500s when at least one terminal became so far from the AP that the data rate was set to a lower value in order to make all mobile stations receive data from the AP. In this case, the data rate has to be set to a completely new value, as in *Adaptive Data-rate*. Thus, when the overall performance of the wireless LAN system is degraded, the data rate should be increased to a higher value, even though some stations may not be able to receive the data.

Figure 5 shows the throughput of the video application transmitted from the AP to the mobile stations in the multicast group. In this graph, the *Adaptive Data-rate* algorithm achieves a higher video throughput than the other algorithm especially after 500s. This indicates that the client application in the mobile station is able to display a greater number of frames per second, with the result that, a greater number of video packets is transmitted from the AP by using *Adaptive Data-rate* algorithm The displayed frames per second is a very important factor related to the QoS. Thus, when the other algorithms (*Fixed Data-rate* or *Maximum Data-rate* algorithm) are used instead of *Adaptive Data-rate*, the QoS of the video application, as well as the overall throughput of the wireless system, is degraded.

Figure 6 shows the throughput of the video application received by the mobile stations and is similar to the Figure 4. It is clear that, after 500s, the throughput obtained by the use of *Adaptive Data-rate* is the highest. Even though some mobile stations may not receive certain packets when *Adaptive Data-rate* is used, the total throughput for all mobile stations is not degraded. Furthermore, the throughput of the receiving mobile stations, as well as the throughput of the AP, is improved.

5   Conclusion

In this study, we proposed a new data rate selection algorithm for multicast packets (*Adaptive Data-rate*), which can provide high system performance in a multi-rate wireless LAN environment. In addition to improve its performance, this algorithm also provides the ubiquitous multimedia service with good service quality, through the efficient use of the high bandwidth available in a multi-rate Wireless LAN environment. The proposed data rate selection algorithm is able to dynamically adjust the data rate of the multicast packet transmission according to the traffic load of the wireless LAN. It consists of two primitive algorithms, a priority queue management algorithm based on packet transmission time, and a data rate selection algorithm for multicast packets based on the average transmission time of all dropped packets. Since wireless LAN supports high data rates, such as the 54Mbps rate defined in IEEE 802.11a, as multimedia traffic using the multicast system increases, the algorithms proposed in this study will become more important. We performed a simulation of the proposed algorithm using the $ns-2$ simulator, and were able to confirm the predicted results. The proposed algorithm contributes to the efficient utilization of resources, as well as increasing the throughput of the wireless LAN and improving the QoS of the multimedia service, by reducing the end-to-end packet delay and increasing the number of frames per second which can be viewed. How to guarantee the fairness in the proposed priority queue management algorithm is one of our future work.

References

[1]   IEEE Computer Society. "802.11: Wireless LAN Medium Access Control (MAC) and Physical Layer (PHY) Specifications". June 1997.
[2]   A. Kamerman and L. Monteban. "WaveLAN II: A high-performance wireless LAN for unlicensed band". Bell Labs Technical Journal, pages 118-133, Summer 1997.
[3]   G. Holland, N. Vaidya and P. Bahl. "A rate-adaptive MAC protocol for multi-hop wireless networks". ACM MOBICOM'01, July 2001.
[4]   B. Sadeghi, V. Kanodia, A. Sabharwal and E. Knightly. "Opportunistic Media Access for Multirate Ad Hoc Networks". ACM MOBICOM'02, September 2002.
[5]   J. Kuri and S. Kasera. "Reliable Multicast in Multi-access Wireless LANs". ACM/Kluwer Wireless Networks Journal, July/August 2001
[6]   Datasheet for ORiNOCO 11 Mbit/s Network Interface Cards, 2001.
    ftp://ftp.orinocowireless.com/pub/docs/ORINOCO/.

Theoretical Throughput/Delay Analysis for Variable Packet Length in the 802.11 MAC Protocol*

Masanori Nakahara

Canon Inc.
3-30-2 Shimomaruko Ohta-ku Tokyo 146-8501, Japan
nakahara.masanori@canon.co.jp

Abstract. This paper presents the results of our analysis of throughput and delay for variable packet length in the IEEE 802.11 MAC protocol. We extend Cali's theoretical model for basic CSMA/CA to derive optional hybrid model and generalize it so that we can incorporate the statistical packet length distribution traced on a real network. We also introduce the delay formula, which allows us to calculate important metrics to evaluate system performance. Our model was validated through simulations which yielded results that are accurate to the 802.11 MAC behavior. In addition by adjusting key parameters in the analytical model, we are able to examine the theoretical performance limit.

1 Introduction

In the IEEE 802.11 Wireless LAN MAC, a *Distributed Coordination Function* (DCF) is the fundamental mode to support a best-effort delivery service. The DCF defines a basic access mechanism for packet transmission: *Carrier Sense Multiple Access with Collision Avoidance* (CSMA/CA) and an optional mechanism: *Request to Send/Clear to Send* (RTS/CTS).

RTS/CTS was originally proposed to combat the *hidden terminal problem* [1]. The function of RTS/CTS is to reserve the channel by transmitting short control frames (RTS and CTS) prior to data packet transmission, so that the data packet of it's node never collides with other packets. Consequently, the use of RTS/CTS is particularly useful for transmitting large data packets in spite of the existence of hidden terminals. While RTS/CTS can decrease the collision cost of data packets, it requires additional transmission time to exchange control frames. Therefore, we need to consider the implied trade-off. The 802.11 standard states that RTS/CTS is used only when a packet size exceeds the *RTSThreshold* parameter, and the actual value of RTSThreshold can be adjusted by a user. In practice, however, it is hard to select an appropriate value for RTSThreshold. The motivation of this paper lies in investigating the benefits of exploiting both CSMA/CA and RTS/CTS to maximize throughput and minimize delay.

---

* This work done at University of Washington as a visiting researcher.

H.-K. Kahng (Ed.): ICOIN 2003, LNCS 2662, pp. 284–294, 2003.
© Springer-Verlag Berlin Heidelberg 2003

Our approach is based on a theoretical analysis of the basic CSMA/CA mechanism [2]. We extend this to derive a *hybrid* mechanism, which combines RTS/CTS and CSMA/CA. We also introduce the analytical formula of the expected transmission delay. We generalize the packet length model so that we can conduct the extensive analysis for any packet length distribution. We compare throughput and delay for a geometric packet distribution, and a statistical distribution obtained by tracing a real network. The geometric distribution is used to analyze general trends in the behavior of the 802.11 standard. We then show the similarity of the trends between the statistical results and the geometric distribution.

We focus on *saturated performance*, which is defined as the performance limit when the offered load increases and the number of users remains constant. The saturated performance is a fundamental figure that shows the potential capability that the system can provide. We aim to analyze this using a mathematical formulation. The key advantage of the mathematical model lies in the fact that it is convenient to manipulate key parameters to optimize system performance. We investigate the theoretical performance limits achieved by tuning two key parameters: RTSThreshold k and transmission probability p.

The DCF mechanism is well studied with simulations [3] [4] and with analytical models [2] . A recent study [5] uses Markov process to analyze the dynamic backoff window behavior of the IEEE 802.11. This also compares CSMA/CA with RTS/CTS. However, it only examines performance at a fixed packet length and does not provide delay analysis. Our work complements [2] and [5] by using variable packet lengths to estimate throughput and delay.

The rest of the paper is organized as follows. Section 2 summarizes the IEEE 802.11 DCF. Section 3 gives an overview of the theoretical analysis of channel capacity in [2] and introduces the hybrid analytical model. Section 4 explains the delay analytical model. Section 5 validates the analytical result. Section 6 investigates the maximum performance with the optimum system parameters and applies a statistical packet length distribution traced on a real network to the analytical model. Section 7 concludes.

2   IEEE 802.11 DCF

This is an overview of the DCF in the 802.11 standard [6]. In the 802.11 Mac protocol, time is divided into slots and a station is allowed to transmit a packet only at the beginning of each slot. In CSMA/CA, a station monitors channel activities during a *Distributed InterFrame Space* (DIFS) period. After a DIFS period is sensed as idle, the station selects a random backoff timer from the *contention window CW*. The backoff timer is decremented while the channel is idle. When it expires, the station transmits a packet. The station will then receive an ACK frame after a *Short InterFrame Space* (SIFS) period. A collision can only occur when more than two stations initiate their transmissions on the same slot. If the station doesn't receive an ACK after a SIFS period or if it detects another carrier on the media, it assumes that the transmission was unsuccessful and selects a backoff timer that is *double* the value of the previous contention window, up to the maximum contention window CW_{max}.

Carrier sense in wireless environment is not perfect and can sometimes fail due to signal fading and attenuation, which in turn can cause the *hidden terminal problem* [1]. RTS/CTS was originally proposed to mitigate this problem, and the basic idea behind it is to reserve the channel by exchanging short control frames (RTS and CTS) prior to data packet transmission. A RTS frame is transmitted after the backoff timer expires. When a destination receives the RTS frame, it responds by sending a CTS frame after waiting a SIFS period. A station is allowed to transmit a data packet only after it successfully receives a CTS frame.

3   Throughput Analysis

First, we will summarize the IEEE 802.11 capacity[1] analysis of the basic access mechanism in [2]. Next, we will introduce an optional hybrid access model.

3.1   Basic Access Model

Our study focuses on saturated analysis: we have a fixed number of stations, each of which will always have a packet ready for transmission. The analytical model in [2] makes three assumptions. First, the channel is error-free without capture. We do not concerned ourselves with the hidden terminal problem. We suppose that the hidden terminal phenomenon can be eliminated in infrastructure networks if the location of base stations is well planned. Second, the backoff timer is chosen from a *geometric distribution* with parameter p. This algorithm is equivalent to a p-persistent protocol and hence differs from the standard 802.11 algorithm. However, it is shown to closely approximate the standard if p is derived from the *average contention window* size. Finally, a station transmits a packet whose length is sampled from a *geometric distribution*.

Instead of the geometric distribution, we adopt a *multinomial distribution* for the packet length model, which is more general. Specifically, a packet has a byte length that is equal to a multiple of the time slots duration. The transmission time of a mean sized packet $\overline{m}$ is calculated as $\overline{m} = t_{slot} \sum_{i=1}^{\infty} i \cdot q_i$, where t_{slot} is a slot time duration and q_i is the probability of transmitting a packet of length equal to the i slot time ($\sum_{i=1}^{\infty} q_i = 1$). To simplify the discussion, we define F_q^j to be the cumulative probability of q_i from slot number 1 to j.

To develop the analytical model, a *virtual transmission time* (t_v) is defined as the period of time between consecutive successful transmissions over the medium, as shown in Figure 1. t_v includes the number of idle slots and collisions before one successful transmission. Let us define the total capacity of the medium $\rho_c = \overline{m}/t_v$. Examining Figure 1 gives us

$$t_v = E[N_c] \cdot \{E[Coll] + \tau + DIFS\} + E[Idle] \cdot (E[N_c] + 1) + E[S] \quad (1)$$

---

[1]   We use the terms capacity and throughput interchangeably.

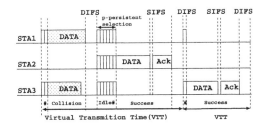

Fig. 1. Virtual transmission time (VTT)

where $E[N_c]$ is average number of collisions in a t_v; $E[Coll]$ is average collision length; $E[S]$ is average successful transmission time; τ is maximum propagation delay between two wireless stations; $E[Idle]$ are the average number of idle slots in a t_v. Each expression in (1) is derived in [2] and we only modify equation (5) as follows to reflect the multinomial packet distribution:

$$E[N_c] = \frac{1-(1-p)^M}{Mp(1-p)^{M-1}} - 1 \tag{2}$$

$$E[Idle] = \frac{(1-p)^M}{1-(1-p)^M} \cdot t_{slot} \tag{3}$$

$$E[S] = MH + \overline{m} + 2 \cdot \delta + SIFS + ACK + DIFS \tag{4}$$

$$E[Coll] = MH + \frac{t_{slot}}{1-(1-p)^M - Mp(1-p)^{M-1}} \cdot$$

$$\left[\sum_{h=1}^{\infty} h \cdot \left\{ (1-p+p \cdot F_q^h)^M - (1-p+p \cdot F_q^{h-1})^M \right. \right.$$

$$\left. \left. - Mp(1-p)^{M-1} \left(F_q^h - F_q^{h-1} \right) \right\} \right] \tag{5}$$

where δ is a packet transmission delay, M is the number of stations that attempt a transmission, and MH is transmission time for the MAC header preamble.

3.2   Hybrid Access Model

For the hybrid analytical model for CSMA/CA and RTS/CTS, the second operates on top of the first only when the payload size of a packet exceeds RTSThreshold. From (2)-(5), $E[N_c]$, and $E[Idle]$ in the hybrid model are same as for CSMA/CA, because both equations depend only on M. $E[S]$ and $E[Coll]$ need to be redefined for the hybrid scheme.

We assume that RTSThreshold is the byte length corresponding to a multiple of the time slot duration (k slot times). A station with a packet whose length is longer than RTSThreshold makes use of RTS/CTS. Otherwise the station will choose CSMA/CA. We define $P_{\leq thresh}$ as the probability that the packet length is less than or equal to RTSThreshold. If all the stations operate using RTS/CTS, the successful transmission time is derived as:

$$E[S_{RTS/CTS}] = MH + \overline{m} + 4 \cdot \delta + RTS + CTS + 3 \cdot SIFS + ACK + DIFS \tag{6}$$

In the general case, where each station transmits a packet of different length, the expectation of a successful transmission time $E[S_{hybrid}]$ results from the combination of CSMA/CA and RTS/CTS: $E[S_{hybrid}] = P_{\leq thresh} \cdot E[S_{CSMA}] + (1 - P_{\leq thresh}) \cdot E[S_{RTS/CTS}]$. Note that $E[S_{CSMA}]$ is $E[S]$ in equation (5).

Let us consider the average collision length $E[Coll]$ for the hybrid mechanism through several examples. When all the stations operate using RTS/CTS, $E[Coll]$ is equal to the transmission time for a RTS frame. If each station employs CSMA/CA and makes use of RTS/CTS only for packets with payload size exceeding RTSThreshold, we can classify the set of colliding packets by two patterns: (1) at least one packet length is less than or equal to k slots; (2) all the packets are RTS frames. Collision length is equivalent to the transmission time for the longest colliding packet. The length of a RTS frame is shorter than that of a MAC header, thus in case (1) we consider data packets as a candidates for longest packet. Let $P_{m_{max}}$ be the probability that the length of m slots ($1 \leq m \leq k$) is longest among colliding packets. Let $P_{RTS_{max}}$ be the probability that all colliding packets are RTS frames. We can represent $E[Coll_{hybrid}]$ as:

$$E[Coll_{hybrid}] = \sum_{m=1}^{k} \{(MH + m \cdot t_{slot}) \cdot P_{m_{max}}\} + RTS \cdot P_{RTS_{max}} \qquad (7)$$

where RTS is the transmission time for a RTS frame. Solving for $P_{m_{max}}$ and $P_{RTS_{max}}$ gives the following;

$$\sum_{m=1}^{k} (MH + m \cdot t_{slot}) \cdot P_{m_{max}} =$$

$$\frac{1}{1 - (1-p)^M - Mp(1-p)^{M-1}} \left[\sum_{h=1}^{k} (MH + h \cdot t_{slot}) \left\{ \left(1 + p \cdot F_q^h - p \cdot F_q^k\right)^M \right. \right.$$
$$\left. \left. - \left(1 + p \cdot F_q^{h-1} - p \cdot F_q^k\right)^M - Mp(1-p)^{M-1}(F_q^h - F_q^{h-1})\right\} \right] \qquad (8)$$

$$P_{RTS_{max}} =$$

$$\frac{1}{1-(1-p)^M - Mp(1-p)^{M-1}} \left\{ (1-p \cdot F_q^k)^M - Mp(1-p)^{M-1}(1-F_q^k) - (1-p)^M \right\}$$
$$\qquad (9)$$

When k is infinity, F_q^k is equal to 1 and we can confirm that equation (8) is exactly equal to equation (5) and $P_{RTS_{max}} = 0$. Also, if $k = 0$, we can easily see $P_{RTS_{max}} = 1$ and $P_{m_{max}} = 0$

4   Delay Analysis

As we stated earlier, a virtual transmission time (t_v) is defined as the time interval between consecutive successful transmissions, not the actual transmission time for a packet. By looking at this from a different point of view, we can consider t_v as the average time between packet departures from a system. If the

Table 1. System Parameters

TIME SLOT	20 μsec	RTS	20 byte
SIFS	10 μsec	CTS	14 byte
DIFS	50 μsec	ACK	14 byte
Bit Rate	2 Mbps	CW_{min}	32
Maximum propagation delay (τ)	2 μsec	CW_{max}	256
MAC Header	34 byte	Packet propagation delay (δ)	1μsec

system is to be stable, this period must equal the average time for new packet generation. Hence, the average arrival rate of packets into the system (λ), can be expressed as: $\lambda = 1/t_v$. According to Little's formula [7], we can derive the expected transmission delay ρ_d as follows:

$$\rho_d = \frac{\Delta}{\lambda} - DIFS = M \cdot t_v - DIFS \qquad (10)$$

where Δ means the average number of packets in the system and Δ is regarded as always equal to the number of stations (M) that attempt to transmit. Although delay is an important metric of system performance, no previous work that we are aware of, validates the formula to estimate the delay.

5   Model Validation

The theoretical models were validated by using the **ns-2** simulator. The parameter values used in the simulation are shown in Table I. We simulated a network topology where all stations were within range of each other, and had the same distance to an AP with propagation delay 1 μsec. Each station generates data traffic directed towards the AP, while the AP only generates ACK and CTS. We estimate throughput by measuring the total number of bytes received at the AP and delay by measuring the time interval from when a packet is generated until it is successfully acknowledged.

We ran 5 simulations for each value of M (from 2 to 50), each simulation with a different random seed, each lasting a total of 120 seconds, and in each case we discarded the first 20 seconds. A packet length is sampled from geometric distribution with six different average lengths (100-1000 bytes). RTSThreshold (k) is set to the average packet length in the hybrid model.

Figures 2 plots capacity and delay for both mechanisms ($M = 5, 20$, and 50), as the average packet length increases. The average values of the 5 simulation results are represented as symbols. The analytical result are shown as lines. We confirm that 95% confidence interval obtained from simulation results are quite small, although they are not shown due to the limitation of space.

6   Performance Analysis

In Section 5, we showed that the analytical model closely approximates the behavior of the 802.11 standard. We decided to use it to examine the maximum

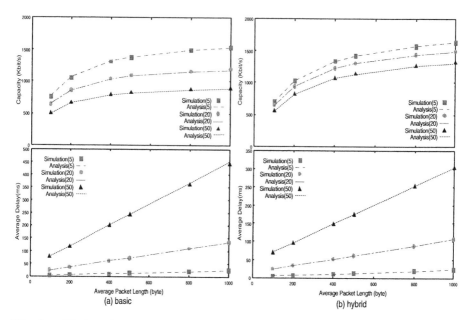

Fig. 2. Notice that the tops show channel capacity and the bottoms show average packet transmission delay for (a) basic and (b) hybrid mechanism. The line (the results by the analytic model) go through symbols (the results by our simulations)

performance limit. Hereafter, we will use the capitalized Hybrid to mean the hybrid model inspired by an optimum RTSThreshold (k), and the term RTS/CTS to refer to the hybrid model with $k=0$.

6.1   The Optimum Parameter Value

According to the equation $\rho_c = \overline{m}/t_v$, the maximum capacity ρ_c is achieved by minimizing t_v. Surprisingly, as expressed in the equation (10), the minimum t_v results in the minimum transmission delay. We can conclude from the analytical model that maximum throughput leads to minimum delay.

Figure 3(a) shows the possible throughput for CSMA/CA, when p varies from 0 to 0.05. We can see that the maximum throughput is almost independent of the number of stations. As the number of station increases, however, the accurate selection of the optimum values for p becomes important. Figure 3(b) plots throughput for hybrid [2] when k varies from 0 to 470. The range of k corresponds to 0-2350 of RTSThreshold. We can see that as the number of station increases, the optimum k gets smaller. k is, however, resistant to the deviations from the optimum value. As examined in [2], the optimum p can be estimated as

---

[2] p is set to the value used in the original CSMA/CA.

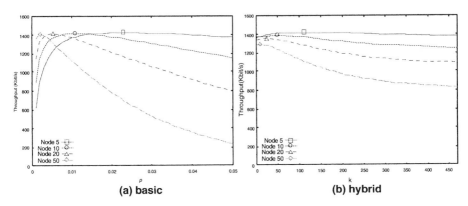

(a) basic   **(b) hybrid**

Fig. 3. Our basic model (a) shows that the maximum throughput for CSMA/CA is almost same for different M. The sensitivity of p, however, increases quickly for larger M. Our hybrid model (b) shows that the optimal k drops as M increases. But, it is resistant to the deviations from the optimum value ($\overline{m} = 500$ bytes)

a function of the number of transmitting stations M. Because a small estimation error in M leads to a dramatic decrease in throughput, we suggest that tuning p to a precisely optimum value will be hard in practice.

To examine the relationship between p and k, Figure 4 plots throughput versus p for several different values of k in the case of $\overline{m}$=500byte. We show only the case for which the number of users M=50, but other number cases showed similar trends. We can observe from Figure 4 that the sensitivity of p is fully dependent on the value of k, and we can see that it becomes less important as k approaches 0. This is reasonable since a larger packet can be more exposed to collisions, as k grows larger, and p is required to be carefully determined.

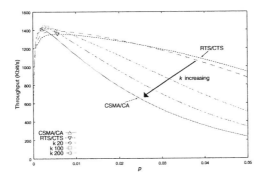

Fig. 4. As k increases, the sensitivity of p is mitigated

6.2   Performance Comparison

We are now ready to compare the performance of five mechanisms: CSMA/CA,
RTS/CTS, Hybrid, CSMA/CA+, and RTS/CTS+. "+" indicates that the evaluation is done at the optimum p value for the original mechanism.

We provide throughput-delay curves for all five cases, in Figure 5. The conventional throughput-delay analysis [8] assumes that each user increases traffic
rate based on a Poisson distribution with either fixed or variable packet size, and
the number of users can be finite or infinite. In this paper we assume that each
station generates packets at the maximum rate. So, we present the throughput
versus delay graphs by increasing packet length with the fixed number of stations
(M=5 and M=50). Each plot shows the performance that is reached for each of
the five mechanisms at the specified packet lengths.

From these graphs, we can make three observations. (1) p is the most effective
parameter that we can use to improve system performance for a large number
of users. Figure 5(b) illustrates how CSMA/CA+ achieves highest throughput
and lowest delay. RTS/CTS+ follows. (2) the performance of RTS/CTS+ approaches that of CSMA/CA+ as packet length grows. From this observation, we
can deduce that the theoretical maximum performance achievable by p-persistent
CSMA/CA is very close to that achieved by p-persistent RTS/CTS. This suggests that the sensitivity of p can be mitigated by decreasing k without incurring performance degradation, when packet size is large. (3) For small numbers
of users, the peak performance we can reach by adjusting k is almost equal to
that achieved by tuning p. We can see that the Hybrid and CSMA/CA+ are
practically superimposed in Figure 5(a).

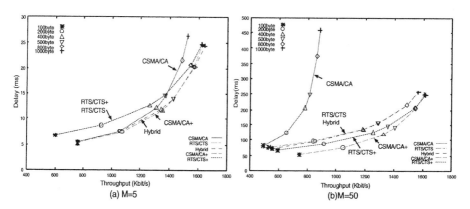

Fig. 5.   Notice that the channel throughput is raised by transmitting larger
packets; yet, the transmission delay will increase proportionally. We can derive
three observations from these graphs (see text)

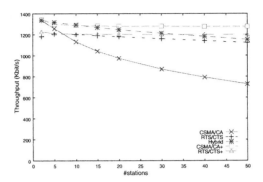

Fig. 6. Our experiments with using packet distribution captured in Internet hold similar trends with Fig. 5

6.3 Performance Analysis on the Current Packet Length Distribution

So far, we have assumed that packet length is sampled from a geometric distribution. In this section, we will show that we will come to the same finding when using a trace of real traffic measured in the current Internet. Our generalized analytical model using a multinomial distribution allows us to examine any packet distribution The trace we used was captured on the MCI backbone in 1997 [9]. Traced data length is divided into 5-byte (a slot time length) bins. The average packet length of the trace data was about 355 bytes.

Figure 6 shows the throughput for all five mechanisms when using the trace data. Unlike the previous graphs, the x-axis represents the number of transmitting stations. We can confirm our findings by comparing the trace result with a geometric distribution: once again, p is the most effective tuning parameter particularly for a large number of users; Hybrid achieves similar performance levels as those of CSMA/CA+ up to 15 stations; since the average packet length is relatively small, we don't observe that throughput for RTS/CTS+ approaches that of CSMA/CA+.

While CSMA/CA+ requires a precise selection in its optimum p, Hybrid is resistant to variations of k. Moreover, the advantage of Hybrid against CSMA/CA+ does not require a modification of the current 802.11 standard. Because of this, we suggest that Hybrid is the most suitable mechanism to employ when the number of stations is not large.

7 Conclusion

Our main contribution in this paper is the extension of the analytical model in [2], so that it is possible to analyze delay and throughput of the 802.11 DCF with a general packet length distribution. Our experiments lead to three findings. First, p is the most effective tunable parameter to improve system performance, but it is very sensitive to deviations from its optimum value. Second, the sensitivity of p can be reduced by decreasing k, without incurring performance degradation when packet length is large. Finally, we observed that the peak performance reached by adjusting k is almost equal to that achieved by tuning p for a small number of users. While p requires to be precisely tuned, k is very resistant to deviations from its optimum point. Moreover, the advantage of adjusting k does not require modifying the current 802.11 standard. We conclude that optimizing k is enough to achieve possible performance limits in networks with a small number of users.

References

[1] Bharghavan, V., Demers, A., Shenker, S., Zhang, L.: "MACAW: A media access protocol for wireless LANs". In: Proceedings, 1994 SIGCOMM Conference, London, UK (1994) 212–225
[2] Cali, F., Conti, M., Gregori, E.: Dynamic Tuning of the IEEE 802.11 Protocol to Achieve a Theoretical Throughput Limit. IEEE/ACM Transactions on Networking 8 (2000) 785–799
[3] Crow, B., Widjaja, I., Kim, J., Sakai, P., the, I.: "Investigation of the IEEE 802.11 Medium Access Control (MAC) Sublayer Functions.". In: Proceedings of IEEE INFOCOM'97., Kobe, Japan (1997)
[4] Weinmiller, J., Woesner, H., Wolisz, A.: "Analyzing and Improving the IEEE 802.11-MAC Protocol for Wireless LANs". In: Proceedings of the Fourth International Workshop on Modeling, Analysis and Simulation of Computer and Telecommunication Systems, San Jose, California, IEEE (1996) 200–206
[5] Bianchi, G.: Performance Analysis of the IEEE 802.11 Distributed Coordination Function. IEEE Journal on Selected Areas in Communications 18 (2000) 785–799
[6] IEEE: "802.11 Draft Standard for Wireless LAN medium Access Control (MAC) and Physical Layer (PHY) Specification" (1997)
[7] Allen, A. O., ed.: "Probability, Statics, and Queueing Theory with Computer Science Applications 2nd ed.". Academic Press (1990)
[8] Kleinrock, L., Tobagi, F.: "Packet switching in radio channels: Part I - Carrier sense multiple-access modes and their throughput-delay characteristics". In: IEEE Tran. Commun. Volume Com-23. (1975) 1400–1416
[9] Paxson, V.: http://www.nlanr.net/NA/Learn/packetsizes.html (1996)

A Bluetooth Scatternet Formation Algorithm for Networks with Heterogeneous Device Capabilities

Dimitri Reading-Picopoulos * and Alhussein A. Abouzeid

Electrical, Computer and Systems Engineering, Rensselaer Polytechnic Institute
Troy, New York 12180-3590
ug99058@ee.ucl.ac.uk
abouzeid@ecse.rpi.edu

Abstract. This paper focuses on Bluetooth, a promising new wireless technology, developed mainly as a cable replacement. We argue that, in practice, Bluetooth devices will have different power capabilities, classifying them as either high-power or low-power nodes. We propose a deterministic, distributed algorithm that accounts for the physical properties of devices, connecting nodes into a scatternet of small diameter. The proposed protocol results in a high effective throughput and allows components to arrive and leave arbitrarily, dynamically updating the cluster formation. Performance is evaluated through extensive *ns-2* simulations.

1 Introduction and Problem Statement

Bluetooth [1, 2, 3] is rapidly emerging as the leading technology in the formation of short-range wireless *ad hoc* networks. The standard provides for low power wireless communication and operates in the 2.4GHz Industrial, Scientific and Medical band. Bluetooth connections are based on a master-slave configuration and employ a frequency hopping spread spectrum.

Bluetooth devices connect into *piconets*, each consisting of a master and up to seven slaves, while master-slave communication is achieved through time division duplexing. The Bluetooth specification [1] facilitates the connection of piconets into *scatternets* through common nodes. These devices can either participate in both piconets as slaves (*bridge device*) or as a master in one and as a slave in the other. The specification, however, does not allude to a specific mechanism by which scatternet formation is to be achieved.

This paper presents a new scatternet formation protocol that takes into account the physical limitations of the devices themselves. The resulting clustering has several attractive features and proves the importance of an efficient clustering algorithm.

Miklós *et al.* [4] take a statistical approach in studying the relationship between scatternet design rules and performance parameters. Raman *et al.* [5] argue

---

* This work was performed while on an exchange program from University College London, UK, at Rensselaer Polytechnic Institute, Troy, NY, USA.

H.-K. Kahng (Ed.): ICOIN 2003, LNCS 2662, pp. 295–305, 2003.
© Springer-Verlag Berlin Heidelberg 2003

for extensive cross-layer optimizations in Bluetooth scatternets, while Salonidis *et al.* [6] discuss the issue of fast connection establishment between Bluetooth devices and propose a symmetric protocol.

Ramachandran *et al.* [7] present two scatternet formation algorithms, but, in contrast to our work, no constraint was imposed on the number of *roles* assumed by the devices. The solutions to efficient scatternet formation presented by [8, 9, 10] are limited by the lack of a mechanism for minimizing the number of piconets in the resulting scatternet. An asynchronous distributed protocol for scatternet construction is presented by Salonidis *et al.* [11], while another scatternet formation algorithm is suggested by Law *et al.* [12], where the scatternet is formed once a leader is elected.

All previous works mentioned above have been performed without taking into consideration the physical limitations of the host units. Not all devices have the required processing power or battery lifetime to sustain a large number of slaves. We propose an algorithm that accounts for such restraints.

Bluetooth link formation is a two-step process with devices having to go through the *inquiry* and *page* states prior to establishing a connection. The purpose of the inquiry procedure is for a master node to obtain the Bluetooth 48-bit MAC address (BD_ADDR) and native clock (CLKN) of devices, which lie within its communication range. Connections are subsequently established through the paging mechanism using the information acquired during the inquiry procedure. Device discovery is potentially a time consuming process, while paging delays are much smaller. The complete state transition sequence, leading to a master-slave connection, is illustrated in Figure 1.

The purpose of any distributed Bluetooth scatternet formation algorithm is the clustering of any group of asynchronous, isolated, Bluetooth devices to

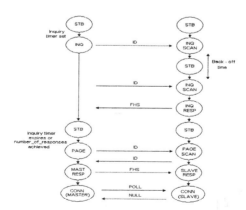

Fig. 1. State transitions leading to connection. STB is the STANDBY (idle) state. ID packets contain an inquiry access code, while Frequency Hop Synchronization (FHS) packets contain the information needed for frequency hop channel synchronization

permit the communication of information between any pair of nodes. Such an algorithm should account for the following.

- Account for the nature of the host device itself, mainly its power limitations and battery lifetime.
- Restrict device participation to, at most, two piconets.
- Not all devices have to be within communication range of one another.
- Maximize the number of devices per piconet, thereby minimizing the number of piconets in the scatternet.
- The resulting network topology should be able to reconfigure, without causing long periods in the loss of connectivity, in the case of nodes arbitrarily joining or leaving the network.
- Path latency – the number of hops between any pair of devices – should be minimized.
- For simplicity, communication loops should be avoided.

The following section outlines our new scatternet formation algorithm. Section 3 introduces the simulation results, while section 4 concludes with some future research possibilities.

2   Scatternet Formation Algorithm

In this section, we present our Bluetooth scatternet formation algorithm based on the properties mentioned above. We differentiate between the following types of device.

High_Power: A host device with high processing power and adequate battery lifetime, capable of connecting, as master, with up to X nodes, of which up to Y are other high_Power nodes acting as slaves and the rest, W, are low_Power devices acting as slaves. Ideally, $Y = 0$ and $X = W$.

Low_Power: Host peer with low processing power, ideally acting as slave. Only under extreme circumstances will such a device connect as master of a piconet, in which case it is limited to Z or less other low_Power nodes acting as its slaves. A special case of a low_Power device is a secure_Device.

Secure_Device: A device that can only send and can neither receive nor forward information. Such a node can only connect as slave and then in a single cluster.

High_ and low_Power nodes assume one of the following roles when connected.

High_Power_Master: A high_Power peer that connects as master of a piconet.

Low_Power_Slave: low_Power node that acts as slave in the resulting topology.

High_Power_Slave: high_Power device forced to assume the role of a slave.

Low_Power_Master: low_Power node forced to act as master.

The algorithm is performed at every node and can be viewed as four stages.

2.1   Stage 1: Piconet Formation

Devices start off isolated, with no prior knowledge of the presence of other nodes. Stage 1 follows the Bluetooth specification [1], but is adapted to the existence of two types of device. By the end of the stage, the n nodes are partitioned into piconets, with a few devices potentially left unconnected. It is assumed that each device knows whether it is a high_ or a low_Power device.

A high_Power peer will enter Stage 1 of the algorithm in the INQUIRY state and wait for an INQUIRY RESPONSE for a period of T_{Inq} seconds. Upon reception of the FHS packet, it will subsequently enter the PAGE state and form the connection with the scanning device as its slave. This cycle is repeated until either T_{Inq} expires or the specified number of responses is achieved.

```
PICONET FORMATION
if (node is a high_Power device) then
  if((number of high_Power_Slaves)<Y and ((number of high_Power_Slaves) +
  (number of low_Power_Slaves))<X) then
    while (forever) perform INQUIRY
      if (INQUIRY RESPONSE received) then
        go to PAGE state →CONNECTION, go to PICONET FORMATION
      if (T_Inq expires or num_Responses achieved) then go to SCATTERNET FORMATION
    endwhile
  else go to SCATTERNET FORMATION
endif
```

Each time the high_Power node enters the INQUIRY state, T_{Inq} gets reset. Devices assigned as low_Power peers enter Stage 1 in the INQUIRY SCAN sub-state. Upon reception of an ID packet, such a device enters the PAGE SCAN state and subsequently connects to the paging unit. In the special case where the node is a secure_Device, the unit will then defer to the COMMUNICATION stage; otherwise it will enter the SCATTERNET FORMATION stage of the protocol. In the case where its clock, $T_{Inq-Scan}$, times out, the device will jump to the SCATTERNET FORMATION stage with secure_Devices returning to the INQUIRY SCAN sub-state.

```
PICONET FORMATION
if (node is a low_Power device) then
  while (forever) perform INQUIRY SCAN
    if (INQUIRY packet received) then
      send INQUIRY RESPONSE, enter PAGE SCAN→CONNECTION
      if (node is a secure_Device) then go to COMMUNICATION
      else go to SCATTERNET FORMATION
    if (T_Inq-Scan expires) then
      if (node is a secure_Device) then go to PICONET FORMATION
      else go to SCATTERNET FORMATION
  endwhile
endif
```

2.2   Stage 2: Scatternet Formation

The aim of Stage 2 is the interconnection of the piconets and the remaining isolated devices into a tree topology that spans the entire n nodes. Following a recommendation put forward by Salonidis et al. [6, 11], devices enter the second stage alternating between the INQUIRY and INQUIRY SCAN sub-states. The amount of time that each unit remains in a particular sub-state dictates T_{I-IS}, the overall time that a node should spend in Stage 2 attempting to connect into a scatternet. A unit alternating between the two states will enter the PAGE procedure upon reception of an INQUIRY RESPONSE sent by a device performing INQUIRY SCAN. The two nodes will then connect with the master device returning to Stage 2. Likewise, upon reception of an ID packet, the node will check for the possibility of the creation of a communication loop. Detecting a loop forces the node to return to the beginning of the second stage without resetting T_{I-IS}, otherwise, an INQUIRY RESPONSE is sent. Upon connection, the unit will reset T_{I-IS} and return to Stage 2, unless it connects as a bridge device, in which case it defers to Stage 3. If T_{I-IS} times out, the device defers to the third stage of the protocol.

```
SCATTERNET FORMATION
if ((high_Power peer and (number of high_Power_Slaves)<Y and ((number
    of high_Power_Slaves) + (number of low_Power_Slaves))<X) or
    (low_Power peer and (number of low_Power_Slaves)<Z)) then
    alternate between INQUIRY and INQUIRY SCAN states
    if (INQUIRY RESPONSE received) then
        enter PAGE state →CONNECTION, go to SCATTERNET FORMATION
    if (INQUIRY packet received) then
        if (loop detected) then do not stop clock, return to SCATTERNET FORMATION
        else send INQUIRY RESPONSE, enter PAGE SCAN→CONNECTION
            if (device becomes bridge) then go to SCATTERNET REORGANIZATION
            else reset clock, go to SCATTERNET FORMATION
    if (clock expires) then go to SCATTERNET REORGANIZATION
else perform INQUIRY SCAN
    if (INQUIRY packet received) then
        if (loop detected) then do not stop clock, return to SCATTERNET FORMATION
        else send INQUIRY RESPONSE, enter PAGE SCAN→CONNECTION
            go to SCATTERNET REORGANIZATION
    if (clock expires) then go to SCATTERNET REORGANIZATION
endelse
```

Loop detection: We propose a mechanism, which makes use of a DIAC (*Dedicated Inquiry Access Code*), which permits only restricted classes of devices to be inquired upon, in the transmission of ID packets by a node in INQUIRY.

Upon the establishment of a master-slave connection, the slave device will form a table, listing the BD_ADDR of its master. If this slave subsequently forms a piconet acting as a master, it will broadcast its table to all of its slaves, which will also append their own master's BD_ADDR. Similarly, each device forming a piconet as a master will request and acquire the tables of each of its slaves combining them to form its own table with the addition of its own BD_ADDR. This procedure ensures that every node in the network holds information on all master devices that it can access. Device table entries are appropriately updated every time a new master enters or leaves the network.

Given the "global" view that each node has of the network, a device in the INQUIRY sub-state will transmit ID packets containing its master's BD_ADDR as a DIAC. A node in INQUIRY SCAN will capture the DIAC and compare it to the entries in its table. Only if a match is not found do the two devices proceed in forming the connection. If a node in INQUIRY does not have a master it uses its own BD_ADDR as the DIAC.

2.3   Stage 3: Scatternet Reorganization

Stage 2 results in a tree topology that covers the entire network of devices and ensures connectivity between any pair of nodes. The resulting clustering, however, is not optimized in terms of minimum number of piconets and minimum path latency, as discussed earlier. Stage 3 seeks to reorganize the clustering in order to maximize network performance. We denote two master devices as x and y. These two units will either be connected directly, with one configured as a slave, or via a third, bridge, device. If both masters are either high_Power_Masters, or both are low_Power_Masters, the master with the highest number of slaves assumes the role of x. In the case where both peers have the same number of children, the master with the highest address becomes x. Similarly, in the situation where the two masters initiated the protocol as devices of opposite class, the master that started off as a low_Power node is configured as y. The low_Power_Slaves of x and y are represented as lp_S(x) and lp_S(y), while their high_Power_Slaves as hp_S(x) and hp_S(y) respectively. Each master device will run the procedure SCATTERNET REORGANIZATION for each master to whom it is directly connected and for every master separated through a bridge.

```
SCATTERNET REORGANIZATION
if (master a high_Power_Master or a low_Power_Master or a
   (single device and a high_Power peer)) then
   if (master has no one hop connection) then go to PICONET FORMATION
   if (one hop devices are not masters nor slaves in another piconet) then
      go to COMMUNICATION
   else go to COMBINE OR TRANSFER
else   if (one hop device is a master) then
           wait for scatternet reorganization, go to COMMUNICATION
       else go to PICONET FORMATION
endelse
```

The COMBINE OR TRANSFER subroutine performs the actual topology reformation and differentiates between the situation where two piconets can be merged into a single cluster and the case where devices can be transferred from one piconet to the other.

```
COMBINE OR TRANSFER
if (x, y are both low_Power_Masters) then
   if ((|lp_S(x)| + |lp_S(y)|)≤Z) then go to COMBINE.A
   else go to TRANSFER.A
else   if ((|lp_S(x)| + |hp_S(x)| + |lp_S(y)| + |hp_S(y)|)≤X) then go to COMBINE.B
       else go to TRANSFER.B
endelse
```

The master y will *convey* a slave device A by terminating their connection, with A entering the PAGE SCAN mode prior to connecting as a slave to master x. If the slave A cannot become a slave of master x (e.g. because it is out of its communication range or because A does not actually exist) then A is deemed *unavailable*. The procedure COMBINE.A is only run in the case where both x and y are low_Power_Masters with a combined number of slaves ≤Z.

is abstract enough as to provide for proficient clustering in wireless networks where link establishment is based on a master-slave relationship. Devices are divided into two categories depending on their power characteristics or other criteria and start off isolated with no prior knowledge of the presence of other nodes within their surroundings. The algorithm connects devices into a single scatternet, while attempting to minimize the number of piconets in the process. No range constraint is imposed and the resulting topology is dynamically updateable to provide for devices arbitrarily leaving or wishing to join the network.

The simulation results show that making the number of high_Power devices adaptive to the network population greatly reduces scatternet connection delays. Network performance and resource utilization are maximized in the ideal situation of a 3:2 ratio between low_ and high_Power nodes and minimized in the case of networks containing 100% low_Power devices. Our future work includes quantitatively assessing the power savings obtained through the implementation of our algorithm and extending the proposed protocol to account for mobility.

References

[1] Specification of the bluetooth system. http://www.bluetooth.com/ [October 2001].

[2] J. Bray and C. F. Sturman. *Bluetooth Connect Without Cables*. Prentice Hall, 2001.

[3] J. C. Haartsen and S. Mattisson. Bluetooth - a new low-power radio interface providing short-range connectivity. In *Proceedings of the IEEE*, volume 88, pages 1651–1661, October 2000.

[4] G. Miklós, A. Rácz, Z. Turányi, A. Valkó, and P. Johansson. Performance aspects of bluetooth scatternet formation. In *Proceedings of The First Annual Workshop on Mobile Ad Hoc Networking and Computing*, pages 147–148, 2000.

[5] B. Raman, P. Bhagwat, and S. Seshan. Arguments for cross-layer optimizations in bluetooth scatternets. In *Proceedings of Symposium on Applications and the Internet*, pages 176–84, 2001.

[6] T. Salonidis, P. Bhagwat, and L. Tassiulas. Proximity awareness and fast connection establishment in bluetooth. In *First Annual Workshop on Mobile and Ad Hoc Networking and Computing*, pages 141–142, 2000.

[7] L. Ramachandran, M. Kapoor, A. Sarkar, and A. Aggarwal. Clustering algorithms for wireless ad hoc networks. In *Proceedings of the 4th International Workshop on Discrete Algorithms and Methods for Mobile Computing and Communications*, pages 54–63, Boston, MA, USA, August 2000.

[8] G. Tan, A. Miu, J. Guttag, and H. Balakrishnan. Forming scatternets from bluetooth personal area networks. Technical Report MIT-LCS-TR-826, MIT, 2001.

[9] Z. Wang, R. J. Thomas, and Z. Haas. Bluenet - a new scatternet formation scheme. In *Proceedings of the 35th Annual Hawaii International Conference on System Sciences (HICSS-35)*, Big Island, HI, USA, January 2002.

[10] G. V. Záruba, S. Basagni, and I. Chlamtac. Bluetrees - scatternet formation to enable bluetooth-based ad hoc networks. In *Proceedings of the IEEE International Conference on Communications*, volume 1, pages 273–277, Helsinki, Finland, 2001.

```
COMBINE.A
if (low_Power_Slave of y available) then
   convey low_Power_Slave from y to x, go to COMBINE.A
else if (x and y connected via bridge device and y available) then
         y disconnects from bridge, y connects to x as slave, go to COMMUNICATION
      else go to COMMUNICATION
endelse
```

For y and x both being low_Power_Masters with a joint sum of low_Power_Slaves $>Z$, a number of slaves will be *conveyed* from y to x with the procedure terminating once x obtains Z low_Power_Slaves.

```
TRANSFER.A
if (|lp_S(x)| < Z) then
   if (x and y connected via bridge device and y available) then
      y disconnects from bridge, y connects to x as slave, go to TRANSFER.A
   else if (low_Power_Slave of y available) then
```

```
         convey low_Power_Slave from y to x, go to TRANSFER.A
      else go to COMMUNICATION
else go to COMMUNICATION
endelse
```

In the case where at least one of the two masters is a high_Power_Master, the equivalent procedures to COMBINE.A and TRANSFER.A are the following.

```
COMBINE.B
if (low_Power_Slave of y available) then
   convey low_Power_Slave from y to x, go to COMBINE.B
else if (high_Power_Slave of y available and |hp_S(x)| < Y) then
         convey high_Power_Slave from y to x, go to COMBINE.B
      else if (x and y connected via bridge device and y available) then
            if ((y is a high_Power_Master and |hp_S(x)| < Y) or
               (y is a low_Power_Master)) then
               y disconnects from bridge, y connects to x as slave, go to COMMUNICATION
            else go to COMMUNICATION
         else go to COMMUNICATION
endelse
```

```
TRANSFER.B
if ((|lp_S(x)| + |hp_S(x)|)<X) then
   if (x and y connected via bridge device and y available) then
      y disconnects from bridge, y connects to x as slave, go to TRANSFER.B
   else if (low_Power_Slave of y available) then convey low_Power_Slave from y to x
            if ((|lp_S(x)| + |hp_S(x)|)<X) then
               if (high_Power_Slave of y available) then
                  if (|hp_S(x)|<Y) then convey high_Power_Slave from y to x, go to TRANSFER.B
                  else     if (low_Power_Slave of y unavailable) then go to COMMUNICATION
                        else go to TRANSFER.B
               else go to TRANSFER.B
            else go to COMMUNICATION
      else if (high_Power_Slave of y available and |hp_S(x)|<Y) then
            convey high_Power_Slave from y to x, go to TRANSFER.B
         else go to COMMUNICATION
else go to COMMUNICATION
endelse
```

Efficient clustering can be achieved through scatternet reorganization provided the following rules are satisfied.

- In order to avoid an excessively high degree of inter-piconet overhead, either one of the two masters, x or y, has to run Stage 3 of the protocol with the complementary unit in COMMUNICATION.

- Piconets must be capable of updating dynamically. When two piconets are reconfigured through Stage 3, all participating devices will be instructed to inform their other masters (if they have any) to enter the third stage.

2.4   Stage 4: Communication

Devices in the COMMUNICATION stage spend their time between communicating information and trying to improve the overall connectivity of the network by periodically returning to Stage 2. The time spent by master devices between transitions, is made inversely proportional to their number of slaves. Equivalently, the inter-procedure switching of slaves is a function of the amount of traffic handled by the node, since idle units will be more willing to connect into new piconets. By permitting devices to return to earlier stages of the protocol, network healing and device assimilation can easily be accounted for.

3   Simulation Results

In this section we evaluate the performance characteristics of our protocol using *Bluehoc*, a Bluetooth performance evaluation tool developed by IBM [13], which provides a Bluetooth extension for *Network Simulator* [14]. The timer values defined in the first stage of our algorithm follow the recommended values given in the Generic Access Profile (GAP) of the Bluetooth specification [1]. When devices alternate between INQUIRY and INQUIRY SCAN, the mean, per state, residence time T_{Res}, based on the results obtained by Salonidis *et al.* [6], was chosen at $T_{Res} = 600msec$. Through simulations we found that a timeout period of 2.60 seconds for Stage 2 would complement the above value of T_{Res}.

The Bluetooth specification restricts $X = 7$. Using typical power consumption specifications we limit $Y = 4$ and $Z = 3$. In all simulations, Bluetooth units arrive uniformly over a *10 second* window and are placed randomly over a given geographical area. Not all devices are within communication range of each other and nodes are assumed to be static or of low mobility for the duration of the protocol. Results are plotted as the average of 10 simulation runs.

Scatternet formation time is dictated by the ratio of low_ to high_Power nodes in a particular geographical region. Devices configured as low_Power spend

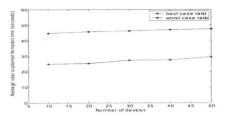
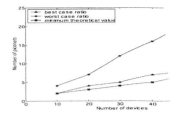

Fig. 3. Time required for scatternet formation, including reconfiguration time. Note that the devices arrive randomly within a 10 second time window, also included in the above results

Fig. 4. Number of piconel resulting scatternets comp the minimum theoretical va

roughly three times more time, compared to high_Power peers, running t stage of the algorithm in order to increase the probability of being disc by an inquiring device. Figure 2 shows the average time taken for a n establish its first connection, for a static number of 10 units, as a func increasing number of high_Power nodes. As expected, connection time i imized when all 10 devices are low_Power and is minimized when 40% nodes are high_Power. It is obvious that making the number of high_ nodes adaptive to the network population can result in a low first conn setup time. In the rest of this section we will refer to the $3:2$ ratio of lc high_Power peers as the *best case ratio* and the *worst case ratio*, the situ where all devices are low_Power.

Figure 3 illustrates the average time taken for scatternet formation, incl reorganization, as a function of the number of devices within considerɛ Both case ratios demonstrate an increase in topology construction time a network size increases. For a given number of nodes, the *worst case ratio* re in a clustering time of roughly twice the duration compared to the time u the *best case ratio*.

Finally, Figure 4 compares the theoretical minimum number of pico with the average number of piconets for the *worst* and *best case ratios*. It be easily shown that the theoretical minimum number of piconets is giver $max(\frac{n}{X+1}, \frac{NHP}{Y+1})$, where NHP is the number of high_Power nodes. Inter-picɛ interference arises from adjacent piconets sharing the same frequency hop] sequence and results in a high degree of packet loss through repetitive c sions. Minimizing the number of piconets per scatternet serves to reduce deterioration in network performance.

4   Conclusion and Future Work

In this paper we present and describe a new scatternet formation protocol. E though the emphasis is put on the Bluetooth technology, the proposed algorit]

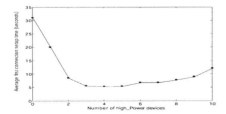

Fig. 2. Establishing the *best* and *worst case ratios*

[11] T. Salonidis, P. Bhagwat, L. Tassiulas, and R. LaMaire. Distributed topology construction of bluetooth personal area networks. In *Proceedings of the 20th Annual Joint Conference of the IEEE Computer and Communications Societies*, 2001.

[12] C. Law, A. K. Mehta, and K.-Y. Siu. Performance of a new bluetooth scatternet formation protocol. In *Proceedings of the ACM Symposium on Mobile Ad Hoc Networking and Computing 2001*, Long Beach, California, USA, October 2001.

[13] Bluehoc: Bluetooth ad-hoc network simulator. `http://oss.software.ibm.com/developerworks/projects/bluehoc/` [October 2001].

[14] The network simulator ns-2. `http://www.isi.edu/nsnam/ns/` [October 2001].

CERA: Cluster-Based Energy Saving Algorithm to Coordinate Routing in Short-Range Wireless Networks

Juan-Carlos Cano[1], Dongkyun Kim[2], and Pietro Manzoni[1]

[1] Department of Computer Engineering, Polytechnic University of Valencia
Camino de Vera s/n, 46071 Valencia, Spain
{jucano,pmanzoni}@disca.upv.es
[2] Department of Computer Engineering, Kyungpook National University
Daegu, Korea
dongkyun@knu.ac.kr

Abstract. We propose a simple protocol that allows mobile nodes to autonomously create clusters to minimize the power consumption and route packets based on the clusters in the mixed wired/wireless network. Our proposed protocol, called *Cluster-based Energy-saving Routing Algorithm* (CERA) is implemented as two separate components: the intra-cluster data-dissemination protocol and the cluster routing protocol. Our protocol focuses on simplicity as well as sufficient robustness and self-recoverability to enable simple and fast node reconfiguration. Simulation results using ns-2 simulator show that, in general, the CERA implementation allows up to 25% of energy saving while keeping the overhead extremely low.

1 Introduction

A *mobile network* consists of a collection of freely movable nodes which communicate each other using wireless links. A new type of wireless networking technologies, the *short-range wireless networks* [1], is becoming more pervasive because of the increased availability of cheap and powerful portable terminals like personal digital assistants, laptops, etc.

We can categorize the short-range wireless networks into two general classes: the *Wireless Personal Area Networks* (WPAN) and the *Wireless Local Area Networks* (WLAN). WPAN technologies (e.g., *Bluetooth*) favor low cost and low power consumption rather than range and peak speed, while WLAN technologies, like the *IEEE 802.11*, pursuit higher speed and longer range rather than cost and power consumption. Most nodes in mobile networks rely on batteries for their lifetime. Energy saving, despite recent advances in extending battery life, is still an important issue and the energy usage optimization can be considered an important design criteria [2].

Various research projects aim to improve interoperability among wireless networks, like: the *Bay Area Research Wireless Access Network* (BARWAN)

H.-K. Kahng (Ed.): ICOIN 2003, LNCS 2662, pp. 306–315, 2003.

Project, the *Cellular IP*, and the *Smart Dust* project [3]. In this paper, we propose a simple protocol that allows mobile nodes to autonomously create clusters to minimize the power consumption and routes packets using the clusters in the mixed wired/wireless network. The protocol, called *Cluster-based Energy-saving Routing Algorithm* (CERA) is implemented as two separate components: the intra-cluster data-dissemination protocol and the cluster routing protocol.

The intra-cluster data-dissemination protocol groups nodes around a special one, called the *cluster leader* (CL). The cluster leader is in charge of cluster's maintenance and communication. We periodically distribute the cluster leader's role among all the nodes inside the cluster in order not to overload a single node. The CL centralizes also the *power management* mechanism and acts as a proxy for data transfer between the cluster and the rest of the nodes. It assumes the burden of buffering data frames for power saving stations and delivering them on stations request, allowing the mobile stations to remain in their power saving state for much longer periods.

The rest of this paper is organized as follows. Section 2 presents some assumptions of our protocol for our overall network architecture. Sections 3 and 4 detail the intra- and inter cluster routing protocols. Section 5 provides us with some discussion points. Section 6 presents the performance of our protocol and Section 7 closes this paper with some concluding remarks.

2   Some Assumptions on the Overall Protocol Architecture

Most cluster-based routing protocols organize the network into groups of nodes, called *clusters*. The main objective is to make a dynamic network appeared "less" dynamic [4]. We concentrated our protocol on the simplicity issue from the perspective of implementation. Our CERA protocol operates between the *Network Interface Layer* (NIL) and the IP layer. The network interface layer includes the physical layer and the medium access control (MAC) sub-layer. This layer deals with the specific physical aspects of each different network technology. Section 3 describes the details of the clustering algorithms. Section 4 describes the details of the routing aspects.

3   The Intra-cluster Data-Dissemination Protocol

We model the overall network as a undirected graph G with a finite nonempty set of vertices $V(G)$ (the mobile nodes) and a nonempty set of edges $E(G)$ (the links). We suppose the set $E(G)$ to vary with time, that is $E(G)(t)$. The CERA intra-cluster data-dissemination protocol is based on a graph-partitioning algorithm. We suppose each domain to be a subgraph D_i induced on G by the nonempty sets of nodes contained in each of the $1 \leq i \leq h$ domains of the overall network. We suppose that D_i is connected, undirected graph with edge weights $w : E(D_i) \rightarrow Z_0^+$. The edge weights represent the power required to send a frame between the nodes connected by the edge (we are supposing symmetrical links).

We consider the spanning connected subgraph D_i' induced on D_i by the edge set J, $\forall e \in J, w(e) \leq w_{max}$. The final goal is to partition D_i' into k, $(D_{i1}', D_{i2}', \ldots, D_{ik}')$, edge-induced sub-graphs, *the clusters*, which are fully connected, that is $\forall v, u \in V(D_{ij}'), v \neq u, \exists e \in E(D_{ij}')$. We require $1 \leq |V(D_{ij}')| \leq B$ where $1 \leq B \leq |V(D_i')|, \forall j \in \{1, 2, \ldots, k\}$.

We use a graph join algorithm [5] to find the various partitions. The join algorithms substitutes two vertices v_1 and v_2 by a unique vertex v_{12}; we do not merge edges from v_1 and v_2 that have the same other endpoint but we create multiple instance of those edges. We define the *join* procedure as follows:

---

procedure $join(H, K)$
    choose randomly $v \in V(K)$
    choose randomly $w \in V(H - K) \mid vw \in E(H)$
    if $(\forall x \in V(K) \, \exists xw \in E(H)) \wedge (|V(K)| < B)$ **then**
        $K \Leftarrow K + \{w\}$
        join(H, K)
    else
        join(H, {w})
    end if

The algorithm is activated in each node by the call: $join(D_i', \{\})$. The algorithm keeps executing autonomously on all nodes thus handling the possible changes in the $E(D_i')$ due to nodes mobility. According to this algorithm, we have two versions of creating the clusters. In the first version, only cluster leaders send HELLO packets to make distributed clusters, while the other requires all nodes to send some information periodically for simplicity of implementation. More details of the protocol on its first version can be found in [6]. In this paper, we therefore present the second version.

We also assume that all nodes in the same cluster should be fully connected in just 1-hop manner. It means that either two nodes selected arbitrarily can reach each other directly in their cluster, or their actual communication should be via a cluster leader (CL). Every node is able to know the list of its 1-hop neighbor nodes by perfoming neighbor discovery procedure.

- All nodes try to create their own clusters periodically every $Period_{CL}$ seconds by broadcasting some information such as ID and the list of their neighbor nodes to their neighbor nodes.
- When node A receives the information broadcasted by node B, node A checks if the list of its neighbor nodes is the subset of that of node B. If so, node A will be contained in node B's cluster whose CL is node B. Otherwise, if the list of node B's neighbor nodes is the subset of that of node A, node B will be contained in node A's cluster whose CL is node A. If the sets are equal, the node whose ID is the lowest value can be the CL of the cluster. If any member which is not contained in each set exists, nodes A and B can be in

individual clusters, further play roles in "gateway (GW)" nodes to connect
different clusters.

- When nodes have only one neighbor node, the nodes create their own clusters
  and they can become CLs of their clusters.
- When nodes detect the change of the list of their neighbor nodes, they repeat
  the above procedure.

We need some investigation of performance comparison according to two
different clustering mechanisms in the future.

4   The Cluster Routing Protocol

The CERA uses an efficient and reactive diffusion protocol to avoid the typical
problems with broadcast protocols based of flooding. Efficiency is achieved by
taking advantage of the energy saving algorithm and by a fast frame discarding
approach. The protocol PDU header allows to discriminate whether a frame
is coming from inside or outside the cluster, and whether the destination is
inside or outside the cluster. Every node knows the other members of its cluster
as mentioned in the above section to create clusters. The DATA type PDUs are
normally handled by CLs only, since the energy saving algorithm forces the other
nodes to be sleeping most of the time. If a node which is not in the Leader state,
receives a DATA type PDU coming from outside the cluster, it must discard it
immediately.

The general behaviour of the cluster protocol is quite straightforward. For
the communication between any two nodes in the same cluster, all nodes become
aware of other nodes' sleeping switch patterns broadcasted in the same cluster.
Hence, the source node can send the packets to the destination when the destination is awake. If the source node doesn't have any knowledge of when the
destination is awake, it should forward the packets to its CL because after all,
the CL can recognize the presence of the node earlier than the source node.

For the communication between two nodes which are in different clusters,
we also have two versions of forwarding the packets to the destination. One is
just to rely on the broadcasting activity in the network for each packet. The
other is to take advantage of the acquired route when the first packet packet is
broadcasted in the network.

For the first version, supposing node a inside cluster A wants to send data
to node b in another cluster B. It sends a DATA frame to its CL by stating that
it is an internally generated frame going outside the cluster. The CL broadcasts
the frame. Whenever a CL receives a frame labeled as coming from outside the
cluster, it first has to check the number of intermediate *hops*. If the packet hopped
more that $\max_{hops}$ times, it is silently discarded. Otherwise, the CL has to check
whether the destination node is inside its cluster or not. If the destination belongs
to its cluster, it is labelled to represent an internal frame and then buffered or
directly sent to the destination node, depending on whether the energy saving
algorithm is active or not. If the destination does not belong to its cluster, the

field of hops is incrementad by 1 and re-broadcasted. We consider that a value for $\max_{hops} = 3$ is a reasonable value to allow early loops detection in relatively small sized domains, e.g., 250×250 meters.

For the second version, they should also rely on the broadcasting capability in the network to find the destination node.

The first data packet will be broadcasted in the network. On receiving the packet, the CL hosting the destination in its cluster can give the source the information on the GWs traversed towards the CL. So, until a route breakage occurs, the path information can be used for further packets transmitted by the CL hosting the source node. This operation can be performed by hop-by-hop as in AODV or by source routing as in DSR.

Therefore, in order to enable the CL to broadcast some message into the network, the CL can achieve some performance improvement if the CL knows which node in its cluster plays a role in functioning as GW. Also, all gateways should solicit their incoming data packets periodically every $Period_{GW}$ seconds to their CLs and other gateways which connect each other.

5   Discussion

In this CERA protocol, we aim to reduce the total power consumption of nodes by switching the nodes except CL nodes into "sleep" mode when they are not in-volved in data communication. However, when some other corresponding nodes want to communicate the nodes in "sleep" mode, we can experience some de-lay until actual communication occurs because each intermediate node should be awake to forward the packets at corresponding time. Furthermore, in order to efficiently utilize the benefits of our protocol, all intermediate nodes includ-ing the source and destination pairs can make some agreement and adjust their "sleep patterns" according to application scenarios. The characteristics of appli-cations can be used as the input data to determine the pattern with overhead for a signaling to enable nodes to change their "sleep patterns".

Besides, as far as the network partition is concerned, we also considered the cases in which all nodes can be fully connected to create distributed clusters. If one of the clusters formed cannot reach others, our routing protocol cannot support the communication between two nodes, which can also be experienced by any other routing protocols with the same network partition. It means that other ways to resolve the partition should be devised.

6   Simulation

The simulation results were obtained using the ns-2 simulator [7]. The reference simulation scenario consists of ten mobile nodes over a 250m × 250m area, spreaded at random place at the beginning of simulation. Each simulation lasts 10 minutes. We considered constant bit rate (CBR) data sources. We used the Random Waypoint model [8] as the mobility model. In this model, each node selects a destination and a speed at random with the maximum value for speed of 10 meters/sec. The node then moves to its selected destination at the selected speed. Once it reaches the destination, it stops for a random pause time. The pause time is uniformly distributed between 0 and PAUSE_TIME. The node eventually selects a new destination and speed combination, and then repeat the movement. Based on the basic scenario, where nodes are initially grouped in two main clusters, we analyzed four different dynamic scenarios trying to repeat activities that are likely to appear in the bounded area.

- In scenario #1, each node alternatively selects the random destination in another cluster.
- In scenario #2, nodes select the random destination inside their cluster.
- In scenario #3, a new activity zone appears. Thus, nodes can randomly choose to behave as in scenario #2, or migrate to the new activity zone, and establish a new cluster.
- Finally, scenario #4 combines the behavior of scenarios #1 and scenario #2.

6.1   The Energy Consumption Model

We supposed that the total transmission power consumption P_{tx} can be decomposed as $P_{tx} = P_e + P_{RF}$ [9]. The P_e factor represents the power consumption of the control logic and the modulator. This factor is technology dependent and cannot be reduced; we considered $P_e = 240mA * v = 1200mW$. The P_{RF} is the Radio Frequency (RF) energy and can vary depending on the transmission range. We approximated the maximum as $P_{RF} = (280mA - 240mA) * v = 200mW$ and considered P_{RF} equal to the radiated power P_t. The NIC we used as a reference specified 250 meters as the maximum range, transmitting at 11 Mbps in a open environment with a bit-error rate (BER) greater than 10^{-5}. We used a simplified formula for the free space propagation model to calculate the power received at the destination: $P_r = \frac{P_t * \lambda^2}{(4\pi)^2 * d^2} = 3.166 \times 10^{-10} W$.

CERA allows cluster leaders to use an appropriate power level to transmit packets inside clusters. Using less power to transmit packets not only saves battery energy but also reduce channel interference thus increasing channel utilization. We assume a typical radius size for a cluster about of $80 - 120$ meters. The correspondent value for P_{RF} can be calculated from the previous formula as: $P_{RF} = P_{t^*} = \frac{P_r * (4\pi)^2 * d^2}{\lambda^2} = 46mW$. Therefore, the energy required to transmit a packet p when using a cluster size of about $80 - 120$ meters is $E_{txcluster}(p) = (P_e + P_{t^*}) * t_p \simeq 1.25 \times 10^{-3} * t_p$ Joules = $w_{max} * t_p$ Joules, where w_{max} is the value used in the clustering algorithm (see Section 3).

6.2   Performance Evaluation

In the simulation process we considered the first version of routing protocol mentioned in Section 4 for simplicity of simulation to enable the communication between nodes in different clusters. We study two different configurations: the $CERA_{S+}$ which uses the $CERA_{s+r+}$ configuration and the $CERA_{S-}$ which uses the $CERA_{s-r+}$ configuration. We evaluate: the energy consumption, the packet error rate and the packet end-to-end delay. We take into consideration the effects of some relevant attributes such as the application sending rate and the node mobility. The comparison is made with the equivalent network configuration but substituting CERA with the DSR [8] protocol without clustering concept. DSR was chosen because while being simple, it is well-known to be one of the most efficient routing protocols, especially in bounded regions [10].

Varying Sending Rate We first analyze the effect of varying the application sending rate. We run simulations varying the packet sending rate with 1.5Mbps, 1Mbps, 0.5Mbps and 0.25Mbps (packet size is 512 bytes; node's speed is 1 m/s). We compare the average energy consumption of the DSR protocol with the average energy consumption of the $CERA_{S-}$ protocol and the $CERA_{S+}$ protocol, respectively. The $CERA_{S+}$ protocol saves up to 20%, 22%, 15% and 18% in scenario #1, scenario #2, scenario #3 and scenario #4 respectively due to the use of the $CERA_{s+r+}$ protocol: as the interaction among clusters increases, like in scenarios #3 and #4, the energy saving decreases. On the contrary, as the interaction decreases, like in scenarios #2, the energy saving increases. Figure 1 shows the energy behaviour in scenario #1.

This result is basically due to the fact that as we increase mobility among different clusters, we reduce the possibility for nodes to use power saving algorithm and therefore to save energy. More stable scenarios like the #2, allow all cluster members to periodically go into the sleep mode, thus saving more energy. The CERA protocol behaves identically to the DSR protocol for all simulations. This means that the CERA protocol does not have any extra overhead with

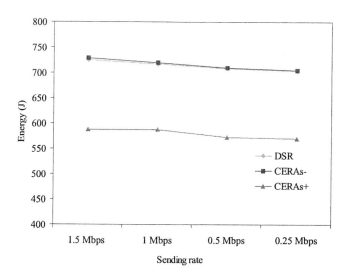

Fig. 1. Energy consumption average. Node speed is 1 m/s

respect to the classical DSR. Relatively to the data packet delivery ratio, the DSR protocol outperforms all the others, even if the difference is quite small. The $CERA_{S-}$ and the $CERA_{S+}$ protocols only differ by 1% and 3% respectively. The $CERA_{S+}$ behaves extremely well in scenario#2, saving up to 22% of the total energy while delivering the same fraction of data packets as in the DSR protocol.

Varying Node Speed We evaluate the effect of increasing mobility from walking speed (1 m/s) to slow vehicle speeds (10 m/s). We fixed the traffic sending rate to 0.25 Mbps, and repeated simulations by varying the maximum node's speed with 1, 2, 5 and 10 m/s. The average energy consumption behaves similarly as seen in Section 6.2. On average, the $CERA_{S+}$ protocol can save energy up to 20%, 21%, 16% and 19% in scenarios #1, #2, #3 and scenario #4, respectively. Node's speed does not significantly affect the average energy consumption.

Figure 2 shows how the average packet loss decreases as node's speed increases since as node's speed increases, the cluster infrastructure stabilizes more quickly and therefore the fraction of data packets successfully delivered increases too. As an example, the $CERA_{S+}$ protocol loses up to 2% of the total data packets in scenario#1 with a node speed of 1m/s, while when increasing the node speed to 10 m/s it only loses 0.3%. In summary, the DSR protocol outperforms again the $CERA_{S-}$ and the $CERA_{S+}$ with a slight percentage: 0.5% and 1%, respectively.

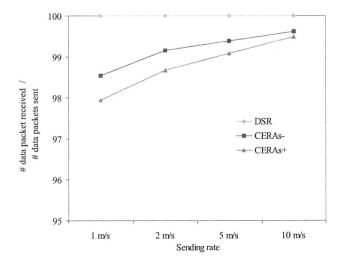

Fig. 2. Fraction of application data packets successfully delivered. Packet sending rate is 0.25Mbps. Scenario #1

The Effect of Radio Transmission Range We evaluate the impact of the radio transmission range on channel utilization by comparing the DSR and the CERA$_{S-}$ protocols. While the DSR protocol uses a tx power of $E_{tx}(p) = 280mA * v * t_p$ Joules, the CERA$_{S-}$ reduces this value to $E_{txcluster}(p) = 1.25 \times 10^{-3} * t_p$ Joules. This power reduction implies a smaller radio transmission range, resulting in reducing channel interference and increasing channel utilization. We evaluate the average end-to-end data packet delay as a measure of channel utilization. We repeated the simulations of Section 6.2 and Section 6.2 by adding a new peer-to-peer constant bit rate (CBR) data stream. The results confirm that for all simulations, the CERA$_{S-}$ protocol significantly reduces the end-to-end data packet delay when compared with the DSR protocol. The explanation can be found in the radio transmission range. While the CERA$_{S-}$ protocol transmits packets from both sources in parallel; the DSR protocol cannot do the same. Therefore, under the DSR protocol, when packets from the first source are being transmitted, packets from the second source must wait in intermediate buffers until the previous packets arrive at the target node due to channel interference.

We also varied the node's speed and observed that the CERA$_{S-}$ protocol reduces the average end-to-end delay by 24%, 69%, 15% and 14% in scenarios #1, #2, #3 and #4. As the clusters interference increases, routes in the CERA$_{S-}$ protocol become longer, therefore increasing not only the channel interference but also the end-to-end delay. Similar results are obtained with varying the sending rate. When using the CERA$_{S+}$ protocol, the end-to-end delay increases. This is because when nodes periodically go into the sleep state, packets must wait in the *Cluster leader* queues until nodes are awake, thus increasing the end-to- end delay.

7   Conclusions

We presented a cluster-based low-complexity routing algorithm for self-organizing networks of mobile nodes called *Cluster-based Energy-saving Routing Algorithm* (CERA). The proposed algorithm allows mobile nodes to autonomously create clusters to minimize power consumption. We considered that simple mobile computing devices should not be overloaded with complex clustering protocols. We therefore maintained our proposal as simple as possible and provided it with sufficient robustness and self-recoverability to enable simple and fast node reconfiguration. CERA is implemented as two separate protocols: the intra-cluster data-dissemination protocol, and the cluster routing protocol. These protocols enable the integration of WLANs and WPANs with wired local area networks, providing mobile devices with access points which operate in a bounded environment.

We simulated the protocols derived from CERA with the *ns* simulator. We evaluated the overhead and stability of the CERA clustering protocol. When adopting the power-saving algorithm, it showed an energy saving of up to the 30%. We then performed a sensitivity analysis of the overall protocol architecture by varying the critical factors related to protocol behavior and showed that, in general, the CERA implementation saves up to 25% of energy while producing an extremely low overhead. We also proposed a possible solution to the general problem of broadcasting data in a wireless environment and showed how it affects the performance of the whole network.

Acknowledgments

This work was supported by the *Oficina de Ciencia y Tecnologia de la Generalitat Valenciana*, Spain, under grant CTIDIB/2002/29 and by the Post-doctoral Fellowship Program of Korea Science & Engineering Foundation (KOSEF).

References

[1] David G. Leeper, "A long-term view of short-range wireless," IEEE Computer, vol. 34, no. 6, pp. 39.44, June 2001.

[2] S. Corson and J. Macker, "Mobile ad hoc networking (MANET): Routing protocol performance issues and evaluation considerations," RFC 2501,January 1999.

[3] J. M. Kahn, R. H. Katz, and K. S. J. Pister, "Next century challenges: Mobile networking for Smart Dust," ACM/IEEE Intl. Conf. on Mobile Computing and Networking (MobiCom 99), August 1999.

[4] C. R. Lin and M. Gerla, "Adaptive clustering for mobile wireless networks," IEEE Journal on Selected Areas in Communications, vol. 15, no. 7, September 1997.

[5] David R. Karger and Clifford Stein, "A new approach to the minimum cut problem," Journal of the ACM, vol. 43, no. 4, pp. 601.640, 1996.

[6] Juan Carlos Cano and Pietro Manzoni, "A low power protocol to broadcast real-time data traffic in a clustered ad hoc network," Proceedings of Globecomm 2001, San Antonio, Texas, USA, November 2001.

[7] K. Fall and K. Varadhan, "ns notes and documents.," The VINT Project. UC Berkeley, LBL, USC/ISI, and Xerox PARC, February 2000, Available at http://www.isi.edu/nsnam/ns/ns-documentation.html.

[8] David B. Johnson, David A. Maltz, Yih-Chun Hu, and Jorjeta G. Jetcheva, "The dynamic source routing protocol for mobile ad hoc networks," Internet Draft, MANET Working Group, draft-ietf-manet-dsr-04.txt, November 2000, Work in progress.

[9] Laura Feeney, "Investigating the energy consumption of an IEEE 802.11 network interface," Swedish Institute of Computer Science, Technical Report n. T99/11, December 1999.

[10] Juan Carlos Cano and Pietro Manzoni, "A performance comparison of energy consumption for mobile ad hoc networks routing protocols," IEEE/ACM MASCOTS 2000: Eighth International Symposium on Modeling, Analysis and Simulation of Computer and Telecommunication Systems, August 2000.

A Mobility Management Strategy for UMTS

Shun-Ren Yang and Yi-Bing Lin

Department of Computer Science and Information Engineering
National Chiao Tung University, Taiwan
{sjyoun,liny}@csie.nctu.edu.tw

Abstract. *Universal Mobile Telecommunications System* (UMTS) utilizes a three-level location management strategy to reduce the net costs
of location update and paging in the packet-switched service domain.
Within a communication session, a mobile station (MS) is tracked at the
cell level during packet transmission. In the idle period of an ongoing
session, the MS is tracked at the UTRAN registration area (URA) level
to avoid frequent cell updates while still keeping the radio connection.
If the MS is not in any communication session, the MS is tracked at the
routing area (RA) level. The inactivity counter mechanism was proposed
in 3GPP 25.331 to determine when to switch between the three location
tracking modes. In this mechanism, two inactivity counters are used to
count the numbers of cell updates and URA updates in an idle period
between two packet transmissions. If the number of cell updates reaches
a threshold K_1, the MS is switched from the cell tracking to the URA
tracking. After that, if the number of URA updates reaches a threshold K_2, the MS is tracked at the RA level. This paper proposes simulation model to investigate the performance of the inactivity counter
mechanism. Our study provides guidelines for K_1 and K_2 selection to
achieve lower net costs of location update and paging.

1 Introduction

Existing second generation (2G) mobile communications systems (such as GSM)
are designed for voice services, which only have limited capabilities for offering
data services. On the other hand, the third generation (3G) systems such as
Universal Mobile Telecommunications System (UMTS) [8] support mobile multimedia applications with high data transmission rates. The UMTS infrastructure includes the *Core Network* (CN) and the *UMTS Terrestrial Radio Access
Network* (UTRAN). The CN is responsible for switching/routing calls and data
connections to the external networks, while the UTRAN handles all radio-related
functionalities. The CN consists of two service domains, the *Circuit-Switched*
(CS) service domain and the *Packet-Switched* (PS) service domain. The CS domain provides the access to the PSTN/ISDN, while the PS domain provides the
access to the IP-based networks. In the remainder of this paper, we will focus on
the UMTS packet switching mechanism. In the PS domain of the CN, the packet
data services of a *Mobile Station* (MS) are provided by the *Serving GPRS Support Node* (SGSN) and the *Gateway GPRS Support Node* (GGSN). The SGSN

H.-K. Kahng (Ed.): ICOIN 2003, LNCS 2662, pp. 316–325, 2003.

connects the MS to the external data network through the GGSN. The UTRAN consists of *Node Bs* (the 3G term for base stations) and *Radio Network Controllers* (RNCs) connected by an ATM network. The MS communicates with Node Bs through the radio interface based on the WCDMA (Wideband CDMA) technology.

The cells (i.e., radio coverages of Node Bs) in a UMTS service area are partitioned into several groups. To deliver services to an MS, the cells in the group covering the MS will page the MS to establish the radio connection. The location change of an MS is detected as follows. The cells periodically broadcast their cell identities. The MS listens to the broadcast cell identity, and compares it with the cell identity stored in the MS's buffer. If the comparison indicates that the location has been changed, then the MS sends the location update message to the network. In the UMTS PS domain, the cells are grouped into *Routing Areas* (RAs). The RA of an MS is tracked by the SGSN. The cells in an RA are further grouped into *UTRAN Registration Areas* (URAs). The URA and the cell of an MS are tracked by the UTRAN.

In UMTS, the mobility management activities for an MS are characterized by two finite state machines: *Mobility Management* (MM) state machine and *Radio Resource Control* (RRC) state machine. The MM state machine for the UMTS PS domain is exercised between the SGSN and the MS for CN-level tracking, while the RRC state machine is executed between the UTRAN and the MS for UTRAN-level tracking. Incomplete state diagrams for these two machines are illustrated in Figure 1. Specifically, the figure only considers the states after the MS has attached to the PS domain. The MM state diagram in the MS is slightly different from that in the SGSN. To simplify the presentation, we only show the common portions of the MS and the SGSN state transitions that are used in this paper. The state diagrams in Figure 1 are described as follows. After an MS is attached to the PS service domain, the MM state machine will be in one of the two states: **PMM-IDLE** and **PMM-CONNECTED**. In the RRC state machine, there are three states: **RRC Idle Mode, RRC Cell Connected Mode** and **RRC URA Connected Mode**. We will briefly elaborate on the MM and RRC state transitions. The readers are referred to [2] and [1] for complete descriptions of the MM and RRC state machines.

When there is no data transmission between the MS and the core network, the MS is in the **PMM-IDLE** state and **RRC Idle Mode**. In this case, UTRAN has no information about the idle MS, and the MS is tracked by the SGSN at the RA level.

When a PS signaling connection is established between the MS and the SGSN (possibly in response to a page from the SGSN), the MS enters the **PMM-CONNECTED** state (see **T1** in Figure 1(a)). Since the establishment of the PS signaling connection triggers the establishment of the RRC connection between the MS and its serving RNC, the RRC state of the MS is switched to **RRC Cell Connected Mode** (see **T1** in Figure 1(b)). In this case, the SGSN tracks the MS with accuracy of the serving RNC, and the serving RNC is responsible

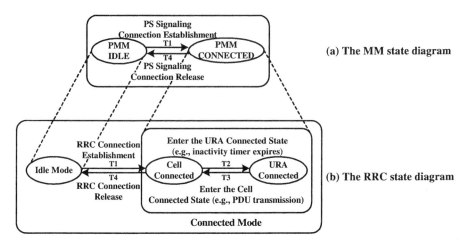

Fig. 1. State diagrams for UMTS mobility management: (a) an incomplete MM state diagram for PS domain; (b) a simplified RRC state diagram

for tracking the cell where the MS resides. Packets can only be delivered in this state.

In the **PMM-CONNECTED/RRC Cell Connected Mode**, if the MS has not transmitted/received packets for a period, the RRC state of the MS is switched to **RRC URA Connected Mode** (see **T2** in Figure 1(b)). In this case, the RRC connection is still maintained while the URA of the MS is tracked by the serving RNC. In this transition, the MM state of the MS remains unchanged; i.e., the state is **PMM-CONNECTED**.

In the **PMM-CONNECTED/RRC URA Connected Mode**, if the MS transmits/receives a packet, the RRC state is moved back to **RRC Cell Connected Mode** (see **T3** in Figure 1(b)). On the other hand, if the PS signaling connection and the RRC connection are released (e.g., a communication session is completed), or if no packet is transmitted for a long time, the RRC state is first switched to **RRC Cell Connected Mode** and then to **RRC Idle Mode** (see **T3** and **T4** in Figure 1(b)). In this case, the MM state is also changed to **PMM-IDLE** (see **T4** in Figure 1(a)).

In the MM and RRC state machines, the mechanism that triggers transitions **T2** and **T4** has significant impacts on the signaling traffic of the UMTS system. This mechanism can be implemented by two approaches. The first approach makes use of two inactivity timers X_1 and X_2. At the end of a packet transmission, timer X_1 is set to a predefined threshold value and is decremented as time elapses. Transition **T2** occurs if the MS does not transmit/receive any packet before timer X_1 expires. When timer X_1 expires, the second timer X_2 is set to a predefined threshold value and is decremented. Timer X_2 is used to determine the time when transition **T4** occurs. In the second approach, two inactivity counters Y_1 and Y_2 are employed. Counter Y_1 counts the number of cell updates in the idle period between two packet transmissions. If the number of cell updates reaches a threshold K_1, then the MS is switched to perform URA updates through transition **T2**. After **T2** has occurred, counter Y_2 is used to count the number of URA updates in the observed idle period. If the number of URA updates reaches a threshold K_2, then the MS is switched to perform RA updates (i.e., transition **T4** occurs).

As pointed out in our previous work [10], the timer approach may have synchronization problem. That is, the peer state machines in the MS and the UTRAN or the SGSN may stay in different states at the same time due to the errors of the clock rates. Besides, the counter approach may significantly outperform the timer approach for the following reason. The timer approach uses two timers of fix-length thresholds. When the mobility rate and/or packet transmission patterns change, the fixed thresholds of the timers do not adapt to the changes. On the other hand, the thresholds K_1 and K_2 of the counter approach always capture the K_1-th cell update and K_2-th URA update of an MS no matter how the mobility rate and packet transmission patterns change. Therefore, this paper will not elaborate on the inactivity timer approach and only focus on the inactivity counter mechanism. In the following sections, we propose simulation model to investigate the performance of the inactivity counter mechanism. Specifically, given any mobility and traffic patterns, we determine the net costs of location update and paging under various K_1 and K_2 threshold values. Our study provides guidelines for K_1 and K_2 selection that results in lower net costs.

2   Simulation Model for Inactivity Counter Mechanism

Discrete event simulations are conducted to study the UMTS inactivity counter mechanism. The simulations adopt the two-dimensional random walk model for user movement. For the demonstration purpose, we consider the hexagonal cell layout in Figure 2. In this configuration, the cells are clustered into several *UTRAN Registration Areas* (URAs), and the URAs are in turn clustered into several *Routing Areas* (RAs). An *n-layer* URA covers $S(n) = 3n^2 - 3n + 1$ cells. Figure 2 illustrates a 3-layer URA. The structure of an n-layer RA is similar to that of an n-layer URA except that the basic elements are URAs instead of cells. Therefore, an n-layer RA covers $S(n)$ URAs. Figure 2 illustrates seven 2-layer RAs (A-G). Each of them consists of seven 3-layer URAs.

We combine the ETSI packet data model [5] with the ON/OFF source model (also known as a packet train model) [3]. As shown in Figure 3, we assume that the packet data traffic consists of communication sessions. Within a communication session, packet traffic is characterized by ON/OFF periods. In an ON-period, a burst of data packets are transmitted. In an OFF-period, no packets are delivered. Other assumptions are summarized as follows.

- The OFF-period t_{p1} is drawn from a Pareto distribution with mean $1/\lambda_{p1}$ and infinite variance. It has been shown that the Pareto distribution with

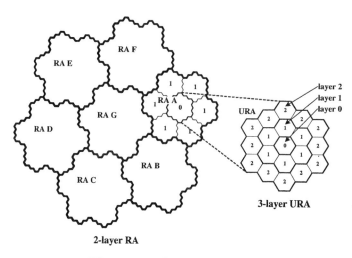

Fig. 2. Cell/URA/RA layout in a UMTS Network

infinite variance can match very well with the actual data traffic measurements [11]. The typical parameter values for OFF-periods obtained in [11] are used in our study.

- The idle period t_{p2} between two consecutive communication sessions has a Gamma distribution with mean $1/\lambda_{p2}$ and variance V_{p2}. It has been shown that the distribution of any positive random variable can be approximated by a mixture of Gamma distributions (see Lemma 3.9 in [9]).
- Following the ETSI packet data model, the number of OFF-periods in a session has a geometric distribution with mean $\alpha/(1-\alpha)$, where $0 \leq \alpha < 1$. In other words, an ON-period is followed by an OFF-period with probability α, and is followed by an inter-session idle period with probability $1 - \alpha$.
- The cell residence times have a Gamma distribution with the mean $1/\lambda_m$ and the variance V_m. The Gamma distribution was employed to model MS movement in many studies [4, 6, 7], and is used in this paper to investigate the impact of variance for cell residence times.

Based on the above input parameters, for specific thresholds K_1 and K_2, we compute the expected number of location updates $E[N_u]$ (including cell updates, URA updates and RA updates) performed in the idle period t_p between two packet transmissions, and the expected number of cells $E[N_v]$ need to be paged for packet delivery. Assume that the cost for performing a location update is U and the cost for paging at one cell is V. Let C_u be the expected location update cost during t_p, then we have

$$C_u = U E[N_u] \tag{1}$$

Let C_v be the expected paging cost during t_p, then we have

$$C_v = V E[N_v] \tag{2}$$

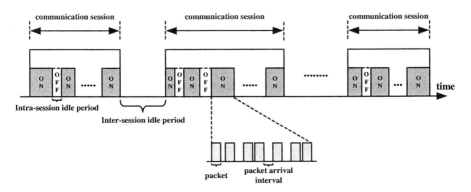

Fig. 3. Packet data traffic

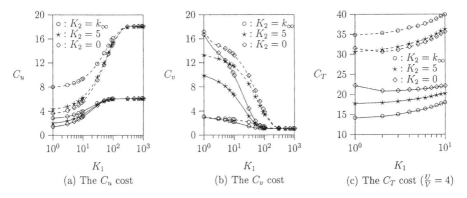

(a) The C_u cost   (b) The C_v cost   (c) The C_T cost ($\frac{U}{V} = 4$)

Fig. 4. Effects of K_1, K_2 and λ_{p2} on C_u, C_v and C_T (solid: $\lambda_{p2} = \frac{1}{200}\lambda_{p1}$; dashed: $\lambda_{p2} = \frac{1}{600}\lambda_{p1}$; $\lambda_m = \frac{1}{10}\lambda_{p1}$, $V_m = \frac{1}{\lambda_m^2}$, $V_{p2} = \frac{1}{\lambda_{p2}^2}$, $\alpha = 0.7$)

From (1) and (2), the net cost C_T for location update and paging during t_p is

$$C_T = C_u + C_v \tag{3}$$

3   Numerical Examples

Based on the simulation model described in the previous section, this section investigates the performance of the UMTS inactivity counter mechanism. To simplify our discussion, we consider the 2-layer URA and 2-layer RA cell layout. The effects of the input parameters are investigated as follows.

Effects of K_1. Figure 4 shows how K_1 affects C_u, C_v and C_T. For a fixed K_2, it is clear that if K_1 increases, the location update (LU) cost C_u increases while the paging cost C_v decreases. When K_1 is very large, the MS always perform cell updates, and no URA or RA update is executed. In this case,

Table 1. The C_u and C_v costs in a t_{p2} period ($\lambda_{p1} = 10\lambda_m$; the costs are normalized by one LU cost)

K_1 and K_2	$\lambda_m = 20\lambda_{p2}$	$\lambda_m = 60\lambda_{p2}$
$K_1 = k_\infty$, $K_2 = k_{any}$	$C_u = 20$, $C_v = 0.25$	$C_u = 60$, $C_v = 0.25$
$K_1 = 1$, $K_2 = k_\infty$	$C_u = 9.1$, $C_v = 1.7$	$C_u = 26.3$, $C_v = 1.7$
$K_1 = 1$, $K_2 = 0$	$C_u = 4.4$, $C_v = 11.7$	$C_u = 11.8$, $C_v = 12.1$

the C_u and C_v costs are not affected by the change of K_1 (see the C_u and C_v curves in Figures 4(a) and (b) where $K_1 \geq 10^2$). In Figure 4(c), the C_T is computed directly from the C_u and C_v using Equation (3), where $\frac{U}{V} = 4$. For the input parameters selected in Figure 4(c), the lowest C_T costs are observed when $K_1 = 1$ or 2. This phenomenon is explained in the following two cases.

Case I. Since the OFF-periods t_{p1} are short (specifically, $E[t_{p1}] = \frac{1}{10\lambda_m}$ in Figure 4), the probability that there are less than 2 cell crossings during t_{p1} is larger than 99%. Therefore, low C_T cost is expected if the MS stays at cell update mode, and only one cell has to be paged for the next packet delivery. In this case, it is reasonable to select $K_1 \geq 2$.

Case II. On the other hand, the MS crosses many cell boundaries during an inter-session idle period t_{p2}. For example, the average number of cell crossings is 20 for $\lambda_m = 20\lambda_{p2}$ and 60 for $\lambda_m = 60\lambda_{p2}$. Consider the case where $\lambda_m = 20\lambda_{p2}$. Table 1 lists C_u and C_v normalized by one location update (LU) cost. When $K_1 = k_\infty$ and $K_2 = k_{any}$, the C_u cost is 20 LUs and the C_v cost is equivalent to 0.25 LUs. Thus, the C_T cost is $20 + 0.25 = 20.25$ LUs. When $K_1 = 1$ and $K_2 = k_\infty$, the C_T cost is $9.1 + 1.7 = 10.8$ LUs. Based on the above discussion, we have

$$C_T(K_1 = 1) < C_T(K_1 = k_\infty) \tag{4}$$

Since Theorem 1 in Appendix A indicates that the lowest C_T value occurs when $K_1 = 1$ or $K_1 = k_\infty$, (4) implies that $K_1 = 1$ is the optimal threshold value in a t_{p2} period.

To obtain low C_T values by considering both Cases I and II, it is appropriate to select $K_1 = 1$ or 2.

Effects of K_2. Figures 4 and 5 show how K_2 affects C_u, C_v and C_T. The effects of K_2 on the location update cost C_u and paging cost C_v are similar to that of K_1: given a K_1 value, the C_u cost is an increasing function of K_2 while the C_v cost is a decreasing function of K_2. Figure 4 also indicates that when K_1 is large, C_u and C_v are not affected by the change of K_2. Figure 5 shows that if λ_{p2} is small, C_T is an increasing function of K_2. On the other hand, if λ_{p2} is large, C_T is a decreasing function of K_2. This phenomenon is explained in the following two cases.

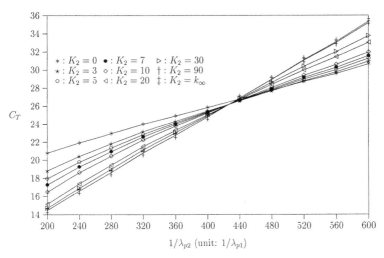

Fig. 5. Effects of K_2 on C_T ($K_1 = 2$, $\lambda_m = \frac{1}{10}\lambda_{p1}$, $V_m = \frac{1}{\lambda_m^2}$, $V_{p2} = \frac{1}{\lambda_{p2}^2}$, $\frac{U}{V} = 4$, $\alpha = 0.7$)

Case I. $\lambda_{p2} = \frac{1}{600}\lambda_{p1}$ (the t_{p2} period is long). When $K_1 = 1$ and $K_2 = k_\infty$, Table 1 indicates that the C_T cost is $26.3 + 1.7 = 28$ LUs. When $K_1 = 1$ and $K_2 = 0$, the C_T cost is $11.8 + 12.1 = 23.9$ LUs. Thus,

$$C_T(K_2 = 0) < C_T(K_2 = k_\infty) \tag{5}$$

From Theorem 1, the lowest C_T value occurs when $K_2 = 0$ or $K_2 = k_\infty$, and (5) implies that the lowest C_T cost is expected when $K_2 = 0$.

Case II. $\lambda_{p2} = \frac{1}{200}\lambda_{p1}$ (the t_{p2} period is short). The C_T cost is $9.1 + 1.7 = 10.8$ LUs when $K_1 = 1$ and $K_2 = k_\infty$, and is $4.4 + 11.7 = 16.1$ LUs when $K_1 = 1$ and $K_2 = 0$. In this case, Theorem 1 indicates that the lowest C_T cost is observed when $K_2 = k_\infty$.

Effects of λ_{p2}. Figures 4(a), (b) and (c) plot the C_u, C_v and C_T curves for $\lambda_{p2} = \frac{1}{200}\lambda_{p1}$ and $\lambda_{p2} = \frac{1}{600}\lambda_{p1}$. The figures show that C_u, C_v and C_T increase as λ_{p2} decreases. A small λ_{p2} implies a long inter-session idle period and more cell movements during this period. Therefore, the location update cost C_u will increase accordingly. For fixed K_1 and K_2 values, increasing the number of cell movements implies increasing the probability that the MS will enter the URA update mode or even the RA update mode when the next packet arrives. Thus, high paging cost is expected. We also notice that when both K_1 and λ_{p2} are large, increasing K_1 only has insignificant effect on C_T. This phenomenon is explained as follows. For large λ_{p2} and K_1, it is likely that K_1 is larger than the number of cell crossings during the idle period. Therefore, the MS will only perform cell updates, and increasing K_1 only insignificantly increases the C_T value.

Effects of α. A smaller α implies more inter-session idle periods. Since more cell crossings are observed in an inter-session idle period than in the OFF-periods of a session, C_T increases as α decreases. Similar to the discussion for the interaction between K_1 and λ_{p2}, C_T is more sensitive to the change of K_1 for a small α than a large α.

Effects of variance V_m. When V_m increases, more short and long cell residence times are observed. Long cell residence times imply small number of cell crossings N_c, which result in a small C_T. On the contrary, short cell residence times imply large number of cell crossings N_c, which increases C_T. We observe that, when $N_c > K_1$, the numbers of URA crossings and RA crossings do not increase as fast as N_c does. The result is that the negative effect of short cell residence times are not as significant as the positive effect of long cell residence times. Therefore, the combined effect is that C_T decreases as V_m increases. We also note that C_T is not sensitive to the change of V_m when $V_m \leq \frac{1}{\lambda_m^2}$.

4   Conclusions

This paper investigated the location management strategy for UMTS PS service domain. When an MS is not in any communication session, the system tracks the RA where the MS resides. Within a communication session, the MS is tracked at the cell level during packet transmission. In the idle period of an ongoing session, the MS is tracked at the URA level to avoid frequent cell updates while still keeping the radio connection. The inactivity counter mechanism was proposed in 3GPP 25.331 [1] to determine when to switch between the three location tracking modes (cell, URA or RA). In this mechanism, two inactivity counters are used to count the numbers of cell updates and URA updates in an idle period between two packet transmissions. If the number of cell updates reaches a threshold K_1, the MS is switched from the cell tracking to the URA tracking. After that, if the number of URA updates reaches a threshold K_2, the MS is tracked at the RA level. We utilized simulation model to investigate the performance of the inactivity counter mechanism. It is clear that as K_1 and K_2 increase, the location update cost increases while the paging cost decreases. There exists optimal K_1 and K_2 that minimize the net cost C_T of location update and paging. For the input parameters considered in this paper, the lowest C_T costs are observed when $K_1 = 1$ or 2. If the inter-session idle periods t_{p2} are long, C_T is an increasing function of K_2. On the other hand, if the t_{p2} are short, C_T is a decreasing function of K_2. We quantitatively showed how C_T increases as inter-session idle periods and user mobility increase. In addition, the variance V_m of cell residence times affects C_T. Our study indicated that as V_m increases, C_T decreases.

Acknowledgment

This work was supported by the MediaTek Fellowship.

References

[1] 3GPP. 3rd Generation Partnership Project; Technical Specification Group Radio Access Network; RRC Protocol Specification for Release 1999. Technical Specification 3G TS 25.331 version 3.5.0 (2000-12), 2000.

[2] 3GPP. 3rd Generation Partnership Project; Technical Specification Group Services and Systems Aspects; General Packet Radio Service (GPRS); Service Descripton; Stage 2. Technical Specification 3G TS 23.060 version 3.6.0 (2001-01), 2000.

[3] Cheng, M. and Chang, L. F. Wireless Dynamic Channel Assignment Performance Under Packet Data Traffic. *IEEE Journal on Selected Areas in Communications*, 17(7):1257–1269, July 1999.

[4] Chlamtac, I., Fang, Y., and Zeng, H. Call Blocking Analysis for PCS Networks under General Cell Residence Time. *IEEE WCNC, New Orleans*, September 1999.

[5] ETSI. UMTS Terrestrial Radio Access (UTRA); Concept Evaluation, Version 3.0.0. Technical Report UMTS 30.06, December 1997.

[6] Fang, Y. and Chlamtac, I. Teletraffic Analysis and Mobility Modeling for PCS Networks. *IEEE Transactions on Communications*, 47(7):1062–1072, July 1999.

[7] Fang, Y., Chlamtac, I., and Fei, H.-B. Analytical Results for Optimal Choice of Location Update Interval for Mobility Database Failure Restoration in PCS networks. *IEEE Transactions on Parallel and Distributed Systems*, 11(6):615–624, June 2000.

[8] Holma, H. and Toskala, A. *WCDMA for UMTS*. John Wiley & Sons, Inc., 2000.

[9] Kelly, F. P. *Reversibility and Stochastic Networks*. John Wiley & Sons, 1979.

[10] Lin, Y.-B. and Yang, S.-R. A Mobility Management Strategy for GPRS. *Accepted and to appear in IEEE Transactions on Wireless Communications*.

[11] Willinger, W., Taqqu, M. S., Sherman, R., and Wilson, D. V. Self-Similarity through High-Variability: Statistical Analysis of Ethernet LAN Traffic at the Source Level. *IEEE/ACM Trans. Networking*, 5(1):71–86, February 1997.

[12] Yang, S.-R. and Lin, Y.-B. A Study on the UMTS Inactivity Counter Mechanism. Technical report, National Chiao Tung University, 2002.

A   Theorem 1

Consider an idle period where no packet is delivered. Let N_c be the number of cell crossings in this idle period. In the inactivity counter algorithm, let $C_T(K_1, K_2)$ be the net cost of location update and paging in the idle period with cell update threshold K_1 and URA update threshold K_2. Let $K^* = (K_1^*, K_2^*)$ be the optimal threshold value pair that minimizes the net cost C_T. Define k_∞ as a number larger than the N_c value in the idle period. Let k_{any} be an arbitrary integer number.

Theorem 1. For an idle period, $K^* = (k_\infty, k_{any})$, $(1, k_\infty)$ or $(1, 0)$.

Proof. Due to space limitation, the proof is not presented in this paper. The readers are referred to [12] for the details of the proof.

A Cluster-Based Router Architecture for Massive and Various Computations in Active Networks *

Young Bae Jang and Jung Wan Cho

Division of Computer Science
Department of Electrical Engineering and Computer Science
Korea Advanced Institute of Science and Technology
373-1, Guseong-dong, Yuseong-gu, Daejeon, Republic of Korea
Tel : +82-42-869-5595 Fax : +82-42-869-3510
ybjang@calab.kaist.ac.kr

Abstract. Traditional network routers are passive in a manner of speaking because they can not manipulate packets but just deliver them. On the contrary, an active network is an innovative approach to the network architecture. Active routers can perform customized computations on the packet flowing through them. They thus can easily adopt or remove protocols and perform operations such as firewall, content-based switching, and multimedia broadcasting. However, they may suffer from the lack of computational power in the near future due to growing active applications. Cluster-based active router can be a solution of this problem. A cluster system is easier to build than a multiprocessor system and brings out similar computing performance. We designed a cluster-based router architecture to share the overall computational load. In our design, packets can be forwarded to other nodes within the cluster using high-speed interconnect network. Therefore, busy node's load can be transferred to other nodes quickly. To test the feasibility of our cluster-based active router design, we modified existing active router daemon, Anetd, so that it can work on the cluster and forward packets inside the cluster. After performing some micro-benchmark, we observed that overall latency is reduced.

1 Introduction

As the expansion of the Internet, the various demands of users have been arisen and many effective protocols for them have been suggested. Traditional routers are passive because they can just deliver packets to the destination and are hard to adopt such new protocols. An active network [18] is an innovative approach to the network architecture. Active packets, or capsule [3] in active network, can change the status of a router or invoke new processes on a router to process other packets. Packets themselves through the active network also can be

---

* This research is supported by KISTEP under the National Research Laboratory program.

H.-K. Kahng (Ed.): ICOIN 2003, LNCS 2662, pp. 326–335, 2003.

modified or replicated. Active routers not only deliver packets but also perform various computations. Thus, they can adopt protocols and perform new router applications such as firewall, multicasting and congestion control [9]. Until now many host-based active router architectures using PCs or workstations have been proposed [5, 15, 16].

However, the additional computation overhead leads to the performance degradation of active routers [13, 11]. Active routers need sufficient computational power for processing massive and various jobs. The easiest solution is simply to forward packets to other routers when the router lacks computational power. In this way, however, some routers may suffer from the concentration of active packets and increase the latency and the drop-rate of packets considerably because routers have no load information of others.

To process many active packets in a single router, the more processors should be added on the router. Several architectures of high-performance active router have been proposed [12], [21]. The essential point of such architectures is how to connect and assign a number of processors. The best solution is to build an entirely new router system to support multiple processors. But this is a very expensive solution because it needs a whole system design to solve the problem of memory bus contention caused by the increased processors [4]. Another solution is adding optional component to take complete charge of active packets [19, 14]. This is easy to construct but can hardly utilize the whole computing resources because the routing resources and the active processing resources are loosely coupled.

A cluster system is the one of the good solutions to get high performance at a relatively low cost. It is built using commodity-off-the-shelf (COTS) hardware components and free or commonly used software components [6]. Nodes consisting a cluster system are connected by low-latency high-bandwidth interconnect network such as Fast Ethernet or Myrinet [7]. Ports and processors can be added on demand because a cluster system is easy to scale and extend. Some cluster-based passive routers have been proposed [17, 11]. But their primary concern is just to deliver packets, without any consideration about active network environment. In active networks, routers manipulate active packets. Thus, the efficient use of processing resources is essential.

In this paper, we propose a cluster-based router architecture to share the overall computational load in active networks. The goal of our router design is forwarding active packets from busy node to other idle node to reduce the processing time per packet and balance resource utilization in the cluster. The rest of this paper is organized as follows. Active networks and cluster-based router architectures are discussed in Sect. 2. A detailed description of our router design is given in Sect. 3. Sect. 4 describes the experimental environment and the results that were taken in the router system. Finally, we will conclude the paper in Sect. 5.

2   Related Work

2.1   Active Network

The active network is a new approach to network architecture. It provides a platform on which network services can be experimented with, developed, and deployed [8]. Nodes, traditionally routers or switches, in active networks can perform customized programs to execute. The functions of active routers are not only the delivery of packets but also the optimization of network protocols and distributed applications by the execution of customized programs. The capability of the active network includes faster deployment of new services, ability to customize services for different applications, and ability to experiment with new network architectures and services on a running network.

The functionality of the active network node is divided among the node operating system (NodeOS), the execution environments (EEs), and the active applications (AAs) [8]. The NodeOS is responsible for allocating and scheduling the node's resources. The NodeOS manages the resources of the active node and mediates the demand for those resources, which include transmission, computing, and storage. Each execution environment presents a programming interface or virtual machine that can be programmed or controlled by incoming packets to it. Users obtain services from the active network via active applications, which program the virtual machine provided by an execution environment to provide an end-to-end service.

The Active Network Encapsulation Protocol (ANEP) [3] provides the way of routing packets to a particular execution environment at a node. The ANEP makes active packets transparent to traditional network nodes that do not support them. Non-active network nodes will simply ignore the active fields. Also, a packet without an ANEP header can be routed traditionally by active nodes. The ANEP can deploy active network elements into the traditional network gradually.

The adaptive power supported by active networks allows services to be tailored to current network environment. The required but not resident services can be automatically and dynamically loaded by code distribution technique. With code distribution it is unnecessary for nodes to have all services. Thus, nodes not requiring certain services do not incur the overhead of storing their code. By code distribution, it is easy to develop, maintain, and upgrade new network services and to spread them throughout an active network. As a consequence, the performance of many network applications has improved such as management [1] and multicast video distribution [2].

2.2   Cluster System

Based on the rapid development of network and microprocessor technologies, the clustering that connects high performance PCs and workstations by high-speed networks have been appeared. A cluster system [6] is a type of parallel or distributed processing system, which consists of a collection of interconnected commodity-off-the-shelf (COTS) computers working together as a single, integrated computing resource. It becomes prevalent to substitute Massively Parallel Processor (MPP) in parallel computing. And it has been used to build internet-based servers.

A computing node can be a single or multiprocessor system with memory, I/O facilities, and an operating system. A cluster generally refers to two or more nodes connected together. An interconnected cluster of computers can appear as a single system to users and applications. Such a system can provide a cost-effective way to gain features and benefits. The network interface hardware acts as a communication processor and is responsible for transmitting and receiving packets of data between nodes via a network and switch. Communication software offers a means of fast and reliable data communication among cluster nodes and to the outside world. The cluster nodes can work collectively, as an integrated computing resource, or they can operate as individual computers. Programming environment can offer portable, efficient, and easy-to-use tools for development of applications. The cluster system offers high performance, expandability and scalability, high throughput, and high availability at a relatively low cost.

2.3   Cluster-Based Routers

Till now some cluster-based routers have been presented. In [12], authors argued Active Network Node (ANN), active network router hardware. It is designed for higher performance and used the combination of a general purpose CPU and an FPGA, called the processing engine, on every port of a switch backplane. The processing engine and a switch backplane are coupled tightly. The majority of non-active packets allow cut-through routing directly through the switch backplane without CPU intervention and packets arrive with minimal overhead through zero-copy direct memory access. The FPGA can be programmed by the CPU on the fly to implement the most performance-critical algorithms in hardware. A packet can first go to the FPGA and then be either passed to the CPU or forwarded straight to the link, depending on whether there is additional processing required. However strictly speaking, this system is not a general cluster-based architecture. Because suggested architecture requires special hardware like FPGA for the performance, the construction of the system is not easy. And we can't have found any information of actual routers based on this architecture.

Suez [11] project built high-performance real-time packet routers with optimized software running on commodity PC hardware. A Suez router is comprised of multiple Pentium-II PCs, connected through a scalable Myrinet switch. One of the PCs is a dedicated control processor, handling all administrative functions such as real-time connection establishment, routing table updates as well as network monitoring and management. Each of the remaining PCs, in addition to being connected to the Myrinet switch, also connects to the outside world, and thus plays the role of input/output link controllers that buffer incoming packets, performing routing-table lookup or real-time packet classification, forward packets to the corresponding output ports, and schedule packets. This architecture is closer to ours than ANN. But Suez router was focused on the way of packet delivery using fast address lookup and real-time packet scheduling.

Software distributed shared memory (DSM) for the active processing engine was suggested in [14]. This design is the extension of a passive router. To perform demanding in-network computations for applications, some sort of attached computation engine must be added to passive routers to offload the computation from the router itself. In that paper, distributed shared memory is proposed as the attached computation engine. Each router must examine packets passing through it to determine whether or not they are active. Non-active packets are routed as usual bypassing the DSM. Active packets are redirected to DSM. This approach of an add-on active engine to traditional routers is good to extend the processing power for active packets but have the bottleneck between the router and the active engine. Moreover, it is hard to balance the utilization of the computation and packet delivery resources.

3   Our Router Architecture

The major difference in architectures between active routers and traditional passive routers is supporting various computations. For efficient packet processing, it is preferred to distribute computing resources per port rather than concentrate them. An active router must be scalable, because the number of active packets will increase much in the near future and computations due to the active packets will increase, too. A cluster-based router architecture meets these requirements. Cluster systems have the generic scalability and each node in the cluster has independent computing resources and ports. Conventional router architecture may suffer from memory interface problem [4], which is caused by the imbalance between memory bandwidth and the number of processors. Cluster systems do not share a bus to communicate with other nodes within the same cluster. Thus, they have no such memory interface problem. Our active router design based on cluster systems is scalable in terms of computation power, memory bandwidth and port capacity. We designed and implemented an active router system by the modification of an active router daemon on a conventional cluster system.

Hardware architecture of our active router is based on a general cluster system. Our system consists of three components, computing nodes, network interface cards, and an interconnect network. Currently, each node has an Intel Pentium III 850MHz processor, 1GB of SDRAM memory, a storage disk and 33MHz/32bit-PCI bus. Network interface cards which connect each node via PCI bus are used both inter- and intra-cluster communications. Through the interconnection network, each nodes share the router information and transfer packets. In our system, nodes are combined by Myrinet interconnect network. Myrinet network interfaces have a 64MHz/32bit-LANai9.1 processor, 4MB of SRAM, and a link speed of 2 * 1.28Gbps. For the communication to the outside cluster, a 100Mbit-ethernet network card is slotted to each node's PCI bus. Fig. 1 (a) shows the outline of our router architecture.

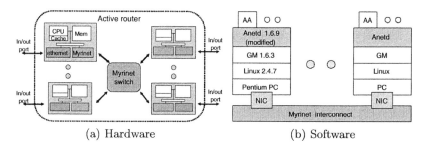

(a) Hardware   (b) Software

Fig. 1. Architecture of our active router

Software architecture is composed of following components. Linux kernel 2.4.7 is used the underlying operating system. To communicate with other nodes, GM 1.6.3 is used. GM is the low-level message-passing system for Myrinet networks. And support for IP over GM is included in GM 1.6.3. Therefore, many existing network applications can run on Myrinet networks without serious modification.

As the NodeOS and execution environment, modified `Anetd` 1.6.9 [20] is running on the host OS of Linux. We have expanded the forwarding mechanism of Anetd supporting cluster environment. Anetd of a node can forward active packets to other node without any assist of running active applications. In this way, packets can be forwarded to an idle node and the load of a busy node can be spread out. There is no centralized management in our router system. It is of great advantage to scale and extend the router. Fig. 1 (b) shows the outline of software architecture and Table 1 summarizes the components of our active router.

Table 1. Components of our active router

Host	Processor	Intel Pentium III 850MHz
	Memory	1Gbytes SDRAM
	I/O	33MHz 32bits PCI bus
Myrinet	Processor	66MHz 64bits LANai9.1
	Memory	4Mbytes SRAM
	Switch	8-port switch
Software	OS	Linux kernel 2.4.7
	Communication	IP over GM 1.6.3
	Router daemon	Modified Anetd 1.6.9

4   Experiment and Evaluation

4.1   Experimental Environment

The main objective of this experiment is how effectively our active router processes packets. We will show the effectiveness by measuring the end-to-end latency of packet delivery. To measure the end-to-end latency, one client node sends an active packet to the active router. Then the router processes the received packet and forwards it to original client node. The client node sends packets 100,000 times, then measures the average latency. For the comparison, original single-node `Anetd` active router was used. In our active router, the overhead of packet forwarding to other node within the same router is added to the latency. For active applications,'`frag`' and '`reed`' in `CommBench` [10] are used. `Frag` is a header-processing program which modifies packet header and computes checksum. `Reed` is a payload-processing program which adds redundancy to data to allow recovery from transmission errors. These applications are downloaded to active router from a local trusted code server. Our experimental environment is illustrated in Fig. 2.

4.2   Results

First, we investigated the transmission latency of packets affected by the load of the processor in a single-processor router. We restricted the availability of processor as 100%, 50% and 25% then measured end-to-end delivery latencies. The active router did not perform any computation on packet except `Anetd` and simply forwarding. Fig. 3 shows the result of end-to-end latencies of 128-byte and 1024-byte packets via a single-processor router. In figures, the x-axis represents router's processor availability (%) and the y-axis is the end-to-end packet forwarding latency (ms). As the availability decreases, the latency increased. Most of increased latency is due to the processing overhead of Anetd router daemon. And the length of packets almost affects only the network latency.

Second, we measured the effect of packet forwarding to idle node of the same cluster in our active router. This experiment was similar to the preceding experiment but performed in a 2-node cluster-based router. The client sends a packet to a host node in the cluster. The host node forwards it to an idle remote node in the same cluster. Then the remote node sends it to the client without any modification. The results of this experiment of a 1024-byte packet are reported in Fig. 4. We give the remote node more CPU availability. In Fig. 4 (a), the host processor is 50% available and the remote is 100% available. When the availability of host processor is 25%, the availability of the remote varies

Fig. 2. Overview of experimental testbed

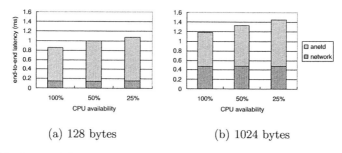

(a) 128 bytes   (b) 1024 bytes

Fig. 3. End-to-end forwarding latency of a packet via single-processor router

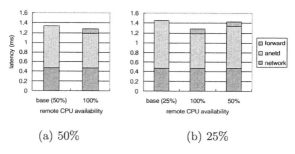

(a) 50%   (b) 25%

Fig. 4. Comparison of latency between single-processor router and cluster-based router

100% and 50% in (b). For the comparison, the latencies of the single-processor router (base) are shown in graphs. Though, there is obvious overhead of packet forwarding to an idle node, overall latencies are decreased. It is incurred by the decreased Anetd processing time and the relatively low latency overhead of packet forwarding.

Finally, we ran real router applications on both active routers. We limited the availability of host processor to 25% and sent a 1024-byte packet to routers. Such results are shown in Fig. 5. Frag is an existing router application whose processing time per packet is very short. Therefore, the result of the execution of frag (Fig. 5 (a)) is very similar to the result in Fig. 4 (b). On the other hand, reed is an active router application (Fig. 5 (b)). The running time per packet is very long and more affected by the CPU availability. The main reason of end-to-end latency is the packet processing time of an active application at a high-loaded router. Packet forwarding in the cluster-based router reduces the overall latency drastically.

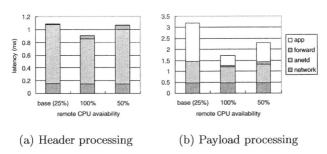

(a) Header processing   (b) Payload processing

Fig. 5. Running Applications

5 Conclusion

In this paper, we proposed a cluster-based router in active networks. We expanded active router software to support general cluster systems. This approach is able to achieve both traditional packet router and active code processing at a low cost. We found the following observations from the experiments. First, the performance of a router depends on the availability of processor. Second, forwarding packets to idle node from busy node in cluster-based router can reduce end-to-end latency. Finally, the latency of active packets is more reduced than of non-active packets in our cluster-based router architecture because active applications require more computational power than traditional network applications. This implies proposed cluster-based router is able to sustain good performance when there are many active packets to process.

With these observations, we will implement load balancing algorithms for cluster-based router. The load balancing of router is quite different from previous load balancing of parallel computer. Because there is little interaction between nodes to process packets in the router, relatively simple load balancing algorithm for the cluster-based router will suffice. Suitable load monitoring must be implemented for the load balancing algorithm. For faster packet treatment, the new version of active router daemon will be tested. Current Anetd router daemon occupies much time in end-to-end forwarding latency. And we will apply the light-weight communication layer to our cluster-based router. It may decrease the overhead of forwarding latency between intra-cluster nodes. The research presented in this paper may be extended. Our approach can be adapted to the clustering of dedicated network processors for fast core routers. The clustering of network processors with our technique can mitigate memory bandwidth problem.

References

[1] Ryutaro Kawamura and Rolf Stadler.: Active Distributed Management for IP Networks. IEEE Communications Magazine (2000)
[2] Ralph Keller, et al.: An Active Router Architecture for Multicast Video Distribution. in Proc. Infocom 2000, Tel Aviv (2000)

[3] D. Alexander, et al.: ANEP: Active network encapsulation protocol. Request for Comments. `http://www.cis.upenn.edu/~switchware/ANEP/docs/ANEP.txt`

[4] Werner Bux, et al.: Technologies and Building Blocks for Fast Packet Forwarding. IEEE Communications Magazine (2001)

[5] Larry Peterson, et. al.: An OS Interface for Active Routers. IEEE Journal on Selected Areas in Communications (2001) 437-487

[6] Mark Baker and Rajkumar Buyya.: Cluster Computing: at a Glance, High Performance Cluster Computing. Vol. 1, 1st edn. Prentice Hall NJ (1999)

[7] Nanette J. Boden, et. al.: Myrinet: A Gigabit-per-second Local Area Network. IEEE Micro (1995) 29-36

[8] K. Calvert.: Architectural Framework for Active Networks version 1.0. (1999)

[9] K. Calvert, et. al.: Directions in Active Networks. IEEE Communications Magazine (1998) 72-78

[10] Tilman Wolf, et. al.: CommBench - a Telecommunications Benchmark for Network Processors. in Proc. IEEE Int. Symp. Performance Analysis Systems Software (2000)

[11] Tzi-Cker Chiueh and Prashant Pradhan.: Suez: A cluster-based scalable real-time packet router. In IEEE ICDCS (2000)

[12] D. Decasper, et. al.: A scalable, High Performance Active Network Node. IEEE Network (1999) 8-19

[13] Simon Walton, et. al.: High-speed Data Paths in Host-based Routers. IEEE Computer (1998) 46-52

[14] P. Graham.: A DSM Cluster Architecture Supporting Aggressive Computation in Active Networks. in Intl. Symp. on Cluster Computing and the Grid (2001)

[15] Patrick Tullmann, et. al.: Janos: A Java-oriented OS for Active Network Nodes. IEEE Journal on Selected Areas in Communications (2001) 501-510

[16] Shashidhar Merugu, et. al.: Bowman: A Node OS for Active Networks. in INFO-COM 2000 (2000) 1127-1136

[17] Larry L. Peterson, Scott Karlin, and Kai Li.: OS support for general-purpose routers. In Workshop on Hot Topics in Operating Systems (1999) 38-43

[18] K. Psounis.: Active networks: Applications, security, safety, and architectures. IEEE Communications Surveys (1999) 1-16

[19] Danny Raz and Yuval Shavitt.: Active networks for efficient distributed network management. IEEE Communications Magazine (2000) 138-143

[20] L. Ricciulli and P. Porras.: An Adaptable Network COntrol and Reporting System (ANCORS). In IFIP/IEEE Intl. Symp. on Integrated Network Management (1999)

[21] T. Wolf and J. Turner.: Design Issues for High-Performance Active Routers. IEEE Journal on Selected Areas in Communications (2001) 404-409

Shortest-Path Mailing Service Using Active Technology

Hyun Joo Kim, Jung C. Na, and Sung W. Sohn

Network Security Department/Information Security Division
Electronics and Telecommunications Research Institute(ETRI)
161 Gajeong-Dong, Yuseong-Gu, Daejeon, 305-350, Korea
{khj63353,njc,swsohn}@etri.re.kr

Abstract. Active network is a new approach of next generation network performing computation and manipulations as injecting programs into the network. Because of these features of active network, active technology is being applied to various areas. Current electronic mail service is useful and popular information transmission media through the Internet. But because it delivers mail through the SMTP mail servers without considering the location of sender or recipient's host, it makes poor performance by generating the redundant network traffic. Therefore in this paper, we suggest Shortest-Path Mailing Service based on active network to solve these problems of traditional mail service. Shortest-Path Mailing Service can reduce the redundant network traffic by transmitting mail to the closest active node from a recipient's host, Active Mail Manager.

1 Introduction

Active network [5][2] is a new approach of next generation network performing computation and manipulations as injecting programs into the network. Because of these features of active network, active technology is applied to various areas. Especially it is being applied to new service development based on active network and many researchers are interested in it more and more. In this paper, as applying active technology to e-mail service widespread with the Internet, we suppose new mail service on active network that can deliver mail to recipient in the shortest path and can integrate and manage some kinds of mail services-Shortest-Path Mailing Service (SPMS) [3]. We first briefly introduce the problems of the traditional mail service and motivation in section 2. Then we explain the SPMS by describing its mechanism and procedures and the architecture of Active Mail Manager in section 3, and discussion about SPMS in section 4. Finally we conclude with our works for the future.

2 Motivation

A major motivation behind SPMS is to intend to solve the following problems over the traditional mail service.

H.-K. Kahng (Ed.): ICOIN 2003, LNCS 2662, pp. 336–345, 2003.

First, in the traditional mail service, a user must send and receive mail through their mail servers (using SMTP) without considering the location of sender or recipient's host. Though the sender's location is near the recipient's, the sender must transmit mail to recipient's mail server via the sender's mail server. But if the sender transmits mail to the closest location from recipient directly, it can reduce the redundant net-work traffic considerably. The explanation about it is detailed in the following section.

Second, in the traditional mail service, a user must connect to each mail server to receive some kinds of mail. Everyday the user must check mailbox of each mail server having user's account. Though it may be possible by using ".forward" file and for-warding mail to one of mail servers, it generates the redundant network traffic when-ever user sends mail.

3   Shortest-Path Mailing Service (SPMS)

SPMS is the new mail service based on active network to provide efficiency, facility, and reliability. To provide them, SPMS supports the shortest mailing path and the three procedures. In this section, we explain the three primary procedures- probing, registration, and authentication and describe the framework of SPMS in comparison to the traditional mail service. Also we discuss Shortest-Path Mailing Protocol (SPMP), the communication protocol used in SPMS and the architecture and function of Active Mail Manager (AMM), the core component.

3.1   Traditional Mail Service Framework

In this part, we discuss problem of the traditional mail service with its framework.

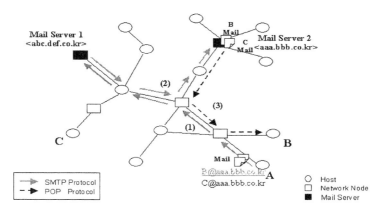

Fig. 1. Framework of traditional mail service

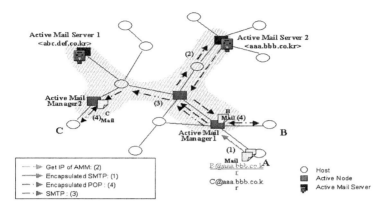

Fig. 2. Framework of SPMS

Fig. 1 shows the framework of the traditional mail service in case user A sends mail to B. In this figure, the SMTP server of A is Mail Server1 and B and C, Mail Server2.

First, A transmits mail to the Mail Server1 through MUA on A's host. Mail Server1 passes it to the Mail Server2 by using the SMTP protocol [4] referring to the 'relay-domains' file. And then, to receive mail, B must be transmitted it from the Mail Server2 although A is near.

3.2   Shortest-Path Mailing Service Framework

We explain the framework of SPMS to provide the shortest path mentioned above. Fig. 2 shows the framework of SPMS on active network of the same case as Fig. 1. Active Mail Manager(AMM) is the closest active node from users, which manages the mailbox.

Sending. (1) A transmits mail to AMM1(A' mail manager) after A's authentication procedure. (2) AMM1 checks the recipient's e-mail address and requests B's AMM information to Active Mail Server2(AMS2, SMTP server of B). (3) AMM1 finds that itself is B's AMM and stores mail in B's mailbox. If A sends mail to C, AMM1 finds that AMM2 is C's AMM and forwards it to AMM2.

Receiving. (4) After B or C is authenticated, he receives mail from each AMM that is the closest location from B or C.

Note that whenever AMM transmits mail, it does not require information of recipient's AMM but use the information stored on cache memory used before. SPMS provides user with speedy and reliable mail transmission by delivering mail to the closest node from the sender and recipient and by having sending authentication procedure. This process is explained in 3.4 Shortest-Path Mailing Service Procedure.

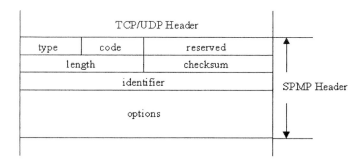

Fig. 3. Header format of SPMP

3.3 Shortest-Path Mailing Protocol (SPMP)

SPMP is the communication protocol between the AMM and AMS in SPMS. This protocol is used in probing, registration, and authentication procedure and Fig. 3 shows the header format of the SPMP.

- type: This field indicates the type value of the packet.
- code: This field means the success/failure of the received packet.
- reserved: It is a field reserved for future use.
- length: This field specifies the length of the SPMP header in 32 bit words. If no options are included in the packet, then its value must be 3. The length of this field is 16 bits.
- checksum: This field is the 32-bit one's complement of the one's complement sum of all 32 bit words in the header.
- identifier: This field includes the value that can uniquely identify the sender and recipient of request/reply packet.
- options: This field is an optional field that includes the active node lists in probing and registration procedure and userID/password in authentication procedure.

3.4 Shortest-Path Mailing Service Procedures

SPMS Procedure is divided into three procedures- probing, registration, and authentication. Probing procedure is the searching mechanism that AMS searches each user's AMM, and registration procedure is the re-registering mechanism that AMS registers the AMM again when user's AMM needs to be modified by change of user's IP or location. Finally authentication procedure provides the sending and receiving authentication for the reliability of SPMS.

Probing Procedure. AMS performs this procedure when a new user is added to AMS. AMS must register not only user's information needed to add user but also additional information, for example, IP addresses of user and AMM. Therefore to search IP address of AMM, this procedure is performed with user's IP address. Fig. 4 illustrates the flow of this procedure.

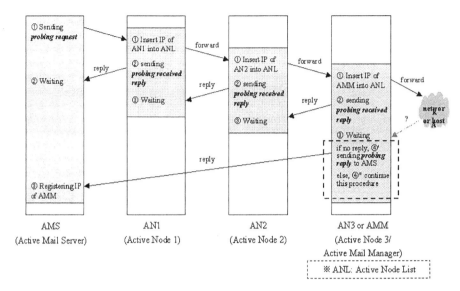

Fig. 4. Probing Procedure

(1) AMS transmits the probing request packet (type=1) whose destination IP address is user's IP address to the network.

(2) If incoming packet is probing packet, active nodes (AN1 and AN2 in Fig. 4) in the path of the destination insert their IP address to the active node list of probing packet and forward it to the next node. At this time, active nodes send probing received reply packet (type=5) to the previous node of active node list and wait for that packet from next node.

(3) If when the active node forwards the probing packet the destination is located in the same network as the active node, the node notifies to AMS that itself is user's AMM. But if not, the active node waits for the probing received reply packet of next active node during the limited period to judge whether it is the closest active node from the destination or not. If there is a different active node in the path of the destination, it will send probing received reply packet. Hence the previous node can receive the packet. But if not, the previous node is the closest active node(AMM) from the destination. Therefore, when it can't receive this packet, the active node inserts own IP address to the active node list of the probing reply (type=2) packet and responds to AMS.

(4) AMS registers the last node of active node list in the received the probing reply packet as user's AMM.

Registration Procedure. This procedure is the mechanism that AMS registers IP addresses of the user' host and AMM again when the user's IP address was changed. In Fig. 2, if user B moves to location of user C, AMM should be changed. Therefore, AMS2 must modify IP address and AMM of B through

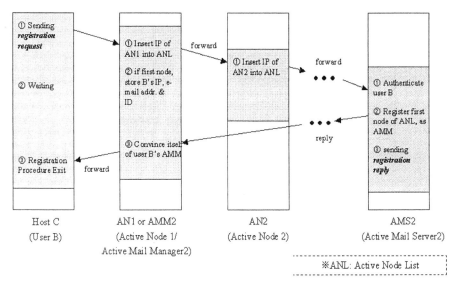

Fig. 5. Registration Procedure

the registration request (type=3) packet transmitted to AMS2 by user B. Fig. 5 below is the flow of this procedure.

(1) B transmits the registration request packet to AMS2 that is a SMTP server of B. The packet includes IP address of B in the source field of IP header.

(2) Each active node (AN1 and AN2 in Fig. 5) in the path of AMS2 inserts own IP address to active node list in SPMP header's option field after receives it, and then forwards this packet to the next node (refer to Fig. 6). Note that the first node of active node list has the possibility that itself can be a new AMM of B, so it must keep B's mail address and identifier of registration request packet.

(3) AMS2 should authenticate the user B using userID/password in option field of registration request packet, and then register a new IP address of B and the first node of active node list as B's AMM and send the registration reply packet to B.

(4) While registration reply packet (type=4) is transmitted to B, AMM2 convinces itself of B's AMM if the identifier of registration request packet corresponds with the identifier of registration reply packet.

(5) After this registration procedure, AMM1 transfers B's mail to AMM2. Since then AMM2 can provide SPMS to B.

Currently many users almost receive mail at home and in the office. Whenever user receives it there, AMM isn't always registered again. That is, user can register secondary AMM as well as primary AMM. So when user connects to secondary AMM, secondary AMM requests user's mail to primary AMM. Then

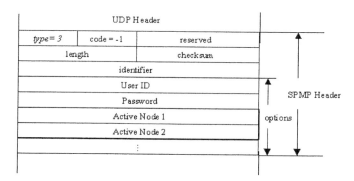

Fig. 6. Format of Registration Request Packet

primary AMM transfers the copy of mail to the secondary. User can select which registration procedure is started.

Authentication Procedure. This procedure decides whether user has the privilege to send and receive mail. SPMS can provide the reliability by sending authentication procedure and optionally user can receive various mails at a time by receiving authentication on AMM.

Sending Authentication: It is performed in AMM when sending mail unlike the traditional mail service. The traditional mail service can make a user who has no account to SMTP server send mail through the server because there is no authentication procedure when sending mail. This can be used as the source of Spam mail and be abused by despiteful user. Also it can cause the security problem that mail server is exposed to despiteful user and others. Accordingly sending authentication is positively necessary.

Receiving Authentication: It is optional. User can choose the way of authentication. First user can select only one receiving authentication. It is performed once in AMM, which makes user receive and manage some kinds of mail at a time. Second user can select the double authentication. It guarantees the more secure service because it is performed in both AMM and AMS.

3.5 Architecture of Active Mail Manager (AMM)

AMM provides the mail service and manages the user's information, cache memory, and mailboxes as well as three procedures. Fig. 7 shows the architecture of AMM. AMM is an active node divided into Execution Environment (EE) and NodeOS [1]. However in this paper we explain the only important components needed to provide SPMS not EE and NodeOS.

- AMM Coordinator: As an AMM's core component, it can arbitrate and manage and control the function of each module. Its detailed function is explained in parts of function description of the relevant components below.

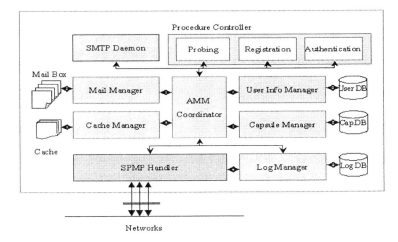

Fig. 7. Architecture of Active Mail Manager

- SPMP Handler: It analyzes the SPMP header and handles the analyzed information to AMM Coordinator. It calculates the header checksum and then checks the type and code field.
- Procedures Controller: It contains the probing, registration, and authentication procedures explained in section 3.4.
- Mail Manager: It manages various mail delivered in AMM. That is, it manages the user's mailbox. And when the capacity of each user mailbox exceeds, it can handle mailbox according to mailbox management policy.
- Cache Manager: It manages the cache memory that stores the IP addresses of mail recipient's AMM temporarily. When sending mail, first AMM Coordinator requires the IP of recipient's AMM of cache manager. If it is not in the cache memory or mail isn't delivered using it in cache memory, AMM Coordinator requests it to AMS and updates cache memory with it.
- User Info Manager: User Info Manager manages information of users provided with SPMS through AMM.
- Log Manager: Log Manager records the specific events, action, and all error that happened in AMM.

4   Discussion

In this section, using Fig. 8 and formulas below we prove that SPMS can deliver mail at the minimum cost. Where, cost means the count of hops passed by when sending and receiving mail. We explain the cost in case that user A sends mail to B and B receives it.

In formula (1), (2), and (6), Cs is the hop count passed by when A sends mail to B and Cr is the hop count when B receives mail. In formula (3), S1 is initial sum of Cs and Cr and is the constant. As the distance of between SMTP

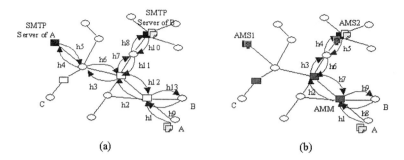

Fig. 8. Total hop count of (a) Traditional Mail Service, (b) SPMS

server and user's host is farther, S1 is greater. In formula (4), S is the total hop count when A sends mail to B at n times, which is the multiple of a1. Here a1 is the constant and changed according to the distance of SMTP server and user's host.

$$C_s = h_1 + h_2 + h_3 + h_4 + h_5 + h_6 + h_7 + h_8 \qquad (C_s > 0) \qquad (1)$$

$$C_r = h_9 + h_2 + h_7 + h_7 + h_{10} + h_{11} + h_{12} + h_{13} \qquad (C_r > 0) \qquad (2)$$

$$S_1 = S_2 = \ldots = S_n = C_s + C_r = 8 + 8 = 16 = a_1 \qquad (3)$$

$$S = \sum_{i=1}^{n} Si = a_1 * n \qquad (n > 0) \qquad (4)$$

In formula (5) and (7), Ci is the hop count passed by when AMM initially requests B's AMM IP address of AMS2 and S1 is the hop count when A first sends mail to B and B receives it. Consequently S is the total hop count that is b1+ b2*n. Here b1 and b1 are constants and less as the distance of between sender and recipient is closer.

$$C_i = h_1 + h_2 + h_3 + h_4 + h_5 + h_6 + h_7 \qquad (C_i > 0) \qquad (5)$$

$$C_s = h_1, \qquad C_r = h_8 + h_9 \qquad (C_s, C_r > 0) \qquad (6)$$

$$S_1 = C_i + C_s + C_r = 7 + 1 + 2 = 10 = b_1 + b_2$$

$$S_2 = S_3 = \ldots = S_n = C_s + C_r = 1 + 2 = 3 = b_2 \qquad (7)$$

$$S = \sum_{i=1}^{n} Si = b_1 + b_2 * n \qquad (n > 0) \qquad (8)$$

In Comparison formula (4) to (8), we know that SPMS has less total cost than traditional mail service. That comparison is more precise as the network is more extensive and the distance of between the SMTP mail server and user'host is farther. SPMS has some advantages. It can reduce the redundant network traffic and response time by shortening the distance of mail transmission and can reduce the server's over-load by distributing the function of the mail service to the network nodes. Also it supports the integrated management of diverse mail service and it guarantees reliability by the sending and receiving authentication.

5   Conclusion and Future Works

In this paper, using active technology, we propose a new and efficient mail service based on active network that solves the problem of the traditional mail service. SPMS can reduce the redundant network traffic happened to pass through the mail servers of the sender and recipient as transmitting mail to the closest active node from the sender's and recipient's host, AMM and can guarantee the reliability by the sending and receiving authentication procedure. Also it can reduce the overload and response time of a mail server and can provide the integrated management of diverse mail service.

In the future we will have the performance evaluation by implementing and simulating it and generalization of path-shortening mechanism of SPMS to apply it to the various services.

References

[1] Node OS Working Group: NodeOS Interface Specification June 11 1999.
[2] S. Bhattacharjee, K. L. Calvert and E. W. Zegura: An Architecture for Active Networking High Performance Networking (HPN'97), White Plains NY, April 1997.
[3] Hyun Joo Kim, Tai M. Chung, Taek Y. Nam, and Sung W. Sohn: Global Mail Service on Active Network The 2002 International Conference on Security And Management (SAM'02) LasVegas, June 2002.
[4] M. Murhammer, O. Atakan, S. Bretz, L. Pugh, K. Suzuki, and D. Wood: TCP/IP Tutorial and Technical Overview International Technical Support Organization Oct. 1998.
[5] D. Wetherall, U. Legedza, and J. Guttag: Introducing New Internet Services: Why and How IEEE Network Magazine, July/August 1998.

Design of Security Enforcement Engine for Active Nodes in Active Networks

Ji-Young Lim[1], Ok-kyeung Kim[1], Yeo-Jin Kim[1], Ga-Jin Na[1], Hyun-Jung Na[1],
Kijoon Chae[1], Young-Soo Kim[2], and Jung-Chan Na[2]

[1] Dept. of CSE
Ewha Womans University, Seoul, Korea
{jylim,kimok,zzin97,nagajin,hjna,kjchae}@ewha.ac.kr
[2] Information Security Technology Division
Electronics and Telecommunications Research Institute, Daejeon, Korea
{blitzkrieg,njc}@etri.re.kr

Abstract. Active networks are a new generation of networks based on
a software-intensive network architecture in which applications are able
to inject new strategies or code the infrastructure to their immediate
needs. Therefore, the secure and safe active node architecture is needed
to give the capability defending an active node against threats that may
be more dynamic and powerful than those in traditional networks. To
secure active networks, the security enforcement engine is proposed in
this paper. We implemented our engine with security, authentication
and authorization modules. Using this engine, it is possible that active
networks are protected from threats of the malicious active node.

1 Introduction

Active Networks[1, 2, 3, 4, 5] introduec a powerful new communication paradigm
within which users inject programs contained in messages into a network capable of performing computations and manipulations on behalf of the user. Active
nodes within the networks facilitate execution of an application or a user specific code that may introduce new protocol elements or process incoming data
streams. Traditional networks expose a fixed set of network services that are
hard to modify and customize. It is difficult to integrate new technologies and
standards into the shared network infrastructure. Even though there have been
tremendous technological gains throughout the years, today's network still suffer
from poor performance due to redundant operations at several protocol layers.

In contrast,active networks seek to create an extensible system by providing
an environment where nearly arbitrary applets, stored in network datagrams,
can be executed, allowing new services to be dynamically created. The execution of active code can be in any active node, including programmable routers,
switches and bridges. Such a framework allows rapid deployment and refinement
of protocols to accommodate the rapid evolution of networking technologies with
the result of dramatic reductions in the development cycle. This in turn enables
intricate customization of services to suit different application.

H.-K. Kahng (Ed.): ICOIN 2003, LNCS 2662, pp. 346–356, 2003.
© Springer-Verlag Berlin Heidelberg 2003

Any networking system faces the threat of invasion for the purpose of criminal activity or exploiting vulnerabilities. Two key areas in security that coexist in modern networks are authentication and authorization. As opposed to traditional networks, active networks allow packets to have direct access to intermediate nodes, including programmable router and switches, which potentially increases their chance of becoming attacked. A solid security architecture is fundamental for the active networks.

We propose the design of the security enforcement engine including two key areas. Our engine consists of the security module for encryption/decryption and message integrity, the authentication module for verification of the valid node/user and the authorization module for verification of the valid access rights to resources. We implement the engine and test the various scenarios to verify our engine. We describe our engine architecture in section 2 and the detail operation of each module in section 3. Finally, the conclusion and future work is described in section 4.

2   The Proposed Architecture

Traditional networks are vulnerable to a variety of threats and exposures and provide limited support to counteract or prevent attacks on the infrastructure. While variable single security systems such as firewall systems and intrusion detection systems[6] have been developed, it is not enough for those systems to deal with various attacks growing complicated. Active networks proposed at DARPA[7] can be introduced to traditional networks to counteract and prevent attacks actively since active networks allow executing a program code at the intermediate routers and modifying the router states based on its result.

However, active networks also have weakness. Packets with malicious codes may attack active nodes in active networks and a malicious active node may disturb normal activities of other active nodes. These may be more powerful attacks than traditional attacks. Therefore, the secure and safe active node architecture is needed to give the capability defending an active node against threats. Figure 1 is our active node architecture including the security enforcement engine for secure communication in active networks. The security enforcement engine is in NodeOS and checks the integrity, encryptes/decryptes and verifies incoming active packets. Normal packets only may be allocated resources and executed in one of EEs (the execution environments). Outgoing packets are done as same as incoming packets by the security enforcement engine.

Figure 2 is the detail architecture of the security enforcement engine designed based on requirements for the secure communication. It consists of a security module, an authentication module and an authorization module. Integrity, secrecy and modification of a message in an active packet are processed in the security module. The user or the node creating or sending a message is certified in the authentication module. The authorization module restricts the access to resources required by the authenticated message. The functions of each module are described next section.

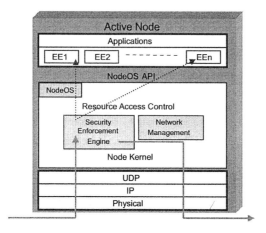

Fig. 1. Active Node Architecture

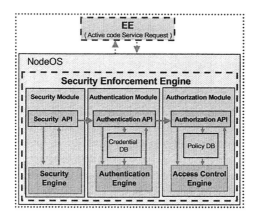

Fig. 2. Security Enforcement Engine Architecture

Figure 3 shows the ANEP packet format [8] used in our engine. IP protocol is used as the network layer protocol and one of TCP and UDP is used in transport layer. ANEP (Active Network Encapsulation Protocol) header and old options are used in ANEP that is the upper layer. Credential, signature, in-line policy and hop-hop integrity are added as new options in ANEP header. A message or a code being delivered to active nodes is contained in an ANEP packet as a payload.

The credential option is used to authenticate the user or the node and may use one of SPKI (Simple Public Key Infrastructure)[9], Kerberos[10] or X.509 format[11]. X.509 format is chosen in our model. The signature is created to verify whether the received message is altered. In-line policy option is used to

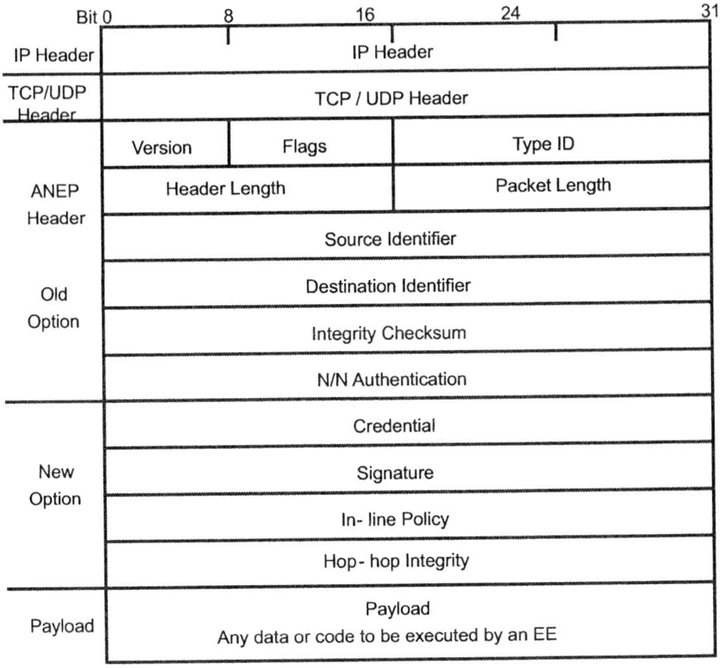

Bit 0	8	16	24	31

Fig. 3. ANEP Packet Format

limit accesses to resources such as the program execution not to spend resources of the active node. Hop-hop integrity means information used to verify the message integrity, authentication, etc. in intermediate nodes. To certify intermediate nodes, credentials issued in intermediate nodes may be added. When an intermediate node modifies the message, the signature issued in it may be added to check the message integrity. These enable to prevent attacks at the intermediate node.

3   The Design of The Security Enforcement Engine

To secure active networks, we designed the security enforcement engine with the security module, the authentication module and the authorization module. In this chapter, we described system environments implemented our engine and the operational procedure of the each module. We assume that communication with active nodes and CA and key exchanges between active nodes is done and not considered in implementation of our engine. Figure 4 shows our system environment to test our implemented engine. Its platform is Linux Red Hat 7.3 version. To emulate NodeOS's function, Anetd (Active Networks Daemon) provided by ABONE was installed. Anetd consists of ABCd (ABone Control dae-

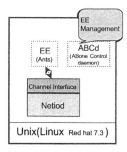

Fig. 4. System Environment

mon) managing EEs and netiod (Network I/O daemon) handling network I/O for UNIX/Linux platform and providing NodeOS Channel interface to EEs[12]. One of Ants (Active Networks Transport System), ASP (Active Signaling Protocol) and PLAN (Packet Language for Active Networks) EEs is possible as EE.

3.1   The Security Module

This module is implemented to satisfy the general security requirements, to provide security service in active networks and to verify integrity and alteration of the received message. We use MD5 algorithm[13]. that limits the length of messages or codes by 128bits for integrity and DES CBC encryption and decryption algorithm[14] for confidentiality. RSA digital signature algorithm[15] is used for digital signature.

Figure 5 shows the operation applied to credential and in-line policy option fields and payload for security services. According to our assumption, the credential, the secret key and the public key for encryption and decryption are given from CA (the Certificate Authority) ((1) and (2) in figure 5). The credential and the in-line policy option fields and the payload are encrypted using DES CBC Encryption/Decryption algorithm not to be exposed to attacker. The encryption procedure uses the public key and the initial vector of the sending node as encryption keys. Before the encryption procedure, MD5 algorithm is applied to three fields to limit the length as 128 bits ((3) in figure 5). Since these encryption keys are also needed at the decryption time, they are encrypted using three 128-bit MD5 message digests and sent to the active node that process a message or a code of the packet ((4) in figure 5). Three fields are encrypted ((5) in figure 5), sent to the next active node (Active Server Node in figure 5) and decrypted at the active server node ((6) in figure 5).

In our module, RSA digital signature is used to verify whether the received packet is modified or not. Figure 6 shows the proposed operation of digital signature. The secret key and public key in figure 6 are as same as those in figure 5. MD5 message digests of in-line policy option field and payload is used

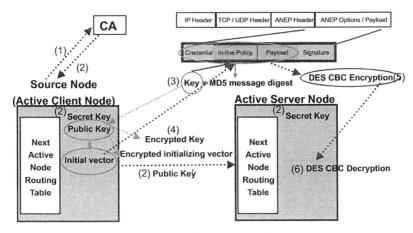

Fig. 5. Flowing of Encryption/Decryption

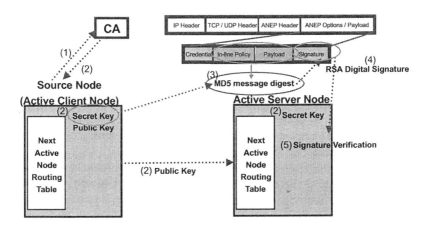

Fig. 6. Flowing of digital signature

to check message integrity ((3) in figure 6). Digital signature is computed from two message digests and secret key ((4) in figure 6), sent to the next node (Active Server Node in figure 6) and verified according to RSA Digital Signature algorithm ((5) in figure 6).

Figure 7 shows the successful process of message digests and encryption at the source active node before sending the packet. If a part of the received packet is lost or modified at another intermediate active node, it is dropped not to be passed to the authentication module at the active server node. Figure 8 shows the result when the failure of the verification results in dropping the packet that is altered.

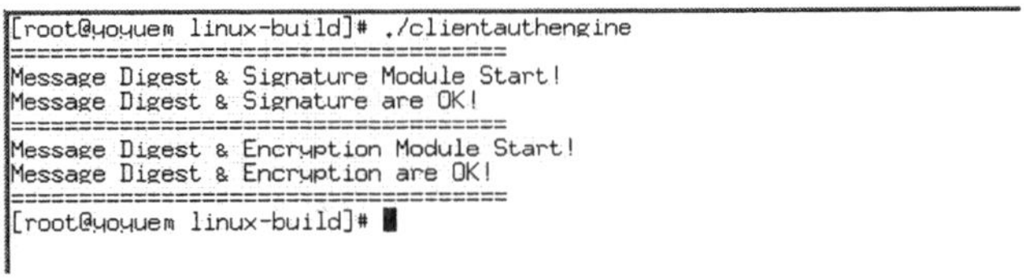

```
[root@yoyuem linux-build]# ./clientauthengine
============================================
Message Digest & Signature Module Start!
Message Digest & Signature are OK!
============================================
Message Digest & Encryption Module Start!
Message Digest & Encryption are OK!
============================================
[root@yoyuem linux-build]# █
```

Fig. 7. Processing result at the Security module in sending active node

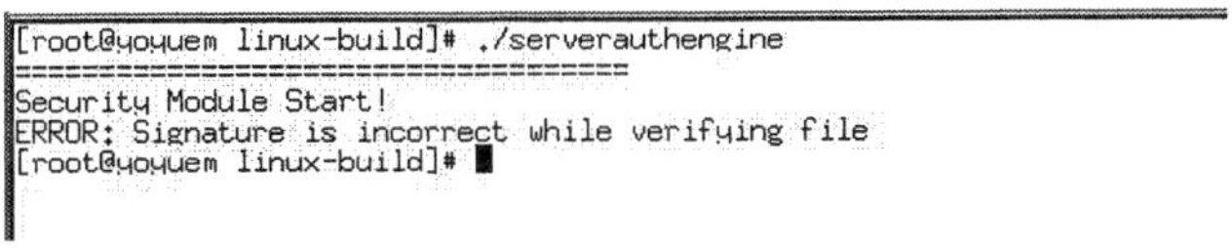

```
[root@yoyuem linux-build]# ./serverauthengine
==============================================
Security Module Start!
ERROR: Signature is incorrect while verifying file
[root@yoyuem linux-build]# █
```

Fig. 8. Processing result at the Security module in receiving active node

3.2   The Authentication Model

It is implemented that after successful verification of message integrity, the sending active node and the user is certified in the authentication module. The packet in the authentication module was already decrypted and verified the message integrity. The authentication module compares the credential extracted from the received packet with the credential given from CA and stored in credential DB. Then it determines whether the received packet is sent by the valid node or the valid user. If the packet is valid, it is passed to the authorization module. Otherwise, it discards.

Figure 9 shows the operational procedure in the authentication module. The node gets credentials of the certifying active nodes from CA and constructs the DB ((1) figure 9). The credential of the received packet is checked in security module ((2) figure 9). It is sent to the authentication module and compared with the credential of DB ((3) in figure 9). It guarantees that nodes or users who creating or modify the packet are valid if two credentials are same. It checks whether the credential is expired or not since the expired credential is prevented from reuse.

Figure 10 shows the result when the packet through the security module discards since its credential is valid but is not in accord with the credential in DB.

3.3   The Authorization Model

The packet in the authorization module finished the verification of the message integrity and the authentication. The authorization module verifies that the packet has the valid access rights for resources and limits its authority for the resource usage. Request for resource allocation is in the in-line policy option field. Each active node has the policy DB in which the policy for each service is

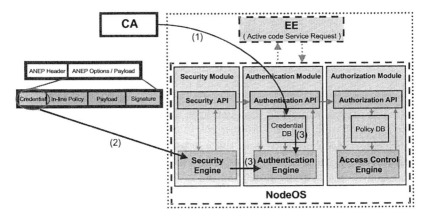

Fig. 9. Processing flows at the Authentication Module

```
[root@yoyuem linux-build]# ./serverauthengine
==========================================
Security Module Start!
Active Packet Signature verified.
==========================================
Authentication Module Start!
It's Valid DATE!!!
no!! issuer name is not match!
[root@yoyuem linux-build]# ▮
```

Fig. 10. Result at the Authentication Module in the receiving node

stored. The policy includes the host, the user, the privilege and time information and its format is in figure 11 and figure 12 [8]. The host and user fields specify the hosts and users accessible to the service. The privilege filed specifies a set of the access modes for the service. Access modes defined in our engine are C (create), M (modify), A (append), R (read), D (delete) and E (execute). The time field specifies time for the service to live in active node. In-line policy in the received packet is compared with policy in DB. If the resource request of the packet is acceptable, it is passed to EE. Otherwise, it discards.

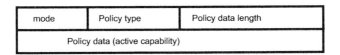

Fig. 11. Policy Format

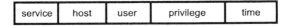

service	host	user	privilege	time

Fig. 12. Policy Data Format

The authorization module also manages the policy DB. The policy in DB may be modified and deleted and new entry is added to DB. The mode field in figure 11 is used to classification of these activities. If the mode is '1', it means the authorization check. If '2', it means addition of a new policy entry or the modification of the existing policy. If '3', it means removal of the policy. The policy type filed in figure 11 means the various policy data formats defined by ANEP Header[8]. The type defined currently is only one and its value is '1' that specifies format in figure 12 . In our current implementation, we consider just one service that is to save the code and to pass it to EE. This service requires a full set of privilege(CMARDE).

Figure 13 shows the result for the verification of the valid access rights of the authenticated packet. The packet is dropped because it has the wrong privileges of the service. Figure 14 shows that the packet through the security and authentication modules is authorized successfully.

4   Conclusion

Active networks are the next generation framework that can solve the many problems for traditional networks. The active code is injected into the networks and executed inside the networks. Therefore, if the code is exposed and modified by the malicious user/node, the threat and attack are more powerful than those occurred in traditional networks. We designed and implemented the security enforcement engine to secure active networks. Our engine has three modules

```
[root@yoyuem linux-build]# ./serverauthengine
=====================================
Security Module Start!
Active Packet Signature verified.
=====================================
Authentication Module Start!
It's Valid DATE!!!
complete match!
Source Node is Authenticated!
=====================================
Authorization Module Start!
Read request information..
Read request information more..
Call authorization policy..
Start authorization function..
Open DeliverToApp DB successfully.
host ok..
user ok..
Checking privilege..
CAN NOT ACCESS!
[root@yoyuem linux-build]# ▊
```

Fig. 13. The result of failure in authorization module

```
[root@yoyuem linux-build]# ./serverauthengine
========================================
Security Module Start!
Active Packet Signature verified.
========================================
Authentication Module Start!
It's Valid DATE!!!
complete match!
Source Node is Authenticated!
========================================
Authorization Module Start!
Read request information..
Read request information more..
Call authorization policy..
Start authorization function..
Open DeliverToApp DB successfully.
host ok..
user ok..
Checking privilege..
Privilege ok..
Checking time..
Date is ok..
Authorization is OK!
========================================
[root@yoyuem linux-build]#
```

Fig. 14. The successful result in authorization module

such as the security, authentication and authorization modules. It can join with existing NodeOSs and EEs.

Active security area is the first step in research and development. We will add functions such as the key exchange and authorization for various services to make our engine more complete.

References

[1] D. L. Tennenhouse, et al., "A Survey of Active Network Research," IEEE Communications Magazine, pp.80-86, Jan, 1997.

[2] K. Psounis, "Active Network: Applications, Security, Safety, and Architecture," IEEE Communications Serveys, 1999.

[3] Security Architecture for Active Nets by AN Security Working Group: 1998, Modified by Seraphim Group: 2000.

[4] R. H. Campbell, et al., "Seraphim: Dynamic Interoperable Security Architecture for Active Networks," IEEE OPENARCH 2000, Tel-Aviv, Israel, Mar. 2000.

[5] Leon Dang, "CANSA (Certificate Active Network Security Architecture)," Basser Department of Computer Science, University of Sydney, 1998.

[6] M. Wood, et al., "Intrusion Detection Message Exchange Requirements:draft-ietf-idwg-requirements-10.txt," October 22, 2002.

[7] Defense Advanced Research Projects Agency,
http://www.darpa.mil/ato/programs/ activenetworks/actnet.htm.

[8] Alexander D. Scot, et al., "Active Network Encapsulation Protocol (ANEP)," Active Network Group Draft, July 1997.

[9] C. Ellison, et al., "SPKI Certificate Theory : rfc2693.txt," Sep. 1999.

[10] B. Clifford Neuman, et al., "Kerberos: An Authentication Service for Computer Networks," IEEE Communications Magazine, Volume 32, Number 9, pages 33-38, Sep. 1994.

[11] R. Housley, et al.,"Internet X.509 Public Key Infrastructure: X.509 Certificate and CRL Profile", RFC 2459, Jan. 1999.

[12] Steve Berson, et al., "Evolution of an Active Networks Testbed", Presentation at DARPA Active Networks Conference and Exposition 2002, San Francisco, CA, 29-30 May 2002.

[13] Rivest, R., "The MD5 Message-Digest Algorithm", RFC 1321, MIT Laboratory for Computer Science and RSA Data Security, Inc., Apr. 1992.

[14] ANSI X3.106-1983, American National Standard for Information Systems - Data Encryption Algorithm - Modes of Operation, American National Standards Institute, Approved 16 May 1983.

[15] C. J. MITCHELL, et al., Digital signature. In Contemporary Cryptology, The Science of Information Integrity, pages 325-378. IEEE Press, 1992.

A Least Number of Stream Used(LSU) Algorithm for Continuous Media Data on the Proxy Server

Seongho Park[1], Seungwon Lee[2], Yongwoon Park[3], Hwasei Lee[4], Yongju Kim[5], and Kidong Chung[2]

[1] Computer Center, Pusan National University, Rep. of Korea
shpark@pusan.ac.kr
[2] Department of Computer Science, Pusan National University, Rep. of Korea
{swlee,kdchung}@melon.cs.pusan.ac.kr
[3] Department of Computer Science, Dongeui Institute of Technology, Rep. of Korea
ywpark@dit.ac.kr
[4] Department of Computer Engineering, Miryang National University, Rep. of Korea
hslee@arang.miryang.ac.kr
[5] Department of Computer Engineering, Silla University, Rep. of Korea
yjkim@silla.ac.kr

Abstract. In this paper, we propose a space replacement algorithm called LSU for the efficient proxy cache management and analyze its performance. For this, we first analyze the log files of one of the major broadcasting system's media server to see what client's characteristics are when they access video objects. As result of this analysis, we generate the probability density function based on each request's system resource usage. We do simulations to check the performance of our proposed algorithm comparing it with LRU and LFU. As result of the simulations, we show that in most cases, LSU works better than LFU and LRU regardless of the size of the replacement time window.

1 Introduction

With the rapid proliferation of the Internet, media streaming service providers such as VoD and advertisement video clip and etc. are increasing rapidly. Along with this, the service providers have made their best efforts to provide as high quality services as possible to clients; due to the development of computing and networking technologies, it is possible to do so[1]. However, the capacity of the Internet cannot keep up with the increasing rate of both clients and objects so that the Internet has been overloaded as time passes[2].

Given the emerging gigabit networking technologies such as Gigabit Ethernet and Fibre Channel, the cost of installing and running a local area gigabit network becomes increasingly cheaper. Therefore, reducing the total bandwidth requirement of the core backbone network should be an important objective in the design of a real time continuous media data delivery system[3]. So, there has been much research about how to minimize the transfer bandwidth required

H.-K. Kahng (Ed.): ICOIN 2003, LNCS 2662, pp. 357–366, 2003.

to transfer objects from the origin server to clients[4,5]. Network caching and CDN(Content Delivery Networks) technology, particularly, exploits the intermediary storage between the origin server and clients so as to intercept network traffic originally destined to the origin server[6,7]. However, current caching algorithms have been designed for the discrete objects such as text and image that they do not go well with continuous media objects. The continuous media objects usually require lots of storage space (ranging from several M bytes to G Bytes) and transfer bandwidth as well as QoS. So, caching in the proxy server must be designed taking these into consideration. Moreover, the discrete objects including text and image are cached in the atomic mode, i.e., they are cached entirely or not; caching policies are oriented to which object is evicted for the newly cached object. So, the replacement metric is usually one of these: recency, frequency and size. In case of streaming service of continuous media objects, it is important to service the requested object without jitter or eviction that the characteristics and the access patterns of continuous media objects should be taken into consideration.

The organization of this paper is as follows. In Section 2, we describe the existing research about proxy server space management policy. In section 3, we analyze the log files of the media servers of one of the major broadcasting corporations in Korea and as a result, show that how the access patterns are changed as time passes. In section 4, we introduce a proxy-caching algorithm based on the result of our analysis described in section 3. We also do simulations to evaluate our proposed algorithm' performance. Finally, in section 5, we draw conclusion and refer to the further research.

2   Related Works

Although many research works have been done to address network-IO bottleneck in terms of servicing continuous media object over the Internet, they are mainly concerned with Web files such as images and text. Continuous media data are used to being accessed similar to text and images using the download and play mode. With a download and play mode of access, data is transferred completely to the client site before display. Due to the large sizes of continuous media objects, this results in large space usage and wait time at the client[8]. Streaming mode of access addresses this problem by enabling the client to initiate display of data with only small start-up latency, without waiting for the entire object to be downloaded. It is not until recently that some proxy caching schemes for continuous media objects are introduced. However, they are focused on reducing the initial latency or smoothing the burstiness of the VBR stream. [8] applies the interval caching to the proxy cache management that each object is decided to be cached based on how much transfer bandwidth and storage space it requires. [3] proposes a proxy cache management policy that some initial portions of the selected objects are cached in the proxy server to smooth the network transfer bandwidth and as a result to provide more qualified service to the clients.

In [9], the initial portion or prefix of each object is also cached to hide the initial latency that occurs when accessing that object from the origin server. As described, most of the proxy caching policies about continuous media objects are focused not on the improvement of the overall performance of the proxy server, but on the minimization of burstiness of servicing VBR streams and the initial latency taken to access the requested object. As time goes by, video streaming service over the Internet will be increasing rapidly that the proxy caching policy based on video streaming service will be considered[10].

In terms of proxy space replacement policy, lots of works have been done so far[11, 12, 13] that the result of these works provide us more efficient proxy space management policy hiding the weakness of the existing proxy space management policy. [11] suggests some proxy design considerations of the conventional data and their combinations in terms of managing proxy space for operational efficiency. [14] proposed the stepwise proxy space management policy where each video is layered-encoded so that caching granularity of each object is decided by its access popularity; the less popular object are serviced directly from the origin server to the clients. [12,13] proposed a proxy space management policy that each object is cached stepwise at the segment-level starting from the initial segment to the entire object so as to reduce both the initial latency and space management overhead. So far, however, none of these works considers how each media object is created and then how their access patterns are changed along with time sequence.

3   The Analysis of Log Trace

In this paper, we analyze the log files of iMBC(http://www.imbc.com) to check out how the access patterns of the objects on the Internet have been changed. In addition, we also analyze the characteristics of users' request patterns to use them as a proxy space replacement metric.

3.1   iMBC Server's Environment

Table 1 shows the iMBC server's specifications. We randomly chose one server among 20 servers and then archived 21 days of the log file of that server. The Windows NT runs on that server and Window Media Server technology is applied for streaming service.

3.2   The Result of our Analysis of our Trace Data

1) The characteristics of media objects

Table 2 shows the characteristics of the objects of the iMBC server. In the iMBC server, 15 ~ 20 objects are created everyday and among all access frequencies, the percentage of access frequencies of the objects created during the monitored period(21 days) is 9.8 The server provides two different transfer modes

Table 1. iMBC server'specification

Parameters	Value
# of servers	1 out of 20 servers
OS	Windows NT
Streaming S/W	Windows Media Server
Log file volumes	2001. 5. 11 -5. 31 (21days)

Table 2. The characteristics of monitored objects

Parameters	Value
# of objects monitored	2,900
# of object created during monitored period	285
Avg. transfer rate	100k / 300k bps
Playback time	5 120 min
Total size of objects	290 G bytes

per object: 100k bps and 300k bps. The playback time of each object ranges from 5 min. to 120 min. and the total size of all objects are 290 Giga Bytes.

2) The users' access patterns

In terms of analyzing the trace log file, we classify each user's request into two modes : Start Access Request and Random Access Request.

– Start Access Request : in case a user want to access his/her requested object from the beginning.
– Random Access Request : in case a user choose his/her requested object's starting point of playback on his/her own at random point in time.

Table 3 shows users'access patterns for objects archived in the iMBC server. The number of overall accesses is 1,887,064. Among this, the percentage of Start Access Request is 26.3the percentage of the number of Start Access Requests is 34.0The number of objects created during the monitored period(21 days) is only 9.8but the percentage of the number of accesses for them reaches 46.6this means the newer the objects, the higher their access ratio.

4   The Replace Policy

Some replacement policies such as LRU or LFU do not work well with continuous media data because continuous media objects, other than convention data type such as text and images, have their remarkable feature as follows.

First, continuous media objects require lots of transfer bandwidth and storage space in terms of proxy caching. For example, an asf-encoded media object of

Table 3. User's access pattern

Parameters	Value	Ratio
All access frequencies	1,887,064	100%
Access frequencies for objects created during monitored period	875,737	46.6%
# of Start Access Request	496,829	26.3%
# of access requests for object created during monitored period	298,063	
# of Random Access Request	1,390,235	73.7%
# of Random Access Requests for object created during monitored period	557,674	

50 minutes of playback time with 300Kbps of transfer rate requires about 112.5 Mbytes storage space; it takes 50 minutes if serviced in streaming mode.

Second, continuous media objects, other than conventional data whose entire data blocks are serviced once they are requested, can be serviced partially. i.e., only part of the object through the interaction with the server. Moreover, users can choose the starting point of his/her requested object at any time.

Third, in streaming mode, continuous media object are transferred in real time mode that system resources including network and disk channel could be reserved during its playback time as much as possible.

Based on these characteristics and the result of our analysis in section 3, we propose a proxy space replacement policy to maximize the efficiency of the network channel's availability and minimize user's initial latency.

4.1 A Replacement Measure

Generally, the most widely known cache space replacement metric is one of the following: frequency, recency and size. They, however, can not work well with continuous media objects because of the following reasons. (Fig.1 shows how a stream is serviced as time passes. $S_i(O_k)$ means a stream for request i of object k or O_k).

First, if continuous media objects are serviced in streaming mode, there are several kind of access frequencies. For example, as shown in Fig.1, there are three kind of access frequencies during Δt : i) frequency about how many streams starts, ii) frequency about how many streams are running at random point in time, iii) frequency about how many streams are played back till the end of its requested object.

Second, the playback time of objects ranges from several minutes to order of ten minutes. So, there is a time gap between stream's starting time and ending time when checking the recency of an object.

Third, a user can stop playing back his/her requested object arbitrarily at his/her own will. For example, as shown in Fig.1, the stream serviced at $t_n + 1$

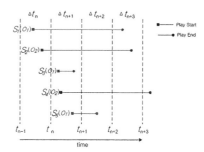

Fig. 1. Example of stream sequence

may not sustain till $t_n + 2$ that the importance of $S_4(O_2)$ is different from $S_5(O_1)$ at $t_n + 1$.

4.2 LSU(Least Number of Streams Used)

In Least number of Streams Used or LSU, the number of active streams for each object at t_n is checked to use it as replacement metric.

1) Replacement function(f_{LSU})

The replacement metric of LSU at t_n for O_k is expressed as function f_{LSU} in Eq. 1.

$$f_{LSU}(O_k, t_n) = Ns(S_i(O_k), t_n) \tag{1}$$

$f_{LSU}(O_k, t_n)$: A Replacement Measure at t_n by LSU algorithm
$Ns(S_i(O_k), t_n)$: The number of streams $S_i(O_k)$ at t_n for Object O_k

$f_{LSU}(O_k, t_n)$ is the number of active streams for O_k at t_n; it can be explained as how popular O_k is at t_n. Each value of f_{LSU} at Fig.1 is given in Table 4 to show you how f_{LSU} is changed as time passes.

2) LSU algorithm

Fig.2 shows how LSU runs. If the requested object O_k does not exist in the proxy cache, the proxy cache selects an object in the proxy cache and then evict

Table 4. The value f_{LSU} in Fig.1

Time Seq.	t_n	$t_n + 1$	$t_n + 2$	$t_n + 3$
$f_{LSU}(O_1, t)$	1	2	1	0
$f_{LSU}(O_2, t)$	1	2	2	1

```
Procedure Replacement_Alg
   Input : f_LSU(O_k), O_k, Rcs
   Output : Whether Object k do replcement
   Begin
      V ← ∅
      While(1)
         if(Size(O_k) < Rcs) return do replacement
         if({∀_j ∈ C_0, j ∉ V, f_LSU(O_j) < f_LSU(O_k)} = ∅ )
            return do not replacement
         Rcs ← Rcs + Size(O_j)
         V ← V ∪ {j}
      Endwhile
   End
```

Fig. 2. LSU replacement algorithm

it to make room for O_k. Rcs and $Size(O_k)$ mean the unallocated cache space and the size of O_k respectively and C_0 is the set of cached objects in the proxy cache.

4.3   Performance Evaluation

In this section, we evaluated our proposed cache replacement policy by comparing it with other existing cache replacement policy: LRU based on the starting point of streams and LFU based on the frequency of streams' starting time. The time interval of setting the value of replacement metric is given 5 minutes.

1) Hit ratios with different cache sizes

Hit ratio in terms of measuring the performance of the proxy server is used to evaluate how much the initial latency is reduced. Fig.3 shows how hit ratios changes with different cache sizes in LRU, LFU and our proposed LSU. As the graphs indicate, LFU and LSU, the frequency-based algorithms, work much better than LRU, the recency-based algorithm. Moreover, LSU shows 2.5% ∼ 8.2% of higher performance gain than LFU does.

2) Byte Hit Ratio with different cache size

Byte Hit Ratio or BHR is used as a performance metric about how much network resource is saved when transferring the requested object form the origin server to the proxy server. Fig.4 shows Byte Hit Ratios with different cache sizes in LRU, LFU and LSU. As with Hit Ratio, LFU and LSU shows better result than LRU does and moreover, LFU shows 3.4% ∼ 7.9% of better performance than LFU.

3) Replacement counts with different cache sizes

Fig.5 compares LSU with LFU and LRU in terms of cache replacement counts. As with other tests, LSU and LFU show much better performance gain

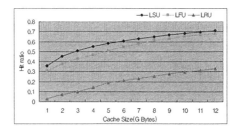

Fig. 3. Hit ratios with different cache sizes

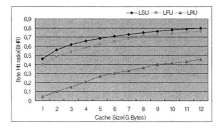

Fig. 4. BHR with different cache sizes

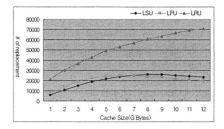

Fig. 5. Replacement counts with different cache sizes

than LRU. LSU is as much similar as LFU; the replacement counts decrease as the cache size gets bigger.

4) Comparison of LSU and LFU

Fig.6 presents how reference ratios change as the cache size changes based on the time interval of setting the value of replacement metric. Compared with other two policies, LSU shows minor changes in terms of reference ratio because in LSU, resetting the value of replacement metric is done at the point in time. Specifically, the reference ratio in LSU changes from 66.9 to 67.7% (about 6.4% of difference) as the time interval of setting the value of replacement metric changes from 5 minutes to 130 minutes.

Fig.7 shows how the number of replacement counts changes as the time interval of setting the value of replacement metric changes from 5 minutes to 130

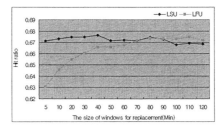

Fig. 6. Reference ratio changes with different time windows of setting the value of replacement measure

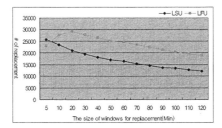

Fig. 7. Replacement count changes with different time windows of setting the value of replacement measure

minutes when the cache size is 9 Giga Bytes. The number of replacement counts in LSU decreases slowly and when the time interval of setting the replacement metric is more than 20 minutes and in this case LFU shows 40% ∼ 57% of more replacement overhead than LSU.

5   Conclusions

In this paper, we proposed and evaluated a new proxy cache replacement policy for the efficient network channel management. More specifically, we analyzed the log files of one of the major broadcasting company's media server to find out what the clients' access patterns are. Based on this analysis, we proposed a cache replacement algorithm called LSU considering the efficient management of the origin server and network channel between the origin server and proxy server. We evaluated our proposed algorithm by comparing it with other two policies in terms of hit raio, byte hit ratio, reference ratio and replacement count.

As the result of performance evaluation, we showed that our proposed algorithm works better than LRU in terms of hit ratio, byte hit ratio and replacement count. Moreover, it shows better performance than LFU in terms of reference ratio, byte hit ratio and replacement count.

References

[1] J. E. PitKow and Colleen M. Kehoe, "The GVU's WWW User Survey," http://www.cc.gatech.edu /gvu/user_surveys, 1998.

[2] Garth A. Gibson, Jeffrey S. Vitter, John Wilkes, "Strategic Directions in Storage I/O Issues in Large-scale Computing." ACM Computing Surveys Vol.28, No. 4, 1996.

[3] Zhi-Li Zhang, S. Dongli, Yuewei Wang, "A Network conscious Approach to End-to-End Video Delivery over Wide Area Networks Using Proxy Servers," Proc. of INFOCOM'98 Page(s) 660-667 vol.2, 1998.

[4] C. C. Aggarwal, J. L. Wolf, P. S. Yu, "Caching on the World Wide Web," IEEE Transactions on Knowledge and Data Engineering, Volume : 11, Jan.-Feb. Page(s) 94-107, 1999.

[5] D. Eager, M. Ferris, and M. Vernon, " Optimized Regional Caching for On Demand Data Delivery," Proc. of the Conference on Multimedia Computing and Networking, Jan. 1999.

[6] C. Maltzahn, K. Richardson and D. Grunwald, " Performance Issues of Enterprise Level Web Proxies," Proc. of the SIGMETRICS Conference on Measurement and Modeling of Computer System, June 1997.

[7] Greg Barish and Katia Obraczka, "World Wide Web Caching: Trends and Techniques," Univ. of Southern California, tech reports, available at ftp://ftp.usc.edu/pub/csinfo/tech-reports/ papers/99-713.ps.Z, 1999.

[8] R. Tewari, H. M. Vin, A. Dan, D. Sitaram, "Resource-based Caching for Web Servers," Proc. of the SPIC/ACM Conference on Multimedia Computing and Networking, Jan. 1998.

[9] S. Sen, J. Rexford, and D. Towsley, "Proxy Prefix Caching for Multimedia Streams," Proc. Of IEEE INFOCOM'99, April 1999.

[10] S. Rexford, J. Towsley, "Proxy Prefix Caching for Multimedia Streams," Proc. of INFOCOM'99. and Eighteenth Annual Joint Conference of the IEEE Computer and Communications Societies, Volume: 3, Page(s) 1310-1319, 1999.

[11] P. S. Sahu and D. Towsley, "Design Considerations for Integrat- ed Proxy Servers," Proc. of IEEE NOSSDAV'99, June 1998.

[12] E. J. Lim, S. H. Park, H. O. Hong, K. D. Chung, "A Proxy Caching Scheme for Continuous Media Streams on the Internet," Proc. of the 15th International Conference On Information Networking, 2001.

[13] S. H. Park, E. J. Lim K. D. Chung, "Popularity-based Partial Caching for VoD Systems using a Proxy Server" Proc. of the 15th International Parallel & Distributed Processing Symposium, 2001.

Enhancing the Quality of DV over RTP with Redundant Audio Transmission

Akimichi Ogawa[1], Kazunori Sugiura[1], Osamu Nakamura[3], and Jun Murai[3]

[1] Graduate School of Media and Governance
Keio University, 5322 Endo, Fujisawa, Kanagawa 252-8520 Japan
[2] CRL (Communications Research Laboratory)
4-2-1 Nukui-Kitamachi, Koganei, Tokyo 184-8795 Japan
[3] Faculty of Environmental Information
Keio University, 5322 Endo, Fujisawa, Kanagawa 252-8520 Japan

Abstract. In this paper, a scheme for reducing the packet loss probability of DV/RTP audio data is introduced. We have created a DV based broadcast quality video transportation tool, DVTS(Digital Video Transport System). Using DVTS, high quality video transportation over the Internet can be realized with low cost. However, DVTS does not have any error correction scheme. Moreover, it does not support reliable data transportation. Thus, a single packet loss can deteriorate the audio quality. In this paper, redundant transportation of DV audio data is intoduced. Using this redundant scheme, the DV audio will be robust. High quality audio can be realized in networks with packet loss.

1 Introduction

Network bandwidth available for the Internet is growing massively. 100Mbps ethernet is very popular for LANs, and 1Gbps class backbone are common. Moreover, 10Gbps class backbone will be common in the future. In such network environments, applications that consume high network bandwidth are reasonable.

We have created a DV[1] based high quality video and audio transportation tool. We have chosen the DV format because it is the most popular format for both consumers and professionals. It is popular because of its small tape media size (6.3 mm, 120min/cassete), full digital recording capability, appropriate cost compared to 8mm camcorder. The video compression of DV format uses only intra frame DCT(Discrete Cosine Transform) and VLC(Variable Length Coding) compression technique at the fixed ratio. Error correction scheme is not included within the DV format.

Unlike MPEG1 and MPEG2, the DV format does not use an inter frame compression technique. This facilitates the editing of the DV format video. Compared to inter frame compression video formats, operations such as fast forward and playing backward is easier. The DV format uses IEEE1394[2] as the linking device. Non-linear DV editing system that uses IEEE1394 are very popular.

H.-K. Kahng (Ed.): ICOIN 2003, LNCS 2662, pp. 367–375, 2003.

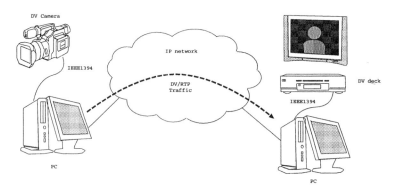

Fig. 1. DVTS overview

The tool we created was named DVTS(Digital Video Transport System)[3, 4]. The overview of DVTS is shown in Fig.1.

DVTS consists of a sender and a receiver. The sender application is called "dvsend" and the receiver application is called "dvrecv". It is assumed that the host using dvsend has an IEEE1394 interface, and a DV device (a DV camera) is connected using the IEEE1394 interface. Dvsend receives DV data via the IEEE1394 interface, encapsulates the DV data using RTP(Realtime Transport Protocol)[5][6][7], and sends the RTP packets to dvrecv using IP. DVTS can be used with both IPv4 and IPv6[8]. The RTP stream consumes about 30Mbps of network bandwidth.

Using DVTS, high quality video transportation over the Internet can be realized with low cost. DVTS does not have any error correction scheme. Moreover, it does not support reliable data transportation. Thus, a single packet loss can deteriorate the audio quality. There is a scheme within dvrecv to deal with loss of DV video data. (Illustration of this scheme is shown in section 2.) However, dvrecv does not have any scheme to deal with audio packet loss. Thus, there is a notable deterioration when a loss of DV audio data exists.

In this paper, a scheme is added within dvsend to reduce the probability of DV audio data loss. Using the scheme introduced within this paper, redundancy can be obtained to prevent the deterioration of audio quality. There is no addition to the dvrecv application. Thus, the scheme added within dvsend retains backward compatibility with the former dvrecv.

The remainder of this paper is organized as follows. Section 2 illustrates DVTS. Section 3 reviews the implementation of the redundant audio mechanism for DVTS. Evaluation of the implemented system is shown in section 4. Finally, section 5 summarizes this paper.

2   DVTS (Digital Video Transport System)

In this section, design and implementation of DVTS is illustrated. The DV format is shown in section 2.1, the dvsend mechanism is shown in section 2.2 and dvrecv mechanism is shown 2.3.

2.1   DV Format

DV format is implemented in its abstraction of framing method. This abstraction observes synchronization issues such as lip synchronization. All data (video, audio and system data) are managed within the units of its video picture frame. The DV digital data stream is composed of three levels of hierarchical structure. The DV format structure is shown in Fig. 2.

A single video frame data in the DV format stream is divided into several "DIF(Digital Interface Format) sequences". A DIF sequence is composed in 150 chunks of 80-bytes length DIF blocks. A DIF block is the primitive unit for all DV stream, and it is common in every DV specifications family. Each 80-byte DIF block contains 3-byte ID header that specifys the type of the DIF block, and its position in the DIF sequence. Five types of DIF blocks are defined in the ID header: and they are, Header, Subcode, Video Auxiliary information (VAUX), Audio data and Video data. Audio DIF block data also consists of audio Auxiliary information and audio data.

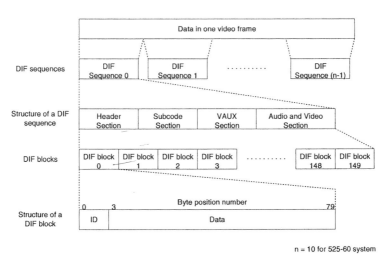

Fig. 2. DV Format Structure

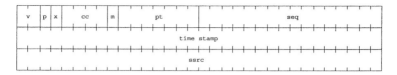

Fig. 3. RTP Header Format

2.2 dvsend

The dvsend application receives DV data via IEEE1394 and encapsulates the DV data into RTP. RTP is designed to accomplish realtime stream transportation using the Internet. RTP provides functions for packet-based realtime communication. The RTP header format is shown in Fig. 3. The "seq" field shown in Fig. 3 is used to show the sequence number of the RTP packet. This value is incremented each time the RTP packet is sent. The "timestamp" field shown in Fig. 3 is used to show the timestamp of the RTP packet. Each RTP packet from the same video frame will include the same value of timestamp.

Every DV stream data is constructed with 80-bytes DIF blocks including 3-bytes ID header. The format of the DV over RTP encoding uses RTP fixed header only, and does not use RTP extension header. The packet format of DV over RTP is shown in Fig. 4. The DV over RTP packet is sent using UDP(User Datagram Protocol)[9].

Any integral number of DIF blocks may be packed into a single RTP packet. The DIF blocks are directly concatenated after the RTP fixed header. All DIF blocks in a single RTP packet must be from the same video frame. Thus, DIF blocks from the next video frame will not be packed into the same RTP packet even if more payload space remains.

Transition from one video frame to the next is indicated by a change in the RTP timestamp. Thus, DV over RTP stream does not rely on particular packets for video frame transition.

2.3 dvrecv

The dvrecv application receives RTP packets sent by dvsend, and sends the reconstructed DV data out from the IEEE1394 interface. The DV data included in the RTP packets are reconstructed into a DV frame. When a transition from

IP header	UDP header	RTP header	DV DIF block	DV DIF block	DV DIF block

Fig. 4. DV RTP Packet

one video frame to the next is indicated, the data of the DV frame is sent out from the IEEE1394 interface. The transition of the video frame to the next is indicated by the change in the RTP timestamp field.

When a DV data of the next DV frame arrives from the RTP stream, it is overwritten in previous buffer of the DV frame. Thus, the former DV data will be used for the area where the DV data does not arrive. When a duplicated data arrives for the same DV data, it is simply overwritten. Using this mechanism, the last DV frame will be displayed when the RTP stream stops.

The DV audio data are flushed every time after it is written to the IEEE1394 interface. Due to the fact that replaing the same audio data multiple times only causes noise. Thus, when a RTP packet that contains a DV audio data is lost, the corresponding part will be left flushed.

3   Design and Implementation of Redundant Audio for DVTS

In this section, a redundant transportation of DV audio data for DVTS is shown. The former DVTS is very weak against loss of RTP packets that contain DV audio data. In the former DVTS stream, each DV data is sent only once. Moreover, RTP is based on UDP and does not support reliable data transportation. Thus, when a RTP packet that contains DV audio data is lost, dvrecv will have to play the DV audio stream with a missing part. The dvrecv application will leave the lost DV audio data blank. Thus, when a RTP packet that contains DV audio data is lost, a remarkable audio noise occurs.

DVTS stream can obtain robustness against a small amount of packet loss using this scheme. In this scheme, DV audio data is sent multiple times. The redundant audio scheme proposed in this paper is shown in Fig.5. The packets shown in the left side of Fig.5 shows the DV data sent by the former DVTS and the right side of Fig.5 shows the DV data sent using the scheme proposed in

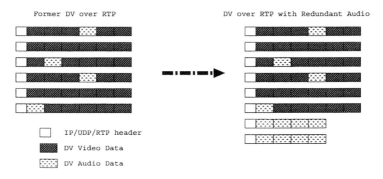

Fig. 5. DV over RTP with Redundant Audio

Table 1. Audio Redundancy and Bandwidth

	No Audio Redundancy	1 Redundant Audio	2 Redundant Audio
525-60 (IPv4)	28.25 Mbps	29.95 Mbps	31.65 Mbps
625-50 (IPv4)	28.27 Mbps	29.98 Mbps	31.68 Mbps
525-60 (IPv6)	28.66 Mbps	30.39 Mbps	32.12 Mbps
625-50 (IPv6)	28.68 Mbps	30.41 Mbps	32.14 Mbps

this paper. Compared to the left side, the right side of Fig.5 has RTP packets that contains DV audio only. By sending DV audio data multiple times, the reachability of the DV audio data to dvrecv will be enhanced. The redundancy level can be configured by changing the number of times the DV audio data is sent. In this paper, redundancy level 0 will describe DVTS stream that does not use the redundant DV audio scheme. Redundancy level 1 describes that each DV audio data is sent twice, level 2 describes that each DV audio data is sent three times, and so on.

The redundant DV audio RTP packets are treated the same way as the normal DV RTP packets. The RTP timestamp value for each redundant DV audio RTP packets are same as the normal RTP packets. Thus, the same DV audio data will be sent using the same RTP timestamp value. The RTP sequence value will be incremented every time when the redundant DV audio RTP packet is sent.

This is to obtain backward compatibility with the former DVTS. In dvrecv, the RTP packet with the same RTP timestamp is treated the same way. Thus, when the same DV data with the same RTP timestamp arrives, it is simply overwritten within the buffer. The former dvrecv uses the RTP sequence number to detect the number of packets lost. If the sequence number does not increment as the RTP packet arrive, it will be assumed that the packet was lost in the intermediate network. The redundant DV audio scheme proposed in this paper will increment the RTP sequnce number, and will have compatibility with the packet loss detection scheme used in the former dvrecv.

Using this DV audio redundancy scheme, robustness can be obtained without any modification within the receiver code. Thus, backward compatibility can be obtained.

4 Evaluation

In this section, evaluation of the redundant audio scheme is shown. The change in usage of network bandwidth by this sheme is shown in section 4.1. Evaluation that shows robustness of this scheme is shown in section 4.2.

4.1 Bandwidth Increasement

In this section, the network bandwidth used for the audio redundancy is shown. Using DV over RTP, multiple 80-bytes long DV DIF blocks can be included within a single RTP packet. The default maximum value of DV DIF blocks included in a single RTP packet used in dvsend is 17. DVTS is usually used with 100baseTX ethernet. The MTU (Maximum Transfer Unit) of the ethernet is 1500 bytes. 17 DV DIF blocks will be the maximum value that can fit within 1500 bytes with both IPv4 and IPv6. The minimum length of a RTP packet, that contains 17 DV DIF blocks, will be 1400 bytes when IPv4 and 1420 bytes with IPv6. The length of the RTP packet will increase when an IP option header for IPv4 is used, or when an IP extension header of IPv6 is used.

The number of DV DIF blocks in one DV frame will be 1500 for 525-60 (NTSC), and 1800 for 625-50 (PAL). For 525-60, 29.97 DV frame will be displayed in one second. For 625-50, 25 DV frame will be displayed in one second.

The DV audio redundancy and the change in network bandwidth is shown in Table 1. Table 1 shows that the increase of network bandwidth is trivial compared to the normal DV RTP traffic.

4.2 Loss of DV Audio Packets

We have done an experiment using the topology shown in Fig.6. A dvsend host is connected to a router on the left side of Fig.6. A dvrecv host is connected to a router on the right side of Fig.6. DV data was sent from dvsend to dvrecv using the intermediate router. We used 525-60 system for dvsend and dvrecv. Thus, 29.97 frames were sent within one second. The RTP packet send from dvsend was dropped at a fixed rate at the intermediate router.

We have sent DV from dvsend to dvrecv with no DV audio redundancy, DV audio redundancy level 1, and with DV audio redundancy level 2. The packet loss rate at the intermediate router was 1%, 5% and 10%. Fig.7 shows the number

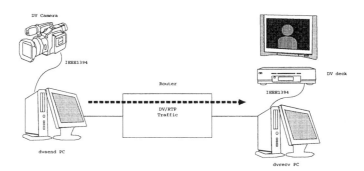

Fig. 6. Topology of Evaluation

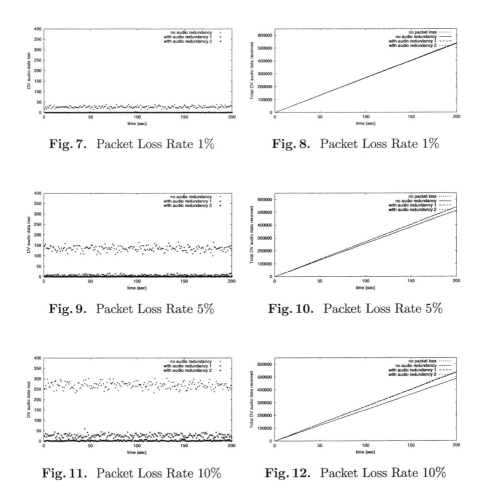

Fig. 7. Packet Loss Rate 1%   **Fig. 8.** Packet Loss Rate 1%

Fig. 9. Packet Loss Rate 5%   **Fig. 10.** Packet Loss Rate 5%

Fig. 11. Packet Loss Rate 10%   **Fig. 12.** Packet Loss Rate 10%

of DV audio data lost at dvrecv, when 1% packet are lost at the intermediate router, Fig.9 and Fig.11 show packet loss correspondingly, 5% and 10%. The x axis of Fig.7, Fig.9 and Fig.11 shows the time in seconds. 29.97 frames were sent within one second. The y axis of Fig.7, Fig.9 and Fig.11 shows the number of DV audio data that was lost.

In each experiment, number of lost DV audio data is the lowest when using audio redundancy level 2. Number of lost DV audio data will be the highest when the audio redundancy scheme is not used. Fig.7, Fig.9, and Fig.11 show that DV audio loss can be reduced using the DV audio redundancy scheme proposed in this paper.

Fig.8, Fig.10 and Fig.12 shows the sum of the DV audio data received at dvrecv. The x axis shows the time in seconds. The y axis shows the total number of DV audio data received at the dvrecv application. Fig.8, Fig.10 and Fig.12

show the result of the same experiment correspondingly Fig.7, Fig.9 and Fig.11. The line "no packet loss" show the estimated total DV audio data to be received when there are no packet loss. Thus, the line that is near the "no packet loss" line will have the highest audio quality.

The evaluation in this section show that the DV audio redundancy scheme can obtain robustness. The robustness will reduce the number of DV audio data loss. Moreover, the robustness can be obtained without much increase of network bandwidth.

5 Conclusion

In this paper, a scheme for reducing the packet loss probability of the DV/RTP audio data is introduced.

We have created a DV based broadcast quality video transportation tool. We have named the tool DVTS(Digital Video Transport System). Using DVTS, high quality video transportation over the Internet can be realized with low cost. DVTS does not have any error correction scheme. Moreover, it does not support reliable data transportation. Thus, a single packet loss can deteriorate the audio quality.

By sending the DV audio data multiple times, robustness of DV audio data can be obtained. The network bandwidth used by the DV audio data is trivial compared to the DV video and system data. Thus, it is scalable to send DV audio data multiple times. Evaluation shows that the system we have implemented can obtain robustness against small amount of packet loss.

References

[1] "Specifications of Consumer-Use Digital VCRś using 6.3mm magnetic tape", HD Digital VCR Conference,1994.
[2] "IEEE Standard for a High Performance Serial Bus", IEEE computer society,1995.
[3] A.Ogawa, "DVTS (Digital Video Transport System) WWW page", URL:http://www.sfc.wide.ad.jp/DVTS/, November 2001.
[4] A.Ogawa and K.Kobayashi and K.Sugiura and O.Nakamura and J.Murai, "Design and Implementation of DV based video over RTP", Packet Video 2000, May 2000.
[5] Audio-Video Transport Working Group and H. Schulzrinne and S. Casner and R. Frederick and V. Jacobson, "RTP: A Transport Protocol for Real-Time Applications", RFC 1889, January 1996.
[6] K. Kobayashi and A. Ogawa and S. Casner and C. Bormann, "RTP Payload Format for DV (IEC 61834) Video", RFC 3189, January 2002.
[7] K. Kobayashi and A. Ogawa and S. Casner and C. Bormann, "RTP Payload Format for 12-bit DAT Audio and 20- and 24-bit Linear Sampled Audio", RFC 3190, January 2002.
[8] S. Deering and R. Hinden, "Internet Protocol, Version 6 (IPv6) Specification", RFC 1883, December 1995.
[9] J. Postel, "User Datagram Protocol", RFC 768, August 1980.

Function Extensible Agent Framework
with Behavior Delegation

Ki-Hwa Lee[1], Eui-Hyun Jung[2], and Yong-Jin Park[1]

[1] Network Computing Lab.
Hanyang University, 17, Haengdang-dong, Sungdong-Ku, Seoul, Korea
{khlee,park}@nclab.hyu.ac.kr
[2] Smart Card Technology Inc.
17, Haengdang-dong, Sungdong-Ku, Seoul, Korea
ehjung@sct.co.kr

Abstract. We suggested an Intelligent Agent Framework that supports agent function extension. Main concern of research in Intelligent Agent has been the enhancement of intelligence. The proposed framework shows a new approach of function extension using the concept of behavior delegation. We define new behavior description language, BDL, which enables users to assemble agent functions without programming. Function configuration and other information are loaded into the framework to create agents dynamically at starting time. All behaviors in agent are executed on external servers using SOAP dynamic binding through Dynamic Invocation Framework. A reference application, the Intelligent Price Finder, is designed and implemented using the proposed framework.

1 Introduction

Rapid growth of the Internet has made an easy way of accessing information and services. However it has also caused the information overload problem [1]. The information overload problem is unlikely to be solved with high performance computing because information consumers have to deal with their information manually. Many researchers have considered the Intelligent Agent as a promising solution to this problem because intelligent agents have an ability to process the information without human intervention [2].

Typical features of the Intelligent Agent include autonomous, adaptability, reactivity, proactivity, and etc [3]. Among these features, adaptability is one of the most important factors because the agent should be able to extend and modify its intelligence and function dynamically according to its environment. Most researches in agent adaptability have focused on the enhancement of intelligence [4].

Intelligence enhancement depends on learning process, whereas agent functions are hard-coded in agent's source code. Thus if users want to modify agent functions, the agent code should be edited and recompiled in the source code level. For this reason, agent functions can't be easily replaced or modified during

H.-K. Kahng (Ed.): ICOIN 2003, LNCS 2662, pp. 376–385, 2003.
© Springer-Verlag Berlin Heidelberg 2003

runtime and the agent function extension has been considered as a hard problem to solve so far [5].

In this paper, we propose an Intelligent Agent Framework in which agents delegate their functions to the external server using Simple Object Access Protocol (SOAP) and eXtensible Markup Language (XML). The proposed framework enables users to assemble their agent functions at starting time or during runtime using external XML documents expressed with proposed Behavior Description Language (BDL). It also provides agents with functional adaptability without complex development process. SinceInt the proposed system, functions are not hard-coded in agents and these can be easily replaced and changed on runtime.

The paper is organized as follows. Section 2 will compare existing binary binding mechanisms such as Remote Method Invocation (RMI) with SOAP to describe the method of extending software without code recompilation. Section 3 addresses proposed system and Behavior Description Language (BDL) defined for agent's decision. Section 4 focuses on the reference application "Intelligent Price Finder" based on the proposed agent framework. Section 5 concludes this paper.

2   Function Extension in Agents

2.1   Legacy Function Extension Methods

In the legacy software engineering, several methods have been suggested to extend running code's functions such as Dynamic Linking Library (DLL). The DLL is the standardized way of extending running code's functions[6]. The DLL is loaded into and connected to calling process when it is required. However, the calling process has to contain the function prototype provided in the DLL to connect it dynamically. This has software extend its functions only through predefined function prototype and it leads to the restriction of function extension. RMI and Common Object Request Broker Architecture (CORBA) have also similar mechanism to extend software's functions and they have the same restriction because these methods are based on the binary binding.

2.2   SOAP and Invocation Framework

The SOAP provides a way of communication between processes on the heterogeneous computing platforms using XML for information exchange [7]. It is somewhat similar to the Internet Inter-ORB Protocol (IIOP) and RMI protocol except binding mechanism. The SOAP has been approved as a promising protocol for the Web Services [8]. In the Web Services, machine readable binding information, called Web Services Description Language (WSDL) files, describes binding information such as parameter, protocol, host address, and etc.

There are two ways of SOAP invocation: static and dynamic. In static invocation, generating stub is important process and it is compiled with calling process code. This kind of invocation is similar to the RMI and has the same limitation. Whereas, dynamic invocation needs no compilation cycle to make stubs. It just takes required parameters described in WSDL files and then calls the SOAP binding using those parameters. To achieve dynamic invocation in the code, Invocation Framework (IF) has been proposed [9][10]. Whenever applications delegate SOAP calling, the IF initiates SOAP-binding after loading given WSDL files and invokes specific operation defined in WSDL files. Using the IF, applications don't need to be recompiled, but just know calling parameters when new WSDL files are added or existing WSDL files are changed.

2.3   Challenges and Design Policy

The IF can be a new method to separate calling codes and called functions in software design. However there are some technical challenges in adopting the IF to the Intelligent Agent because the IF is just a mechanism to call functions. First of all, agents don't know why and when they call SOAP message. Although binding information is described in the WSDL file, there is no information about agent's action. Therefore additional information should be provided for the agent to decide its actions. Second, typical agent's task can't be performed with simply SOAP call, but may be logically composed of several SOAP calls. This structural information is not described in WSDL files.

   To cope with these problems, the proposed framework should provide a number of features. First, the agent framework should separate agent functions from agent using SOAP binding. Second, new information schema should be designed for agent's action. At last, the designed schema should have a flow control mechanism that can express several functions for one task.

3   Architecture

3.1   System Component

The components and internal action flows in the proposed agent framework are shown in Fig. 1.

Agent Core. Each agent in the agent framework is a form of Agent Core, which loads BDL files and decides when and how to invoke functions according to agent's environment. Since the Agent Core is the base class for all agent applications, it is customized for each application's need.

BDL Document. Each function is modeled as a "behavior" that describes the information for execution and binding information. These behavioral descriptions and flow information are grouped as a "solution". Typically one BDL document for agent may contain several solutions.

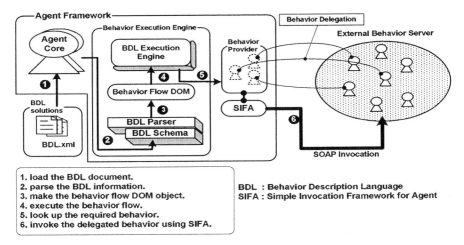

Fig. 1. Agent Framework Architecture

Behavior Execution Engine. Agent Core has its own Behavior Execution Engine (BEE) that parses agent's BDL documents and composes behavioral flow Document Object Model (DOM) and executes behaviors according to behavioral flows described in the BDL document.

Behavior Provider. When the BEE needs to invoke real functions, it sends the function request to Behavior Provider (BP) using a Simple Invocation Framework for Agent (SIFA), we implemented it to support dynamic SOAP binding for connecting external Behavior Server (BS). The BS maintains real functions corresponding to behaviors and can service SOAP requests from the BP.

3.2   BDL Schema

BDL is a description language to express the delegated behavior and behavioral flow information for the proposed agent framework. Although delegated behaviors are executed on external servers and described in WSDL document, agent system requires information about when the delegated behaviors are called and how the delegated behaviors are composed. This requirement is accomplished by BDL description. Designed schema has four types of components. These are Port components, Behavior component, I/O components, and Flow Control components.

Port Components. Port components are attached modules. They present additional I/O information for other BDL component. There are "InPort", "OutPort", "ExInPort", and "ExOutPort" in Port components symbols as shown in Fig. 2. InPort and OutPort define parameter information used for connecting BDL components in BDL description. InPort accepts data and delivers it to its

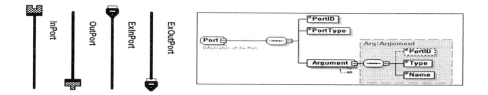

Fig. 2. Port components' symbols and its schema diagram

attached BDL component. OutPort publishes data acquired from its attached BDL component to other BDL component's InPort. ExInPort and ExOutPort provide the communication path between the agent framework and external computing actors such as I/O devices. Since these ports provide the inter-operability between the agent framework and external I/O, agent developers can make the customized external I/O as long as they satisfy the corresponding interface.

Behavior Component. Behavior component is a basic component to express one behavioral unit. As seen as Fig. 3, Behavior component has one InPort and one OutPort to connect other components in the BDL document. Real function of the Behavior component is delegated to external servers using SOAP binding. Since delegated services are accomplished by the Web Services, Behavior component has the information about the WSDL location and Web Services operation name for the delegated behavior. At the Behavior execution time, InPort converts input data to the request message of the Web Services operation and then the BP invokes the delegated behavior using the SIFA. After completion of behavior execution, OutPort collects the response message from the Web Services operation and delivers these to other components' InPort.

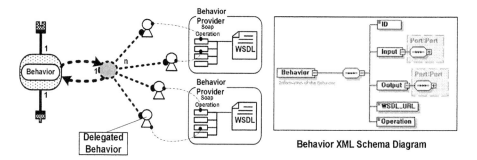

Fig. 3. Behavior component's symbol and its schema diagram

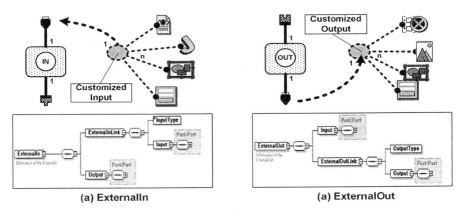

Fig. 4. I/O components's symbols and their schema diagrams

I/O Components. I/O components connect external I/O to BDL components. "ExternalIn" and "ExternalOut" provide the channel between agent and external computing environment. ExternalIn component converts an external input data into a compatible data, which can be used in BDL components. Fig. 4-(a) shows ExternalIn symbol and schema diagram. It has one ExInPort for the external data and one OutPort to deliver converted data to other BDL components.

ExternalOut component has the opposite role of ExternalIn. Fig. 4-(b) shows ExternalOut symbol and schema diagram. It also has mapping mechanism of how the inner data in the agent is mapped onto external data.

Flow Control Components. Flow Control components describe the flow of behaviors in agents. There are "SourceBar", "SyncBar", and "Branch" in the flow control components. SourceBar component launches new execution path. It delivers one input data to many output ports for parallelism. Fig. 5-(a) shows SourceBar symbol and schema diagram. It has one InPort and two or more OutPorts. After executing SourceBar, execution paths are divided as the number of OutPorts.

SyncBar component has the opposite role of SourceBar. It aggregates multiple execution paths using attached multiple InPorts and synchronized these paths. Fig. 5-(b) shows SyncBar symbol and schema diagram. SyncBar has one OutPort and two or more InPorts. Whenever a BDL component needs multiple inputs from different execution paths, SyncBar synchronize executions until all data are arrived.

Branch component provides behavioral flow such as loop or branches using its condition. It seems like switch expression in programming language. Fig. 5-(c) shows Branch symbol and schema diagram. It has one InPort and two or more OutPorts. When InPort receives data, the Branch component judges the matched condition for input data and selects appropriate flow. Branch can only

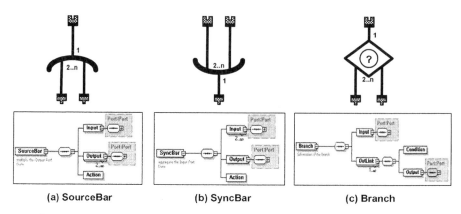

(a) SourceBar   (b) SyncBar   (c) Branch

Fig. 5. Flow Control components' symbols and their schema diagrams

select one OutPort at a time depending upon the internal condition unlike other flow components.

4 Reference Application

4.1 Intelligent Price Finder

In this section, we consider a reference application, "Intelligent Price Finder", to show the use of the BDL components and the agent framework. There have been many search engines to find products in the Web. However, the main concern of legacy search engines is exact search results, not providing additional intelligence such as language translation of the product name or currency exchange of the product price. Suggested Intelligent Price Finder performs multi-locale search using product name and price with given locale.

The Intelligent Price Finder has the internal structure of BDL components as shown in Fig. 6. There are three behaviors, "Language Translation", "Currency Exchange", and "Product Search" in the given BDL document. After loading the BDL document, the agent takes the initial product information through the customized UI. ExternalIn component converts the external data into internal BDL data and delivers it through the Port1.

At source bar, the execution paths are divided to Port2 and Port3. At Port2, data from Port1 flows into Language Translation behavior. Port3 injects the same data into Currency Exchange behavior. Language Translation behavior translates the product name into target locale name and delivers it to Port4. Currency Exchange behavior also executes currency exchange for target locale and delivers it to Port5. SyncBar aggregates each data from the Port4 and Port5 and delivers it to Port6. Product Search behavior searches product information using input data from Port6. Finally, ExternalOut translates data from Port7 and delivers it to external customized UI.

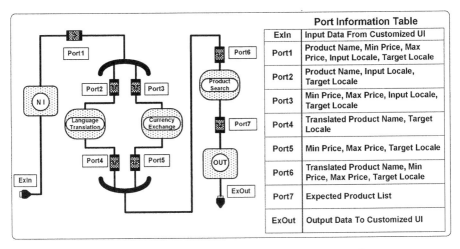

Port Information Table	
ExIn	Input Data From Customized UI
Port1	Product Name, Min Price, Max Price, Input Locale, Target Locale
Port2	Product Name, Input Locale, Target Locale
Port3	Min Price, Max Price, Input Locale, Target Locale
Port4	Translated Product Name, Target Locale
Port5	Min Price, Max Price, Target Locale
Port6	Translated Product Name, Min Price, Max Price, Target Locale
Port7	Expected Product List
ExOut	Output Data To Customized UI

Fig. 6. BDL document of Intelligent Price Finder

4.2 Evaluation

Developing process of agent applications using the proposed framework is quite simple. As seen as Intelligent Price Finder reference application, new application can be easily assembled with existing functions and simple BDL document. For Intelligent Price Finder, users write BDL document for the application using normal XML editor. After completion of the BDL document, the framework loads it and starts to make new agent instance of Intelligent Price Finder with the document.

To test the function extension that we assume, a new behavior is added to the existing BDL document. Added behavior is "Sorting" that provides sorting function using user's preference. This behavior is inserted into Port7. The modified Intelligent Price Finder is well performed without recompiling and modifying its agent's code. Fig. 7. shows the result when the Sorting behavior is inserted into the BDL document. While the previous Intelligent Price Finder has gotten the unsorted result, new one shows the sorted result in the price order.

The proposed framework has provided a new approach of function extension in software agent system. Besides, the framework can easily integrate various functions made by other developer without knowing of programming languages. It also contributes to increase the reusability of the agent function.

5   Conclusion and Future Work

The function extension of the Intelligent Agent has been considered a difficult problem because typical software should be edited and recompiled if it needs function modification or replacement after launching. To extend agent functions dynamically, these functions should be separated from the agent's hard-coded

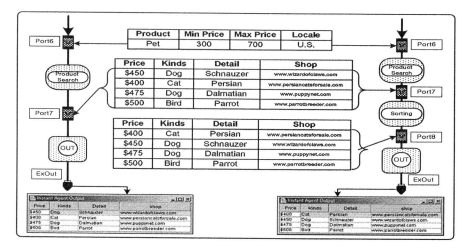

Fig. 7. Inserting the Sorting Behavior

source. We adopt SOAP dynamic binding mechanism to solve this issue and implemented a SIFA to make a SOAP dynamic binding in the agent's code. The SIFA makes the delegated function be loosely coupled with the agent system. In this architecture, all agents' functions are delegated and executed on external servers through SOAP.

BDL schema is designed to describe agent's behavioral information and flows. It can hold all information related to agent execution. Using BDL schema, agent developers simply make their own agent application without programming. Another advantage of the framework is function reusability.

Dynamic Function Locating (DFL) using Universal Description, Discovery and Integration (UDDI) [8][11] and graphical BDL Editor deserves the focus. The DFL provides agents with finding proper functions automatically. The BDL Editor can provide more convenient graphical environment with users to easily assemble and maintain functions.

References

[1] P. Mates: Agents that Reduce Work and Information Overload, CACM, vol. 37, no. 7, pp.31-40, Jul. (1994)
[2] M. Wooldridge: Agent-based software engineering, IEEE Proc. on Software Engineering, vol. 144, no. 1, pp.26-37, (1997)
[3] Björn Hermans: Intelligent Software Agents on the Internet: an inventory of currently offered functionality in the information society and prediction of (near-) future developments, Thesis, Tilburg University, Tilburg, The Netherlands, (1996)
[4] Aaron Sloman: What sort of architecture is required for a human-like agent?, Cognitive Modeling Workshop, Aug. (1996)

[5] Nicholas R. Jennings, Katia Sycara, Michael Wooldridge: A Roadmap of Agent Research and Development, Autonomous Agents and Multi-Agent Systems, pp. 7-38, (1998)

[6] Introducing Dynamic Link Libraries, http://webclub.kcom.ne.jp/ma/colinp/win32/dll/ intro.html

[7] Jepsen T., "SOAP cleans up interoperability problems on the Web." , IT professional, v.3 no.1, pp.52-55, 2001

[8] Curbera F., Duftler M., Khalaf R., Nagy W., Mukhi N. and Weerawarana S., "Unraveling the Web Services Web: an introduction to SOAP, WSDL and UDDI.", IEEE Internet computing, vol.6 no.2, pp.86-93, (2002)

[9] Matthew J. Duftler, Nirmal K. Mukhi, Aleksander Slominski and Sanjiva Weerawarana. ,"Web Services Invocation Framework (WSIF)", OOPSLA 2001 workshop on Object-Oriented Web Services

[10] Harshal Deo, "The need for a dynamic invocation framework", http://www.webservices.org/index.php/article /articleview/469

[11] UDDI White Page, http://www.uddi.org/whitepapers.html

Robust Audio Streaming over Lossy Packet-Switched Networks

Jari Korhonen

Nokia Research Center
Speech and Audio Systems laboratory, P.O. Box 100, 33721 Tampere, Finland
jari.ta.korhonen@nokia.com

Abstract. Multimedia streaming is getting more and more important class of applications in IP networking. As real-time streaming sets contradictory requirements for network latency and error robustness, practical implementations have to trade-off between interactivity and quality experienced by the end user. In this paper a transport scheme for streaming perceptually coded audio over an unreliable packet-switched network is presented. The scheme is based on selective retransmissions and it provides improved robustness against packet loss and efficient utilization of network resources. The scheme is implemented into streaming software and it is evaluated by simulating an unreliable network environment.

1 Introduction

Evolution of IP networking is leading towards fast connections allowing real-time multimedia streaming even in mobile terminals via wireless LANs and next generation mobile networks. From IP networking perspective Internet telephony and teleconferencing applications are especially challenging, because high level of interactivity is an essential requirement. Especially in telephony long transport and buffering delays may severely disturb communication, which sets quite strict limits for buffering and usage of retransmissions.

Different audio- and video-on-demand services form another significant class of multimedia streaming applications. They could be compared to CD or DVD players, the difference being that the content is physically located separate from the end user and streamed via network to the receiver in real-time while playing. This kind of applications can tolerate buffering delays up to a few seconds. In contrast, the quality requirements are typically much higher than for IP telephony or teleconferencing. That is why different techniques providing optimal trade-off between network latency and packet loss robustness are important for practical multimedia streaming implementations.

In this paper an error resilient transport scheme for streaming high quality audio is proposed. The scheme is based on data element shuffling among multiple transport packets and selective retransmissions. Data element shuffling is needed to turn a set of individually decodable audio frames into a set of packets with different priorities: each frame is divided into smaller elements with different

H.-K. Kahng (Ed.): ICOIN 2003, LNCS 2662, pp. 386–395, 2003.

priorities and the elements are shuffled among different packets with respective priority. This arrangement is needed to take the full benefit of the unequal error recovery scheme using selective retransmissions. In addition, the scheme allows each frame to be partially reconstructed in case of packet loss, which improves the performance of error concealment. The scheme is tested by running a practical test implementation capable to simulate lossy network environment and the results are discussed accordingly.

2   Multimedia Streaming

To avoid unnecessary utilization of the digital transmission channel resources efficient compression methods are essential in multimedia streaming. In advanced multimedia compression both lossy and lossless compression schemes are typically used in parallel. The demanding nature of real-time data transport sets special requirements for the network protocols as well. In the world of the Internet Real-time Transport Protocol (RTP) [1] is the de facto standard for carrying real-time multimedia content. In this chapter general aspects of multimedia coding and real-time delivery of multimedia content are addressed.

2.1   Multimedia Coding Principles

Advanced video compression used with MPEG standards is based on removing the spatial and temporal redundancies. There is typically a lot of redundant information in consecutive video frames. Hence a sequence of images to be divided into three classes based on their dependence on other pictures. In intra-picture coding mode encoded I-pictures are not predicted from any other picture, and therefore they can be decoded independently. In contrast, inter-pictures are predicted from other pictures. They are called P-pictures if the prediction takes place from one picture or B-pictures if from two pictures, respectively. The B-pictures can be dropped from the video sequence without affecting any other frame [2].

One-dimensional audio signal does not contain such obvious redundancies as video data and thus a different approach is required for compression. Low bit-rate speech codecs utilize the knowledge about waveforms that are typical for speech, but they do not work well with other types of audio, such as instrumental music. That is why perceptual audio coding is commonly used for compressing audio for advanced multimedia applications. Perceptual audio coding is based on psychoacoustic principles: sensitivity of human ear depends on the frequency, and a loud tone masks other tones near to the dominating tone. The masking effect applies both in spectral and temporal domain. In perceptual audio codecs sound signal is transformed into spectral domain and perceptually irrelevant frequency bands are omitted or coded with lower accuracy [3].

Typically perceptually encoded audio frames contain three different types of data: 1) Critical data that includes the vital information for decoding the actual audio data, for example Huffman codebook indices. Without this part the whole frame is useless. 2) Scalefactors defining the range of encoded spectral samples. If there are missing or damaged scalefactors, decoding process can continue, but audio output may be significantly distorted for the damaged frame. 3) The encoded spectral audio samples. To improve coding efficiency, entropy coding is usually applied for scalefactors and spectral samples. In general, this makes perceptual audio codecs quite vulnerable to bit errors.

2.2   Real-Time Multimedia Transport

In data transport point of view the video coding based on different types of pictures is not very robust against data loss: a missing I-picture makes also the dependent P- and B-pictures useless or corrupted. In addition, typically I-pictures have to be fragmented into more several transport packets than P- and B-pictures, because intra-pictures contain more data than inter-pictures. This leads to a dependency between transport packets as shown in Fig. 1 a). Obviously the significance of the packets containing I-picture data is higher for the overall quality than of those containing inter-picture data.

On the contrary, audio frames are typically individually decodable, equal in priority and relatively constant in size. There is usually no need to fragment audio frames among multiple transmission unit either. However, perceptually coded audio frames typically use overlapping filtering windows and optional prediction tools, which make the quality of each frame to be at least partially dependent on the previous frame(s). Fig. 1 b) illustrates a dependency graph for audio transport packets.

In packet-switched networking RTP is commonly used for transporting real-time data, such as video pictures or audio frames. RTP provides sequence numbers and timestamps for synchronization and keeping packets in right order. Unreliable underlying protocols, such as UDP, are usually utilized for carrying data. Because the original emphasis in RTP design was especially in multicast conferencing applications, retransmission mechanisms were not included into the RTP specifications. However, Real-time Control Protocol (RTCP) is intended to convey feedback information between RTP session participants about network conditions. According to the initial RTP design philosophy, RTP packets should carry timely consistent data units, such as individually decodable video or audio

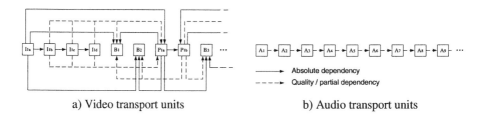

a) Video transport units   b) Audio transport units

Fig. 1. Dependency graphs for video and audio transport units (packets)

frames or fragments of them. This approach minimizes latency and allows fast interaction.

However, following the original RTP philosophy does not provide optimal error robustness in all conditions. For example, in circuit-switched digital speech transmission, as in mobile cellular networks, adjacent speech samples are often interleaved or shuffled over longer period to alleviate the quality degradation caused by bursty transmission errors. Individual missing samples are typically much easier to replace with different error concealment techniques than a longer set of missing adjacent samples [4]. Another problem with RTP is the lack of retransmission mechanism. Because of the interdependencies between packets as explained above, losing one packet may cause significant quality degradation or even make many other packets useless. That is why frameworks for selective RTP retransmissions have been proposed to enable more reliable transport for the most critical data units, such as I-pictures in a video sequence [5] [6].

3   Transport Scheme for Perceptually Coded Audio

Because perceptually coded audio frames are typically individually decodable units with similar priority, RTP retransmissions are not as obviously advantageous to audio streaming as for video streaming. However, it is possible to take a set of audio frames, tear the individual data elements in each frame apart from each other and shuffle them among multiple different transport packets. This scheme allows each frame to be partially reconstructed even if some of the packets are lost. In case of packet loss audio output is significantly less distorted, because error concealment techniques can be used more efficiently for missing individual spectral samples or scalefactors than for missing entire frames [7].

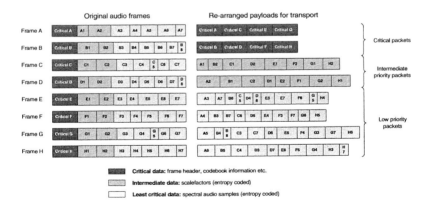

Fig. 2. Turning a set of audio frames into transport packet payloads

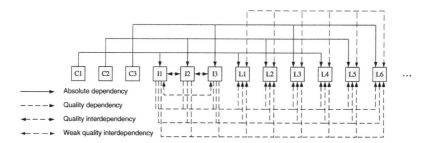

Fig. 3. Dependency graph of the transport units in the proposed scheme. C stands for critical, I for intermediate, and L for low priority packets

Fig. 2 illustrates the basic principle for turning a set of audio frames into high priority packets with the critical data sections, intermediate priority packets with the scalefactors and low priority packets with spectral samples. Fig. 3 shows the dependency graph for transport packets generated by the proposed technique. Obviously, high priority packets have to be transported with higher reliability. In [7] two different approaches to deal with priority distinction are discussed: redundant transmissions for critical data sections and establishing separate streams using different transport protocols for data packets with different priorities. In this paper a transport scheme based on selective RTP retransmissions is proposed.

Proper ordering of packets is essential to provide efficient utilization of retransmissions. After a set of frames has been turned into a sequence of packets as described above, the packets are arranged into descending order in terms of priority allowing the most critical (class A) packets to be sent first. This ordering allocates more time for class A packet retransmissions. When the receiver gets the first packet that belongs to class B, it sends the first compound retransmission request (negative acknowledgement, NAK) for the missing class A packets, if necessary. It is assumed that the sender gets the NAK while still sending class B packets of the same sequence, and retransmits the requested class A packets before continuing the normal transmissions. More compound NAKs can be generated later, if needed, containing retransmission requests for all the missing class B and C packets in addition to the requests for still missing class A packets. In this way different number of retransmission attempts can be allocated for different packets, depending on the priority class. When the sender receives a NAK, it resends the requested higher priority packets before it continues transmitting the lower priority packets. The optimal length of the sequence depends on user requirements and network conditions: naturally a long sequence allows more retransmissions and improves error robustness by spreading data elements more efficiently, but requires a large buffer and a long buffering delay. Optimal number of NAKs per sequence and the intervals between positions where NAKs are sent depend on the round-trip time: new retransmission request should not be sent before all the packets requested last time should have been most likely received if not lost afresh.

The scheme allows efficient utilization of the network bandwidth, because the critical packets are always sent before the lower priority packets, and the bandwidth usage can be controlled by the application using the retransmission mechanism. The bandwidth utilization in different conditions is illustrated in Fig. 4: a) shows the optimal network conditions without any packet loss, b) with reasonable packet loss rate and c) with heavy packet loss rate. Gaps between packets illustrate the bandwidth reserved for retransmissions. Average time slot reserved for each sequence of packets is T and it cannot be extended more than T_a. However, exact timing between packets is an implementation issue, and it does not need to follow precisely the scheme presented in the figure. In practice, the slot size can be defined as a maximum amount of data allowed to be sent during the slot rather than in time units. This approach allows simple application level control over the network resource usage: if the upper limit for data to be sent during the slot is reached before all the packets in the sequence have been sent, sender discards all the rest unsent packets and starts transmitting the packets of the next sequence. This is not fatal, because the lowest priority packets are the last in the sequence.

The proportion of critical data, scalefactors and spectral samples varies depending on the codec and coding attributes. As a very rough estimate for AAC, scalefactors take typically about 10% of the frame size and the critical data

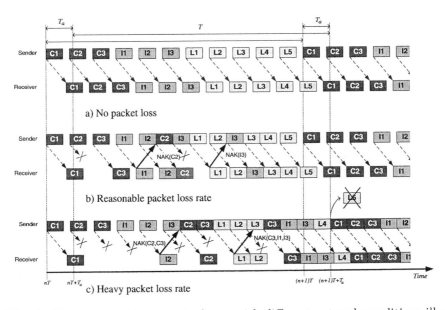

Fig. 4. The proposed transport scheme with different network conditions illustrated. C stands for critical, I for intermediate and L for low priority

slightly less. Usually more than 80% of the data consists of spectral samples. In this case packet sizes can be made relatively close to each other by selecting the proportion of different priority packets to be approximately 1:1:8.

4   Simulations

If we assume the packet loss probability to be constant p for all packets, including NAKs, the residual packet loss rate p_{res} follows (1), where r is the maximum number of retransmission attempts:

$$p_{res}(p, r) = p(2p - p^2)^r \tag{1}$$

Thus the theoretical residual packet error rate increases exponentially after each retransmission point. Correspondingly, the network traffic overhead t_{ovr} caused by retransmissions can be theoretically estimated by (2):

$$t_{ovr}(p, r) = \sum_{n=1}^{r} p^{n-1}(1 - p - (1 - p)^{n+1}) \tag{2}$$

We implemented the explained scheme for transport and packetization in a streaming server and client software to test the concept. AAC was used as codec and RTP as transport protocol with RTCP messages to convey NAKs from the receiver to the sender. In the test implementation the maximum number of NAKs is m and each retransmission point is defined as a sequence number n_{arq} (compound NAK for all the packets with sequence number lower than n_{arq} is sent when a packet with sequence number equal or higher than n_{arq} arrives). Relative numbers of high, intermediate and low priority packets as well as the retransmission points are configurable. Random packet losses and transport delays were produced in the streaming software to simulate an unreliable network environment.

We used AAC [8] stereo bitstreams with sample rate of 44.1 kHz and bitrate 128 kbit/s using main profile to test the concept. The average frame size of the test streams was about 365 bytes, with about 5% high priority data, 10% scalefactors and 85% spectral samples. As each decoded frame produces 1024 time samples, duration of each frame is 23 ms, which makes approximately 43 frames per second.

In the first test a sequence consisted of 64 packets carrying 64 frames, divided into different priority class packets in proportion 8:8:48. The first retransmission point was set at packet number 8, the second at packet number 16, and the third at packet number 32. No limitations were set for the total retransmission overhead. With this configuration the class A packets were protected by three retransmission attempts (NAKs), class B packets by two and the first third of the class C packets by one NAK. Because no transport delay was applied, resulting residual packet loss rates and retransmission overhead for different priority classes should follow the theoretical model. This is also the case, as depicted in Fig. 5.

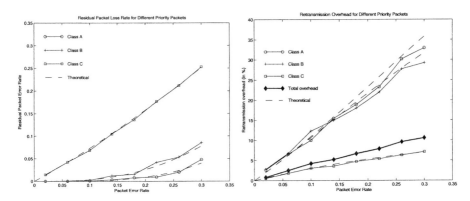

Fig. 5. Residual packet error rates and network overhead caused by retransmissions

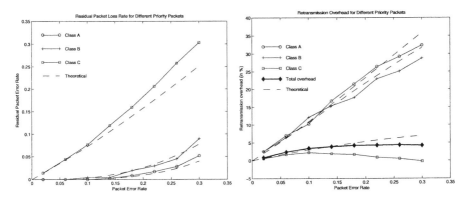

Fig. 6. Residual packet error rates and network overhead when the peak retransmission overhead is restricted to be less than 5%

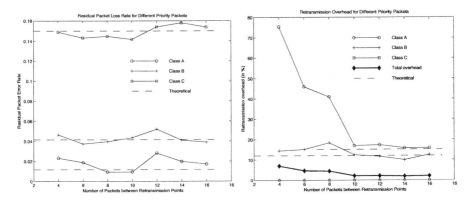

Fig. 7. Residual packet error rates and the network overhead with different time gaps between retransmission points

For the second test a similar configuration was used, but the maximum re-transmission overhead for each sequence was limited to 5%. This allows the server application to control the bandwidth resource utilization. The simulation results are shown in Fig. 6. As expected, the residual packet error rate gets higher for class C packets in comparison to the first test when packet loss rate increases. However, the total retransmission overhead can be efficiently restricted, and the packet loss rate for the class A and B packets does not increase.

If the retransmission points are too close to each other, the network delay causes unnecessary retransmissions leading to waste of network resources. In the third test this was simulated by adding a randomly selected transport delay between 40 and 60 ms, resulting to round-trip times between 80 and 120 ms. Packet loss rate was kept constant (0.15) while the number of class A and class B packets varies from 4 to 16. Retransmission points were located correspondingly at the boundaries between different priority classes allocating two retransmission attempts for class A packets and one attempt for class B packets. The result is illustrated in Fig. 7. As we can see, the number of unnecessary retransmissions for class A packets grows when the NAKs are sent too close to each other. This effect, however, does not in practice influence the residual packet loss rate.

5   Conclusions

In this paper a transport scheme for streaming perceptually coded audio over a lossy packet-switched network is explained. The scheme allows selective re-transmissions to be used most efficiently by rearranging the internal data elements of audio frames into transport packets with different priorities. By arranging a set of transport units into descending order in terms of priority more retransmission attempts can be allocated for the most critical packets. This approach provides efficient utilization of network resources.

We have implemented the scheme in a streaming software and tested it by simulating packet losses and network latency. The results indicate that the total retransmission overhead can be efficiently controlled by the application and it is still possible to achieve high robustness against packet loss by using the scheme. The basic scheme explained in this paper can be modified relatively easily to cope with different network conditions.

References

[1] Schultzrinne, H., Casner, S., Frederick, R., Jacobsen, V.: RTP: A Transport Protocol for Real-Time Applications. RFC 1889, 1996.
[2] Sikora, T.: MPEG Digital Video-Coding Standards. IEEE Signal Processing Magazine, vol. 14 no. 5, pp. 82-100, September 1997.
[3] Painter, T., Spanias, A.: Perceptual Coding of Digital Audio. Proceedings of the IEEE, vol. 88 no. 4, pp. 451-515, April 2000.
[4] Wah, B. W., Su, X., Lin, D.: A Survey of Error-Concealment Schemes for Real-Time Audio and Video Transmissions over the Internet. Proc. of IEEE International Symposium on Multimedia Software Engineering, Taipei, Taiwan, pp. 17-24, December 2000.

[5] Varsa, V., Leon, D.: RTP Retransmission Framework. Internet draft, June 2002.

[6] Miyazaki, A., Fukushima, H., Hata. K., Wiebke, T., Hakenberg, R., Burmeister, C., Takatori, N., Okumura, S., Ohno, T.: RTP Payload Formats to Enable Multiple Selective Re-transmissions. Internet draft, June 2002.

[7] Korhonen, J.: Error Robustness Scheme for Perceptually Coded Audio Based on Interframe Shuffling of Samples. Proc. of International Conference on Acoustics, Speech and Signal Processing 2002, Orlando, Florida, pp. 2053-2056, May 2002.

[8] ISO/IEC: Generic coding of moving pictures and associated audio information - Part 7: Advanced Audio Coding (AAC). International Standard 13818-7, 1997.

An XML-Based Mediation Framework for Seamless Access to Heterogeneous Internet Resources

Seong-Joon Yoo[1], Kangchan Lee[2], and Kyuchul Lee[2]

[1] School of Computer Engineering
Sejong University, Seoul, 143-747, Korea
`sjyoo@sejong.ac.kr`
[2] Department of Computer Engineering
Chungnam National University, Daejon, 305-764, Korea

Abstract. This paper proposes an XML-based mediation framework for seamless access to the Internet information resources that are distributed, autonomous, and heterogeneous. The proposed method is faster in query processing than previously proposed methods since it integrates data in advance while the previous work integrates data as a result of query processing. There have been several approaches for seamless access to heterogeneous Internet resources. While other methods are applicable to integration of structured heterogeneous data that are usually stored in DBMS, mediation approach is appropriate to integration of unstructured, semi-structured, and structured data. XML is introduced to describe a common data model for the mediation by supporting query processing of information resources which have their own query methods, data representation, and schema structures.

1 Introduction

The Internet emerges as the largest database. However, Internet contents are stored in various format according to the type of information resources that they maintain and to the interface that they provide. For example, biologists often access several single nucleotide prototype(SNP) databases through the world wide web where these data are written in XML, flat file format, MySql, or ASN.1 format[1]. The main problem in accessing these contents is heterogeneity. Since each site has its own semantic, structure, schema, and data model and access to these sites requires laborious manual job. To remedy this problem and allow seamless access to internet resources in heterogeneous format, it needs to integrate information resources and support query in a single format.

There have been several integration methods such as universal DBMS method[10, 11], federated databases[10, 11, 12], data warehouse[10, 11, 13], multi-databases[14, 15], and mediator method[10, 16, 17, 18, 19, 20, 21]. While other methods are applicable to integration of structured heterogeneous data which is usually stored in DBMS, mediation approach is appropriate to integration of unstructured, semi-structured, and structured data.

H.-K. Kahng (Ed.): ICOIN 2003, LNCS 2662, pp. 396–405, 2003.

Since recent integration requirements on the Internet are extended to integrate databases as well as the data produced by Internet applications, the mediator approach should be chosen to solve the problem of accessing seamlessly to internet information resources.

Each integration method requires a data representation model. XML is introduced as a common data model for the mediation by supporting query processing of information resources which have their own query methods, data representation, and schema structures. The XML based common data model is superior to previous data models such as object oriented data model[17, 18], rule-based data model[19, 21, 22] and semi-structured data model[20] since XML is easy to model, highly readable, needs non-redundant processing of encoding or decoding in document transmission and natural in the Internet environment.

In this paper, we propose a new approach named XMF(XML-based Mediation Framework)[3, 4] for integrating Internet information resources. XMF adopts a mediator-wrapper architecture[3] to provide end users with an integrated view of the underlying information sources.

XMF is similar to MIX[2] in that it focuses on wrapper-mediator systems which employ XML as a means for information modeling, as well as interchange, across heterogeneous information sources. The wrapper associated with each source exports an XML view of the information at that source. The mediator is responsible for selecting, restructuring, and merging information from autonomous sources and for providing an integrated XML view of the information. However, XMF is faster in query processing than MIX since XMF integrates data in advance while MIX integrates data as a result of query processing. XMF is more compatible to the Internet environment since it uses a standard query language Xpath whereas MIX uses its own query language called XMAS.

XMF has three characteristics. First, XMF describes the information resources and mapping rules in XML[5]. XMF solves the problem of schematic conflict with XMF mapping rules. XMF supports the integration of various kinds of information resources and dynamic management of global schema information in XML. Second, XMF's wrappers support well-known protocols such as HTTP and JDBC. Third, XMF uses Xpath[6] as a query language because integrated result of XMF is an XML document.

In the remaining part of this paper, we introduce features and architecture of XMF in Section 2. Section 3 describes XMF mediation rules. Section 4 describes implementations of an example system. Finally, we present conclusion and future works in Section 5.

2   XMF Architecture

Figure 1 illustrates the layered architecture of XMF. XMF is composed of three layers, i.e. *Application Layer, Mediation Layer*, and *Resource Layer*. Various Internet information resources reside in the Resource Layer. DBMSs, search engines, file systems, and applications on WWW are the examples of the Internet information resources.

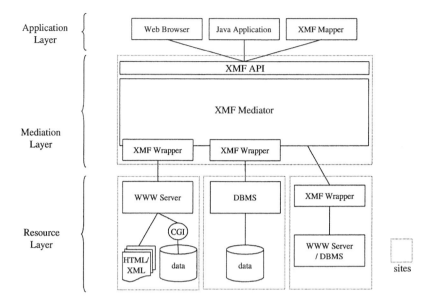

Fig. 1. Layered Architecture of XMF

The XMF wrapper is a module for extracting data from each Internet information resource and converting them into XML instances. Each wrapper has one-to-one mapping relationship with each information resource in the resource layer. The Wrapper does not need to run on the same platform with the mediator.

The mediator in the mediation layer is a module for information resource integration. The mediator controls each wrapper. The mediator manages its own program library, which supports API of XMF. Programmers can make their own client programs using the XMF library with Java Language.

The application layer is at the top of XMF architecture. Web browsers and any kind of user interfaces of XMF can be built on the XMF library in the application layer. The XMF mapper which generates translation rules between global schema and local schema is also at the application layer

2.1   Mediator

The XMF mediator is composed of *a query processor, a result integrator*, and *an* XMR(XMF Mediation Rules) *handler*. The query processor receives, analyzes, and decomposes global queries into deliverable forms for wrappers. At first, the query processor tests user's query whether it conforms with a stored global schema. After that, the query processor generates subqueries, and distributes the subqueries to appropriate wrappers.

Table 1. Wrapper types of XMF

	Type1	Type2	Type3	Type4	Type5
Information Resources	WWW	WWW	DBMS	DBMS	WWW
Protocol	HTTP	HTTP	JDBC	JDBC	HTTP
Query Type	Query String	Query String	SQL	SQL	XPath
Result	HTML	XML	Database Result	XML	XML

The result integrator receives XML data which are the results from each sub-query, constructs XML trees, integrates the trees with the rule processor, and eliminates duplicated nodes. And then, the result handler checks the validity of the result with global schema, and return the result to the application layer.

The XMR handler is a module for metadata management. The XMR handler includes wrapper location information, user ids, passwords, global and local schemas, and the relationship between global schemas and local schemas. Resource Map Table, TypeID Table, and Function Table are major tables for managing metadata. If users want to use XMF, users construct map information using the XMF mapper which is a visualization tool for generating map information. Once map information is generated, the XMR handler manages the map information for the query processor, the result integrator, and user applications.

2.2 Wrapper

The main function of the XMF wrapper is translating sub-queries into local queries according to its information resource management system and translating the local result into the XML data format. The XMF wrapper is composed of a query translator, result translator, and an XMR handler.

The role of the query translator is receiving queries from the XMF mediator, generating local queries for local information resource management systems, and transmitting the queries to local resource management systems. The local information resource management system returns query results with their proprietary format. The result translator in the wrapper harmonizes the result in the XML format. XMF supports various wrapper types since there are several types of information systems on the Internet and each information system has its own characteristics in query usage, result format, and connection method.

XMF supports five types of wrappers as shown in Table 1. Type 1, 2, and 5 are wrappers for integrating WWW information resources, and Type 3, and 4 are wrappers for integrating DBMS information resources.

3 XMF Mediation Rules

XMF uses XML as a tool for integrating the heterogeneity since XML provides a common data modeling framework of heterogeneous information integration,

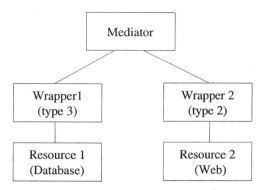

Fig. 2. XMF Usage Environment

and interoperability between applications. In addition, XML supports the separation of content, structure, and presentation. The mediator based on XML in XMF can easily integrate data, and users of XMF can access information transparently without having to know what software and hardware platform are resided and without having to know where it is located.

In this section, we show an example of integrating distributed, heterogeneous, and autonomous on-line computer stores on the Internet. Suppose that there are two on-line computer stores. For convenience, we name each computer store R1(Resource1) and R2(Resource2). Although R1 and R2 contain computer information, data structures and units are different. Let's assume that the evaluation information of R1 is represented by 'A', 'B', and 'C', and the evaluation information of R2 is divided into 'good', 'average', and 'poor', and the price unit of R1 is Dollar($) and R2 is Korean Won. And more, R1 has image thumbnails of products, whereas R2 does not.

In XMF, a wrapper makes it possible that each on-line computer store can be considered as an XML repository. Each wrapper of computer stores understands the XMF queries and returns the results in XML format as shown in Figure 3, which shows the query results of each wrapper and the results are similar but the structures are different.

For the integration of the result of the wrapper, XMF utilizes XMR, which defines mapping information between global schemas and local schemas. XMR consists of three consecutive blocks: (i) The *data registry* block contains access method (resource name, location information, ID, password) description for local information systems, user-defined functions for resolving the semantic conflict, (ii) The *schema definition* block has the information of global/local schema represented in DTD, XML schema, and (iii) relationship information between global schemas and local schemas are in the *mapping rule* block.

Figure 4 is an example of XMF mediation rule for integrating the Internet information resources in Figure 3. XMR consists of a main element, 'xmr', and the subelements 'registry', 'schemas', and 'maps'. The 'registry' subelement in

Wrapper1 (Resource1)

```
<Products>
<Product Id = "ABP3BF">
<Product Name>ASUS P38-F</Product_Name>
<Description>Supports Intel...</Description>
<Manufacturer>Asus</Manufacturer>
<Price>110</Price>
<Evaluation>A</Evaluation>
<Thumbnail>1.gif</Thumbnail>
</Product>
......
</Products>
```

Wrapper2 (Resource2)

```
<Products>
<Item No ="103">
<Name>D815EPEA(815E/Socket/Sound)<Name>
<Description>Supports Intel...</Description>
<Type>Main Board</Type>
<Company>Intel</Company>
<Price unit ="\">159000</Price>
<Evaluation>Good</Evaluation>
</Product>
......
</Products>
```

Fig. 3. An XML Representation of the Computer Store Information Resources Translated by XMF Wrappers

turn contains other subelements such as 'integrate', 'source', and 'dest'. Integrating resources are defined using the 'integrate' element. Remarkable point of the 'integrate' element is that it has the operation attribute. According to the value of operation attribute, the XMF mediator integrates the Internet information resources. Available operations are 'merge', 'union', and 'join'. The name, connection, and type attribute in the 'source' element indicate the wrapper name, location, and type of integrating information resources, respectively, and UDF(user defined function) for resolving semantic conflict is registered in function element.

The global and local schema is defined in 'schema' element. The name attribute is correspondence with name attribute in 'source' and 'dest' element, and the value of 'ref' attribute is global and local schema filenames.

XMF can analyze both DTD and XML Schema[7] for schema description of each information resource. And 'map' elements in mapping rule division are the relationship between local schemas and global schemas. The 'map' element has 'dest' and 'source' attributes. Each attribute has the value,

```
<?xml version="1.0" encoding="euc-kr" ?>
- <xmr>
- <registry>
  - <integrate operation="union">
  - <source name="ls1" connection="REMOTE" method="POST" type="3">
    <url>http://211.255.252.141/wrapper2/wrapper2Servlet</url>
    <query name="query" />
    </source>
  - <source name="ls2" connection="REMOTE" method="POST" type="2">
    <url>http://211.255.252.142/wrapper3/wrapper3Servlet</url>
    <query name="query" />
    </source>
    </integrate>
  - <dest name="gs" />
    <func name="arithmetic" sclass="division" dclass="multiplication" />
    <func name="representation1" sclass="InUnit1" dclass="ConvertedUnit1" />
    <func name="representation2" sclass="InUnit2" dclass="ConvertedUnit2" />
    <func name="representation3" sclass="InUnit3" dclass="ConvertedUnit3" />
  </registry>
- <schemas>
  <schema name="gs" ref="/export/home/xmf/mediator/gs.dtd" />
  <schema name="ls1" ref="/export/home/xmf/mediator/ls1.dtd" />
  <schema name="ls2" ref="/export/home/xmf/mediator/ls2.dtd" />
  </schemas>
- <maps>
  <map source="ls1:/results" dest="gs:/Product_List" />
  <map source="ls2:/Catalog" dest="gs:/Product_List" />
  <map source="ls1:/results/result" dest="gs:/Product_List/product" />
  <map source="ls2:/Catalog/item" dest="gs:/Product_List/product" />
  <map source="ls1:/results/result/pname/@id" dest="gs:/Product_List/product/@id" />
  <map source="ls2:/Catalog/item/itemno" dest="gs:/Product_List/product/@id" />
  <map source="ls2:/results/result/classification" dest="gs:/Product_List/product/category" />
  <map source="ls2:/Catalog/item/groupname" dest="gs:/Product_List/product/category" />
  <map source="ls2:/Catalog/item/company" dest="gs:/Product_List/product/manufacturer" />
  <map source="ls1:/results/result/pname" dest="gs:/Product_List/product/product_name" />
  <map source="ls2:/Catalog/item/name" dest="gs:/Product_List/product/product_name" />
  <map source="ls2:/results/result/summary" dest="gs:/Product_List/product/description" />
  <map source="ls2:/Catalog/item/illust" dest="gs:/Product_List/product/description" />
  <map source="arithmetic(1, ls1:/results/result/price)" dest="gs:/Product_List/product/price" />
  <map source="ls2:/Catalog/item/value" dest="gs:/Product_List/product/price" />
  <map source="ls1:/results/result/thumbnail/@src" dest="gs:/Product_List/product/image" />
  <map source="ls1:/results/result/capable_of_buying" dest="gs:/Product_List/product/availability" />
  <map source="ls2:/Catalog/item/buyable" dest="gs:/Product_List/product/availability" />
  <map source="ls1:/results/result/delivery" dest="gs:/Product_List/product/deliveryterm" />
  <map source="ls2:/Catalog/item/receivingterm" dest="gs:/Product_List/product/deliveryterm" />
  <map source="representation2(1, ls1:/results/result/evaluation)" dest="gs:/Product_List/product/estimation" />
  <map source="representation3(1, ls2:/Catalog/item/estimate)" dest="gs:/Product_List/product/estimation" />
  </maps>
  </xmr>
```

Fig. 4. An XMF Mediation Rule for the Example Computer Store Information Resources

```
<Products>
<Product Id = "201">
<Name > ASUS P38-F</ Name>
<Description>Supports Intel...</Description>
<Manufacturer>Asus</Manufacturer>
<Category>Mother Board</Category>
<Price> 132,000  </Price>
<Evaluation>A</Evaluation>
<Thumbnail>1.gif</Thumbnail>
</Product>
<Products>
<Product Id = "103">
<Name>D815EPEA(815E/Socket/Sound)</Name>
<Description>Supports Intel...</Description>
<Manufacturer>Intel</Manufacturer>
<Category >Mother Board</Category>
<Price >159,000</Price>
<Evaluation> A </Evaluation>
<Thumbnail></Thumbnail>
</Product>
......
</Products>
```

Fig. 5. An XML mapper for the Computer Store Databases

which consists of a global/local schema name and an XPath expression like "gs:/catalog/book/@ISBN". If the value of a 'source' attribute has a form of function call, it means that the value of a 'source' attribute is the result of UDF, which is defined in 'function' element. And more, setting the default value is available like "Computer".

An XMF administrator should understand all of the information resources and write out the XMR. Because the XMR is the instance of XML, any one who knows the syntax of XML can write the XMR files easily with a text editor. However, manual writing is a tedious job. To resolve this problem we provide an XMF mapper which automatically generates an XMR in Java. Figure 4 shows an example of an XML mapper automatically generated with XMF mediator for the computer store databases.

After the construction of XMR, users can use XMF query. XMF supports the query language which is the subset of XPath. The noble aspect of XMF query is that the mediator and the wrapper have the same form of query language with execution model and the result format. So, if there are several wrappers, a user executes the query using hierarchical XML architecture. Figure 5 shows integrated form of the example computer store information resources represented in XML shown in Figure 3.

4 Implementation

Currently the prototype system consists of three resources, three wrappers on three computer systems and one mediation server system. We implemented type 3 and 4 wrappers for ORACLE, MySQL and MDB databases.

5 Conclusion

In this paper, we propose and implement a new integration framework, XMF, which provides uniform views over large number of Internet information resources by using only XML and Internet. XML provides a self-describing modeling method for capturing the meanings of heterogeneous information resources, and the Internet protocol supports common data communication mechanism. The features of XMF are integrating various kinds of information sources and its application on the Internet, supporting common data model and run-time integration of information resources by using its mediation mechanism and query language. In consequence, XMF supports common architecture and query language for integrating the Internet information resources and users can easily access XMF with a uniform method. Furthermore, XMF can be easily implemented with current Internet technology and XML-related software similarly to the experimental prototype described in the previous section.

Acknowledgement

We acknowledge with thanks for the support of professor Taekyoung Kwon, and graduate students Hyunchul Jang and Junghwan Lim in evaluating an example system.

References

[1] ASN.1 Project Homepage(see http://www.itu.int/ITU-T/asn1/).
[2] Chaitanya K. Baru, Amarnath Gupta, Bertram Ludascher, Richard Marciano, Yannis Papakonstantinou, Pavel Velikhov, Vincent Chu, "XML-Based Information Mediation with MIX," Proceeding of International Conference on ACM SIGMOD, pp. 597-599, 1999.
[3] Kangchan Lee and Kyuchul Lee, "XML-based Mediation Framework(XMF) for Integration of Internet Information Resources," Proc. of World Multiconference on Systemics, Cybernetics and Informatics, Vol. 14, pp.191-195, 2001-07, SCI2001.
[4] Kangchan Lee, Jaehong Min, and Kishik Park, "A Design and Implementation of XML-based Mediation Framework(XMF) for Integration of Internet Information Resources", Proc. of Hawaii International Conference on System Sciences, 35, January 2002.
[5] Tim Bray and C. M. Sperberg-McQueen, "Extensible Markup Language (XML): Part I. Syntax", World Wide Web Consortium Recommendations, February 1998, Available at http://www.w3.org/TR/REC-xml.

[6] James Clark, Steve DeRose, "XML Path Language (XPath) Version 1.0", World Wide Web Consortium Recommendations, Nov. 1999, Available at http://www.w3.org/TR/xpath.

[7] David C. Fallside, "XML Schema Part 0: Primer", World Wide Web Consortium Candidate Recommendation, October 2000, Available at http://www.w3.org/TR/xmlschema-0/.

[8] Hector Garcia-Molina, Yannis Papakonstantinou, Dallan Quass, Anand Rajaraman, Yehoshua Sagiv, Jeffrey D. Ullman, Vasilis Vassalos, Jennifer Widom, "The TSIMMIS Approach to Mediation: Data Models and Languages," Journal of Intelligent Information Systems, Vol. 8, No. 2, pp. 117-132, 1997.

[9] B. Panchapagesan, J. Hui, G. Wiederhold, S. Erickson, L. Dean, "The INEEL Data Integration Mediation System White Paper", Available at http://id.inel.gov/idim/paper.html.

[10] Ruxandra Domenig, Klaus R. Dittrich, "An Overview and Classification of Mediated, Query System", SIGMOD Record,, 28(3), pp. 63-72, 1999.

[11] Won Kim, "Introduction to Part 2: Technology for Interoperaing Legacy Databases," Modern Databases Systems, pp. 515-520, 1995.

[12] A. P. Sheth and J. A. Larson, "Federated Databases for managing distributed, heterogeneous, and autonomous databases" Computing Surveys 22:3, pp. 183-236, 1990.

[13] S. Chaudhuri and U. Dayal, "An overview of data warehousing and OLAP technology", ACM SIGMOD Record 26:1, pp. 65-74, 1994.

[14] Won Kim, Injun Choi, Sunit K. Gala, Mark Scheevel, On Resolving Schematic Heterogenetity in Multidatabases Systems, Modern Database Systems, pp. 512-550, 1995.

[15] UniSQL/M Manual, UniSQL Inc, 1994.

[16] Gio Wiederhold, "Mediators in the Architecture of Future Information Systems," The IEEE Computer Magazine, 25(3):38-49, March 1992.

[17] V.S Subrahmanian, Sibel Adali, AnneBrink, Ross Emerym J. Lu Adil Rajput, J. Rogers Robert Ross, Charles Ward, "HERMES : A heterogeneous Reasoning and Mediator System", ARPA, 1995.

[18] Thomas Kirk, Alon Y. Levy, Y. Sagiv, D. Srivastava, "The Information Manifold", In Proceedings of the AAAL Spring SymPosium on Information Gathering in DisTributed Heterogeneous Environments, March, 1995.

[19] H. Garicia-Molina, Y. Papakonstantinou, D. Quass, A. Rajaraman, Y. Sagiv, V. Vassalos, J. D.Ullman, J. Widom, "The TSIMMIS approach to mediation : data models and languages", J. Intelligent Information System 8:2, pp. 117-132, 1997.

[20] C. Baru, A. Gupta, B. Ludascher, R. Marciano, Y. Papakonstantinou, and P. Velikhov. XML-Based Information Mediation with MIX, In Proceeding of ACM-SIGMOD, 1999.

[21] Y. Papakonstantinou and P. Velikhov. Enhancing Semistructured Data Mediators with Document Type Definitions. In Proceeding of Data Engineering (ICDE), Australia, 1999.

[22] Y. Papakonstantinou, H. Garcia-Molina, J. Widom, "Object Exchange across Heterogeneous Information Sources", In Proceeding of Data Engineering, pp. 251-260, 1995.

Internet Camera Selections in Response to Location-Based Requests from Multiple Users

Akira Hiyamizu[1], Koji Okamura[2], and Masatoshi Arikawa[3]

[1] Graduate School of Information Science and Electrical Engineering
Kyushu University, Japan
hiyami@dontaku.csce.kyushu-u.ac.jp
http://dontaku.csce.kyushu-u.ac.jp/~hiyami/
[2] Computing and Communications Center
Kyushu University, Japan
[3] Center for Spatial Information Science
University of Tokyo, Japan

Abstract. Mobile communications have become popular in recent years, and the mobile users can connect to the Internet at anyplace. More and more mobile users will inquire about their locations in the future. When we provide information about a certain place, it is one of the effective methods to use images taken by fixed-point observation cameras, because the camera images can explicitly tell us about the situation of the place. Moreover, using multiple cameras is more effective because we can see many places from many angles. There are, however, some problems to be solved when we use multiple cameras. One problem is that it is difficult to select one target camera from a great number of cameras. In our research, we suggest an algorithm to select the optimal camera that suits for user requests based on positional information. We can use multiple cameras with unique parameters by using our proposed method.

1 Background and Goal of this Research

Mobile communications have become popular in recent years, and the users have been able to connect to the Internet at anyplace. Therefore, an increasing number of mobile users search information about their locations using the Internet, and many applications that provide such information have been developed[1]. When we provide the information about a certain place, it is one of the effective methods to use images taken by fixed-point observation cameras, because the camera images can explicitly tell us about the situation of the place. Many cameras are connected to the Internet and set up in various places at present, and the camera images from such Internet cameras are provided to users. In fact, many cameras, such as Web cameras[2], which we can access via Internet are already available. Development of mobile communications and the spread of broadband networks also enable us to set up cameras in various places where setting up of any Internet cameras was impossible before, and it is expected that

H.-K. Kahng (Ed.): ICOIN 2003, LNCS 2662, pp. 406–415, 2003.

more fixed point observation cameras are set up. When we provide information about a location, using multiple cameras is more effective than using a single camera because we can see many places from many angles. There are, however, some problems to be solved when we use multiple cameras. One problem is that we must select one camera, from multiple cameras, which can provide images of the place we want to see. When a great number of cameras exist, we cannot select the target camera easily. In addition, when we set up cameras, we cannot set them up regularly due to physical restrictions. We cannot therefore select the target camera by a simple method, such as changing a camera one by one. We need a certain reference value or method in order to select an optimal camera.

In our research, we unify multiple Internet cameras based on positional information about cameras in order to solve the problems in using multiple cameras. We show an outline of our research in Fig. 1. We assume that a great number of Internet cameras exist. Those cameras are managed by a camera server and we can control a camera through a camera server. The server, called "Union-Camera server", collects the information about cameras and coordinates multiple cameras. When the Union-Camera server receives a request from a user, the server selects an optimal camera that is suitable for the request and provides the camera image. In this way, the user can control multiple cameras with unique parameters that do not depend on each camera, and get image. In addition, we can easily develop an application that uses multiple Internet cameras.

In this paper, we propose an algorithm for selecting the optimal camera that is suitable for a request of a user from multiple Internet cameras. We select the optimal camera based on the positional information about cameras. We adopt a concept of "Photo vector" in the camera selection algorithm. In Section 1.2, relations between the photo vector and the camera selection algorithm are described. We can coordinate multiple cameras by using the algorithm we propose, and can control multiple cameras with photo vector as unique parameters.

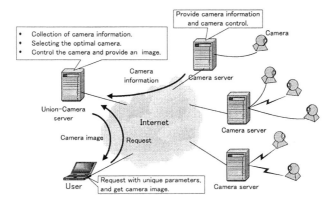

Fig. 1. Outline of our research

Fig. 2. Direction of the movement

1.1 Cameras with a Controllable Camera Platform

There are a large variety of cameras that can be used as a fixed-point observation camera. Our research concerns cameras with a controllable camera platform[3]. In this section, we describe a camera with a controllable camera platform. The camera is controlled by three camera parameters: pan, tilt and zoom. A pan value specifies the angle of horizontal direction, and a tilt value specifies the angle of vertical direction. Fig. 2 is one of the cameras with a camera platform, which is a Canon VC-C3[4]. We show the direction of movement that we can control with the pan and tilt. In this research, we assume that a great number of such cameras are set up and connected with the Internet.

1.2 Photo Vector and Selection of a Camera

In this section, we describe the concept of photo vectors in order to explain the purpose of our research plainly. A photo vector is a vector which has a point of a camera as the starting point and has a point of the object that is taken in the center of a picture as an end point. If one photo vector is determined, one optimal picture suitable for the photo vector can be determined. Originally a photo vector is a 3-dimensional vector. We can easily obtain 2-dimensional information about a position by using the information received from GPS or a map, etc. However, it is difficult to obtain the information on height. When a user requires information by using a photo vector, it is difficult to specify the height. Thus, we select a camera by using 2-dimensional position information. The photo vector has the information about a camera point and an object that a user requests. The required camera point is the starting point of the photo vector, and the required object point is the end point of the photo vector. The optimal camera is a camera that can take the end point from the starting point. Selecting of the optimal camera with photo vector consists of two processes. The first process is to select cameras that can take the required end point. The next process is to select one optimal camera about required starting point from the selected cameras. In our research, we propose the selection algorithm of a camera which has the photo vector that meets a request of a user based on the positional information about cameras.

2   Camera Selection Algorithm

2.1   Information Required for Camera Selection Algorithm

In this section, we describe the information required for camera selection algorithm and preconditions. We select the optimal camera suitable for requests based on positional information, camera capability information, and information of objects. We defined the two preconditions for the algorithm proposed in this paper.

1. An object(obstacle,building, etc) is defined as a polygon, and it is possible to obtain the information on each edge and each peak.
2. The information about each camera is obtained.

In our previous research[5], selection of a camera was possible when there were no obstacles. In fact, however, there are unavoidable obstacles to any objects. We need to consider them when we select a camera. Therefore, the information of objects is needed, and the definition of obstacles is given as a polygon in order to simplify calculation.

Next, we describe the information about a camera. The following information is necessary to select the optimal camera.

- Height of a camera
- Coordinates of a camera position
- Direction that a camera faces to
- Capability of a camera

The height value is a height from the ground. The coordinates of the camera position is expressed as a relative value. We determine one reference point and axis of coordinates and measure a camera position based on them. It is easy to convert relative coordinates to absolute ones, because the conversion of a relative value to absolute value is possible by using a conversion function, and vice versa. Therefore, it is easy to convert a global position, latitude and longitude, to relative coordinates. In addition, both the information of cameras and the information of objects need to be in the same coordinate system. If they are in the different coordinate systems, it is necessary to carry out coordinate conversion. The direction of the camera is an angle that a camera faces to when it faces the front. The direction is expressed in the same coordinate system. The information about capability of cameras is also needed, because we use a camera with a controllable camera platform. The capability information of cameras consists of degree range of pan and tilt, and range of zoom that a camera has. The information is used for judging whether a camera can take images of a place. In addition, some other information is needed. They are the name of camera device and information about a camera server. Such information is needed when we control a camera. In our research, information about a camera server is needed because we assume that a camera is controlled through the Internet.

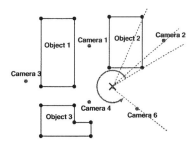

Fig. 3. Scanning peaks of objects and points of cameras

2.2   Algorithm for the End Point

When we look at an object with a camera, we think that the object to see is more important than the place from which we see the object. Thereby, we start with selecting the camera which can take the point of the object, in other words, the end point of photo vector. First of all, a set of cameras which can be seen from an end point of required photo vector is computed. This is the same as computing the camera which can take an end point. In our proposed method, as it is shown in Fig. 3, we scan peaks of objects and points of cameras with a focus on an end point, and obtain a set of cameras which can take an end point.

1. Points of cameras and peaks of objects are sorted in order of declination in the polar-coordinate system that takes the end point for origin of the coordinate system. A sorting sequence is $(v_0, ..., v_n)$.
2. Segments intersect a ray whose declination is 0 is set to a balanced binary search tree S, and elements of the set are saved in the order of the distance from the origin.
3. Each point $V_i(i = 1, ..., n)$ is scanned in order. If V_i is a peak of an object, when two segments that have V_i as the end point are contained in S, they are removed. And if not contained, they will be added to S.
   If V_i is a point of a camera, V_i is compared with the minimum element of S. The minimum element here means an element that has the smallest distance with the origin. If V_i is smaller than the minimum element of S, the camera that has V_i can take image of the end point. At the same time, the capability of a camera is checked.

In this algorithm, when the scanning line enters in and comes out a segment of an object, S is updated. The set of segments which intersects a scanning line is always stored in S. When a scanning line arrives at the point of a camera position, if the camera point is closer to the end point of required photo vector than the minimum element of S, we assume that there is no obstacle between the end point and the camera. At the same time, it is necessary to consider a capability of a camera. We judge whether the camera can take the point from the camera information. We can compute the pan and tilt values of a camera with positional information about an end point and a camera point, and we can judge whether a camera can take the point by comparing the capability of the camera with the pan and tilt values. The method to compute pan and tilt is described in Section 3.

2.3   Algorithm for the Starting Point

We can get a set of camera that can take the end point of a required photo vector by using algorithm described in Section 2.2. After that, we need to select one optimal camera with the information about starting point. When we select a camera, we may be able to simply select a camera located in the nearest place from the required starting point. However, we need to take into account that the camera is connected to the Internet and is controllable by zoom. Therefore we judge the optimal camera comprehensively using the distance, zoom, number of users and bandwidth of a camera server.

Zoom and Distance. When zoom is used, an angle of view will narrow and an object is taken widely. It is the same as the case when we approach an object and take it without zoom. Thus, we convert zoom into distance and use it for selection of a camera. We can obtain the relational expression of zoom and angle of view by measuring an angle of view in each zoom. Fig. 4 shows the results of actual measurement, from which we obtained the formula (1). Moreover, the ratio of the distance when we shoot an object with high zoom and when we approach an object and shoot without zoom can be expressed in the ratio of angle of view. We can thus compute an angle of view in each zoom by using the formula (1), and we can convert zoom into distance by using the ratio of them.

$$\tan \frac{\theta}{2} = 0.004312z^2 - 0.134z + 1.05 \tag{1}$$

Number of Users and Bandwidth of a Camera Server. We assume that a camera is connected to the Internet. It is important to consider the load of

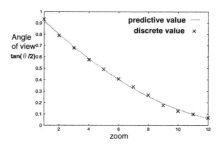

Fig. 4. Relation between angle of view and zoom

a network since many people may use a camera simultaneously. We need to consider the number of users. Camera servers manage the information about the number of users. When the number of users changes, a camera server sends the information, and we use it. A camera server obtains the number of users from connection to the server and disconnection to the server. In addition, the maximal value of the bandwidth is set up static, and we guess the bandwidth currently used from the number of users. When the maximal value of the bandwidth is given as B_{max}, the number of users is given as n, and the amount of average used per user is given as B_{avg}. The rate of use B is expressed in a formula (2). In our research, we use the rate of use for camera selection also.

$$B = \frac{n \cdot B_{avg}}{B_{max}} \qquad (2)$$

The Camera Evaluation Method. In our research, we convert the value of each element into the evaluation value of 0 to 1 in order to select the optimal camera comprehensively. We select a camera that has the largest total of evaluation values as the optimal camera. When the total is the same among some cameras, we select one that has the highest evaluation value in order of distance, the number of users and bandwidth. When we convert each element to the evaluation value, we use the conversion function which is designed based on a selection policy. In this research, we select a camera that is near the required vector and has a few users. Based on this camera selection policy, we architect formula (3) to convert distance to an evaluation value, and formula (4) to convert the number of users. A constant is appropriately chosen according to the situation. We assume that the same network is shared among all camera servers, and the evaluation value of bandwidth is constant.

$$f(x) = e^{\frac{-x^2}{C1}} \qquad (3)$$

$$f(x) = \frac{C2}{(x+1)} \qquad (4)$$

2.4   Consideration about Camera Selection Algorithm

First, we consider the algorithm for end points. An operation of a balanced binary search tree, such as extraction of the minimum element, addition and deletion of an element, requires the time of $O(\log k)$. k is a number of elements. In our proposed algorithm, the number of elements of a binary search tree is in proportion with the number of objects, and the number of times of operating a balanced binary search tree is in proportion with the sum of the number of cameras and the number of objects. Therefore, the amount of calculation depends on the number of cameras and edges of an object. If the number of cameras is set to m and the number of edges set to n, the amount of calculation will be set to $O(n + m) \log n$. If n and m become large, calculation time may pose a problem. However, the domain in which we need to search can be divided into the small cells, and the amount of calculation can be cut down. Additionally, the information on a camera and the information on an object are separable. An object such as a building remains in a fixed location for a long time. However, a camera may be moved frequently. When a camera is moved, we need only to update the information on the camera. Therefore a quick reaction can be expected.

Next, we consider the selection of a starting point. We select the optimal camera considering the distance, the load of a network and the number of users. Thereby, we can select a camera in consideration of the performance of a camera not only of a demand of a user.

3   Camera Control and Conversion of Parameters

We have described the camera selection algorithm in Section 2. We need to control a camera and get images after the optimal camera is selected. When we control a camera, we must convert from photo vector to individual camera parameters, because we cannot control a camera with photo vector as it is. Camera parameters here mean pan and tilt. In this section, we describe the method to convert the positional information to pan and tilt.

3.1   Conversion from Positional Information to Pan

The pan value can be converted by vector operation. When the camera position is given as (x_1, y_1) and the object position is given (x_2, y_2), the absolute value of θ in Fig. 5 can be calculated by using formula (5). At the same time, the direction of rotation is decided by comparison of camera position with required position. If y_1 is smaller than y_2, θ is $-|\theta|$. If y_1 is bigger than y_2, θ is $|\theta|$. The pan value is computed by direction of a camera and θ. If the direction of a camera is given as θ_c, the pan value is $\theta - \theta_c$.

$$|\theta| = \cos^{-1}\left(\frac{x_2 - x_1}{\sqrt{(x_2 - x_1)^2 + (y_2 - y_1)^2}}\right) \tag{5}$$

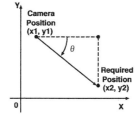

Fig. 5. Calculate the pan valuel

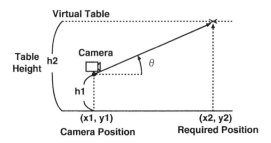

Fig. 6. Conversion from positional information to tilt

3.2 Conversion from Positional Information to Tilt

When we calculate tilt, the height of a camera is important. However, we assume that the photo vector is 2-dimensional. For this reason, when we calculate tilt, we set the height used as a reference value. We make a virtual table used as a reference value by the relation between the height of an object and the height of a camera. We projects a required end point on the table and calculates a tilt value. When the camera is set on a high place, it looks down. Therefore, we set the table on the ground. When the camera is set on a low place, we adjust the table higher. When the required end point, the camera position, the camera height and the table height are given as Fig. 6, the tilt value is calculated by formula (6).

$$\theta = \tan^{-1}\left(\frac{h_2 - h_1}{\sqrt{(x_2 - x_1)^2 + (y_2 - y_1)^2}}\right) \tag{6}$$

4 Implementation

We implemented an application using our proposed camera selection algorithm. We present the overview of the application in Fig. 7. When we click arbitrary two points on a map and specify a photo vector, the optimal camera image will be displayed. We can select the optimal camera that meets the user's requests from multiple cameras by using our proposed algorithm. In addition, coordination between map and camera images can be easily achieved by using a photo vector as an interface.

5 Conclusion and Future Work

In this paper, we proposed and described the camera selection algorithm based on positional information about cameras and camera control methods. We can select the optimal camera, from multiple cameras, which suits users' requests by using our proposed algorithm. If we adapt our method for multiple network cameras,

Fig. 7. Overview of the application

we can easily get the picture that we want to see. In addition, coordination between spatial information and camera images can be easily achieved by using a photo vector as an interface. We assume the environment where a great number of cameras exist. When there are only a few cameras, the picture which surely suits a request of a user may not be obtained. However, the optimal camera is selected by best effort, and we can also respond to a request by carrying out image processing or increasing the number of cameras. Future work will include improving our application as well as evaluating its algorithm.

References

[1] Xiaofang Zhou, Joseph D. Yates and Guihai Chen "Using visual spatial search interface for WWW applications", Information Systems, Vol. 26, Issue 2, April 2001, pp 61-74
[2] WebView World Home Page, http://www.x-zone.canon.co.jp/WebView
[3] Akira Hiyamizu, Koji Okamura and Masatoshi Arikawa, "Design and Implementation of Union-Camera", Proceedings of Multimedia, Distributed, Cooperative and Mobile Symposium, June 2001, pp. 241-246 (In Japanese)
[4] Canon Inc. Home Page, http://www.canon.com
[5] Akira Hiyamizu, Koji Okamura and Masatoshi Arikawa, "Implementation of virtual Internet camera for mapping positional and spatial information by advanced method", Proceedings of Joint Workshop on System Development, April 2002

Implementation of Third Party Based Call Control Using Parlay Network API in SIP Environment*

Hyoung-min Kim[1], Hwa-sung Kim[1], Kwang-sue Chung[1], and Hyuk-joon Lee[2]

[1] School of Electronic Eng., KwangWoon Univ., Korea
meruru98@kw.ac.kr
{hwkim,kchung}@daisy.kw.ac.kr
[2] Department of Computer Eng., KwangWoon Univ., Korea
hlee@daisy.gwu.ac.kr

Abstract. The communication network is in a transition toward the NGN (Next Generation Networks), which can support both of voice and data at the same time, according to the explosive demand of new services. The NGN allows the third-party application provisioning by defining the networks as layers of Services, Distributed Processing Environment and Transport. The Service layer can further be divided into Application and Service Component layer. In order to realize the third-party application provisioning, the Parlay Group has adopted an open Parlay API approach as an interface between the Application and the Service Component layer. Using Parlay API, the third parties may build and deploy the new applications at the Application layer exploiting the service components within network operators' domain. In this paper, we presents the implementation details about the two-party call setup and release using the third party service logic based on Parlay API Rel. 3.0., where SIP is used as a signaling protocol in Transport layer.

Keywords: Parlay API, NGN, MSF, MSS, SIP

1 Introduction

The communication network is in a rapid transition not only in terms of technical respects, but also in terms of user requirements for higher bandwidth and new kind of high quality services. Especially, the timely provision of the new customized services is becoming crucial to the network operators to fulfill the user requirements. This transition of communication environment is resulted in the buildup of new communication network infrastructure, namely NGN, which allows the third-party application provisioning by opening the network capability. On the contrary, the traditional communication services have been exclusively under the control of network operators.

* This work was supported by Grant No.R01-2002-000-00179-0(2002) from the Korea Science & Engineering Foundation and the Research Grant from Kwangwoon University in 2002

H.-K. Kahng (Ed.): ICOIN 2003, LNCS 2662, pp. 416–425, 2003.

The third party application provisioning can be achieved by defining an open application programming Interface (API) that resides between the application layer and the service component layer, and makes it possible to program the network by exposing the functionalities of network element outside world through well-defined API [1, 2]. The Parlay API [2] is an example of the open programmable interfaces that control the voice, video and data networks in the same manner. In this paper, we describe the implementation details about the two-party call setup and release process using the third party service logic based on Parlay API Rel. 3.0.

2   Open Networking Architecture

MSF (Multi-service Switching Forum) defines the reference architecture of MSS (Multi-Service Switching Systems) for an open networking architecture [3]. Fig. 1 shows the reference architecture of MSS, which is the network node that enables the integrated provision of versatile network services. MSS reference architecture consists of various function blocks as illustrated in Fig. 1. Also, the inter-block interfaces are defined using various reference points. Roughly speaking, MSS consists of MGs (Media Gateway) and MGCs (Media Gateway Controller), which is characterized by distributed switching systems based on packet switching. MSS aims to provide voice, video, data services including ATM, frame relay, IP in the same manner. On the other hand, MSS adopts the multi-plane system model, which consists of application, control, switching, adaptation and management planes as shown in Fig. 1. MSS pursues that each plane can be implemented independent of specific technologies. For example, it should be possible for the adaptation plane can include diverse interfaces and protocols. The implementation of the control and the switching plane should also be possible using diverse technologies.

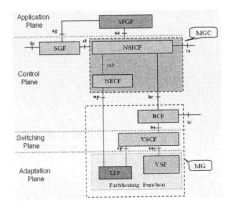

Fig. 1. MSS(MG/MGC) Architecture [3]

Especially, the application plane can provide diverse services by implementing the service logic through several sub-layers. These services may be directly accessed by the application plane or can be triggered by the control plane. SFGF (Service Feature Gateway Function) is a function block that enables the access to the IN services or other applications. Also, it enables the access to the functions of control plane for the application requested by application plane directly through signaling.

3   Parlay API

The architecture of Parlay API, which can be used at sg+ sa reference points of MSS reference architecture, is presented in Fig. 2. The Parlay API consists of two kinds of open interfaces as follows: [2]

– Service Interfaces - enables for the application to access the capabilities and information provided by networks. Generic Call Control Service, INAP1 Call Control Service, Generic Messaging Service, Generic User Interaction Service and Call User Interaction Service (Voice prompt to user, DTMF input from user) are the examples of these interfaces.
– Framework Interfaces - provide the subsidiary functions, which make the service interfaces operated and maintained in secure, robust fashion. Authentication, Discovery, Event Notification, Integrity Management, Operation, Administration and Maintenance are the example of this interfaces.

The Parlay API, which provides the open and technology independent API specification, enables the independent third party software vendors to develop and offer the value added advanced telecommunication services to network operators or network service providers. Since the Parlay API is designed to be simple to be used and extensible, the independent third party software vendors can develop the applications easily using Parlay API, which can be applied to between different types of networks and services. Using the Parlay API, the network

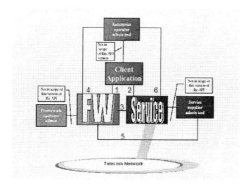

Fig. 2. The architecture of Parlay interface [2]

service developers can develop and test the new customized services without interrupting existing services. The API also tries to provide the secure solutions so that the network operators would not worry about the integrity, performance and security. The integrity and security issues are the principle obstacles for the network operators to willingly open their networks to third party vendors through API. In order to achieve these goals, an object-oriented approach to specifying the API has been adopted. The main specification is defined in the technology independent Unified Modeling Language (UML) [1, 2].

4   Third-Party Based Call Control

This paper shows how the third-party based call setup and release are achieved when voice service is provided over packet networks through parlay API. The basic building blocks for interconnection between Public Switched Telephony Network (PSTN) and the packet network have been identified in several functional components such as Media Gateway (MG), Media Gateway Controller (MGC). Media gateway is classified into Residential Gateway (RGW), Access Gateway (AGW), and Trunking Gateway (TGW) according to their functions. A TGW performs an interface between PSTN trunks, typically connected to a local exchange, and a packet network. A TGW has no subscriber interfaces. Now if a packet network such as IP/ATM network interconnects these components, the result is a network similar to one shown in Fig. 3 [4].

MGC that consists of software-based functions is equivalent to the unbundling of call control functionality inherent in a central office switch. It allows interworking between PSTN and packet network with different signaling systems and media transports. On the other hand, MGs provide conversion of streamed media formats such as voice or video, and interface between the PSTN telephone network and the IP/ATM networks. Such gateways typically manage a large number of digital virtual circuits and TDM bearer circuits with signaling carried on a separate path. The links from the gateway to the network

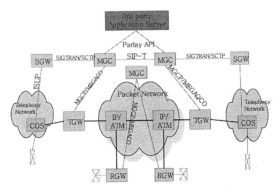

Fig. 3. Net composition for VoP

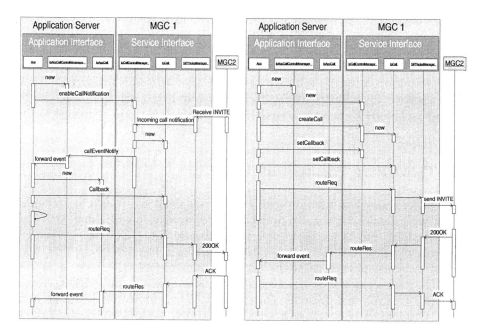

Fig. 4. Call Setup based on Parlay API (incoming call setup)

Fig. 5. Outgoing call setup

are IP/ATM links. A Signaling Gateway (SG) is responsible for termination of Switched Circuit Network (SCN) signaling (typically SS7) and transport of signaling messages to MGC.

The third party based call setup procedure is illustrated in Fig. 4. The left side rectangular box in sequence chart represents the Application Server, which performs the Application layer functions of NGN architecture and operated by the third party service provider. The right side rectangular box represents MGC that performs the Service Component layer functions of NGN architecture and controls MGs. The inner rectangular boxes that reside in Application Server and MGC represent the class objects, which are the instances of the parlay interface classes. The objects are physically distributed across the Application Server and MGCs. In terms of the parlay API architecture in Fig. 2, an Application Server takes the role of the parlay client and executes the application logic provided by the third party vender. On the other hand, MGC takes the role of the parlay server by implementing parlay APIs. The functions of parlay server may be located in MGC or in a separate system such as a parlay gateway. In this paper, we assume that the function of service interface is located in MGC for simplicity. The interface between the Application Server and the MGC is achieved through parlay API.

The Service Interface implemented in MGC consists of two interface classes: IpCall and IpCallControlManager. On the other hand, the Application Interface implemented in Application Server consists of two interface classes: IpAppCall and IpAppCallControlManager. We implemented both interfaces based on Java environment such as Java 2 ORB IDL compiler.

Fig. 4 is a modified version of the sequence diagram that is presented by the parlay API 3.0 specification. We modified the sequence diagram because it does not include the mapping relation with the SIP messages [5] especially in MGC side. We implemented the Incoming Call Routing service based on the sequence diagram shown in Fig. 4. The sequence diagram shows the procedure of Incoming Call Routing, in which the following modules are involved:

- IpCallControlManager: This module creates and manages IpCall. This is registered to CORBA when the server is started or created.
- IpAppCallControlManager: This module provides the application call control management functions to the generic call control service. This performs the similar function to IpCallControlManager.
- IpCall: This module performs the call routing function according to the information delivered from application server.
- IpAppCall: The client application developer implements this module and it is used to handle call request responses and state reports.

In the following, the sequence of events in Fig. 4 is explained in detail:

1. The application logic registers itself to MGC by calling enableCallNotification() in order to receive the notification from MGC when a new incoming call is occurred.
2. A new INVITE request is received by SIP PacketManager implemented in MGC and the SIP PacketManager notifies IpCallControlManager.
3. IpCallControlManager creates an IpCall object for the new call and notifies IpAppCallControlManager through CallEventNotify() that is a callback.
4. IpAppCallControlManager creates a new IpAppCall object and notifies the request of new call to the application logic.
5. Performing the data translation, the application logic invokes routeReq() that informs IpCall of the routing information of the called party.
6. IpCall requests the SIP PacketManager to send SIP 200 OK message to other server destined to the given called route.
7. When called party answers, SIP PacketManager notifies IpCall using SIP ACK message, IpCall forwards to IpAppCall using callback called routeRes(). Finally, this result is sent to the application logic.

The Application Server can use the specific services, which is provided by the MGC, through CORBA naming service and then synthesize the application logic using those services. On the other hand, the MGC has no idea of the existence of the Application Server. Therefore, the MGC (IpCallControlManager) waits until the Application Server sends the registration information (enableCallNotification()) as described in step 1 of sequence of events in Fig. 4. Once receiving the registration request from the Application Server, the MGC (IpCallControlManager) returns the assignment ID to the application as shown in Fig. 4. If the MGC (IpCallControlManager) detects the occurrence of new call as in step 2, then the MGC notifies the Application Server that the new call is occurred using the callback mechanism as in step 3. When more than one application is registered, the MGC (IpCallControlManager) may be able to distinguish the different application through assignment ID value.

The other sequence diagram shown in Fig. 5 presents the procedure of Outgoing Call Setup. The outgoing call setup is different from incoming call in terms of the order that the interface objects are created. In the case of incoming call, IpCall is created earlier than IpAppCall. But, outgoing call has opposite order. In Fig. 4, the SIP Packet Manager in MGC receives the invite message and then the MGC notifies IpCallControlManager of this event. But IpCall in Fig. 4 receives all messages from other MGC.

In addition to the interface classes defined in parlay API 3.0 specification, we need to define the several additional classes for the actual implementation of the service interface. The packet manager class that is shown in Fig. 4 is such an example. The other example is the class, namely P-Server, which triggers the service to begin the operation, but is not shown in Fig. 4. This class will register IpCallControlManager to CORBA. Particularly, this class is very important because it is the place that the service interface starts for the first time. But Parlay API specification does not mention about these classes. Also we defined the other kind of class, which performs the parsing function when the SIP packet manager receives SIP Packet, and which creates a SIP Packet when message sending is needed.

5   Synthesis of Third Party Application

5.1   Mapping between Parlay API and SIP

When implementing the service interface in service component layer, the translation of methods mentioned in section 4 into SIP requests or translation of the SIP responses into methods implemented in Parlay API interface should also be included as illustrated in Fig. 6. The method, routeReq () is used when creating a SIP message such as INVITE. On the contrary, routeRes() method is used when it receives a SIP message such as 200 OK or CANCEL. It is important to

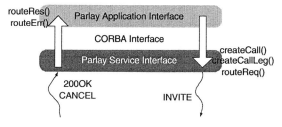

Fig. 6. Mapping between Parlay API and SIP

Table 1. Mapping between Parlay and SIP

SIP	Parlay
Receive INVITE	callEventNotify()
Send INVITE	routeReq()
Receive ACK	routeRes()
Send ACK	routeReq()
Receive 200OK	routeRes()
Send 200OK	routeReq()
BYE	deassignCall() ,routeRes()
UA A	IpAppCall 1, IpCall 1
UA B	IpAppCall 2, IpCall 2

map the Parlay API to SIP message when implementing the Parlay API in SIP environment. The mapping described above is summarized in Table 1.

5.2   Case Study: Two-Party Call Setup & Release

We implemented the two-party call setup and release procedure in CORBA/Java environment as an example of synthesizing the third party application. As explained in previous sections, the third party call control refers to the general ability to manipulate calls outside the transport network. In our implementation, the third party controller will trigger to set up and manage a communication relationship between two other parties.

Fig. 7 illustrates the SIP sequence diagram of the two party call control [6]. In Fig. 7, the controller means a third party call control block or Application Server whose role is to establish a session between two other user agents namely participants A and B. The session establishment is orchestrated by third party call control block. On the other hand, Fig. 8 representing the two-party call setup and Fig. 9 representing the two-party call release show the detail sequence diagram based on Parlay Generic Call Control service [2], in which the internal

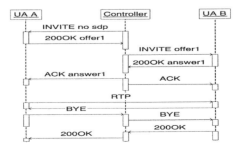

Fig. 7. Third party based Two Party call control flow [6]

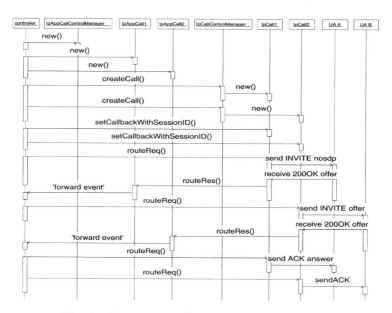

Fig. 8. Two party call setup using parlay API

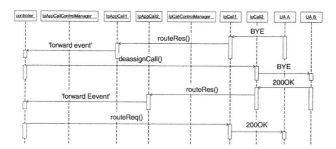

Fig. 9. Two party call release using parlay API

function of the controller is described. Both figures represent the mapping process between Parlay API and SIP messages in addition to the interaction process through SIP messages. In order to implement the third party applications, the SIP interaction process like Fig. 7 should be translated into the sequence diagram that includes the mapping between Parlay API and SIP messages.

Two-party call setup process is different from the incoming call setup process described in section 4 in that the Application Server and the MGC (IpCallControlManager) are connected each other by calling createCall() method of IpCallControlManager and createCall() method calls user agent A and B in turn. The procedures after this connection take the same sequence of those described in section 4. It is also possible to implement the third party based call control shown in Fig. 7 using multiparty call control and call leg other than generic call control. The detail of this implementation is going to be introduced in another paper.

6 Conclusion

The Parlay API is a nice candidate for the interface between the application layer and service component layer. The Parlay API has the potential to revolutionize how to offer the value-added services in an open networking architecture. In the case of current switching systems, the call control functions are performed only through the signaling interaction between the switching nodes that has the all the necessary embedded service logic in them, which makes it difficult to change the service logic and to provide the new services promptly. Using the parlay API, however, the timely provision of new network services becomes feasible by the diverse service logic providers.

In this paper, we described the implementation details of the interaction between the Application Server (application layer) and the Service Interface implemented in MGC (service component layer) during call setup process using Parlay API on CORBA/Java platform. In this case, the translation between Parlay API and SIP should be carefully implemented because the mapping between Parlay API and SIP is not always straightforward. Furthermore, we illustrated the entire call control flow using the service logic provided by the third party through Parlay API when the call is connected and tear down between two independent parties.

References

[1] Parlay Group, Parlay API Business Benefits White Paper, June 1999
[2] Parlay Group, Parlay API Spec. 3, Dec. 2001
[3] MSF, MSF-ARCH-001.00-FINAL 1A, System Architecture Imple-mentation Agreement, May 2000
[4] Abdi R. Modarresi, Seshadri Mohan, "Control and Management in Next-Generation Networks: Challebges and Opertunities", IEEE Computer Magazine, Oct. 2000
[5] IETF, draft-ietf-sip-rfc2543bis-03.txt, "SIP: Session Initiation Proto-col", May 2001
[6] IETF, draft-ietf-sipping-3pcc-02.txt, "Best Current Practices for Third Party Call Control in the Session Initiation Protocol", June 5, 2002

A Location Transparent Multi-agent System with Terminal Mobility Support

Akira Hosokawa[1], Kazuhiko Kinoshita[1],
Nariyoshi Yamai[2], and Koso Murakami[1]

[1] Department of Information Networking
Graduate School of Information Science and Technology, Osaka University
2-1 Yamadaoka, Suita, Osaka 565-0871, Japan
{akira,kazuhiko,murakami}@ist.osaka-u.ac.jp
[2] Computer Center, Okayama University
3-1-1 Tsushimanaka, Okayama, Okayama 700-8530, Japan
yamai@cc.okayama-u.ac.jp

Abstract. On a multi-agent system, an agent cooperates with others to complete its task. Such a system has good flexibility and scalability for network management and service control so that it is expected to be a unified platform for next generation network architecture. However, since each agent migrates from a node to another node autonomously, it is difficult for an agent to know the location of others. Thus, it is important to support location transparent communications. In addition, it is also useful for mobile terminals to use mobile agent technology. However, it is usually difficult for agents to know the network location of the terminal when the terminal moves. In such a case, therefore, agents cannot communicate with the terminal. Practically, there are no systems that support the location transparency with mobile terminals. In this paper, we propose a location transparent multi-agent system with terminal mobility support.

1 Introduction

Recently, as information technologies make progress quickly, computer networks such as the Internet widely spread. Accordingly, the amount of information and the number of services over networks increase more and more. The more information and services there are over the Internet, however, the more difficult the users get the contents they want. For example, we can know many content servers on a network. However, we cannot often select the most suitable one among them because we do not know which server is most appropriate. In such a case, we have to check a server one by one whether it is suitable or not.

To overcome such a situation, mobile agent technology receives much attention [1]. Mobile agent is a kind of software that autonomously migrates from computers to computers on a network while performing a given task using resources on the network. Mobile agent technology is developed actively, and many agent systems are proposed so far, such as Telescript [2], Aglets [3], AgentSpace [4], Voyager [5] and so on.

H.-K. Kahng (Ed.): ICOIN 2003, LNCS 2662, pp. 426–435, 2003.

An agent system where multiple mobile agents perform a common task in cooperation with others is called a multi-agent system. On a kind of multi-agent system, when a user lets his/her agent visit to many nodes, the agent produces its clone(s) autonomously if necessary. Moreover, on such a system, an agent communicates with others to complete its task.

On the other hand, mobile agent technology is also fit for distributed network management by virtue of its autonomy [7]. For example, when a network administrator asks an agent to manage the network, it autonomously migrates to each node on the network and complete its task, even if there are many nodes on the network. Thus, mobile agent technology is expected to be a unified platform for next generation network architecture.

However, it is difficult for an agent to know the network location of other agents, because each agent migrates from a node to another node on the network autonomously. In such a case, it is difficult for agents to communicate with other agents, and it is hard for an agent to cooperate with others. Because of the same reason, it is also difficult for a user to know the network location of agents and it is hard for him/her to send messages to his/her agents. Thus, it is important to support the location transparency on a multi-agent system. However, most of existing multi-agent systems do not realize it.

By the way, users have come to get access to computer networks with mobile terminals such as a laptop computer, a PDA and a cellular phone. However, in general, bandwidth of a wireless network is narrower than that of a wired network. Thus, when users get the contents via a wireless network, it takes more costs and longer time than via a wired network to complete the communication till they get ones. Mobile agent technology can also overcome such a problem because a user has to communicate via a network only when he/she sends his/her agents and receives the result. However, the user terminal may move while the agents are in action. It is usually difficult for agents to know the new network location of the terminal in that situation. In such a case, agents cannot communicate with their user. In other words, the user cannot receive the result of his/her task from his/her agents.

Practically, there are no systems that support the location transparency with mobile terminals. In this paper, we propose a new multi-agent system with the location transparency. The proposed system is also designed to support the terminal mobility.

2   Location Transparent Communications

Location transparent communications are that one can communicate with others without awareness of network location of them. On a multi-agent system, multiple agents perform their tasks efficiently in cooperation with others on other nodes. However, it is difficult for a user and agents to know the network locations of others, because they migrate autonomously. It is a critical issue for multi-agent system.

2.1   Location Transparency with CORBA

A few existing multi-agent systems support the location transparency, and most of them use the Common Object Request Broaker Architecture (CORBA) [6].

CORBA is a basic technology to support communications among objects on distributed computing environments. It works well regardless of operating system and programming language. It supports location transparency by the following ways.

The Object Request Broker (ORB), which is one of the principal elements of CORBA, mediates all communications between a server object and a client object. When the client object wants to communicate with a server object, it first sends a message to the ORB. The ORB sends the message to the server object, when the server object is on the local host. In case there is the server object on a remote host, the ORB sends the message to the ORB on that host. Then, CORBA naming service are used to know the network location of the remote object.

Naming service, which is one of CORBA services, is a service to search an object. An object is registered when it is used as a distributed object with naming service. Each object can communicate with a remote object by getting the reference to it with naming service.

On a multi-agent system with CORBA, location transparent communication is also supported. In this case, a user and agents can get the network locations of all agents with naming service. However, each agent has to register its network location to the naming server whenever it migrates. So such a system does not work well in a large scale network.

2.2   A New Multi-agent System

We propose a new multi-agent system which supports location transparency. Moreover, the proposed system also provides high scalability.

Outline of the Proposed System. This system is installed on each node where agents are allowed to execute. Each node has a history file that records the activities such as arrival, execution and migration, of agents. On the proposed system, the agent migrates to a neighbor node with recording its activity in the history file of each node. A history file also records the footprints of messages from an agent and/or a user, such as an arrival time, the name of the neighbor node which the message was sent from/to, and so on. The proposed system provides the location transparency with these history files. We explain it in detail in the following paragraphs.

Agent Migration. On the proposed system, an agent visits to all nodes designated by a user based on the flooding algorithm [8]. Because it is guaranteed that an agent visits to all the nodes through the shortest paths. That is, when an agent visit to the neighbor nodes, the node where the agent exists refers to

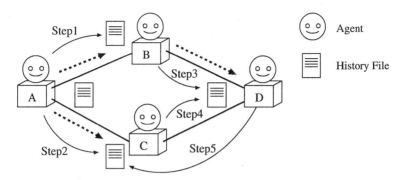

Fig. 1. Agent Migration

the history files of all neighbor nodes except the node from which the agent migrates. If there is only one neighbor node where the agent has not arrived yet, the agent migrates to the neighbor node. On the other hand, if there are two or more neighbor nodes where the agent has not arrived yet, the node where the agent exists produces clone(s) of the agent, and then, the agent migrates to one of the neighbor nodes and the clone(s) migrate to the other neighbor node(s). When an agent arrives at a node, the node records its arrival to the history file.

In these ways, an agent visits to all the designated nodes with producing clone agent(s) if necessary. Note here that the set of the links used for an agent migration forms a spanning tree, and the root of the spanning tree is the node from which the agent originates.

Figure 1 shows how an agent visits to all the nodes in a network with the nodes A, B, C and D. Suppose that an agent visits to all the nodes from the node A. When the agent finishes its task on the node A, the node A refers to the history files of the neighbor nodes, the nodes B and C. Since the agent has not arrived at these nodes yet, the node A produces a clone of the agent, and then, the original agent migrates to one of the nodes B and C and the clone agent migrates to the other node (Step1, Step2). Then, suppose that the agent on the node B finishes its task earlier than the agent on the node C. Here, since the agent on the node B has migrated from the node A, the node B refers to the history file of the node D. The history file of the node D has no record of the agent, therefore, the agent migrates to the node D (Step3). Then, the node C refers to the history file of the node D in the same manner as the node B did, when the agent finishes its tasks on the node C (Step4). Since there is a record of the agent on the node D, the agent on the node C does not migrate. In the same way, the agent on the node D does not migrate to any neighbor nodes either (Step5).

User-Agent Communication. All message forwarding for user-agent communications is performed with referring to the history file on each node. As mentioned before, the set of the links used for an agent migration forms a span-

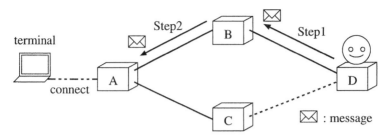

Fig. 2. User-Agent Communication

ning tree. So, all user-agent communication messages are delivered along the tree. When a node receiving a message from a user and/or an agent, the node refers to the history file, and then the node duplicates the messages if necessary and forwards the messages to its neighbor nodes. Moreover, all inter-agent communications are performed in the same way.

Figure 2 shows an example of user-agent communication. Suppose that the agent on the node D is active after the agent visits to all the nodes as shown in Figure 1, and that the agent tries to communicate with the user connected to the node A. Firstly, the node D refers to its history file, and then it knows the parent node, that is the node B. Consequently, the node D sends the message to the node B, and then the node B accepts it (Step1). Secondly the node B refers to its history file, and then knows that the agent has migrated from the node A. Consequently, the node B sends the message to the node A, and the node A accepts it (Step2). Finally, the node A forwards the message to the user terminal.

3   Terminal Mobility on a Multi-agent System

When a user terminal moves, an agent is unable to communicate with the terminal because the agent cannot know the new location of the user terminal. For example, in Figure 2, suppose that the terminal leaves the node A and then connects to another node. Then, the message cannot be forwarded to the terminal. To overcome this problem, we propose a new method to support the terminal mobility on the proposed agent system.

3.1   Existing Terminal Mobility Supporting Method

Mobile IP technology enables a mobile terminal to continue to communicate without awareness of movement when the terminal moves over IP networks [9]. When we apply the concept of Mobile IP technology to the proposed system, agents are able to communicate with a terminal after the terminal leaves a node and connects to another node. In this method, the node to which a user sends an agent at first acts as Home Agent which manages the network location of the

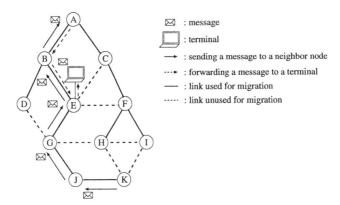

Fig. 3. Existing Terminal Mobility Supporting Method with Mobile IP

terminal. When the terminal moves and changes the location on the network, a message to the user is received by HA at first, and then it is forwarded to the terminal.

Figure 3 shows an example of the terminal mobility with such a method. Suppose that a user terminal is connected to the node A, and he/she sends an agent to visit to all nodes. After the agent visits all the nodes, the links used for migrations of agents are shown by solid lines and other links are shown by dotted lines in the figure. Next, suppose that the terminal leaves the node A and then connects to the node E, and that there is an active agent on the node K and the agent sends a message to the user. The message is forwarded to the root node, that is the node A, via the nodes J, G, E and B in turn with referring to their history files. Then, since the node A knows the network location of the terminal, the node A forwards the message to the terminal.

In this method, however, the message is forwarded via the root node regardless of the location of the terminal. Thus, the amount of traffic increases according to the distance between the root node and the node to which the terminal connects after movement.

3.2   Proposed Method

Then, in this paper, we propose a new method to overcome the problem. It supports the terminal mobility on the proposed multi-agent system with less increase of the amount of traffic.

In this method, at first, when the terminal leaves a node and connects to another node, the terminal sends a request to the node to which the terminal connects after movement. The request is used for modifying the parent-child relationship into reverse in the history file.

When a node receives the request, the node refers to its history file. Then, if there is an arrival record of the agent, it modifies the parent-child relationship into reverse in its history file and forwards the request to the parent node.

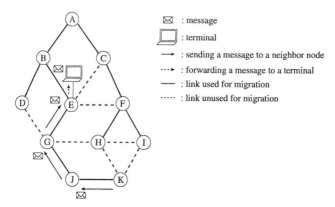

Fig. 4. Proposed Method

After several times of the same procedure, the request is arrived at the node to which the terminal connected before movement and the request is discarded. As a result, the proposed method changes the root of the spanning tree from the node to which the terminal connects before movement to the node to which the terminal connects after movement.

On the other hand, when the history file of the node to which the terminal connects after movement has no arrival records of the agent, the node sends a message to the terminal in order to report that the agent has not arrived yet. Then, when the terminal receives the message, it sends another request to the node to which it has connected before movement. After the node to which the terminal has connected before movement receives the request, when a message from an agent is arrived at the node, the node sends the message to the terminal.

This procedure continues until the node receives the request to modify the parent-child relationship into reverse in the history file from its child node. In such ways, the terminal can receive the messages from the agents, even if the agent has not arrived at the node to which the terminal connects after movement yet.

Figure 4 shows an example of the terminal mobility support with the proposed method. We assume the same situation as shown in Figure 3. When the terminal connects to the node E, the terminal sends a request to modify the history file about his/her agent. When the node E receives the request, the node E refers to its history file. Here, suppose that the agent has already arrived at this node. In such a case, the node E knows that the parent node of the node E is the node B. Then, the node E modifies its history file in order that the node E turns into the parent node of the node B, and it forwards the request to the node B. Then, the request is forwarded to the node A, and the node B modifies its history file in the same manner. When the node A receives the request, it discards the request because it is the original root node. Here, suppose that there is an active agent on the node K, and the agent sends a message to the user. The message is forwarded to the node E via the node J and G in turn.

4   Performance Evaluation

In this section, we evaluated the performance of the proposed multi-agent system to show its efficiency.

4.1   Evaluation of Location Transparent Communication

We provided a lattice network as a network model. The number of nodes on the network was 25, and the agent system was supposed to be installed on all the nodes.

At first, we selected one of the nodes to which a terminal connected at random, and then we let an agent visit from the node to all the other nodes. Then the terminal sent a message to all the agents. We compared the proposed system with the existing system with CORBA. On the existing system, whenever each agent migrates to a node, it sends a message for registration of its location to the naming server. Here, suppose that the naming server was installed on the center node of the network. We measured the cumulative sum of the number of links that all messages have passed as the amount of traffic for the following procedures because the volume of each message is almost the same.

Registration of the Agent Location: When an agent migrates to a node, the agent sends a message for registration of its location to the naming server.

User-Agent Communication When a terminal communicates with all agents, the terminal sends a messege to them.

Table 1 shows the amount of traffic on average over 10,000 simulation experiments. It is shown that the proposed system achieves location transparent communications with much less traffic than the existing system. On the existing system, since each node has to send a message to naming server to register its location every agent migration, the amount of traffic increases more on a larger scale network. To the contrary, on the proposed system, agent migration is recorded only on a local node (history file). Moreover, on the proposed system, the amount of traffic for user-agent communications is also less than on the existing method. Note here that the occurrence ratio of registration of the agent location to user-agent communication depends on a use of the agent system. However, this simulation result shows that the proposed system is always much better than the existing system.

4.2   Evaluation of User-Agent Communication
with Terminal Mobility

We provided lattice networks as network models. The number of nodes on these networks was a variable parameter of the experiments, and the proposed system was supposed to be installed on all the nodes.

Table 1. Evaluation of the Location Transparent Communication

	Registration of the agent location	User-Agent Communication
Proposed System	0	24.0
Existing System with CORBA	60.0	40.6

(messages)

At first, we selected one of the nodes to which a terminal connected at random, and then we let an agent visit from the node to all the other nodes. Next, we randomly selected one of the nodes, and the terminal moved there. After that, all the active agents sent a message to the terminal. We compared the proposed method with the existing terminal mobility supporting method with Mobile IP on the proposed system. We measured the cumulative sum of the number of links that all messages have passed as the amount of traffic for user-agent communications.

The results of simulation experiments are shown in Figure 5. It shows the amount of traffic on average over 10,000 simulation experiments as a function of the number of nodes.

It is shown that the proposed method supports the terminal mobility with much less traffic than the existing method. When the number of nodes increases, the difference in the amount of traffic between the proposed method and the existing method becomes larger. Thus, the proposed method is particularly fit for large scale networks.

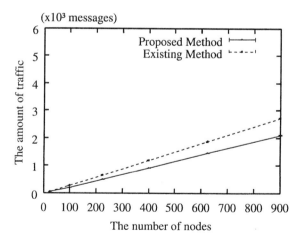

Fig. 5. User-Agent Communication with Terminal Mobility

5   Conclusion

In this paper, we described that a multi-agent system had good potential to be a unified platform for network management and service control on the next generation network architecture. We also insisted that location transparent communications were important for efficient use of multi-agent systems. In addition, we explained the problem of the terminal mobility.

Consequently, we proposed a new multi-agent system with location transparency. It provides location transparent user-agent communications by distributed location management with history files.

Moreover, we also proposed a method to support the terminal mobility by modification of history files.

Performance evaluation results showed that the proposed system provided location transparent communications scalably and supported the terminal mobility with much less increase of traffic. In particular, it was shown that the proposed system had better performance in larger scale networks. It comes to the conclusion that the proposed multi-agent system has enough potential to be a platform for next generation network architecture.

As further studies, we try to enhance the security of the proposed system and to implement a prototype system.

References

[1] V. A. Pham and A. Karmouch: "Mobile Software Agents: An Overview," *IEEE Communications Magazine*, Vol. 36, No. 7, pp. 26-37 (July 1998).
[2] J. E. White: "Telescript Technology: in Mobile Agents," in *Software Agents*, J. Bradshaw (Ed.), MIT Press, Massachusetts (1996).
[3] IBM: IBM Aglets Software Development Kit Home Page,
http://www.trl.ibm.com/aglets/index.html.
[4] I. Satoh: Agent Space,
http://www.research.nii.ac.jp/ ichiro/agent/agentspace.html.
[5] ObjectSpace Inc.: Voyager,
http://www.recursionsw.com/products/voyager/voyager.asp.
[6] Object Management Group, Inc: Complete CORBA 3.0 specification with change bars, http://www.omg.org/cgi-bin/doc?formal/02-06-01.pdf.
[7] Andrzej Bieszczad, Bernard Pagurek, Tony White: "Mobile Agents for Network Management," IEEE Communications Surveys (1998).
[8] Andrew S. Tanenbaum: Computer Networks Third Edition, Prentice-Hall, New Jersey (1996).
[9] Charles E. Perkins, David B. Johnson: "Mobility Support in IPv6," Proceedings of the Second Annual International Conference on Mobile Computing and Networking (Nov. 1996).

Efficient Handover Scheme with Prior Process for Heterogeneous Networking Environment

Hyunseung Choo[1], DaeKyu Choi[1], and Keecheon Kim[2]

[1] School of Information and Communication Engineering
Sungkyunkwan University, 440-746, Suwon, Korea
{choo,eunpiri}@ece.skku.ac.kr
+82-31-290-7145
[2] Department of Computer Science and Engineering
Konkuk University, Seoul, Korea
kckim@konkuk.ac.kr

Abstract. Recently real-time and QoS guaranteed services for mobile terminals between two adjacent heterogeneous networks are investigated based on the cellular structure of small cells in the literature. An efficient handoff is a very important factor for those services and it is called inter-system handoff(ISHO). The time required to complete the handoff can vary and depends on the network structure. And also the transmission of additional signals can increase the probability of failure for ISHO. The proposed scheme significantly reduces the ISHO failure rate up to about 16% compared to the existing one. This can be done by alleviating workloads of boundary base stations and distributing them to predetermined neighboring cells. The overhead of the proposed scheme is considered as well.

1 Introduction

Due to the need of high QoS mobile computing, PCS has been designed primarily to be capable of forwarding and receiving expected data whenever users request. The third generation systems represented by ITU IMT-2000 and ETSI UMTS services include services such as audio visual multimedia, roaming, virtual home, billing and security in order that various service configurations would be possible[1].

Recently real-time and QoS guaranteed services for mobile terminals in cellular networks are investigated in the literature. Handoff[2, 3] is very important to guarantee the quality of service. It is the mechanism that transfers an ongoing call from one cell to another as a mobile moves at the coverage area of the cellular system. Intra-system handoff occurs when the mobile user moves within the area of same system(or cell). Meanwhile, inter-system handoff(ISHO) occurs when the user moves into an adjacent cell which is in the different system that entire mobile connections must be transferred to the new system. For a successful ISHO, several important issues must be handled and additional new techniques must be required[4, 5].

H.-K. Kahng (Ed.): ICOIN 2003, LNCS 2662, pp. 436–445, 2003.

We focus on the cellular structure of small cells which are required for the high density of population and the handoff scheme implemented between the two heterogeneous systems. The ISHO requires several additional processes such as path rerouting and data transformation. Rerouting process can be implemented in a relatively short period of time at the same network. The recent study [4] shows that it takes some time to finish the handoff, since boundary cell base stations (BBSs) handle every ISHO process. Here, we employ Sub-BBSs which are located near BBSs at the same location area(LA). As we know, LA is a service area where several clustered cells are managed by a mobile switching center. BBSs are located at boundary cells of LA, and Sub-BBSs are adjacent to BBSs but located inner side of BBSs in LA. When the mobile terminal(MT) enters Sub-BBS area, the network starts finding a new route, and after entering BBS area, the MT initiates the transformation process. The proposed scheme significantly reduces the ISHO failure rate up to about 16% compared to the existing one which is the most recent and known to be effective.

In the following section, the basic structure of cellular networks and a recent scheme for handoff is discussed. The proposed scheme is described in section 3. The performance of the proposed scheme in terms of the probability of failure for ISHO is evaluated in section 4 comparing to the previous one in which only BBSs work for the ISHO. Finally, we conclude our paper with the future direction.

2   Background

Due to the increase of user demands on better user mobility, the next-generation wireless network supports terminal mobility, personal mobility, and service provider portability. Terminal mobility means the ability of the network to deliver calls to the MT regardless of its location to the network. Personal mobility refers to the ability of the user to access her personal services independent of the her location to the network or terminal[6]. Service provider portability means that the user or the MT moves beyond regional mobile networks, and the user will be able to receive their personalized services regardless of their current network[1]. IMT-2000 users are provided their prescribed network services regardless of their roaming situation[1, 6]. However, there are some limitations according to their resident network. Inter-system handoff is an essential technique supporting the global roaming.

The typical three-stage process for handoff is first initiated by the user, a network agent, or changed network conditions to identify the need for handoff. Second, new connection is established that the network must find new resources for the handoff connection and perform any additional routing operations. Finally, data-flow control for the delivery of the data from the old connection path to the new one is maintained according to agreed-upon service guarantees.[1]

A set of signaling messages needed to communicate between two BBSs(SWs) has been developed at ISHO schemes in the literature. Basically ISHO procedure is divided in three phases. 1) when MT arrives at boundary cell(BC) of a network, rerouting to the other network is performed. 2) MT performs a format transformation at the same cell. 3) MT hands over to the other network using the handoff scheme prepared by the other one. For a small region with large population, the probability of failure for handoff is highly likely because MT reconfiguration and path rerouting, etc. are performed within the boundary cell. Especially for two heterogeneous networks environment, the time for ISHO takes even more, thus the probability of failure for ISHO is even higher than that of intra-system handoff. Therefore ISHO problem should be carefully investigated. The delay time due to inefficient handoff procedure may result in communication failure. This is the reason why we need this study. By path rerouting in advance at Sub_BBS, we can considerably reduce the probability of failure for ISHO especially for real-time services or QoS guarantee required services.

3   The Proposed Scheme

The proposed scheme discussed in this paper assumes several conditions. At first, MT can compare strength of signals from BBSs of two adjacent cells. Secondly, there are many processes for handoff. Finally, MT can control every procedure for handoff(MCHO). Extremely special cases for handoff procedure are out of the concern in this paper.

Figure 1 shows the border of the cellular network in which smaller sized cells are used for high density of mobile users. In this case, transformation process and path rerouting procedure are required additionally. Path rerouting may take very short period of time during handoff procedure in the same network. However in ISHO, the elapsed time for path rerouting depends on the network configuration. The additional signaling for the two processes must be an important for deciding factor ISHO failure due to increased total time for handling them. This is the reason why Sub-BC(Sub-BBS area called Sub-BC) is employed in our scheme to reduce the probability of failure. When MT moves to Sub-BC from outer cells, the reroute process is launched. And when MT comes to BC, it prepares

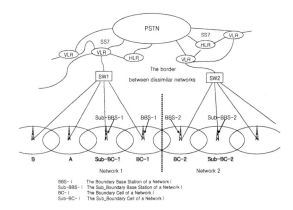

Fig. 1. Cellular structure of wireless network with Sub_BBS

for the transformation. If MT goes back to cell A, the information for the path rerouting is maintained. And if MT goes to cell B, the information is abandoned refer to Figure 1. Let us suppose that MT moves into Sub_BC-1 and possibly the user moves into BC-1. Then the network expects the user may move into BC-2. The entire ISHO procedure is divided into three phases. The first phase starts when MT enters Sub_BBS-1. At this moment, MT informs SW1 that MT may need ISHO into the adjacent network. After a certain moment, the MT may enter BC-1. While the MT resides in BC-1, it faces ISHO at last and requires mode transformation to be adapted to the new system. The last phase for ISHO is implemented when the MT passes through the boundary between two different networks. During ISHO, MT changes its mode that is reconfigured by the transformation which makes MT to be able to communicate in the new system continuously.

Phase 1: Path Rerouting. If MT moves into Sub_BC-1, the current network perceives that the MT may need ISHO into the other network. First, MT sends ISHO warning message to Sub_BBS-1 as soon as finishing intra-system handoff into Sub_BC-1. The message must contain an identification of the MT being used and information for the previous base station. Then Sub_BBS-1 sends an acknowledgement message to the MT. The warning message for ISHO is also sent to SW1. After finding the root switch of BBS-1, SW1 sets a path to BBS-1. It means that SW1 sends BBS-1 some information for the path from BBS-1 to SW1. While the MT resides in Sub_BC-1, SW1 and SW2 find new route for the MT.

Phase 2: Transformation. The second phase for ISHO is implemented when the MT resides in the BC-1. In this phase, the MT obtains an information for the protocol utilized in the other network and transforms data based on the current networking protocol to be adapted to the other system with information received. After moving into BC-1, BBS-1 sends compare-message to the MT. If the MT receives this message from the BBS-1, it keeps comparing the strength of two signals from BBS-1 and BBS-2. While the MT resides within BC-1, BBS-1 sends Ready-Mode-Change message to the MT. After receiving Ready-Mode-Change message from BBS-1, the MT starts to prepare for ISHO. If the MT is multi-mode terminal, it needs only mode change process. However if the MT is not multi-mode one, the MT begins to download information for the protocol used in network 2 from BBS-1. As soon as the MT finishes changing the mode, it sends BBS-1 Ready-Mode-Change-Done message. If SW1 finishes finding a new path, it sends reroute-done message to BBS-1. If the MT finishes preparing mode change and SW1 sends the new path, the entire steps for phase 2 are completed.

Phase 3: Into the New System. While the MT resides within BC-1, it compares signal powers from both networks continuously. If the MT estimates that the signal from network 2 is surpassing than that from the current network,

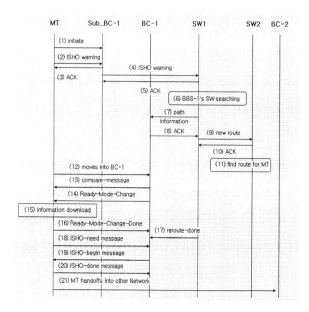

Fig. 2. Signal transmissions in all phases

it launches the ISHO phase 3. If the MT decides the ISHO is needed, it sends ISHO-need message to BBS-1. After receiving ISHO-need message, the BBS-1 sends ISHO-begin message to the MT as an acknowledgement. And the MT sends ISHO-done message to the BBS-1. With all these steps, the MT can handoff into the other network and is served even in the other network continuously. Figure 2 shows every message flow from phase 1 to phase 3.

4   Performance Evaluation

For a successful ISHO between two different networking systems, the MT must perform steps explained in section 3 before it enters different system. The minimum residency time in boundary cells and the minimum boundary cell size should be given for calculating the chain of steps. The following parameters are used in this section. T_s: additional signal processing time at ISHO; T_i: execution time for step i; T_{req}: required time at Sub-BBS for format transformation; α_i: message transmission time; β_i: message propagation time; γ_i: processing time for the control message; N_f: the number of wireless link failures; q: probability of wireless link failure; V: the expected velocity of mobile terminal; λ: the arrival rate of mobile terminal into boundary cells; S: the boundary cell area; L: the one hexagonal perimeter of the boundary cell; l: the length of the side of the boundary cell.

4.1   Additional Signal Processing Time

Inter-system handoff additional signal processing time T_s is the entire time accomplishing all steps from downloading the information for the other network protocol to mode conversion in order that the MT can communicate in the new network. We obtain ISHO time by adding all time for each step mentioned in section 3 and focus on additional time at BC-1 to evaluate the system performance. As we know there is enough time to finish steps (1) through (12) at Sub_BC-1 for path rerouting. Hence those are excluded from the additional signal processing time. $T_s = \sum_{i=13}^{21} T_i$. At each step the time to send a message is composed of transmission time, propagation time and processing time. $M_i = \alpha_i + \beta_i + \gamma_i$ $i = 13, \ldots, 21$. The transmission time α_i is computed by the size of the control message in bits(b_i) over the bit rate of the link on which the message is sent(B). $\alpha_i = \frac{b_i}{B}$. Propagation time β_i may vary depending on the medium of environment. Time γ_i is a parameter that has the same value at switches, base stations and mobile terminals. In steps (13), (16), (17), (19), (20) message retransmission is not needed so signal processing time is same as message sending time. $T_i = M_i$ $i = 13, 16, 17, 19, 20$. At steps (14) and (18), messages are sent in wireless environment and possibly signals can be lost. Therefore, the control signal is retransmitted until ACK message is received. By considering the number of link failures in wireless environment N_f and the probability of link failure, we obtain additional signal processing time at these steps. $T_i = \sum_{N_f=0}^{\infty} T_i(N_f) \cdot Prob(N_f \; failures \; and \; 1 \; success)$. In wireless environment ACK signal may not be received for T_w after the request signal is sent and then we assume that the message is lost and control signal must be retransmitted. If there are N_f failures, then T_w and message transmissions occur N_f times. So $T_i(N_f)$ is induced as follows. $T_i(N_f) = M_i + N_f(T_w + M_i)$ $i = 14, 18$. Therefore signal processing time of requiring retransmission steps is same as below. $T_i = \sum_{N_f=0}^{\infty} \{M_i + N_f(T_w + M_i)\} \cdot Prob(N_f \; failures \; and \; 1 \; success) = M_i + (T_w + M_i) \sum_{N_f=0}^{\infty} N_f \cdot Prob(N_f \; failures \; and \; 1 \; success)$ $i = 14, 18$. Here $\sum_{N_f=0}^{\infty} N_f \cdot Prob(N_f \; failures \; and \; 1 \; success)$ is induced by infinite geometric progression. Link failure probability, q, is smaller than 1. So, $\sum_{N_f=0}^{\infty} N_f \cdot Prob(N_f \; failures \; and \; 1 \; success) = \frac{q}{1-q}$. Generally q has value of 0.5. So T_i is $T_i = M_i + (T_w + M_i)\frac{q}{1-q} = M_i + (T_w + M_i)\frac{0.5}{0.5} = 2M_i + T_w$ $i = 14, 18$. At step (15) MT performs only mode transformation process when MT is multi-mode. Meanwhile, when MT is not multi-mode, MT must receive the protocol information for network2 thereafter MT changes its mode. Here T_{15} is the downloading time that is measured by system parameter in the given environment.

4.2   The Probalility of ISHO Failure

In the previous subsection, we have discussed two factors for ISHO failure. One is the signal loss in wireless environment. When ACK signal is not received in T_w after request signal is sent, link failure is determined. If link failure occurs more than two times then the time required increases for ISHO. The other factor is the case that MT's velocity is too fast. In this case MT doesn't stay in BC-1 for going through the series of steps to change mode. T_{req} is the time required that MT stays in the boundary cell. The parameter P_f is the probability of the threshold value and T is a random variable that takes on values of the time to the next handoff after the MT's arrival into the boundary cell. T is exponentially distributed. So it is induced as follows. $P = Prob(T < T_{req}) = 1 - \exp(-\lambda T_{req}) < P_f$. Here λ is the arrival rate of MT into the boundary cell and MT's movement direction is uniformly distributed on the interval $[0, 2\pi)$. So λ is calculated as follows. $\lambda = \frac{VL}{\pi S}$. The V is the expected velocity for MT that varies in the given environment. The L is the perimeter. Generally, a cell is a hexagonal shape. Six sides of the hexagon have a length of l each. Area of boundary cells, S, is $3l^2 sin(\pi/3)$. Then λ becomes $\lambda_{MT} = \frac{2V}{\pi l sin(\pi/3)}$. To reduce the probability of ISHO failure, the minimum boundary cell size is necessary with expected velocity for MT staying in the boundary cell. So l is induced like equation as follows. $l > \frac{2V T_{req}}{\pi log(1/(1-P_f)) sin(\pi/3)}$. Hence, the boundary cell size is determined. $A^{cell} > 3l^2 sin(\pi/3)$.

4.3   Numerical Results

To obtain additional signal processing time we must consider each parameter. Especially, an important thing is the deliberation of transmission environment. Table 1 shows parameters used to analyze the system in the proposed ISHO scheme.

Transmission environment is divided into wired environment and wireless one. At step (17), control signal is transmitted in wired environment. At the remaining steps message is sent in wireless environment. Table 2 shows the additional signal processing time, nominal handoff time and total handoff time in the proposed ISHO scheme.

Figure 3 shows results for ISHO failure probability comparison between the existing scheme and the proposal one. And we can find that the smaller cell

Table 1. System parameters

Bit Rates B	Values	Propagation Times β	Values
Wireline Link	155 Mbps	Wired Link	500 msec
Wireless Link		Wireless Link	2 msec
Low Mobility	2 Mbps	**MT Velocity** (V)	**Values**
Medium Mobility	384 Kbps	Low Mobility	3 Km/hr
Vehicular Mobility	144 Kbps	Vehicular Mobility	10-100 Km/hr
High Mobility	64 Kbps	High Mobility	300 Km/hr
Processing Times λ	**Values**	**Nominal Handoff Times** (T_{17})	**Values**
Switch	0.5 msec	PACS	20 msec
Base Station	0.5 msec	DECT	50 msec
Mobile Terminal	0.5 msec	GSM	1 sec
the others	**Values**		
Number of Hops (N)	3		
Download Time (T_{15})	10 msec		
Link failure Probability (q)	0.5		

Table 2. Time required for accomplishing ISHO in a given cell

Variables	Indoor	Pedestrian	Vehicular	High Speed
MT Velocities (V)	3 Km/hr	3 Km/hr	10-100 Km/hr	300 Km/hr
Bit Rates (B)	2 Mbps	384 Kbps	144 Kbps	64 Kbps
Additional Signaling Time (T_s)	34.60 msec	41.00 msec	53.60 msec	82.60 msec
Nominal Handoff Times (T_{21})	20 msec	50 msec	1 sec	1 sec
Total handoff Time	54.60 msec	91.00 msec	1053.60 msec	1082.60 msec

area gets exponentially high probability of ISHO failure. First, the case of pico cell and the nominal values used in PACS(North American Personal Access Communication System). In Figure 3 if we assume acceptable probability of ISHO failure is 2%, a minimum boundary cell size is about $7m^2$. Second the case of micro cell, the numerical values for DECT(European Digital European Cordless Telecommunication) one used. So total handoff time is 91 msec. The last case for macro cell has nominal values for GSM(European Global System for Mobile Communication). This case has the total handoff time of 1.08 sec. In wireless environment communication service provider's goal is to provide the better quality of service to more persons. Therefore cell size should be smaller.

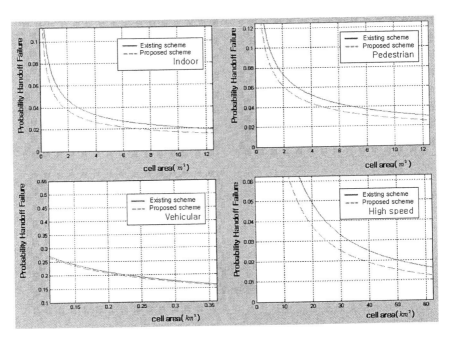

Fig. 3. The probability of ISHO failure with cell area

In that sense, the proposed ISHO scheme has a merit that provides the better quality of service.

4.4 Overhead

When MT arrives to the boundary cell of two systems, it performs path rerouting and data transformation in sequence. If MT dose not perform ISHO after preparing for the handoff, the time spent for the preparation becomes the total waste and thus the overhead for the scheme. The overhead is acquired as $F_{existing}(p) = (1-p)T$. We assume the time spent for MT in preparing ISHO is time T, rerouting time + transformation time, and thus the overhead is calculated by multiplying the probability of non-ISHO, $(1-p)$, the time T. ISHO process for the proposed scheme is shown in Figure 4.

The time spent for the preparation is T_1, hence the overhead for the corresponding step is $f_1(p) = (1-p)T_1$. The MT which has experienced the path rerouting arrives to BC-1 and then handovers to BC-2 with ISHO probability p. Then MT does not perform ISHO to BC-2 with the probability $(1-p)$. The time required for the data transformation is T_2 and thus the overhead here is calculated as $f_2(p) = p(1-p)T_2$. The probability that MT from other cells is $1-p$. Therefore, such overhead is denoted by f_3, i.e. $f_3(p) = (1-p)T_2$. The duplicated part is represented by f_4 and must be extracted from the overall overhead, i.e. $f_4(p) = p(1-p)^2T_2$. Therefore, $F_{proposed}(p) = \sum_{i=1}^{3} f_i - f_4$. On calculating the overhead in percentage, $F_{proposed}(p) - F_{existing}(p)$ is the total time of preparation, where $F_{existing}(p)$ is obtained similarly by the case without prior process. The value of p is approximately 0.67 when $F_{proposed}(p) - F_{existing}(p)$ is maximized.

$$\{F_{proposed}(0.67) - F_{existing}(0.67)\}/T \tag{1}$$

The probability p that MT moves from Sub_BC-1 to BC-1 is 1/6 in hexagonal cellular environments. Hence the general overhead is as below. Refer to Table 3.

$$\{F_{proposed}(1/6) - F_{existing}(1/6)\}/T \tag{2}$$

The overhead becomes 12.7% when maximized. On the other hand, in typical hexagonal case, the overhead takes just up to 2%. Therefore, we conclude that

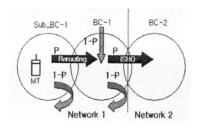

Fig. 4. ISHO of the proposed scheme

Table 3. Overhead percentage

	Indoor	**Pedestrian**	**Vehicular**	**High Speed**
Maximum case(1)	10.5%	10.3%	6.2%	12.7%
Hexagonal environment(2)	1.6%	1.6%	0.9%	1.9%

the overhead is relatively small comparing to the performance improve done by the proposal scheme.

5   Conclusion

In this paper, we propose the new handoff scheme that reduces the probability of failure for inter-system handoff. Since every communication with MT is under the wireless environment, every signal can be lost at any time. Thus after waiting for ACK message in a time period, MT and BBS resend the message. The other is very high speed for MT. MT cannot have enough time for mode change. Therefore MT must reside within the boundary cell for the time required to prepare for the mode change. As we all know, the technology must be developed for the convenience of users and QoS must be concerned when the new technology is applied. The proposed scheme handles ISHO reasonably well and it shows better performance in terms of the probability of failure for ISHO comparing to the existing one. In the future study, we present the way to pursue more stable networks.

References

[1] I. F. Akyildiz, et al., "Mobility management in next generation wireless systems," Proceedings of the IEEE, vol. 87, no. 8, 1999.
[2] A. Bora and D. Cox "Rerouting for Handoff in a Wireless ATM Network," IEEE Personal Communications, pp.26-33, 1996.
[3] M. A. Marsan and C.-F. Chiasserini, "Local and global handovers for mobility management in wireless ATM networks," IEEE Personal Communications, vol.4, no. 5, pp.16-24, 1997.
[4] J. McNair, I. F. Akyildiz, and M. D. Bender, "An inter-system handoff technique for the IMT-2000 system," IEEE INFOCOM 2000, vol. 1, pp.208-216, 2000.
[5] C.-K. Toh, "A unifying methodology for handovers of heterogeneous connections in wireless ATM networks," ACM Comp. Comm. Rev, vol. 27, no. 1, pp.12-30, 1997.
[6] R. Pandya, et al., "IMT-2000 standards: Net-work aspects," IEEE Personal Communications, pp.20-29, 1997.

Personal Server Model
for Real-Space Networking

Wakayama Shirou[1], Kawakita Yuusuke[1], Kunishi Mitsunobu[2],
Hada Hisakazu[3], and Murai Jun[1]

[1] Graduate School of Media and Governance, Keio University, Japan
{shirou,kwktjun}@sfc.wide.ad.jp
[2] Graduate School of Science and Technology, Keio University, Japan
kunishi@tokoro-lab.org
[3] Research Division, Digital Library Nara Institute of Science and Technology, Japan
hisaka-h@wide.ad.jp

Abstract. As the Internet population grows, not only computers but
household electric appliances, such as VCRs and a refrigerators, will
become connected to a network. The demand of connecting appliances
to a network is increasing, however, other cheap objects are difficult
because of the cost or technical problems.

In this research, we focus on connecting people to the Internet. We propose the *Personal Server Model* that sets up a personal agent on the
Internet. An agent, we called *Personal Server*, obtains person's information by using sensors. Personal Server also provides information via
Internet. With this Personal Server, any person can become an Internet
node virtually. By treating person as an Internet node, applications that
use the personal information can be developed easily like conventional
Internet applications.

Keywords: Ubiquitous, Context Based Computing, Sensor Network,
Real-Space Networking.

1 Introduction

Ubiquitous computing environment is said to be embedding a computer at various things, and offers convenience to a user. In ubiquitous computing environment, not only electric appliances but also even homes and vehicles have connected to Internet[1]. Moreover these things, it is thought that there are many
advantages of connecting other cheap and common object like a coffee-cup to
the Internet in Real-Space[1]. However, there cheap object can not be embeded
because of the cost or technical problems.

The Real-Space Networking aims to connecting these objects also to the
Internet by using *sensors*. And any network application can obtain information
about these object via Internet.

The target object in this research has the following features.

---

[1] In this paper, "Real-Space" is human sensible space. Opposite word is "Virtual-
Space".

H.-K. Kahng (Ed.): ICOIN 2003, LNCS 2662, pp. 446–459, 2003.

– No Internet connectivity because of various reasons like cost, a size.
– Mobile. Active or Passive is not concerned.
– Elemental substance. In other words, multistage composition is not target.

Among the objects with such features, we especially focus on the "person". By connecting person to Internet, various kinds of information can be obtained.

There are numerous kinds of information that are personal information. For instance, schedules, contact addresses and context information such as walking, sitting, attending a meeting, driving a car and so on. The files in the person's home directory may be included. All personal information is important. It can become more convenient by using this information. Though, we will concentrate on the personal location information in this research.

Next, the two scenarios of the application using person's location information will be described. and after scenarios, we introduce related works in section 2. In section 3, we propose *Personal Server Model* that manages personal sensor information. In section 4, we describe detail of Personal Server Model . In section 5 and 6, we introduces our experimentation and evaluation. Section 6 concludes the paper with discussions and future work.

Health-Care. A patient does not appear in the hospital at a scheduled time for a medicine, and cannot contact. In this case, an ambulance can be dispatched to the place where the patient's is by grasping it's position. However, in the case with people, moving to places where cannot be grasped by GPS, such as underground and inside buildings, must also be considered. In addition, person can be use heterogeneous sensor to obtain various kinds of information.

Alarm Clock. Even in a case where an alarm clock goes off, and the person is still in bed, the alarm clock can detect from the location information that the person is still sleeping. An alarm clock would then continue to ring until necessary.

2   Related Works

The Active Bat system [3] is the Location-aware Computing system. Active Bat uses an ultrasound in order to detect location. In Active Bat, users and objects have cheap and small tags which receive ultrasound in order to detect location.

Cricket Location Support System[4] uses ultrasound and cheap receivers in object in order to detect the location of object. Cricket use radio frequency signals additionally in order to time measurements. In Cricket, location is decided by triangulation relative to the beacons.

Recently, studies based on IEEE 802.11b has became major like robot location [5]. Since IEEE802.11b becomes an wireless network method that has wide usage, becomes cheap and small device. However, if target object has no electronic power like cheap object, system can not detect its location.

Geographical Location Information (GLI)[6] is able to map a mobile entity on the Internet to a geographical location. The GLI system consists of home and area servers. These servers are managed based in a hierarchical server structure toward highly scalable system However, GLI system assumed absolute coordinate like GPS.

3   Personal Server Model

This paper focuses especially on person. In this section, we describe our personal server model that manages personal information from the sensor at each own personal server.

3.1   Assumptions

Sensors are required in order to obtain information from Real-Space. However, there are various kinds of device type of sensor like data set. In addition, the movement of person who are targeted in this paper becomes problem. And, in order to discriminate person uniquely, it is necessary to have Globally Unique ID. Next, we will describe about detail of these assumptions.

Heterogeneous Sensors Infrastructure. To realize Real-Space Networking environment, sensors are needed to obtain personal location information. However, contrary to related works, we use not only the sensor which the user holds, but also the sensor infrastructure which is installed in the environment. Because, one sensor can not cover all location. For example, GPS can cover above ground, however, GPS can not detect under ground or in building.

Since that, we need various kinds of sensors to detect person's location. In figure 1 illustrates these sensor environment. There are various kinds of sensor infrastructures. Sensor infrastructure A in figure 1 has many sensors and these sensors connects each other like most Sensor-Network. Sensor infrastructure B and C has many sensors and send information to sink-node(Data-Base) like Active-Bat. Sensor infrastructure D has only one sensor device and that sensor can cover like GPS. User moves these sensor infrastructure and sensors nearby user obtain user's current information.

In addition, each sensor infrastructure has Administrator. The administrator manages and maintain these sensors. Consequently, administrator can get user's information. Therefore, security and privacy become problem.

All person has Globally Unique ID(GUID). In order to identify the user from sensors, we assume every persons have GUID and sensors can get these GUID using Auto-ID[7] or other method. GUID is used SDP described later.

We assume only uniqueness about GUID. However, structure of ID space and ID development is out of Target.

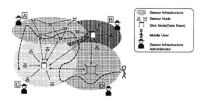

Fig. 1. Supposed Environment

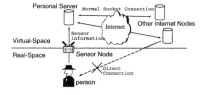

Fig. 2. Personal Server Model

3.2 Personal Server Model

In heterogeneous sensor infrastructure as mentioned above, there are two problems. One is where manages sensor information and the other is privacy.

We propose the Personal Server Model in order to resolve these problems. In this model, each person have the *Personal Server* that stores personal information like Location Information. Figure 2 shows this model.

Sensors sense personal information, and send it to his/her Personal Server. Personal Server accumulates the personal information. The other Internet nodes can access to the Personal Server in order to get that personal information in stead of direct access to person who is not connected to the Internet. In other words, Personal Server will serve as proxy or agent in the person's virtual-space.

In Personal Server Model, we can say the Personal Server is mapped object on Internet from person. According to that, when other Internet node obtains personal information, connect with other ordinary Internet manner like socket connection. Eventually, Internet Application that use personal information can be developed easily.

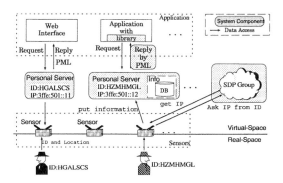

Fig. 3. System Component

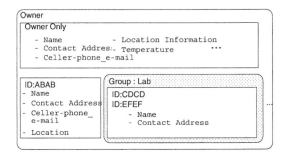

Fig. 4. Access Control in the Personal Server

4 System Architecture

In this section, we will describe the system architecture based on Personal Server Model. Figure 3 shows the system components which is half-tone dot meshing. We then describe about these system components.

4.1 Sensors

As mentioned, we assume using Heterogeneous Sensors in order to obtain user's personal information. We classify sensors into the following two kinds.

– Self-directed type sensor
– Sensor Infrastructure

Self-directed type sensor is a sensor where the sensor and sensing target is same. For example, GPS sensor receives the electric wave from a satellite and obtains its position. Therefore, sensing target is GPS sensor itself. Although this type has high accuracy, in many cases, these devices tend to be quite expensive.

Sensor Infrastructure are installed sensor as infrastructure. Many kinds of these sensor infrastructure sense the target when the target enter within the coverage of sensing. For instance, there are many studies to detect cheap target devices by using RF(Radio Frequency) or other method and obtain target owner's location. Though sensor infrastructure itself is expensive, target devices are cheap. Consequently, this kind of sensor is useful to operate as infrastructure. In addition, it must be noted that, when we use this kind of sensor as a location sensor, obtained information becomes relative location. For example, GPS uses latitude and longitude and altitude. However, RF sensor cannot get this absolute coordinate, but only labels such as room name.

As mentioned above, this research assumed on the Heterogeneous Sensor Infrastructure. Hence, it is necessary to accommodate to two kinds of sensors. However, Users do not always have sensor devices in Heterogeneous Sensor Infrastructure.

4.2 Personal Server

Personal Server is the Internet node that holds the user's information in Real-Space. The other Internet node can obtain personal information such as location information of the owner of that Personal Server by accessing to the Personal Server.

Internet nodes that wants to obtain user's information sends a request message to the Personal Server which has the GUID of the target person. Request message must contain the GUID of the person whom is using that Internet node. Personal Server returns user's information described by PML to the request node. Basically, Personal Server has this function.

It may be setup the Personal Server anywhere on the Internet. Moreover, the Personal Server always needs to be connected to the Internet. However, as long as it is a short time, the Personal Server may move. It becomes impossible to obtain the person's information in the meantime.

Privacy Protection. Personal Server may be set up anywhere on Internet and many Internet nodes may connect to the Personal Server. Since Personal Server has the personal information like location information, privacy protection is required. Moreover, only one Personal Server has managed his personal information. Since the personal information managed unitary on one Personal Server, when performing privacy protection, it is necessary to protect only one Personal Server.

We prepare an access control mechanism in order to realize privacy protection. When requesting personal information, Internet node must include GUID in a request message. The access control mechanism inside the Personal Server changes sending personal information to according to GUID contained in a request message. Personal Server may return nothing if request is invalid. Thereby, information cannot be passed to a request from an unexpected partner or the partner who does not want to give. Figure 4 shows access control list in Personal Server. Access control mechanism is moved according to this access control list. In this figure, personal server may send name, contact address and location information to the user which has ID:ABAB. However, personal server may send only name and contact address to the user(ID:CDCD) and user(ID:EFEF) because they are belong to Group:Lab. The group is maintained only one personal server. For that reason, though the user which is belongs to Group A in someone's access control list, the same user do not necessarily belongs to Group A in another user's access control list.

4.3 User's Information in Real-Space

We propose *Person Markup Language(PML)* in order to describe the user's information in Real-Space. When Personal Server exchange personal information with other Internet node, PML is used.

```
<d2:D2
  xmlns:d2=''http://www.sfc.wide.ad.jp/~shirou/Work/d2/d2/''
  xmlns:person=''http://www.sfc.wide.ad.jp/~shirou/Work/d2/person/''>
    <d2:core>
    <d2:id>HZMHMGL</d2:id>
    <d2:location>Meeting Room 1</d2:location>
    <d2:timestamp>2002-09-10 18:24:49</d2:timestamp>
    </d2:core>

    <person:person>
    <person:name>WAKAYAMA Shirou</person:name>
    <person:nickname>Shirou</person:nickname>
    <person:e-mail_addr>shirou@sfc.wide.ad.jp</person:e-mail_addr>
    <person:organization>KEIO SFC</person:organization>
    </person:person>
</d2:D2>
```

Fig. 5. An example of user's Information described by PML

PML. PML is XML[8] based language for describing personal information. PML must contain a core part. A core part contains GUID, Location Information, and Timestamp that obtained the last Location Information. Because every object have these information.

Figure 5 shows an example of a simple description of one person.

The core section of PML example in Figure 5 shows also that this person has ID:HZMHMGL and that his actual location is in the room called *Meeting Room 1*, and the Timestamp from which the Location was last obtained.

Moreover, PML can describe various kinds of information in addition to the core section. Figure 5 shows also person name and e-mail address. Thus, PML is extensible so that various kinds of information may be expressed. User's information described by PML is updated on the fly by the sensors or user himself.

4.4 Server Discovery Protocol (SDP)

Other Internet nodes and sensor nodes need to know the Personal Server IP address in order to access to the certain person's Personal Server. In this paper, as mentioned above, it is assumed on all persons holding GUID.

Server Discovery Protocol (SDP) changes person's GUID into that person's Personal Server IP address. We can say that SDP changes GUID that is independent of location into IP address depending on a location.

Group of nodes using SDP handles Server Discovery Protocol. These SDP groups are overlay network on Internet. There connect each other, and they exchange the conversion table to Personal Server IP address from GUID. Node that has participated in SDP Group can convert the GUID to that Personal Server IP address. SDP mechanism is designed based on pure P2P network like Gnutella[9].

However, it is not desirable on privacy protection that other unnecessary nodes hold the conversion table to IP address from GUID. Therefore, Conversion table has no GUID itself but one-way hashed ID such as MD5. Only node that knows right GUID can map certain IP address. Thereby, even if Personal Server IP address understands the middle node of SDP group, it can not know whose GUID it is. In addition, even if it connect the Personal Server IP address, it will be denied in the access control mechanism of Personal Server.

5   Experimentation with Some Hundreds of People

The experimentation based on Personal Server Model was conducted during the WIDE-camp in September 2002. WIDE-camp is research conference held at a Hotel where member of WIDE Project[10] gather and open BoFs to have discussions. 274 people participated in this experiment during 3 nights and 4 days.

In this experimentation, in order to create many Personal Servers, IPv6 was used. Next, we then describe about each part of system component detail.

Figure 6 shows the Map. Each room are labeled with a name like "BOF1". Location information of Participants is expressed using this labels. In this experiment, labeling was performed manually. The white circle in figure 6 shows the place in which RF-Reader where setup as mentioned below. Note that figure 6 will be used at section 6.

For the Location detection by LIN6 mentioned later, segments is divided for each and every room to show a prefix for each room. Furthermore, in this experiment, since the host for many Personal Servers was not able to be prepared, one host was attached two or more IPv6 addresses as Alias. Then, Personal Servers for all the members was prepared virtually.

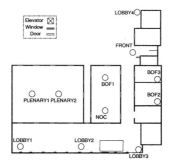

Fig. 6. Hotel Map

5.1   Sensors

We used two kinds of devices as Location Sensor. One is RF-Reader and the other is LIN6-MA. RF-Reader is human location sensor in Real-Space, and LIN6-MA can provide location information from network prefix.

RF-Reader. We use the RF-Code[12], which is a system for detecting a human location.

The cost of sensor construction is one of reasons for having chosen RF-Code. Although other systems, such as Active Bat, have very high accuracy, however, huge cost is required. In this experimentation, it is enough if personal location is known per room. Therefore, cost becomes important rather than accuracy.

Therefore, the Self-directed type sensor that needs a high cost device for an individual user is unsuitable. We selected RF-Code as an Environmental installation type sensor, and all users always carry a cheap and small tag. The tag is shown in Figure 7. The right tobacco is used to compare the size.

This RF-ID tag is a tag carrying its own battery, each with a Globally Unique ID, and sending the ID by using 300MHz radio frequency. We used this RF-ID GUID as GUID of Personal Server Model. Tag sends this GUID once every 3 seconds, and when RF-BaseStation catches this GUID, it detects that a user is in detection within a sensing area of that RF-BaseStation. Figure 8 shows RF-BaseStation.

Furthermore, we prepared three kinds of the antenna attached to RF-BaseStation. Since this sensitivity different, the antenna was properly used according to the size of the each room. Thereby, one RF-BaseStation area can cover mostly equivalent to one room. Moreover, two or more leaders are installed in the large room.

When two RF-BaseStation and the person enter between them, it becomes a most important problem when choosing its location.

Moreover, Hotel used at WIDE-camp had thin walls, and radio waves passed through them easily. Therefore, a RF-reader BaseStation will receive an ID rang-

Fig. 7. RF-Code Tag

Fig. 8. RF-BaseStation

ing over the room. Consequently, person may seem to have moved even though they didn't.

However, many BoFs and presentation are made at WIDE camp. In this circumstances, we can assume that once people goes into the room, they will not leave that place for a certain period of time.

Therefore, the position diagnosis algorithm in consideration of hysteresis was adopted. This algorithm can reduce miss detection of movement.

We used the Mobile Gear[11] for connecting network and RF-BaseStation. The Mobile Gear connects to network by using Wireless LAN; added to this, RF-BaseStation and Mobile Gear have a battery. RF-Reader set is completely movable by wire free.

Mobile Gear connect to RF-BaseStation by using RS-232C, and sends GUID sensed by RF-BaseStation via network to Personal Server of owner known by GUID. Mobile Gear sends the label of RF-Reader set additionally. According to this label, location information is determined.

Moreover, the Mobile Gear is operated by NetBSD/hpcmips. Reliability is increasing by using not WindowsCE but NetBSD. In addition, it is manageable from remote host.

LIN6-MA. In LIN6[13], unique ID of a meaning called LIN6ID is assigned to each host. After a host can assign network prefix, LIN6ID and network prefix will be compounded and an IPv6 address will be generated automatically. By registering this IPv6 address into Mapping Agent (MA), a host notifies a communication partner of in which network it exists, and realizes Location Independency.

If a user connects Note PC with the network, LIN6ID and network prefix will be compounded and an IPv6 address will be generated automatically . This IPv6 address is sent to MA with LIN6ID. MA, which received the IPv6 address, transmits an IPv6 address and LIN6ID to the process called NLX (Network Location eXchanger). NLX can know owner of the note PC by LIN6ID, and the network by network prefix. As shown below in subsection??, the network that has different prefix for every room in this experiment, it is possible to get to know the location of Note PC by network prefix.

In this experimentation, we used LIN6 as another location sensor. Every 16 LIN6ID was assigned every participant in order to support when a participant has two or more PCs,

5.2   User Interface

Two kinds of interfaces were prepared as an interface to Personal Server. One is interface for human, the other is used to develop.

Web Interface. We prepared the web interface to connecting Personal Server and obtaining location information. Users can get the following things through this web interface.

– Search location information
  By specifying a location, it is possible to get the list of users who are near that location. From this list, the situation of that location and how many participants are gathering can also be known via network.
– Search information from location
  By specifying a location, you can get list of users who exist near that location. From this list, the situation of that location and how many users are gathering can be known also via network.
– Tracking Note-PC
  We are able to find out the location where Note PC is put based on LIN6ID which described below.
– Acquisition of Personal Server IP address
  It is possible to perform SDP(Server Discovery Protocol) through a web interface. Thereby this interface, it is also possible to get Personal Server IP address and to obtain information from Personal Server directly.

Library. Moreover, we prepared library of the C language for accessing Personal Server so that it might be easy to develop the application which uses location information. This library contains the functions, which can get the IPv6 address of Personal Server using SDP, and connect to Personal Server and acquires PML. In addition, SDP command line program was also created.

6   Evaluation

In this section, we will describe about Evaluation of this system. First, we will start about accuracy of sensor. The three kinds of evaluation will be shown that are Human Location Tracking, Human behavior and General Movement. Next, we will discuss about Personal Server Model.

6.1   Human Location Tracking

At first, we compared tracked one person movement and actual movement. Figure 6.1 shows sensed location and actual location while 24 hours (actually, 18 hours).

6.2   Human Behavior Observation

We collected every user's location from each personal server in order to obtain statistics human behavior data. From this results, we select at two rooms it show the characteristic motion of people and these graphs is shown in Figure 10 and11. These figures, which mean how many people, are in the room for every time. As for Figure 10, BOF2, people are beginning to be detected from the 11:30 time by which BOF was started. Moreover, it is not detected at all midnight. However, in Lobby3 which is a passage, it is always detected except for early morning. In addition, all graphs has decline pattern. It is possible that people move all together at the beginning and end of Meeting. A behavior of people can be grasped from these figures.

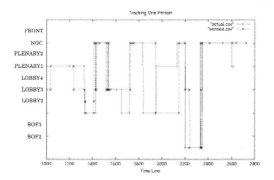

Fig. 9. Sesnsed Movement and Actural Movement

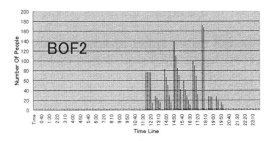

Fig. 10. Number of People of room "BOF2"

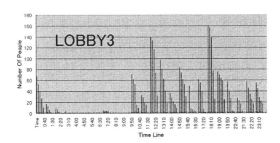

Fig. 11. Number of People of room "LOBBY3"

General Movement. Matrix of table 1 shows movement of people about "from where" and "to where" in one day. This table shows whether Location-Modeling could be performed by combining the map of figure 6. The still more efficient location detection technique can be used now like this experiment not by the short period of time but by applying in the long term.

Table 1. Movement Table

From \ To	BOF1	BOF2	BOF3	PLENARY1	PLENARY2	LOBBY1	LOBBY2	LOBBY3	LOBBY4	NOC	FRONT	Total
BOF1	-	24	0	14	20	11	28	54	0	18	6	175
BOF2	18	-	12	24	28	24	37	51	0	13	6	213
BOF3	1	13	-	1	1	1	5	9	0	0	7	38
PLENARY1	36	32	6	-	188	83	134	173	4	43	21	720
PLENARY2	20	29	2	160	-	47	115	157	2	43	23	598
LOBBY1	11	28	1	46	61	-	129	61	0	26	10	373
LOBBY2	23	39	4	76	117	133	-	102	2	34	14	544
LOBBY3	55	65	11	78	153	59	108	-	5	113	47	694
LOBBY4	1	0	0	0	5	3	1	3	-	0	1	14
NOC	18	19	0	22	41	24	33	104	0	-	5	266
FRONT	9	5	2	10	25	9	13	46	0	5	-	126
Total	192	254	38	431	639	394	603	760	13	295	141	3760

7   Conclusion

This research aims being connecting the person whom it did not connecting network to Internet.

We proposed Personal Server Model that uses an agent on the Internet as a communication node, named "Personal Server". Sensors sense personal information like location, and send to his Personal Server. Then Personal Server holds this personal information. If other Internet node wants to obtain personal information, connect to the Personal Server, and can get personal information.

According to this Personal Server Model, durability, low cost and privacy protection were performed. In addition, Internet applications, which use the personal information on Real-Space, can be developed by the same technique as usual Internet application.

7.1   Experimentation

In order to establish Personal Server Model, the actual proof experiment was conducted. In this experiment, RF-ID was used for Location-Sensing, 274 participants' movement was detected, and it held to each Personal Server.

Moreover, web Interface for using this location information and library for using from the C language were prepared. Thereby, application that uses personal location information can be developed easily.

7.2   Future Work

Present Personal Server has only a function, which is Access Control and holding information described by PML. However, Personal Server holds various kinds of personal information. Therefore, more convenient environment will be able to be offered if it becomes possible to extend the function of Personal Server easily if needed.

In addition, we use two kinds of sensor devices in this implementation. We cannot say heterogeneous sensor environment by using only two kinds of devices. Since that, we will develop more sensor devices and more applications.

Although especially Personal Server Model has focused to person, it is applicable also to the simple device that does not have Internet connection nature by the reasons of cost and so on. Therefore, it is necessary to extend the target range of this research as a Real-Space network will spread from now on. It will be necessary to specifically perform further extension of PML.

References

[1] K.Uehara, Y.Watanabe, H.Sunahara, O.Nakamura, J.Murai, "InternetCAR - Internet Connected Automobiles-", *Proc. of INET'98*, Jul 1998

[2] B. Schilit and N. Adams.'Context-aware computing applications *In Proceedings Workshop on Mobile Computing Systems and Applications. IEEE*, December 1994

[3] Mike Addlesee, Rupert Curwen, Steve Hodges, Joe Newman, Pete Steggles, Andy Ward, Andy Hopper. Implementing a Sentient Computing System. *IEEE Computer Magazine*, Vol.34, No. 8, August 2001, pp. 50-56

[4] N. Priyantha, A.Miu, H. Balakrishman, and S.Teller. The Cricket compass fo context-aware mobile applications. In *Proc. of the 7th Annual ACM/IEEE International Conference on Mobile Computing and Networking(MOBICOM 2000)*,pages 1-14, Rome,Italy,July 2001

[5] Andrew M. Ladd,Kostas E. Bekris,Algis Rudys,Guillaume Marceau,Lydia E. Kavraki and Dan S. Wallach,"Robotics-Based Location Sensing using Wireless Ethernet" In *Proc. of the 7th Annual ACM/IEEE International Conference on Mobile Computing and Networking(MOBICOM 2002)*

[6] Yasuhito Watanabe, Atsushi Shionozaki, Fumio Teraoka, Jun Murai, "The design and implementation of the geographical location information system.", Proc. of INET'96. Internet Society, June 1996

[7] AutoIDCenter, *http://www.autoidcenter.org*

[8] XML: eXtensible Markup Language `http://www.w3c.org`

[9] Gnutella, *http://gnutella.wego.com.*

[10] WIDE Project *http://www.wide.ad.jp*

[11] Mobile Gear MC/R550 NEC Corporation.
`http://www.nec.co.jp/press/en/9803/1101.html`

[12] RF-Code *http://www.rf-code.com*

[13] Mitsunobu Kunishi, Masahiro Ishiyama, Keisuke Uehara, Hiroaki Esaki, Fumio Teraoka, LIN6: A New Approach to Mobility Support in IPv6, *International Symposium on Wireless Personal Multimedia Communication*, 2000

Socket Level Implementation
of MCS Conferencing System in IPv6

Balan Sinniah[1], Gopinath Rao Sinniah[2], and Sureswaran Ramadass[3]

[1] KDU College Sdn.Bhd., A Paramount Corporation Company
32 Jalan Anson, 10400 Penang, Malaysia
sbalan@kdu.edu.my
http://www.kdupg.edu.my/
[2] Asian Institute of Medicine, Science and Technology (AIMST)
2 Persiaran Cempaka, Amanjaya, 08000 Sungai Petani, Kedah, Malaysia
gopi@nrg.cs.usm.my
http://www.aimst.edu.my/
[3] Network Research Group, School Of Computer Science
Universiti Sains Malaysia, 11800 Minden, Penang, Malaysia
sures@cs.usm.my
http://nrg.cs.usm.my/

Abstract. The technology of application development is currently shifting to use IPv6 [5, 6, 9] as a main protocol instead of IPv4. This includes multimedia streaming, which has been an important tool for communication, locally and internationally. The need for multimedia based application in IPv6 is further supported with the use of IPv6 in 3G implementation. Even though a lot of conferencing systems available for current use, there are very few that supports IPv6. This paper discusses the design for multimedia (audio and video) streaming in IPv6 using point to point and multicast technology. The main topic in this paper is socket level programming that has been defined in the API.

1 Introduction

IPv6 has been in the Internet for over 10 years and has been evolving since then. It was started with proposals for few standards that to be used by Internet community. Now it is the time for application to be implemented using the standards that has been identified and defined.

The trend is now moving towards having conferencing system for all purposes by all the sectors, i.e., for home and business users, government and non-governmental use. In simple word, conferencing system will be used daily as a mean of communication tool. The wide implementation of this conferencing system will be easier with the use of IPv6.

A conferencing system can use the features that have been defined in the IPv6 protocol. The flow label to control the flow of the packets will make the conferencing run smoothly. This is achieved by marking the packets as belonging to a stream of data that need special handling by the router [1]. Beside that,

H.-K. Kahng (Ed.): ICOIN 2003, LNCS 2662, pp. 460–469, 2003.
© Springer-Verlag Berlin Heidelberg 2003

the scope of the conferencing can also be defined with the use of IPv6 multicast header. In IPv6, the multicast is compulsory. This means the multicast packet can be routed in the Internet with the use of multicast algorithm that will be deployed. This is a big advantage over IPv4. The flow label and the traffic class fields in the IPv6 header are to define the quality of service [8, 10] needed by an application.

The main issue when designing a network application is the socket programming. It is important to use the correct header file in the programming. Beside that the appropriate functions are needed when developing a program in different operating system. It is important to tackle these entire problems before going deep into programming of a network application.

The objective of this paper is to outline the socket programming method used in developing a point to point and multicast conferencing system. This paper is divided into few sections. Section 2 discusses on the current conferencing systems. This includes MCSv4 that will be used in the implementation. In Section 3, the design and implementation of point to point multimedia streaming in IPv6 will be discussed. This is followed by IPv6 multicast streaming [5, 6]. The paper will then be concluded with the summary and future plans.

2   Conferencing Systems

There are a lot of conferencing systems available in IPv4 but only few of them supports IPv6. Three of the systems will be described in the following section.

2.1   VIC

The UCB/LBNL video tool, vic, is a real-time, multimedia application for video conferencing over the Internet. Vic was designed with a flexible and extensible architecture to support heterogeneous environments and configurations. For example, in high bandwidth settings, multi-megabit full-motion JPEG streams can be sourced using hardware assisted compression, while in low bandwidth environments like the Internet, aggressive low bit-rate coding can be carried out in software.

Vic is based on the Draft Internet Standard Real-time Transport Protocol (RTP) developed by the IETF Audio/Video Transport working group. RTP is an application-level protocol implemented entirely within vic. Although vic can be run point-to-point using standard unicast IP addresses, it is primarily intended as a multiparty conferencing application. To make use of the conferencing capabilities, your system must support IP Multicast, and ideally, your network should be connected to the IP Multicast Backbone (MBone). Vic also runs over RTIP, the experimental real-time networking protocols from U.C. Berkeley's Tenet group and over ATM using Fore's SPANS API.

Vic provides only the video portion of a multimedia conference; audio, whiteboard, and session control tools are implemented as separate applications.

The Intra-H.261 encoder combines the advantages of nv's block-based conditional replenishment scheme (i.e., robustness to loss) with those of H.261 (i.e., higher compression gain and compatibility with hardware codecs). For a fixed bit rate, the H.261 coder achieves frame rates typically 2-4 times that of the nv coding format.

Vic has several dithering algorithms for representing continuous-tone color video streams on color-mapped displays. The user can trade of run-time complexity with quality.

Video streams can be displayed simultaneously on a workstation display and on an external video output port, provided your hardware supports external output. This allows you to render a single stream to a full-sized NTSC/PAL analog video signal, which when viewed on an external monitor (or video projector) generally provides a more comfortable picture compared to video displayed in a small X window.

2.2   Intel Proshare

The ProShare Video, manufactured by Intel PC and Networking Products, is a desktop videoconferencing system. The system is based on the PC format and Windows must be installed

This component is made up of hardware and software that changes analog signals into compressed digital signals to transmit over digital networks, while simultaneously performing the reverse process from the distant codec. (Kind of similar to what modems do, except they modulate and demodulate digital signals and translate them to analog). The codec is responsible for digitizing, compressing and transmitting video and audio signals at speeds from 56 to 112 kilobits per second (kbps), constantly refreshing parts of the video transmission that change.

The full duplex audio feature incorporated into the main unit enables both sites to talk simultaneously without experiencing a 'clipping' effect. Additionally, it provides clarity and fidelity without echo, feedback or distortion.

The Proshare supports both H.261 and H.263 video coding. During the H.323 connections the system will negotiate the highest bitrate up to 400Kbps, whereas for audio, G.728 and G.711 are supported algorithm for ISDN calls. For IP conferencing, the user can connect at G.711 or the aH.323 algorithm, G.723.

2.3   MCSv4

MCS or Multimedia Conferencing System [2] developed by Network Research Group is one of the versatile multipoint video conferencing systems. MCS currently runs on IPv4. The thorough research through out the development of the MCS have result into a dynamic availability and compatibility of this system in any existing LAN or WAN.

There are 4 entities in MCSv4; client, server, mlic and compression.

1. **Client.** As MCS support distributed network entities, this entity would provide GUI at the users PC. It captures the video and audio packet and transmits it.
2. **Server.** The server entity maintains and controls the conference via RSW control criteria.
3. **MLIC.** MLIC or Multiple LAN IP Converter allows interconnected different LANs to join a conference. MLIC would convert multicast packets to unicast before it transmits these packets to another LAN. The other side of MLIC would convert the unicast packet back to multicast packet.
4. **Compression.** Its an entity which would enhance the performance of real time multimedia streaming over the LAN and WAN when there is a limitation in bandwidth availability. It does the compression of audio/video packets. There are few compression algorithms that can be used in the implementation such as VDO Wave, H.323 and MPEG4.

Bandwidth constraint is always an issue in any multimedia conferencing system across the different LANs. Multicasting would be an essential architecture of IPv6 which could be use to reduce this bandwidth problem. Although MCS is able to transmit multicast packets to a different LAN using a specific mechanism (MLIC), it is just going to be an additional overhead to the overall architecture of MCS.

Thus, MCS would be an ideal aiding system for research and development of IPv6 multicast multimedia streaming as performance and constraints are concerned. Additionally this would be a part of enhancement of the current MCS to support IPv6.

MCSv4 to MCSv6 Enhancement of MCSv4 [1] to support IPv6 is not an easy task to be accomplished within a day. Methodological approach such as programming aspects and techniques, and use of new version of Winsock APIs need to be considered before implementation.

Multicasting is our major enhancement to MCSv4 as this would eliminate the MLIC entity. MCS without MLIC would reduce the conversion workload from multicast packet to unicast and vice versa.

Even the basic notion of multicasting is similar between IPv4 and IPv6; few new features have been introduced based on result collected in IPv4 multicasting. Fixed address field explicitly limits the scope of multicast addresses in IPv6 where as in IPv4, the TTL (time to live) field specified the scope [12]. Thus, no straightforward approach for implementing IPv6 multicasting is possible. Original approach has to be introduced in implementing IPv6 multicasting.

Implementing IPv4 multicast can be done in very straight forward manner whereby it uses unicast addresses to identify a network interface. However, this approach is not applicable in IPv6 as IPv6 node maybe assigned with multiple addresses on a single interface which include link local address. Consequently it may not identify the address that need to be used. User has to specify the interface index as well as the address in such case.

The implementation of IPv6 multicasting in current MCS has to be done in stages. The initial stage of development should focus on point to point to multimedia streaming. The conversion of Winsock API [7] from older version to newer version should be done in a very careful manner as IPv6 supports only Winsock 2 [7] and above. New data structures would be used to hold IPv6 address as the address is 4 times larger than IPv4 address. This data structure must be large enough to hold the 128 bits IPv6 address. Conversion of string to network address has to be done using specific function which supports Windows 2000 or XP. Modification need to be done to video and audio capturing part as well. As IPv6 only supports Windows 2000 or Windows XP, the original video/audio capturing module that was developed in different operating system need to undergo modification.

Upon successful completion of point to point video streaming, this could be extended to implementing IPv6 multicasting in MCS. A direct enhancement from point to point streaming could force to have a fixed multicast address and not from the server as in the current version of MCS. All the clients would use this address as a multicast address to join in and starts the multimedia packet transmission. As mentioned earlier there would be successive steps in complete implementation of IPv6 multicasting in multimedia streaming.

3   IPv6 Point to Point Streaming

The development of point to point system was done in stages. It is divided into 2 main sections, audio and video. For each section, the transmission of the packets will be in two stages; first it is developed in simplex and later in full duplex method. The client is developed in Windows 2000 using Visual C++ version 6 as the programming language.

3.1   Audio

The design and implementation in the first phase is to stream an audio unicast packet from sender to receiver.

The client was further enhanced to send and receive the packets. This full duplex transmission is shown in the figure 2 below.

3.2   Video

The video transmission method is same as audio. First it is developed in simplex and later in full duplex method. Figure 3 below shows the full duplex video

Fig. 1. Simplex Audio Transmission. The client can either send or receive the audio at one time

Fig. 2. Full duplex audio transmission. The client able to send and receive the audio at the same time

Fig. 3. Full duplex video transmission

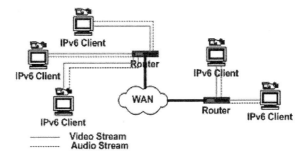

Fig. 4. The overall design of point to point conferencing system using IPv6

transmission. The video packets are captured and send in 16 bit true color. This will allow transmission of real life colors as it is captured. It is able to transmit 64K colors. This leads to high quality (sharp and vibrant) picture. Both in audio and video, the packets are transmitted as raw packets without being compressed.

3.3   Winsock Optimization

Implementation of an IPv6 application using socket in Windows platform is similar to Unix. Only Winsock 2.0 and above in Windows supports IPv6. In windows platform the standards for implementing IPv6 using Winsock is clearly defined in RFC2133 which was later superceded by RFC2533. For the implementation of IPv6 multimedia streaming using Visual C++, the use of WSASocket() as the predefined socket objects is compulsory. This is because CAsynSocket and CSocket do not support IPv6. Thus, the fragment of code below is necessary to initialize the socket version and trigger Winsock 2.0 [3, 4] [7, 11].

```
wVersionRequested = MAKEWORD( 2, 2 );
WSAStartup( wVersionRequested, &wsaData );
```

To create a WSASocket with the new address family name, the following fragment of code need to be used.

```
m_hSocket = WSASocket (AF_INET6, SOCK_DGRAM, IPPROTO_UDP, NULL, 0, 0);
```

The AF_INET6 definition distinguishes between the original **sockaddr_in** address data structure, and the new **sockaddr_in6** data structure.

A number of new socket options are defined for IPv6. All of these new options are at the **IPPROTO_IPV6** level. That is, the "level" parameter in the **setsockopt()** calls is **IPPROTO_IPV6** when using these options. The constant name prefix IPV6_ is used in all of the new socket options. This is to clearly identify these options as applying to IPv6.

The declaration for IPPROTO_IPV6, the new IPv6 socket options, and related constants defined in this section are obtained by including the header <ws2tcpip.h> and <tpipv6.h>.

```
if(setsockopt (m_hSocket, IPPROTO_IPV6, IPV6_UNICAST_HOPS, (char *) &hoplimit,
sizeof( hoplimit)) == SOCKET_ERROR)
```

As stated above, incompatibility of CSocket and CAsynSocket with IPv6 has lead to use of WSASocket. Continuous receiving of audio/video frames in a real time mode, need a receiving event which runs on thread together with adequate buffer to hold received data. Thus additional optimization has to be done to WSASocket to cater for this operation. This could be done as stated below.

```
if(WSAAsyncSelect (m_hSocket, mesg_ptr->m_hWnd, WM_SOCKET_EVENT, lEvent ) ==
SOCKET_ERROR)
```

The IPv6 address which is in the form of string has to be converted to network address before sending the audio/video packets. This conversion is slightly different in Windows platform with the Unix environment. We cannot use the function as shown below for converting the string to network address.

```
int inet_pton(int af, const char *src, void *dst);
```

So, in order to convert the string to network address in Windows, WSASocket need to be used. This can be done using the function below.

```
WSAStringToAddress    (sndBuf,AF_INET6,    NULL,    (LPSOCKADDR)&m_sendtoadd,
&add_size)
```

The best approach to ensure that the size of the data structures is properly defined is to use the **SOCKADDR_STORAGE** structure. This single code as mentioned below can be used to store either/both the IPv4 and IPv6 address.

```
SOCKADDR_STORAGE m_sendtoaddr
```

The audio and video frames can be transmitted using the function as declared below.

```
sendto (m_sndSocket, buffer, sendsize, 0, (LPSOCKADDR) &m_sendtoadd, add_size)
```

The 5th argument is a pointer to const sockaddr. So the IPv6 address has to be type cast with LPSOCKADDR in order to avoid unexpected results.

4   IPv6 Multicast Streaming

Unicast and broadcast are the main source of network traffic. In the context of unicast, a separate copy of the data is sent from the source to each client that requests it whereas in broadcasting a single copy of the data is sent to all clients on the network. Both of these method is going to waste the network bandwidth

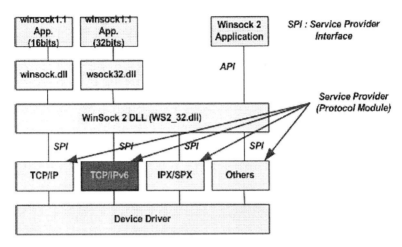

Fig. 5. Winsock 2 architecture that is used in developing the system

when the same data needs to be sent to only a portion of the clients on the network

Multicasting sends a single copy of the data to those clients who request it. Multiple copies of data and unintended recipient are avoided in multicasting. Multicasting allows the deployment of multimedia applications on the network while minimizing their demand for bandwidth.

4.1 Proposed Design

As for the initial stage of multicast streaming testing, a fixed multicast address will be used. Audio or video packet will be sent to this fixed multicast address. The IPv6 nodes that joined this multicast address will be able to receive the streamed (audio/video) data.

Implementing multicasting in IPv4 requires unicast addresses to identify a network interface. Usually a structure is used when adding and deleting an interface for multicast routing. One of this structures members specifies an IPv4 address which serves as an interface identifier.

This concept cannot be used in IPv6 as IPv6 might be assigning more than one addresses on a single interface, could lead to configuration mismatch. On the other hand, a link-local address is not MUST be unique within a node. This leads to complexity in identifying a single interface. So in IPv6, a user must specify the interface index and the address as well.

4.2 Winsock Optimization

Multicast socket in IPv6 can be created as in Unicast mode, WSASocket since the mode of transmission can be set thru one of the parameter in socket option.

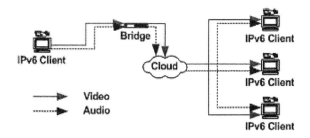

Fig. 6. The proposed multicast streaming in IPv6

IPv6 applications may send UDP multicast packets by simply specifying an IPv6 multicast address in the address argument of the sendto() function [3, 4, 7, 11].

```
sendto(m_sndSocket, buffer, sendsize, 0, (LPSOCKADDR) &m_sendtoadd, add_size)
```

m_sendtoadd would be a fixed multicast address (FF01:0:0:0:0:0:0:2) as this would be a testing stage. The socket options at the IPPROTO_IPV6 layer control some of the parameters for sending multicast packets. The setsockopt() options for controlling the sending of multicast packets are summarized below.

```
setsockopt(m_hSocket, IPPROTO_IPV6, IPV6_MULTICAST_IF (char *) &hoplimit,
sizeof(hoplimit))
```

IPV6_MULTICAST_IF is to set the interface index for sending out the multicast packets. The argument is a type of **unsigned int**. In order for an IPV6 node to receive these multicast packets, the node has to join this fixed multicast address. This can be done immediately after the WSASocket is created, as shown below.

```
struct ipv6_mreq imr6;
imr6.ipv6mr_interface = 0;
setsockopt(m_hSocket,IPPROTO_IPV6,IPV6_JOIN_GROUP,(char*) &imr6, sizeof(imr6));
```

This sends an ICMPv6 Multicast Listener Discovery (MLD) message to the multicast group.

The **ipv6mr_interface** should be either zero to choose the default multicast interface or the interface index of a particular interface if the host is multihomed. When an IPv6 node leaves the group, it calls setsockopt with **IPv6_LEAVE_GROUP** as shown below.

```
struct ipv6_mreq imr6;
setsockopt(s, IPPROTO_IPV6, IPV6_LEAVE_GROUP, (char *)&imr6, sizeof(imr6))
```

The **imr6** parameter contains the same values as used to add the membership.

5   Conclusion and Future Work

The socket level programming in IPv6 takes huge amount of time. This is mainly on troubleshooting the errors that are due to either header file of wrong functions being used. After the socket programming has been settled, the other functionalities of the conferencing system are adopted from MCSv4 with slight changes on the architecture to suit IPv6.

For future work, the conferencing system will be enhanced to support better compression for both the audio and video. This is important for transmission over low bandwidth. The multicasting that has been proposed will be tested in WAN with the use of available multicast algorithms.

The other important aspect that needs to be looked into is quality of service (QoS) [8, 10] . Flow label and traffic class field in the IPv6 header has been defined for this purpose [5]. It will be better if both this fields are used to increase the quality of the audio and video streams. The method of using these fields is still new, so some research must be started for this purpose. With the use of all this technology the conferencing system can also be adopted to use with 3G and other upcoming technology.

References

[1] Gopinath Rao, S., Ettikan Kandasamy, K., Sureswaran, R.: Migration Issues of MCSv4 to MCSv6. Proceedings Internet Workshop 2000, Tsukuba, Japan (14-18 February 2000).

[2] Mlabs Sdn. Bhd.: MCS Ver. 4.0 Technical White paper.
http://www.mlabsglobal.com/techdetail.html

[3] Stevens, W., Thomas, M.: Advanced Sockets API for IPv6. RFC2292, IETF (February 1998).

[4] Gilligan, R., Thomson, S., Bound, J., Stevens, W.: Basic Socket Interface Extensions for IPv6. RFC2553, IETF (March 1999).

[5] Deering, S., Hinden, R.: Internet Protocol, Version 6 (IPv6) Specifications, RFC2460, IETF (December 1998).

[6] Hinden, R.: IP Next Generation Overview. Communication of the ACM Vol. 39 (June 1996) page 61-71.

[7] Windows Socket 2, Protocol - Specific, Annex, Revision 2.0.3 (May 10, 1996)

[8] Schmid, S., Scott, A., David, H., Konrad, F.: QoS based Real Time Audio Streaming on IPv6 Networks, University, U. K., University of Ulm, Germany.

[9] Goncalves, M., Niles, K.: IPv6 Networks, McGraw Hill (1998).

[10] Martin, H.,: QoS, Multicast & IPv6: Technologies & Challenges,
http://www.ipv6forum.com/navbar/globalsummit/slides/html/martin.hall/tsld001.htm

[11] Crawford, M.: Transmission of IPv6 Packets over Ethernet Networks, RFC2464, IETF (December 1998).

[12] Tatuya Jinmei: Implementation and Deployment of IPv6 Multicasting, Toshiba Corporation, Japan

Part III

QOS in Internet

An Extended Multimedia Messaging Service Architecture for Efficiently Providing Streaming Services[*]

Hocheol Sung, Juhee Hong, and Sunyoung Han

Department of Computer Science and Engineering
Konkuk University 1 Hwayangdong, Kwangin-gu, Seoul, 143-701, Korea
{bullyboy,jhhong,syhan}@cclab.konkuk.ac.kr

Abstract. Streaming multimedia may contain much more information than other data and should be processed steadily and continuously. Existing Multimedia Messaging Service(MMS) environment does not consider these characteristics of multimedia stream. Therefore, this may carry a heavy burden on the whole network and MMS system and deteriorate the overall service quality. For this reason, 'an extended MMS architecture' is proposed. Two new elements, the MMS proxy and the MMS Proxy Selection Agent(PSA), are added for transferring multimedia stream efficiently. Consequently, the overall load of the entire MMS core network and MMS system can be reduced.

1 Introduction

Recently streaming technologies are becoming increasingly important with the growth of the Internet[7, 8]. With streaming, user's application can start playing the multimedia data before the entire file has been transmitted. Therefore, users do not need to download large multimedia files and can save time. For this reason, streaming service is suitable for wireless handset like PDA, which has only poor storage and low network bandwidth. Besides it is expected to core service in the wireless networks such as 3G and 4G networks[3, 4, 6].

For streaming service to work, a lot of multimedia data should be processed and transmitted continuously. This means that significant system and network resources are needed for delivering streamed multimedia. However, existing MMS environment has no regard for these characteristics of streaming multimedia[1, 2, 5]. When demands of many users occur in centralized existing MMS environment, MMS relay/server and a streaming server have overload to process a number of user's requests. More over it requires high network bandwidth to transmit continuously streaming data to all users. In conclusion, existing MMS environment cannot provide the optimal streaming service to users because of overload of MMS system and lowering of network efficiency.

---

[*] This work is supported by NCA(National Computerization Agency) of Korea for Next Generation Internet Infrastructure Development Project 2002.

H.-K. Kahng (Ed.): ICOIN 2003, LNCS 2662, pp. 473–482, 2003.

Therefore, a MMS proxy-based extended MMS architecture is proposed, which is for guarantee of efficient transmission of multimedia streaming. The MMS proxy extends the function of streaming data transmission over the basic function of the MMS relay. It retransmits the unicast or multicast streaming data to end user from streaming server. What is more, the MMS proxy has extra functions for multicasting service and a message caching. Clients should use a PSA to select the optimal MMS proxy. The PSA is located in each of user networks and performs the function that informs end users of the optimal MMS proxy among locally distributed MMS proxies. Consequently, the overload of the entire MMS core networks and MMS system can be reduced since requests of users and responded multimedia message would be processed through some distributed MMS proxies.

The rest of this paper is organized as follows. In section 2, a proposed MMS system architecture and operation is described. In section 3, detail functions and required protocols about each system component are described. In section 4, the result of numerical evaluation is presented. In section 5, we conclude this paper.

2   Architecture

The architecture proposed in this paper consists of three elements. Those are MMS Center(MMSC), the MMS proxy and the PSA. MMSC is a set of systems that compose MMS environments defined by 3GPP and WAP Forum [1, 5]. It includes the existing MMS elements, such as MMS relay/server, and contains an additional streaming server that transfers multimedia streaming [8].

The following Fig.1 shows the overall system architecture proposed in this paper.

2.1   MMS Proxy

MMS proxies are distributed over the MMS core network and literally play the role as a "proxy", which relays multimedia streaming to users. The MMS proxy provides two types of relay as follows.

Unicast Relay The MMS proxy requests and receives that message from MMSC and retransmits the message to the client and caches it into its local storage. If a client requests the same message, MMS proxy sends the cached message to user without requesting it from MMSC.

Multicast Relay The MMS proxy manages clients who request the service as a virtual group. If the MMS proxy is not a member of the requested multicast group, it joins that multicast group. After that, the MMS proxy relays the multimedia streaming delivered in multicast to all users in the virtual group in unicast.

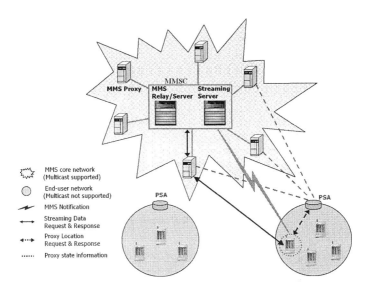

Fig. 1. An extended MMS system architecture

2.2 PSA

The PSA is located in each user's network and provides the function that informs the optimal MMS proxy to users. The PSA make a list and enrolls MMS proxies that notified their information in the list. In order to maintain information about the optimal MMS proxy, the PSA has to check the state of MMS proxies registered in the list and rebuild the list periodically. Also, the PSA broadcasts its own information on the network for all clients on that network to be indicated it. Because the optimal MMS proxy may be different according to physical location of networks, clients on a different network may receive multimedia streaming from a different MMS proxy.

3 Protocol Design

3.1 Basic Sequence

The sequence about the operation of whole MMS system proposed in this paper is described in Fig.2 and Fig.3.

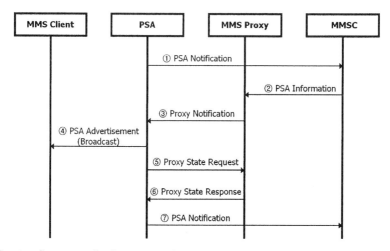

Fig. 2. Sequence for keeping information about the optimal MMS proxy

Sequence for PSA to Keep Information about the Optimal MMS Proxy

① When a PSA boots up, it sends a notification to MMSC and waits for notifications from MMS proxies for a while.
② MMSC that receives a notification from the PSA notifies all MMS proxies in MMS environment of information about the notified PSA.
③ The PSA creates a list in which MMS proxies from which it receives a notification within some time are registered.
④ The PSA broadcasts periodically its own information on the network.
⑤ The PSA requests periodically state information from all MMS proxies registered in the list.
⑥ The PSA updates the list according to a response form each MMS proxy.
⑦ The PSA may send a notification to MMSC to rebuild the list. Operations listed above shall be repeated.

Sequence for a MMS Client to Request and Receive Multimedia Streaming

① A MMS client gets a notification or an advertisement for multimedia streaming from MMSC.
② In order to receive multimedia streaming, The MMS client requests information about the optimal MMS proxy from the PSA.
③ The PSA responds with information about the optimal MMS proxy.

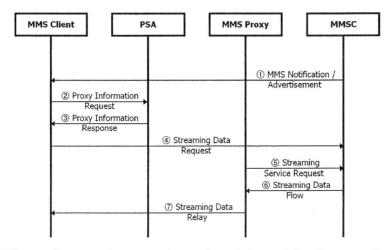

Fig. 3. Sequence for requesting and receiving multimedia streaming

Table 1. Information element in the PSA Notification message

$Information Elements$	$Descriptions$
Message Type	Identifies this message as PSA Notification.

④ The MMS client requests streaming data from the selected MMS proxy. If necessary, the MMS proxy may create a member list for multicast relay and join the multicast group.

⑤ The MMS proxy may request streaming data requested by the MMS client from MMSC. If the MMS proxy keeps the data in its cache, it may transmit directly the data.

⑥ The MMS proxy receives the requested streaming data from MMSC.

⑦ The MMS proxy transmits the data to MMS client(s) and may store the data in its cache.

3.2   Messages

For these operations among components, several messages are defined. Involved abstract messages are outlined as follows:

PSA Notification It is a message to inform MMSC that a PSA is booting up.

PSA Information MMSC that receives a PSA Notification message shall sends PSA Information message to MMS proxies in order to notify information on the PSA.

Table 2. Information elements in the PSA Information message

InformationElements	Descriptions
Message Type	Identifies this message as PSA Information.
PSA Information	The address of the PSA that sent a PSA Notification.

Table 3. Information element in the Proxy Notification message

InformationElements	Descriptions
Message Type	Identifies this message as Proxy Notification.

Table 4. Information element in the PSA Advertisement message

InformationElements	Descriptions
Message Type	Identifies this message as PSA Advertisements.

Table 5. Information elements in the Proxy State Request message

InformationElements	Descriptions
Message Type	Identifies this message as Proxy State Request.
Flag	If this flag is set, it indicates that the Load Test module must be run

Table 6. Information element in the Proxy State Response message

InformationElements	Descriptions
Message Type	Identifies this message as Proxy State Request.

Proxy Notification MMS proxies that receive a PSA Information message from MMSC shall send Proxy Notification message to the PSA in order to notify their own information.

PSA Advertisements A PSA broadcasts a PSA Advertisements message on its local network in order to notify all clients on the network of its own information.

Proxy State Request A PSA sends a Proxy State Request message to a MMS proxy in order to estimate RTT between the PSA and the MMS Proxy.

Proxy State Response A MMS Proxy that receives a Proxy State Request message from a PSA shall respond with a Proxy State Response message to the PSA.

Table 7. Information elements in the MMS Notification/Advertisement message

InformationElements		Descriptions
MM1_notification.REQ		
Flag	S	If this flag is set, it indicates that this message contains streaming data.
	B	If this flag is set, it indicates that it is broadcast data.

Table 8. Information element in the Proxy Information Request message

InformationElements	Descriptions
Message Type	Identifies this message as Proxy Information Requet.

Table 9. Information elements in the Proxy Information Response message

InformationElements	Descriptions
Message Type	Identifies this message as Proxy Information Response.
Proxy Information	The address of the optimal MMS proxy.

MMS Notification/Advertisement MMSC sends MMS Notification or MMS Advertisement message to clients in order to notify them that they have a new multimedia message. In this paper, the MM1_notification.REQ message is extended, which is used in MM1 interface defined by 3GPP. Two new fields are added to it.

Proxy Information Request A client sends a Proxy Information Request message to a PSA in order to know information on the optimal MMS Proxy.

Proxy Information Response A PSA that receives a Proxy Information Request message form a client shall responds with a Proxy Information Response message to the client in order to notify the client of information on the optimal MMS proxy

Streaming Data Request A client sends a Streaming Data Request message to a MMS proxy in order to request a streaming data to be delivered.

Streaming Service Request/Response Depending on the protocol used in MMSC for the streaming service, many types of a Streaming Service Request/Response message format may exist. Also, MMS proxies have to support the protocol. According to implementations, the MM1_retrieve.REQ and the MM1_retrieve.RES messages used in MM1 interface may be extended to use.

Table 10. Information elements in the Streaming Data Request message

InformationElements		Descriptions
Message Type		Identifies this message as Streaming Date Request.
Message Reference		URI for the multimedia message.
Flag	S	If this flag is set, it indicates that MMS proxy should send a request message to the streaming server in MMSC.
	B	If this flag is set, it indicates that MMS proxy should join the multicast group to receive the broadcasting data.
Retransmission Flag		If this flag is set, it indicates that MMS proxy should retransmit the first packet which contains the header of streaming data

Table 11. Information elements in the Streaming Data Relay message

InformationElements	Descriptions
Message Type	Identifies this message as Streaming Data Relay.
Message Reference	URI for the multimedia message.(only for unicast relay)
Sequence Number	It may be used by the MMS client to request retransmission the header of streaming data to the MMS proxy (only for unicast relay)
Data	The content of multimedia message. e.g., streaming data

Streaming Data Relay A MMS proxy sends a Streaming Data Relay message to clients in order to retransmit the streaming data received from MMSC.

4   Numerical Evaluation

Lets assume that K is a default load rate of a streaming server and M is a load rate added to the total load of a streaming server when a client requests to it. In case MMS proxies are not used, the total load rate of a streaming server is expressed as follows.

$$Load_{stream} = \frac{K}{(1 - M)}$$

In case MMS proxies are used, the total load rate of a streaming server is expressed as follows.

$$Load_{stream} = \frac{K}{\left(1 - \frac{M}{L}\right)}$$

where L means the number of clients per a MMS proxy. In case MMS proxies are used, the average cost value of a MMS proxy is expressed in the following way.

$$Load_{proxy} = \frac{\sum_{i=0}^{L} KM^i}{N}$$

where N is the average number of MMS proxy. However, the performance evaluated when the optimal MMS proxy is used may be different from that evaluated when the non-optimal MMS proxy is used because of network latency between a MMS proxy and a client. Assume that d_p is the average value of network latency between MMS proxies and clients when each client uses the non-optima MMS proxy. Also, assume that D_p is the least value of network latency between MMS proxies and clients when each client uses the optimal MMS proxy, and let Q be the number of clients. An inequality below can be expressed.

$$\alpha \times Load_{proxy} + \beta d_p \geq \alpha \times + \beta d_p - \beta D_p \left(\frac{N}{Q} \right)$$

where α is the weight of the load of a MMS proxy, β is the weight of network latency and the number of the MMS proxy is less than that of the client. The total load rate of a streaming server and the cost value of a MMS proxy are shown in Fig.4. This value is calculated with expressions above. Where we assume that the value of K, M and L are 10%, 0.02% and 10 respectively.

5   Conclusion

In this paper, an extended MMS architecture is proposed, which is for providing streaming service efficiently. MMS proxies distributed physically in MMS environment are the interface to both user's network and MMS environment. Therefore, using the optimal MMS proxy can reduce the burden of MMSC and networks and then provide the optimal service to users. In order to using the optimal MMS proxy, clients should request information about the optimal MMS proxy from a PSA on their network. Another advantage obtained by using MMS proxies is to enable to provide services between heterogeneous networks. In the proposed architecture, because MMS proxies can relay multicast data, clients on networks not supporting multicast service are able to receive data transmitted in multicast.

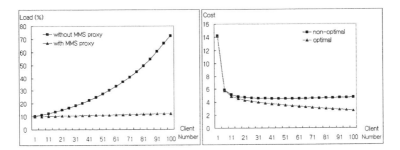

Fig. 4. The total load rate of a streaming server and the cost value of a MMS proxy

MMS proxies may perform various additional functions between user's networks and MMSE. For examples, MMS proxies may set up parameters for supporting QoS or authenticate users for security facilities between user's network and MMSE. The functional extension of a MMS proxy would be a topic of our future research.

References

[1] 3rd Generation Partnership Project.: Technical Specification Group Services and System Aspects; Service aspects; Stage 1 Multimedia Messaging Service (Release 2000), 3G TS 22.140 v.5.2.0 (2002-06)

[2] 3rd Generation Partnership Project.: Technical Specification Group Terminals; Multimedia Messaging Service (MMS); Functional description; Stage 2(Release 4), 3G TS 23.140 v.5.3.0 (2002-06)

[3] 3rd Generation Partnership Project 2.: Technical Specification Group Services and System Aspects; Multimedia Streaming Services Stage 1, 3GPP2 S.R0021 v2.0 (2002-4)

[4] 3rd Generation Partnership Project 2.: Technical Specification Group Services and System Aspects; Broadcast/Multicast Services Stage 1, 3GPP2 S.R0030 v1.0 (2001-7)

[5] Wireless Application Protocol Forum.: WAP MMS Architecture Overview, WAP-205-MmsArchOverview, Draft version 01-Jun-2000

[6] I. Elsen, F. Hartung, U. Horn, M. Kampmann, L. Peters. Streaming technology in 3G Mobile Communication Systems, IEEE Computer, pp. 46-52, September 2001

[7] Real Networks, http://realforum.real.com/cgi-bin/realforum/wwwthreads.pl

[8] MS Windows Media Home,
http://www.microsoft.com/windows/windowsmedia/default.asp

A Look-Ahead Scheduler to Provide Proportional Delay Differentiation in Wireless Network with a Multi-state Link

Yuan-Cheng Lai and Arthur Chang

Department of Information Management
National Taiwan University of Science and Technology
laiyc@cs.ntust.edu.tw, arthur@mail.sjsmit.edu.tw

Abstract. The Proportional Delay Differentiated Model was proposed to provide predictable and controllable queueing delay for different classes of connections. However, most of related works focused on providing this model in a wired network. This paper proposes a novel scheduler to provide the proportional delay differentiation in a wireless network, in which a multi-state channel exists. This scheduler, Look-ahead Waiting-Time Priority (LWTP), can achieve proportional delay differentiation and low queueing delay, by adapting with the location-dependent capacity of a wireless link and conquering the head-of-line blocking problem. The simulation results demonstrate that the LWTP scheduler actually achieves much closer to the desired delay proportion between classes and induces smaller queueing delays, compared with the past scheduler.

1 Introduction

Providing the guarantee of the diverse Quality of Services (QoS) in the network has been an emerging issue in recent years. Currently, the Differentiated Services (DiffServ) has evolved in two directions: absolute differentiated services and relative differentiated services. In terms of further taxonomy of service differentiation, the absolute differentiated services have two branches: premium services [1] and assured services [2], and the relative differentiated services have five models: strict prioritization, price differentiation, capacity differentiation, additive differentiation, and proportional differentiation [3]. Our paper focuses on the proportional differentiation model for it can provide controllable and predictable service differentiation.

In the proportional differentiation model, the performance metrics is often regarded as packet queueing delay or packet loss rate. Dovrolis et al. proposed the Waiting-Time Priority (WTP) scheduler to approximate the desired proportional delay differentiation under heavy traffic load [3]. Bodamer proposed the weighted earliest due date (WEDD) algorithm to provide tunable delay differentiation for applications with a mechanism that has not only different delay bounds but also different deadline violation probabilities [4].

H.-K. Kahng (Ed.): ICOIN 2003, LNCS 2662, pp. 483–492, 2003.

These studies do have remarkable contributions, however, only in the wired environment. Those approaches designed for a wired network are not directly applicable in a wireless environment due to some specific characteristics of a wireless link, namely, 1) high error rate and burst errors; 2) location-dependent and time-varying capacity; 3) scarce bandwidth [5]. These characteristics, if without being specially and carefully treated in designing a scheduler, often cause the HOL blocking problem and low channel utilization.

This paper proposes a novel scheduler, which can provide the proportional delay differentiation in a wireless network with a multi-state channel. Our proposed LWTP scheduler, modified from the WTP scheduler, tries to transmit packets to a mobile host which has a high-capacity channel and maintain the proportional delay differentiation simultaneously. Thus the LWTP scheduler can obtain the following characteristics: 1) to provide proportional delay differentiation; 2) to offer lower queueing delay; 3) to conquer head-of-line (HOL) blocking problem.

2   Background

2.1   Proportional Differentiation Model

The proportional differentiation model was proposed by Dovrolis et al. to provide relative quality spacing between classes [3]. This model has two objectives: to be controllable and to be predictable. To be controllable indicates that the network manager has a means to control and adjust the quality spacing for different classes, based on his/her criteria. To be predictable indicates that whatever the traffic load of classes is, traffic belonging to higher priority classes should always receive better performance (at least no worse) levels than that belonging to lower ones.

Considering N classes of services, each class has a dedicated queue. The proportional differentiation scheduler is responsible for scheduling packets from the N classes and providing differentiated services. Let $\overline{d_i}$ denote the average queueing delay of class-i packets. Assume class j is expected to have shorter average queueing delay than class i does, where $i < j$. The proportional delay differentiation model has the following constraint for any pair of classes:

$$\frac{\overline{d_i}}{\overline{d_j}} = \frac{\delta_i}{\delta_j} \quad (1 \le i, j \le N) \tag{1}$$

where $\delta_1 > \delta_2 > ... > \delta_N > 0$, and δ_i is the delay differentiation parameter (DDP) of class i.

2.2   WTP

The WTP scheduler was originally from Kleinrock's Time-Dependent-Priorities algorithm, a non-preemptive packet scheduling algorithm that provides a set of control variables to manipulate the instantaneous priority of any packet [6].

In WTP, the priority of a packet increases in proportion to its waiting time. Let $P_i^k(t)$ denote the priority of the k-th packet of the class i at time t, and $W_i^k(t)$ be its waiting time. Then, the priority is set as

$$P_i^k(t) = \frac{W_i^k(t)}{\delta_i} \quad (1 \leq i \leq N) \tag{2}$$

where $\delta_1 > \delta_2 > ... > \delta_N > 0$ is the DDP to control the priority increasing rate for certain class.

Among each class, the HOL (i.e., k=1) packet has the longest waiting time than any packet else behind it, so only the HOL packet of each class has to be considered for scheduling. Thus Eq. (3) is re-expressed as follows,

$$P_i(t) = \frac{W(H_i(t))}{\delta_i} \quad (1 \leq i \leq N) \tag{3}$$

where $W(H_i(t))$ is the waiting time of the HOL packet of class i at time t.

WTP is examined to successfully approach the targeted delay proportion under heavy load, but does not achieve it under light load. Dovrolis et al. later proposed a hybrid proportional delay (HPD) scheduler, which combines WTP and PAD, to provide proportional delay differentiation regardless of class load distribution [7]. Saito et al. proposed a local optimal proportional delay packet scheduler, which reaches an optimal decision when no further packets are arriving and induces more accurate approximations of delay ratio than WTP does [8]. Leung et al. proposed an adaptive control algorithm by adjusting some control parameters in response to the variable system load [9].

2.3   Problems of Scheduling over Wireless Network

Figure 1 illustrates the proportional differentiation model in a wireless network. In this model, all mobile hosts share one link. Because each host could be located at different locations, different capacities appear when this scheduler transmits data to different mobile hosts via the same wireless link. Also as the mobile host moves, the capacity to this host varies. Thus a wireless link has a location-dependent and time-varying capacity. For clear description, a logical channel (for simplicity, we call "channel" in this paper) can be regarded as a wireless link to a mobile host. Note that more than two channels are concurrently used is impossible because only one wireless link actually exists. Let $B_j(t)$ denote the channel capacity when the scheduler transmits packets to a mobile host j at time t.

In a TDMA network, a wireless channel has either full capacity when it is error-free or zero capacity when it is error-prone, that is, $B_j(t) = L$ or $B_j(t) = 0$, where L is full capacity. Thus an individual channel can be modeled by employing a two-state Markov chain, which including error-free and error-prone states. When any channel has zero capacity $(B_j(t) = 0)$, a HOL blocking problem may happen in this channel, especially when many input flows are aggregated into a small number of classes. HOL blocking is that more than two packets

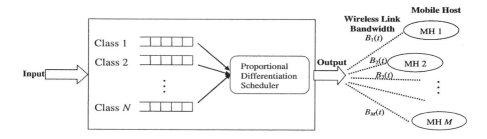

Fig. 1. The proportional differentiation model in a wireless network

are in the same queue and the HOL packet cannot be served because of its bad destined channel, then all packets behind this HOL packet in the same queue can not be served even their destined channels are good. So, subtle considerations need to be put into a new wireless approach, just like WWTP in later subsection.

However, a channel in a CDMA network contains multiple states other than just good or bad state, and each state has its own capacity (i.e., $B_j = \alpha_i L$, where $0 \leq \alpha_i \leq 1$). Thus a HOL blocking problem still happens in a CDMA network when a channel has zero capacity ($\alpha_i = 0$). Besides, a serious problem of low throughput should be considered. When a channel capacity is scarce, i.e. α_i is close yet not equal to zero, transmitting packets on this channel would take a long time, which results in a low throughput and a long packet queuing delay. Thus, a scheduler should note the channel condition to obtain the high throughput in this CDMA environment.

2.4   WWTP

Due to the current schedulers designed for wired communications are inapplicable for wireless networks, Jeong et al. modified the WTP scheduler and proposed a Wireless WTP (WWTP) scheduler to provide relative delay differentiation [10] in the wireless network in which wireless links have only two states (i.e. error-free or error-prone). They assume that the scheduler has full information of each link so it can avoid transmitting packets on an error-prone channel. WWTP serves a packet that has the highest priority and error-free destined channel at each scheduling time. The steps are taken as the followings.

1) Choose HOL packets of each non-empty queue initially.
2) Calculate the priority value, $P_i(t) = \frac{W(H_i(t))}{\delta_i}$, for each HOL packet.
3) Compare the calculated priority values of all chosen packets. Selects the packet from the class with the largest priority value, and then checks the destined channel state of the selected packet.
4) If the destined channel state is error-free, transmit the selected packet. Otherwise, exclude the selected packet from scheduling and choose the subsequent packet of the same class instead. Re-calculate the priority value of this new packet. Go to step 3.

WWTP can avoid the HOL blocking problem by excluding those packets suffering channel errors from transmission. Once those packets' destined channels recover from bad to good, those packets would get compensated, since their waiting times accumulate and the priority values increase during the channel error durations.

3   The LWTP Scheduler

Assume the scheduler can fully identify the channel status of every mobile host. Each channel has multiple states, not just good or bad state. Our proposed LWTP scheduler has the following goals in a multi-state wireless network: 1) to provide proportional delay differentiation, 2) to offer low queueing delay, and 3) to conquer the HOL blocking problem.

LWTP does scheduling in two phases: avoiding HOL blocking in phase I and then increasing throughput and keeping the proportional delay differentiation in phase II. This scheduler prefers transmitting packets on channels with better capacity to reduce the overall waiting time and increase throughput.

In phase I, LWTP finds out exact one candidate, rather than a HOL packet, in each non-empty queue. A packet will be selected as a candidate only when its destined wireless channel has any capacity except zero. Initially, candidate packets are the HOL packets from all non-empty queues. If any candidate is blocked, then this packet is substituted with the subsequent one of the same queue until a packet with non-zero destined channel capacity is encountered or the end of the queue is reached. Let $C_j(t)$ be the candidate of class i at time t. It may be NULL if all packets in the queue are blocked or in an empty queue.

In phase II, LWTP employs a pseudo-service technique, which virtually transmits the candidate packet in turn and evaluates the virtual new waiting times of all candidate packets after that has been transmitted. Let $W(C_j(t)), PS(C_j(t))$, and $B(C_j(t))$ denote the waiting time, packet size, and destined channel capacity of the candidate $C_j(t)$ at time t, respectively. When the pseudo-served packet belongs to class i, LWTP calculates the virtual normalized waiting time of class $V_j^i(t)$, and obtain the maximum proportion, $MP_i(t)$, as,

$$V_j^i(t) = \frac{W(C_j(t)) + X_i}{\delta_j}, \text{ where } X_i = \begin{cases} 0, \text{ if } j = i; \\ \frac{PS(C_i(t))}{B(C_i(t))}, \text{ if } j \neq i. \end{cases} \quad (4)$$

$$MP_i(t) = \max_{1 \leq i \leq N} \{V_j^i(t)\}. \quad (5)$$

X_i is the extra waiting time caused by transmitting the candidate packet of class i. For each class i, its corresponding $MP_i(t)$ is calculated. Then, the minimum value of all $MP_i(t)$, and the corresponding index are determined by,

$$MMP(t) = \min_{1 \leq i \leq N} \{MP_i(t)\}; \quad (6)$$

Table 1. An example for the LWTP algorithm

	$j=1$	$j=2$	$j=3$
Pseudo-service of class-1 packet (i=1)	0.5	0.733	0.417
Pseudo-service of class-2 packet (i=2)	0.667	0.4	0.292
Pseudo-service of class-3 packet (i=3)	0.833	0.567	0.25

$$S(t) = \underset{1 \leq i \leq N}{\operatorname{argmin}} \{MP_i(t)\}. \tag{7}$$

The LWTP scheduler chooses the candidate packet of class $S(t)$ and actually transmits it. When more than one $MP_i(t)$ equals $MMP(t)$, LWTP randomly selects one of them.

The fundamental concept of LWTP is to preview the influence brought by a packet selected for transmission. By adopting the pseudo-service of candidate packets, LWTP compares and reduces the difference between the normalized waiting times of all candidates, and then transmits the packets which yield the most accurate proportion. When pseudo-servicing a packet of class i, the obtained maximum proportion $MP_i(t)$ implies the most over-proportional value. A larger $MP_i(t)$ corresponds to a poorer proportion. Thus LWTP selects the packet with the minimum $MP_i(t)$, that is, the most correct proportion.

Consider an example of three classes ($N = 3$) and let the current queuing delays of three candidate packets be $W(C_1(t)) = 0.5$, $W(C_2(t)) = 0.8$, and $W(C_3(t)) = 1.0$ time units. Their packet size are $PS(C_1(t)) = PS(C_2(t)) = PS(C_3(t)) = 441$ bytes, and current destined channel capacity are $B(C_1(t)) = 1323$, $B(C_2(t)) = 2646$, and $B(C_3(t)) = 661.5$ bytes/sec. The DDPs for the three classes are $\delta_1=1$, $\delta_2=2$, and $\delta_3=4$. For WWTP, the priorities of the candidate packets at time t are $P_1(t) = 0.6$, $P_2(t) = 0.4$, and $P_3(t) = 0.25$; consequently, the WWTP algorithm delivers the candidate packet of class 1. By employing Eqs. (5) and (6), given in Table 1, the LWTP algorithm calculates $MP_1(t)=0.733$, $MP_2(t)=0.667$, and $MP_3(t)=0.833$, represented as underlined, and gives $MMP(t)=0.667$ and $S(t)=2$, marked by an oval. Thus LWTP selects the candidate packet of class 2 to transmit. The selections of the WWTP scheduler (class 1) and the LWTP scheduler (class 2) are clearly different.

In particular, the LWTP scheduler exhibits the following characteristics. Due to the space limitation, the detailed explanation can be found in [11]

- *No HOL blocking occurs.*
- *Delay proportion is maintained.*
- *Packet destined to the channel with higher capacity is preferred.*
- *Packet of the class with higher priority is preferred.*

4   Simulation Study

The simulation study evaluates the WWTP and LWTP packet schedulers in the context of the proportional delay differentiation model over a wireless link, which has a time-varying, location-dependent, and multi-state capacity. The simulations are conducted to investigate the effects of packet arrival rate, mean channel capacity, and coefficient of variation of channel capacity on the *delay ratio* and *delay improvement*, whose definitions are described later.

The model we simulate is depicted as in Fig. 1. There are three service classes ($N=3$). The DDPs are set as $\delta_1=1$, $\delta_2=2$, and $\delta_3=4$. Packet arrival follows a Poisson process and its mean arrival rate is $\lambda=0.9$ packets/sec. The packet size is fixed at 441 bytes for all classes. The buffer size for each class is infinite, i.e., no packet loss, and the full wireless channel capacity is 2646 bytes/sec. The wireless channel, which is modeled by a multi-state Markov chain, has five states with the capacity varying among 100%, 75%, 50%, 25%, and 0% of the full capacity. The Markov chain transition matrix of channel capacity is set as

$$
\begin{array}{c c c c c c}
 & 100\% & 75\% & 50\% & 25\% & 0\% \\
100\% & a & (1\text{-}a)p_1 & (1\text{-}a)p_1^2 & (1\text{-}a)p_1^3 & (1\text{-}a)p_1^4 \\
75\% & (1\text{-}a)p_2 & a & (1\text{-}a)p_2 & (1\text{-}a)p_2^2 & (1\text{-}a)p_2^3 \\
50\% & (1\text{-}a)p_3^2 & (1\text{-}a)p_3 & a & (1\text{-}a)p_3 & (1\text{-}a)p_3^2 \\
25\% & (1\text{-}a)p_4^3 & (1\text{-}a)p_4^2 & (1\text{-}a)p_4 & a & (1\text{-}a)p_4 \\
0\% & (1\text{-}a)p_5^4 & (1\text{-}a)p_5^3 & (1\text{-}a)p_5^2 & (1\text{-}a)p_5 & a
\end{array}
$$

The default value of a is 0.8, and the values of p_1, p_2, p_3, p_4, and p_5 can be calculated by letting the sum of each row equal to 1. In each simulation, at least 500,000 packets for each class are generated for the sake of stability.

The average queueing delay of class i, $\overline{d}_i$, is obtained by averaging the measured delays of all serviced class-i packets. The performance metrics used in this paper for comparing WWTP with LWTP are delay ratio and delay improvement. The delay ratio of class i over class j is defined as $\overline{d}_i/\overline{d}_j$ and the delay improvement of class i is defined as $(\overline{d}_i^W - \overline{d}_i^L)/\overline{D}_i^W$, where $\overline{d}_i^W$ and $\overline{d}_i^L$ are the average queueing delays of class i, made by WWTP and LWTP, respectively.

4.1   Packet Arrival Rate

Figure 2(a) shows that the delay ratios achieved by both WWTP and LWTP reach the target ratios when the arrival rate exceeds 1.0 packets/sec, which is heavy traffic load. At high packet arrival rate, the packet waiting time greatly exceeds the packet transmission time, and thus LWTP and WWTP make the decision quite similar to each other. Nevertheless, when traffic load is moderate, neither WWTP nor LWTP approaches the desired delay proportion. However, LWTP has delay ratios higher than WWTP has, since the former prefers the packets of higher-priority class (herein, class-1 packets) than does the latter. At low packet arrival rate, LWTP has few choices different from WWTP, thus both have similar behaviors. Figure 2(b) reveals that the queueing delay improved by

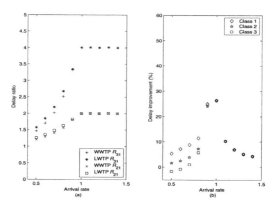

Fig. 2. The effect of packet arrival rates

LWTP increases as packet arrival rate increases, but to an extent, the improvement starts to decrease. The reason is the same as discussed above. Another fact to notice is that under moderate traffic arrival rate, the delay improvements of each class by LWTP are $class1 > class2 > class3$, since LWTP prefers packets of higher-priority class.

4.2   Mean Channel Capacity

The channel capacity is absolutely associated with the channel state. The mean channel capacity is varied, with the coefficient of variation fixed at 0.6375, to observe the behaviors of WWTP and LWTP for the achieved delay ratios and delay improvement. Figure 3(a) reveals that when channel capacity is small, both WWTP and LWTP achieve the target ratios. But as channel capacity increases, both schedulers' ratios slide down. The reason is that we fix the arrival rate for each class at 0.9 packets/sec, so the traffic becomes heavy-loaded when channel capacity is small, while moderate-loaded when channel capacity is large. Thus the consequence comes in the same way as the previous discussion of packet arrival rate. Still, LWTP has closer ratios to the target than WWTP. In Fig. 3(b), the queueing delay improved by LWTP increases as mean channel capacity increases, but to an extent, the improvement starts to decrease.

4.3   Coefficient of Variation of Channel Capacity

The coefficient of variation (CV) of channel capacity is varied, with the mean fixed at 1323 bytes/sec, to see the effect of CV on the achieved delay ratios and delay improvement. Figure 4(a) shows that when the CV is large, both WWTP and LWTP approximate the target ratios, but when the CV is small, both schedulers cannot maintain their desired ratios. Smaller CV represents more number of states near the mean, while larger CV represents more number of states far away from the mean. Figure 4(b) reveals that the queueing delay

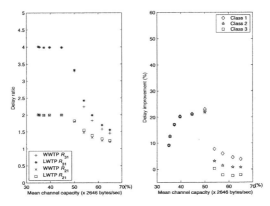

Fig. 3. The effect of different mean channel capacity (with the fixed coefficient of variation 0.6375)

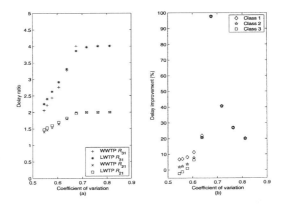

Fig. 4. The effect of coefficient of variation of channel capacity (with the fixed mean 1323 bytes/sec)

improved by LWTP increases as the CV of channel state increases, but to an extent, the improvement starts to decrease. When most of the channels are at the mean state, LWTP has few choices different from WWTP, so the improved delay ratio is small.

5   Conclusions

The characteristics of time-varying channel capacity and location-dependent error exhibited in wireless communications make packet scheduling be a challenge. We proposed the LWTP scheduler to provide the proportional delay differentiation for the wireless environment with the following goals: 1) to provide proportional delay differentiation; 2) to offer low queueing delay; and 3) to conquer the HOL blocking problem.

Simulations are conducted to investigate the effects of the traffic load, mean channel capacity, and coefficient of variation of channel capacity on the delay ratios and delay improvement. From the simulation results, LWTP is examined to deal with location-dependent channel capacity well and does provide more accurate (or at least no worse) proportional delay differentiation than WWTP does under different circumstances. Moreover, LWTP is able to reduce the average queueing delay, compared with WWTP, and thus provide better service performance.

Our future works include developing a wireless proportional scheduler to provide both delay and loss differentiation, and adapting the LWTP scheduler in a CDMA environment where a base station will transmit multiple packets concurrently.

References

[1] Nichols, K., Jacobson, V., Poduri, K.: Expedited Forwarding PHB Group, IETF RFC 2598, June 1999.
[2] Clark, D., Fang, W.: Explicit Allocation of Best Effort Packet Delivery Service, IEEE ACM Transactions on Networking, Vol. 6, pp.362-373, Aug. 1998.
[3] Dovrolis, C., Stiliadis, D., Ramanathan, P.: Proportional Differentiated Services: Delay Differentiation and Packet Scheduling, ACM SIGCOMM 1999.
[4] Bodamer, S.: A New Scheduling Mechanism to Provide Relative Differentiation for Real-Time IP Traffic, Proc. IEEE GLOBECOM 2000, pp. 646-650, Nov. 2000.
[5] Cao, Y., Li, V. O. K.: Scheduling Algorithms in Broad-Band Wireless Networks, IEEE Proceedings of The IEEE, Vol. 89, No.1, pp.76-87, Jan. 2001.
[6] Kleinrock, L.: Queueing Systems, Volume 2: Computer Applications, Wiley-Interscience, 1976.
[7] Dovrolis, C., Stiliadis, D., Ramanathan, P.: Proportional Differentiated Services: Delay Differentiation and Packet Scheduling, IEEE/ACM Transactions in Networking, Feb. 2002.
[8] Saito, H., Lukovszki, C., Moldovan, I.: Local Optimal Proportional Differentiation Scheduler for Relative Differentiated Services, Proc. IEEE ICCCN, Nov. 2000.
[9] Leung, M. K. H., Lui, J. C. S., Yau, D. K. Y.: Adaptive Proportional Delay Differentiated Services: Characterization and Performance Evaluation, IEEE/ACM Transactions on Networking, 2001.
[10] Jeong, M. R., Kakami K., Morikawa H., Aoyama T.: Wireless Scheduler Providing Relative Delay Differentiation, WPMC'00, pp.1067-1072, Nov. 2000.
[11] Lai, Y. C, Chang, A.: A Look-Ahead Scheduler to Provide Proportional Delay Differentiation in Wireless Network with a Multi-state Link, http://lai.cs.ntust.edu.tw/publication/lwtp.ps

A Practical QoS Network Management System Considering Load Balancing of Resources

Heeyeol Yu, Shirshanka Das, and Mario Gerla

Computer Science Department
University of California, Los Angeles
Los Angeles, CA 90095-1596

Abstract. This paper presents a load balancing scheme for Multipath QoS Routing in QMulator which was created by UCLA and utilized for QoS network. With this multipath routing, QMulator has showed that it was ready to address the varied characteristics of traffic and can assure a high degree of adherence to the quality of service demand as well as the load balancing in the sense of scattering packets. But it did not consider the whole network resources, such as link utilization or load balancing. The objective of this paper is to verifty the original benefits gotten from the multipath scheme and to use specific paths among several paths which are marked as lowest path utilization, the average of each link utilization in the path. The extensive experiments in QMulator showed that the proposed path utilization scheme gave the less variance values in each link where random traffics and flows were delivered and worked as load balacing metric in QMulator.

1 Introduction

Network systems for providing QoS guarantees may consist of various building components such as *resource assurance, service differentiation*, and *QoS routing*. Among them QoS routing, especially *multiple QoS constraints* is deemed to be the most fundamental groundwork for the other QoS components, and major researches about the aspect were started with [1, 2] which showed a possible use of the Bellman-Ford algorithm in the link state environment for QoS routing. These works clearly pointed that a set of multiple QoS constraints must be satisfied at the same time for QoS-sensitive applications, and they formed meaningful specifications for practical approaches [3]. Along with these foundations, [4] showed the practicality of the basic QoS routing algorithms by applying them to a network environment for IP Telephony. QoS routing issues were further accelerated by the development of the Multiprotocol Label Switching Protocol (MPLS) [5] which deploys fast packet-forwarding mechanisms in IP core networks.

Furthermore a new additional issue has been highlighted and the conventional QoS routing approaches are lacking for the issue; robust QoS guarantees in unreliable network environment [6, 7]. This is very important for mission-critical applications which require not only reasonable QoS guarantees but also transparent reliability from underlying network. For this purpose, recently a novel approach

H.-K. Kahng (Ed.): ICOIN 2003, LNCS 2662, pp. 493–503, 2003.

for reliable QoS support was introduced in [8], which effectively computes *multiple QoS paths* with minimally overlapped links. By provisioning multiple QoS paths, the network system can provide backup paths when one or more paths are detected corrupted. Besides, spreading network traffic over the provisioned multiple QoS paths in parallel has the network resources more evenly utilized. All these benefits of provisioning multiple QoS paths are sufficiently proved and demonstrated in [8, 10]. With this purpose, HPI Lab in UCLA has built a QoS network and showed that with multiple QoS path in the Qmulator[11, 12] gave the fault tolerance of the network and found that the multiple paths provides much greater degree of robustness.

However, the multiple QoS paths may require more network bandwidth resources than the single shortest path. So we must consider the network resource as a whole. In this respect, load balancing is an important aspect of Traffic Engineering. It can use some mechanisms to map a part of traffic in the overutilized routes to less underutilized routes to avoid congestion in a certain path, and to promote total network throughput and network resource utilization. In this paper, we present a load balancing scheme with the practical QoS system equipped not only with the conventional QoS routing but also with an effective QoS routing algorithm for multipath and fault tolerance.

In Section 2, the core QoS routing algorithms including a load balancing are briefly reviewed. Section 3 depicts the entire network system architecture and Section 4 presents experiment results obtained with the proposed QoS scheme and proves its and effective performance.

2   Multipath QoS Routing Algorithms

The proposed and implemented network system in the previous paper [12] has QoS routing mechanisms ready for QoS applications in both conventional and fault-tolerant ways. The mechanisms are associated with a certain Call Admission Control(CAC). When a QoS application comes in for its corresponding QoS services, it consults the underlying routing algorithm for feasible paths. If no feasible path is found, the connection is rejected and the application exits. Thus, the QoS routing path computation algorithm not only provides the capability of finding QoS paths but also plays an important role in CAC. Several papers have been presented to show the benefit of Multipath QoS routing. In [13], the dynamic routing algorithm for MPLS networks is proposed where the path for each request is selected to prevent the interface among paths for the future demands. It considers only single path routing for simplicity and does not include the constraint. [14] proposes an adaptive traffic assignment method to multiple apths with measurement information for load blalacning. However, how to find the appropriate multiple paths is not covered. [15] showed that resource-based static or dynamic load balancing algorithm only for single path which is better than the traditional shortest path aglorithm in MPLS. As previously mentioned and discussed in [8], we adopted the two different QoS routing algorithms in our system; the conventional single path algorithm and the newly introduced multiple path algorithm. In addition, among the resulting multiple paths we decide how many flows are to be assigned to each path. Then we compared the load balancing metric of each case in the practical QoS system, Qmulator.

2.1   Finding Multiple QoS Path

The originally proposed multiple QoS path algorithm[8] is a heuristic solution. We do not limit ourselves to strictly "path disjoint" solutions. Rather, the algorithm searches for multiple, maximally disjoint paths such that the failure of a link in any of the paths will still leave one or more of the other paths operational. All the previously generated paths are kept into account in the next path computation. The detailed descriptions of the multiple path algorithms are shown in [8] [9] as well as Alg. 2.1.

Algorithm 2.1: MULTIPLE PATH COMPUTATION(s, d, n)

s : source node	d : destination node
n : maximum number of paths	G : graph (V, E) for network
S : path set	r : path
P, T, T' : QoS descriptor set	$N(x)$: neighbor nodes of node x
$D_p(i)$: preceding node of i	

procedure COMPUTEPATH()
  $P \leftarrow \emptyset$ $T \leftarrow D(s)$ from G
  while $D(d) \notin P$ **and** $T \neq \emptyset$
  $\begin{cases} T' \leftarrow \emptyset \\ \textbf{for each } D(i) \in T \\ \quad \begin{cases} \textbf{if } f_C(D(i)) = \textbf{true} \textbf{ and } D(i) \notin P \\ \quad \textbf{then } T' \leftarrow T' + D(i) \\ \quad \textbf{else discard } D(i) \end{cases} \\ \textsc{NextHop}() \end{cases}$
  build r from P
  return (r)

procedure NEXTHOP()
  $T \leftarrow \emptyset$
  for each $D(i) \in T'$
  $\begin{cases} \textbf{for each } D(j) \in N(i) \text{ from } G \\ \quad \begin{cases} \textbf{if } j \neq D_p(i) \\ \quad \textbf{then } T \leftarrow T + F'(D(i), D(i, j)) \end{cases} \end{cases}$

main
  repeat
  $\begin{cases} r \leftarrow \textsc{ComputePath}() \\ \textbf{if } r \text{ is valid} \\ \quad \textbf{then } \begin{cases} S \leftarrow S + r \\ \text{remove all } E \in r \text{ from } G \end{cases} \end{cases}$
  until r is invalid **or** $|S| = n$
  return (S)

This multiple QoS path computation algorithm searches for maximally disjoint paths through the nodes satisfying given QoS constraints yet minimizing hop counts. This satisfies our requirement for fault tolerance with multiple QoS path provisioning.

2.2   Splitting Bandwidth among Paths

After finding N multiple paths through the previous step, application which requested the multiple QoS paths now asks the path utilization of each paths returned from OSPF Daemon (OSPFD). The definitions of path utilization and link utilization are $\sum_i$ *link util. on path*$/N$ and *Used BD/ BD of link*, respectively.

Based on this path utilzation metric, we can decide which path has the lowest path utilization, then a bundle of flows whose traffic demand is requested from the application are divided into N paths in quantum way and in proportion to the ratio of each path utilization as shown in Alg. 2.2

Algorithm 2.2: SPLITTING FLOWS(P, R)

P : Path set satisfied with QoS R : Ratio set among traffic flows
N : Num. of elements in P

main
 Measure link utilization and path utilization.
 Sort set P in the increasing order.
 Make the ratio set R of traffic division in decreasing order
     among the satisfied paths set P.
 for $i \leftarrow 1$ **to** N
   do $P_i \leftarrow R_i$

---

3   System Architecture for M-QoS w/ Load Balance

Qmulator consists of PCs running Linux, and all the QoS-capable features are embedded in the Linux kernel. Each of the machines has a few modules running on it, namely the *link emulator, metric collector, OSPF daemon, MPLS forwarding* and the *applications*, and the entire system architecture is depicted in Fig. 1.

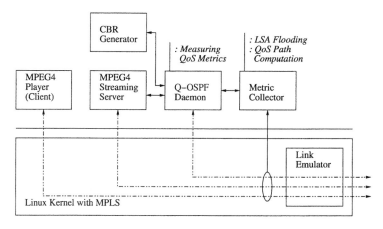

Fig. 1. The entire QoS system architecture

LS age		Options	LS type 10	Link state header
Opaque type		Opaque ID		
Advertising router				
LS sequence number				
LS checksum		Length		
Type = 2		Length = 32		Link TLV
Type = 1		Length = 1		Link type
All 0			P2P = 1	sub TLV
Type = 4		Length = 4		Local interface
Local interface address				IP address sub TLV
Type = 32768		Length = 4		Available bandwidth
Available bandwidth (bps) IEEE floating point				sub TLV
Type = 32769		Length = 4		Link delay
Delay (sec) IEEE floating point				sub TLV
Type = 32770		Length = 4		Link utilization
Utilization ratio IEEE floating point				sub TLV

Fig. 2. Opaque LSA format

Using this Qmulator several papers showed that the multiple QoS paths gave the good performance in terms of fault tolerance and load balancing [8, 11, 12]. In this paper we are going to present some specific parts needed for load balancing. The rest parts of Qmulator are the same as described in [11, 12]

To provide QoS, reserving bandwidth and bounding on delay for a connection, we require the knowledge of link characteristics at all times through *metric collector*. This information is needed by the QoS allocation module which looks at the current usage of the network and figures out if the new request can be satisfied or not. So the link metric collection module is an integral part of any routing scheme implementation providing the measurement-based QoS routing.

To propagate QoS metrics among all routers in the domain, we need to use Interior Gateway Protocol (IGP). OSPF is one of the major IGPs and significant researches have been recently made on OSPF with traffic engineering extensions. We selected the open source OSPF daemon (OSPFD) to implement our QoS routing scheme. [16] defines Opaque LSA for OSPF nodes to distribute user-specific information. Likewise, we extended our specific Opaque LSA entries by assigning new type value for link utilization in the following Opaque LSA format as described in Fig.2.

In addition to LSA flooding, OSPFD exchanges router LSAs to build a full network topology. Router LSAs originate at each router and contain information about all router links such as interface addresses and neighbor addresses. We bind the link metrics that we are interested in, viz. bandwidth and delay to the opaque LSA specified by the link interface address.

One of the key assumptions for Q-OSPF to be effective is the capability of setting up explicit paths for all packets of a stream to use. For this purpose

MPLS has given its efficient support of expplicit routing through the use of Label Switched Paths(LSP). We took advantage of these features which are implemeted in Qmulator[12] to evalute our new QoS scheme.

4 Experiments

We performed three kinds of experiments for this paper. The first one deals with the QoS routing scheme with the result of distribution of traffic among some parts of network resource to verify the original multipath is better in terms of our load balancing metric. The second experiment involves comparing the performance of load balancing with respect to CBR traffic in a simple triangle topology. The last set involves comparing the performance of load balancing with respect to CBR using multiple paths in whole QMulator. So we performed experiments in certain representative scenarios where multiple path with load balancing scheme is expected to give better results as well as in totally general scenarios.

In each network topology each unidirectional link is set to 1.5 or 1 Mpbs link capacity. Furthermore to measure the load balancing of the whole network, we use the link and path utilization which are previously defined and calculate the metric defined as $\sum_i \sigma$ *of link util.* for load balancing of whole network to evalute our load balancing scheme.

4.1   A Simple Topology w/ QoS and w/o LBQOS

The first set of experiments shows the QoS capability of the implemented system. Fig. 3 (a) shows the network topology for the first set of experiments. 4 nodes are connected directly to each other through 1.5 Mbps links. The traffic model which is injected into the test bed is a CBR (Constant Bit Rate) traffic with 1000-byte UDP packets. There are two extra connections functioning as interfering background traffic; 1000 Kbps connection from QoSR1 to QoSR4 and 500 Kbps connection from QoSR2 to QoSR3.

We introduce three new connections (800, 300, 500 Kbps) from QoSR1 to QoSR 3. Firstly, we examine the performance of the conventional IP routing which does not provide any QoS capability. In this case, packets of all the connections are routed over the shortest path (QoSR1, QoSR2, QoSR3). Thus, all three connections share the same path and cause network congestion on the link from QoSR2 to QoSR3, while other links are empty and unused. This is an inevitable condition due to the lack of the QoS capability in the current networks.

Fig. 3 (b) shows the throughput of the connections as a function of time when the conventional IP routing is used. It can be seen from the graph that the throughput are proportionally decreased after the connections are inserted and their traffic exceeds the link capacity between QoSR2 and QoSR3.

In contrast, the enabled QoS routing feature in the system shows the capability of routing the given connections over the paths with sufficient resources.

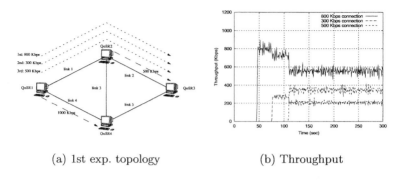

(a) 1st exp. topology   (b) Throughput

Fig. 3. Simple topology w/o QoS

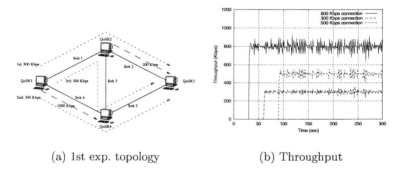

(a) 1st exp. topology   (b) Throughput

Fig. 4. Simple topology w/ QoS

Fig. 4 (a) depicts the path selections for the given connections with the QoS routing.

The path computation process in the extended OSPF daemon of the system computes three different paths as expected for the incoming connections. Fig. 4 (b) shows the throughput of the connections when the QoS routing feature becomes effective. All the connections completely meet their requirements by avoiding congestion. This experiment proves that appropriate paths are selected by the QoS routing capability with given QoS constraints. In addition, as you can see in the table 1 QoS routing gives better performance in terms of load balancing scheme.

4.2   A Simple Triangle Topology w/ QoS and LBQOS

Compared to the first, this 2nd experiment involved sending several random streams of CBR traffic from QoSR1 to QoSR3 with multiple QoS paths. During this experiment the three links from QoSR1 to QoSR2, QoSR2 to QoSR3, and

Table 1. The standard deviation of link utilization in each link

	w/ QOS	w/o QOS
Avg. Util. of Link 1	0.864	0.998
Avg. Util. of Link 2	0.866	0.980
Avg. Util. of Link 3	0.333	0
Avg. Util. of Link 4	0.865	0.652
Avg. Util. of Link 5	0.533	0
σ	$\sqrt{0.049}$	$\sqrt{0.203}$

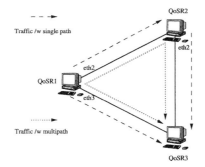

Fig. 5. The second experiment topology and corresponding network traffic

Table 2. The sum of standard deviation of each link

	LBQOS	NO-LBQOS
σ of QoSR1-eth2	0.185	0.198
σ of QoSR1-eth3	0.199	0.210
σ of QoSR2-eth2	0.176	0.196
Total sum	0.562	0.605

QoSR1 to QoSR3 are used for other traffics randomly as shown in Fig. 5. For those traffic only single QoS routing is used to fill up the some bandwidth of these links. While sending some traffics, we generated several traffics which can be grouped and separated using multiple QoS paths. With these traffics we used the load balancing scheme at first time. In the experiment we only considered two multiple QoS paths because that is the only possibility. After that with the same set of traffic, we didn't use the load blancing scheme. In each case we observed the link utilization of three links then calculated the average of the link utilization and the standard deviation.

The table 2 gives the result of the 2nd experiment for Load Balanced QoS(LBQOS) and NO-LBQOS cases proving that with LBQOS scheme the simple triangle topology give the less standard deviation values. The differenc is around 0.04 and this means that overall network resources are more safe in terms of load balancing.

4.3   Whole QMulator Topology w/ QoS and LBQOS

The experiment in this set was a totally general scenario, in which random connections were set up, and the traffic characteristics were measured. We first make Qmulator busy with random CBR traffic and then make dedicated CBR traffic from QoSR5 to QoSR6 in multiple path. Owing to performance bounds of the server and the CPUs of the nodes, we reduced the bandwidth emulated on each link to 1Mbps. The total number of connections generated in CBR case was 25 x 9 = 225. These connection were generated using randomly chosen source and destination from the 9 QOS routers. And the duration and turnaround of experiment was 5 minutes. During this experiment the traffic characteristics measured are the usual throughput profile on the links.

The Fig. 7 shows the overall summation of the standard deviation of each link in Qmulator. In this case the number of multiple QoS path is 2 and for this case 20,30,40,60,70,80, and 90 percentage of traffic demand is divided and assigned to the lowest of these path utilization according to the retuned measure from OSPFD. As you can see in the Fig. 7, as the portion increases passing by 50 the summation value for LBQOS is less than that for NO-LBQOS. In case of 50, these two values are back and forth, so we didn't consider. But as the portion goes from 40 to 60, LBQOS wins over NO-LBQOS.

5   Conclusion and Future Work

This paper provides an insight into the interaction of a measurement based QoS algorithm and a load balancing scheme in multiple QoS paths. We present our results based on an evaluation over a real network set up in the form of a testbed, Qmulator, where MPLS integrated into it with our QoS scheme. The first experiment gives the basic proof that QoS routing is better in terms of LBQOS. The second experiment proves that using the lowest path while sending traffic demand gives the smaller value of standard deviation. The third experiment shows that in general scenario of Qmulator load balancing scheme gives less total value of standard deviation as the portion for dividing increases.

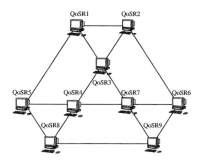

Fig. 6. The third experiment topology

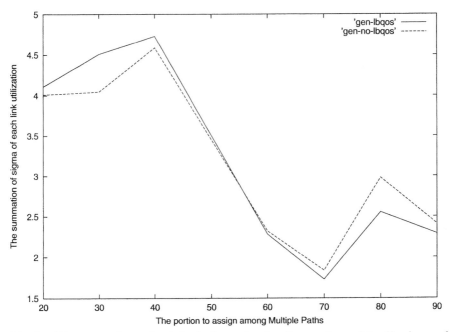

Fig. 7. The sum of standard deviation value by the portion of traffic demand

It is evident from the results that our measurement based load balancing QoS scheme gives a good performance. The emulation environment described in the paper is in a very active stage, with new changes happening every week. The current goals include setting up mechanisms for label assignment, distribution and signalling, so that applications can transparently use the network. To implement our load balancing QoS scheme, each ingress LSP must keep the status of division information of each bundle of traffic flows in granularity way. So it may cause a scalability problem to cope with for an ingress and egress LSP.

References

[1] R. Guerin, A. Orda, and D. Williams, "QoS routing mechanisms and OSPF extensions," in *Proc. of Global Internet (Globecom)*, Phoenix, Arizona, Nov. 1997.

[2] Dirceu Cavendish and Mario Gerla, "Internet QoS Routing using the Bellman-Ford Algorithm," in *IFIP Conference on High Performance Networking*, 1998.

[3] G. Apostolopoulos, S. Kama, D. Williams, R. Guerin, A. Orda, and T. Przygienda, "QoS routing mechanisms and OSPF extensions," Request for Comments 2676, Internet Engineering Task Force, Aug. 1999.

[4] Alex Dubrovsky, Mario Gerla, Scott Seongwook Lee, and Dirceu Cavendish, "Internet QoS Routing with IP Telephony and TCP Traffic," in *Proc. of ICC*, June 2000.

[5] E. Rosen, A. Viswanathan, and R. Callon, "Multiprotocol label switching architecture," Request for Comments 3031, Internet Engineering Task Force, Jan. 2001.

[6] Shigang Chen and Klara Nahrstedt, "An overview of quality of service routing for next-generation high-speed networks: Problems and solutions," vol. 12, no. 6, pp. 64–79, Nov. 1998.

[7] Henning Schulzrinne, "Keynote: Quality of service - 20 years old and ready to get a job?," *Lecture Notes in Computer Science*, vol. 2092, pp. 1, June 2001, International Workshop on Quality of Service (IWQoS).

[8] Scott Seongwook Lee and Mario Gerla, "Fault tolerance and load balancing in QoS provisioning with multiple MPLS paths," *Lecture Notes in Computer Science*, vol. 2092, pp. 155–, 2001, International Workshop on Quality of Service (IWQoS).

[9] S. S. Lee, S. Das, H. Yu, K. Yamada, G. Pau, and M. Gerla "Practical QoS Network System with Fault Tolerance," *Computer Communications Journal*, to be published in May 2003.

[10] Scott Seongwook Lee and Giovanni Pau, "Hierarchical approach for low cost and fast qos provisioning," in *Proc. of IEEE Global Communications Conference (GLOBECOM)*, Nov. 2001.

[11] S. Das, K. Yamada, H. Yu, S. Lee, and M. Gerla, "A QoS Network Management System for Robust and Reliable Multimedia Services," in *IFIP/IEEE International Conference on Management of Multimedia Networks and Services*, 2002.

[12] S. S. Lee G. Pau K. Yamada S. Das, M. Gerla and H. Yu, "Practical QoS Network System with Fault Tolerance," in *2002 International Symposium on Performance Evaluation of Computer and Telecommunication Systems*, 2002.

[13] M. Kodialam and T. V. Lakshman, "Minimum interference routing with applications to mpls traffic engineering," in *INFOCOM*. ACM, 2000, also in "Computer Communication Review" 20 (4), Oct. 1990.

[14] A. Elwalid, Low C. Jin, and I. Widjaja, "Mate: Mpls adaptive traffic engineering," in *INFOCOM*, 2001.

[15] Zhongshan Zhang Keping Long and Shiduan Cheng, "Load balancing algorithm in mpls traffic engineering," in *IEEE workshop on High Performance Switching and Routing 2001*, 2001, pp. 175–179.

[16] R. Coltun, "The OSPF Opaque LSA Option," RFC2370

A QoS-Aware Framework for Resource Configuration and Reservation in Ubiquitous Computing Environments

Wonjun Lee[1,*] and Bikash Sabata[2,**]

[1] Department of Computer Science and Engineering, Korea University
Seoul, Korea
[2] Department of Computer Science, Stanford University
Stanford, CA, USA

Abstract. In this paper we present a new admission control algorithm that exploits the degradability property of applications to improve the performance of the system in pervasive networking environments. The algorithm is based on setting aside a portion of the resources as reserves and managing it intelligently so that the total utility of the system is maximized. We also applied our Soft-QoS framework to the admission controlling and resource scheduling for ubiquitous multimedia devices, such as PDA or VOD servers, where multimedia applications can generally tolerate certain variations on QoS parameters. It exploits the findings about human tolerance to degradation in quality of multimedia streams.

Keywords: Quality of Service, Pervasive and Ubiquitous Computing, Admission Control.

1 Introduction

This paper presents a protocol that allows such integrated admission control and negotiation with graceful adaptation. To start with, we assume that we have a single resource where admission control is being performed. We will extend this to multiple resources next. We use the constructs of resource demand function and benefit functions developed in [8]. These constructs simplifies the mechanisms for graceful adaptation of the applications. An application running at a degraded QoS uses a lower amount of resources. The constructs of benefit functions and the resource demand functions tell the system the cost of such a tradeoff for the application [5]. This paper specifically deals with the policies the system adopts when a new application arrives. We refer to this as the application admission and the negotiation process. The system answers the questions

---

* Corresponding author. This work was supported by grant No. R01-2002-000-00141-0 from the Basic Research Program of the Korea Science & Engineering Foundation.

** Early version of this work was in part funded by DARPA through the SPAWARSYSCEN under Contract Number N66001-97-C-8525.

H.-K. Kahng (Ed.): ICOIN 2003, LNCS 2662, pp. 504–514, 2003.

"can the application be admitted?" and if admitted "at what QoS level?" We also address the mechanisms the system invokes when an application is completed and hands back the resource to the system. We refer to this as the resource reclamation and redistribution process.

The paper first presents the admission control and negotiation process. Section 3 describes the two applications used for experimental evaluation. In section ??, we present the experimental evaluations of our admission control algorithm on video distribution application and on disk access, followed by the details of the relevant previous work in admission control and negotiation in section 5. Finally we conclude in section 6. The mathematical definitions and details about the resource demand functions and benefit functions are not described in this paper for brevity. summarized

2   Threshold Negotiation: Admission Control and QoS Adaptation

When there is a shortage of resources the admission control process can still service some applications that accept degraded services. The admission process can negotiate down the resource allocation to the application. To enable the process of admission and negotiation, applications arrive with a request for a certain amount of resources. In our formulation of the problem this information is provided to the system via the **resource demand function**. The function presents the resource demand for the different QoS settings of the application.

In addition to the resource demand function, all the applications specify a **benefit function** that describes the relative benefit between the different QoS parameter settings for the application. This function is useful to tradeoff between the QoS parameters when there are constraints on the resources [?]. The goal of the admission control process is to maximize the total benefit over all the applications. The total system benefit (B_{total}) can be defined as a weighted sum of the benefit of each application. The term *benefit* (or *utility*) may take on the meaning of users' satisfaction or of pleasure, depending on the context. For example, the function of *CPU utilization* in a queuing system can be a benefit function to be maximized. In another case, *frame loss probability* or *total waiting time* can be benefit functions which should be minimized. The details about the resource demand funciton and benefit function are described in Appendix.

As described in section 5, there have been several admission control algorithms in the past. Much of the work in soft-real time systems such as continuous media servers, has focused on either conservative approaches based on bandwidth peaks or statistical methods which model arrival rates and stream bandwidths with probability distributions and determine a satisfactory level of performance by the disk and network subsystems, either separately or as a system [7]. Most of the existing admission control approaches are purely greedy strategy in the sense that a new application can be accepted only if the server could give the client all the requested resources. We are proposing a new admission control scheme which is a mixture of greedy and non-greedy strategy. We start with a brief description

Table 1. Attributes used in Algorithm

Attributes	Descriptions
s_i	stream (application) i
r_i	resource demand of s_i
R_{total}	total available resources initially given
$R_{reserve}$	amount of resources assigned to the reserve
R_{free_res}	available resources in Reserve : $$\begin{cases} R_{total} - R_{used} \text{ if } R_{used} > R_{total} - R_{reserve} \\ 0 \qquad\qquad \text{otherwise} \end{cases}$$
$R_{free\_non\_res}$	available resources outside Reserve : $$\begin{cases} R_{total} - R_{reserve} - R_{used} \text{ if } R_{used} \leq R_{total} - R_{reserve} \\ 0 \qquad\qquad\qquad\quad \text{otherwise} \end{cases}$$
R_{alloc}	allocated resources to the requesting stream
R_{avail}	available resources to assign : $(= R_{total} - R_{used})$
R_{used}	total resources currently used : $\sum_i v_i$, where $s_i \in \mathcal{RS}$
$\Psi(R_{free\_res})$	heuristic function for reserve assignment in *Admission Test* : e.g. $\frac{R_{free\_res}}{k}$, where $k = k_1^{(R_{used} - R_{reserve}) \cdot k_2}$
$congestBit$	bit flag to indicate congested/uncongested state : $$\begin{cases} 1 \text{ if } R_{used} > R_{total} - R_{reserve} \\ 0 \text{ otherwise} \end{cases}$$

about the typical greedy approach and then discuss the mixture strategy which is the focus of the paper. Next, we will illustrate the two different applications of video distribution and disk access. In the following sections, we will describe experimental evaluations of the performance of our approach using the examples explained in section 3. Table 1 describes the attributes used in the algorithms.

The natural and naive approach to admission control is to use a greedy strategy where applications are admitted as long as there are resources [4]. On application arrival the policy admits the application if the resources available is greater than the resources requested. On application departure the policy just adds the released resources from the application to the resources available.

The key advantage of the greedy strategy is that it is simple. There is no need to estimate the relative benefit of the applications and the computations are fast. Usually the difficult part is the problem of estimating the resource utilization of the application. Since in long-lived applications this varies with time, their goal is to find the best strategy so that the application can run continuously and the total resource utilization is high.

In some sense admission control tries to make an estimate of what the future holds and makes the current decision by trading off future gains for current gains. The greedy strategy ignores the future and accepts what ever comes in first. However, the greedy strategy may cause an important application that arrives a few moments later than a low-priority application to be rejected (we are assuming that there is no pre-emption).

Reserve-Based Strategy: The main idea of our approach is that we assign a portion of the resources as reserved. This reserve is specified as a threshold on the resource utilization (e.g., above 80% utilization). We say that a resource is congested if the resource is opening in its reserve. The admission policy AD-MIT_POLICY, is the protocol that is invoked when an application arrives and requests for resources. The ADMIT_POLICY changes in the mixed strategy when the resources operate in the reserves. This is different from the greedy strategy where the policy is fixed viz. "admit if sufficient are resources available." The admission control process starts with R_{total} amount of initially available resources, and $R_{reserve}$ is the amount of resources assigned to the reserve. At any instance of the run-time applications have been allocated certain fraction of the resources, this is represented by R_{used}. R_{free_res} indicates available resources in the reserve. A new application arrives with a request for r_i resources and R_{alloc} is resource that is actually allocated to the application. Note that these were the same in Greedy policy. The middle-ware admits the applications and assigns resources to the application using the ADMIT_POLICY. When the resource is not congested then the ADMIT_POLICY uses the greedy strategy and the application agent just uses the maximum QoS values for the application as the QoS assignment When the resource is congested, the ADMIT_POLICY changes to a more conservative policy where smaller and smaller amounts of the resources are allocated to the applications. We use the function $\Psi(R_{free\_res})$ to represent the different policies for reserve assignments. For example, one of the implemented policies is that when the resource is congested, we do not assign more than half of the remaining resources to an incoming application. After the admission control process allocates a certain amount of resources to the application, the application agents compute how best to use the resources for maximizing the application benefit. If the amount of resources assigned to the application is less than the amount originally requested by the application then the application has to run with a degraded set of QoS parameters. The application agents tradeoff between the different QoS dimensions of the application and compute the QoS assignment for the application. This tradeoff is computed using the benefit function and the resource demand function of the application (*Get_degraded_QoS_level*(.). The ADMIT_POLICY holds the key to how the current application benefit is traded off with the future benefit that the resource can generate for potentially more important applications. The policy should be simple for the purpose of making a real-time decision for admission. There are a few admission policies that we have tried: **Fixed Policy, Linear Variable Policy**, and **Exponential Variable Policy**.

Dynamic Threshold Adaptation: When applications depart the application agents return the resources back to the system. These resources are reclaimed by the system and redistributed to the applications that were admitted with degraded quality. There are different strategies for the redistribution (implemented by *redistribute_resources*: **First-come-first-served, Prioritized queue,** and **Maximization of the benefit** policy).

3   Streaming Applications in Pervasive and Ubiquitous Computing Environments

In this section, we will describe two multimedia applications which can be considered for our simulation studies.

For a video distribution application, we use the application benefit functions generated for real video distribution applications [?]. The video applications with soft-QoS requirements arrive at the system in an arbitrary order. The applications also come with a *resource demand* that is a function of the QoS parameters. For example we may have a set of video distribution applications that provide streaming video with different types of content to the remote users. Each application has a different *benefit function* and depending on the user profile and the data content, the different applications are ranked by relative importance of the applications.

As for the second application, we consider a disk access problem in video server system as another application. For the admission control process, we check the following two constraints: (1) disk I/O bandwidth constraint, and (2) available buffer constraint. Each continuous media request (e.g., video stream) arrives with a certain amount of *rate* value (e.g., play-back rate: frames per second). Every time either (1) a new application request arrives; (2) a rate control operation (such as *Set_Rate, Pause, Resume,* and *Fast_Forward*) is received; (3) the playing of a running stream is over; or (4) when the resources in the reserves are not used for some time, the following two constraints would be checked for admission control requirements. The admission controller in continuous media server determines whether or not to admit the stream. The details of the two constraints with respect to admission controlling are not described in detail for brevity.

For the *resource reclamation (adaptation) policy* in our reserves-based admission control algorithm (RAC), we consider the following two heuristics which are based on *done_ratio* (or *remaining_ratio*).

- *Dynamic Reserve Adaptation (DRA)* : this is invoked only if there are applications that need resources and do not have them. On departure of streams, DRA returns the assigned resources to the leaving streams (q_i) back to the available resource pool and re-calculate the new R_{avail} using the total available resource (i.e. old $R_{avail} + q_i$). Among the requests that are not fully serviced yet, we select a request (s_{min}) from the queue (DS), whose *done_ratio* is the smallest first, and assign the proper resource to the request. The **maximum** $\Phi(R_{free\_res})$**-rule** applies here.

  Assign $R_{alloc} = \min (\Phi(R_{free\_res}), \text{remaining\_rate}_{min})$ to the selected request (s_{min}).
  The loop continues until there is no more available resource to assign or R_{alloc} is too small to assign.

- *Reclamation within Returned Reserve (RWR)* : this is a similar method to Dynamic Reserve Adaptation method, but the difference is that it redistributes the only resources returned from the leaving streams (v_i). In $\mathcal{DRA}$, we used R_{avail} (instead of v_i) for reclamation. R_{alloc} per stream is calculated according to the ratio of *remaining_ratio* to *sum_remaining_ratio*.

$$R_{alloc} = R_{avail} * (remaining\_ratio_k / sum\_remaining\_ratio)$$

The key idea here is not to touch the reserves, but to utilize the only returned resource due to departure of streams.

We will validate the performance of the proposed policy in terms of video streaming distribution applications and disk access application in pervasive ubiquitous multimedia environments.

4 Experimental Evaluation of Disk Access

Here we present the performance results obtained from simulations for admission control based on the disk I/O bandwidth constraint under the various load conditions.

In disk access simulation, QoS parameter is *frame rate of playout (f_i)* and *resource demand* (returned value of $R(.)$) is *disk bandwidth*. That is, resource demand function is described as $R(f_i)$. Though various Benefit Function ($B(.)$) can be considered, we consider *done_ratio* as the major benefit function in our experiment. The *done_ratio* is the ratio of serviced resource to requested resource for a stream. We assume that all the resource demand and benefit function values are normalized. We simulate situation where the applications with soft-QoS requirements arrive at the system in an arbitrary order. The applications also come with a play-out rate (which maps into an amount of required resource) depending on the user profile and the data content, which is a value of the data rate parameters. For example, we may have a set of video distribution applications that provide streaming video with different types of content to the remote users. Ideally, the application load should be characterized using experimental data. But for the simulations, we assume the following The applications arrive based on a Poisson process with average inter-arrival rate (λ). The duration of the applications is described using a Gaussian process with the mean (μ_d) and the standard deviation (σ_d) of the distribution. The experiments are run by fixing the generating process parameters and then generating many application traffic traces. The admission control and scheduling algorithm is tried on each one of the application traffic traces. We measure the following:

1. *Accumulated number of admitted streams over time*: Streams are admitted if there are resources minimally available ($> MIN\_FRACTION_i$). The basic greedy algorithm and the two policies of $\mathcal{RAC}$ are compared by evaluating the effect on the number of admitted streams.

2. *Total quality:* The total quality is computed as *Total system utilization* divided by the number of current running streams :

$$total\_quality = (\sum_{s_k \in \mathcal{RS}} done\_ratio_k)/number\_of\_current\_running\_streams$$

We assume the data rates of input streams which change in time as follows. *Heavy traffic* is generated by setting $\lambda = 0.333$, $\mu_d = 90$, and $\sigma_d = 3.5$. *Medium traffic* is generated by having $\lambda = 0.125$, $\mu_d = 60$, and $\sigma_d = 3.5$. The values on the *x*-axis are normalized. Under the heavy traffic, the resource demands arrive more frequently than under the medium traffic.

For brevity, we show an experimental result on *Total Quality* in this artile.

The major advantage of the $\mathcal{RAC}$ is its guaranteeing more numbers of streams to be admitted and to run simultaneously with tolerable degradation of quality. In this section, we compare the efficiency of $\mathcal{RAC}(\mathcal{RWR}$ and $\mathcal{DRA})$ ĩgorithm and greedy algorithm from the aspect of accumulated number of admitted streams. Under traffic loads that demand more than the available resources, the accumulated occurrences of admission decreases (rejection increases) both in the basic method and in our strategy. Overall, $\mathcal{RAC}$ achieves better performance than basic greedy algorithm in terms of admission rate and in particular $\mathcal{RWR}$ mode results in the best performance. The heavier the traffic is, the better performance $\mathcal{RAC}$ algorithm achieves compared to the basic greedy algorithm.

To demonstrate that the $\mathcal{RWR}$ and $\mathcal{DRA}$ algorithms achieve the improvement in the total quality, ??, we plot the curve. Figure 1 illustrates the total quality which is quantified by dividing the sum of each application's done_ratio by the number of current running streams. It is observed that the total quality achieved by $\mathcal{RWR}$ algorithm is almost the same as that achieved by the greedy algorithm. Only in the initial part of simulation, it is occasionally below 1.0, but still over 0.85 which is fairly good. However, the quality gets to be almost 1.0 after that. In contrast, the performance of $\mathcal{DRA}$ algorithm in total quality is quite low most of time. This arises due to the fact that the $\mathcal{DRA}$ algorithm holds off the reserves for future streams which arrive soon before some of current running streams depart. Under medium traffic, though the performance of $\mathcal{DRA}$ gets better, it is still too low. Here we could conclude that we should be more careful to dip into the reserves for resource reclamation. In $\mathcal{RWR}$ mode, we hold off the reserves by reclaiming the only returned resources.

5   Related Work

Pervasive or ubiquitous computing has evolved and come to us with promoting the proliferation of various contemporary distributed multimedia network environments, such as ad-hoc wireless, mobile and embedded devices [3]. With respect to quality-sensitive applications in pervasive multimedia computing, a challenging problem is to provide dynamic Quality of Service (QoS) provisioning, admission control, and efficient resource reservation framework with guaranteed

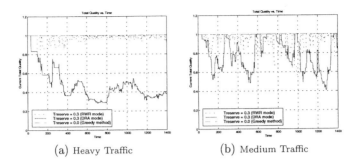

(a) Heavy Traffic   (b) Medium Traffic

Fig. 1. Total Quality over Time

end-to-end QoS. There have been several existing admission control algorithms in which the number of blocks of a media strand retrieved during a round is dependent on its playback rate requirement. Ensuring continuous retrieval of each strand requires that the service time not exceed the minimum of playback durations of the blocks retrieved for each strand during a round, which is a typical greedy admission control strategy [12, 9]. The simple admission control decisions are based on the worst case scenario. The worst case policy is based on the observation that multimedia traffic is characterized by its bursty nature, therefore, if sufficient resources are not available for the worst case scenario, then the applications may fail [10]. These worst-case assumptions that characterize deterministic admission control algorithms cause severe under-utilization of server resources due to excessive constraints and strict performance guarantees. To improve the resource utilization, predictive (observation-based) admission control algorithm for clients have been suggested. These methods simply use the average amount of time spent in retrieving a media block instead of the worst-case assumptions [11]. Vin et al. [12] described a statistical admission control algorithm in which new clients are admitted for service as long as the statistical estimation of the aggregate data rate requirement, rather than the corresponding peak data rate requirement, can be met by the server. They improve the utilization of server resources by exploiting the variation in the disk access times of media blocks, as well as the variation in playback rate requirement by variable rate compression techniques. These methods improve resource utilization at the risk of application failure. Statistically they are safe but in the worst case it may cause applications to fail because of resource congestion. In some recent research on pervasive computing, dynamic QoS adaptation frameworks within middle-ware instead of application-level or O/S-level, have been examined. In [6, 2], they describe a control-based middle-ware framework for their modeling QoS adaptations by dynamic digital control and reconfigurations to the internal functionalities of a distributed multimedia application. In [1], a middle-ware application is proposed, which abstracts O/S interfaces for QoS scheduling such as dynamic QoS allocations to applications during execution.

The QoS resource manager in their prototype reads *application execution statistics* including CPU usage and determines, using several policies, what execution levels will be selected for each application given the current system resources. They use each application's user-specified benefit information and application-specified maximum CPU usage to determine a QoS allocation of CPU resources that maximizes overall user benefit. The above mentioned schemes lack in their theoretical basis to maximize system benefit (utility): e.g., using a fixed (and even small) number of execution level information and application state information may not be proper to negotiate system's dynamic fluctuation efficiently and optimally.

6   Concluding Remarks and Future Work

In this paper we have proposed a threshold (reserves)-based admission control algorithm that exploits the degradability properties of soft-real time applications in pervasive computing environments. We have performed extensive simulation studies of the approach and compared it to naive approaches to admission control. For the simulation studies, we illustrate two different applications: video distribution application and disk access application. We presented a subset of our results related to the performance evaluation of our admission control scheme with respect to a set of metrics such as total system benefit and rejection ratio. In the simulations on video distribution application, to evaluate the performance, we measured the accumulated number of input application rejections, maximum and actual observed total system benefit, system benefit with different reserve values and with different assignment policies. We found that our new admission control scheme reduces the rejection ratio significantly and increases the total system benefit. The performance improvement is marked in particular when the system traffic gets to be heavy. We also figured that the exponential variable policy achieves the best performance in terms of total system benefit, compared to the other two policies such as linear variable policy and fixed policy. We next apply our algorithm to the disk access experiment, i.e. admission control with respect to disk bandwidth control. The performance results in terms of several metrics designed to measure the admission ratio and total quality in CM servers, are also presented. It was observed that under heavy traffic, our algorithm achieves much better performance than the greedy algorithm. Using our scheme, we could expect that more streams could be running with an acceptable range of data quality in a given system resource. We are currently continuing the study of different strategies for admission control and negotiation. In particular we are studying the dependence of the different parameters of the algorithm on the traffic characteristics of the applications. We are also setting a testbed to evaluate the performance on real application data. In addition to the experimental evaluations, we are transitioning the algorithms into working systems.

In another vein, we are applying an *Markov Decision Process (MDP)* approach to our QoS and resource adaptation model because we figured out that our framework can be formulated and described mathematically using the well-known MDP theories. We could define a policy (or decision) rule π be a mapping from a set of states X to a set of actions A and let Π be the set of all possible policies. For a given policy $\pi \in \Pi$, starting with a given initial state $X_0 = x$, we follow the policy π over time such that a particular system path is given as follows: $(X_0 = x_0, a_0, X_1 = x_1, a_1, ..., X_t = x_t, a_t, X_{t+1} = x_{t+1}, a_{t+1}, ...)$ with a_t being the action taken at time t and $a_t = \pi(x_t)$ with $x_t \in X$, where $X_t, t = 0, 1, 2...$ be a random variable that denotes a state at time t. Over the system path, the system accumulates the discounted rewards of

$$\sum_{t=0}^{\infty} \gamma^t R(x_t, \pi(x_t)), 0 < \gamma < 1,$$

and the probability that the system has the particular path over time can be obtained as $\prod_{t=0}^{\infty} P(x_{t+1}|x_t, \pi(x_t))$. Here, R denotes a *reward function* that maps $X \times A$ to a real number, which is analogized with *benefit* in our QoS framework. P is a state transition function that describes how the system evolves over time. Therefore, we can take the expectation of the discounted rewards over all possible system paths with those probabilities. Finally, the next phase of this work will extend the admission control and negotiation process over multiple resources as well.

Acknowledgements

This work was supported by a Korea University Grant in 2002.

References

[1] Brandt, Scott and Nutt, Gary and Berk, Toby and Humphrey, Marty: Soft Real-time Application Execution with Dynamic Quality of Service Assurance. Proceedings of 6th International Workshop on Quality of Service, 1998,May, Napa, CA

[2] Yi Cui and Dongyan Xu and Klara Nahrstedt: SMART: A Scalable Middleware Solution for Ubiquitous Multimedia Service Delivery. Proceedings of IEEE International Conference on Multimedia and Expo,2001,August

[3] Xiaohui Gu and Klara Nahrstedt: Dynamic QoS-Aware Multimedia Service Configuraiton in Ubiquitous Computing Environments. Proceedings of IEEE 22nd International Conference on Distributed Computing Systems (ICDCS),2002,July

[4] Lau, S. W. and Lui, John C. S:A Novel Video-On-Demand Storage Architecture for Supporting Constant Frame Rate with Variable Bit Rate Retrieval. 5th International Workshop on Network and Operating Systems Support for Digital Audio and Video,1995,April,Durham, N. H.

[5] Lee, Wonjun and Srivastava, Jaideep:An Algebraic QoS-based Resource Management Model for Competitive Multimedia Applications. Multimedia Tools and Applications, Kluwer Academic Publishers,2001

[6] Li, Baochun and Nahrstedt, Klara:A Control Theoretical Model for Quality of Service Adaptations. Proceedings of 6th International Workshop on Quality of Service,1998,May, Napa, CA

[7] Makaroff, Dwight and Neufeld, Gerald and Hutchinson, Norman:An Evaluation of VBR Disk Admission Algorithms for Continuous media File Servers. Proceedings of the ACM Multimedia Conference,1997,Dec., Seattle, Wa

[8] B. Sabata and S. Chatterjee and M. Davis and J. Sydir and T. Lawrence:Taxonomy for QoS Specifications. Proceedings of IEEE Computer Society 3rd International Workshop on Object-oriented Real-time Dependable Systems(WORDS '97),1997,Feb., Newport Beach, California

[9] Shenoy, Prashant J. and Goyal, Pawan and Rao, Sriram S. and Vin, Harrick M.:Symphony: An Integrated Multimedia File System. Proceedings of SPIE/ACM Conference on Multimedia Computing and Networking (MMCN'98),1998,January,San Jose, CA

[10] S. V.Raghavan, Satish K. Tripathi:Networked Multimedia Systems: Concepts, Architecture, and Design. Prentice Hall,1998

[11] Vin, Harrick M. and Goyal, Alok and and Goyal, Pawan and Goyal, Anshuman:An Observation-Based Approach for Designing Multimedia Servers.Proceedings of the IEEE International Conference on Multimedia Computing and Systems,1994,May,Boston, MA

[12] Vin, Harrick M. and Goyal, Pawan and Alok Goyal and Goyal, Anshuman:A Statistical Admission Control Algorithm for Multimedia Servers. Proceedings of ACM Multimedia '94, 1994,October,San Francisco

SAP: On Designing Simple AQM Algorithm for Supporting TCP Flows

Miao Zhang, Jianping Wu, Chuang Lin, Ke Xu, and Mingwei Xu

Department of Computer Science and Technology
Tsinghua University, Beijing, P.R.C.
{zm,clin,xmw,xuke}@csnet1.cs.tsinghua.edu.cn, jianping@cernet.edu.cn

Abstract. In this paper we propose a new AQM scheme, Simple Adaptive Proportional (SAP). The key idea of SAP is to add adaptive mechanism to the proportional controller. SAP has several advantages. With the property of proportional controller, SAP is responsive and stable. Compared with doing update at each packet arrival, the periodic update method in SAP requires much less processing overhead. The adaptive mechanism in SAP makes it work well in a wide range of network environments without reconfiguration. Both the implementation and configuration of SAP are very simple.

1 Introduction

Congestion control algorithms have been extensively studied in recent years. Active Queue Management (AQM) is an important topic in congestion control research. Its goal is to operate the network at high throughput with negligible loss and queuing delay. RED (Random Early Detection)[1] is one of the most famous AQM schemes. Its basic idea is to warn sources of incipient congestion by probabilistically marking packets. RED shows better performance than DropTail, but it has two major shortcomings [2]: RED is extremely sensitive to parameter setting; RED fails to prevent buffer overflow as sources increase.

In [3], the property of RED is analyzed with control theory. It is pointed out that the problem of RED arises because of the coupling between the (average) queue size and the marking probability. RED has a *steady state error*, which is dependent on network parameters. With the introduction of integral factor, PI (Proportional and Integral) controller [4] can make this *steady state error* down to zero. But the advantage of PI controller is not gotten free. Though the *steady state error* is eliminated, the response speed is slowed down. It is the inherent property of PI controller [4]. The traffic pattern of the Internet is suggested to be very bursty [5]. A slow-responsive controller will inevitably result in long queuing delay and lots of packet loss. Apart from PI controller proposed in [3], REM [2] and AVQ [6]also use integral factor in their algorithms. They have similar disadvantage as PI controller on response speed.

From control theory, proportional controller is more responsive than PI controller. It is also very simple to implement and easy to configure. The design of

H.-K. Kahng (Ed.): ICOIN 2003, LNCS 2662, pp. 515–524, 2003.

RED is in fact a proportional controller plus low pass filter. It is very interesting to study why RED can't work well. It is suggested that there are four main reasons. First, the averaging filter used in RED is a cause of sluggishness of the response. It can lead to instability and low frequency oscillation in the regulated output [3]. Second, the update method used in RED is not appropriate [3, 7, 8]. Doing update on every packet arrival is unnecessary. It also makes the weight used in averaging filter very sensitive to different packet arrival rate. Third, the maximum and minimum thresholds used in RED are unnecessary and harmful to the performance. They are a source of discontinuity of RED control [7]. Fourth, the fixed maximum probability in RED can only work well in a limited range of network environments.

With the lessons learned above, we design a new AQM algorithm, Simple Adaptive Proportional (SAP). SAP is an attempt to design self-configured AQM algorithm with proportional controller. There are three key points in the design of SAP. First, the marking probability is calculated with a proportional controller and *instantaneous* (instead of *average*) queue size is used in the calculation. Second, periodic update method is applied in SAP and a clear guideline for setting the update interval is provided. Third, a simple adaptive mechanism is designed for adjusting the coefficient used in the proportional controller.

The rest of this paper is organized as follows: in Section 2, we present the design of SAP. In Section 3, we show the simulation results with network simulator ns-2. Section 4 summarizes the results and discusses the future work.

2   The SAP Algorithm

Two parts compose the SAP algorithm: *the basic algorithm* and *the adaptive algorithm*. The basic algorithm is a proportional controller. It calculates marking probability with the instantaneous queue size. The adaptive algorithm is used to adjust avg_p, which is one coefficient used in the proportional controller. Figure 1 depicts the relationship between multiple function blocks inside SAP. In the following parts of this section, we will describe the design of SAP in detail.

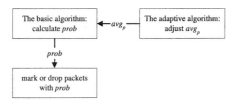

Fig. 1. The diagram of SAP

2.1   The Proportional Controller in SAP

In the basic algorithm of SAP, the marking probability is calculated as follows:

$$prob = avg_p * q/target \qquad (1)$$

Here, q is the instantaneous queue size. It is sampled every $interval\_s$. $target$ is the destination for average queue size. avg_p is the average value of $prob$. It is adjusted by *the adaptive algorithm* of SAP. Though avg_p is not a constant, the update frequency of it is much lower than the update frequency of $prob$. So essentially SAP is a proportional controller. Instantaneous queue size is used in Equation (1). It is different from the design of RED, which uses average queue size calculated with EWMA. Averaging filter can filter out transient bursts, but the weight (w_q) used in averaging filter is shown sensitive in different network environments [9]. This problem is related to the update method used in RED and we will make a discussion on it later. In SAP the troublesome setting for w_q is avoided.

2.2   Periodic Update Method

In RED, probability calculation is triggered by packet arrival. Such method is not applicable for high-speed link where the interval between packet arrivals is very short. Also, the maximum packet arrival rate is not same for different bandwidth. The fixed w_q used in RED can't fit different packet arrival rates.

The purpose of doing update at every packet arrival epoch is to "fully observe" the dynamic of queue size. It is suggested by experiment results that such small granularity is unnecessary. Before the congestion signal is conveyed to the end system, there is a feedback delay about 1 RTT. This feedback delay will decrease the benefit of "fully observation".

In SAP marking probability is calculated periodically for every $interval\_s$. Such method doesn't have the disadvantages mentioned above. It is suggested by simulation results that using an update interval one order of magnitude less than RTT is appropriate when RTT is larger than 100ms. It is also observed that SAP with an update interval of 0.01 second works well when RTT is less than 100 ms, even when RTT is less than 10ms. Now it is recommended to set $interval\_s$ to 0.01 second.

2.3   The Adaptive Algorithm of SAP

The fixed value of max_p in RED can only work well in a limited range of network environments. In [9] and [10] adaptive mechanisms are proposed to adjust the value of max_p. We borrow some good ideas from them in designing the adaptive algorithm of SAP.

The adaptive algorithm of SAP is shown in Figure 2. A threshold $target$ is used in the adaptive algorithm. The basic idea is: if the average queue size (avg) is above $target$, increase avg_p; if avg is below $target$, decrease avg_p. It

```
for every interval_l:
    calculate the average queue size avg;
    if (avg < target)
        avg_p = avg_p / α; // decrease avg_p
    elseif (avg > target)
        avg_p = avg_p * α; // increase avg_p
```

Fig. 2. The adaptive algorithm of SAP

is similar to the proposal in [9]. There are five further considerations for the adaptive algorithm of SAP.

First, avg is not calculated with EWMA in SAP. It is desirable that avg can reflect the queue size situation in $interval\_l$. With EWMA, how long the queue size situation can be reflected by avg actually depends on the value of w_q. So we use following form to calculate avg (n is the number of samples in $interval\_l$):

$$avg = \sum q/n \qquad (2)$$

Second, two rules are proposed on setting target: $target > 10.0$; $target < B/2$, B is buffer size. The first rule comes from the observation to experiment results. It is observed that the performance of SAP is not good when target is small. It is suggested to be a problem of the adjusting granularity. In SAP, the granularity for adjusting p is:

$$\Delta p = avg_p/target \qquad (3)$$

Assume that queue size oscillates near $target$, the relative granularity for adjusting p (i.e., $\Delta p/p$) will be very large ($> 10\%$) when $target$ is very small (e.g., 3.0). Such adjustment can lead to wild oscillation at end systems and thus degrade the performance. With large $target$ (e.g., 40), such oscillation can be mitigated. The intent of the second rule is to avoid severe packet loss when the scope of queue oscillation is beyond the buffer size.

Third, avg_p is constrained to remain within the range $[0.0001, 0.5]$ to avoid over-adjustment of avg_p. This scope is not very strict. The minimum limitation of 0.0001 is for the situation when there are no packet arrivals. Without this limitation, avg_p may be decreased to near zero. A maximum limitation of 0.5 is also applied for both avg_p and $prob$ in SAP. 0.5 is big enough for all conditions.

Fourth, two guidelines are proposed for setting $interval\_l$. $interval\_l$ should be larger than RTT. Severe performance degradation is observed when $interval\_l$ is less than RTT. $interval\_l$ should also be small enough to keep the adaptive mechanism responsive. Consider these two factors, the default value for $interval\_l$ is 1.0 second.

Fifth, the setting for α depends on the network traffic pattern. α decides the adjustment step of avg_p. From intuition, if the congestion level always changes very aggressively, a larger value of α (such as 5.0) should be used. But too large α

may cause system oscillation. From the experiment experience, the performance will degrade if α is too large. It is also observed that large value of α (e.g., 5.0) doesn't have much advantage over small one (e.g., 2.0) in response time. It's recommended to use the default setting of $\alpha=2.0$.

3   Simulations

In this section, we will use ns-2 to simulate SAP. Experiments 1 to 5 are used to verify the performance of SAP and the design rules described in Section II. Experiment 6 is used to compare the performance between SAP and other well-known AQM schemes. For the limitation of space, we only show part of the experiments we have done in this paper. More experiment results can be found in [11].

3.1   Simulation Setup

For most of the experiments in this section, we consider a single link of capacity 64Mbps that marks or drops packets according to some AQM schemes. The link is shared by N (N = 20, 40, 60, ... , 160) Reno users. They are all activated at time 0s and stopped at time 200s. The packet size is 1000 bytes. Each TCP connection has a round-trip propagation delay of 140ms. For SAP, the default settings are: $interval\_s = 0.01$ second, $interval\_l = 1.0$ second, $target = 40.0$, $\alpha = 2.0$. The adaptive mechanism of SAP is enabled by default. We use packet dropping in all experiments. The default buffer size is 800 packets.

3.2   Experiment 1

In this experiment, we verify the rules given in Section 2 on choosing $interval\_s$. We use RTT = 140ms and the settings for $interval\_s$ are 0.1, 0.01 and 0.001 second. To focus on the effect of different $interval\_s$, the adaptive mechanism in SAP is disabled in this experiment.

The experiment results are shown in Figure 3. The performance with $interval\_s = 0.01$ is better than that with $interval\_s = 0.1$. But there is little improvement when adjusting $interval\_s$ from 0.01 to 0.001. In Experiment 1, the packet arrival rate is about 8000 pkt/s. With $interval\_s = 0.01$ and 0.001, the frequency of update is much lower than the frequency of packet arrival and the performance is still very good.

3.3   Experiment 2

In this experiment, we verify the guideline given in Section 2 on choosing $interval\_l$. The settings for $interval\_l$ are 0.05, 0.2, 0.5 and 1.0 second. The experiment results are shown in Figure 4. There is significant performance degradation when $interval\_l$ is 0.05. But for other settings of $interval\_l$, the performance is quite similar.

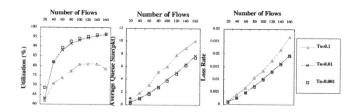

Fig. 3. Exp.1. Performance with different *interval_s* (Ts)

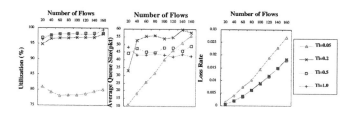

Fig. 4. Exp.2. Performance with different *interval_l* (Tl)

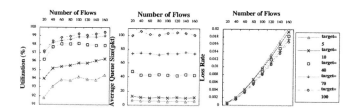

Fig. 5. Exp.3. Performance with different *target*

3.4 Experiment 3

In this experiment, we verify the guidelines given in Section 2 on choosing *target*. The settings for *target* are 5, 10, 40, 70 and 100.

The experiment results are shown in Figure 5. The performance is improved with the increase of *target*, but the "step" of improvement is decreased. There is little difference in performance with *target* = 70 and 100. With *target* larger than 100, there is little performance improvement, for there is little improvement in adjusting granularity. The average queue size is very stable with different number of flows. The difference between round trip time for *target* = 10 and *target* = 100 is very little, for the average queuing delay is less than 10 % of RTT even with *target* = 100.

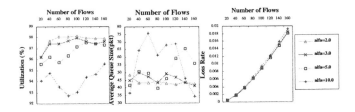

Fig. 6. Exp.4. Performance with different α (alfa)

3.5 Experiment 4

In this experiment, we verify the performance of SAP with different value of α. The settings for α are 2.0, 3.0, 5.0 and 10.0. The goal of this experiment is to test the performance of SAP in steady state with different value of α.

The experiment results are shown in Figure 6. The performance is similar when changing α from 2.0 to 5.0. When α =10.0, there is evident performance degradation. It is because of too large oscillation in adjusting avg_p. It is suggested that setting α to 2.0 or 3.0 is proper.

3.6 Experiment 5

In this experiment, we verify the performance of SAP in presence of traffic oscillation. The bandwidth of shared link is 15 Mbps and the buffer size is 375 packets. The packet size is 500 bytes. 200 FTP flows are created at time 0s. At time 50s, 160 FTP flows drop out, and at time 100s they start back again. The experiments are done with $\alpha = 2.0$ and $\alpha = 5.0$. Figure 7(a) shows the experiment results with $\alpha = 2.0$. The adjustment can be completed in about 2.0 seconds. The average link utilization is 99.0% and the average queue size is 40.2. Figure 7(b) shows the experiment results with $\alpha = 5.0$. The adjustment can be completed in less than 2.0 seconds. The average link utilization is 98.8% and the average queue size is 58.5.

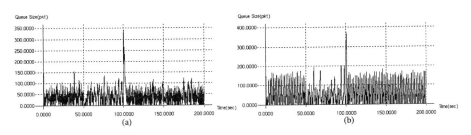

Fig. 7. Exp.5 Dynamics of queue size with SAP in presence of traffic oscillation

3.7   Experiment 6

In this experiment, we compare the performance of SAP with some AQM schemes that have been proposed. The network configuration is same to that used in Experiment 5. We make comparison in two simulation scenarios. In the first scenario, the number of flows is constant. The link is shared by N (N = 20, 40, 60, ... , 160) Reno users. They are all activated at time 0s and stopped at time 200s. In the second scenario, the number of flows is fluctuating. 200 FTP flows are created at time 0s. At time 50s, 160 FTP flows drop out, and at time 100s they start back again.

The AQM schemes to be compared with are:

1. Adaptive RED proposed in [9]. In our experiments, we use the "gentle" version of RED. The parameters are chosen as recommended in [11]: *interval* = 0.5, β = 0.9. *target* is set to 40.0. We use the automatic mode provided by Adaptive RED to set max_{th}, min_{th} and w_q, resulting in max_{th} = 60, min_{th} = 20, w_q = 0.000267.
2. Adaptive Virtual Queue (AVQ) proposed in [6]. We let γ, the desired utilization, be 0.98. The damping factor α is set to 0.15.
3. PI controller proposed in [3]. We set the parameters of PI according to the design rules given in [3]. The target queue size is 40.0.
4. Random Exponential Marking (REM) proposed in [2]. The parameters of REM are chosen as recommended in [2]. The target queue size is 40.0.

Figure 8 shows the results in the first experiment scenario. The link utilization is similar for all AQM schemes in the experiment. In most of the experiments, the link utilization is above 90%. The average queue size is about 40.0 with SAP, ARED, PI and REM. But with AVQ, the average queue size is much larger than 40.0. There is no way to directly control queue size in AVQ. The loss rate with SAP is higher than with other AQM schemes. In summary, the performance of SAP is at the same level with other AQM schemes in this scenario.

Figure 9 shows the dynamics of queue size in the second experiment scenario. We can compare the results in Figure 9 with the performance of SAP shown in Figure 7. We get the response speed by observing the queue size dynamics near

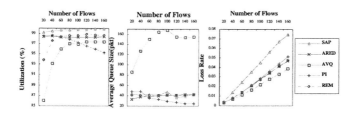

Fig. 8. Exp.6. Performance with SAP, Adaptive RED (ARED), AQV, PI and REM

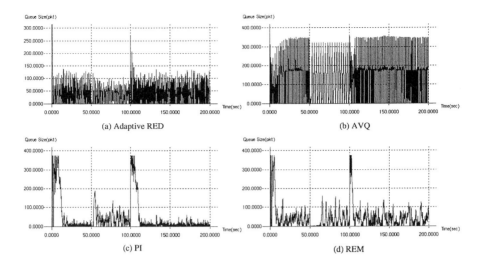

Fig. 9. Exp.6. Dynamics of queue size with ARED, AQV, PI and REM

time 50s and 100s. With Adaptive RED the queue size can complete response in about 3.0 seconds. With PI, the response time is about 10 seconds. With REM, the response time is longer than 10 seconds. With AVQ, though there is no peak after 100s, the queue size hasn't got stable until 107s. It is suggested that AVQ, PI and REM are not applicable in the Internet for their slow responsiveness.

4   Conclusion

In this paper, a simple AQM scheme - SAP is proposed. The basic idea is to add adaptive mechanism to the simple proportional controller. SAP has three advantages. First, it is very easy to implement. By using periodic update method, the processing overhead is decreased. The adaptive mechanism of SAP is also very simple. Second, it is very easy to configure. A clear guideline is provided in this paper to set the four parameters of SAP. The experiment results also suggest that SAP can work well in a wide range of network environments with the recommended configuration.

The work in this paper is based on some assumptions about the Internet traffic pattern. It is assumed that the traffic is relatively stable in several seconds, which makes it possible for the system to make response. Internet traffic pattern is very important to the design and success of AQM. The flow statistics of small time granularity deserves further study.

Acknowledgements

This work was supported by the National Natural Science Foundation of China under Grant No.90104002 and No.60203025, the National High Technology Development 863 Program of China under Grant No.2001AA121013 and the National Grand Fundamental Research 973 Program of China under Grant No. G1999032707.

References

[1] Floyd, S., Jacobson, V.: Random Early Detection Gateways for Congestion Avoidance. IEEE/ACM Transactions on Networking, 1993, 1(4): 397–413

[2] Athuraliya, S., Li, V. H., Low, S. H., Yin, Q.: REM: Active Queue Management. IEEE Network, 2001, 15(3): 48–53

[3] Hollot, C. V., Misra, V., Towsley, D., Gong, W.: On Designing Improved Controllers for AQM Routers Supporting TCP Flows. In: Sengupta, B. ed. Proceedings of IEEE INFOCOM. Anchorage, Alaska , USA: IEEE Communications Society, 2001. 1726–1734

[4] Franklin, G. F., Powell, J. D., Emami-Naeini, A.: Feedback Control of Dynamic Systems. Addison-Wesley, 1995

[5] Ryu, R., Cheney, D., Braun, H. W.: Internet flow characterization: adaptive timeout strategy and statistical modeling. In: Proc Passive and Active Measurement Workshop, Amsterdam, Netherlands, 2001, 94–105

[6] Kunniyur, S., Srikant, R.: Analysis and Design of an Adaptive Virtual Queue (AVQ) Algorithm for Active Queue Management. ACM Computer Communication Review, 2001, 31(4): 123–134

[7] Firoiu, V., Borden, M.: A Study of Active Queue Management for Congestion Control. In: Sidi, M. ed. Proceedings of IEEE INFOCOM. Tel Avlv, Israel: IEEE Communications Society, 2000. 1435-1444

[8] Misra, V., Gong, W., Towsley, D.: A Fluid-based Analysis of a Network of AQM Routers Supporting TCP Flows with an Application to RED. ACM Computer Communication Review, 2000, 30(4): 151–160

[9] Floyd, S., Gummadi, R., Shenker, S.: Adaptive RED: An Algorithm for Increasing the Robustness of RED's Active Queue Management. http://www.aciri.org/floyd/papers.html, August 2001

[10] Feng, W., Kandlur, D., Saha, D., Shin, K.: A Self-Configuring RED Gateway. In: Doshi, B. ed. Proceedings of IEEE INFOCOM. New York, USA: IEEE Communications Society, 1999. 1320–1328

[11] Zhang, M., Wu, J. P., Lin, C., Xu, M. W., Xu, K.: SAP: On Designing Simple AQM Algorithm for Supporting TCP Flows. Technical report. http://netlab.cs.tsinghua.edu.cn/~zm/publication/sap.pdf

A VoIP Traffic Modeling for the Differentiated Services Network Architecture*

Daein Jeong

Hankuk University of Foreign Studies, Korea
djeong@hufs.ac.kr

Abstract. We propose a traffic modeling, (σ, ρ, ξ)-model, which is suitable for QoS provisioning with hard bound on performance degradation. ξ in (σ, ρ, ξ)-model defines the portion of data that fails to conform to the (σ, ρ)-constraint. An exact analysis of a discrete-time non-buffered token bucket is presented to introduce the (σ, ρ, ξ)-model. The analysis is verified through simulations, and is shown to be more accurate than a fluid model approach. The applicability of the (σ, ρ, ξ)-model is examined in the assessment of the required resources for aggregate VoIP traffic. It is observed that the allowance of bounded degradation enhances the resource utilization significantly. We also observe that the buffer adjustment is preferable to the bandwidth adjustment in accommodating a variety of traffic burstiness.

1 Introduction

The continuing and massive advances in communication technologies are bringing the variety of service integrations into the Internet. Especially, delivering voice traffic over the Internet is increasingly deployed over the enterprise networks as well as the residential environment, while introducing lots of challenging issues[1]. The voice traffic requires strict performance on both end-to-end delay and delay variations, which are beyond the capability of today's Internet. It also requires that the portion of data which is not available for replay at the receiver must be limited within a hard bound. In this sense, the EF(Expedited Forwarding) PHB(Per-Hop Behavior)[2] in the DS(Differentiated Services)[3] domain is a promising service for the voice traffic. Due to the strictness in service, the appropriate traffic description of the EF class is crucial for efficient resource allocation. The LBAP(Linear Bounded Arrival Process) traffic characterization[4] is most widely preferred for delivering the strictness in QoS guarantee. Instead, however, it is convenient to model the voice traffic by a stochastic on-off source[5], described by two-state Markov process. Transformation of the on-off source model into the LBAP traffic characterization, therefore, is fairly helpful in accommodating the voice traffic via EF PHB.

* This work was supported by grant No.R01-2000-00280 from the Korea Science and Engineering Foundation.

H.-K. Kahng (Ed.): ICOIN 2003, LNCS 2662, pp. 525–534, 2003.

The LBAP traffic characterization is represented by the (σ, ρ)-constraint model introduced in [4]. A source conforms to the (σ, ρ)-constraint if, in any interval of time τ, the amount of generated bits is no larger than $\sigma + \rho\tau$. σ and ρ are two parameters that characterize the source. Considering the loss-tolerant nature of the voice traffic, however, it is possible to alleviate the strictness the (σ, ρ)-constraint model imposes. We propose the following traffic model.

Definition 1 *A source conforms to the (σ, ρ, ξ)-model if it satisfies*

$$P_{rob}\{A(t,\ t+\tau) > \sigma + \rho\tau\} \leq \xi \tag{1}$$

for all $t, \tau \geq 0$, where $A(t, t+\tau)$ denotes the amount of data generated from the source during time interval of $[t, t + \tau]$, $P_{rob}\{event\}$ denotes the probability of the event occurrences, and ξ is any value in the range of $0 \leq \xi \leq 1$.

Notice that if $\xi = 0$ in (1), the (σ, ρ, ξ)-model turns out to be the $(\sigma,\ \rho)$-constraint model. For any interval of length τ, the (σ, ρ, ξ)-model allows the source traffic to violate the envelope curve, $\sigma + \rho\tau$, within the ratio of ξ. The applicability of the (σ, ρ, ξ)-model can be found in the assessment of the required resources for aggregate VoIP traffic served via EF PHB. We observe that the required resources for aggregate VoIP traffic is far less than the linear sum of individual requirements, which in turn shows that the allowance of violation enables one to enhance the resource utilization significantly.

The traffic profile of the (σ, ρ, ξ)-model is efficiently configurable by the token bucket algorithm. An analysis of the token bucket mechanism, thus, is needed to calculate the parameters σ and ρ considering the stochastic nature of the on-off source. A rigorous analysis is presented in this paper. For exact results, a discrete-time on-off traffic is considered rather than a fluid one assumed in [6], or a Poisson process considered in [7]. Simulation results demonstrate the validity of the analysis, and also show that our approach is more accurate than the previous work[6] which is based on the fluid approximation.

Section 2 describes the token bucket mechanism including the mathematical modeling of the discrete-time on-off source offered to the token bucket. Based on this modeling, the token bucket is exactly analyzed in section 3. In section 4, simulation results are examined including a demonstration of the accuracy of the analysis. Conclusions are in section 5.

2   Modeling the Policing Device

2.1   Discrete-Time On-Off Source

Consider a traffic source described by the discrete-time on-off source model. The packet size is assumed to be fixed, and time is scaled in unit of one packet transmission time. The source alternates between on and off states. We characterize the source by the utilization γ (i.e., the probability in on state), and the average number of packets generated during an on period ℓ. Then, the state transition rate from off to on state is $\frac{\gamma}{\ell(1-\gamma)}$, and from on to off state is $\frac{1}{\ell}$. For convenience, we denote those transition rates by p and q, respectively. If the source is in the off state, it generates no data, while in the on state, it generates a packet per time slot. The sojourn times in each state last for a geometrically distributed number of time slots.

2.2 Policing Device Mechanism

Consider a token bucket and a discrete-time on-off source. The token pool size is σ, and the tokens are generated with fixed rate ρ. In each D $(= \frac{1}{\rho})$ time slots, a token is generated and stored in the pool if the pool contains less than σ tokens. Otherwise, the generated token is discarded. An arriving packet that finds the token pool non-empty enter the network immediately, and one token is removed from the token pool. When the arriving packet sees the token pool empty, it is *marked* to be distinguishable from others within the network, and is allowed to enter the network. We refer to that mechanism the policing device in the rest of this paper.

2.3 Arrival Description

Consider an interval of D time slots, and assume a new token is generated at the beginning of each interval. For convenience of analysis, we assume D is an integer. Notice that the number of packet arrivals during any interval of time slots is correlated with the source status just prior to the beginning of that interval. So, we define a conditional probability, $P_{S^-,S}^{n,j}$, as follows.

$$P_{S^-,S}^{n,j} \triangleq P_{rob}\{j \text{ packets are generated during interval of } n \text{ time slots,}$$
$$\text{and end with state } S \mid \text{source state was } S^- \text{ just prior}$$
$$\text{to the interval}\} \tag{2}$$

in which $S^-, S = \{on, off\}$, $n \geq 1$ *and* $0 \leq j \leq n$. For the given interval of D time slots, we need values of

$$P_{S^-,S}^{D,j}, \quad for \quad S^-, S = \{on, off\} \quad and \quad 0 \leq j \leq D.$$

According to the source characteristics, we have the following a priori known values:

$$P_{on,off}^{1,0} = q \ , \ P_{on,on}^{1,1} = 1 - q \ , P_{off,off}^{1,0} = 1 - p \ , \ P_{off,on}^{1,1} = p \ , \tag{3}$$

and

$$P_{on,off}^{1,1} = P_{off,off}^{1,1} = P_{on,on}^{1,0} = P_{off,on}^{1,0} = 0 \ . \tag{4}$$

Additionally, we have, for $S^- = \{on, off\}$ and $1 \leq n \leq D$,

$$P_{S^-,on}^{n,0} = P_{S^-,off}^{D,D} = 0 \ . \tag{5}$$

The first term in (5) is obviously 0 because the source cannot become on state at the last time slot of the interval with no packet generation during that interval. Similar reasoning gives the second term in (5).

When we apply (3) and (4), which represent the dynamics of the on-off source in one time slot, we can get the following recursive equations for $0 \leq j \leq n \leq D$:

$$P_{on,on}^{n,j} = pP_{on,off}^{n-1,j-1} + (1-q)P_{on,on}^{n-1,j-1} \tag{6}$$

$$P_{on,off}^{n,j} = (1-p)P_{on,off}^{n-1,j} + qP_{on,on}^{n-1,j} \tag{7}$$

$$P_{off,on}^{n,j} = pP_{off,off}^{n-1,j-1} + (1-q)P_{off,on}^{n-1,j-1} \tag{8}$$

$$P_{off,off}^{n,j} = (1-p)P_{off,off}^{n-1,j} + qP_{off,on}^{n-1,j} \tag{9}$$

With a given value of D, the values of $P_{S^-,S}^{D,j}$, for $0 \leq j \leq D$ and $S^-, S = \{on, off\}$, can be computed iteratively using recursive equations (6) through (9), with the initial values given in (3), (4) and (5), starting from $n = 2$ to $n = D$.

3   Analysis of the Policing Device

Let $P_S^t(j)$ be the joint probability that the source status is in S and the available number of tokens is $\sigma - j$ at time t ($t = 0, D, 2D, \cdots$), just prior to the token generation instances. It is obvious that the Markov chain $P_S^t(j)$ is ergodic, which assures the existence of the steady state probability denoted by $P_S(j)$, such that, for $0 \leq j \leq \sigma$ and $S = \{on, off\}$,

$$P_S(j) = \lim_{t\to\infty} P_S^t(j) = P_{rob}\{source\ is\ in\ S, \tag{10}$$

$$and\ the\ number\ of\ tokens\ is\ \sigma - j\}.$$

Obviously, $P_S(j) = 0$ for $j > \sigma$ regardless of S. The reason we defined the joint probability is that the number of packets generated during a token generation interval depends upon the source status at just prior to the beginning of that interval. The steady-state equations of the policing device are given in two cases depending on the values D and σ. This is because the number of new arrivals are limited up to D packets during a token generation interval.

1) case 1 : $\sigma > D$

In this case, the steady-state equations consist of two valid regions depending on the token pool occupancy as follows:

If $0 \leq j \leq D$,

$$\begin{cases} P_{off}(j) \\ = P_{off}(0)P_{off,off}^{D,j} + \sum_{k=0}^{j}[P_{off}(k+1)P_{off,off}^{D,j-k} + P_{on}(k+1)P_{on,off}^{D,j-k}] \\ \\ P_{on}(j) \\ = P_{off}(0)P_{off,on}^{D,j} + \sum_{k=0}^{j}[P_{off}(k+1)P_{off,on}^{D,j-k} + P_{on}(k+1)P_{on,on}^{D,j-k}] \end{cases} \tag{11}$$

and, if $D + 1 \leq j \leq \sigma - 1$,

$$
\begin{cases}
P_{off}(j) \\
\quad = \displaystyle\sum_{k=0}^{D} [P_{off}(j - k + 1) P^{D,k}_{off,off} + P_{on}(j - k + 1) P^{D,k}_{on,off}] \\
\\
P_{on}(j) \\
\quad = \displaystyle\sum_{k=0}^{D} [P_{off}(j - k + 1) P^{D,k}_{off,on} + P_{on}(j - k + 1) P^{D,k}_{on,on}]
\end{cases}
\tag{12}
$$

$P_S(j)$ for $S = \{on, off\}$ and $0 \leq j \leq \sigma$ can be obtained by some rearrangement and iterative computations using (11) and (12). For an example, consider an integer $j(< D)$. The second equation in (11) gives the description of $P_{on}(j)$ with $P_{off}(k)$ for $0 \leq k \leq j$. If we substitute this description for $P_{on}(j - 1)$ in the first equation in (11), we get the description of $P_{off}(j)$ in terms of $P_{on}(k)$ and $P_{off}(k)$ for $0 \leq k \leq j - 1$. By substituting that description for $P_{off}(j)$ in the second equation in (11), we get the description of $P_{on}(j)$ in terms of $P_{on}(k)$ and $P_{off}(k)$ for $0 \leq k \leq j - 1$.

Accordingly, we get total of 2σ equations, while there are $2\sigma + 2$ variables, $P_S(j)$, $j = 0, \cdots, \sigma$ and $S = \{on, off\}$. Among those variables, we know that $P_{on}(0) = 0$, because the token pool cannot be full of tokens when the source is *on* state. Then, all other $P_S(j)$, $j = 1, \cdots, \sigma$ and $S = \{on, off\}$, can be expressed with $P_{off}(0)$. So, at first, we assume a value for $P_{off}(0)$, and solve $P_S(j)$'s for $j = 1, \cdots, \sigma$ and $S = \{on, off\}$, iteratively. Then, all quantities are normalized according to

$$
\sum_{j=0}^{\sigma} [P_{on}(j) + P_{off}(j)] = 1 \ .
$$

We define the throughput (T) of the policing device as the average number of unmarked packets during a token generation interval. T is given as follows:

$$
\begin{aligned}
T = \sum_{j=0}^{\sigma-D+1} & [P_{on}(j) \sum_{k=0}^{D} k(P^{D,k}_{on,on} + P^{D,k}_{on,off}) \\
& + P_{off}(j) \sum_{k=0}^{D} k(P^{D,k}_{off,on} + P^{D,k}_{off,off})] \\
& + \sum_{j=\sigma-D+2}^{\sigma} [P_{on}(j)\{ \sum_{k=0}^{\sigma-j+1} k(P^{D,k}_{on,on} + P^{D,k}_{on,off}) \\
& + (\sigma - j + 1)(1 - \sum_{k=0}^{\sigma-j+1} (P^{D,k}_{on,on} + P^{D,k}_{on,off}))\} \\
& + P_{off}(j)\{ \sum_{k=0}^{\sigma-j+1} k(P^{D,k}_{off,on} + P^{D,k}_{off,off})
\end{aligned}
$$

$$+(\sigma - j + 1)(1 - \sum_{k=0}^{\sigma-j+1}(P_{off,on}^{D,k} + P_{off,off}^{D,k}))\}] \tag{13}$$

The first square bracket is for the case when there are at least D available tokens in the token pool at the beginning of the interval, while the second square bracket is for the other case. Notice that one token which is to be generated during the interval of interest is taken into account. Applying the throughput T, we get the marking probability denoted by P_{mark} of the policing device as follows:

$$P_{mark} = 1 - \frac{T}{\gamma D}. \tag{14}$$

2) case 2 : $\sigma \leq D$

In this case, we have, for $0 \leq j \leq \sigma - 1$:

$$\begin{cases} P_{off}(j) \\ = P_{off}(0)P_{off,off}^{D,j} + \sum_{k=0}^{j}[P_{off}(k+1)P_{off,off}^{D,j-k} + P_{on}(k+1)P_{on,off}^{D,j-k}] \\ \\ P_{on}(j) \\ = P_{off}(0)P_{off,on}^{D,j} + \sum_{k=0}^{j-1}[P_{off}(k+1)P_{off,on}^{D,j-k} + P_{on}(k+1)P_{on,on}^{D,j-k}] \end{cases} \tag{15}$$

The throughput is obtainable through the same procedure as that of case 1.

With a given required bound on the ratio of performance degradation ξ, we are looking for the values of σ and ρ which give $\hat{T}$ such that

$$\hat{T} = (1 - \xi)\gamma D. \tag{16}$$

Let us denote them as $\hat{\sigma}$ and $\hat{\rho}$, respectively. Notice that they are tunable parameters towards $\hat{T}$, and are closely related to the available network resources, the buffer space and the bandwidth, respectively. With the given relationship given in (13), those two parameters can not be determined uniquely. Depending on the resource availability, one of them is assigned at first with a specific value, and then the remainder is determinable through repeated applications of (13). This enables an efficient usage of the network resources depending on their availability. Regarding the selection of ρ, the feasible range is $\gamma < \rho < 1$.

4 Simulation Results

Simulations are performed to show the validity and the applicability of the proposed traffic modeling. At first, the accuracy of the analysis is examined. The on-off traffic is assumed to have $\gamma = 0.01$, and $\ell = 5$ or 10. Defining the token bucket load as $\lambda = \frac{\gamma}{\rho}$, we let $\lambda = 80\%$. With the corresponding ρ, we apply the analysis given in this paper to determine the token pool sizes(σ) that yield the

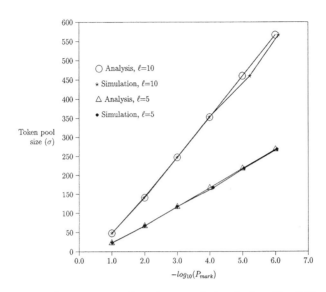

Fig. 1. Comparison of analysis and simulation (λ=80%, γ=0.01)

marking probabilities of 10^{-1} through 10^{-6}, respectively. Then, token buckets configured with those σ and ρ are simulated with the offered on-off source of $\gamma = 0.01$, and $\ell = 5$ or 10. Fig. 1 demonstrates the accuracy of the analysis for each ξ value with both $\ell = 5$ and 10.

In the following, performance of the analysis is compared with a fluid traffic approach addressed in [6]. With $\xi = 10^{-5}$, $\gamma = 0.01$, and $\lambda = 80\%$, we compute the σ values for ℓ of 1 through 20. Letting σ_f and σ_d be the results for fluid and discrete-time approach, respectively, we plot in Fig. 2 (a) both $\sigma_f - \sigma_d$ and $\frac{\sigma_f - \sigma_d}{\sigma_d}$ versus ℓ. Additionally, with $\xi = 10^{-5}$, $\gamma = 0.01$, and $\ell = 10$, we also compute the σ values versus λ of 50% through 90%, and plot in Fig. 2 (b) both the resultant $\sigma_f - \sigma_d$ and $\frac{\sigma_f - \sigma_d}{\sigma_d}$. In Fig. 2 (a), an almost constant number(i.e., around 25) of over-estimation is observable, while an almost constant ratio(i.e., about 5%) of over-estimation is observable in Fig. 2 (b), both in the fluid traffic approximation. The discrete-time analysis given in this paper, therefore, is proved to be more close to the optimality than the fluid one in terms of resource utilization.

The effects of aggregate (σ, ρ, ξ)-model traffic are examined next. We at first compute σ of an on-off source with $\ell = 5$, $\gamma = 0.01$, $\xi = 10^{-5}$, and $\lambda = 80\%$. Then, aggregation of $n(2 \leq n \leq 20)$ homogeneous on-off sources(i.e., $\ell = 5$ and $\gamma = 0.01$ for each) is offered to a token bucket which is configured with that σ, and $\lambda = 80\%$ for the aggregate offered load $n\gamma$. Remarkably, the resultant aggregate marking probability stays at the vicinity of 10^{-5} independent of n. Similar experiments with $\ell = 10$ and $\ell = 15$ yield the same results, which are all shown in Table 1. As far as $\lambda = 80\%$ is maintained, the aggregate burstiness seems to have no significant effect on the marking probability even with token

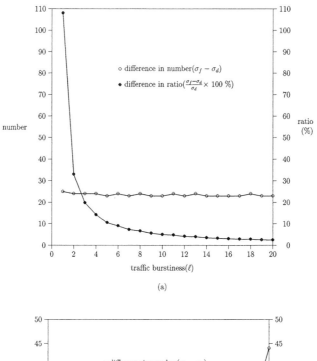

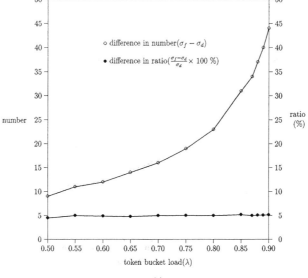

Fig. 2. Comparison between the discrete-time and fluid analysis. (a) with $\xi = 10^{-5}$, $\gamma = 0.01$, $\lambda = 0.8$ versus ℓ, and (b) with $\xi = 10^{-5}$, $\gamma = 0.01$, $\ell = 10$ versus γ. σ_f and σ_d denote the resultant σ for fluid and discrete-time analysis, respectively. From [6], $\sigma_f = \dfrac{\ell(1-\gamma)(\lambda-\gamma)}{\lambda(1-\lambda)} \ln \left\{ \dfrac{(\lambda-\gamma)(1-\lambda(1-\xi))}{\lambda(1-\gamma)\xi} \right\}$

Table 1. Aggregate marking probability in token bucket with parameters of $\lambda = 80\%$, and σ chosen to yield $\xi = 10^{-5}$ for one flow

	number of homogeneous flows in aggregate						
ℓ	1	4	8	12	16	20	σ
5	0.9×10^{-5}	0.3×10^{-5}	0.9×10^{-5}	1.1×10^{-5}	1.1×10^{-5}	1.1×10^{-5}	218
10	0.6×10^{-5}	0.5×10^{-5}	1.1×10^{-5}	0.9×10^{-5}	1.0×10^{-5}	0.9×10^{-5}	460
15	0.8×10^{-5}	0.7×10^{-5}	0.5×10^{-5}	1.2×10^{-5}	0.6×10^{-5}	0.9×10^{-5}	702

pool size scaled for only one flow. This observation is fairly encouraging when the aggregate voice traffic for EF PHB is considered. Once a voice traffic is described by $\hat{\sigma}$ and $\hat{\rho}$ computed for $\hat{\xi}$ and $\hat{\lambda}$, the aggregation of homogeneous voice traffic is also describable by the same values, $\hat{\sigma}$ and $\hat{\xi}$, with ρ tuned to meet $\lambda = \hat{\lambda}$. The required resources(i.e., buffer and bandwidth) for a specified $\hat{\xi}$ can be easily determined from that aggregate description.

Finally, we investigate the efficiency of parameter adjustment to increment of traffic burstiness. Assume a σ computed for specified values of γ, ℓ, λ, and ξ. When ℓ increases, there can be two ways of maintaining ξ constant: σ or λ (that is, ρ) adjustment. We examine the adjustment efficiency by measuring the resultant increasing ratio of σ and ρ. For σ adjustment, we at first compute σ for $\gamma = 0.01$, $\ell = 2$, $\lambda = 80\%$, and ξ of 10^{-1} through 10^{-5}, and refer to them as $\sigma_{\ell=2}$. Then, we compute σ for $\ell=8$(and 10) needed to maintain each ξ constant, and denote them by $\sigma_{\ell=8}(\sigma_{\ell=10})$. Meanwhile, for ρ adjustment, we at first compute $\rho_{\ell=2}$ determined from $\gamma = 0.01$, $\ell = 2$, $\lambda = 80\%$ for ξ of 10^{-1} through 10^{-5}. Then, we compute ρ for $\ell = 8$(and 10) required to maintain each ξ constant, and denote them by $\rho_{\ell=8}(\rho_{\ell=10})$. The resultant increasing ratio $\frac{\sigma_{\ell=8}}{\sigma_{\ell=2}}(\frac{\sigma_{\ell=10}}{\sigma_{\ell=2}})$ and $\frac{\rho_{\ell=8}}{\rho_{\ell=2}}(\frac{\rho_{\ell=10}}{\rho_{\ell=2}})$ are plotted in Fig. 3, which clearly shows that σ adjustment(i.e., buffer control) is far more efficient than ρ adjustment(i.e., bandwidth control). This property is desirable when we compare the cost scales of memory and bandwidth.

5   Conclusions

We propose a traffic description, (σ, ρ, ξ)-model, which is applicable for the services that require hard bound on the performance degradation, such as voice traffic. The accuracy as well as the validity of the derivation of that model are examined through simulations. It is also observed that the buffer control outperforms the bandwidth control coping with increment of traffic burstiness. When a call admission control scheme is collaborated, the (σ, ρ, ξ)-model is capable of adaptation in resource management by tuning σ and ρ depending on the resource availability. Practically, the envelope curve built with non-zero ξ is by far close to the actual arrival curve owing to the drastically lowered σ than that for $\xi = 0$. This shows the appropriateness of the (σ, ρ, ξ)-model for provisioning the

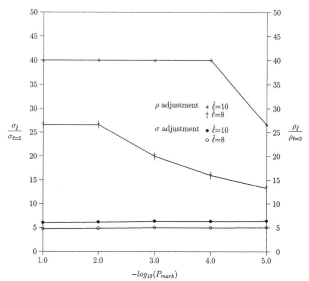

Fig. 3. Parameters increasing ratio for maintaining the marking probability over increment of traffic burstiness from $\ell = 2$ to $\ell = 8$ and 10 (σ vs. ρ adjustment)

voice services in packet networks overcoming the low utilization problem of the deterministic approach.

An arithmetic analysis of the aggregate voice traffic is remained for further study. In the Differentiated Services domain, the class-level traffic management including the EF class has significant effects on the efficiency of the per-hop behavior. The further study also includes the packet-level selective treatment in network nodes towards the service guarantee for the unmarked packets.

References

[1] W. J. Goralski, and M. C. Kolon : IP Telephony. McGraw Hill, 2000.
[2] B. Davie, A. Charny, J. C. R. Bennett, K. Benson, J. Y. Le Boudec, W. Courtney, S. Davari, V. Firoiu, and D. Stiliadis : An Expedited Forwarding PHB(Per-Hop Behavior). RFC 3246, Mar. 2002.
[3] S. Blake, D. Black, M. Carlson, E. Davie, Z. Wang, and W. Weiss : An Architecture for Differentiated Services. RFC 2475, Dec. 1998.
[4] R. L. Cruz : A Calculus for Network Delay, Part I:Network Elements in Isolation. *IEEE Trans. Information Theory*, Vol. 37, No. 1, pp. 114-131, Jan. 1991.
[5] R. G. Tucker : Accurate Method for Analysis of a Packet-Speech Multiplexer with Limited Delay. *IEEE Trans. Commun.*, vol. COM-36, pp. 479-483, Apr. 1988.
[6] M. G. Hluchyj, and N. Yin : On the Queueing Behavior of Multiplexed Leaky Bucket Regulated Sources. in *Proc. IEEE INFOCOM'93*, pp. 672-679, 1993.
[7] M. Sidi, W. Z. Liu, I. Cidon, and I. Gopal : Congestion Control Through Input Rate Regulation. *IEEE Trans. Commun.*, Vol. 41, No. 3, pp. 471-477, Mar. 1993.

CORBA Extensions
to Support QoS-Aware Distributed Systems

Jose Rodriguez[1], Zoubir Mammeri[1], and Pascal Lorenz[2]

[1] IRIT, Toulouse, France
{mammeri,rodrigue}@irit.fr
[2] University of Haute Alsace, Colmar, France
lorenz@ieee.org

Abstract. The CORBA specification is the most complete architecture for developing Object-Oriented systems. QoS-based systems are developed with CORBA using the Messaging and the Notification Services. However, some features are missing in the CORBA specification in order to make it capable of guaranteeing end-to-end QoS (quality of service). Some of these features are proposed in several projects. The objective of this paper is to review and analyze the most important proposed approaches to provide QoS guarantees with CORBA architecture.

1 Introduction

Distributed systems are becoming more and more popular. With the advent of multimedia applications, new requirements for servers, network and clients are appearing. Consider, for example the case where several clients make requests to the same server; in this case, policies, levels of service, scheduling and dispatching functions must be implemented in order to avoid overflow of communication and execution in network and server side. To solve this problem, it is necessary to meet some Quality of Service (QoS) requirements. Meeting QoS in a distributed system is not a trivial task. Several considerations must be taken into account: scheduling and dispatching policies must be implemented, QoS parameters for application and communication must be specified and mapped at several layers wide and along the system, resource reservation and administration must be assured. Communication, negotiation, and monitoring are the major functions concerned with QoS parameters administration.

The Common Object Request Broker Architecture CORBA [1] is the most advanced specification for implementing distributed systems. Although CORBA Messaging [2] and Notification Service [3] have been specified to include QoS features, some considerations must be taken into account for developing QoS-based distributed systems using CORBA as middleware.

In [4] the main functions of QoS-based distributed systems are analyzed: QoS parameters specification, QoS mechanisms for control, provision, and management are presented. A general analysis of QoS mechanisms of some generic QoS-based distributed system is given, and we think that the components of this generic architecture can be applied to CORBA-based systems.

H.-K. Kahng (Ed.): ICOIN 2003, LNCS 2662, pp. 535–542, 2003.

We have analyzed several frameworks for QoS-based distributed systems. Every framework makes a contribution to some particular aspects of interest. In this paper we analyze those contributions and we propose a way to include those components in a general CORBA-based QoS framework. In section 2, the parameter specification models are presented. In section 3, the mapping of these parameters at different layers is analyzed. In section 4, the problems of communication, resource reservation and routing are presented. Different clients may be connected to a server at the same time, every one with different needs of service, so control mechanisms are needed to accept or reject some requests. These aspects are presented in section 5.

2   QoS Parameters Specification

QoS specification is concerned with capturing QoS requirements. Specification is different at each system layer and is used to configure and maintain QoS conditions in the endsystems and the network [4]. QoS specification is application-specific, and specification formats are tailored for targeted application domains. Thus, translation from the original application-level notation into the system-level parameters is needed [5].

QoS specification includes requirements for:

− *Performance:* expected performance characteristics, parameters to establish resource commitments.
− *Level of service:* specifies the degree of resources commitment required to maintain performance guarantees (deterministic, predictive and best effort are the resource commitment levels of service).
− *Policy:* a policy dictates a number of conditions that must be met before a specified action can be taken.
− *Cost of service:* specifies the cost the user will incur for the level of service.
− *Synchronization:* specifies the degree of synchronization required between related flows, services and events.
− *Reliability:* availability, fault tolerance or data integrity.
− *Security:* request protection and partner authentication.
− *QoS management:* the degree of QoS adaptation that can be tolerated and the actions to be taken if the contracted QoS cannot be met.

Specification of QoS parameters listed above and assignation to entities can be established with different approaches [6]:

− *IDL:* QoS parameters are defined by IDL types or CORBA services, in this approach the original CORBA is left untouched.
− *QoS language definitions:* new language for QoS definitions and assignment to entities is defined, in addition to IDL; the original IDL rest untouched, but a mapping must be done between the IDL and the QoS language.
− *Implicit IDL enhanced:* no changes to the IDL are defined, instead special reserved identifiers in IDL are parsed by an enhanced IDL compiler that

treats those identifiers as QoS definitions. In this approach QoS definitions are known in all the system, but the problem is the identification of QoS related definitions and original IDL ones.

- *Explicit IDL enhanced:* IDL is enhanced with special keywords and structures to express QoS definitions and the assignment to entities. The advantage of this approach is that QoS definitions and assignments are made explicit and easy to be recognized, but objects may loose their portability.

The primary objective in specifying QoS parameters with IDL is to keep the notion of separation between interfaces and their implementations. One way to specify QoS parameters in IDL is through the *Any* CORBA data type and the use of IDL structures [7, 8]. Approaches of this type are kinds of *implicit IDL enhanced* mentioned above. It is necessary that IDL allows specifying different QoS constraints, application types, type of requests, execution times, deadlines, dependability between request, etc. [9].

3   Parameter Mapping

QoS mapping performs the translation between representations of QoS at different system levels. For example, QoS mapping can be used to derive the scheduler QoS parameters. QoS requirements listed above are used to derive resource requirements such as computation times, communication, memory, etc. QoS parameters are mapped into various system layers to give quantitative QoS parameters; these quantitative parameters may be oriented towards [10]:

- Performance, sequential or parallel processing.
- Format, transfer rate, data format, compression schema, image resolution, etc.
- Synchronization, synchronous or asynchronous, loosely or tightly coupled.
- Cost, platform rates, connection and data transmission rates.
- User, quality of images, sound, response time.

Although several projects have been developed, there are few papers that make reference to how to map parameters between layers: user parameters into application parameters, application parameters or end system into connection parameters, etc. Parameters translation is not a trivial task, Nahrstedt [11] presented a translation relationship between Media quality parameters and Connection quality parameters; for example packet rate is evaluated as the character size divided by the packet size multiplied by the sample rate.

In [12] QoS parameters are divided into generic application-level and generic-system level. Parameter mapping is done in a direct way; for example, the end-to-end delay is evaluated as network delay plus operating system delay, and the application cost is mapped into network cost plus operating system cost, etc.

In CORBA Architecture QoS mapping must be done at several layers, through application, domain specific and common middleware services, operating systems and protocols, and hardware devices. Fig. 1 presents the layers where QoS mapping have to be done.

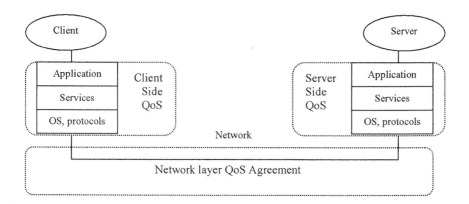

Fig. 1. Layers concerned with QoS mapping

Developing distributed QoS-aware systems with CORBA, implies mapping between the CORBA layers, from top to bottom. Stubs and skeletons capable to map QoS parameters into the format adapted to the ORB are needed. Mappings are needed too, in the ORB services, mainly Name, Scheduling, Notification, and Messaging. Mapping from ORB into operating system is needed too; for example, in real-time operating systems a mapping could be needed to use the scheduling policy of the operating system. Mapping is needed too at the endsystem boundary; protocols must be capable of translating QoS parameters into a format adapted to the network to only transport the data or to make resource reservation along the request path.

4   Communication

In distributed multimedia systems, meeting QoS guarantees from application to application is a fundamental issue. QoS must be configurable, predictable, and maintainable, from end system devices to communication subsystems and networks [4] and all end-to-end elements in the distributed architecture must work with the same efficiency to guarantee the required QoS level.

Several specific protocols have been implemented. In general minimizing context switching, system calls, data copying, and preempting are the main properties dealt with in these implementations. Mainly extensions in the transport layer, application layer, and middleware layer are implemented. Since several layers are implied, some protocols have been implemented incorporating several types of communication, layer-to-layer, layer-to-OS, peer-to-peer [13, 14, 15].

Internet Protocols (IP)-based networks provide best effort level of service by default. In this kind of networks, the complexity resides in the end-host, so the network remains as simple as possible. As more hosts are connected more resources are required; eventually, system resources capacity may be exceeded, system may start to degrade gracefully.

Some QoS protocols and algorithms have been specified to provide some level of quantitative and qualitative services to IP-based network services. Adding some 'intelligence' in the network layer to distinguish types of traffics is the main philosophy of QoS parameters. QoS-aware protocols are applied at network layer; the end-to-end QoS provisioning principle is the primary objective of this type of protocols. Providing different levels of service for different kinds of clients is one of the primary challenges for this architecture [16]. Several QoS-aware protocols and algorithms have been specified to take into account this type of requirements:

- Resource Reservation Protocol (RSVP), a signaling protocol to enable network resources reservation.
- Differentiated Services, used to categorize and prioritize network flows.
- Multi Protocol Label Switching, a bandwidth management via network routing control according to label in the packet headers.
- Subnet Bandwidth Management, categorization and prioritization at data-link layer on shared switched IEEE 802 networks.

Efficient mapping between layers in endsystems and network resources administration are necessary to obtain end-to-end predictability, guarantees, and efficiency in QoS-distributed systems.

In QoS-based routing, flow's paths could be determined based on some knowledge of resources availability in the network, as well as the QoS requirements of flows. QoS-based routing has some primary tasks: dynamic determination of feasible paths, optimization of resource usage, graceful performance degradation [17]. Internet is focused on connectivity and best-effort service. Extending current routing paradigm to allow QoS routing implies:

- Support traffic using integrated services, multiple paths between node pairs must be calculated. If QoS metrics change frequently, routing updates can become more frequent.
- Internet's opportunistic routing will shift traffic from one path to another as soon as a better path is found, event if the existing path can meet the service requirements, frequently changing routing can increase the variation in the delay and the jitter.
- Routing algorithms do not support alternate routing.

While resource reservation protocols provide network resources reservation, they do not provide mechanisms for determining a network path that has adequate resources to accommodate the requested QoS. Conversely, QoS routing allows the determination of a path that has a good chance to accommodate the requested QoS, but it does not include a mechanism to reserve the required resources. QoS-based routing may be used in conjunction with resource reservation protocols; combining both technologies allows fine control over the route and the resources.

The OMG has defined the General Inter-ORB Protocol (GIOP) with the objectives of availability, simplicity, scalability, low cost, generality, and architectural neutrality in mind [1]. The GIOP can be mapped onto any connection oriented transport protocol. The CORBA messaging specifies a routing algorithm to solve the problem when the target is not active, but this routing is not based on QoS parameters.

The CORBA specification has considered neither the resource reservation nor the QoS-based routing. The OMG GIOP is an abstract protocol framework that defines the major components necessary for clients and servers to communicate, whereas the Internet Inter-Operable Protocol (IIOP) is a mapping of GIOP in TCP/IP. To turn GIOP in a concrete protocol specification, it is necessary to specify the encoding of the Inter-operable Object Reference (IOR).

It is clear that in order to specify resource reservation protocols and QoS-based IP routing is not sufficient. To guarantee QoS properties, CORBA applications must be capable to interact with other technologies. As CORBA communication is based on IP protocol, one approach to integrate IP and RSVP is by using the Type Of Service field in the IP header; another approach is by placing a protocol on top of IP occupying the place of a transport protocol in the protocol stack [18]. Developing CORBA QoS-distributed systems needs the participation of different types of protocols, transport, signaling, and routing. Minimizing data copies, context switching, and system calls is the responsibility of endsystem protocols, but optimizing network resources is the responsibility of resource reservation and QoS-routing protocols.

5 Negotiation, Admission Testing, and Monitoring

In QoS-based distributed systems, clients and servers may be added at anytime, so the general conditions of system may change at execution time. In consequence, negotiation, admission test and monitoring the resource status are necessary. Developing of adaptable distributed systems is not an easy task, some implementations that have been developed use QoS policies, contracts, and control functions.

QoS-negotiation begins with the evaluation of resources requirements arising from the request against the available resources in the system. If resource availability is guaranteed and resource management policies are respected the request can be accommodated; if admission testing is successful resources are reserved. If after admission test the request is rejected, the endsystem application may change its resources needs.

QoS-monitoring allows tracking the status of QoS levels at different layers in the system. QoS-monitoring allows too the administration of QoS status, availability, degradation, and maintenance of resource entities. Mechanisms to deal with QoS monitoring have been proposed. Particularly, resource adaptors component and configurators are proposed like entities neutral to applications and specific to each resource type [5]. The resource adaptors control all concurrent applications sharing the same resource in the same host. Configurators perform QoS adaptation by deleting, replacing, or adding application components. Contracts and SysCon objects are presented in [19]. Contracts are used to specify the level of service desired by the client, the level of service that the object expects to provide, QoS measures, and actions to take if QoS level changes. SysCond objects are used to measure and control QoS.

Mechanisms for admission testing, monitoring, and negotiation, could be implemented in CORBA using the interceptors. Interceptors are objects in the ORB that are used to invoke ORB services. Interceptors are interposed in the invocation path between a client and the server. In order to allow application negotiation and admission control, the CORBA interceptors could be extended. At creation time, interceptors may get the policies that apply to the client and the target, binding is performed by interceptors. Binding uses information about policies to apply, security mechanisms and protocols supported, and execution context. QoS-Interceptors and binding could use information about the request and policies in order to make an access control decision; binding may reserve resources in order to guarantee QoS.

6 Conclusions

CORBA is the most complete middleware architecture to implement object-oriented distributed systems. In the beginning, CORBA has mainly addressed client-server applications, but with the development of systems that need more control over resources and predictable behavior of applications, some extensions to the CORBA architecture have been necessary.

Vertical and horizontal mapping of QoS parameters is needed. The vertical mapping is between the application layer and the operating system and network. The horizontal mapping is needed to guarantee that network resources will be available for the requests when needed. An integral mapping including endsystem layers and network resources is needed.

There are two ways to develop QoS-based distributed systems, the first one is based on priority mechanisms and preemption, the second on resource reservation. We think that for building reliable QoS applications, it is necessary to combine the best provided mechanisms by both philosophies.

Making CORBA adaptable for QoS-aware applications implies optimal functioning and extensions of several mechanisms, stubs and skeletons, CORBA services, protocol switching contexts and data copies. Adapting those components to the CORBA architecture may not be easy. CORBA has been in constant evolution; components and services for message and event oriented applications have been added to CORBA. The main objective of CORBA specification is to alleviate developers of problems related to network and communication, but in developing QoS-distributed systems communication and resources control is an important requirement, that is what CORBA try to hide them to developers.

All the mechanisms analyzed in this paper could be used in CORBA in some case, news capabilities are needed, in other cases extending actual mechanisms of CORBA is sufficient, but it is clear that the CORBA specification is still in evolution.

References

[1] Object Management Group: The Common Object Request Broker: Architecture and specification 2.4. OMG (2000)

[2] Object Management Group: CORBA Messaging. Document orbos/98-05-05. OMG (1998)

[3] Object Management Group: CORBA Notification Service. Document telecom/99-07-01. OMG (1999)

[4] Aurrecoechea, C., Campbell, T., Hauw, L.: A Survey of QoS Architectures. Multimedia Systems **6** (1998) 138-151

[5] Nahrstedt, K., Dongyan, X., Duangdao, W., Baochun, L.: QoS-Aware Middleware for Ubiquitous and Heterogeneous Environments. IEEE Communications Magazine (2001) 140-147

[6] Becker, C., Geihs, K.: Generic QoS Specification for CORBA. Proceedings of kommunikation in verteilten systems (KIVS '99) Darmstadt/Germany (1999)

[7] Becker, C., Geihs, K.: Quality of Service- Aspects of Distributed Programs. International workshop on Aspect-Oriented programming, ICSE'98, Kyoto Japan (1998)

[8] Campbell, A., Coulson, G., Hutchinson, D.: A Quality of Service Architecture. ACM SIGCOMM **24(2)** (1994) 6-27

[9] Mammeri, Z., Rodríguez, J., Lorenz, P.: Framework for CORBA Extensions to Support Real-Time Object-Oriented Applications. Telecommunication Systems **19(3)** (2002) 361-376

[10] Hansen, G.: Quality of Service. (1997) www.objs.com/survey/QoS.htm

[11] Nahrstedt, K., Smith, J.: The QoS Broker. IEEE Multimedia Spring **2(1)** (1995) 53-67

[12] Siqueira, F.: The Design of a Generic QoS Architecture for Open Systems. www.cs.usyd.edu.au/~mingli/QoS-architecture.html

[13] Becker, C., Geihs K.: MAQS-Management for Adaptive QoS-enabled Services. IEEE Workshop in Middleware for Distributed Real-Time Systems and Services, San Francisco USA (1997)

[14] Gopalakrishnan, R., Parulkar, G.: A Framework for QoS Guarantees for Multimedia Applications within an Endsystem (1997) www.objs.com/surveys.QoS.htm

[15] Schmidt, D., Levine, D., Mungee, S.: The design of TAO real-time object request broker. Computer Communications **21** (1998) 294-324

[16] Stardust: White Paper-QoS Protocols & Architectures. (1999) www.qosforum.com

[17] Crawley, E., Nair, R., Rajagopalan, B., Sandick, H.: A Framework for QoS-based Routing in the Internet. IETF, RFC 2386 (1998)

[18] Braden, R., Zhang, L., Berson, S., Herzog, S., Jamin, S.: Resource Reservation Protocol -version 1 Functional Specification. RFC 2205, IETF (1997)

[19] Loyall, J., Schantz, R., Zinky, J., Bakken, D.: Specifying and Measuring Quality of Service in Distributed Objects Systems. Proceedings of ISORC'98, Kyoto, Japan (1998)

Class-Based Fair Intelligent Admission Control over an Enhanced Differentiated Service Network

Ming Li, Doan B. Hoang, and Andrew J. Simmonds

Faculty of Information Technology
University of Technology Sydney, Sydney, NSW 2007, Australia
{mingli,dhoang,simmonds}@it.uts.edu.au

Abstract. Integrated Service (IntServ) can provide powerful QoS on a per flow basis but it requires routers to perform per flow admission control and maintain per flow state. Differentiated Service (DiffServ) model is more scalable, but cannot provide service that is comparable to IntServ. To achieve scalability and a strong service model, we believe that DiffServ equipped with scalable admission control is a possible solution. This paper examines some fair intelligent admission control schemes over the enhanced DiffServ [1]. Specifically, we propose two admission control schemes: random early dropping admission control and random early re-marking admission control. We apply these admission control schemes to address traditional DiffServ problem concerning fairness. The simulation results demonstrate our schemes can offer a class-based fair resource sharing. Such admission control schemes can also help enforce the desired service assurances.

1 Introduction

Internet only provides best-effort service for all traffic. Traffic is processed as quickly as possible but there is no guarantee as to timeliness or even successful delivery. The network makes no attempt to differentiate its service response between the traffic streams generated by concurrent users of the network. This means that the network is not able to guarantee the level of service required by an application that demands more stringent response in terms of delay, jitter, bandwidth, etc. Work on Quality of Service (QoS) -enabled IP networks has led to two distinct approaches: the Integrated Services (IntServ) and the Differentiated Service (DiffServ).

The goal of IntServ is to provide end-to-end QoS [2]. The IntServ architecture needs an explicit setup mechanism to convey information to routers so that they can provide the requested services to flows. The resource requirements for running per-flow resource reservations on routers increase in direct proportion to the number of separate reservations that need to be accommodated. The use of per-flow state and per-flow processing is thus not feasible across the high-speed core of a network.

H.-K. Kahng (Ed.): ICOIN 2003, LNCS 2662, pp. 543–552, 2003.
© Springer-Verlag Berlin Heidelberg 2003

In contrast, the DiffServ architecture achieves scalability by limiting QoS functionalities to class-based priority mechanisms. The DiffServ architecture is composed of a number of functional elements: packet classifier, traffic conditioner and per-hop forwarding behaviors (PHB) [3]. The PHB determines the priority and the maximum delay in the transmission queues. There are two types of PHBs besides the default best-effort service: expedited forwarding (EF) PHB and assured forwarding (AF) PHB. EF PHB is a high-priority behavior, for e.g. network control traffic, while AF forwards packets based on some assigned level of queuing resources and three-drop precedence. However, without per-flow admission control, such an approach weakens the service model as compared to IntServ.

To achieve scalability and a strong service model, we have devised class-based fair intelligent admission control over DiffServ architecture. These admission control schemes are not only based on the traffic service level agreement but also based on the network resource capability. Due to DiffServ's lack of knowledge of network resource capability, we developed an enhanced DiffServ model, Fair Intelligent Resource Discovery DiffServ (FIRD-DiffServ) [1], which can provide network resource capability dynamically. The admission control decisions are made solely at the ingress edge router; per-flow state is not maintained in the network core router, and there is no coordination of state with core nodes. Therefore, admission control is performed in a scalable way.

It has been observed that the traditional DiffServ in some cases can hardly achieve the desired quality of service and result in unfair resource sharing [4]. We apply our admission control schemes to address above problem. The simulation results show our admission control over enhanced DiffServ can offer better fair resource sharing.

The paper is organized as follows. Section 2 puts our work in context. Section 3 presents our fair intelligent admission control schemes. In section 4, we present a set of simulation experiments to assess the performance of our fair intelligent admission control schemes against traditional DiffServ model. Section 5 concludes with suggestions for future work.

2   Context

Over the last two years, several research efforts have been made to find an appropriate QoS architecture, which combines the advantages of DiffServ and IntServ.

De Meer et al [5] provided an analysis of existing IP quality of service solutions and the implied signaling issues. It is pointed out that an improvement to the QoS DiffServ architecture could be achieved by providing congestion signaling from within a DiffServ domain to the boundary between the two administrative domains. It is also believed that feedback information and signaling is needed in the next generation of a DiffServ architecture that delivers its specified classes of service by a combination of resource provisioning and cooperation with the subscribers. Our proposal addresses these issues.

Jeong, et al [6] proposed a set of router-based QoS mechanisms including queue policy, resource reservation and metering using the enforcement of traffic profile. The proposed queue policy is to ensure that UDP flows get required bandwidth and TCP flows are protected from unresponsive UDP flows. The proposal only considered simple buffer partitions for allocating bandwidth. Our project uses class-based fair intelligent admission control mechanisms at the edge router to dynamically adjust incoming class traffic.

Gerla, et al [7] considered bandwidth feedback control of TCP and real time sources in the Internet. The end hosts use this information to adjust their congestion window. However, their scheme needs to modify current TCP implementation by adding one state variable to store the round trip propagation delay and the available bandwidth-delay product. Our scheme is transparent to TCP, requires no modifica-tions to current TCP implementations and also can be applied to UDP traffic.

The Endpoint Admission Control scheme [8] conveys the congestion status of network nodes to the end-points. The idea is sound but may not be adequate to control the connection's QoS. Our approach employs feedback information explicitly from both the endpoints and the core routers. We also introduce an acceptance region to replace a single threshold to avoid thrashing.

Kumar, et al [9] proposed an intelligent marker, which relies on an Explicit Congestion Notification (ECN) -like feedback control mechanism. The marker uses a congestion factor provided by the control mechanism to calculate marking probability. The proposal only considers congestion status as 1 or 0 without indicating congestion degree. Our project applies the class-based fair intelligent admission control mechanism with an accurate level to control incoming class traffic at ingress router.

3   Fair Intelligent Admission Control over Enhanced DiffServ

The key idea of enhanced DiffServ, FIRD-DiffServ, is to calculate class-based network resource capability via closed-loop feedback. Refer to [1] for detailed description. Besides the normal functions of ingress edge, FIRD-DiffServ ingress edge router is responsible for generating Resource Discovery (RD) packet, which is used to collect networking resource capability state. The RD packet is generated proportionally to the class traffic rate. Each core router along the path consults its QoS state for that class and modifies the parameters of the RD packets it can support, then forwards the RD packets to the next router till to the destination edge router. All intermediate core routers perform exactly the same algorithms to calculate mean fair share rate, avail-able buffer and congestion condition for the class traffic.

In this paper we focus on the admission control model over enhanced Diff-Serv. Fig. 1 illustrates our framework: the explicit feedback control loop operates between a destination edge router and a source edge router. In particular, we

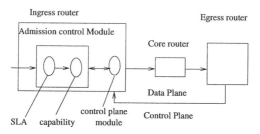

Fig. 1. The enhanced DiffServ framework

will illustrate the key design issues that must be addressed for scalable admission control.

The role of admission control is to check whether there is enough network resources in the data path to accommodate the class-based traffic requirements. Pure DiffServ architecture only considers the class-based traffic requirements via service level agreement (SLA) in a static way; it doesn't consider dynamic network resource capability. Once a flow is allowed to enter a DiffServ domain, it is usually policed or conditioned at the ingress node according to its agreement profile even when the network is congested. Our admission control not only consider the traffic requirement, it also considers the network capability.

The admission control mechanism introduces a new admission control module to an ingress router, which decides to accept or reject the class-based traffic.

In the following subsections, we introduce two different admission control mechanisms: dropping with probability and remarking with probability.

Our design consists of two inter-related components: (1) the enhanced Diff-Serv, FIRD-DiffServ, including control plane and original DiffServ data plane, (2) the admission control module which accepts or rejects incoming class-based traffic.

Fig. 1 depicts the relationship among these system components. The admission control module has two components: SLA (service level agreement) component and capability component, which decides admission or rejection based on the network capability state provided by FIRD-DiffServ. When a packet arrives an ingress edge router, the SLA component exercises the normal DiffServ classifying function according to the service level agreement and then forwards the packet to the capability component. The capability component will decide on admission or rejection based on the network capability state. Meanwhile, the control plane of the enhanced DiffServ periodically sends a resource discovery packet to the network to detect class-based network resource capability. The admission control module at ingress edge router uses this information in making its decision.

In our implementation, we set maximum admission control threshold (max_th) as the explicit capability rate and set the minimum threshold (min_th) as the 90% of the explicit capability rate.

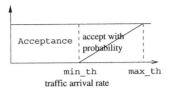

Fig. 2. Acceptance region

If the incoming traffic rate is below the minimum threshold, the packet will be forwarded to the network. If the incoming traffic rate is beyond the maximum threshold, the packet will be dropped or remarking (downgrade). If the incoming traffic rate is between the minimum threshold and maximum threshold, the packet will be dropped or remarked with probability β. This is similar to the RED [10] idea.

3.1 Random Early Dropping Admission Control Scheme

Armed with FIRD-DiffServ, the admission control module has the network capability knowledge, explicit capability rate. According to this explicit capability rate, the random early dropping admission control converts the network capability state to packet dropping probability b. This random early dropping admission control drops the packets with probability at the edge router in order to avoid congestion in core routers. The algorithm is described in following algorithm.

Random early dropping admission control algorithm

```
set max_th = network capability rate; //by FIRD-DiffServ
set min_th = 0.9 * max_th;
set factor = 0.95; //adjust factor

When a~new packet arrives,
if the class traffic arrival rate <= min_th
   forward the packet into network
else if min_th <= the class traffic arrival rate <= max_th
   drop_prob =
      factor * (class traffic rate - min_th)/(max_th - min_th)
   drop the packet with the drop_prob probability
else if the class traffic arrival rate > max_th
   drop the packet
When backward RD packet arrives,
     update the network capability rate
```

The dropping scheme works as follows: if the incoming class traffic arrival rate is below the minimum threshold, the admission control will forward the packet to the network. If the incoming packet rate is between the minimum threshold and maximum threshold, the admission control will drop the incoming packet with the probability β. If the incoming packet is beyond the maximum threshold, the admission control scheme will drop the incoming packet.

3.2   Random Early Remarking Admission Control Scheme

The idea of the remarking probability algorithm is similar to the dropping scheme. The remarking admission control scheme only works on IN-profile packet.

We set the maximum admission control threshold as the network capability rate provided by FIRD-DiffServ and set the minimum threshold as the 90% value of the maximum threshold. The remarking with probability algorithm and edge router behavior is described in the following algorithm.

Random early remarking admission control algorithm

```
set max_th = network capability rate; //by FIRD-DiffServ
set min_th = 0.9 * max_th;
set alpha = 0.95; //adjust factor

When a~new packet arrives,
if the class traffic arrival rate <= min_th
   forward the packet into network
else if min_th <= the class traffic arrival rate <= max_th
   remark_prob =
      alpha * (class traffic rate - min_th)/(max_th - min_th)
   downgrade IN packet with the remark_prob probability
else if the class traffic arrival rate > max_th
   downgrade IN packet
When backward RD packet arrives,
   update the network capability rate
```

The remarking admission control scheme works as follows: if the incoming class traffic arrival rate is below the minimum threshold, the admission control scheme will forward the incoming packet to the network. If the incoming packet rate is between the minimum threshold and maximum threshold, the admission control scheme will re-mark or downgrade the IN-profile packet to the OUT-of-profile packet with probability b. If the incoming packet is beyond the maximum threshold, the admission control scheme will downgrade the IN-profile packet to an OUT-of-profile packet.

4   Fairness Evaluations for Admission Control over FIRD-DiffServ

In this section, we apply our admission control scheme to address traditional DiffServ problem: unfairness resource sharing [11]. We classify the unfairness problems into two categories: (a) fairness among TCP traffic classes with different round trip times, the traditional DiffServ is biased against long round trip time TCP traffic. (b) Fairness among TCP/UDP traffic mixes, the UDP traffic has advantage over the TCP traffic in traditional DiffServ.

We studied the simplified traditional AF DiffServ model [12], RED with In and Out (RIO). The basis of the RIO mechanism is RED-based differentiated dropping of packets during congestion at the router. In RIO, traffic profiles for end-user are maintained at the edge of the network. When user traffic exceeds the contracted target rate, their packets are marked OUT-of-profile. Otherwise, packets are marked IN-profile.

4.1 Topology Configuration

For the standard DiffServ simulation in ns-2 [13], we use software developed at Nortel Networks [14]. We have developed several new modules - the FIRD-DiffServ, and the admission control - and incorporated them into ns-2.

Fig. 3 shows the configuration topology. There are two class traffics sending data to the destinations through a "DiffServ" region. A class may be an aggregation of TCP or UDP flows. The TCP version used in our simulations was TCP Reno, which includes fast retransmission and fast recovery. The packet size is 512 bytes. The receiver's advertised window is configured large enough so that it is never a limit on the TCP sender's window. We use a constant bit rate, CBR, to model non-responsive sources UDPs since a CBR source does not have a congestion control mechanism. The sending rate of CBR is 10 Mbps and the achieved throughput of the CBR connection is calculated at the receiver and calculated the received packets over time. We compare the performance of our different admission control mechanisms and the DiffServ scheme (Time Sliding Window (TSW) DiffServ scheme by Fang and Clark [12]). In particular we want to investigate the throughput performance with respect to RTT. The Time Sliding Window (TSW) tagging algorithm runs on the edge routers that tag packets as IN or OUT according to specific service profiles.

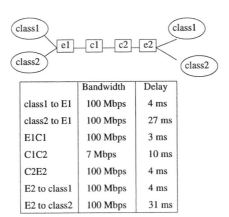

	Bandwidth	Delay
class1 to E1	100 Mbps	4 ms
class2 to E1	100 Mbps	27 ms
E1C1	100 Mbps	3 ms
C1C2	7 Mbps	10 ms
C2E2	100 Mbps	4 ms
E2 to class1	100 Mbps	4 ms
E2 to class2	100 Mbps	31 ms

Fig. 3. Scenario topology and configuration

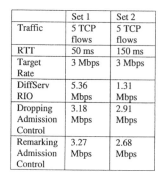
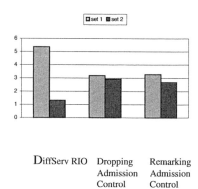

	Set 1	Set 2
Traffic	5 TCP flows	5 TCP flows
RTT	50 ms	150 ms
Target Rate	3 Mbps	3 Mbps
DiffServ RIO	5.36 Mbps	1.31 Mbps
Dropping Admission Control	3.18 Mbps	2.91 Mbps
Remarking Admission Control	3.27 Mbps	2.68 Mbps

Fig. 4. Throughput (in Mbps) for different schemes

The simulation is running for 200 seconds. We assume that the system is in the steady state (no slow start), and all TCP flows have unlimited data to send. We also assume that the system is under-subscribed, i.e., there is a surplus of bandwidth that can be allocated to each flow.

4.2 Experiment 1:
Class-Based TCP Traffic with Different Round Trip Time

The goal of this experiment is to study the impact of round trip time on through-put. In this scenario, there are two sets of traffic. Each set of traffic has 5 TCP flows and has target rate of 3 Mbps. One set of traffic's round trip time, RTT, is 50 ms while the RTT for another set of traffic is 150 ms.

Fig. 4 displays the throughput of 2 traffic sets, they have same target rate but different round trip time. Traditional DiffServ, RIO scheme, does not solve the unfairness of TCP protocol. Traffic set 1 with short round trip time TCP flows, claims the most bandwidth and traffic set 2 with longer round trip time TCP flows, claims the least. Fig. 4 compares the throughput performance of our admission control schemes with DiffServ. It is seen that our admission control schemes with FIRD-DiffServ provide much better fairness to all traffic sets irrespective of round trip times.

4.3 Experiment 2:
TCP and UDP Traffic Co-exist Environment

The second scenario is to study the impact of flows aggregated with different traffic types. One of traffic sets has 5 TCP flows with the same round trip time. Another traffic set has 5 UDP flows with the same sending rate, 10 Mbps.

Fig. 5 displays the throughput of two traffic sets all with the same propagation time and same target rate, the traffic set 1 contains 5 TCP flows and

	Set 1	Set 2
Traffic	5 TCP flows	5 UDP flows
Sending rate		10 Mbps/flow
Target Rate	3 Mbps	3 Mbps
DiffServ RIO	0.2 Mbps	6.72 Mbps
Dropping Admission Control	2.71 Mbps	3.17 Mbps
Remarking Admission Control	2.6 Mbps	3.02 Mbps

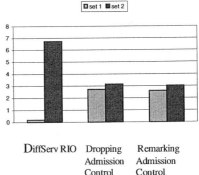

Fig. 5. Throughput (in Mbps) for different traffic type

the traffic set 2 contains 5 UDP traffic flows. The bottleneck link capacity is configured as 7 Mbps. Clearly, without being restrained by the TCP congestion control algorithms, the set 2, with UDP traffic flows, claims a larger amount of bandwidth. This may not be acceptable to a subscriber who would like the two traffic types to be treated equally. In this graph, the desired service with the class-based priority is not realized. However, because the TCP connection reduces its own rate when the packet discard occurs, the TCP throughput of set 1 is little bit lower than UDP set 2.

5   Discussions and Conclusion

Our admission control schemes rely on FIRD-DiffServ that provides the network capability dynamically. The idea behind our admission control schemes is that our admission control is not only based on the class traffic service level agreement but also based on the network capability state.

In this simulation there is little to choose between our two admission control schemes. This is because in the simulation our resource discovery packet and feedback loop ensures the ingress router has an accurate knowledge of the network state or capability, hence whether packets are dropped immediately (Random Early Drop-ping), or marked as OUT-of-profile (Random Early Remarking) has little effect, since remarked packets will soon be dropped by the network anyway. There would be difference with a less accurate knowledge of the network capability state.

We apply our admission control schemes over FIRD-DiffServ to address the fundamental problem of traditional DiffServ: fairness resource sharing. The simulations results show our admission control schemes have better performance than traditional DiffServ. Comparing to the remarking admission control, the dropping admission control may cause TCP sources to back off when their packets are dropped. In case of mixed TCP and UDP traffic, they protect the TCP traffic more efficiently from the greedy UDP traffic than traditional DiffServ scheme. They also can help reduce traffic bursts.

With these two admission control schemes we have shown that RD and feedback works so that the ingress edge router has an accurate view of the capability of the network.

Currently, we only study the admission control schemes within an intra-DiffServ domain. To provide connections' end-to-end QoS, we plan to apply these admission control schemes to multiple domains situation.

References

[1] Hoang, D. B., Li, M.: Resource Discovery and Control Mechanism for DiffServ. In: Proceedings of IEEE ICC'03, Anchorage, Alaska (2003)

[2] Braden, R., Clark, D., Shenker, S.: Integrated Services in the Internet architecture: an overview. IETF RFC1633. (1994)

[3] Blake, S., Black, D., Carlson, M., Davies, E., Wang, Z., Weiss, W.: An architecture for Differentiated Service. IETF RFC2475. (1998)

[4] Ibanez, J., Nichols, K.: Preliminary simulation evaluation of an assured service. Internet Draft, draft-ibznez-diffserv-assured-eval-00.txt. (1998)

[5] De Meer, H., O'Hanlon, P., Feher, G., Blefari-Melazzi, N., Tschofenig, H., Karagiannis, G., Partain, D., Rexhepi, V., Westberg, L.: Analysis of Existing QoS Solution. Internet Draft, draft-demeer-nsis-analysis-02.txt. (2002)

[6] Jeong, S. H., Owen, H., Copeland, J., Sokol, J.: QoS support for UDP/TCP based networks. Computer Communications, Vol. 24. (2001) 66–77

[7] Gerla, M., Weng, W., Cigno, R. L.: Bandwidth feedback control of TCP and real time sources in the Internet. In: Proceedings of IEEE GLOBECOM. San Francisco, CA, USA (2000)

[8] Breslau, L., Knightly, E. W., Shenker, S., Stoica, I., Zhang, H.: Endpoint Admission Control: Architecture Issues and Performance. In: ACM SIGCOMM. Stockholm, Sweden (2000)

[9] Kumar, K. R. R., Ananda, A. L., Jacob, L.: Using edge-to-edge feedback control to make assured service more assured in DiffServ networks. In: Proceedings LCN 2001. (2001)

[10] Floyd, S., Jacobson, V.: Random early detection gateways for congestion avoidance. IEEE/ACM Transactions on Networking, Vol. 1. (1993) 397–413

[11] Seddigh, N., Nandy, B., Pieda, P.: Bandwidth assurance issues for TCP flows in a differentiated services network. In: Proceedings of IEEE GLOBECOM. (1999)

[12] Clark, D., Fang, W.: Explicit allocation of best-effort packet delivery service. IEEE/ACM Transactions on Networking, Vol. 6. (1998) 362–373

[13] VINT-Project: Networking Simulator version 2. LBL. (2000)

[14] Shallwani, F., Ethridge, J., Pieda, P., Baines, M.: Diff-Serv implementation for ns. http://www7.nortel.com:8080/CTL/software. (2000)

Optimal QoS Routing Based on Extended Simulated Annealing[*]

Yong Cui, Ke Xu, Mingwei Xu, and Jianping Wu

Department of Computer Science, Tsinghua University
Beijing, P.R.China, 100084
{cy,xuke,xmw}@csnet1.cs.tsinghua.edu.cn, jianping@cernet.edu.cn

Abstract. Quality-of-service routing (QoSR) is to find an optimal path that satisfies multiple constraints simultaneously. As an NPC problem, it is a challenge for the next-generation networks. In this paper, we propose a novel heuristic SA_MCOP to the general multi-constrained optimal path problem by extending simulated annealing into Dijkstra's algorithm. The heuristic first translates multiple QoS weights into a single metric and then seeks to find a feasible path by simulated annealing. Once a feasible path is found, it optimizes the cost without losing the feasibility. Extensive simulations demonstrate that SA_MCOP has the following advantages: (1) High performance in both success ratio and cost optimization. (2) High scalability regarding both network size and the number (k) of QoS constraints. (3) Insensitivity to the distribution of QoS constraints.

1 Introduction

Providing different quality-of-service (QoS) support for different applications in the Internet is a challenging issue [1], in which QoS Routing (QoSR) is one of the most pivotal problems [2] [3]. The main function of QoSR is to find an optimal path that satisfies multiple constraints for QoS applications. For the NP-completeness of QoSR problem [4] [5], many heuristics have been proposed. However, these algorithms have some or all of the following limitations [2]: (1) Most of the heuristics only focus on a branch of the QoSR problem. (2) High computation complexity prevents their practical applications; (3) Low performance sometimes leads to fail to find a feasible path even when one does exist. (4) Some algorithms only work for a specific network.

Simulated annealing is a meta-heuristic method for combinational optimization [6]. Based on an initial solution, it repeatedly iterates to a new solution. For the Multi-Constrained Optimal Path (MCOP) problem, we extend simulated annealing to take an end-to-end path as a solution in routing computation, and propose a novel heuristic, SA_MCOP (Simulated Annealing for MCOP), by extending simulated annealing [6] to Dijkstra's algorithm.

[*] Supported by: (1) the National Natural Science Foundation of China (No. 90104002; No. 69725003); (2) the National High Technology Research and Development Plan of China (No. 2002AA103067).

H.-K. Kahng (Ed.): ICOIN 2003, LNCS 2662, pp. 553–562, 2003.
© Springer-Verlag Berlin Heidelberg 2003

When a QoS connection request arrives at a router, the router uses our SA_MCOP to compute an optimal feasible end-to-end path (or the next hop) based on the network state information it maintains. This algorithm uses a non-linear energy function to translate multiple QoS constraints into a single metric. It first computes a complete shortest path tree (SPT) with respect to the traditional cost as the initial solution for simulated annealing by Dijkstra's algorithm. If the path along the current SPT is not feasible, it marks all of the nodes in the graph according to the current SPT. Then a new SPT is created in simulated annealing mode by our improved Dijkstra's algorithm with a nonzero probability $P(T)$ to select a non-optimal path, where T is the temperature for simulated annealing. If the path along the new SPT is not feasible yet, the algorithm then marks each node again based on the current SPT and computes a new SPT with a lower temperature T iteratively. When T decreases to $T \to 0$, we have $\lim_{T \to 0} P(T) = 0$. If a path is feasible, the algorithm optimizes the cost at last. Based on the theory about simulated annealing, SA_MCOP guarantees to find a feasible path when one exists. Extensive simulations also show that SA_MCOP performs well.

The rest of this paper is organized as follows. In Sect. 2 we analyze how to translate multiple weights to a single metric and summarize the simulated annealing. SA_MCOP is proposed in Sect. 3, and extensive simulations show the performance evaluation in Sect. 4. Finally, conclusions appear in Sect. 5.

2   Background information

2.1   Problem Formulation

A network is represented by a directed graph $G(V, E)$. V is the node set and an element $v \in V$ is called a node representing a router. E is the set of edges representing links, which connect the routers. An element $e_{ij} \in E$ represents an edge $e = v_i \to v_j$ of G. In QoSR, each link has a group of independent weights $(w_0(e), w_1(e), \cdots, w_{k-1}(e))$, which is also called QoS metric. For a path $p = v_0 \to v_1 \to \cdots \to v_n$ and $0 \le l < k$, the weight $w_l \in R^+$ satisfies the additive property if $w_l(p) = \sum_{i=1}^{n} w_l(v_{i-1} \to v_i)$.

Definition 1. *Multi-Constrained Optimal Path (MCOP): For a given graph $G(V, E)$ with source node s, destination node t, and a constraint vector $c = (c_0, c_1, \cdots, c_{k-1})$, when $k \ge 2$, the path p from s to t is called a multi-constrained optimal path, if (1) $w_l(p) \le c_l$ for each $0 \le l \le k - 1$ (we write $w(p) \le c$ in brief), and (2) $cost(p) \le cost(p')$ for any p' satisfying $w(p') \le c$.*

Note that both $w(e)$ and c are k-dimensional vectors. For a given QoS request and its constraint c, QoSR seeks to find the path p with optimal cost based on the network state information, where p satisfies $w(p) \le c$. Dijkstra proposed the shortest path tree (SPT) algorithm, which has a low computation complexity [8]. However, QoSR problem is related to multiple weights simultaneously. Thus the problem is changed to an NPC one that the original Dijkstra's algorithm cannot solve in polynomial time. Therefore, one feasible method is to translate the multiple weights into a single metric, as follows:

Definition 2. *Energy Function:* $g(p) = \max_{l=0}^{k-1}\{w_l(p)/c_l\}$ *is called the energy function of path p, where* $c = (c_0, c_1, \cdots, c_{k-1})$ *is the constraint vector of a QoS application.*

2.2   Simulated Annealing

Research on statistical mechanics shows that in temperature T, the probability for a molecule of substance to stay in the state r satisfies Boltzmann's distribution:

$$Pr\{\hat{E} = E(r)\} = \frac{1}{Z(T)}exp(-\frac{E(r)}{k_B T}) \qquad (1)$$

$\hat{E}$ is the stochastic variable representing the energy of a molecule. $E(r)$ is the energy of a molecule that stays in the state r. T is the temperature. k_B is Boltzmann's constant and $Z(T)$ is the normalized factor.

Annealing is a physical process. After a metal body is heated, when it cools down slowly, the molecules of the body stay in different states with different probabilities, which satisfy Boltzmann's distribution. Annealing usually requires the following two conditions:

(1) The initial temperature is high enough so that the probabilities for a molecule to stay in arbitrary states are approximately equal. If we have

$$T_0 \gg E(r)/k_B \qquad (2)$$

then $E(r)/k_B T_0 \approx 0$. As a result, $Pr\{\hat{E} = E(r)\} \approx 1/Z(T_0)$, i.e. the probabilities are approximately equal.

(2) When it cools down to $T = 0$, all of the molecules will stay in the least-energy state with the probability being one. If r^* presents the least energy state, when $T \rightarrow 0$, we have

$$Pr\{\hat{E} = E(r)\} = \begin{cases} 1, & r = r^*; \\ 0, & others; \end{cases} \qquad (3)$$

The idea of simulated annealing was first proposed by Metropolis [9] and was applied to combinational optimization successfully in 1983 [6]. In its basic form, it first generates an initial solution as the current solution. It then selects another solution in the neighborhood of the current solution and replaces the current solution with the new one. The same process continues iteratively for many times. Although the goal is to find an optimal solution, it selects a non-optimal solution with a non-zero probability $P(T)$ to avoid being stuck in a local optimization. When the temperature decreases, $P(T)$ also decreases. When $T \rightarrow 0$, it is guaranteed to find an optimal solution since the probability $P(T)$ is zero to select a non-optimal solution.

3   MCOP Based on Simulated Annealing

3.1   The Idea of SA_MCOP

The key issues to use metaheuristics in QoSR include (1) how to express a solution, and (2) how to iterate. We extend simulated annealing to take an end-to-end path as a solution and use Dijkstra's algorithm to guarantee that a new solution is still an end-to-end path in iterations. When we compute the SPT by Dijkstra's algorithm, we select with probability $P(T)$ a node that is not the current optimal node. Therefore, our SA_MCOP can overcome the local optimization problem that all heuristics face.

For a given QoS request from s to t, node s first uses Dijkstra's algorithm to calculate the least-cost SPT rooted by s and marks each node in the network. Then it uses an improved Dijkstra's algorithm to compute new labels for each node iteratively based on the old labels computed last time. With different probabilities $P(T)$, it selects different links including non-optimal links, where $\lim_{T\to 0} P(T)$ satisfies Eq. 3. After each iteration, the temperature T decreases. When the algorithm iterates enough times, we guarantee $T \to 0$. In order to optimize the cost, when multiple feasible paths are found, the heuristic will choose the path that has the least cost.

3.2   Pseudo-code Description

Fig. 1 shows the pseudo-code of the algorithm, where SA_MCOP is the main function. The input of SA_MCOP includes a given graph with multiple QoS weights, a QoS request from s to t and a constraint vector $c = (c_0, c_1, \cdots, c_{k-1})$. In addition, we can configure the initial temperature (T_0), the gradient $(grad)$ for cooling down the temperature and the iteration times (I). If the k-dimensional weight $d[t]$ of the forward least energy path from s to t satisfies the constraint c, the algorithm returns the path successfully. Otherwise, it refuses the request. Table 1 shows the notations used in the pseudo-code.

1. **Function SA_MCOP** In function SA_MCOP, we first use Dijkstra's algorithm to compute the least-cost SPT rooted by s (Line 2), where the initial solution is the path along the SPT from s to t. We then compute the new SPT by simulated annealing (SA_Dijkstra) backwards and forwards iteratively, including (1) computing the SPT rooted by t (Line 5); (2) computing the SPT rooted by s (Line 8). After the complete SPT is computed each time by Dijkstra's algorithm or SA_Dijkstra, $d[.]$ is updated to save the newly computed weights from each node to the root of the new SPT. On the other hand, SA_Dijkstra computes a new SPT based on the $d[.]$ updated last time (Line 1 in function SA_Relax). In addition, after a new SPT is constructed, the path along this SPT from s to t is checked to see whether it satisfies constraint c (Line 3, 6 and 9). Once it does (i.e. a feasible path is found), the algorithm seeks a least-cost path by OPT_Dijkstra. If it is not feasible, we then change the temperature T for simulated annealing to construct a new SPT by SA_Dijkstra iteratively (Line 5, 7, 8 and 10).

```
SA_MCOP(G = (V, E), s, t, c, T_0, grad, I)        SA_Cheapest()
1. T = T_0                                         1. g* = min_{v∈NB} max_{l=0}^{k-1}(r_l[v] + d_l[v])/c_l
2. Dijkstra(G,s); //label with least cost          2. FOR each node v in NB
3. IF (d[t] < c) RETURN OPT_Dijkstra(G,t)          3.   E(v) = max_{l=0}^{k-1}(r_l[v] + d_l[v])/c_l − g*
4. FOR( i = 0; i < I; i + +)                        4. Z = Σ_{v∈NB} exp(−E(v)/T)
5.    SA_Dijkstra(G, t, T); // min g(p)             5. x = Z * uniform(0, 1) //Randomicity
6.    IF (d[s] < c)                                 6. sum = 0
..       RETURN OPT_Dijkstra(G, s)                  7. FOR each u in NB
7.    T = T/grad                                    8.   sum = sum + exp(−E(v)/T)
8.    SA_Dijkstra(G, s, T); //min g(p)              9.   IF sum >= x RETURN u
9.    IF (d[t] < c)
..       RETURN OPT_Dijkstra(G, t)                 SA_AddNode(u)
10.   T = T/grad                                    1. NB = NB − u // remove u from NB
11.RETURN failure                                   2. SPT = SPT + u // add u to SPT
                                                    3. FOR each node v in u's neighbor
                                                    4.   IF v is not in SPT
OPT_Cheapest()                                      5.     NB = NB + v // add u's neighbor
1. ret = INFINITY;
2. minCost = INFINITY;
3. FOR each node v in NB
4.   IF r[v] + d[v] < c //feasibility              SA_Relax(u, v)
5.     IF minCost > cost[v] //optimize cost         1. tmp = max_{l=0}^{k-1}(r_l[v] + w_l(u, v) + d_l[v])/c_l
6.       ret = v                                    2. IF g[v] > tmp // relax v to be u's child
7. IF ret = INFINITY RETURN no node                 3.   g[v] = tmp
8. ELSE RETURN ret                                  4.   r[v] = r[u] + w(u, v)
                                                    5.   cost[v] = cost[u] + cost(u, v)
OPT_Relax(u, v)                                     6.   Pr[v] = u
1. IF r[u] + w(u, v) + d[v] < c
2.   IF cost[u] + cost(u, v) < cost[v]
3.     relax v to u's child
```

(a) SA_MCOP   (b) Sub-functions

Fig. 1. Pseudo-code of the proposed heuristic

2. **Function SA_Cheapest** This function presents the idea of simulated annealing: a non-optimal node will be selected with a certain probability and the probability decreases to zero when temperature T decreases enough. In the first line of SA_Cheapest, $\max_{l=0}^{k-1}(r_l[v] + d_l[v])/c_l$ is the energy of a path defined by Def. 1. $r_l[v]$ is the forward weight, i.e. the l'^{th} weight of the path from the root of the current SPT to node v. $d_l[v]$ is the backward weight, i.e. the l'^{th} weight of the path from node v to the root of the old SPT calculated last time. The backward weight is saved when the old SPT is computed last time (Line 12-13 in function SA_Dijkstra). SA_Cheapest first selects the least energy g^* of the neighbors of the current SPT (Line 1). It then computes the energy $E(v)$ for simulated annealing (Line 2-3), which guarantees the least energy to be 0. Then the normal factor $Z(T)$ in Eq. 1 is calculated (Line 4). Finally, a node u, which will be added to the partially created SPT, is selected according to the probability distribution in Eq. 1 (Line 5-9).

3. **Function SA_AddNode**
   Similar to the original Dijkstra's algorithm, when node u is added to the partially created SPT, we use this function to change the set NB, which is the neighborhood of node u. This includes two parts: deleting node u from NB (Line 1), and adding u's neighbors that are not in the current SPT to NB (Line 3-5).

Table 1. Notations in the pseudo-code of SA_MCOP

Symbol	Meanings	Symbol	Meanings
T_0	initial temperature	I	maximum number of iterations
E(v)	the energy of node v in formula (i)	$Pr[v]$	the precedent node of node v
grad	gradient for decreasing tempera-ture	c	k-dimensional constraints of a QoS request
OPT_Dijkstra (G,s)	proposed heuristic finding the least-cost paths rooted by s	$d[u]$	backward weights of the path along the old SPT from its root to u
Dijkstra (G,s)	standard Dijkstra's algorithm for SPT rooted by s	NB	the set of the neighbors of the current SPT
u	an intermediate node	$g[u]$	the energy of node u
SA_Dijkstra (G,s,T)	proposed heuristic for SPT rooted by s based on simulated annealing	$r[u]$	forward weights of the path along the current SPT from its root to u
Z	the normal factor in formula (i)	v	a child node of node u
g*	a locally minimal energy	SPT	a partially created SPT

4. **Function SA_Relax** We relax node v via v's parent u. The energy of v via u is computed (Line 1). If this new energy of v is smaller than the old one (Line 2), node v will be relaxed to use the new energy (Line 3), the forward weight (Line 4), the cost (Line 5) and the precedent node (Line 6).

5. **Function OPT_Cheapest**
   OPT_Relax In order to guarantee the feasibility of newly found optimal path, when OPT_Dijkstra chooses the least-cost node and relaxes node, it has to check the feasibility of the new node. Only when the new potential paths satisfy the constraint, the node on such paths can be chose or relaxed. In OPT_Cheapest, when the feasibility is guaranteed (Line 4), least-cost node is added to the optimal SPT (Line 5-6). It should be note that, the optimal SPT may not be a complete one that connects all of the nodes in the graph. Instead, we only consider feasible ones marked by above SA_Dijkstra or Dijkstra (Line 7). If there is no more feasible node, OPT_Dijkstra returns the current optimal path as the least-cost one.

3.3   Complexity and Parameters

Since the computation complexity of an improved Dijkstra's algorithm only considering the cost is $O(m + nlogn)$, the complexity of SA_Dijkstra and OPT_Dijkstra is $O(km + knlogn)$, respectively. As a result, including the iteration, the overall computation complexity of SA_MCOP is $O(Ik(m + nlogn))$, where I is the maximum number of iterations. Because the feasibility of a path newly found is checked before the next iteration, when most of the QoS requests are feasible, only some of them need to iterate for multiple times. Therefore, the above complexity is the worst-case one. In fact, the running time of our SA_MCOP is almost independent of the maximum number of iterations.

Simulated annealing requires a new solution selected randomly enough at the beginning, i.e. initial temperature T_0 should be high enough. Because the energy $E(v)$ is often much less than one in SA_MCOP, it suffices to set $T_0 = 1$ to satisfy Eq. 2 . In addition, simulated annealing also requires that when the temperature

$T \rightarrow 0$, all of the molecules stay in the state with the least energy. Thus, in order to decrease the temperature quickly, we select $grad = 10$ according to the geometric proportion. In this way, after $2I$ times of iteration, the temperature $T(2I) = 10^{-2I} \ll E(v)$ satisfies Eq. 3. The following simulations show that such parameters perform well.

4   Performance Evaluation

We simulate purely random network graphs with N nodes [10] and generate k weights for each link, where $w_l(e) \in uniform[1, 1000]$ for $l = 0, 1, \cdots, k - 1$ and $w_l(e)$ has no correlation for different e or l. We simulate 10 graphs with N being 50, 100 and 200, respectively. In each graph, we select the source-destination node pair (s, t) 100 times (a particular node may be selected more than once), where we guarantee that the minimum hop is not less than three. Each source node s uses SA_MCOP to compute the least energy path for different numbers of iterations respectively. For performance evaluation, we use the success ratio (SR), which is defined as the ratio of the number of requests satisfied using a heuristic algorithm and the total number of requests generated. We first get SR of the 100 (s, t) pairs in one graph, and then calculate the average SR of 10 graphs with same number of nodes.

4.1   The Performance with Two Constraints

The evaluation depends heavily on the generated constraints of the requests, e.g. the distribution of constraints. Therefore, based on the normalized weights in the whole graph, for a given request pair (s, t), we use the method of weighted ratio simulation to generate the constraints. First, we assume that each QoS application concerns the weight w_l to a_l degree. Then we use Dijkstra's algorithm to find the path $p(s, t)$ that minimizes the linear energy $\sum_{l=0}^{k-1} a_l w_l(s, t)$. Finally, we take the weights of the path $p(s, t)$ as the QoS constraints of the pair , i.e. $c(s, t) = w(p(s, t))$.

In the case of two dimensions, we let $a_1 \in [0, 1]$ and $a_0 = 1 - a_1$ for simplicity. Because different QoS applications concern a weight to different degrees, we use the following three methods to generate a. (1) NORMAL: $a \in normal(0.5, 0.16)$; (2) UNIFORM: $a \in uniform(0, 1)$; (3) AB_NORMAL: $a \in normal(0, 0.16)$ and $a \in [0, 0.5]$. In order to guarantee that the difference between a_1 and the expectation are less than 0.5 with the probability of 99.7%, we set the standard deviation to be 0.16 in NORMAL and AB_NORMAL distributions.

The relation between the maximum number of iterations and the performance of our SA_MCOP is shown in Fig. 2 against H_MCOP. The x-axis is the method to generate QoS constraints, and the y-axis is the success ratio (SR) representing the performance of heuristic routing. With only a few iterations (e.g. $I = 1$), SA_MCOP does not perform well. The main reason is that $T_0 = 1$ is much greater than energy $E(v)$ and the strong randomicity cannot guarantee an optimal path. With more iteration times, the performance of SA_MCOP increases rapidly and

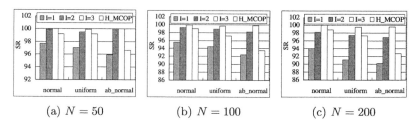

(a) $N = 50$ (b) $N = 100$ (c) $N = 200$

Fig. 2. Performance evaluation with two constraints

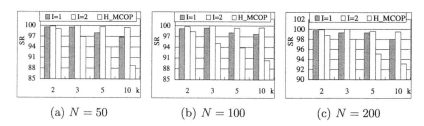

(a) $N = 50$ (b) $N = 100$ (c) $N = 200$

Fig. 3. Performance evaluation with multiple constraints

reaches almost 100%. This shows that the simulated annealing can increase the performance of QoSR greatly.

H_MCOP has different performance with different QoS constraints. The reason is that when it computes the SPT for the first time, it concerns the two weights to the same degree. Therefore, when applications concern the two weights to the same degree (normal distribution in Fig. 2), H_MCOP performs well; otherwise, it will degrade, especially in the ab_normal distribution. On the contrary, our SA_MCOP performs well in all conditions, including different distributions of QoS constraints and different network scales.

4.2   Performance with Multiple Constraints

In order to study the relation between the maximum number of iterations and the performance of SA_MCOP, we use the following method to generate the constraints for a given (s, t) pair. We first take the random number $b_l \in uniform(0, 1)$ for $l = 0, 1, \cdots, k$ and calculate $a_l = b_l / \sum_{l=0}^{k-1} b_l$. We then construct the least energy path from s to t according to the energy function $g_1'(p) = \sum_{l=0}^{k-1} (a_l w_l(p))$ and take the weights of the path as the constraints of the given (s, t), i.e. $c(s, t) = w(p_i)$.

Fig. 3 shows the performance for multiple constraints, where the x-axis is the number of constraints and the y-axis is SR. SA_MCOP performs well for large k, while H_MCOP does not. Furthermore, our SA_MCOP has good scalability on the size of network with multiple constraints.

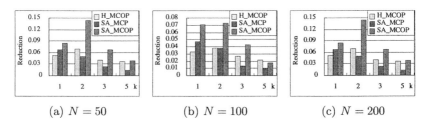

(a) $N = 50$ (b) $N = 100$ (c) $N = 200$

Fig. 4. Performance evaluation with two constraints

4.3 Optimization of Traditional Cost

The rigorous constraints that we generate for a given (s, t) pair in the above experiments restrict the number of feasible paths, even to only a single one. In order to represent the performance of optimization, we use Dijkstra's algorithm to find the shortest path p_l w.r.t. w_l for each $l = 0, 1, \cdots, k - 1$. Then we take random $c_{l+1} \in uniform(0.8w_{l+1}(p_l), 1.2w_{l+1}(p_l))$ as one element of the constraint vector c.

We compare four heuristics in this section: H_MCP, H_MCOP [7], SA_MCP and SA_MCOP. H_MCP is a variation of H_MCOP by removing the optimization parts from H_MCOP [7]. SA_MCP is a variation of our SA_MCOP without optimization (i.e. in line 3, 6 and 9 SA_MCOP just returns the feasible path along the current SPT rather than compute optimal path by OPT_Dijkstra). Furthermore, SA_MCP computes the shortest path w.r.t. the hop number as the initial solution instead of the one w.r.t. cost in SA_MCOP. SR relies on the number of iteration in SA_MCP and SA_MCOP. Because the comparison of optimization is meaningful with same or close SR, we adjust iteration times to keep a close SR. In our experiments, most iteration times are one or two.

Fig. 4 shows the average performance of cost optimization. The y-axis represents the percentage of the cost reduction by other heuristics compared with H_MCP. This figure demonstrates the four points: (1) SA_MCOP performs better than H_MCOP for more reduction of cost. (2) SA_MCOP is much better than SA_MCP, so the optimization parts in SA_MCOP algorithm are necessary and efficient. (3) When $k = 1$, the method generating constraint $c_0 \in uniform(0.8w_0(p_0), 1.2w_0(p_0))$ is so strict that there are not many feasible paths as candidates. Therefore, for the limitation of the generation method, this figure does not show a high reduction with $k = 1$. If we omit the part with $k = 1$, we will find that when k increases (i.e. more weights and constraints), for the short of feasible paths, the reduction decreases. (4) The larger the network, the more the reduction. The reason is that in larger networks, there may be more feasible path, where the importance of optimization is exhibited better.

5 Conclusion

For the NP-completeness of the multi-constrained optimal QoSR problem, there is no efficient algorithm up to now. In the paper we summarize simulated annealing and propose a novel heuristic SA_MCOP to the general QoSR problem based on simulated annealing. SA_MCOP first takes the least-cost SPT as the initial solution and marks all of the nodes in the network. It then computes a new SPT and marks the nodes again in simulated annealing mode iteratively, until the path along the new SPT is feasible or maximum iteration time is reached. If a feasible path is found, the heuristic then starts to optimize the cost, where the feasibility is still guaranteed. Extensive simulations show that SA_MCOP achieves high performance with respect to both success ratio and cost optimization. It is also scalable in both network scale and constraint number k. Furthermore, it is insensitive to the distribution of QoS constraints. In addition, although the worst-case computation complexity is $O(Ik(m+nlogn))$, which is proportional to the maximum iteration time I, the practical running time is almost independent of I.

References

[1] Xiao, X., Ni, L. M.: Internet QoS: A big picture, IEEE Network, vol.13, no.2 (1999) pp.8–18

[2] Cui, Y., Wu, J. P., Xu, K., Xu, M. W.: Research on Internetwork QoS Routing Algorithms: A Survey. Chinese Journal of Software, vol.13, no.11 (2002) pp.2065–2075

[3] Cui, Y., Xu, K., Wu, J. P.: Precomputation for Multi-constrained QoS Routing in High-speed Networks. Proc. of IEEE INFOCOM'03 (2003)

[4] Wang, Z., Crowcroft, J.: Quality-of-service routing for supporting multimedia applications. IEEE Journal on Selected Areas in Communications, vol.14, no.7 (1996) pp.1228–1234

[5] Garey, M. S., Johnson, D. S.: Computers and Intractability: A Guide to the Theory of NP-Completeness. W. H.Freeman, New York (1979)

[6] Kirkpatrick, S., Gelatt, J. C. D., Vecchi, M. P.: Optimization by simulated annealing. Science, vol.220 (1983) pp.671–680

[7] Korkmaz, T., Krunz, M.: Multi-Constrained Optimal Path Selection. Proc. of IEEE INFOCOM'01, vol.2 (2001) pp.834–843

[8] Dijkstra, E.: A Note on Two Problems in Connexion with Graphs. Numerische Mathematik, vol.1 (1959) pp.269–271

[9] Metropolis, N., Rosenbluth, A., Rosenbluth, M., et al.: Equation of state calculations by fast computing machines. Journal of Chemical Physics, vol.21 (1953) pp. 1087–1092

[10] Zegura, E. W., Calvert, K. L., Donahoo, M. J.: A Quantitative Comparison of Graph-based Models for Internet Topology. IEEE/ACM Transactions on Networking, vol.5, no.6 (1997) pp.770–783

Adaptive Playout Algorithm
Using Packet Expansion for the VoIP

Jae-Hyun Nam[1], Won-Joo Hwang[2], Jong-Gyu Kim[3], Soong-Hee Lee[2],
Jong-Wook Jang[4], Kyo-Hong Jin[4], and Jung-Tae Lee[5]

[1] Silla University, Division of Computer and Information Engineering
1-1 Gwaebop-dong, Sasang-gu, Busan 617-736, Korea
jhnam@silla.ac.kr
[2] Inje University, School of Electronics and Telecommunications Engineering
607 Obang-dong, Gimhae, Gyungnam 621-749, Korea
{ichwang,icshlee}@inje.ac.kr
[3] Youngjin College, Division of Computer information Technology
218 Bokhyun-dong, Buk-gu, Daegu 702-721, Korea
jkkiim@yjc.ac.kr
[4] Dongeui University, Division of Computer Application Engineering
24 Gaya-dong, Busanjin-gu, Busan 614-714, Korea
{jwjang,khjin}@dongeui.ac.kr
[5] Pusan National University, School of Electrical and Computer Engineering
30 Jangjeon-dong, Geumjeong-gu, Busan 609-735, Korea
jtlee@pusan.ac.kr

Abstract. Internet telephony means placing telephone call over the
Internet which providing "best effort" service only instead of Public
Switched telephone networks. Because of this, Internet telephony cannot guarantee QoS. In this paper, we propose a new algorithm to reduce
the packet loss and to compensate the jitter which are factors affection
QoS called the Frame Expansion for Adaptive Playout Time algorithm.
This algorithm is suitable for soft real-time applications which can tolerate a certain amount of packet loss and delay. This is because it has
low packet loss and delay regardless of some playout delay caused by
expanding the received frame size using Synchronized Overlap and Add
algorithm in receiver side. In order to analyze and evaluate the performance of proposed algorithm, we carry out a simulation of delayed packet
loss rate and playout delay, which is computed by taking the difference
between the playout time and the arrival time on receiver side. From the
simulation, our proposed algorithm can reduce the probability of packet
loss considerably, and improve the quality of the voice compared with
existing playout buffering algorithm.

1   Introduction

The Internet is more and more used not only for data traffics, but also for real-
time traffic such as audio, voice and video transmission. In particular, Internet
telephony service has emerged as an important service because it is much cheaper

H.-K. Kahng (Ed.): ICOIN 2003, LNCS 2662, pp. 563–572, 2003.
© Springer-Verlag Berlin Heidelberg 2003

and easier to introduce the value-added service than the plain old telephone service (POTS). Internet telephony service means that the voice traffics are transmitted over the IP based network such as the Internet instead of PSTN. However, the Internet typically offers only a "best effort" service, which does not make commitment about a required minimum bit-rate or a allowed maximum delay. Consequently, when the network gets congested, real-time packets may arrive too late at the receiver or may be dropped at routers due to buffer overflow. A number of studies have attempted to measure packet loss rates and delays [1, 2]. The results indicate substantial variability, with typical packet loss rates of 0 to 20% and one-way delays of 5ms to 500ms. Unfortunately, when packet loss rates exceed 10% and one-way delays exceed 150ms, voice quality can be quite poor [3].

To alleviate this problem, some works mainly focus on issues such as packet loss recovery techniques [4, 5, 6] and adaptive playout buffering technique [7]. The packet loss recovery techniques are for concealment of lost voice packets. The adaptive playout buffering technique is for smoothing such delay and jitter. However, the former are considerably degraded the quality of voice in case of consecutive packet losses, though they were compensated the lost packets. The latter need an extra memory for buffering that is reason of raising the cost.

In this paper, we propose the Frame Expansion for Adaptive Playout Time (FEAPT) algorithm. It expands the received frame size using the Synchronized Overlap and Add (SOLA) algorithm [8] in the receiver side, which is a technique for time scale modification of audio signal, and uses it to reduce packets loss caused by deadline miss[1]. Our proposed algorithm is suitable for soft real-time applications like Internet telephony which must be played back consecutively but may delay or loss a certain amount of packets is permissible. This is because it is applied considering the tradeoff between loss and playout time.

The remainder of this paper is organized as follows. In section 1, we describe related works on packet loss recovery techniques for voice data. Section 3 describes FEAPT algorithm that is used in this paper. Section 4 presents simulation and experimental results to illustrate the performance of our algorithm. Section 5 presents conclusion.

2   Related Works

In this section, we briefly describe the related works dealing with compensating the loss, delay and jitter of a voice packet. These works mainly focus on issues such as loss recovery technique and adaptive playout technique [4, 5, 6, 7].

---

[1] The deadline constraint simply states that every voice frame must be arrived (a_i) before its playout time (p_i). Namely, when an a_i-p_i is less than zero, we call this deadline miss.

2.1   Packet Loss Recovery Techniques

Sender-Based Repair Technique: It may be split into two major classes: active retransmission and passive channel coding. It is further possible to subdivide the set of channel coding techniques with following two schemes.

Forward error correction (FEC) schemes rely on the addition of repair data to a stream, from which the contents of lost packets may be recovered.

Interleaving-based schemes disperse the effect of packet losses. Units are resequenced before transmission so that originally adjacent units are separated by a guaranteed distance in the transmitted stream and returned to their original order at the receiver.

Receiver-Based Repair Technique: It is of use when sender-based repair techniques fail to correct all loss, or when the sender of stream is unable to participate in the recovery. They rely on producing a replacement for a lost packet which is similar to the original. This is possible since audio signals, in particular voice, exhibit large amounts of short-term self-similarity. These schemes spilt into three categories.

Insertion-based schemes repeat lost packets by inserting a fill-in packet such as silence, noise or repetition of the previous packet. They features that the signal is not used to aid reconstruction. Because of this, they are simple to implementation but have poor performance.

Interpolation-based schemes repair lost packet by interpolation accounting for the changing characteristics of signal. Comparing with insertion-based schemes, they have better performance, but are more difficult to implement and require more processing time. The time scale modification (TSM) like the SOLA algorithm adopted in this paper belongs to the category.

Regeneration-based scheme derive the codec's state information from neighbors of the lost packets and regenerate a replacement for the lost packet.

2.2   Adaptive Playout Buffering Technique

Packets experience the delay and jitter while traversing the network, especially in the Internet. Ramjee et al. proposed an adaptive playout buffering algorithm to smooth such delay and jitter [7]. It was designed to dynamically adjust the talkspurt's playout delays to the traffic conditions of the underlying network without assuming neither the existence of an external mechanism for maintaining accurate clock synchronization between the sender and the receiver during the voice communication, nor a specific distribution of the voice packet transmission delay.

3   FEAPT Algorithm

We presented some techniques for compensating or smoothing the loss, delay and jitter of a voice packet. Among the earlier works, TSM which one of interpolation-based schemes is very useful technique because it allows listeners to throttle the rate of audio-information in much the same way a reader controls his reading rate by moving his eyes across a page. TSM refers to the process of compressing or expanding the time scale of an audio segment. A properly time scale modified signal maintains the properties of the original signal such as local pitch period, speaker identity, and intelligibility. When this is performed on a voice signal, the resulting signal sounds as though the same person is talking faster or slower in the same voice. A number of algorithms for TSM have been proposed; SOLA [8], short-time fourier transform (STFT) [9], pitch overlap and add (PLOLA) [10]. Of all existing TSM, we consider the SOLA, which is suitable for the VoIP because of simple implementation, low complexity and high voice quality relatively.

SOLA extracts some signal segments of a voice signal to create a new one. The segments are extracted every S_a samples from input signal then overlap and add every S_a samples. These S_a and S_s are determined according to the time-scale modification factor $\alpha = \frac{S_a}{S_s}$ [2]. If S_s is different from S_a, the length of the output signal is different too. To avoid phase mismatches when a new segment is add to output signal, they propose to adjust the value of S_s at each addition to ensure synchronization. The shift is determined by the maximization of the cross correlation between the output signal and the segment. This technique provides a locally optimal match between successive packets' combining the packets in this manner tends to preserve the time-dependency pitch, magnitude, and phase of a signal.

For reducing the packet loss caused by deadline miss, we propose the FEAPT algorithm in Fig. 1 which expands the received frame size using SOLA algorithm in receiver, and uses it to reduce packet loss due to delay. The receiver decides whether or not to apply the FEAPT algorithm to the received packet according to the condition of playout buffering time. FEAPT algorithm operates as follows:

- Receiver examines whether the ith packet arrives (a_i) before its playout (p_i) or not.
- If the packet arrives later than the playout time $(a_i \leq p_i)$, it is considered lost. Otherwise, it is expanded its size using SOLA, depending on satisfying whether or not the condition of playout buffering time $(p_i \geq 2 \times \Delta_{max})$[3]

Let us consider the simple example of the playout mechanism at the receiver (Fig. 2). We first assume that every packet experiences some packet delay time in the networks, and the SOLA computation time is δ_{SOLA}[4].

In the case 1, it schedules playout time to play back immediately, when the packets arrive at the receiver. This leads to the dropping of 5th, 9th, and 10th

---

[2] If $\alpha > 1$, the signal is expanded time scale. Otherwise, the signal is compressed time scale.

[3] In this paper, we set the playout buffering time to the maximum value $(2 \times \Delta_{max})$ considering playout buffer underflow and over flow [1, 11]. This is because of reducing the packet loss by delay to the utmost.

[4] For example, the 1st packet transmitting form sender on $1T$ arrives at receives at receiver on $6T$. Namely, the 1st packet experiences $5T$ packet delay in the network.

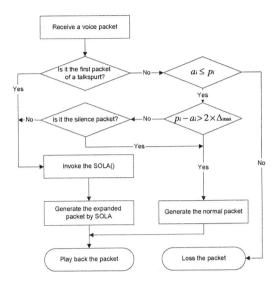

Fig. 1. FEAPT algorithm using SOLA

packet, because they cannot arrive until scheduled playout time (i.e. deadline miss). However, if the playout for the 1st packet is delayed during the $2T$, there is no lost packet but it causes the additional delay of $2T$ between end-to-ends (case2). If the receiver applies the SOLA to all received packets, the length of each packet is excessively expanded as compared with the original signal (case 3). This leads to longer end-to-end delays than the original signal; the long delays are intolerable for interactive communication system like Internet telephony. However, in the case of FEAPT algorithm (case 4), it decides on whether the SOLA is applied to the received packets or not according to the amount of buffering time. Because it is not apply the SOLA to all received packets, this improves the quality of the voice by reducing the probability of packet loss end-to-end delay considerably.

In adaptive playout buffering scheme, it computes optimal playout time using end-to-end delay and variation of previous packets to satisfy the deadline constraint. After determining the optimal playout time, it delays the playout time to the first packet in a talkspurt period, and a playout time for any subsequent packet in a talkspurt is computed as an displacement from the time when the first packet was played back. Accordingly, after the playout time of the first packet in a talkspurt is determined, if the delay for any subsequent packet in a talkspurt is very large, it cannot play back and will be lost. Besides, this scheme may be degraded the QoS of voice due to increasing the end-to-end delay. However, FEAPT algorithm expands the received frames under threshold to satisfy the deadline constraint for each frame. If it is applied the received frame to the SOLA which has the time scale factor $\alpha = 1.25$, the size of the received

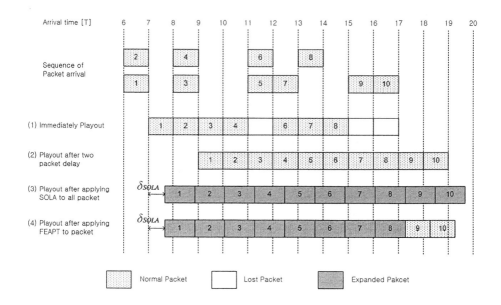

Fig. 2. Example of playout mechanisms

frame (N) becomes expand $N \times 1.25$. For example, we suppose that the size of sending frame is 40ms and the size of the received frame, which is applied to the SOLA, is about 50ms. In the case of the receiver, the playout time of the expanded frame is $p_j = p_i + 50ms$. This provides the extra buffering time (i.e. 10ms) for each frame. Even if a packet cannot play back due to long delay, it can play back the packet at a scheduled time in the system adopting FEAPT algorithm.

4   Performance Measurements and Analysis

We organize the empirical environment to evaluate the performance of FEAPT algorithm in Fig. 3.

Sender transmits the frame that was sampled by the frequency of 8000Hz to the traffic monitor, approximately every 40ms. Receiver expands the received frame size using SOLA algorithm which has $\alpha = 1.25$ to reduce packet loss caused by deadline miss: The size of the input frame (M) is 320 bytes and sampled input packet size (S_a) is $\frac{M}{4}$. We used the Shunra Software Ltd.'s Cloud which for emulating a network and set the network parameters such as average delay and packet loss rate according to [12]: average delay of 20ms and packet loss 0.03% in domestic network; and average delay of 300ms and packet loss of 8.36% in North America network; and average delay of 420ms and packet loss of 12.2% in Asia/Pacific network; and an average delay of 200ms and packet loss of 13.4%

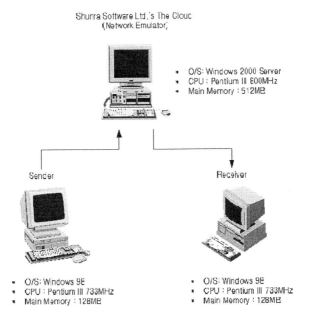

Fig. 3. Empirical Environment

Table 1. Packet loss rate in each empirical environment

Region	Packet loss rate [%]		
	FEAPT	Adaptive PBA	Non-PBA
Domestic	1.21	1.19	3.28
North America	9.27	9.11	13.82
Asia/Pacific	14.1	13.78	28.7
Europe	15.7	15.21	22.11

PBA=Playout Buffering Algorithm

in Europe network. We conduct the performance comparison among FEAPT algorithm, adaptive playout buffering algorithm and Non-playout buffering algorithm in two ways: packet loss rate by deadline miss and playout delay by expanding the frames or buffering.

We first examine the packet loss rate by deadline miss at the receiver. Table 1 presents that FEAPT algorithm covers much less packet loss rate in receiver than Non-playout buffering algorithm. However, in spite of any buffering scheme, it has similar packet loss rate comparing with adaptive playout buffering algorithm.

Next, we verify the probability density functions (PDF) as the playout delay, which was computed by taking the difference between the playout time of frame and the arrival time in receiver. Fig. 4, 5, 6 and 7 show that FEAPT algorithm outperforms adaptive playout buffering algorithm over a number of measured playout delay. In the case of adaptive playout buffering algorithm, average playout delay is ranged from 62.09 to 347.72ms, but in FEAPT algorithm, it is ranged from 55.48 to 97.98ms. Note that, in the results of playout delay, Non-playout buffering algorithm have not playout delay because there is not any buffering or frames expanding mechanism.

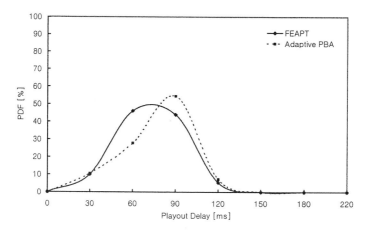

Fig. 4. Playout delay in Domestic network environment

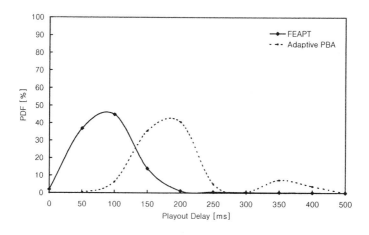

Fig. 5. Playout delay in North America network environment

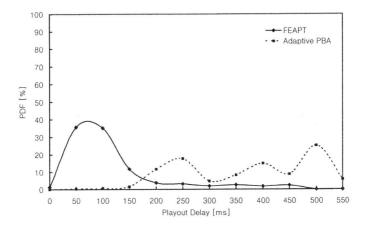

Fig. 6. Playout delay in Asia/Pacific network environment

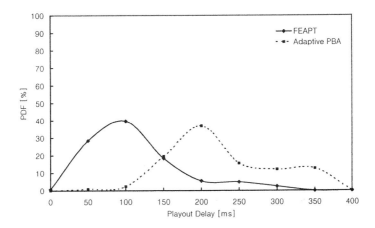

Fig. 7. Playout delay in Europe network environment

5   Conclusions

The VoIP like Internet telephony service becomes an increasingly important application. However, best effort property of the Internet, the Internet telephony cannot guarantee the QoS.

In order to reduce the packet loss and compensate the jitter which are factors affecting QoS, we proposed FEAPT algorithm. It is suitable for real-time applications which demand low delay but can tolerate or conceal a small amount of delayed packets. This is because it has high performances of packet loss and jitter regardless of some playout delay caused by expanding the received frame size using SOLA algorithm in receiver.

From the results, we first know that the FEAPT algorithm has less packet loss rate than Non-playout buffering algorithm. Furthermore, in spite of any Buffing scheme, it has similar packet loss rate comparing with adaptive playout buffering algorithm. Second, the FEAPT algorithm has much better performance of playout delay than the Adaptive playout buffering algorithm over a number of measured playout delay.

Our further efforts are focus on applying a variable time scale factor to SOLA according to network traffic. We are also investigating an adaptive silence deletion algorithm in SOLA, which will more reduce the end-to-end delay than our algorithm.

References

[1] J. C.Bolot: Characterizing End-to-End Packet Delay and Loss in the Internet, Journal of High-Speed Networks, Vol.2, No.3, pp. 305-323, Dec. 1993.

[2] Vern Paxson: End-to-end internet packet dynamics, Procedings of SIGCOMM Symposium on Communications Architectures and Protocols, Cannes, France. Sept. 1997.

[3] J. Rosenberg, L. Qiu, and H. Schulzrinne: Integrating Packet FEC into Adaptive Voice Playout Buffering Algorithms in the Internet, Procedings of IEEE INFO-COM 2000, Tel Aviv, Israel, pp 1705-1714, Mar. 2000.

[4] J. Suzuki and M. Taka: Missing packet recovery techniques for low-bit-rate coded speech, IEEE Journal on Selected Areas in Communications, vol SAC-7, no. 5, pp 707-717, June 1989.

[5] L. A. DaSilva, D. W. Petr, and V. S. Frost: A class-oriented replacement technique for lost speech packets, IEEE Transactions on Acoustics, Speech, and Signal Processing, vol. ASSP-37, no 10, pp 1597-1600, Oct 1989.

[6] C. S. Perkins, O. Hodson and V. Hardman: A Survey of Packet Loss Recovery Techniques for Streaming Audio, IEEE Network Magazine, PP. 40-48, Sep./Oct. 1998.

[7] R. Ramjee, J. Kurose, D. Towaley and H. Schulzrinne: Adaptive Playout Mechanisms for Packetized Audio Applications in Wide-Area Networks, Procedings of IEE INFOCOM 1994, Toronto, Canada, pp. 680-688, June 1994.

[8] S. Roucos and A. M. Wilgud: High Quality Time-Scale Modification for Speech, Proceeding of ICASSP, pp. 493-496, Apr. 1986.

[9] M. R. Portnoff: Time-scale Modification of Speech Based on Shorttime Fourier Analysis, IEEE Transactions on Acoustics, Speech, and Signal Processing, vol. ASSP-29, no. 3, pp. 374-390, Jun. 1981.

[10] E. Moullines and F. Charpentier: Pitch Synchronous Waveform, Proceeding for Text-to Speech Synthesis Using Diphones, Speech Communications, vol. 9, no. 5/6, pp. 453-467, 1990.

[11] M. Roccetti, V. Ghini, G. Pau, P. Salomoni and M. E. Bonfigli: Design and Experimental Evaluation of an Adaptive Playout Delay Control Mechanism for Pactetized Audio for Use over the Internet, Multimedia Tools and Applications, an International Journal, Kluwer Academic Publishers, vol. 14, no. 1, pp. 23-53, May. 2001.

[12] W. Hwang and J. Lee: Internet Phone Technology, The magazine of the IEEK, IEEK, vol. 26, no. 8, Aug. 1999.

Dynamic Quality of Service on IP Networks

Tippyarat Tansupasiri and Kanchana Kanchanasut

Computer Science and Information Management Program
Asian Institute of Technology
Pathumthani, 12120, Thailand
{fon,kk}@cs.ait.ac.th

Abstract. In this paper, we propose a dynamic QoS, D-QoS, model which allows the QoS requirements be reconfigured dynamically. A privilege user can request a network interruption to guarantee its own smooth traffic flow at the expense of possibly interruptions or blockages of other traffic flows. Based on the concept of active IP network, an interruption can be triggered by sending an active packet requesting for an interruption to the network. Under normal circumstances, our model provides DiffServ based on CBQ and, in case of an interruption, priority queue is employed. The two queuing mechanisms operate alternately in response to the programs sent via active packets. A prototype of the model has been successfully implemented where the two operation modes have been shown to operate satisfactorily. Details of the design and implementation of D-QoS and its evaluation are discussed in this paper.

1 Introduction

Since the introduction of time sensitive applications on the Internet, the need for quality of service (QoS) has been raised and focused on. Among the proposed QoS models, one which can scale well and is suitable as the core network QoS model is Differentiated Services (DiffServ) model [1], initiated by the IETF's Differentiated Services Charter [2]. It provides a simple and coarse way to offer differentiated classes of service on flow aggregates, where the complexity on packet classification marking, policing and shaping functions are needed only at network boundaries. Packets are classified into predefined classes based on the values, called codepoints, specified in their Differentiated Services (DS) fields of the IP headers [3]. Different traffic classes may receive different packet forwarding treatments on DiffServ routers.

In the case of internetworking among network domains with different service policies, any adjacent domains should establish the agreement, Service Level Agreement (SLA), to provide the same level of service for a particular codepoint. However, DiffServ model itself operates on the static predefined SLAs in which human intervention for reconfiguration is needed when there are changes on the predefined QoS settings. As Internet traffic could be randomly generated at any time and possibly from anywhere, these reconfigurations are definitely necessary. For example, one may wish to send bulky telemedicine traffic which requires high precision of data transmission over the net during an emergency situation.

H.-K. Kahng (Ed.): ICOIN 2003, LNCS 2662, pp. 573–582, 2003.

This paper proposes a dynamic QoS system based on the concept of active network. This concept was introduced in 1994 by Defense Advanced Research Projects (DARPA) research community as a future direction of networking system [4]. It replaces normal packets with active packets containing small programs and possibly data. Network nodes are substituted by active nodes capable of performing user- or application-specific functions on packets passing through them. For the IP networks, Murphy [5] has proposed an active IP network that allows the IP network to perform customized computations on user data, as additional processing to the traditional IP forwarding mechanism. His work has demonstrated how to construct an active IP node, and how to seamlessly integrate an active IP node into IP network. Hence, both normal IP nodes and active IP nodes can cooperate and coexist in a single IP network.

As the feasibility of the active network implementation on IP networks has been shown, a model that provides dynamic QoS based on this concept can also be constructed. With our proposed QoS model, Dynamic Quality of Service (D-QoS), the IP network could be reconfigured without human intervention. It aims to provide dynamic QoS upon receiving the interruption requests from authorized users to get highest possible network bandwidth. However, our prototype implementation does not fully implement an active IP node, but only emulates the active IP node with restricted features related to providing dynamic QoS.

Details of the D-QoS model and its implementation are explained in Section 2. Section 3 provides the experimental results from several tests based on different network scenarios. Related works to this paper are discussed in Section 4. Finally, the last section presents the discussion and conclusion.

2 Dynamic Quality of Service (D-QoS)

Unlike DiffServ in which a predefined portion of the bandwidth is shared within each traffic class, the proposed model is a reconfigurable system that allows multi-level interruptions upon the requests of prioritized flows. We refer to the prioritized flow as a super user flow which has all the privilege over other flows, in the same manner as an operating system super user. Default QoS set up of the system is configured as DiffServ service. A super user can send a request for a specific level interruption level prior to the actual flow transmission. Upon receiving a request, the interruptions are generated on all the nodes along the communication path and normal DiffServ is then suspended where the network output queuing mechanism on each node is automatically reconfigured. This allows high priority telemedicine traffic to flow through with the highest possible bandwidth at the expense of other lower priority traffic. Normal DiffServ will resume after the prioritized flow has completed. D-QoS allows multi-level of interruptions where a number of super users may request for the interruptions at the same time. All interruption requests are kept in a stack where previous QoS configuration is resumed after an interruption has finished. D-QoS also requires that the active IP network consists of a contiguous set of active IP nodes, which is not the case in [5], otherwise, unexpected result may occur since normal IP nodes do not provide the requested QoS to the prioritized flow.

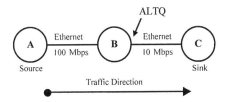

Fig. 1. The D-QoS Prototype

2.1   An Implementation of D-QoS

A prototype of D-QoS consisting of three D-QoS nodes has been implemented to demonstrate the functioning of the model, as shown in Fig. 1. Each D-QoS node is a Pentium 100 MHz PC running FreeBSD. It emulates an active IP node with restricted primitives for queuing manipulation related to interruption handling, i.e., making and removing an interruption. Active packets carrying programs are transmitted through an opened UDP socket where their contents are examined and the appropriate actions are performed. These program packets contain the information used to identify a particular super user flow and a required interruption level as the parameters for each primitive. A flow is identified by its source and destination addresses, source and destination port numbers and protocol. Data packets are represented as normal IP packets carrying the flow payload.

Upon receiving an interruption request packet, the program carried in the packet's content is executed. This program periodically forwards the request packet to the next neighbor towards the destination until a confirmation message is received on time out. Hence, only one program packet is required to trigger all the nodes along the path. At the end of the flow transmission, another program packet is transmitted to remove the interruptions and recover the nodes back to the previous state according to the information in the stacks.

Since traditional FreeBSD provides only FIFO, other queuing mechanisms are presented by Alternate Queuing (ALTQ) which was introduced by Cho [6]. In normal circumstances, the system provides DiffServ service with Assured Forwarding (AF) and Expedited Forwarding (EF) as specified in RFC 2597 [7] and RFC 3246 [8] respectively. DiffServ implementation operates on a Class-Based Queuing (CBQ) proposed by Floyd and Jacobson [9]. CBQ is a hierarchical structure of classes representing their bandwidth partitioning and sharing. Each class has its own queue with a predefined bandwidth partition. In this case, the CBQ structure for DiffServ consists of four AF classes and one EF class, as illustrated in Fig. 2. Each of the four AF classes applies an extended version of Random Early Detection with In and Out (RIO) to provide three drop precedence levels. An AF class may receive excess bandwidth if available up to 100% of the link bandwidth, while the EF class has a limitation on its maximum bandwidth acquired at 75% of the total link bandwidth. As EF class requires low-loss, low-delay and low-jitter, this class is configured to have higher priority. Unknown traffic is classified into AF3 Class.

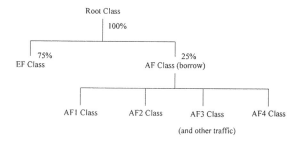

Fig. 2. Class-Based Queuing (CBQ) Structure for DiffServ Implementation on D-QoS Prototype

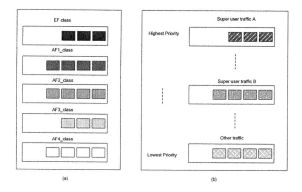

Fig. 3. (a) Queuing Structure for DiffServ in D-QoS (b) Queuing Structure for Multi-level Interruption in D-QoS

When a node receives an interruption request, this CBQ structure for DiffServ is automatically substituted by the Priority Queue (PQ). The requested interruption level is examined, if that level is not available, a lower priority is given to that requesting flow. Our PQ offers 16 interruption levels ranging from 0-15. When changing the queuing mechanism, the queue content of the previous mechanism is discarded. The super user flow is given the specified priority level, while other traffic flows are treated as having the lowest priority (priority 0). Fig. 3(a) demonstrates the network output queuing structure of CBQ for DiffServ. The queuing structure of PQ in Fig. 3(b) illustrates a particular case when there are two interruptions by super user A, with the highest priority, and super user B, with lower priority, at the same time.

3   Experimental Results

Our focus is mainly on the case where the available bandwidth is not large enough for the total bandwidth requirement of flows. Hence, the two Ethernet links connecting D-QoS nodes have been configured with different speeds, 100

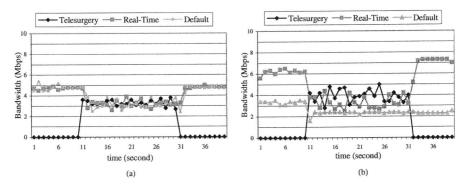

Fig. 4. (a) Result from Best-Effort IP Network (b) Result from DiffServ Service

Mbps and 10 Mbps. Each flow was generated, by a traffic generator program called Iperf [10], as a unidirectional UDP flow from node A to node C. The middle node, Node B, acted as the bottleneck where all flows competed for bandwidth. ALTQ was enabled on node B's interface connecting to node C to provide appropriate queuing mechanisms on traffic passing through it.

The experiments aimed to measure the transmission quality of a particular telesurgery flow, as an example of telemedicine flows, which needs high precision of data transmission and also has high bandwidth requirement. Similar to the transmission of other time sensitive streams where the transmission rate is preferred over reliability, the flow was transmitted over UDP rather than TCP. The flow was presented as a constant rate UDP stream of size 6 Mbps which consumed 60% of the 10 Mbps link. It has been found that when generating a traffic flow with 6 Mbps bandwidth, the bandwidth achievement result may be up to 6.3 Mbps as there are overheads from packet headers. The maximum bandwidth that could be achieved from the 10 Mbps link is 9.6 Mbps. Each experimental result is provided as a bandwidth achievement graph.

To validate the correctness of our prototype implementation, two experiments were conducted with two different settings, the traditional best-effort IP network and the DiffServ service in D-QoS. In both experiments, two traffic flows representing the real-time traffic and the default Internet traffic were first generated, with the same size of 6 Mbps for 40 seconds. After the first 10 seconds, the telesurgery flow was generated and lasted for 20 seconds.

From Fig. 4(a) which is the result obtained from the traditional best-effort IP network, during the first 10 seconds, the real-time traffic and the default Internet traffic flows acquired equal bandwidth of about 4.7 Mbps in average. In the next 20 seconds where three traffic flows were generated, total link bandwidth was equally shared among them, each flow gained 3.3 Mbps in average.

Experimental result from DiffServ service based on the CBQ structure in Fig. 2 is shown in Fig. 4(b). The real-time traffic flow was classified into EF class with higher priority compared to the default Internet traffic flow placed in AF3 class. During the first 10 seconds, the real-time traffic flow gained the required bandwidth of 6.2 Mbps while the default Internet traffic flow got 3.3 Mbps bandwidth. Then, in the period where three traffic flows were generated, 75% of link bandwidth or 7.5 Mbps assigned to EF class was shared by both the telesurgery and the real-time traffic flows, while the default Internet traffic flow obtained the rest available bandwidth of 2.3 Mbps. The packet loss percentages experienced by each flow in this period were 42% for the telesurgery traffic flow, 44% for the real-time traffic flow and 63% for the default Internet traffic flow. In the last 10 seconds, there was a period where the real-time traffic flow obtained about 7.2 Mbps which was higher than the generated rate and nearly reached the maximum bandwidth limitation for EF class. The reason is because there were some packets left in the buffer queue during the congestion time and they were forwarded out when the bandwidth became available.

3.1   Single Interruption

Fig. 5 demonstrates the bandwidth achievement results in the case where there was an interruption. Two experiments were conducted with different sizes of telesurgery flows, 6 Mbps and 10 Mbps. Other two flows were generated in the same manner as the previous experiments. Fig. 5(a) shows the result when the telesurgery flow was 6 Mbps and Fig. 5(b) is the case when it was 10 Mbps. The results in the first 10 seconds were based on DiffServ service and yielded the same result as that shown in Fig. 4(b). Then, at the time when the telesurgery traffic flow was generated, an interruption on the output queuing mechanism was created and normal DiffServ service with CBQ was automatically substituted by PQ providing two priority levels. The telesurgery flow was treated as the higher priority flow, other flows were classified into the lowest priority class. When the telesurgery flow completed, normal DiffServ on CBQ was resumed.

During the interruption period, the telesurgery flow acquired 6.3 Mbps with no packet losses. This is obviously different from the case of DiffServ service where it experienced 42% of packet losses. The rest bandwidth was shared by both the real-time and the default traffic flows with equal bandwidth of 1.6

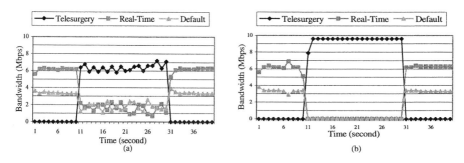

Fig. 5.   (a) Result from Single Interruption of Telesurgery Flow of Size 6 Mbps (b) Result from Single Interruption of Telesurgery Flow of Size 10 Mbps

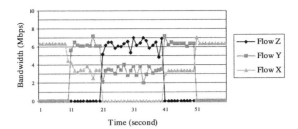

Fig. 6. Result from Multi-level Interruptions

Mbps with packet loss percentages of 62.3% and 59.8% respectively. The graph on the last 10 seconds shows the result when the interruption was completed and normal DiffServ resumed, which was the same as in the first 10 seconds.

It is also possible that the interrupting flow may solely occupy the whole link bandwidth, as shown by the experimental result in Fig. 5(b). The telesurgery flow was generated with the bandwidth of 10 Mbps. During the interruption period, the telesurgery flow automatically obtained 9.6 Mbps which was the maximum bandwidth that could be achieved from the link while other flows got less than 0.1 Mbps. However, even the telesurgery traffic flow had higher priority, its packet loss percentage was 2.5%, due to the limitation on link bandwidth, while the other two flows had 99.6% and 98.7% of packet loss. Comparing to the result obtained in case of the DiffServ service, the interruption mechanism provided in D-QoS has shown that the telesurgery flow could be delivered with much better quality based on the available link bandwidth.

3.2   Multi-level Interruption

This section discusses on the experimental result obtained from the D-QoS prototype when there were a number of interruptions with different priority levels, as shown in Fig. 6. There were two interruption requests from super user flow Y and Z, with the priority of 10 and 15 respectively, on an existing flow X. All flows had the same size of 6 Mbps. The total time of the experiment was divided into periods of 10 seconds each. In the first period, there was only one flow, flow X, and the network was based on DiffServ service with CBQ. This CBQ was then changed to PQ with two priority levels when an interruption of flow Y was received. Flow Y, which would last for 40 seconds, was given higher priority and got the bandwidth of 6.2 Mbps, while flow X only got 3.4 Mbps. When another super user flow, flow Z, was generated with an interruption level 15, the previous PQ with two priority levels was substituted by a PQ providing three priority levels. Flow Z received 6.2 Mbps of bandwidth during its transmission of 20 seconds. Flow Y with lower priority received 3.3 Mbps. Flow X received less than 0.1 Mbps as it had the lowest priority. Upon completion of flow Z, the PQ resumed with only two priority levels. Finally, normal DiffServ was resumed.

Table 1. Priority Level in Different Time Periods

	t=1	t=11	t=21	t=31	t=41	t=51	t=61	t=71
Flow X	5	5	10	10	10	15	5	5
Flow Y	10	10	5	10	15	15	15	10
Flow Z	-	15	15	15	15	10	10	-

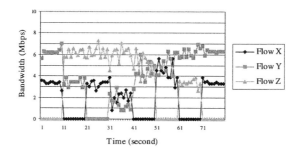

Fig. 7. Result from Dynamically Priority Changing

To validate that we can switch our priority settings upon each interruption, we ran an experiment with multi-level interruptions whose levels changed dynamically during the transmission. Three flows were generated with the same size of 6 Mbps and their priority levels within each period are shown in Table 1.

The result, illustrated in Fig. 7, has shown that the prototype correctly reacted to different levels of the interruption requests. The bandwidth acquired by each flow depended on its priority at that time. The first highest priority flow got 6.2 Mbps while the rest of the link available bandwidth, 3.3 Mbps, was given to the second highest priority flow and the lowest priority flow got less than 0.1 Mbps. In the case where two flows had the same priority, they shared the available bandwidth equally based on their priorities compared to another flow; 10 Mbps bandwidth was shared if they both had the highest priority, or the remaining bandwidth of 3.4 Mbps was shared if they both had lower priority.

4 Related Works

Dynamic QoS may also be accomplished with another technique, i.e., with the interoperation of Integrated Service (IntServ) based on Resource Reservation Protocol (RSVP) [11] and DiffServ as presented in RFC 2998 [12] and Detti et al. [13]. The main idea of those works is to have an additional network function that acts as an admission control agent or a bandwidth broker within DiffServ networks that monitors its resource usage and communicates with other IntServ nodes outside DiffServ networks. However, the interoperation requires that DiffServ network must be able to provide support for the standard IntServ services and that the admission control information must be provided.

The idea of giving privilege to some important flows in our system is also similar to the work proposed by Quadros et al. [14]. Nevertheless, their work is based on another technique where a DiffServ packet handling mechanism, called *Dynamic Degradation Distribution (D3)*, was introduced. D3 dynamically distributes network resources, transmission capacity and memory, to DiffServ classes according to their loss and delay sensitivities. In case of congestion, more sensitive classes experience less performance degradation. On the other hand, in D-QoS, the flows are transmitted according to their priorities. Therefore, the super user flow with the highest priority receives the best network service.

5   Discussion and Conclusion

The proposed D-QoS model has been successfully implemented and evaluated through our D-QoS prototype which offers dynamic QoS reconfiguration upon receiving the authorized requests. The DiffServ service with CBQ is dynamically substituted by PQ when an interruption request is received. Available network bandwidth is always given to the privilege flow such as telemedicine traffic flow. Telemedicine flow which needs high precision of data transmission is transmitted with the best service the network can offer. Compared to the static QoS configuration of DiffServ model, the super user flow can gain more bandwidth as required based on the network capacity. This also leads to the dramatically reduction of packet loss percentages. In the case where the prioritized flow needs the bandwidth of equal or more than the network link capacity, that link is then dedicated to the flow and other traffic is completely discarded. However, the privilege traffic may experience some packet losses in this case.

Additionally, in the case of multi-level interruption where super users may request for interruptions at different interruption levels at the same time, the prototype can dynamically switch its queuing mechanisms according to the requests and correctly services the flows according to their priorities. Upon the completion of each interruption, the previous QoS state is automatically resumed.

Even though the experiments were performed on a small D-QoS network with only three machines, it is sufficient to demonstrate the transmission quality experienced by a number of flows through a network bottleneck. More network nodes may be added to the prototype, but it would not impact the experimental results. However, our future work is to evaluate the D-QoS model by network simulation software (ns-2). With ns-2, the model can be applied to more complicated networks with a larger number of nodes, various types of network links and variety of network traffic patterns which is closer to the real Internet characteristics, where the scalability issue of D-QoS can be examined.

Acknowledgements

The authors would like to thank the Royal Thai Government for providing scholarship for the first author.

References

[1] Blake, S., Black, D., Carlson, M., Davies, E., Wang, Z. and W. Weiss, 'An Architecture for Differentiated Services', RFC 2475, December 1998.

[2] IETF, Differentiated Services (DiffServ) Charter, http://www.ietf.org/html.charters/diffserv-charter.html

[3] Nichols, K., Blake, S., Baker, F. and D. Black, 'Definition of the Differentiated Services Field (DS Field) in the IPv4 and IPv6 Headers', RFC 2474, December 1998.

[4] Tennenhouse, D. L., Smith, J. M., Sincoskie, W. D., Wetherall, D. J. and G. J. Minden, 'A Survey of Active Network Research', IEEE Communications Magazine, 1997.

[5] Murphy, D. M., 'Building an Active Node on the Internet', Master's thesis, MIT, 1997.

[6] Cho, K., 'A Framework for Alternate Queueing: Towards Traffic Management by PC-UNIX Based Routers', Proceedings of USENIX 1998 Annual Technical Conference, 1998.

[7] Heinanen, J., Baker, F., Weiss, W. and J. Wroclawski, 'Assured Forwarding PHB Group', RFC 2597, June 1999.

[8] Davie, B., Charny, A., Bennett, J. C. R., Benson, K., Le Boudec, J. Y., Courtney, W., Davari, S., Firoiu, V. and D. Stiliadis, 'An Expedited Forwarding PHB (Per-Hop Behavior)', RFC 3246, March 2002.

[9] Floyd, S. and V. Jacobson, 'Link-Sharing and Resource Management Models for Packet Network', IEEE/ACM Transactions on Networking, Vol. 3 No. 4, August 1995.

[10] Tirumala, A., Qin, F., Dugan, J. and J. Ferguson, 'Iperf', May 2002, http://dast.nlanr.net/Projects/Iperf/

[11] Braden, B., Zhand, L., Berson, S., Herzog, S. and S. Jamin, 'Resource Reservation Protocol (RSVP) - Version 1 Functional Specification', RFC 2205, September 1997.

[12] Bernet, Y., Yavatkar, R., Baker, F., Zhang, L., Speer, M., Braden, R., Davie, B., Wroclawski, J. and Felstaine E., FRC 2998, 'A Framework for Integrated Services Operation over Diffserv Networks', November 2000.

[13] Detti, A., Listanti, M., Salsano, S. and L. Veltri, 'Supporting RSVP in a Differentiated Service Domain: an Architectural Framework and a Scalability Analysis', ICC'99, June 1999.

[14] Quadros, G., Alves, A., Monteiro, E. and F. Boavida, 'An Approach to Support Traffic Classes in IP Networks', QoFIS'2000, Berlin, Germany, September 2000.

Part IV

Mobile Internet

An Efficient Wireless Internet Access Scheme*

Jen-Chu Liu and Wen-Tsuen Chen

Department of Computer Science
National Tsing Hua University, Hsin-Chu, Taiwan 30043 R.O.C.
{dr888301,wtchen}@cs.nthu.edu.tw

Abstract. Many popular services provided on wired Internet, such as WWW and FTP, are also used in wireless communication environments. Wireless networks usually have some disadvantages such as lower and variable bandwidth, higher transmission error and connection broken probability, etc. These disadvantages make the wireless Internet access performance inefficient. Furthermore, if a mobile host (MH) move frequently, how to guarantee the reasonable data accessing efficiency is important. This paper proposed a proper architecture and access schemes for wireless Internet accessing, named Smart Follower Cache (SFC). We choose WLAN to be the simulation platform and designate an access point (AP) as a Wireless LAN proxy. The proxy maintains two kinds of cache, one is public cache, and the other is private cache. Every service, like WWW and FTP, has its own corresponding public cache and each MH has its own private cache in the proxy. By caching necessary data in public cache, the proxy can prefetch the necessary data into the public cache and retransmit the data from the public cache immediately while transmission error occurs. Recording the user profile and several personalized data, which were stored in private cache, can make proxy providing different data qualities to different type of MH easily. Moreover, this architecture can accelerate data accessing. When a MH handoff, the original proxy may automatically transmit the MHs relative cache data that are located in the public cache and its private cache to the new proxy. Therefore the MHs can usually access their wanted data from the new proxy more quickly and efficiently.

1   Introduction

The Internet and wireless communications networks revolutionized communications in recent years. The convergence of these two technologies leads naturally to the wireless Internet. Many popular services provided on wired Internet, such as World Wide Web (WWW) and File Transfer Protocol (FTP), are also used in wireless communication environments. However, the protocols that those services used in wired line are not proper for wireless networks and make the wireless Internet access performance inefficient. For example, wireless networks usually have some drawbacks such as lower and unfixed bandwidth, higher transmission

---

* This research was supported by the Ministry of Education, Taiwan, R.O.C. under Grant 89-E-FA04-1-4.

error and connection broken probability, etc. On the other hand, the difference of functionality and processing ability in different Mobile Hosts (MHs) is significant. For example, some MHs, like laptop computer or notebook, have better processing ability, larger screen and storage size than the others, like personal digital assistant (PDA). These hardware differences made the design of proper scheme to achieve good performance for different kind of MHs more difficult. Many researches have proposed several kinds of scheme to relieve the performance degradation that those drawbacks caused [1,6,11-16]. [11,12] describe a PDA-based wireless browser. These schemes use a proxy to fetch, parse, and separate HyperText Markup Language (HTML) documents into multiple screens. [13,14] describe a mobile system, named Wit system, that partition applications and manage communication. The Wit system supports caching, prefetching, data reduction, and other performance-enhancing features. [1,15] have developed a new protocol, respectively, to replace the HyperText Transfer Protocol (HTTP) used between access point (AP) and MH. And designed a wireless WWW proxy to interconnect the wireless and wired users.

The proxy prefetches documents to clients based on clients cache content. It also reduces the resolution of large bitmaps when the wireless link is slow in order to speed up document fetches. All these schemes use a proxy to deal with the communication between wired and wireless networks, but there are some disadvantages. For example, [13,14] include prefetching and content reduction by outlining, but it does not use resolution reduction and it is not clear whether inline images for prefetched documents are sent to client. The approaches proposed in [11,12] are geared towards PDAs and not suit for other types of MH. [8] uses a proxy to convert the content quality according different display ability among different MHs. [1,15] have proposed new protocols, but both the MH and AP should be modified, it is too complicate. Therefore, this paper wants to design a proper architecture and access scheme for wireless Internet accessing, especially focus on improving the bandwidth utilization, dealing with the channel transmission error problem, and accelerating the data accessing. How to deploy a proxy in wireless networks? The simplest way is designating a base station (BS) or an AP as a proxy in wireless networks. Wireless local area network (WLAN), for example, usually maintains several APs to handle the interconnection between wired and wireless networks. Each AP creates a basic service set (BSS) and the mobile hosts (MH), which are in the same BSS, can access the Internet through the AP. Furthermore, the MH may be a notebook or PDA-like handset, and the difference among different hardware functionality maybe significant. For example, the CPU processing ability and storage capacity of a notebook are nearly the same as PC, but PDA-like handsets usually have lower CPU processing ability and storage capacity than PC. It is necessary that using a proxy to deal with the connection service according to different hardware functionality.

This paper proposes a proper architecture and access schemes for wireless Internet accessing, named Smart Follower Cache (SFC). We choose WLAN to be the simulation platform because it is easy to implement. First, we designate an access point as a Wireless LAN proxy. The proxy contains two kind of cache, one is public cache, and the other is private cache. The proxy assigns a public cache for each service (like WWW and FTP) and the data will be cached in its corresponding public cache. Therefore, the proxy can prefetch the necessary data into the public cache and retransmit the data form the public cache immediately while transmission error occurs. On the other hand, the proxy also maintains a private cache for each MH. The private cache caches the link information that is referred to the public cache data. The user profiles and several personalized data are also recorded in its corresponding private cache. The proxy can provide different data qualities to different type of MH more easily according the information resides in private cache. Moreover, this architecture can accelerate data accessing. When a MH move to another BSS, the original proxy will automatically transmit the MHs relative cache data that are located in the public cache and its private cache. Therefore, the MH can always hit the data in the nearest cache and achieving higher data accessing performance. In brief, this paper prose a proper architecture and access schemes for wireless Internet accessing. These schemes can improve the bandwidth utilization, deal with the channel transmission error problem, and accelerate the data accessing.

This paper is organized as follows. Section 2 describes our architecture and schemes. The simulation results and analysis are reported in section 3, and section 4 concludes the paper.

2 The Proposed Schemes

In our architecture, we designate a base station (BS) or an access point (AP) as a proxy in wireless networks. The proxy can handle the connection status between source and destination. In other words, the proxy is a bridge between wired and wireless networks. The idea is a common sense on previous researches [1,6,11-16], but how to use this architecture to increase the wireless Internet access performance is still an open issue, especially including the considerations about mobility. Our main idea is maintaining two kind of cache in the proxy, one is public cache, and the other is private cache. The proxy assigns a public cache for each service (like WWW and FTP) and the data will be cached in its corresponding public cache. On the other hand, the private cache caches the link information that is referred to the public cache data. The user profile and several personalized data are also recorded in its corresponding private cache. When a MH moves to another BSS, its corresponding cached data may be transmitted automatically to another proxy. Therefore, we name our scheme Smart Follower Cache (SFC).The detail of cache maintenance is described as follows:

2.1   Cache Maintenance Scheme

The cache architecture is shown as Figure 1. Each service, like WWW and FTP, has its corresponding public cache and each MH has its own private cache in the proxy. Any type of MH can cache useful data in the public cache, retrieve these data according to its private cache in the proxy, regardless MHs hardware limit.

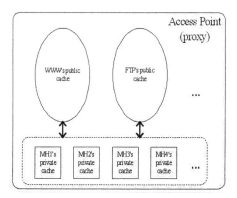

Fig. 1. The proposed cache architecture

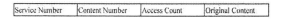

Fig. 2. Adding three kinds of information for the Original cache data

Public Cache Each service has its corresponding public cache in proxy. The content cached in the public cache will be adding three kinds of information: service number, content number and access count (see Figure 2).

- Service number: the service number can be use to indicate the data belong to which kind of service.
- Content number: the content number is a sequential number and can be used by private cache to accessing cache data.
- Access Count: the access count can represent the popularity of the data.

This scheme only needs to add little information before cache content and can cooperate with other cache maintenance policy conveniently.

Private Cache Each MH has its own private cache in corresponding proxy. The private cache is a table which recording the information referred to the public cache data. The table format is shown as Figure 3.

- MH_{ID}: Indicate the cache belong to which MH.
- AP_{ID}: A metric to decide that when the corresponding public cache data should be automatically transmitted to the new proxy.
- Service Number: Indicate the corresponding service type.
- Content Number: Indicate the corresponding content number in public cache.
- AP_{IP}: To point out which machine stored the corresponding public cache content.
- Hop Count: The distance between old and new proxy is called 1 hop. Its also a metric to decide that when the corresponding public cache data should be automatically transmitted to the new proxy.

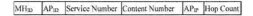

Fig. 3. The table format of private cache

When a MH moves to another BSS, its corresponding private cache will be transferred to the new proxy automatically. But its corresponding public cache data will not be transferred except the hop count larger than 5 or the AP_{ID} are different. The APs that are in the same domain will exchange connection information with each other periodically. The decision of the AP_{ID} is according to the IP domain which the AP belong. For example, there are three APs connecting with networks (see Figure 4), the IP of AP_1 and AP_2 are 140.114.xxx.xxx, and AP_3 is 140.113.xxx.xxx. The value of AP_{ID} in AP_1 and AP_2 will be the same. In this example we set the AP_{ID} is1, and the AP_{ID} of AP_3 will be 2. When a MH move form AP_1 to AP_2, its corresponding private cache content will be transferred to AP_2 immediately, because the AP_{ID} of AP_1 and AP_2 are the same, the referenced public cache content does not need move automatically. If the MH needs to retrieve the public cache content in AP_1, it can find the information on its private cache in AP_2. On the other hand, if the hop count is larger than 5, its corresponding public cache data will be transferred automatically. The detail cache-movement decision algorithm is shown in Figure 5. In summary, when a MH handoff, its corresponding private cache will be move to new AP, but the movement of its corresponding public cache data is depend on the AP_{ID} or the hop count.

2.2 Connection Monitor

To handle the connection status more easily, we implement a connection monitor to deal with all connection situations. The monitor will record a set of information for every connection and its corresponding cache. The format of the connection information table is described as follows:

- IP + port num.: This field records the IP and port number of the connection.
- Access count: This field records the data accessed times. It is also an important factor to decide data replacement priority.
- File size: This field indicates the data size. This information is helpful for data retransmission.
- AC flag: Access complete flag, to indicate whether the data had been transferred complete.
- Connection status: We divide the connection status into two levels; they are good, and noisy respectively. If the connection status is noisy, the data retransmission and prefetching policy need more aggressive.
- Priority: It is an important factor to decide whether to prefetch or retransmit data automatically. The connection monitor is helpful for a proxy to make some decisions, like cache replacement and data transmission policy. Moreover, the administrator of proxy can use these data to administer the whole wireless connection status.

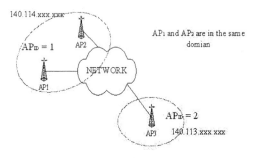

Fig. 4. An example of assigning AP_{ID}

```
function IfNeedMove
    AP_old.private_cache move to AP_new.private_cache
    if AP_old.ID == AP_new.ID then
        if hop_count <= 5 then
            AP_old.public_cache do not need move
        Else
            AP_old.public_cache move to AP_new.public_cache
            Reassign AP_new.public_cache.content_number
            Reassign AP_new.private_cache.content_number
            AP_new.private_cache.AP_IP = AP_new 's IP address
        End if
    else
        AP_old.public_cache move to AP_new.public_cache
        Reassign AP_new.public_cache.content_number
        Reassign AP_new.private_cache.content_number
        AP_new.private_cache.AP_IP = AP_new 's IP address
    End if
End function
```

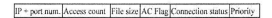

Fig. 5. The detail cache-movement decision algorithm

IP + port num.	Access count	File size	AC Flag	Connection status	Priority

Fig. 6. The connection information table

3 Simulation Results and Analyses

As this paper mentioned above, we choose WLAN to be the simulation platform because it is easy to implement. We use some normal personal computers (PCs) to construct Access Points. The reason is that the disk size is large enough to be a cache of a WLAN. Furthermore, we implement our public cache and private cache on these proxies.

3.1 Simulation Environment

We extended the environment, which was described in Figure 4 . The movement of MH can be separated into two parts, one is the movement in a same AP_{ID}, and the other is in different AP_{ID}.

Table 1. Traffic source type (Web accessing)

Script number	No. of document	Total size	Content category
1	20	3 MB	Commercial WebPages
2	12	2.6 MB	Photo gallery pages
3	20	2.4 MB	Normal WebPages
4	20	1.6 MB	Technical documents WebPages

3.2 Web Accessing Performance

We deployed four MHs in our environment. Every MH will execute four kinds of script (see Table 1) to simulate the Web accessing scenario, and each page will be fetched sequentially.

The transmission speeds are 11Mbps, 2Mbps, and variable speed respectively. We compare the total page loading time with the HTTP/1.1[7] and the MHSP scheme [1]. The simulation result is shown in figure 7. We notice that when a MH handoff between the AP with same AP_{ID}, our scheme has better performance than MHSP and HTTP/1.1, especially in the variable transmission rate scenario. The reason is our scheme will record each connection status. The connection can be automatically resumed when transmission error occurs. Moreover, we notice that when MH1 access these web pages, the contents will be cached in the public cache. After that, MH2 can find these pages in the public cache soon. The total loading time of MH2 is shorter than MH1s loading time. Therefore, in our architecture, the successive users can shorten their total page loading time easily. We continue our simulation with another consideration, when a MH handoff between the AP with different AP_{ID}. The simulation result is shown in figure 8.

We notice that when a MH handoff between the AP with different AP_{ID}, our scheme has much better performance than MHSP and HTTP/1.1. Our SFC scheme can store the needed data of MH, and these data may be transferred with MH. Therefore, in our architecture, the successive users can shorten their total page loading time more easily. In summary, when a MH handoff between the AP with different AP_{ID}, the difference in performance between SFC and MHSP is significant. Our scheme is suitable for large scale WLAN environment on Web accessing.

3.3 File Transfer Performance

In the file transfer scenario, we prepare several files with large size for long period transfer (see table 2). The link speed is 11Mbps, and the simulation result is shown as table 3 and 4.

The total transfer time in table 3 and 4 refer to the first time when a MH accesses those files in table 2. We notice that our scheme can avoid connection broken while MH handoff. Although the average transfer speed is lower than the scenario without mobility, but the difference is slight. Therefore, our scheme still achieves good performance in FTP scenario.

	SN	1.1 Mbps			2Mbps			Variable		
		HTTP/1.1	MHSP	SFC	HTTP/1.1	MHSP	SFC	HTTP/1.1	MHSP	SFC
MH1	1	10.2	5.6	4.2	78.1	55.1	40.2	213.5	151.1	121.6
	2	11.3	4.3	2.6	70.5	49.1	37.5	220.5	144.2	102.2
	3	10.5	4.1	2.3	74.3	45.8	26.1	231.7	140.6	97.9
	4	10.1	3.6	1.9	72.6	41.2	20.7	219.8	131.7	72.2
MH2	1	10.6	5.3	3.9	77.0	59.1	33.7	256.8	194.1	103.3
	2	11.5	4.6	2.2	87.1	47.2	31.2	243.3	162.7	93.7
	3	11.1	4.3	2.0	72.5	46.1	24.9	238.6	153.3	88.9
	4	12.3	3.5	1.6	71.2	45.6	20.2	238.5	144.1	64.1
MH3	1	13.1	7.6	4.0	78.3	66.9	36.1	246.1	172.3	99.6
	2	11.1	5.6	2.3	77.6	52.5	31.1	247.3	166.7	92.2
	3	10.5	4.9	2.0	75.5	46.6	22.6	242.5	142.4	85.7
	4	9.8	4.1	1.7	75.8	40.5	19.7	237.3	139.1	61.6
MH4	1	12.5	5.1	3.7	76.7	50.7	34.5	266.5	186.3	93.2
	2	11.7	4.9	2.4	75.6	45.8	30.6	253.7	175.8	82.2
	3	10.8	4.3	2.2	75.8	41.1	25.1	258.1	170.2	78.3
	4	11.6	3.9	1.9	73.1	36.5	16.5	236.0	149.5	61.1

(unit: seconds)

Fig. 7. Total document loading time (between same AP_{ID})

	SN	1.1 Mbps			2Mbps			Variable		
		HTTP/1.1	MHSP	SFC	HTTP/1.1	MHSP	SFC	HTTP/1.1	MHSP	SFC
MH1	1	31.1	23.6	6.6	118.7	111.2	52.2	282.1	263.0	134.5
	2	31.3	30.1	5.8	117.1	97.7	43.7	301.7	234.1	111.7
	3	33.6	23.7	5.1	124.5	96.9	41.1	313.6	257.7	97.5
	4	30.7	19.4	3.9	119.3	84.3	36.2	297.0	212.3	81.1
MH2	1	31.6	25.3	6.2	117.0	124.6	50.5	331.1	311.2	126.2
	2	30.1	24.7	4.7	117.1	101.2	42.8	326.6	268.5	108.1
	3	32.5	20.1	3.5	122.5	99.3	38.4	313.3	273.7	95.6
	4	32.3	13.5	3.1	121.4	90.6	32.5	306.2	226.1	74.3
MH3	1	33.7	28.1	6.5	118.6	119.9	50.1	311.7	275.2	119.2
	2	31.2	23.1	4.5	117.2	104.6	41.1	307.4	261.5	102.0
	3	30.8	20.5	3.8	125.8	100.1	38.6	292.6	245.9	90.9
	4	31.1	17.7	3.5	125.5	95.2	30.7	299.1	237.1	77.8
MH4	1	32.0	25.9	4.9	126.7	114.3	49.5	331.0	299.2	111.1
	2	31.6	26.4	4.1	115.2	102.8	40.6	328.5	261.4	100.7
	3	30.2	22.2	3.4	115.3	90.8	37.1	332.3	256.0	90.5
	4	31.1	17.7	3.0	123.7	81.2	31.5	319.0	222.8	73.2

(unit: seconds)

Fig. 8. Total document loading time (between different AP_{ID})

Table 2. The file size for FTP simulation

File name	File size
Film1	600 MB
Film2	300 MB
Film3	550 MB
Film4	650 MB

4   Conclusion

This paper proposes a proper access schemes for wireless Internet accessing, named SFC. The proxy maintains two kinds of cache, one is public cache, and the other is private cache. The proxy can cache the necessary data in the public cache and retransmit the data from the public cache to MH immediately while

Table 3. The simulation results of FTP performance (with traditional protocol)

	Withput Mobility	With Mobility
Average Transfer Speed	4.11 Mbps	0.867 Mbps
Total Transfer Time	4120(s)	19378(s)
Connection Broken while MH handoff	NO	Sometimes

Table 4. The simulation results of FTP performance (with SFC)

	Withput Mobility	With Mobility (SFC)
Average Transfer Speed	4.11 Mbps	3.87 Mbps
Total Transfer Time	4120(s)	4345(s)
Connection Broken while MH handoff	NO	NO

transmission error happened. This architecture can accelerate data accessing. When a MH handoff, its corresponding data may be transmitted to the new proxy automatically. According the simulation results, under our architecture, the MHs can access their wanted data more quickly and efficiently.

References

[1] T. B. Fleming, S. F. Midkiff, and N. J. Davis: Improving the Performance of the World Wide Web over Wireless Networks, GLOBECOM 97.

[2] J. C. Mogul: Squeezing More Bits Out of HTTP Caches, IEEE Network Magazine, June 2000.

[3] L. Fan et al. : Summary Cache: A Scalable Wide-Area Web Cache Sharing Protocol, IEEE/ACM Transactions on Networking, June 2000, pp. 281-293.

[4] G. Voelker et al. : On the Scale and Performance of Cooperative Web Proxy Caching, Proc. 17th SOSP, Kiawah Island, SC, December 1999, pp. 16-31.

[5] V. Padmanabhan and J. C. Mogul: Using Predictive Prefetching to Improve World Wide Web Latency, Comp. Commun. Rev., vol. 26, no. 3, 1996, pp. 22-36.

[6] M. Srivastava and P. P. Mishra: On Quality of Service in Mobile Wireless Networks, Proceedings of the IEEE 7th International Workshop on Network and Operating System Support for Digital Audio and Video, 1997, pp. 147 V 158.

[7] R. Fielding, J. Gettys, J. C. Mogul, H. Frystyk and T.Berners-Lee: HyperText Transfer Protocol V HTTP/1.1, RFC 2068, Network Working Group, Jan. 1997.

[8] H. Bharadvaj, A. Joshi, and S. Auephanwiriyakul: An Active Transcoding Proxy to Support Mobile Web Access, Seventeenth IEEE Symposium on Reliable Distributed Systems, Oct. 1998.

[9] A. Joshi, S. Weerawarana, and E. N. Houstis: On Disconnected Browsing of Distributed Information, Seventh International Workshop on Research Issues in Data Engineering, April 1997.

[10] G. Anastasi and L. Lenzini: QoS provided by the IEEE 802.11 wireless LAN to advanced data applications: a simulation analysis, ACM Journal on Wireless Networks, 2000, pp. 99-108.

[11] J. F. Bartlett: W4 V the Wireless World Wide Web, Workshop on Mobile Computing Systems and Applications, Dec. 1994, pp. 176-178.

[12] S. Gessler and A. Kotulla: PDAs as Mobile WWW Browsers, Electronic Proc. Second World Wide Web Conference 94(Mosaic and the Web), Oct.1994.

[13] T. Watson: Application Design for Wireless Computing, Workshop on Mobile Computing Systems and Applications, Dec. 1994, pp. 91-94.

[14] T. Watson: Effective Wireless Communication through Application Partitioning, Proc. Fifth Workshop on Hot Topics in Operating Systems, May 1995, pp. 24-27.

[15] M. Liljeberg, H. Helin, M. Kojo, and K. Raatikainen: MOWGLI WWW Software: Improved Usability of WWW in Mobile WAN Environments, IEEE Global Internet, Nov. 1996.

[16] K. Kim, H. Lee, and H. Chung: A Distributed Proxy Server System for Wireless Mobile Web Service, ICOIN-15, February 2001.

An Experimental Performance Evaluation of the Stream Control Transmission Protocol for Transaction Processing in Wireless Networks

Yoonsuk Choi[1], Kyungshik Lim[1], H.-K. Kahng[2], and I. Chong[3]

[1] Computer Science Department, Kyungpook National University
#1370 Sankyuk-dong Buk-gu, Daegu, 702-701, Korea
{yoonsuk,kslim@ccmc.knu.ac.kr}
[2] Electronics Information Engineering Department, Korea University
#208 Suchang-dong, Chochiwon, Chungnam, 339-700, Korea
kahng@tiger.korea.ac.kr
[3] Information and Communications Engineering Dept., Hankuk University of FS
#270 Imun-dong, Dongdaemun-Gu, Seoul, 130-790, Korea
iychong@hufs.ac.kr

Abstract. In this paper, we show that the Stream Control Transmission Protocol(SCTP) could be an efficient alternative to the Transmission Control Protocol(TCP) for processing wireless multimedia transaction with multiple in-line objects in a transaction. This comes from three improvements of SCTP over TCP: the standard feature of selective acknowledgment, a different set of the initial and maximal values for retransmission counter and timer, and the multi-streaming feature. Among them, the multi-streaming feature of SCTP has the most profound impact on the overall performance in terms of throughput and delay in wireless networks with high probability of burst errors.

1 Introduction

Over the years, the Transmission Control Protocol(TCP) has been widely used for reliable transaction processing in both wired and wireless IP networks. However, as a demand on supporting a range of commercial applications such as real-time multimedia and telephony applications grows, the limitations and shortcomings of TCP have motivated the development of the Stream Control Transmission Protocol(SCTP) by the Signaling Transport(SIGTRAN) working group in the IETF[1].

Like TCP, SCTP provides a reliable end-to-end message transportation service over IP-based networks. It inherits many features from TCP such as connection management, congestion control, error control, and flow control mechanisms. However, two major SCTP enhancements over TCP include multi-homing and multi-streaming. Multi-homing enables an SCTP host to establish an association with another SCTP host over multiple links or paths, achieving fast link/path recovery with little interruption to the data transfer service. Multi-streaming

H.-K. Kahng (Ed.): ICOIN 2003, LNCS 2662, pp. 595–603, 2003.

makes it possible to define subflows in an overall SCTP message flow and enforce message ordering only within each subflow, solving the head-of-the-line blocking problem of TCP.

There have been several research efforts on the evaluation of the SCTP performance for various applications in different environments. Andreas Jungmaeir, et al.[2] shows that SCTP shares network resources fairly with TCP in a wide area networks so that it can be properly used for the transport of telecommunication signaling messages in an engineered IP-based signaling network. However, this paper does not show how the protocol features of SCTP such as selective acknowledgements, fast retransmission and out-of-order delivery affect the high performance requirements in a telecommunication signaling system. P. Conrad, et al.[3] shows that the multi-streaming feature of SCTP can give great benefits in reducing latency for multimedia streaming services in high loss environments. J. Iyengar, et al.[4] proposes the Rhein algorithm that solves some problems of SCTP such as unnecessary retransmission and congestion window overgrowth during changeover. A. Caro, et al.[5] decouples failure detection from the recovery process and makes it possible to begin the recovery process during early signs of a possible failure in order to improve overall throughput.

In this paper, we focus on the effect of the multi-streaming feature of SCTP on wireless multimedia transaction processing when multiple in-line objects are transferred in a transaction. Furthermore, we also investigate what other features of SCTP could be used for the same purpose. This paper is organized as follows. Section 2 presents the experimental network setup such as experimental network architecture, wireless error model, and transaction generation model. In Section 3, experimental results are given and experimental analyses are discussed. Finally, the conclusion and future works are presented in Section 4.

2   Experimental Network Setup

2.1   Experimental Network Architecture

Fig. 1 shows our experimental network architecture where two 100Mbps Ethernets are interconnected via the SHUNRA/Cloud WAN emulator on Windows NT. The emulator makes it possible to set various network parameters such as latency parameters, packet loss, packet effects, link faults, and congestion manually and to record latency and packet loss externally. We use this emulator to simulate a contemporary IS-95C CDMA network of 144Kbps. A Linux implementation of the SCTP client and server is from Siemens AG and the Essen University. To compare the performance of SCTP with that of TCP, we also used a Linux implementation of the TCP Reno. We assume that there are 20 SCTP/TCP clients and 1 SCTP/TCP server.

2.2   Wireless Error Model

We use the Gilbert model as an error generation model to take an advantage of representing both burst or non-burst error characteristics of wireless networks

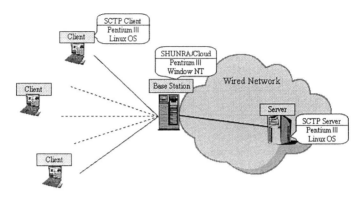

Fig. 1. Experimental Network Architecture

in an easy manner[6, 7]. Let p and q denote the transition probabilities of going from state 0 to state 1 and the reverse, respectively, where state 0 means that packet is delivered and state 1 represents that packet is lost. Let X_n denote the probability variable which is set to 1 if packet n is lost, and 0 otherwise. Then, the transition probabilities p and q are presented using X_n as follows.

$$P = Pr\left[X_{n+1} = 1 | X_n = 0\right], Q = Pr\left[X_{n+1} = 0 | X_n = 1\right] \tag{1}$$

The steady state properties of the chain are given by $Pr[X_n = 0]$ and $Pr[X_n = 1]$ to represent that the channel is in state 0 and state 1, respectively[6, 8], as follows.

$$Pr\left[X_n = 0\right] = \frac{q}{p+q}, Pr\left[X_n = 1\right] = \frac{p}{p+q} \tag{2}$$

We assume that packet loss rate is less than 50% and the average number of consecutive packet loss is from 2 to 10 packets in mobile computing environments. Then, p and q are calculated using the Mean Residence Time(MRT) of state 0 and 1 in the Gilbert model, respectively. Since MRT of state 1 is given by $1/q$ and the average number of consecutive packet loss is 2-10 by the assumption, q is $0.1 \leq q \leq 0.5$[6]. Thus, we experiment the performance of the SCTP and TCP protocols when FER is 1-50%, and q is 0.1, 0.3 and 0.5.

2.3 Transaction Generation Model

Fig. 2 shows the relationship between transaction and session, where a session is a set of transactions and a transaction is a pair of a request and its multiple responses(i.e., in-line objects). In Table 1, we do not consider parsing time and object viewing time. The user behavior is modeled as an on-off client model. Transaction inter arrival time and session duration are determined by the truncated pareto distribution, and session inter arrival time is determined by the exponential distribution[9]. Response object size and the number of in-line objects of server are determined by truncated pareto distribution[9, 10, 11].

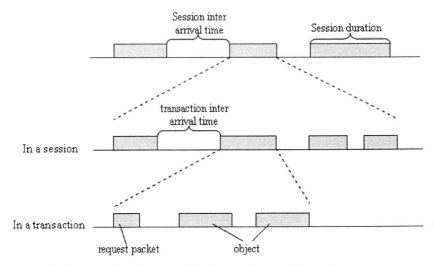

Fig. 2. The relationship between transaction and session

Table 1. Transaction generation model with multiple in-line objects

Process	Parameters	Distribution
Request size	min = 36bytes max = 80bytes	uniform
Object size	shape = 1.2 scale = 101bytes max = 1500bytes	truncated pareto
Number of in-line objects	Shape = 1.4 scale = 5 max = 2	truncated pareto
Transaction inter arrival time	shape = 1.5 scale = 17sec max = 46800sec	truncated pareto
Session inter arrival time	Lambda = 0.011 (mean = 90sec)	exponential
Session duration	shape = 1.5 scale = 260sec max = 900sec	truncated pareto

3   Experimental Analysis

In the experiment, we consider an IS-95C CDMA wireless network with a bandwidth of 144kbps and an RTT of 200ms. Over 20 clients generate transaction traffics to a server going through the internet emulator that simulate the above network condition. We are interested in both normalized overall throughput and transaction processing delay as Frame Error Rate(FER) increases in conventional wireless environments. The main goal of the experiment is to figure out how the multi-streaming feature of SCTP affects the overall performance of transaction processing in wireless networks. Fig. 3 shows the normalized throughput and transaction processing delay of TCP and SCTP when the number of in-line objects is one. In other words, we measure the transaction success rate and the average execution time of transactions over wireless networks in a burst error and non-burst error condition. When FER is between 30% and 50% and q is between 0.1 and 0.3, SCTP shows better throughput than TCP. In other cases, SCTP is similar to TCP in terms of throughput. This means that SCTP performs better than TCP in case of high error rate and high probability of burst errors. This improvement is due to the SACK feature of SCTP which is a mandatory feature of SCTP, but not available in TCP Reno. The details can be explained as follows. With the limited information inherent to cumulative acknowledgments, a TCP sender can only learn about a single lost packet per round trip time. This might force the sender to either wait a roundtrip time to find out each lost packet, or to unnecessarily retransmit segments which might have been correctly received. However, with SCAK, the data receiver can inform the sender about all segments that have arrived successfully, so the sender needs retransmit only the segments that have actually been lost[12].

Another interesting point is that the transaction processing delay of SCTP is about one tenth times shorter than that of TCP. We believe that this comes from the fact that SCTP uses suitable initial and maximal parameter values for retransmission counter and RTO over TCP in high burst error conditions. Table 2 compares the initial and maximal values for retransmission counter and RTO of SCTP and TCP. By a simple calculation based on Table 1, we can see that the maximal data sending time of SCTP is twenty times shorter than that of TCP even if the connection setup time of SCTP is three times longer than that of TCP.

Fig. 4 shows the normalized throughput and transaction processing delay of TCP and SCTP when the number of in-line objects is more than one. Compared to the Fig. 3, Fig. 4 shows that SCTP performs much better than TCP in both throughput and delay. TCP usually uses multiple connections for receiving more than one object in a transaction and restores the correct sequence of objects in case of errors on some connections, resulting in a long delay of in-order delivery of multiple objects. However, the multi-streaming feature of SCTP makes it possible to transfer multiple in-line objects independently, eliminating the TCP's head-of-line blocking delay that might be frequently occurred in case of high error rate of wireless links. For that reason, SCTP shows much higher throughput than

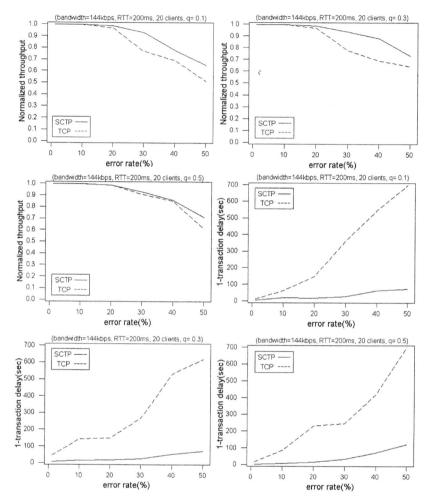

Fig. 3. The comparison of the normalized throughput and transaction processing delay of TCP and SCTP(when one in-line object in a transaction is used)

Table 2. The comparison of retransmission counter and RTO of SCTP and
TCP

		During connection setup	During sending data
SCTP	Retransmission counter	9	6
	RTO	Initial value : 3sec	Initial value : 3sec
		MAX : 60sec	MAX : 48sec
TCP	Retransmission counter	5	15
	RTO	Initial value : 3sec	Initial value : 1sec
		MAX : 24sec	MAX : 120sec

TCP for the transfer of multiple in-line objects in a transaction over wireless network with high error rate.

4 Conclusion and Future Works

SCTP has been developed as an efficient transport layer protocol for the transfer of call control signaling message over IP networks by IETF. In addition, there

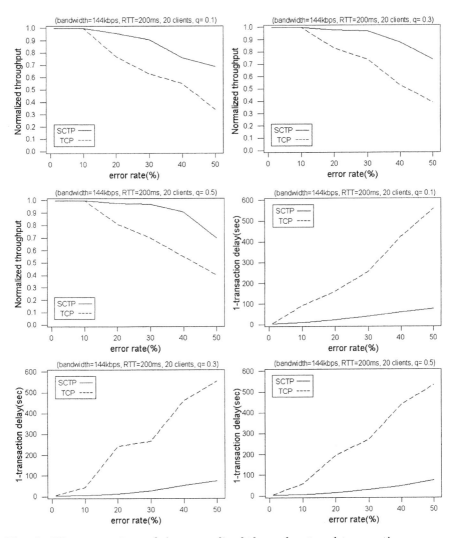

Fig. 4. The comparison of the normalized throughput and transaction processing delay of TCP and SCTP (when multiple in-line objects in a transaction are used)

have been a number of research efforts to take advantage of SCTP to overcome the limitations of TCP for the support of various multimedia applications. As such an effort, in this paper, we conducted the use of SCTP for transaction processing in wireless environments. Traditionally, TCP has been widely used for transaction processing in wired IP networks. But, as the demand on wireless transaction processing increases, alternative protocols such as TCP/T, a variant of TCP, and Wireless Transaction Protocol have been investigated. In this paper, we show that SCTP could be an efficient alternative over TCP for wireless multimedia transaction processing with multiple in-line objects in a transaction. The standard feature of selective acknowledgment in SCTP takes an important role to improve the overall throughput in wireless networks with high probability of burst errors. In addition, a different set of the initial and maximal values for retransmission counter and RTO of SCTP over TCP might result in that the transaction processing delay of SCTP is around ten times shorter than TCP. Lastly, most importantly, the multi-streaming feature of SCTP could greatly improve the overall performance in terms of throughput and delay for the processing of wireless transactions in case of high probability of burst errors. In the future researches, we are supposed to conduct the use of multi-homing feature of SCTP for host mobility management in combination of how to handle multiple bundles of multiple streams between two end hosts.

References

[1] R. Stewart, et al.: Stream Control Transmission Protocol. RFC 2960. Internet Engineering Task Force. October 2000
[2] Andreas Jungmaeir, Michael Schopp, Michael Tuxen: Performance Evaluation of the Stream Control Transmission Protocol. Proceedings of the IEEE Conference on High Performance Switching and Routing, Heidelberg, Germany. June 2000
[3] P. Conrad, G. Heinz, A. Caro, P. Amer, J. Fiore: SCTP in Battlefield Network. Proceeding of MILCOM 2001, Washington. October 2001
[4] J. Iyengar, A. Caro, P. Amer, G. Heinz, R. Stewart: SCTP Congestion Window Overgrowth During Changover. Proceedings of the 6th World Multiconference on Systemics, Cybernet-ics and Informatics, Orlando, Florida. July 2002
[5] A. Caro, J. Iyengar, P. Amer, G. Heinz, R. Stewart: A Two-level Threshold Recovery Mechanism for SCTP. Proceedings of the 6th World Multiconference on Systemics, Cybernetics and Informatics, Orlando, Florida. July 2002
[6] Yeunjoo Oh, Nakhoon Baek, Hyeongho Lee, Kyungshik Lim: An adaptive loss recovery algorithm based on FEC using probability models in mobile computing environments. SK Telecommunications Review, Vol. 10, No. 6, pp. 1193-1208. December 2000
[7] Jean-Chrysostome Bolot, Hugues Crepin, Andres Vega Garcia: Analysis of Audio Packet loss in Internet. Proceedings of Network and Operating System Support for Digital Audio and Video NOSSADV'95, pp. 163-174, Durham, NH. April 1995
[8] J-C. Bolot, S. Fosse-Parisis, D. Towsley: Adaptive FEC-Based Error Control for Internet Telephony. Proceedings of IEEE INFOCOM'99, New York. March 1999
[9] Yingxin Zhou and Zhanhong Lu: Utilizing OPNET to Design and Implement WAP Ser-vice. OPNETWORK 2001. August 2001

[10] Thomas Kunz, et al.: WAP traffic: Description and Comparison to WWW traffic. Proceedings of the 3rd ACM international workshop on Modeling, analysis and simulation of wire less and mobile systems, pp. 11-19, Boston. August 2000

[11] B. Bellalta, M. Oliver, D. Rincón: Análisis de tráfico y capacidad para servicios WAP sobre redes GSM/GPRS. XI Jornadas de I+D en Telecomunicaciones, Madrid, Spain. 2001

[12] M. Mathis, et al.: TCP Selective Acknowledgment Options. RFC 2018. Internet Engineering Task Force. October 1996

TCP Recovering Multiple Packet Losses over Wireless Links Using Nonce*

Chang Hyun Kim[1] and Myungwhan Choi[2]

[1] LG Electronics Institute of Technology, Seoul 137-724, Korea
kevinch@lge.com
[2] Dept. of Computer Science and Eng., Sogang University, Seoul 121-742, Korea
mchoi@ccs.sogang.ac.kr

Abstract. In the current TCP, the packet losses over the wireless link are considered as the indication of the congestion and the TCP congestion control is invoked, which may severely degrade the performance of TCP. We propose a new method to effectively overcome this type of TCP performance degradation using nonce. The proposed scheme running at the base station (BS) can identify the lost segments which can be quickly retransmitted. It is shown that the proposed scheme is robust to the consecutive packet losses. It has another advantage that the nonce in the proposed scheme can be used to implement the strategy to block the attacks by the misbehaving receiver.

1 Introduction

In the wireless part of the network accommodating the wireless/mobile users, the packet loss occurs due to the relatively high bit error rate (BER) over the wireless link and the user mobility. The current TCP is designed under the assumption that the underlying network is rather reliable. So the packet losses over the wireless link are considered as the indication of the congestion. The TCP congestion control invoked thereby may severely degrade the performance of TCP over the wireless link.

To hide the packet loss due to the transmission error over the wireless link from the sender in the wired part of the network, the BS in the boundary of the wired network and the wireless network plays a key role. The initial approach was to split the TCP connection at the BS and two separate TCP connections are maintained over the wired network and wireless link, respectively [1]. This approach, however, does not keep the end-to-end TCP semantics. M-TCP [2] also falls in this category. Snoop module approach [3] and WTCP [4] cache the arriving packets at the BS to be transferred to the mobile host (MH). The BS transmits acknowledgements only for those packets successfully received by the MH. How quickly and efficiently the BS can detect the packet losses and retransmit the lost packets differs for different proposals. Pure link layer approach [5]

---

* This work was supported by the Institute for Applied Science and Technology, Sogang University.

H.-K. Kahng (Ed.): ICOIN 2003, LNCS 2662, pp. 604–613, 2003.

tried to minimize the effect of the packet loss over the wireless link on the performance of the TCP by using the techniques such as adaptive forward error correction, fully-reliable and semi-reliable link layer ARQ protocols. To overcome the TCP performance degradation during handoff period, fast retransmit approach was proposed [6].

In parallel to the effort to improve the TCP performance over the wireless networks, it is also important to notice that the end-to-end congestion control mechanisms, such as those used in TCP, implicitly rely on the assumption that both endpoints cooperate in determining the proper rate at which to send data. If a sender misbehaves, then it can send data more quickly than well-behaved hosts. A·misbehaving receiver can achieve the same result [7].

The current TCP congestion control mechanism will operate properly only when the duplicate ACKs are generated by the successful reception of the distinct packets transmitted by the sender. To defend against the attacks such as duplicate ACK spoofing and optimistic ACKing by misbehaving receiver, it is necessary to identify the data segment that led to the generation of each duplicate ACK and to guarantee that it took each ACK with valid nonce a full round-trip time to return to the sender. It is shown that the utilization of nonce makes this possible [7].

In this paper, we propose a nonce TCP (N-TCP) which, by using nonce, can effectively overcome the TCP performance degradation when the multiple consecutive packet transmission errors occur over the wireless link in the wired/wireless networks. The nonce used in the N-TCP can be utilized to block the attacks by the misbehaving receiver as shown in [7]. The use of nonce in the Internet community is not new. In [8], the incorporation of explicit congestion notification (ECN) to TCP and IP is specified. The authors in [9] describe the ECN-nonce, as optional addition to ECN that prevents receivers from exploiting ECN to gain an unfair share of network bandwidth.

In Sect. 2, the operation of the proposed N-TCP is described. The simulation results are shown in Sect. 3. The conclusion follows in Sect. 4.

2   The Operation of N-TCP

N-TCP resides in the BS as shown in Fig. 1 and it stores the packets delivered from the fixed host (FH) before it transmits the received packets to the MH.

In the N-TCP, nonce is used to quickly recognize the multiple packet losses without receiving duplicate ACK and those lost packets are retransmitted, which prevents the congestion control by the sender side TCP from being invoked. We assume that the mobile IP is used in the network layers of the BS and the MH. In this work, however, mobility is not considered. Considering that the TCP performance degradation typically occurs in ignorance of the existence of the wireless link, we limit our description of the N-TCP operation to the packet flow from the FH to MH.

Two additional fields - nonce and nonce reply - need to be added to the current TCP header format to run the N-TCP. When a packet with sequence

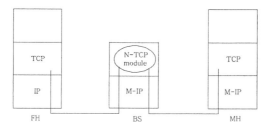

Fig. 1. N-TCP connection

number s arrives at the BS, the packet PKT(s, nonce) is stored at the BS buffer before it is transmitted to the MH, where nonce is the random number unique to that packet. When the receiver replies for the correctly received packet, it transmits ACK(r, nonce reply) to the BS where r is the expected packet sequence number and the nonce reply is the copy of the nonce value of the correctly received packet. Therefore, the BS can determine whether there were any packet losses or not by comparing the ACK number r with the sequence number of the packet corresponding to the value of nonce reply field of the received ACK. For example, assume that ACK(3, 32) is received by the BS meaning that the MH has successfully received packets with sequence numbers up to 2. Therefore, if the sequence number of the packet corresponding to the nonce value 32 is 2, it means that there was no packet loss. If it is 7, it means that packets with sequence numbers 3, 4, 5, and 6 are lost.

2.1 Circular Queue Management

N-TCP stores the PKT(s, nonce) when a packet with sequence number s arrives from the FH and transmits it to the MH. When the ACK(r, nonce reply) is received, the N-TCP removes the packets from the BS buffer which are verified to be received by the MH and the ACK r is transmitted to the FH. For this operation, circular queue is implemented and two indices, buftail_ and bufhead_, are maintained.

buftail_ indicates the location of the packet with the smallest sequence number among packets stored in the buffer. bufhead_ indicates the next empty space to the location of the packet with the largest sequence number among packets in the buffer. When the queue is either full or empty, the buftail_ and bufhead_ indicate the same location. Therefore, a flag is used to indicate the full state. If a packet arrives when the buffer is full, buftail_ is incremented by one and the arriving packet is stored in the location being indicated by the bufhead_. In this case, the retransmission timeout may occur at the sender side for the removed packet. This event, however, will rarely occur if the buffer size is large enough to store full congestion window size worth of packets. The packet arriving in-sequence is stored in the location indicated by the bufhead_. When the packet arrives out-of-sequence, the rearrangement of the stored packets in the buffer is performed so that the arriving packet can be stored in the buffer in-sequence.

The arriving packet whose sequence number is larger than that of the packet stored at bufhead_−1 is said arriving in-sequence.

Assume that q is the sequence number of the packet whose corresponding nonce value is p. When the ACK(k, p) arrives at the BS, the packets whose sequence numbers are smaller than k are removed from the buffer. Now buftail_ indicates the location where the packet whose sequence number is k is stored. If there was no packet loss, q will coincide with $k − 1$. It there were some packet losses, q will be larger than $k − 1$.

2.2   Processing of Data Packet Arriving at the BS from the FH

The arriving packets are grouped into the following three cases and the processing required for each case is described below.

Case a) in-sequence packet arrival (arrival of a new packet). The arriving packet is stored at the location indicated by the bufhead_ along with the nonce value. The PKT(sequence number, nonce) is transmitted to the MH. bufhead_ is incremented by one.

Case b) arrival of out-of-sequence packet which is already in the buffer. This happens when the arriving packet is retransmitted by the FH. The nonce value for the packet in the buffer is replaced by the new nonce value and the packet with new nonce value is transmitted to the MH along with its sequence number.

Case c) arrival of out-of-sequence packet which is not in the buffer. This can be categorized into the following three cases: case c.1) when the sequence number of the arriving packet is smaller than that of the packet stored at the location indicated by the buftail_, case c.2) when the sequence number of the arriving packet falls somewhere between the sequence number of the packet indicated by buftail_ and that of the packet indicated by bufhead_−1, case c.3) when the ACK for the arriving packet is already sent to the FH and the corresponding previous packet was removed from the buffer.

For the case c.1), the buftail_ is decremented by one and the arriving packet is stored in the location indicated by new buftail_ along with newly generated nonce value. The packet with nonce value is transmitted to the MH. For the case c.2), it is necessary to find the location to store the arriving packet. Assuming that the appropriate location is buftail_+k (meaning that the sequence number of the packet stored at the location buftail_+k−1 < the sequence number of the arriving packet < the sequence number of the packet stored at the location buftail_+k), the arriving packet is inserted at the location buftail_+k along with the nonce value and transmitted to the MH. For this operation, the packets which have been residing at locations buftail_+k through bufhead_−1 should be relocated at new locations buftail_+k + 1 through bufhead_. After the completion of this process, the bufhead_ is incremented by one. The ACK loss on the way from BS to the FH will cause the event coming under the case c.3). Therefore, for the case c.3), the BS sends the ACK for the packet with the largest sequence number among packets received by MH for sure and the arriving packet is discarded.

2.3   Processing of ACK Arriving at the BS from the MH

Assume that $ACK(k, p)$ arrives at the BS from the MH. Let last_ACK denote the largest ACK number among the acknowledgement numbers transmitted by the MH to the BS. Also assume that q is the sequence number of the packet whose corresponding nonce value is p. Then the arriving ACKs can be grouped into the following three cases and the processing required for each case is described below.

Case a) Arrival of ACK with a new acknowledgement number ($k >$ last_ACK). A new $ACK(k, p)$ has arrived. So, all the packets in the buffer whose sequence numbers are smaller than k are removed and the ACK k is transmitted to the FH. Let $L = q - (k - 1)$. Then $l = \max\{L, 0\}$ packets are lost and packet q now is residing at buftail_+L. The nonce values of the packets stored at buftail_ through buftail_+$L - 1$ are replaced by the newly generated nonce values and those packets are retransmitted to the MH. Finally, the packet stored at buftail_+L is removed.

Case b) ACK number of the arriving ACK is smaller than the acknowledgement number transmitted to the FH ($k <$ last_ACK). The arriving ACK is spurious and discarded. This won't happen during the normal operation.

Case c) Arrival of duplicate $ACK(k, p)$ ($k =$ last_ACK). Let $L = q - (k - 1)$. Then $l = \max\{L, 0\}$ packets are lost and packet q now is residing at buftail_+L. Among the packets stored at buftail_ through buftail_+$L - 1$, only the packets which have never been retransmitted or for which more than twice the round trip time between the BS and the MH since last transmission elapsed, are retransmitted. Those retransmitted packets carry the newly generated nonce values. The packet q is removed from the buffer. Notice that the ACK is not transmitted to the FH in this case.

2.4   Timer

Two timers, round-trip timer and persist timer, are used in N-TCP. Round-trip timeout interval is computed using smoothed round-trip time (srtt) mechanism, $srtt = (1 - \alpha) \times srtt_{old} + \alpha \times rtt_{curr}$, where $\alpha = 0.25$. In order to get out of the deadlock situation between the BS and the MH, persist timer is used. When there are some packets remaining in the buffer and there is no communications occuring between the BS and the MH for longer than 100 ms, the persist timer expires, which enables the retransmission of the packets awaiting ACK at the buffer.

2.5   Recovery of the Multiple Packet Losses in N-TCP

N-TCP can effectively recover the multiple packet losses. To show this and its features, the operation of the N-TCP is compared with the Snoop module. The snoop module is widely known as the baseline system to overcome the TCP performance degradation in the wireless networks. Readers are referred to [3], [10] for detailed description of the snoop module.

Figure 2 illustrates the key differences between these two schemes. For this illustration, congestion window size is 7 and delayed ACK is used. Using the delayed ACK, ACK is generated for every two correctly received packets. When there are packet losses, the ACK is generated by the next correctly arriving packet.

We briefly explain the operation of the N-TCP in this example. When ACK(2, 15) is received, packet 1 is removed from the buffer. At this time, packet 2 resides at location buftail_. The packet 5 whose nonce value is 15 resides at location buftail_+3. Therefore, we know that three ((buftail_+3)− buftail_ = 3) packets are lost and these packets are retransmitted each carrying new nonce value. Since

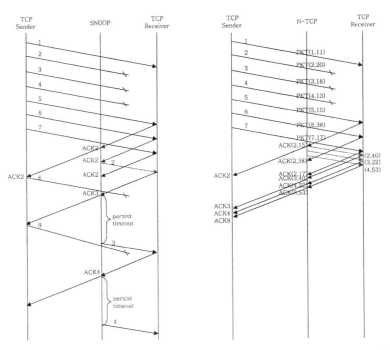

Fig. 2. Multiple loss recovery of snoop and N-TCP (Delayed ACK)

packet 5 is successfully received by the MH, it is removed from the buffer. When ACK(2, 38) arrives, packet 6 whose corresponding nonce value is 38 resides at the location buftail_+3. Therefore, we know that three packets are lost. This time, however, those packets are not retransmitted because those packets are already retransmitted not long ago (before the twice the srtt elapses). Notice that in this example, the snoop module requires the reception of the duplicate ACK to retransmit the lost packet and retransmission of only one packet for duplicate ACKs is allowed. The additional delay caused by this operation further increases the possibility for the persist timeout as shown in this example. In contrast, the feature of the N-TCP that the lost packets identified by the reception of an ACK can be retransmitted allows the fast recovery from the multiple packet losses.

3 Experimental Evaluation

3.1 Simulation Model

We assume that the FH and the BS are interconnected by the 10 Mbps Ethernet. The BS and the MH are connected using the 2 Mbps IEEE 802.11 WLAN protocol. It is also assumed that the packet transmission delays from the FH to the BS and from the BS to the MH are 200 msec and 10 msec, respectively. The TCP-Reno is used on the FH and the MH. We built the simulation model using network simulator v.2.1b8a [11].

Four bytes long nonce and nonce reply fields are defined as option fields in the TCP header. Snoop supported in the network simulator is used as the simulation system for the snoop module. Payload size of the TCP segment is 536 bytes long. Maximum window size is 35 segments. Minimum interpacket generating time is assumed 0.4 msec. The BER over the wired link is assumed zero. We assigned the buffer in the BS large enough so that any packet loss on the path from the BS to the MH is caused by the transmission error over the wireless link only.

We modeled the bursty wireless link transmission characteristics as two-state Markov model using multi-state error model . The BERs at good and bad states are 10^{-6} and 10^{-2}, respectively. The time duration at the good state is exponentially distributed with mean 1 sec. The time duration at the bad state is also exponentially distributed with mean x msec. In the simulation, we varied the mean bad period x from 1 msec up to 256 msec to obtain the average BERs ranging from 10^{-5} to 2×10^{-3}.

3.2 Simulation Results

We ran simulations for the cases where the FH sends a large file of size 1 Mbytes to the MH for various mean bad periods over the wirless link.

Figure 3 shows the sequence number vs. time curve at the MH when the mean bad period is 8 msec (BER = 8×10^{-5}). As illustrated in Fig. 2, the N-TCP retransmits all the lost packets right after receiving the first ACK after any packet losses while the snoop module can retransmit a packet only after receiving

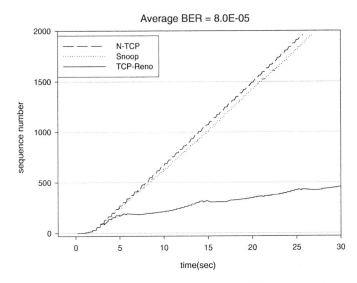

Fig. 3. Sequence number vs. time at MH, $0 \leq t \leq 30\,\text{sec}$

a duplicate ACK. Therefore, the N-TCP gives better performance than the snoop module as shown in Fig. 3.

When the mean bad period is 32 msec (BER $= 3 \times 10^{-4}$), the number of consecutive packet errors increases compared with the previous case. Therefore, it is more important to have the ability to retransmit lost packets quickly. It is observed that the file transmission completion time has increased by a negligible amount of time when the N-TCP is used. Meanwhile, it has increased by a large amount of time when the snoop module is used (in this experiment, it has changed from 27.03 second to 42.13 second).

When the mean bad period increases to 256 msec (BER $= 2 \times 10^{-3}$), the number of consecutive packet errors further increases and even the retransmitted packets are sometimes lost. So, the performance of both then N-TCP and the snoop severely deteriorates (Fig. 4).

In the more detailed sequence number vs. time curve (not shown here for space limitation), for example at around 6 second, it is observed that multiple retransmissions are required for the MH to receive a packet successfully. It can also be observed that when the snoop is used, it takes much more time to send the lost packet successfully to the MH compared with the case of N-TCP and this causes retransmission timeout at the FH. This invokes the TCP congestion control and further degrades the TCP performance.

For the TCP-Reno, the packet loss over the wireless link invokes the congestion control and poor performance is observed.

Figure 5 shows the time taken to complete the transmission of 1 Mbytes of data for various BERs.

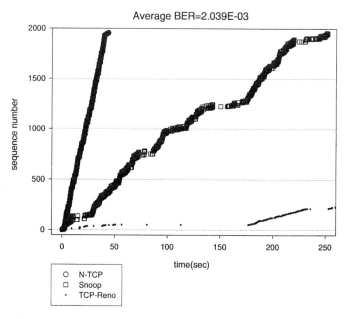

Fig. 4. Sequence number vs. time at MH. $0 \leq t \leq 250\,\text{sec}$

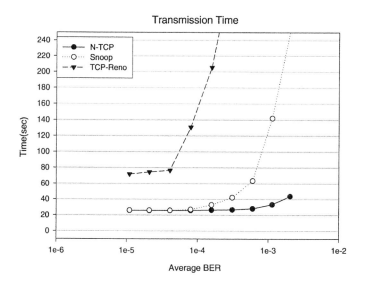

Fig. 5. Comparison of file transfer times

4   Conclusion

This work is motivated by [7] which proved that the TCP is open to the attacks of the misbehaving receivers such as the duplicate ACK spoofing and those attacks can be effectively blocked using nonce over the wireless link. In this paper, we have shown that the nonce can also be utilized to overcome the TCP performance degradation due to the consecutive packet losses over the wireless link. For this, however, it is also required to modify the TCP: new fields (nonce and nonce reply) must be either added in the TCP header or defined as option.

Even though it is unlikely that the N-TCP will be deployed in the near future, we strongly believe it is important to consider the potential of using nonce over the wireless segment of the network. The proposed N-TCP may be used to effectively recover the TCP performance degradation due to the packet errors over the wireless link and to block the attacks of the misbehaving receivers.

References

[1] Baker, A., Badrinath, B. R.: I-TCP : Indirect TCP for Mobile Hosts. Proc. 15th Intl. Conf. on Distributed Computing Systems (1995) 136–143

[2] Brown, K., Singh, S.: M-TCP: TCP for Mobile Cellular Networks. ACM Computer Communications Review, Vol. 27, No. 5 (1997) 19–43

[3] Balakrishnan, H.: Challenges to Reliable Data Transport over Heterogeneous Wireless Networks. Ph.D. Thesis, Univ. of California at Berkeley (1998)

[4] Ratnam, K., Matta, I.: WTCP : An Efficient Mechanism for Improving TCP Performance over Wireless Links. Proc. Third IEEE Symposium on Computer and Communications (1998)

[5] Ayanoglu, E., Paul, S., LaPorta, T. F., Sabnani, K. K., Gitlin, R. D.: AIRMAIL: A Link-Layer Protocol for Wireless Networks. ACM/Baltzer Wireless Networks J., Vol. 1, No. 1 (1995) 47–60

[6] Caceres, R., Iftode, L.: Improving the Performance of Reliable Transport Protocols in Mobile Computing Environments. IEEE J. on Selected Areas in Communications, Vol. 13, No. 5 (1994) 850–857

[7] Savage, S., Cardwell, N., Wetherall, D., Anderson, T.: TCP Congestion Control with a Misbehaving Receiver. ACM Computer Communications Review, Vol. 29, No. 5 (1999) 71–78

[8] Ramakrishnan, K., Floyd, S., Black, D.: The Addition of Explicit Congestion Notification (ECN) to IP. RFC 3168 (2001)

[9] Spring, N., Wetherall, D., Ely, D.: Robust ECN Signaling with Nonces. Internet-drafts draft-ietf-tsvwg-tcp-nonce-04.txt. work in progress (2002)

[10] Balakrishnan, H., Seshan, S., Katz, R. H.: Improving Reliable Transport and Handoff Performance in Cellular Wireless Networks. ACM Wireless Networks J., Vol. 1, No. 4 (1995) 469–481

[11] McCanne, S., Floyd, S.: ns - Network Simulator. http://www.isi.edu/nsnam/ns/

UMTS Implementation Planning on GSM Network

Tutun Juhana and Maiyusril

Telematics Laboratory, The Department of Electrical Engineering
Institute of Technology Bandung
Jalan Ganesha 10 Bandung - Indonesia
`tutun@telecom.ee.itb.ac.id`

Abstract. Nowadays, GSM Network has been deployed all over the world, including in Indonesia. Some GSM operators in Indonesia get prepared to develop their GSM networks toward GPRS. After GPRS, the next generation technology is Universal Mobile Telecommunication Service (UMTS). UMTS is a third generation of mobile telecommunication technology that is defined in ITU specification as International Mobile Telecommunications-2000 (IMT-2000). GSM/GPRS operators that want to deploy UMTS will wish to reuse the components of its existing network to the greatest extent possible. There is a significant opportunity to reuse existing equipment because of the fact that the core network of UMTS is essentially the same core network as is used for GSM/GPRS. 3GPP Release 1999 network architecture reuses a great deal of GPRS functionality for its packet data network. In this paper, we design an implementation planning of UMTS on existing GSM network. We use some network data of PT Telkomsel in Batam Area as our case study.

1 Introduction

Improvement of human mobility has caused the need for communication that not depends on place, time and situation. One of the solutions is by using mobile communication systems that is called cellular technology. The most popular of this technology is GSM.

Nowadays, GSM Network has been deployed all over the world, including in Indonesia. Some GSM operators in Indonesia get prepared to develop their GSM networks toward GPRS. After GPRS, the next generation technology is Universal Mobile Telecommunication Service (UMTS). UMTS is a third generation of mobile telecommunication technology that is defined in ITU specification as International Mobile Telecommunications-2000 (IMT-2000).

GSM/GPRS operators that want to deploy UMTS will wish to reuse the components of the GSM/GPRS network to the greatest extent possible, such as MSC, SGSN, HLR, etc. There is a significant opportunity to reuse existing equipment because of the fact that the core network of UMTS is essentially the same core network as is used for GSM/GPRS. 3GPP Release 1999 network architecture reuses a great deal of GPRS functionality for its packet data network.

H.-K. Kahng (Ed.): ICOIN 2003, LNCS 2662, pp. 614–622, 2003.

The implementation of UMTS network on GSM/GPRS network infrastructure can combine telecommunication service for voice, data and video in a mobile phone terminal. In this case, there is a dual mode system between UMTS and GSM. This dual mode property can give a flexible solution for GSM/GPRS operator by sharing a new frequency spectrum. Old GSM/GPRS network infrastructure is used for voice and low speed data services, while new network infrastructure is used for video high speed data services.

In this paper, we design an implementation planning of UMTS on existing GSM network. We use GSM network of PT Telkomsel in Batam Area as an example of case. After implementation process, the existing network can support both of GSM/GPRS services and UMTS services

2   Basic Theory

2.1   Global Systems for Mobile Communication (GSM)

GSM is a cellular mobile phone system that uses digital technology with SIM (Subscriber Identity Module) card as a personal identity. This system makes using of cellular phone become easier, such as car mounted, transportable, or handheld. Appearing of GSM is influenced by incompatible equipment problem and to make a global communication and a fair competition for equipment vendor.

The GSM system architecture consists of three major interconnected subsystems that interact between themselves and with the user through certain network interfaces. The subsystems are the Base Station Subsystem (BSS), Network and Switching Subsystem (NSS), and the Operation Support Subsystem (OSS). The Mobile Station (MS) is also a subsystem, but is usually considered to be part of the BSS for architecture purposes.

The BSS provides and manages radio transmission paths between MS and Mobile Switching Center (MSC). The BSS also manages the radio interface between MS and all other subsystems of GSM. Each BSS consists of many Base Transceiver Stations (BTS). BSC (Base Station Controller) control some BTS.

The NSS manages the switching functions of the systems and allows the MSC to communicate with other network, such as PSTN, ISDN, etc. The MSC (Mobile Services Switching Centre) is the central unit in the NSS and controls the traffic among all of the BSC. In the NSS, there three different databases called HLR (Home Location Register), VLR (Visitor Location Register), and Authentication Center (AuC). The HLR is a database which contains subscriber information and location information for each user who resides in the same city as the MSC. The VLR is a database which temporarily stores customer information for each roaming subscriber who is visiting the coverage area of a particular MSC. The AuC is a strongly protected database which handles the authentication and encryption keys for every single subscriber in HLR and VLR.

The OSS supports one or several Operation Maintenance Centers (OMC) which are used to monitor and maintain the performance of each MS, BS, BSC, and MSC within a GSM system. The OSS has three main function, which are: to maintain all telecommunications hardware and network operations with a particular market, manage all charging and billing procedures, and manage all mobile equipment in the system. Within each GSM system, an OMC is dedicated to each of these tasks and has provisions for adjusting all base station parameters and billing procedures, as well as for providing system operators with the ability to determine the performance and integrity of each piece of subscriber equipment in the system.

In the telecommunication network, the establishment of a connection needs some specific signal to control the connection, especially when establish a call setup. In the GSM system, this control function works through some specific channel outside traffic channel (TCH). These channels consist of three groups, BCCH (Broadcast Control Channel), CCCH (Common Control Channel), and DCCH (Dedicated Control Channel).

Frequency spectrum of GSM consists of two sub-bands with 25 MHz bandwidth, from 890 to 915 MHz and from 935 to 960 MHz. The lower sub-band is used as uplink frequency, for transmission from MS to BTS. While, the upper sub-band is used as downlink frequency from BTS to MS. Both of sub-bands are divided into several channels. Each channel has its couple on another sub-band. Moreover, each of sub-bands are divided into 124 channels and numbered as ARFCN (Absolute Radio Frequency Channel Number). In GSM, frequency ranges inter a coupled sub-band (uplink and downlink) is always 45 MHz. Inter channels range is 200 kHz, and in the beginning of every sub-band is allocated as guard band. Therefore, there are 124 ARFCNs for different GSM operator.

2.2   General Packet Radio Service (GPRS)

GPRS is designed to provide packet data services at higher speeds than those available with standard GSM circuit-switched data services. The greatest advantage of GPRS is the fact that it is a packet-switching technology. This means that a given user consumes RF resources only when sending or receiving data. If a user is not sending data at a given instant, then the timeslots on the air interface can be used by another user.

The GPRS air interface is built upon the same foundation as the GSM air interface GSM, the same 200 kHz RF carrier and the same eight timeslots per carrier. This allows GSM and GPRS to share the same RF resources.

Similar to GSM, GPRS requires a number of control channels. GPRS control channels are Packet Common Control Channel (PCCCH), Packet Broadcast Control Channel (PBCCH), and Dedicated Control Channels (DCCH).

Figure 1 shows the GPRS network architecture. One can see a number of new network elements and interfaces. In particular, we find the Packet Control Unit (PCU), Serving GPRS Support Node (SGSN), Gateway GPRS Support Node (GGSN), and Charging Gateway Function (CGF).

The PCU is a logical network element that is responsible for a number of GPRS-related functions such as the air interface access control, packet scheduling on the air interface, and packet assembly and re-assembly. SGSN is analogous

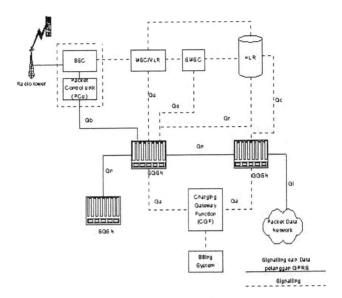

Fig. 1. GPRS Network Architecture

to the MSC/VLR in the circuit-switched domain. Its functions include mobility management, security, and access control functions. A GGSN is the point of interface with external packet data network, such as Internet.

2.3 Universal Mobile Telecommunication Service

Third generation mobile telecommunication technology (3G) is defined in the ITU specification International Mobile Telecommunications-2000 (IMT-2000). This specification is called UMTS. UMTS specifications define four service classes; they are conversational services, interactive services, streaming services, and background services.

The radio access for UMTS is known as Universal Terrestrial Radio Access (UTRA). This is a WCDMA-based radio solution with 5 MHZ nominal bandwidth. WCDMA supports multiple simultaneous user data channels in the downlink, so that a single user can achieve rates of over two Mbps.

There are several specifications about UMTS that is released by 3GPP (3G Partnership Project). Each of specification has different network architecture. The first specification is 3GPP Release 1999 that the core network uses the same basic architecture as that of GSM/GPRS. Therefore, this specification will be used to deploy on GSM network, so that we can reuse the components of GSM network to the greatest extent of possible.

3GPP Release 1999 focuses mainly on the access network and the changes needed to the core network to support that access network. The next specification, 3GPP Release 4, focuses more on changes to the architecture of the core network. 3GPP Release 5 introduces a new call model, which means changes to user terminals, changes to the core network, and some changes to the access network; although the fundamentals of the air interface, remain the same.

3   UMTS Implementation Planning on GSM Network

In this planning, we deploy UMTS on GSM network by implementing GPRS for the first step. GSM operator can see the market penetration before deciding to deploy UMTS. UMTS network (3GPP Release 1999 network architecture) reuse GPRS functionality for its packet data network. Thereby, GSM/GPRS operators that want to deploy UMTS can reuse the entities of their GSM/GPRS network as many as possible.

3.1   Optimization of GSM Network

Before implementation of GPRS systems, we need to optimize the existing GSM network. The objective is checking whether the existing capacity can support generated traffic. Therefore, subscriber will get satisfying quality of service. The first step of this optimization is measuring the latest traffic on the cells that will be optimized. Then, we compare it with the existing capacity. If the measured traffic is greater than the existing capacity, then that cells will be overload. We do optimization by upgrading the existing TRX or adding new cell near the overload cells. Optimization results can be seen in Table 1.

3.2   Implementation of GPRS System

The deployment of GPRS system on GSM network infrastructure is done by adding some new entities and by upgrading some existing entities, especially on access network, core network, transmission network.

In access network, we need to dimension the air interface channels that will be allocated for GPRS. GPRS air interface is same as GSM air interface, with the same 200 KHz carrier and the same 8 timeslot per carrier. It allows GSM and GPRS can use the same radio resources commonly. However, this channels need to be allocated, which channels will be used for voice (GSM) and which channels will be used for data (GPRS). Besides that, we need to determine whether the allocation is dynamic or not. In this planning, from calculating result by using some assumption, we get that the number of channels for being allocated in Batam area are 7 channels. This allocation property is dynamic.

In the core network, the entities that need to be upgraded are Serving GPRS Support Node (SGSN), Gateway GPRS Support Node (GGSN), and Packet Control Unit. SGSN is analogous to the MSC/VLR in the circuit-switched domain. Its functions include mobility management, security, and access control functions. A GGSN is the point of interface with external packet data network, such as Internet. The PCU is a logical network element that is responsible for a number of GPRS-related functions such as the air interface access control, packet scheduling on the air interface, and packet assembly and re-assembly.

Table 1. Optimized Cell

Site	Initial Configuration	Optimized Configuration
BTM010	4/4/4	4/7/5
BTM011	4/5/4	4/7/4
BTM015	3/3/3	4/3/3
BTM018	3/3/3	3/3/4
BTM019	3/3/3	3/4/3
BTM023	3/3/3	4/3/4
BTM028	4/4/4	4/5/4
BTM032	3/3/3	8/3/3
BTM033	3/3/3	3/4/3
BTM038	4/4/4	5/8/4
BTM062	4/4/4/4	4/5/6/4
BTM077	4/4/4	6/5/4
BTM078	4/4/4	7/6/6
BTM089	4/4/4	5/6/4
BTM094	3/3/3	6/7/7
BTM205	3/4	4/4
BTM306	2/2	5/5
BTM401	2	3
BTM402	2	2/1
BTM404	2	4
BTM048	3/4/4	3/9/5

In this planning, we use one entity each for SGSN, GGSN and PCU. Besides that, we also need other supporting entities, such as Charging Gateway (CG), Border Gateway (BG), Lawful Interception Gateway (LIG), Domain Name Server, etc. Then, some existing entities also need to be upgraded, such as Base Station Sub-system (BSS), Home Location Register (HLR), etc.

There is a little change in transmission network. The existing GSM transmission network is still used. The existing Abis interface is used for both of voice and data traffic. Because of adding some new entities in core network, we need some new interfaces in transmission network. For example, we use Gb interface for transmission between SGSN and BSS, Gn interface for transmission between SGSN and GGSN, Gi interface for transmission between GGSN and external network, etc.

Thereby, we have finish GPRS implementation process. We can see GPRS network architecture from this implementation process in figure 2. This architecture is based on Nokia GPRS network structure.

3.3 Implementation of UMTS Network

Implementation of UMTS network in GSM/GPRS network infrastructure allows combining of voice communication services, data and video and a mobile phone.

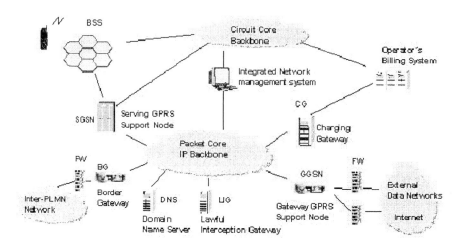

Fig. 2. Nokia GPRS Structure

In this case, we use 3GPP Release 1999 network architecture. 3GPP Release 1999 is the first specification of UMTS. Its core network uses the same basic architecture as that of GSM/GPRS by re-using GPRS functionality for packet data network. Therefore, the existing core network can support the new radio access technology. This can save the network implementation cost. The existing GSM/GPRS operator can reuse the components of GSM network to the greatest extent of possible.

In supporting UMTS services, we add some new entities in GSM/GPRS network infrastructure, such as Node B and Radio Network Controller (RNC). Node B function is identical with BTS function in GSM/GPRS network, while RNC is identical with BSC in GSM/GPRS network. Besides that, the existing GSM/GPRS network entities need to be upgraded to support both of GSM/GPRS services and of UMTS services, such as MSC/VLR, HLR, SGSN, and GGSN. Several vendors have designed their GSM/GPRS base station in order to be able to support both of GSM/GPRS services and of UMTS services. Then, we add an interface to connect GSM BSS with W-CDMA network. It causes a dual mode at W-CDMA/GSM terminal. This dual mode property can give a flexible solution for GSM/GPRS operator. The old GSM/GPRS network architecture is used for voice and low speed data services, while the new infrastructure is used for video and high-speed data service.

In transmission network, we use some new interface, such as Iub interface, Iur interface, Iu-CS interface, Iu-PS interface, etc. Besides that, we still use the existing interface.

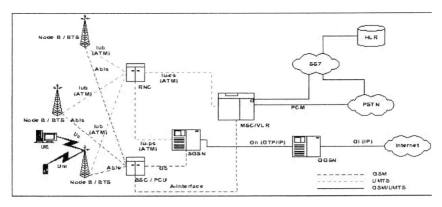

Fig. 3. Combined Infrastructure of GSM/GPRS and UMTS Network

Thereby, we have finished the implementation of UMTS on GSM network. We can see the combined infrastructure of GSM/GPRS and UMTS network in Figure 3.

4   Conclusion

We have studied detail of UMTS Implementation on GSM network. In this case, we use some cells in GSM network of Telkomsel at Batam area. We can get some conclusion, that is:

1. Some overloads cells need to be optimized to offer better quality of service for subscriber by upgrading the capacity oh the cells.
2. In supporting GPRS services, we need to allocate 7 channels for each Telkomsel's cells in Batam area. In the core network, we add some entities, such as SGSN, GGSN, and PCU. While in the transmission network we add some new interface. However, we still use the existing GSM entities and transmission network by upgrading it to adapt with GPRS technology.
3. In UMTS implementation, we use the existing BTS and add one new RNC, while the existing entities need to be adapted in supporting both of UMTS and GSM services. In transmission network, we do the dimensioning process.

References

[1] Clint Smith, Daniel Collins, 3G Wireless Network, New York, McGraw-Hill, 2002
[2] Ericsson Team, GSM Network Planning - Switching, Ericsson Radio Systems AB, 2000
[3] ETSI, GSM 03.60 version 6.1.1 release 1997, Digital cellular telecommunication system (Phase2+); General Packet Radio Service (GPRS); Service Description; Stage 2, Paris, ETSI, 1998

[4] Maiyusril. Perencanaan Sel-sel Ericsson di Area Batam pada Jaringan Seluler Telkomsel, Student Report, Departemen Teknik Elektro - ITB, 2001

[5] Mouly M., Pautet M., The GSM System for Mobile Communication, New York, McGraw-Hill, 1992

[6] Ramjee Prasad, Werner Mohr, Walter Konhauser. Third Generation Mobile Communication Systems, London , Artech House, 2000

[7] Tektronix Team, Whitepaper : UMTS Protocols and Protocol Testing., Tektronix, 2000

[8] Vijay K. Garg, Wireless Network Evolution, New Jersey, Prentice Hall PTR, 2002

[9] Wisnu A. S., General Packet Radio Service Nokia, Student Report, Departemen Teknik Elektro - ITB, 2001

[10] Y. S. Rao, Prof. M. N. Roy., Mixed Voice and Data Capacity Estimation for 3G Wireless Network"., 2001

A Mobile-Node-Assisted Localized Mobility Support Scheme for IPv6*

Miae Woo

Dept. of Information and Communications Eng.
Sejong University
98 Kunja-Dong, Kwangjin-Ku, Seoul, Korea
`mawoo@sejong.ac.kr`

Abstract. As the number of portable devices roaming across the Internet increases, the problem of routing packets to mobile hosts generates increasing research and commercial interest. This paper presents a mobile-node-assisted scheme to support localized mobility in the Internet in a distributed fashion. In the proposed scheme, a mobile node is enhanced to utilize an access router in a foreign domain as an anchor point in the domain. On the other hand, entities in the network-side are only required to provide Mobile IPv6 functionalities in this scheme. Our scheme has three main benefits; reducing the load on the backbone of the Internet, scalability, and interoperability.

1 Introduction

As 3GPP adopts IPv6 as a protocol for all-IP wireless network [1], future wireless networks will include large number of IP-enabled mobile devices moving among wireless cells while the devices maintain connections with others using TCP/IP protocol suite. Also, to accommodate new bandwidth intensive multimedia applications, the cell size will be reduced to enlarge the system capacity. As some cellular network operators reported that the processing capacity of the mobile switching center consumed by mobility signaling traffic in a high density urban environment has been as high as 40% [2], reducing the mobility management overheads becomes an important aspect to the IP-level mobility support in the cellular network. Subsequently, an effective mobility management scheme that generates minimum signaling is needed.

Mobile IPv6 [3] is proposed to support mobility of IP-enabled devices in IPv6 network. However, adoption of Mobile IPv6 in a cellular network environment imposes a significant burden in terms of the amount of signaling generated, processing required, and bandwidth consumed. As solutions to the above problems, several proposals [4]–[6] have been proposed. These schemes are aim to reduce the amount of signaling to the Home Agent (HA) in the mobile's home subnet and to improve the performance of Mobile IPv6. Hierarchical Mobile IPv6 [4] introduces a new functional entity in the visiting domain whose function is similar to the

---

* This research was supported by University IT Research Center Project.

H.-K. Kahng (Ed.): ICOIN 2003, LNCS 2662, pp. 623–632, 2003.

foreign agent in Mobile IPv4 to support mobility management. Multicast-based schemes[5, 6] usually employ sparse-mode multicast routing, thus requiring add-on functionality on the routers in the visited domain. To adopt schemes proposed in [4]–[6], functionalities on the network entities such as routers in the foreign domain need to be enhanced. Also, functionalities of Mobile Nodes (MNs) are required to be tailored to the specific schemes in order to understand, process, and generate specialized messages used.

In this paper, we propose a new mobility management scheme to provide localized mobility support. The proposed scheme is called a MObile-Node-Assisted localized MobilitY support Scheme, namely MONAMYS. Compared to the other proposals given in [4]–[6], our proposal only needs a slight modification at the mobile nodes while functionalities of the entities in the network side remain the same as those in Mobile IPv6. Our scheme distributes mobility management overheads over the Access Routers (ARs) in the visited domain, by employing a new concept, namely "Localized Mobility Agent (LMA)". A LMA is an ordinary Mobile IPv6-enabled AR in the foreign subnet. It is used as an anchor point for the MN in the foreign domain to provide localized mobility management service. In other word, a MN uses a LMA as its HA in the foreign domain and updates its location with the LMA when it roams inside the foreign domain. The performance of the proposed MONAMYS is evaluated in terms of signaling rates to various network entities. Based on the observation, our scheme can provide three main benefits. First, it localizes the signaling messages for location update in the access network and therefore reduces the load on the backbone of the Internet. Second, it is highly scalable since it operates in a distributed fashion. Third, our scheme is applicable to any access network if it provides the basic Mobile IPv6 functionality.

The remainder of the paper is organized as follows. We first overview Mobile IPv6[3] and Hierarchical Mobile IPv6[4] briefly in Section 2. In Section 3, the proposed scheme is presented. The rates of signaling messages generated using Mobile IPv6, Hierarchical Mobile IPv6, and the proposed scheme are analytically derived in Section 4. In Section 5, we investigate the results of Section 4 by applying numerical examples. Section 6 concludes this paper.

2   Related Works

2.1   Mobile IPv6

Mobile IPv6 (MIPv6) [3] allows a MN to move from one link to another without changing its IP address. In Mobile IPv6, each IPv6 MN has at least two addresses per interfaces; home address and the Care-of Address (CoA). The home address is an IP address that is permanent to the MN, while the CoA provides information about the MN's current location in a foreign subnet.

Each time a MN moves from one subnet to another, it gets a new CoA either by stateless or by stateful address autoconfiguration. It then registers its binding which is an association between the MN's home address and the CoA with a router in its home subnet, requesting this router to act as its HA. The HA registers this binding in its binding cache, and serves as a proxy for the MN until the MN's binding entry expires. Any packet addressed to the MN's home address are intercepted and tunnelled to the mobile's CoA using IPv6 encapsulation by the HA. When a MN connects to a new link and forms a new CoA, it may establish forwarding of packets from the previous CoA to this new CoA by sending a Binding Update (BU) to any home agent on the link on which the previous CoA is located. In this sense, all the ARs in the foreign domain are assumed to have ability to behave as HAs on their links throughout this paper.

The MN may send a BU to its correspondent nodes (CNs), which can then learn and cache the new mobile's CoA. As a result of this mechanism, the CN can send any packets destined for the MN directly.

2.2   Hierarchical Mobile IPv6

Hierarchical Mobile IPv6(HMIPv6) [4] introduces the hierarchical mobility management model to enhance the performance of Mobile IPv6. It requires some extensions for Mobile IPv6 and neighbor discovery. It also introduces a new node called the Mobility Anchor Point (MAP) to provide a local entity to assist handoffs and to reduce mobility signaling with external networks. A MAP can be located at any level in a hierarchical network of routers.

Though there are two different MAP modes proposed, we only present the basic mode of HMIPv6 here. The basic mode of HMIPv6 operates as follows. Upon the arrival in a foreign network, a MN discovers the global address of the MAP in the router advertisement provided by the AR. It forms its own global CoA (RCoA) on the MAP's subnet while roaming within a MAP domain. The MN then registers with a MAP by sending a BU containing its RCoA and on-link address (LCoA). The MAP stores this information in its binding cache, and acts essentially as a local home agent for the MN, receiving its packets and tunneling them to the MN's LCoA. After receiving the binding acknowledgement from the MAP, the MN registers the RCoA with the HA and the CNs if applicable.

The boundary of a MAP domain is defined by the ARs' advertising the MAP information to the attached MNs. If the MN changes its current address within a local MAP domain, it only needs to register the new LCoA with the MAP. Hence, the RCoA registered with the CNs and the HA does not change.

3   MONAMYS

The proposed MONAMYS uses IPv6 neighbor discovery for the movement detection and address auto-configuration protocols for acquisition of CoA procedures as in Mobile IPv6. For the localized mobility support, a MN takes different actions depending on whether it changes a domain or not during a handoff. When a MN transits to a new foreign domain, it sends an inter-domain binding update to notify it's HA about its movement. It may also establishe forwarding from the previous CoA as is in MIPv6. In addition to that, the MN creates a record for

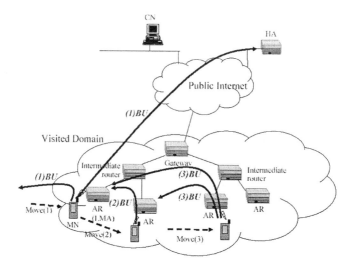

Fig. 1. Binding update procedure in MONAMYS

the LMA in the visiting domain. The record contains a global unicast address of the AR in the newly attached subnet. When a MN handoffs inside the foreign domain, it sends a BU to its LMA. It may send a BU to the AR in the previous foreign subnet optionally. Fig. 1 shows the procedure for the proposed scheme.

If the lifetime of the BU to the HA expires while a MN resides in the same visiting domain, the MN updates its binding information with the HA and revises its LMA as the AR in the currently attached link. The procedure for handling mobility in a MN is listed in the following.

MN procedure

1 Detect movement
  1.1 If it enters a new foreign domain,
    1.1.1 Sends a BU to the HA.
    1.1.2 Saves the AR in the current subnet as a LMA.
    1.1.3 May send a BU to the AR in the previous subnet optionally.
  1.2 If it moves to an adjacent subnet in the same foreign domain,
    1.2.1 Sends a BU to the LMA.
    1.2.2 If the AR in the previous subnet is different from the LMA, may send a BU to the AR in the previous subnet optionally.
2 If the binding with the HA expires,
  2.1 Sends a BU to the HA.
  2.2 Saves the AR in the current subnet as a LMA.

Compared to MIPv6, MONAMYS localizes binding updates inside the foreign domain, thus reducing mobility signaling in the external networks like HMIPv6. In MONAMYS, any AR in the foreign domain can be act as a LMA for MNs, and MNs may use different ARs for their LMAs. In that sense, MON-AMYS distributes the localized mobility management overheads over the ARs in the foreign domain while HMIPv6 puts all the overheads on the MAPs.

4   Performance Analysis

To evaluate the performance of MONAMYS and to compare it with those of Mobile IPv6 and HMIPv6, we analytically derive the rates of binding messages to the HA, MAP and ARs. There are two events that trigger binding messages; one is MN's crossing the registration area boundary and the other is the expiration of the lifetime of binding update that is established with a HA. In this paper, we use a *binding update* to refer the message generated by the former event and a *binding refresh* to refer the message generated by the latter event, respectively.

For the analysis, we introduces several parameters and assumptions. For the network architecture in the visiting domain, it is assumed that there are $\bullet_D$ square-shaped subnets and the perimeter of the subnet is $\bullet_B$. It is also assumed that the shape of a foreign domain is square too. For the analysis of the binding traffic, we assume a simple fluid flow mobility model [7] for MNs. The model assumes that MNs are moving at an average velocity of $\bullet$, and their direction of movement is uniformly distributed over $[0 \cdot 2\bullet]$. Also it is assumed that MNs are uniformly populated with a density of $\bullet$.

If the registration area boundary for binding updates is of length $\bullet_u$, then the rate of binding updates, $\bullet_u$, is given by the following equation [7].

$$\bullet_u(\bullet_u) = \frac{\bullet\bullet\bullet_u}{\bullet} \tag{1}$$

A binding refresh is generated only when a MN stays in the same registration area boundary for binding refresh for longer than the binding lifetime $\bullet_B$. Let $\bullet_r$ be the length of registration area boundary for binding refresh. Then, the mean stay time of a MN in the registration area is $\bullet\bullet_r \cdot (16\bullet)$. Consequently, the rate of binding refreshes, $\bullet_r$, is

$$\bullet_r(\bullet_r) = \frac{\bullet\bullet\bullet_r}{\bullet} \left\lfloor \frac{\bullet\bullet_r}{16\bullet\bullet_B} \right\rfloor \bullet \tag{2}$$

The total rate of binding messages to a specific network entity, $\bullet$, is in general

$$\bullet = \bullet_u \bullet_u(\bullet_u) + \bullet_r \bullet_r(\bullet_r)\bullet \tag{3}$$

where $\bullet_u$ and $\bullet_r$ is the number of registration areas for binding updates and refreshes from where a network entity is responsible for mobility management.

4.1   Mobile IPv6

When Mobile IPv6 is used, we are interested in the rate of binding updates and refreshes delivered to the HA at the home subnet. In Mobile IPv6, the registration area boundary is a subnet area for both binding updates and refreshes.

Therefore there are $\bullet_D$ registration areas in the foreign domain. Thus, the rate of binding messages to the HA, $\bullet^{HA}_{MIPv6}$, is as follows:

$$\bullet^{HA}_{MIPv6} = \frac{\bullet\bullet\bullet B}{\bullet}\left(\bullet_D + \bullet_D\left\lfloor\frac{\bullet\bullet B}{16\bullet\bullet_B}\right\rfloor\right) \qquad (4)$$

4.2 HMIPv6

In HMIPv6, a MN updates its binding to the MAP while it resides inside a MAP domain. It registers its location with the HA only if it crosses a MAP domain. Though HMIPv6 considers a distributed-MAPs environment, it is recommended to choose the furthest MAP to avoid frequent re-registration[4]. Based on the recommendation, we choose the gateway as a MAP. As a result, the boundary of a MAP domain is the same as that of the visited domain. Therefore, binding updates with the HA are performed when MNs cross the foreign domain boundary. Also, binding refreshes are generated if MNs stay in the domain longer than $\bullet_B$. As a result, the rate of binding messages to the HA, $\bullet^{HA}_{HMIPv6}$, is

$$\bullet^{HA}_{HMIPv6} = \frac{\bullet\bullet\bullet B}{\bullet}\left(\sqrt{\bullet_D} + \sqrt{\bullet_D}\left\lfloor\frac{\bullet\bullet B\sqrt{\bullet_D}}{16\bullet\bullet_B}\right\rfloor\right)\bullet \qquad (5)$$

On the other hand, the rate of binding messages to the MAP, $\bullet^{MAP}_{HMIPv6}$, is becomes

$$\bullet^{MAP}_{HMIPv6} = \frac{\bullet\bullet\bullet B}{\bullet}\left(\bullet_D + \sqrt{\bullet_D}\left\lfloor\frac{\bullet\bullet B\sqrt{\bullet_D}}{16\bullet\bullet_B}\right\rfloor\right)\bullet \qquad (6)$$

4.3 MONAMYS

For MONAMYS, the rate of binding messages to the HA, $\bullet^{HA}_{MONAMYS}$, is same as that of HMIPv6 as shown in Eq. (5).

Since localized mobility management in MONAMYS is achieved by bindings with ARs in the foreign domain, we analyze the rate of binding messages to ARs. As explained in Section 3, binding updates are delivered to an AR if it is a LMA. Binding updates to the AR in the previous subnet may generated optionally. On the other hand, no binding refresh is generated for ARs. The rate of binding updates to the AR in the previous subnet is $\bullet_u(\bullet_B)$, and it is same to all the ARs in the foreign domain. If an AR is serving as a LMA for MNs, the rate of binding updates varies depending on the location of the AR in the domain. The location of the ARs can be classified as follows; corner of the domain, boundary but not at the corner of the domain, and the inside of the domain. For an AR, dependency of the binding rates on the location comes from the probability of becoming a LMA for MNs in the domain. After the expiration of binding lifetime for a MN in the foreign domain, all the ARs are equally probable to be a LMA for the MN. However, before the first expiration of binding lifetime, a MN uses the AR in the firstly visited subnet as its LMA. Consequently, ARs in the corner and boundary of the domain serve more MNs as LMAs than those inside the doamin.

The rate of binding updates to the ARs that are located inside the domain, $\bullet_{MONAMYS}^{AR_i}$, is given in the following equation.

$$\bullet_{MONAMYS}^{AR_i} = \frac{\cdots B}{\bullet}\left(1 + \frac{(\sqrt{\bullet\,D}-1)(\bullet\,D-1)\left\lfloor\frac{\pi L_B\sqrt{B_D}}{16vT_B}\right\rfloor}{\bullet\,D\sqrt{\bullet\,D}\left(\sqrt{\bullet\,D}+\left\lfloor\frac{\pi L_B\sqrt{B_D}}{16vT_B}\right\rfloor\right)}\right) \tag{7}$$

The rate of binding updates to the ARs that are located at the corner of the domain, $\bullet_{MONAMYS}^{AR_c}$, is

$$\bullet_{MONAMYS}^{AR_c} = \frac{\cdots B}{\bullet}\left(1 + \frac{(\sqrt{\bullet\,D}-1)\left(2\bullet\frac{2}{D}-\bullet\,D+4(\bullet\,D-1)\left\lfloor\frac{\pi L_B\sqrt{B_D}}{16vT_B}\right\rfloor\right)}{4\bullet\,D\sqrt{\bullet\,D}\left(\sqrt{\bullet\,D}+\left\lfloor\frac{\pi L_B\sqrt{B_D}}{16vT_B}\right\rfloor\right)}\right)\bullet \tag{8}$$

The rate of binding updates to the ARs that are located on the boundary but not the corner of the domain, $\bullet_{MONAMYS}^{AR_b}$, is given as follows:

$$\bullet_{MONAMYS}^{AR_b} = \frac{\cdots B}{\bullet}\left(1 + \frac{(\sqrt{\bullet\,D}-1)\left(4\bullet\frac{2}{D}-3\bullet\,D+16(\bullet\,D-1)\left\lfloor\frac{\pi L_B\sqrt{B_D}}{16vT_B}\right\rfloor\right)}{16\bullet\,D\sqrt{\bullet\,D}\left(\sqrt{\bullet\,D}+\left\lfloor\frac{\pi L_B\sqrt{B_D}}{16vT_B}\right\rfloor\right)}\right) \tag{9}$$

5   Numerical Examples

In this section, numerical examples for the equations derived in Section 4 are presented to illustrate the advantages of MONAMYS over Mobile IPv6 and HMIPv6. For that purpose, we investigate the effect of various parameters, such as user density, user speed, and binding lifetime, on the rate of binding messages. The values given in Table 1 are used as default values in the examples.

The results are illustrated in Fig. 2 – Fig. 4. For the proposed MONAMYS, the rate of binding messages delivered to an AR in the corner of the foreign domain is only illustrated in the figures, since it is the biggest value among the rates of binding messages to the ARs. Overall, the rates of binding messages delivered to the various entities in the network increase as the user density and user velocity increase. They decrease as the binding lifetime increases. One noticeable result is that the rate of binding messages delivered to the MAP in

Table 1. Parameter values

Item	Type	Value
ρ	User density	$200/\text{Km}^2$
v	User speed	10 Km/hr
L_B	Perimeter of a subnet	10 Km
B_D	Number of subnets in a domain	256
T_B	Binding lifetime	3600 sec

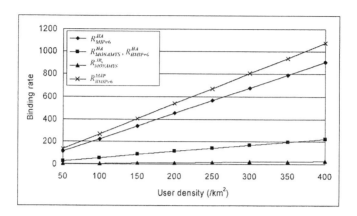

Fig. 2. Binding update rates depending on user density

HMIPv6 is about 1.2 times of the rate of binding messages delivered to the HA in Mobile IPv6. The reason for giving such results is that more binding refreshes are generated in the HMIPv6 than that in the MIPv6. On the other hand, the rate of binding messages delivered to an AR in the proposed scheme is the least one. By employing the localized movement management schemes in the foreign domain, binding rates delivered to HA, $\bullet_{MONAMYS}^{HA}$ and $\bullet_{HMIPv6}^{HA}$, are reduced to $1/4$ of $\bullet_{MIPv6}^{HA}$. $\bullet_{MONAMYS}^{AR_c}$ is about $1/40$ of $\bullet_{HMIPv6}^{MAP}$ and $1/34$ of $\bullet_{MIPv6}^{HA}$ respectively. Considering that there are 256 ARs in the foreign domain, the rate reduction is not proportion to the number of ARs. However such a reduction requires far less processing overheads at the network entities, and provides more scalability. If we compare the overall binding rates generated in the foreign domain, MONAMYS results about 1.7 times of binding messages than HMIPv6.

The result shown in Fig. 2 reveals that the rates of binding messages delivered to the various entities in the network is linearly increased as the users are densely populated. In Fig. 3, $\bullet_{MONAMYS}^{HA}$ and $\bullet_{HMIPv6}^{HA}$ do not increase proportionally as the user velocity increases. Such a situation occurs because as the MN changes its point of attachment frequently as it moves faster, the rate of binding refreshes decreases. In Fig. 4, $\bullet_{MIPv6}^{HA}$ remains the same if the binding lifetime is longer than 1200 seconds since a MN moves to an adjacent subnet before the binding lifetime expires. Thus, extending the binding lifetime does not reduce the binding rates for the Mobile IPv6 after a certain value of the binding lifetime. $\bullet_{MONAMYS}^{HA}$, $\bullet_{HMIPv6}^{HA}$ and $\bullet_{HMIPv6}^{MAP}$ decrease as the binding lifetime increases because of the reduction of the rates of binding refreshes. On the other hand, $\bullet_{MONAMYS}^{AR_c}$ increases as binding lifetime gets longer. Such a situation occurs because if there is no binding refresh occurred, then the LMA for a MN remains the same as it roams around in the foreign domain.

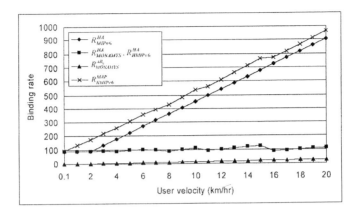

Fig. 3. Binding update rates depending on user velocity

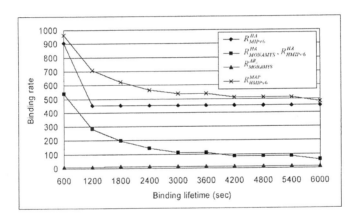

Fig. 4. Binding update rates depending on binding lifetime

6   Conclusion

In this paper, we propose a new localized mobility support scheme that can work with Mobile IPv6. The proposed scheme localizes the binding messages in the visiting domain when MNs move inside the visiting domain. The most beneficial feature of our scheme over the other proposals is that it is fully interoperable among the different domains because it only requires enhancements on the MN. Subsequently, a MN can utilize a localized mobility support in any domain regardless of to where it roams. In order to evaluate the performance of the proposed scheme, we derive analytically the rate of binding updates that are delivered to the HA and compare it with that of Mobile IPv6. From the analysis, it is observed that the proposed scheme reduces the signaling load to the Internet backbone by decreasing the binding update rate to the HA. We also derive analytically the rates of binding updates delivered to the entities inside the foreign domain. For them, a comparison is made with HMIPv6. The result shows that the proposed MONAMYS provides better scalability that is invariant to various parameters such as user density, user velocity and binding lifetime.

References

[1] G. Patel and S. Dennett, "The 3GPP and 3GPP2 Movements Toward an All-IP Mobile Network," IEEE Personal Communications, pp. 62–64, Aug. 2000.

[2] Tabbane, S., "Modelling the MSC/VLR processing load due to mobility management," in ICUPC '98, pp. 741–744, 1998.

[3] D. B. Johnson and C. Perkins, "Mobility Support in IPv6," Internet draft, draft-ietf-mobileip-ipv6-18.txt, Jun. 2002.

[4] H. Soliman, C. Castelluccia, K. El-Malki, and L. Bellier, "Hierarchical MIPv6 Mobility Management (HMIPv6)," Internet draft, draft-ietf-mobileip-hmipv6-05.txt, Jul. 2001.

[5] C. Castelluccia, "A hierarchical mobility management scheme for IPv6," Proceedings in ISCC '98, pp. 305–309, 1998.

[6] A. Mihailovic, M. Shabeer and A. H. Aghvami, "Multicast for Mobility Protocol (MMP) for Emerging Internet Networks," Proceedings in PIMRC 2000, Vol. 1, pp. 327-333, 2000.

[7] S. Mohan and R. Jain, "Two User Location Strategies for Personal Communications Services," IEEE Personal Communications, Vol. 1, No. 1, pp. 42-50, 1994.

The In-vehicle Router System
to Support Network Mobility

Koshiro Mitsuya, Keisuke Uehara, and Jun Murai

Shonan Fujisawa Campus, Keio University
5322 Endo, Fujisawa, Kanagawa, 252-5322, Japan
mitsuya@sfc.wide.ad.jp, kei@wide.ad.jp, jun@wide.ad.jp
http://www.sfc.wide.ad.jp/~mitsuya/

Abstract. This paper proposes a communication system which ensures Internet connectivity and network transparency to a group of nodes with several network interface devices. We also implement this system as In-vehicle Router System. A vehicle consists of a group of nodes such as sensor nodes and devices held by passengers, is connected to the Internet through several network interface devices. It is typical for buses, trains, and airplanes to have such Mobile Internet environment, and the nodes in such vehicles have to change the attachment point of the Internet frequently. But the nodes would then not be able to maintain transport and higher-layer connections if it changes the attachment point. Thus, it is important to provide mobility support like Mobile IP. These nodes include low cost network appliances with only limited space for extra functions. It is preferred that a solution has no impact on these low cost nodes. Therefore, existing Host Mobility Protocol is not suitable for this situation. In this article, we propose the In-vehicle Router System as a solution to this situation by combining network mobility protocol with Interface switching system. We also implement and evaluate our system on the InternetITS testbed. We have confirmed that our system provides enough functionality to satisfy the requirements of Mobile Internet vehicles.

1   Introduction

Automobiles contain a group of nodes, consist of sensor nodes and devices held by passengers. Passengers can also be considered as a group of nodes with PDA's and cellular phones. These groups of nodes are connected to the Internet using several communication devices such as cellular phones and Wireless LANs.

This paper proposes a communication system which ensures Internet connectivity and network transparency to a group of nodes with several network interface devices. Based on the suggested model, we have designed and implemented the In-vehicle Router System for automobiles connecting to the Internet.

The InternetITS Project [1] aims at building the ITS infrastructure, and ultimately promoting ITS-related industries and businesses. ITS (Intelligent Transport Systems) is a system which is designed to promote the advance of car navigation technology, to ensure that the Electronic Toll Collection System (ETC) is

H.-K. Kahng (Ed.): ICOIN 2003, LNCS 2662, pp. 633–642, 2003.

effective for tolling and to support safe driving. With this system, people, roads, and vehicles mesh using the latest information communications technology.

The InternetITS Project carries on three pilot programs, Nagoya Pilot Program: services for use by taxi companies, drivers and passengers; Tokyo Pilot Program: services for general car drivers; High Function Prototype Car Program: future services assumed. This paper is a part of High Function Prototype Car program. A High Function Prototype Car is equipped with functions we expect for future InternetITS, safe driving assistance, management of drivers' physical conditions, group multi-media communications, vehicle status monitoring and so on.

This paper explores the requirements of the prototype car, and proposes the design of the In-vehicle Router System which satisfies the requirements. In order to verify the validity of this system, we also evaluate this system.

2   Assumed Environment for Our System

The purpose of this study is to propose a communication system for the High Function Prototype Car. In order to arrange the requirements for the In-vehicle Router System, we explain the assumed environment, some applications and features of the prototype car in this section. The applications include a probe information service, a group multi-media communication, vehicle monitoring service.

As we mentioned in section 1, we assume that automobiles contain a group of nodes and these groups of nodes are connected to the Internet using several communication devices such as cellular phones and Wireless LANs. Figure 1 shows the Mobile Internet environment currently assumed.

A probe information service is an application of the High Function Prototype Car. The probe information service utilizes automobiles as roving sensors scattered all over. The center system collects desired informations from each automobiles. New information is produced by processing collected data. The information will be used for operation management of vehicles, public service and etc. Many different types of computers are carried in the present cars, and those computers are processing more than 100 different types of information. However, these data collected by a car are currently used only by the car itself and not

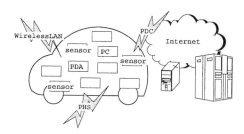

Fig. 1. Mobile Internet environment assumed

exported. In the probe information service, data are collected and processed by the information center of vehicles, and then, the processed information is sent back to the vehicles. This model could be applied to traffic congestion information service, road information service, weather information service, parking lot usage information service, etc. efficiently.

The prototype car has touch-sensitive displays, video cameras, and microphones on each seat. We can perform a group multi-media communication with movies and audios with those devices using the existing applications and products. The video stream, VIC/RAT [2] over IP, are delivered by Explicit Multicast [3] to the Correspondent Nodes. VIC and RAT are widely deployed video conference applications.

We can also monitor the current status of the car at anytime by using custommade applications. The status includes the current location, speed, wiper and lights status and so on. These status information are collected through SNMP using the InternetITS Management Information Base (MIB). The specification of the MIB is available from [1].

Other featured services include portal web sites, individual authentication by IC card, access by center-type voice recognition, electric money reconciliation settlement, push-type content delivery service.

3 Requirements for Our System

In this section, we summarize the requirements to design the communication system in the environment described in section 2.

End-to-End Communication, Permanent Connectivity, Reaching via Fixed IP Address: As a car moves around, the attachment point of the Internet could change frequently. For example, within a few seconds, a car communicating by IEEE802.11b could be out of the cell range for the cell size of IEEE802.11b is about 300 meters. To keep stable connectivity, we have to switch the network devices in time, because there are no network devices which covers all over the world. When a node changes its attachment point to the Internet, the IP address changes. Thus nodes inside the car can not have fixed identity on the Internet. It is not preferable to have an extra naming mechanism on the session layer or more upper layer such as TCP migration [4] or I-TCP [5], because it has an impact to the existing applications. Thus we assume End-to-End communication and fixed IP address for each in-vehicle node.

Migration Transparency, Transparency from Application: The changes of IP address also trigger disconnection of the session between Corresponding Nodes and nodes inside a car. To keep the session connected with the existing applications, we need to introduce migration transparency into our system. The solution dose not affect the existing applications and products.

No Impact to Nodes Inside a Car: We have Low Cost Network Appliance(LCNA) such as sensor nodes inside a vehicle. We can not extend LCNA, thus the solution should support those nodes and have no impact to nodes inside the vehicle. This allows us to use existing products without any changes to nodes inside a car.

Desirable Bandwidth: Our applications, the probe information service and vehicle monitoring service, should be in operation at all times. To collect information which are used by those applications, we use SNMP. The MIB of InternetITS is 788 bytes in total. To obtain the information every 1 second, we need at least 6,304 bps. Additionally, the group communication application requires the following bandwidth with h263 codec (320 x 240 size); if the object is moving and the frame rate is 1.3 fps, it requires 25 Kbps, if the object stays still and the frame rate is 15 fps, it requires 14 Kbps. Therefore, about 20-31 Kbps is desirable in total.

Small Round Trip Time: We have the video conference application which requires real-time communication. According to [6], we need less than 200 ms to get good performance with this applications. Thus, it is preferable to keep the latency within 200 ms, between Correspondent Node and node inside a car. Of course, the round trip time depends on the topology and Layer 2 technologies.

4 Related Works

We use Mobile IPv6, Interface Switching and Prefix Scope Binding Update as components of the proposed communication system.

Mobile IPv6 [7] provides host mobility on the network layer. Each mobile node is always identified by its home address, regardless of its current point of attachment to the Internet. In this case, a mobile node has a *fixed IP address as Home Address*. While situated away from its home, a mobile node is also associated with a care-of address, which provides information about the mobile node's current location. IPv6 packets addressed to a mobile node's Home Address are transparently routed to its care-of address. The protocol enables IPv6 nodes to cache the binding of a mobile node's Home Address with its care-of address, and then send any packets destined for the mobile node directly to this care-of address. Thus Mobile IPv6 can provide *End-to-End communication* between Mobile Node and Correspondent Node as well as *Migration Transparency*.

Interface Switching using Care-of Address provide a *permanent connectivity*, a mobile node has to optimize the data-link media according to the situation. "Multiple Network Interface Support by Policy-Based Routing on Mobile IPv6" [8] is a framework for optimum interface selection with respect to each network connection. It provides persistent Internet connectivity, better throughput and efficient utilization of multiple network interfaces.

Prefix Scope Binding Update (PSBU) [9] is one of the solution of the network mobility. In contrast with host mobility, the network mobility support is considered where an entire network changes its point of attachment to the Internet and thus its topology changes. PSBU is an enhanced Mobile IPv6 Binding Update which associates a care-of address with a network prefix instead of a single IP address to provide network mobility. Network mobility can provide *the mobility for nodes inside a car* even if we can not extend the nodes.

5   Design of Our System

We choose the combination of those three technologies, Mobile IPv6, Interface Switching and Prefix Scope Binding Update, which are mentioned in section 4 to satisfy the requirements. Each function alone can not satisfy all the requirements. However by integrating these functions in to a system, we can satisfy the requirements. In this section, we show how those three technologies work as one system. Refer to [7], [9] and [8] about the detailed actions.

We describe the entire picture of the In-vehicle Router System in Figure 2. There are Mobile Router(MR), Home Agent(HA) and Correspondent Node(CN). MR is a router which provides mobility support for a group of nodes. We assume that the group of nodes is addressed on a subnet behind the HA. HA is a router which knows the current care-of address of MR. The care-of address is the current IP address of the external interface of MR. CN is a node which communicates with the group of nodes.

When the CN sends packets to the group of nodes, the packets are forwarded toward the HA according to the routing of the Internet. Assuming that the MR is not behind HA, the packets will become unreachable. However, the HA knows the current IP address of MR enabling HA to re-route the packet which destined to the group of nodes to the current location.

The In-vehicle Router System is designed as an extension to SFC-MIP6[10], which is our own implementation of Mobile IPv6. Our Mobile IPv6 stack is comprised of Header processing part, care-of address management part and home address selection. The packets sent from Mobile Node are destined to the CN by

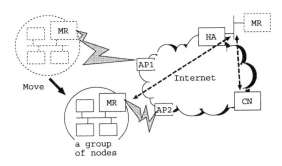

Fig. 2. System image of In-vehicle Router System

using care-of address as source IP address. CN can know that the packets are sent from the Mobile Node, because the packets also include the home address as an IPv6 destination header. Thus the header processing part can provide Transparency to the upper layer. CN has to manage the binding between home address and care-of address as Binding Cache. Mobile IPv6 also defines the management of Binding. For details, refer to [10].

The distinctions between Mobile IPv6 and PSBU are the entity of Binding Cache and signaling messages of Binding Update. CN has to hold the network prefix corresponding to the network behind MR, in this case the network inside a car, in the Binding Cache. In order to transmit the information of the network prefix, we also have to extend the signaling between CN and MR. Thus PSBU is designed by extending the header processing part. To perform interface switching, we have to decide the care-of address depending on the situation. In our design, each network interface has a priority. The interface is switched depending on the priority and link status of the interface. For example, assume having two interfaces, A and B; A has priority 1 and B has priority 2. When interface A is down and interface B is up, the system uses interface B. However the system switches to the interface A if interface A comes up. This function works as an extension of the care-of address management parts. In this way, these functions are carried out.

CN and MR in the vehicle are based on this design, but CN have only Header Processing Part and Binding Cache. Nodes inside a car keep the normal specification of IPv6.

6 Implementation

This implementation is an extension of SFC-MIP. We implemented this system on NetBSD 1.5.2 and on FreeBSD 4.4. In this section, we explain about basic data structures, the interface switching module and the extension of signaling for Binding.

The implementation has data structures shown in Figure 3. Structure mobileip_data includes the pointer of all data for this implementation. Structure mobileip_softc includes the information only related to the function of Mobile Router. When the system works as CN or HA, this structure isn't referred. Structure mip_if includes the information of pseudo device, virtual device, which has a home address. In this implementation, each home address is held on each pseudo device. Thereby, it becomes possible to have multiple home addresses and switch the home address for the purpose of use. Structure coa_list includes the list of care-of addressees(CoA) which is managed by CoA selection parts. The care-of addresses are stored with a priority, the care-of address which have the highest priority is used as the primary care-of address. Structure home_agent_list includes a list of HAs. Structure binding_update_list includes a list of CNs to which Biding Update should be sent. All nodes have the Binding Cache as structure binding_cache and the corresponding network prefix as Structure mobile_network_cache . We divided the mobile network cache from the

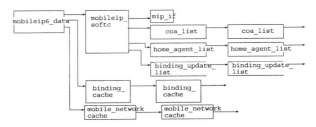

Fig. 3. The Basic Data Structures of our implementation

binding cache, because it is less cost to implement. When CN sends a packet to the nodes inside a car, CN looks up the Binding which corresponds to the destination of the packet from the mobile network cache. If it matches, the corresponding care-of address is searched from Binding Cache. Thereby, nodes can know the corresponding care-of address of nodes inside the car.

When nodes inside the car communicate with CN for the first time or when the timer of an entry in Binding Cache has expired, MR send the Binding Update to the CN. The Binding Update includes the network prefix inside the router as an optional header defined in the specification of Mobile IPv6.

All functions except interface switching daemon are implemented in the kernel. Such a native implementation is needed, because many functions are processed by IPv6 option header. Meanwhile, we have to select the best network interface by using the link status and the priority. It's easy to collect those information from the user space, because various APIs are available. We used these APIs to collect the link status and the priority, and implemented a API to advertise the care-of address from the user space to the kernel.

The interface switching daemon selects the primary care-of address from the priority and link status of the network interfaces. We set the priorities in a configuration file. The daemon monitors the status of all the interfaces, and tries to enable the network interfaces as long as possible. For example, if a Wireless LAN is down, the daemon sets the new ESS-ID and the WEP Keys registered beforehand to search for a new access point. If PPP link is down, the daemon carries out the negotiation of PPP once again. The daemon compares the priority of the network interface with other interfaces, and selects the care-of Address for the current situation.

This implementation was performed to support the environment among illustrated in Figure 4. The nodes in the vehicle behind the MR, 5 are FreeBSD, 4 are Mircosoft Windows 2000, and 1 is Linux. They are connected to the MR via Fast Ethernet. The MR has 5 access links. 2 are Personal Digital Cellular(PDC) (PDC-P 9.6Kbps, PacketOne 28.8Kpbs), 2 are Personal Handy-Phone System(PHS) (AirH" 32Kbps, P-in 64Kbps) and 1 is IEEE802.11b. PDC and PHS currently only provide IPv4 connectivity. Thus, we use *IPv6 over IPv4 tunneling* and *Dynamic Tunnel Configuration Protocol*. A Dedicated Short Range Communication (DSRC) is provided from another router which is running Linux

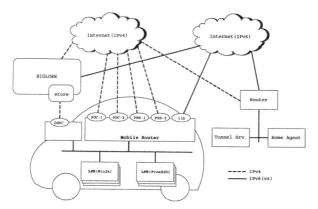

Fig. 4. The Testbed of the High Function Prototype Car: This network is deployed in one of the pilot program. A Correspondent Node, the HA and a tunnel server are located on the Shonan Fujisawa Campus Keio University, Japan. The vehicle establishes Internet access via distinct access technologies

under IPv4. The DSRC connection is not performed by the same MR because there is no DSRC driver available in NetBSD. The CN is running a SNMP client under FreeBSD 4.4 or NetBSD 1.5.2. The MR, CN and HA are running our implementation.

7   Evaluation

We verify the validity of the In-vehicle Router System. There are functional requirements and the performance requirements describe in section 3.

We used the ping program to verify that the operations of this system satisfy the functional requirements. We check that CN and nodes inside the car are still communicating even when the router changes the network interface. Fixed IP addresses are assigned to nodes inside a car, so that there is no impact for the nodes. Therefore this system was verified to meet the requirements: End to End communication, Reaching via fixed IP address, Migration transparency, No impact for nodes inside a car, Permanent connectivity.

The results of this performance evaluation of the performance issue are shown in Table 1. To evaluate the throughput, we transfered a file with size of 10M bytes from CN to nodes inside a car via Wireless LAN, and a file whose size is 1M bytes via PHS. We repeated this 100 times. Wireless LAN is 11M bps mode of IEEE 802.111b, PHS is 9600 bps.

We obtained enough throughput mentioned in section 3 when using Wireless LAN, which means that all applications can work fine. We can not obtain enough bandwidth to perform the group communication application when using PHS. However, it's enough for other applications if we do not use the Group Communication Application described in section 2. Thus, we may need an approach which gives a priority to the traffic, such as Quality of Service.

Table 1. The throughput and RTT with Wireless LAN and PHS

Throughput(bps)		RTT(ms)			
		min	avg	max	std-dev
Wireless LAN	228,572	4.009	4.631	43.104	1.704
PHS	41,734	125.923	168.188	323.396	29.423

We also evaluated the Round Trip Time (RTT) from CN to nodes inside the car by using the ping program. The results are described in Table 1. The result shows that we obtained the desired value of RTT mentioned in section 3 when using Wireless LAN, which means that all applications can work fine. We can not obtain the desired value to perform the group communication application when using PHS. However, it will be difficult to improve this, because the RTT strongly depend on the topology and layer 2 technologies.

Additionally, we investigated the relation between the condition and the packet size in this testbed. Figure 5 shows the changes of RTT when the router changes the network interface between Wireless LAN and PHS. We pinged from CN to nodes inside the car at a rate of one packet per 0.15 ms with 56 bytes in packet size. Figure 6 shows the change when the packet size is 768 bytes and sent every 1 second.

When the packet size is 56 bytes, there is no packet loss when we performed vertical handoff. At vertical handoff, the packet loss can not occur easily, because two or more interfaces can be used at the same time. In contrast, it could not

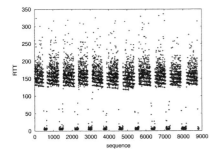

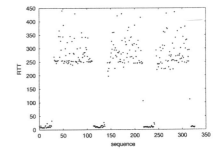

Fig. 5. The change of RTT during MR changes its interface Wireless LAN and PHS, packet size is 56 (40 + 8 + 8) bytes, per 0.15 second

Fig. 6. the change of RTT, packet size is 816 (40+8+768) bytes, every 1 second communicate at all when the packet size is more then 816 bytes. We may have to consider about the maximum transfer unit to obtain better condition while communicating.

8   Conclusion

We proposed a communication system which ensures Internet connectivity and network transparency to a group of nodes with several network interface devices. We also implemented it as In-vehicle Router System.

We summarized the requirements for the communication system of the prototype car from the applications running on the prototype car. We proposed the combination of Mobile IPv6, Network Mobility and Interface Switching to satisfy those requirements. The PSBU is one of the solutions to provide Network Mobility, the Interface Switching using Care-of Address provides strong Internet connectivity. Those are designed as an extension of Mobile IPv6. Therefore, We proposed the design of this system based on Mobile IPv6.

We implemented In-vehicle router system based on the design, and evaluated the system using the InternetITS testbed. We verified that our system satisfies the functional requirements. We also verified that this system can not satisfy the performance requirements with the current cellular phone. This problem will be solved if better Layer 2 technologies appear.

In conclusion, we have confirmed that our system provides enough functionality to satisfy the requirements of Mobile Internet vehicles. We believe that this research will bring the further possibility to the Mobile Internet.

References

[1] The InternetITS Project. *InternetITS*, 2002. http://www.internetits.org/.
[2] The UCL Multimedia and Networking group. *Videoconferencing Tool and Robust Audio Tool.*
[3] XCAST Project. *Explicit Multicast.* http://www.xcast.jp/.
[4] Alex C. Snoeren and Hari Balakrishnan. An end-to-end approach to host mobility. In *Proc. 6th International Conference on Mobile Computing and Networking (MobiCom)*, 2000.
[5] Ajay Bakre and B. R. Badrinath. I-TCP: Indirect TCP for mobile hosts. *15th International Conference on Distributed Computing Systems*, 1995.
[6] R. Cole and J. Rosenbluth. Voice over ip performance monitoring. In *ACM Computer Communication Review*, page 31(2), 2000.
[7] Charles E. Perkins and David B. Johnson. Mobility support in ipv6. In *Mobile Computing and Networking*, pages 27–37, 1996.
[8] Ryuji Wakikawa, Keisuke Uehara, and Jun Murai. Multiple network interfaces support by policy-based routing on mobile ipv6. In *ICWN 2002*, 2002.
[9] Thierry Ernst, Alexis Olivereau, and Hong-Yon Lach. *Mobile Networks Support in Mobile IPv6 (Prefix Scope Binding Update)*, February 2002. Work in Progress.
[10] InternetCAR Project in Keio University. *Mobile IPv6 stack from InternetCAR.* http://www.sfc.wide.ad.jp/MIP6/.

Independent Zone Setup Scheme for Re-configurable Wireless Network

Jae-Pil Yoo and Kee-cheon Kim

School of Computer Science & Engineering
Konkuk University, Seoul, Korea
Tel: +82-2-450-3518

Abstract. ZRP (Zone Routing Protocol), is especially considered to be suitable for moves of dynamic nodes and scalability of RWN. In configuring the ZRP zone, if we remove redundant function of a node, we can get performance improvement in routing. This improved performance of routing is usually from configuring a ZRP zone in which a node doest not have functioning zone head, relaying (internal) node, and peripheral node simultaneously. In such cases, internal routing information size based on link state can be significantly reduced. Hop count is also reduced when routing is reduced and routing loop prevention steps are taken. In this paper we introduce a zone setup scheme based on ZRP in which a node does not have overlapping function as head and relay, and border. Also we introduce a routing scheme for proposed new zone setup scheme.

1 Introduction

RWN (Re-configurable Wireless network) is a scalable network in which a number of nodes can moves around dynamically. A node itself has a sending or receiving wireless interface and with these interfaces nodes can communication each other. When they communicate each other, each node needs to relay some traffic for others without relying on fixed transmission infrastructure. We assume that almost all the nodes in RWN have limited wireless transmission radius and power supply.

Routing protocol that reflects features mentioned above is classified into two types: proactive scheme and reactive scheme. Proactive is a scheme that all the nodes constantly and periodically exchange network path information prior to sending packets such as OSPF [1]. On the other hand, reactive scheme sends path setup signal only when it is needed. It is also called 'on-demand' scheme [2, 3]. But it needs packet buffering. Since all the nodes using proactive scheme already know their destination path information before they send data traffic, they are able to send data traffic without delay or buffering. In spite of its merits, the use of proactive scheme heavily depends on the number of nodes in RWN. Therefore, it has a scalability problem. In order to overcome this problem, various clustering (grouping) scheme were introduced [4, 5, 6, 7, 8]. Clustering scheme is to grouping local nodes into a cluster and then separates inter-cluster routing information and intra-cluster routing information for routing efficiency.

H.-K. Kahng (Ed.): ICOIN 2003, LNCS 2662, pp. 643–650, 2003.

One of above scheme is ZRP, ZRP uses hybrid scheme; proactive method for intra-zone routing such as link-state and on-demand based reactive method for inter-zone routing for dynamic RWN

2   ZRP and Independent Zone Setup Scheme for RWN

ZRP is a hybrid of reactive and proactive scheme [9, 10]. A zone is established as setting the zone center node as the central point of zone and the zone center node has n-hop radius within the zone, in which nodes make maximum n-hop path. Intra-zone routing protocol such as IARP monitors intra-zone routing information and keeps routing table constantly. Inter-zone routing protocol such as IERP finds paths using so-called border-casting. If the zone center node named zone head needs to send some packets to the destination, it sends query packets to its peripheral nodes using border-casting, which groups the same zone. At the same time, one of peripheral nodes that receives query packet also functions as a center node of a certain zone. This node checks whether the destination node resides in its zone or not. If the destination exists, that node sends reply packet to the source node in the reverse order. If not, it repeats border-cast to its peripheral nodes. When a query packet travels around zone's center nodes and peripheral nodes, it adds node's unique ID to its packet header. Finally, when a source node receives a reply packet, it sets reverse path of the reply packet as a routing path. In case of ZRP routing, all the nodes have their zones and also belong to other zones. Each node has functions of zone head, relaying (internal) nodes, and peripheral nodes simultaneously. The number of zones matches exactly with the number of nodes. It means that the number of different intra-zone routing information is set on each zone. Figure 1 Depicts ZRP zone topology. Solid line circles (cells) are needed to route from the source 'node S' to the destination 'node D'. Dotted line circles (cells) also represent each zone, but it does not participate in routing from 'node S' to 'node D'

However, if we possibly prevent a zone area from overlapping each other, and we only permit overlapping peripheral nodes, each node in RWN should function as a center node, internal node, or peripheral node. A node will function as a center node, peripheral node or a relay node when setting up a zone. This removal of overlapping zones results in simplification of node function and routing efficiency and it also reduce the traffic that is used to prevent the routing loop. In case of the existing ZRP, each node belongs to many different zones depending on n-hop radius. It means that if a node belongs to ten different zones, it must monitor different ten routing information. But if we keep zones not to be overlapped, a node does not have to function as a center, internal, and peripheral node at the same time, all the nodes in a zone only need to keep its internal routing information.

At this time, the only problem is to configure the initial zone that does not overlap and keeps the zone without overlaps constantly. If we just adopt previous research result for setting up a simple cluster or hierarchical cluster [5, 6], it is not easy to reflect zone radius of ZRP. It results in imbalance of each

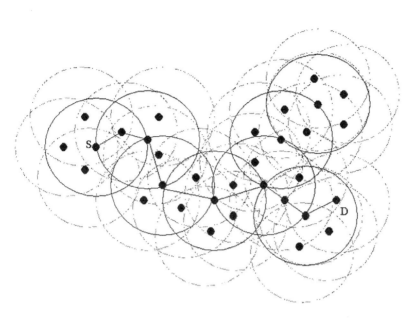

Fig. 1. Zone Topology of ZRP and an example of its IERP operation

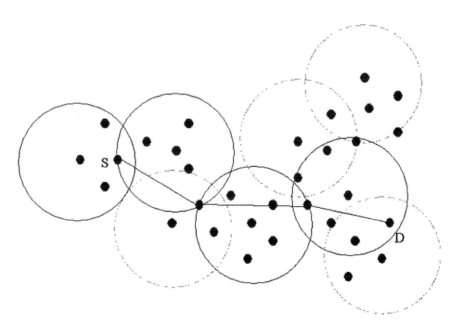

Fig. 2. Newly configured Independent Zone Topology and an example of its IERP operation

zone. After all, we need new algorithm for a zone configuration that does not overlap that named independent zone. We also need a scheme that maintains overlap-free zones even when the node moves and zones with modified routing schemes. Figure 2 depicts a newly configured independent zone topology and an example of its IERP operation. Solid line circles (cells) are needed to route from the source 'node S' to the destination 'node D'. Dotted line circles (cells) also mean each zone, but it does not participate in routing from 'node S' to 'node D'. The details of its operation will be explained in the coming sections of this paper.

3   Algorithms for Independent Zone Scheme

3.1   Independent Zone Configuration

Algorithm as follows, tries to make a zone without overlaps during the initial zone configuration allowing as many nodes as possible in a zone radius. Once a zone is configured, it starts making another zone and shares a peripheral node that belongs to each zone to make a connection from one to others.

a) Each node finds out the number of the neighbor node contained within suitable signal strength. Such signal strength also becomes an n-hop radius of independent zone.

b) Each node exchanges neighbor counts with its neighbors. A node that has maximum neighbor node count becomes the center node of an independent zone. It informs their neighbors that it is the center node of the zone for them to join a specific zone. It implicitly tells that all the nodes once joined to a specific zone are not to respond to a zone configuration request except for the peripheral nodes. (Center node for zone X: cX, internal node for zone X: iX, peripheral node for zone X: pX)

c) This step finds candidate zones. A center node for zone X, cX, selects one of its peripheral nodes, pX, and delegates pX to configure the zone by sending request message. pX temporarily becomes a center node, cA, of the virtual zone A. And then, cA (=pX) tries to find out the peripheral node p'A that has the largest neighbor count in its 'virtual zone A'. Undoubtedly, neighbor count means the number of nodes that does not configure the zone at that time. To find out a node that has the largest number of neighbors, we need to just apply step a). If the nodes are uniformly distributed, the peripheral node p'A with the largest neighbor count would be located in a straight line from cX via pX, that is to say, p'A will be the farthest peripheral node belongs to the virtual zone A from cX. This scheme makes each zone not to be overlapped each other and allows a zone to have the maximum number of nodes as possible. Fig.3 depicts this step.

d) p'A with the largest neighbor count configures a new zone Y. At the same time, p'A becomes a center node cY of zone Y. Step b) can be applied in configuring the zone. At this time, pX naturally belong to the zone Y. so pX also becomes pXY, it then acts as a peripheral node of zone X and zone Y at the same time.

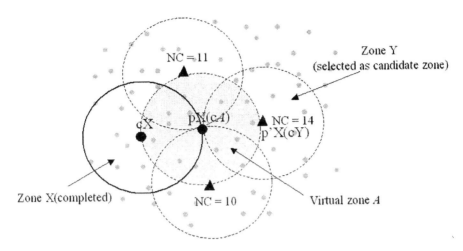

Fig. 3. Finding next candidate zone

e) Repeat step c) and d) to extend the zone coverage. If a node heard requesting messages to configure a zone that already belongs to a certain zone, it just discards the requesting message.
f) During the step 1 through 5, if some nodes could not configure a zone for the critical time value, they just make zones with them. Also, step b) can be applied in configuring the zones. However, a configured zone needs to have at least on minimum peripheral node sharing with other zones to constitute a routing path.

3.2 Independent Zone Maintenance

a) Movement of a node within a zone does not affect the external routings. It only affects the internal routing information. So it easily fixes up the internal routing information changes
b) If a peripheral node of a zone moves to another location, and if the peripheral node just breaks routing path between two zones, one of the center node of two zones needs to strengthen the signal power to find out the substitute peripheral node to connect the two broken zones.
c) If the center node of a zone moves to another location, and if the nodes within the zone cannot communicate each other, the zone should be divided into two zones to restore the connection. Some time later, if some nodes reach a location that connects the split zones into one, they can be merged.

3.3 Independent Zone Routing

A source node s of a zone X, sX, needs to follow the following steps to find a routable path to the destination node d of zone Z, dZ.

a) sX tries to find out dZ in its routing table belongs to zone X. if dZ is found
   in the internal routing table of sX, sX can send packets to dZ directly .
b) If sX could not find dZ, sX border-casts query message to all the peripheral
   nodes pX in the zone X.
c) One of the peripheral node pX in zone X may also belongs to other zones
   and also has many routing tables depends on the number of zones it belongs
   to. So, pX that receives a query message tries to find out the destination
   node dZ in its routing table except for the zone from which query message
   arrived.
d) If dZ is not found in each routing table on the peripheral node, pX border-
   casts the query message to the rest of the peripheral nodes except for the
   zone from which query message came.
e) Repeats c), d) until source node sX finds the destination node dZ.

4   Performance Evaluation

The results of our simulation are presented in the following graphs. We use
the assumption that a network of 200 nodes moves in an area of 1000x1000m
Squares. In this model every node can select a random location and moves there
at a uniformly chosen speed of between 0 and 1m/s. We ignored border-casting
delay and IARP retrieving delay setting as 0ms respectively. A node can send
query packets by maximum of ten times in one second.

Figure 4 illustrates the transmitted packets of basic ZRP and the proposed
scheme supporting the independent zone with IERP operation. The number of
transmitted packets depends on the zone radius. But, in most cases, the proposed
scheme shows less IERP operation packets in a network. IERP Operation is

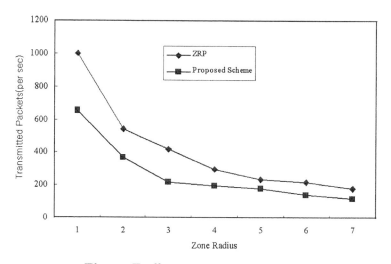

Fig. 4. Traffic per Route Query after initial setup

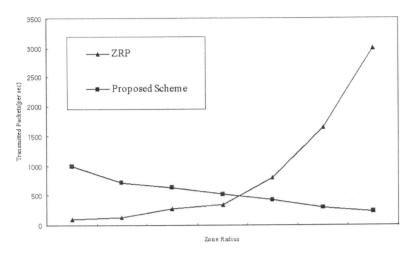

Fig. 5. Control Traffic variation due to mobile

mostly depends on the number of zones. So it is natural that the proposed scheme is more efficient than the basic ZRP, because the proposed scheme has by far less number of zones than that of the basic ZRP. Besides, the proposed scheme just needs nearly half the number of zones than the basic ZRP with IERP operation.

Figure 5 illustrates the transmitted packets of basic ZRP and the proposed scheme supporting the independent zone with re-configuration due to the movement of a node. Basic ZRP shows when a zone radius grows larger, much more control traffic flows on the network. In case of the proposed scheme, when its zone radius is small, it uses much traffic to re-configure the zone than the basic ZRP. However, as the zone radius grows larger, it uses less traffic to re-configure the zone in comparison with the basic ZRP. We expect that intra-zone movement and inter-zone movement ratio causes this result. With the smaller zone radius, almost all the movement of node are considered to be the inter-zone movement. So it needs to re-configure the zone frequently. But the larger zone radius means more inter-zone movement. So it will have less chance to re-configure the zones.

5   Conclusion

In this paper, we proposed a new independent zone configuration scheme for ZRP in which a node does not have multiple functions as head, internal, and peripheral node. We also introduced a modified routing scheme for the newly established zone scheme. This scheme tries to make a zone without overlaps during the initial zone configuration and allows as many nodes as possible in a zone radius. This new independent zone configuration scheme has the following advantages as shown in the performance evaluation. Internal routing information size based on the link state can be significantly reduced. A hop count is reduced and the routing loop prevention steps are possible with less cost. Out simulation results tells us that our independent zone scheme is more suitable for less dynamic large RWN.

Zone re-configuration algorithm of the proposed scheme is more complex than ZRP since It needs to keep the zone not to be overlapped all the time. This produces negative performance especially for dynamically moving node in narrow radius zone. Since the performance of RWN is affected by such parameters as node movement and zone radius, as a future research, we will research on the dynamically adaptive RWN to such parameters to enhance the performance.

References

[1] J. Moy., OSPF version 2. RFC 2178, Internet Engineering Task Force, July 1997.
[2] C. E. Perkins and E. M. Royer., Ad-Hoc On-Demand Distance Vector Routing. In Proceedings of the Second Annual IEEE Workshop on Wireless and Mobile Computing Systems and Applications, February 1999.
[3] J. J.Garcia-Luna-Aceves and M.Spohn, Source-Tree Routing in Wireless Networks, In Proc, IEEE ICNP 99, 7th Intl, Conf. On Network Protocols, Toronto, Canada, Oct 1999.
[4] D. J. Baker and A. Ephremides., A Distributed Algorithm for Organizing Mobile Radio Telecommunication Networks. In Proceedings of the Second International Conference on Distributed Computer Systems, April 1981.
[5] C.-C. Chiang., Routing in Clustered Multi-hop, Mobile Wireless Networks. In Proceeding of ICOIN Nobemver, 1996.
[6] M. Gerla and J. T.-C. Tsai, Multi-cluster, Mobile, Multimedia Radio Network, Wireless Networks, October 1995.
[7] R. Krishnan, R. Ramanathan, and M. Steenstrup., Optimization Algorithms for Large Self-Structuring Networks. In Proceedings of IEEE INFOCOM '99, March 1999.
[8] J. Zavgren., NTDR Mobility Management Protocols and Procedures. In Proceedings of the IEEE Military Communications Conference, November 1997.
[9] Z. J. Hass and M. R. Pearlman., The Performance of Query Control Schemes for the Zone Routing Protocol. In Proceedings of SIGCOMM '98, September 1998.
[10] M. R. Pearlman and Z. J. Haas., Determining the Optimal Configuration of the Zone Routing Protocol. IEEE Journal on Selected Areas of Communications, August 1999.

Information-Theoretic Bounds for Mobile Ad-hoc Networks Routing Protocols

Nianjun Zhou and Alhussein A. Abouzeid*

Dept of Electrical, Computer and Systems Engineering
Rensselaer Polytechnic Institute
Troy, New York 12180, USA
{zhoun,abouza}@rpi.edu

Abstract. In this paper, we define the *routing overhead* as the amount of information needed to describe the changes in a network topology. We derive a universal lower bound on the routing overhead in a mobile ad-hoc network. We also consider a *prediction-based* routing protocol that attempts to minimize the routing overhead by predicting the changes in the network topology from the previous mobility pattern of the nodes. We apply our approach to a mobile ad-hoc network that employs a dynamic clustering algorithm, and derive the optimal cluster size that minimizes the routing overhead, with and without mobility prediction. We believe that this work is a fundamental and essential step towards the rigorous modeling, design and performance comparisons of protocols for ad-hoc wireless networks by providing a universal reference performance curve against which the overhead of different routing protocols can be compared.

1 Introduction

Research on distributed multi-hop wireless networks, also known as wireless ad-hoc networks [1] has evolved from DARPA packet radio program during the early 1970's [2]. Since wireless ad-hoc networks can be deployed rapidly in a non-organized (i.e. ad-hoc) fashion without requiring any existing infrastructure, they are expected to find applications in a number of diverse settings [3, 4].

Much of the research in the area of ad-hoc networks has focused on developing routing protocols. *Proactive* routing protocols (e.g. [5, 6, 7]) attempt to compute paths in advance and determine them continuously so that a route is readily available when a packet needs to be forwarded. *Reactive* routing protocols (e.g. [8, 9, 10]) are based on a source initiated query/reply process and typically rely on the flooding of queries for route discovery. A network with a few relatively fast moving nodes favors reactive protocols while a network with many slowly moving nodes favors proactive protocols.

The two strategies are combined in routing scheme. *Hierarchical* routing schemes based on the formation of a virtual backbone are developed in [12, 13].

---

* Correspondence Author: Alhussein A. Abouzeid, Rensselaer Polytechnic Institute, 110 Eighth St, JEC-6038, Troy NY 12180 Tel: (518)276-6534, Fax: (518)276-4403

H.-K. Kahng (Ed.): ICOIN 2003, LNCS 2662, pp. 651–661, 2003.

The main advantage of hierarchal routing is that it overcomes scalability problems by designating a node within a group of nodes to be responsible for routing, and thus only these nodes need to maintain routing information about the rest of the network. The main disadvantage is the overhead in the cluster maintenance operations (i.e. joining and leaving the clusters) which mainly depend on the degree of mobility of the nodes. This paper considers *proactive hierarchal* routing protocols.

Considerable effort has been directed towards evaluating routing protocols [14, 15, 16, 17, 18, 19, 20, 21]. The procedure of evaluation relies on simulations of the protocol and comparing its performance against other existing protocols.

There is a need for a *universal reference performance curve* that can tell us how good (or bad) the performance of a specific protocol really is. In coding theory, a channel coding algorithm is good if it achieves the Shannon capacity [22]. Similarly, we seek to derive a universal curve against which we can measure how good a routing protocol performs, in terms of minimizing the *routing overhead*, which is the amount of information needed to describe the changes in a dynamic network topology. Up to our knowledge, this paper is the first that attempts to derive a theoretical lower bound on the routing overhead of ad-hoc network protocols. Related work in [23, 24] do not consider the routing overhead in deriving the capacity of wireless networks.

2   Network Model

2.1   Assumptions

We make the following assumptions about the network: **(A1)** The mobile nodes are identical; **(A2)** The mobile nodes are distinguishable;**(A3)** The number of mobile nodes in the network is fixed, denoted by N; **(A4)** The communication region of interest is fixed and bounded. A1 means that all the nodes have the same physical characteristics. A2 means that each node is addressable by a node unique identifier - which we denote as NUI. Furthermore, We assume that the statistical node mobility patterns of individual nodes are independent and known. In this paper, we allow any type of mobility model as long as it allows a formulation using Markov chains.

Regarding the self-organization of the network and the geographic distribution of the nodes, the model can be viewed as a two-level abstraction. We assume that nodes move freely within a bounded region of space. The whole region is divided into sub-regions. For this preliminary work, we assume the number of sub-regions is fixed. There are two levels of hierarchy for nodes. Each region has its unique identifier - RUI. All the nodes of a sub-region form a cluster and a cluster head, which we call *routing node*. Routing node is selected at random from the nodes within the sub-region. It is possible that a sub-region becomes empty and thus will not have a cluster or a cluster head. The nodes belonging to the same sub-region communicate directly. The nodes belonging to different sub-regions communicate through the routing nodes. Thus, only two modes of *direct* (i.e. *single-hop*) communication are permissible: nodes within the same sub-region, and routing nodes of neighboring sub-regions (See Figure 1).

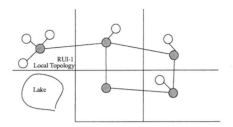

Fig. 1. A snapshot of the network topology. Blue (dark) nodes are routing nodes. There is one routing node in every sub-region. Sub-regions do not have to be rectangular or identical

2.2 Definitions

For clarity of presentation, we provide the formal definition of the terms used in the paper as follows. **Routing Node:** The cluster head for a sub-region, assumed to be selected randomly from the set of nodes that belong to the sub-region. There is only one routing node for a sub-region. If a routing node leaves a sub-region, a new routing node is selected from the remaining nodes in the same sub-region. Each routing node has an RUI associated with it; **Regular Node:** a node that is not currently a routing node. Any regular node can be selected as a routing node; **Topology:** is the connectivity relationship of the nodes as depicted in Figure 1. Due to the dynamic behavior of the topology, it can be described as an instance of a random graph. An edge exists between two arbitrary nodes if single-hop communication between those two nodes is admissible, according to the rules presented earlier. Due to the mobility of the nodes, the edge of two nodes can be established or torn down randomly; **Global Topology:** The topology of the whole network. **Local Topology:** The topology of a sub-region; **Topologies Cardinality:** The total number of the topologies. The cardinality of the network topologies is used to calculate the minimum amount of information needed to describe a network snapshot without any extra knowledge; **Information Overhead:** The bits needed to distinguish one topology from another.

3   Analysis

Any algorithm for solving some problem must do some minimal amount of work. The most useful principle of this kind is that the outcome of a comparison between two items contains at most one bit of information. Hence, if there are m possible input strings, and an algorithm purports to identify which one it was given solely on the basis of comparisons between input symbols, then $\lceil \log_2 m \rceil$ comparisons are needed (all logarithms in this paper are base 2 and hence we will shortly write log instead of $\log_2$). This is because $\lceil \log m \rceil$ bits are necessary to specify one of the m possibilities [25].

3.1 Outline and Notation

In this section, we apply the above principle to derive a lower bound on the overhead of proactive hierarchal routing protocols with fixed number of clusters in a mobile wireless ad-hoc network. The steps followed are as follows: (a) Compute the number of different topologies possible for each local sub-region and for the whole network. These are termed local topologies cardinality and global topologies cardinality, and are denoted by denoted $\mathbf{L}$ and $\mathbf{G}$, respectively; (b) Derive expressions for h, the average holding time for a node within a sub-region, and for H, the average holding time before a topology change; (c) Compute the minimum amount of information needed to keep track of the topology changes using the individual local information from the different sub-regions $\mathbf{I_L}$ and from the global view of the network $\mathbf{I_G}$. (d) Finally, consider the effect of predicting the future locations of the nodes based on their mobility on the local $\mathbf{I_L^P}$ and global $\mathbf{I_G^P}$ routing overhead.

3.2 Cardinalities of Global and Local Network Topologies

Theorem 1: Consider N mobile nodes randomly distributed over M sub-regions. Choose a routing node for each non-empty sub-region randomly from the set of nodes belonging to the sub-region. The total number of possible topologies is given by:

$$G = \sum_{i=1}^{\min(N,M)} \frac{N!M!}{i!(M-i)!(N-i)!} i^{N-i} \tag{1}$$

Proof [1]: See Appendix.

Theorem 2: Consider N mobile nodes randomly distributed over M sub-regions, then the total number of local topologies for a given sub-region is:

$$L = (2^{N-1})N + 1 \tag{2}$$

Proof: See Appendix.

3.3 Inferring Network Topology Dynamics from Individual Node Mobility

We consider the following simple Markov mobility model. For a given time-step τ, we assume that[2] the probability that a mobile node will stay in the same sub-region is $0 \le p_0 \le 1$, and leave its current sub-region is $1 - p_0$. We further assume that a node moves to any of the K neighboring nodes with equal

---

[1] The result can be verified using a simple example with $N = 4$ and $M = 2$, where it is easy to show using manual calculation that the total combination is 56, which is the same as the result from (1).

[2] This model follows from discretizing a continuous-time two-dimensional Markov process description of the individual node mobility.

probability (random walk) $\frac{1-p_0}{K}$. Thus, the probability distribution function of the node holding time at the same sub-region is geometric. For a given node, the mean holding time $h = \frac{p_0}{1-p_0}$.

Since the individual nodes move independently, The probability p_1 that a specific topology does not change is equal to the probability that all nodes do not change their sub-regions. Assuming independent node mobility, $p_1 = p_0^N$. Hence, the average holding time for a topology is thus

$$H = \sum_{i=1}^{\infty} i(1 - p_1)p_1^i = \frac{p_1}{1 - p_1} = \frac{p_0^N}{1 - p_0^N} \tag{3}$$

3.4 Routing Overhead without Mobility Prediction

The information needed to identify a specific network topology from the set of all the possible topologies is the entropy of the set.

Global Topology without Prediction: From (1)

$$I_G = \log(G) \tag{4}$$

Since we do not have any knowledge of the topology probability distribution, we use the maximum entropy method [26] to infer it. From the proof of Theorem 1, we have shown that for a distribution $r = (r_1, r_2, \ldots, r_M)$ of nodes for each sub-region, the total number of possible topologies is

$$n(r, N) = \frac{N!}{r_1! r_2! \ldots r_M!} \prod_{i=1}^{M} g(r_i) \text{ with } \sum_{i=1}^{M} r_i = N \tag{5}$$

We have also shown that $\prod_{i=1}^{M} g(r_i)$ reaches its maximum value when $r = (r_1, r_2, \ldots, r_M)$ has a uniform distribution. In this case

$$n(r, N) \le \frac{N!}{r_1! r_2! \ldots r_M!} \left[\frac{N}{M} \right]^M \tag{6}$$

It is easy to show after some algebraic manipulations that

$$M^N \le G \le \left[\frac{N}{M} \right]^M M^N \tag{7}$$

and equivalently

$$N \log M \le I_G \le N \log M + M \log \left\lceil \frac{N}{M} \right\rceil \tag{8}$$

The result (8) can be interpreted as follows. The lower bound in (8) is the minimum information needed to describe the network without identifying which nodes are the routing nodes. The introduction of routing nodes increases the complexity of the network by $M \log \left\lceil \frac{N}{M} \right\rceil$.

Local Topology without Prediction: The entropy of a specific sub-region is $\log L$. From (2), the *sum* of the information overhead required to maintain each of the local topologies, I_L, is ,

$$I_L = M \log(2^{N-1}N + 1) \tag{9}$$

where the "M" coefficient comes from the fact that there are M local topologies.

For large N,

$$\lim_{N \to \infty} \frac{\log(2^{N-1}N + 1)}{N} = 1 \tag{10}$$

and hence

$$I_L \approx MN \ ; N \gg 1 \tag{11}$$

From (8) and (11),

$$\frac{I_L}{I_G} \approx \frac{M}{\log M} \ ; N \gg 1 \tag{12}$$

A comment on the above result (12) is in order. It states that, in the limit $N \to \infty$, $I_L > I_G$ (i.e. the sum of the entropies of the local topologies is larger than the entropy computed directly from the global network topology). The reason is that I_G assumes the knowledge of N and M when computing the different possible topologies, while I_L computes the local topology information locally (independent of the node distribution over the rest of the sub-regions) even though the local topologies of different sub-regions are *not* independent (a node may not exist in two different sub-regions).

3.5 Routing Overhead with Mobility Prediction

Global topology with Prediction: The information needed to update the new location of a node (i.e. which sub-region it belongs to) given the knowledge of the current location is

$$I = -1\left(p_0 \log p_0 + (1 - p_0) \log(\frac{1 - p_0}{K})\right) \tag{13}$$

and hence

$$NI \leq I_G^P \leq NI + M \log\lceil \frac{N}{M} \rceil \tag{14}$$

When $p_0 = 1/(K+1)$, I reaches its maximum value. From (8) and (14),

$$\frac{I_G^P}{I_G} \leq \frac{\log(K+1)}{\log M} \ ; N \gg 1 \tag{15}$$

For many practical scenarios, even the largest number of neighboring sub-regions is usually less than the total number of sub-regions. In this case, for large N, the above result states that using prediction to update the topology information will result in large savings in the routing overhead.

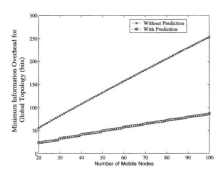

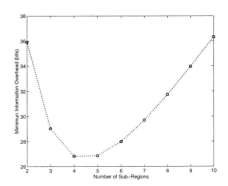

Fig. 2. Minimum information overhead with and without prediction. $M = 5$, $p_0 = 0.9$, $k = 3$

Fig. 3. Minimum information overhead, with prediction. $N = 100$, $q = 0.5$, $p_0 = 0.9$ and $k = 3$. Optimal number of sub-regions is 4

Local Topology with Prediction: Assuming each node is equally likely to belong to any of the sub-regions with probability $\frac{1}{M}$,

$$I_L^P \leq -MN \left(\frac{1}{M} \log \frac{1}{M} + (1 - \frac{1}{M}) \log(1 - \frac{1}{M}) \right) + M \log N \qquad (16)$$

Note that $M \log N$ bits are needed to specify the routing node for the cluster.

For large M, it is easy to show after some mathematical manipulations that,

$$I_L^P \leq N \log M + M \log N \; ; M \gg 1 \qquad (17)$$

From (11) and (17),

$$\frac{I_L^P}{I_L} \leq \frac{N \log M + M \log N}{NM} \; ; M \gg 1 \; ; N \gg 1 \qquad (18)$$

4   Numerical Results and Applications

In this section, we show some numerical results from the bounds derived in the previous section as well as an application of these bounds in finding the optimal number of sub-regions that minimizes the information overhead.

Figure 2 shows a numerical example for the lower bound on the routing overhead with and without prediction. Clearly, large amount of savings is achieved with prediction.

Let q denote the probability that two arbitrary communicating nodes belong to the same cluster. Let O (O^P) denote the minimum routing overhead without(with) prediction. Then,

$$O = q\frac{I_L}{M} + (1 - q)I_G \qquad (19)$$

without prediction and

$$O^P = q\frac{I_L^P}{M} + (1 - q)I_G^P \tag{20}$$

Figure 3 depicts the results for an arbitrary network. The figure shows the optimal number of sub-regions which is the one that requires the least amount of information exchange. Decreasing the number of sub-regions M increases the information overhead in maintaining the local cluster topology (more nodes per cluster) while increasing the number of sub-regions increases the overhead in maintaining the inter-cluster (global) topology, and the optimal point represents the balance between those two opposing factors.

5 Conclusion

This paper presented an information-theoretic approach to the analysis of routing protocols. We applied the framework to derive lower bounds on the routing overhead in a mobile ad-hoc network, with and without mobility prediction. We quantified the amount of overhead needed for clustering, and also the amount of savings. By combining both, we find the optimum number of clusters that minimize the routing overhead. A number of future avenues of work remain. Due to the distributed nature of the network, the local topology information will be exchanged and used to infer the global network topology. Therefore, the mechanism in which the local topology information will be exchanged is an important topic that has not been addressed in this paper. Including this aspect may result in a tighter lower bound (i.e. a curve that is "above" the one derived in this paper).

References

[1] Z. J. Haas et al. eds. Wireless ad hoc networks. *IEEE Journal on Selected Areas in Communications*, 17(8), August 1999.

[2] J. Jubin and J. D. Tornow. The DARPA packet radio network protocols. *Proceedings of the IEEE*, 75(1), 1987.

[3] D. L. Estrin et al. *Embedded Everywhere: A research agenda for networked systems of embedded computers*. National Research Council, 2001. http://www.nap.edu.

[4] I. F. Akyildiz, W. Su, Y. Sankarasubramaniam, and E. Cayirci. Wireless sensor networks: A survey. *Computer Networks (Elsevier)*, 38(4):393–422, March 2002.

[5] C. E. Perkins and P. Bhagwat. Highly dynamic Destination-Sequenced Distance-Vector routing (DSDV) for mobile computers. In *Proceedings of ACM SIGCOMM'94*, pages 234–44, London, United Kingdom, October 1994.

[6] P. Jacquet, P. Muhlethaler, and A. Qayyum. Optimized link state routing protocol. IETF MANET, Internet Draft, November 1998.

[7] S. Murthy and J. J. Garcia-Luna-Aceves. A routing protocol for packet radio networks. In *Proceedings of ACM International Conference on Mobile Computing and Networking (MOBICOM '95)*, pages 86–94, Berkeley, CA, USA, December 1995.

[8] V. D. Park and M. S. Corson. A highly adaptive distributed routing algorithm for mobile wireless networks. In *Proceedings of IEEE INFOCOM'97*, volume 3, pages 1405–13, Kobe, Japan, April 1997.

[9] D. B. Johnson and D. A. Maltz. Dynamic source routing in ad hoc wireless networking. In T. Imielinski and H. Korth, editors, *Mobile Computing*. Kluwer, 1996.

[10] C. E. Perkins and E. M. Royer. Ad hoc on-demand distance vector routing. In *Proceedings of IEEE Workshop on Mobile Computing Systems and Applications (WMCSA'99)*, pages 90–100, New Orleans, LA, USA, February 1999.

[11] M. R. Pearlman and Z. J. Haas. Determining the optimal configuration for the zone routing protocol. *IEEE Journal on Selected Areas in Communications*, 17(8):1395–414, August 1999.

[12] B. Das, R. Sivakumar, , and V. Bharghavan. Routing in ad hoc networks using a spine. In *Proceedings of International Conference on Computer Communications and Networks*, pages 34–9, Las Vegas, NV, USA, September 1997.

[13] B. Das, R. Sivakumar, , and V. Bharghavan. Routing in ad-hoc networks using minimum connected dominating sets. In *Proceedings of International Conference on Communications (ICC'97)*, volume 2, pages 765–9, Montreal, Canada, September 1997.

[14] S. Ramanathan and M. Steenstrup. A survey of routing techniques for mobile communications networks. *Journal of Special Topics in Mobile Networks and Applications (MONET)*, 1(2):89–104, October 1996.

[15] E. Royer and C.-K. Toh. A review of current routing protocols for ad-hoc mobile wireless networks. *IEEE Personal Communications Magazine*, pages 46–55, April 1999.

[16] J.-C. Cano and P. Manzoni. A performance comparison of energy consumption for mobile ad hoc network routing protocols. In *Proceedings of 8th International Symposium on Modeling, Analysis and Simulation of Computer and Telecommunication Systems (MASCOTS)*, pages 57–64, San Francisco, CA, USA, August 2000.

[17] S. R. Das, R. Castaneda, and Yan Jiangtao. Simulation-based performance evaluation of routing protocols for mobile ad hoc networks. *Mobile Networks and Applications*, 5(3):179–89, 2000.

[18] C. E. Perkins, E. M. Royer, S. R. Das, and M. K. Marina. Performance comparison of two on-demand routing protocols for ad hoc networks. *IEEE Personal Communications*, 8(1):16–28, February 2001.

[19] A. Boukerche. A simulation based study of on-demand routing protocols for ad hoc wireless networks. In *Proceedings of IEEE 34th Annual Simulation Symposium*, pages 85–92, Seattle, WA, USA, April 2001.

[20] C.-K. Toh, H. Cobb, and D. A. Scott. Performance evaluation of battery-life-aware routing schemes for wireless ad hoc networks. In *Proceedings of IEEE International Conference on Communications (ICC)*, volume 9, pages 2824–9, Helsinki, Finland, June 2001.

[21] A. Nasipuri, R. Burleson, B. Hughes, and J. Roberts. Performance of a hybrid routing protocol for mobile ad hoc networks. In *Proceedings Tenth International Conference on Computer Communications and Networks*, pages 296–302, Scottsdale, AZ, USA, October 2001.

[22] C. E. Shannon. A mathematical theory of communication (2 parts). *Bell System Technical Journal*, 27:379–423, 623–56, July and October 1948. http://cm.bell-labs.com/cm/ms/what/shannonday/shannon1948.pdf.

[23] P. Gupta and P. R. Kumar. The capacity of wireless networks. *IEEE Transactions on Information Theory*, 46(2):388–404, March 2000.

[24] M. Grossglauser and J. Bolot. On the relevance of long range dependence in network traffic. *IEEE/ACM Transactions on Networking*, 7(5):629–40, October 1999.

[25] B. Weide. A survey of analysis techniques for discrete algorithms. *ACM Computing Surveys*, 9(4):291–313, December 1977.

[26] A. Papoulis and S. U. Pillai. *Probability, Random Variables and Stochastic Processes*. Cambrige studies in advanced mathematics. McGraw-Hill, 4th. edition, 2002.

Appendix: Proof of Theorems

Proof of Theorem1 First, consider the case $N \geq M$. Let $r = (r_1, r_2, \ldots, r_M)$ denote a specific organization of the nodes over the sub-regions. Then the possible topologies for this case is

$$n(r, N) = \frac{N!}{r_1! r_2! \ldots r_M!} \prod_{i=1}^{M} g(r_i)$$

where $g(r) = r \forall r > 0$ and $g(0) = 1$. The function $g(r)$ follows from the fact that we have r different ways of selecting a routing node for each subregion.

Let R denote the set of all possible organizations of the nodes. Let

$$f(x_1, x_2, \ldots, x_M) = (x_1 + x_2 + \cdots + x_M)^N = \sum_{r \in R} \frac{N!}{r_1! r_2! \ldots r_M!} x_1^{r_1} x_2^{r_2} \ldots x_M^{r_M} \tag{21}$$

The total number of topologies in which there are no empty sub-regions can be calculated by taking the partial derivative of (21) w.r.t x_i and then setting $x_i = 1$, which yields

$$\left(\frac{N!}{(N-M)!} \right) M^{N-M} \tag{22}$$

Consider now that there is a single empty sub-region s. Then the total number of topologies can be calculated by taking the partial derivative of (21) w.r.t x_i and then set $x_i = 1 \forall i \neq s$ and $x_s = 0$, which yields

$$\left(\frac{N!}{(N-M+1)!} \right) M^{N-M+1} \tag{23}$$

And since there are $\binom{M}{1}$ ways of having a single empty sub-region, the total number of topologies with a single sub-region is thus

$$\left(\frac{M!}{1!(M-1)!} \right) \left(\frac{N!}{(N-M+1)!} \right) M^{N-M+1} \tag{24}$$

By induction, the number of topologies with exactly i empty sub-regions is

$$\left(\frac{M!}{i!(M-i)!} \right) \left(\frac{N!}{(N-M+i)!} \right) M^{N-M+i} \; ; 0 \leq i \leq M-1 \tag{25}$$

Summing over all i yields the result. Similar derivation applies for the case $N < M$ (but notice in this case that the number of empty sub-regions will range from $M - N$ to $M - 1$).

Proof of Theorem2 Let r be the number of nodes in the sub-region. The total number of possible topologies is thus

$$L = \sum_{r=0}^{N} \frac{N!}{r!(N-r)!} g(r) \tag{26}$$

where $g(r)$ has been defined in the proof of Theorem 1 (above). Let

$$f(x_1, x_2) = (x_1 + x_2)^N = \sum_{r=0}^{N} \frac{N!}{r!(N-r)!} x_1^r x_2^{N-r}$$

Taking the partial derivative of both sides w.r.t x_1, then setting $x_1 = x_2 = 1$ yields

$$N2^{N-1} = \sum_{r=1}^{N} \frac{N!}{r!(N-r)!} g(r)$$

For the trivial case $r = 0$ (no nodes in the sub-region, there is only one possible topology.

History-Aware Multi-path Routing
in Mobile Ad-Hoc Networks

Sangkyung Kim[1] and Sunshin An[2]

[1] Service Development Lab., KT
137-792 Seoul, Korea
skkim98@kt.co.kr
[2] Department of Electronic Engineering, Korea University
136-701 Seoul, Korea
sunshin@dsys.korea.ac.kr

Abstract. This paper presents a new on-demand ad hoc routing protocol, History-Aware Multi-path Routing (HAMR) protocol, which introduces a session history as one of routing metrics. A session history implies how many times and how much duration a node is involved in communication sessions between mobile nodes in a network. The motivation of HAMR is that if a node's session history is higher, the node will be more stable than nodes with lower histories. HAMR supports the establishment of multiple paths, which are selected in consideration for their session histories. HAMR's approach sets the goal at reducing control traffic overhead and route reconfiguration time by decreasing frequent route re-starts due to route failures. We have evaluated the performance of HAMR through a series of simulations using the Network Simulator 2 (ns-2).

1 Introduction

A mobile ad hoc network [3] is a temporary network that consists of mobile nodes that can communicate with each other without relying on any infrastructure. Thus, a mobile ad hoc network is inherently quite dynamic due to frequent and unpredictable node mobility. This leads to the problem of keeping track of network connectivity. Routing in mobile ad hoc networks should be highly adaptive to rapid changing topology.

Traditional routing protocols [6, 7] known from wired networks do not work efficiently because they have not been designed with the unstable network topology in ad hoc networks [5]. Therefore, numerous routing protocols have been developed for ad hoc networks [1, 2, 3, 8, 9, 11]. Most of existing ad hoc routing protocols are concerned about reducing control overhead and providing a stable route from source to destination, and adopts on-demand routing schemes, which creates routes only when desired by the source node using a route discovery process [1, 9, 11]. Dynamic source routing (DSR) [1, 4] is one of typical on-demand routing protocols. DSR is based on source routing and uses the shortest path as the routing metric. Ad hoc on-demand distance vector (AODV) [9]

H.-K. Kahng (Ed.): ICOIN 2003, LNCS 2662, pp. 662–671, 2003.
© Springer-Verlag Berlin Heidelberg 2003

routing protocol improves the performance characteristics of DSDV [8] in the creation and maintenance of routing information. AODV minimizes the number of broadcasts by creating routes by on-demand basis and does neither maintain a complete list of routing information nor participate in periodic routing table exchanges as DSDV does. Associativity-based routing (ABR) [11] is another on-demand routing protocol that uses associativity state as a new routing metric. Associativity state implies periods of link stability and it is measured by recording the number of control beacons received by a node from its neighbors.

In a dynamic ad hoc network, route discoveries due to frequent link changes increase largely the amount of control messages through the network. In addition, this may cause the increase of packet loss and packet transmission delay. Hence, it is quite required to select relatively a stable route and maintain multiple paths to reduce the number of route re-discoveries and route reconfiguration time. Fast route reconfiguration can improve performance measures, such as packet delivery fraction and average end-to-end packet delay.

This paper presents a new on-demand ad hoc routing protocol, History-Aware Multi-path Routing (HAMR), which uses 'session history' as one of route selection criteria. Session history means how many times and how long a node is involved in communication sessions within a certain time. The motivation behind history-aware routing is that a node with high session history will be in low movement compared with nodes having low session history, and a route consisting of such nodes is likely to be long-lived and avoid frequent route re-starts. HAMR provides multiple paths in consideration of session history from source to destination. HAMR's approach sets the goal at reducing control traffic overhead and route reconfiguration time by decreasing frequent route re-starts due to route failures.

The remainder of this paper is organized as follows. The following section gives descriptions about detailed algorithms and procedures of HAMR. Sect. 3 presents the results of the analysis of HAMR using simulations. Finally, we present our conclusions and directions for future research in Sect. 4.

2 History-Aware Multi-path Routing

2.1 Communication Session History

To enable history-aware routing, session history should be maintained at each node. During a route discovery process, a route query message transports to a destination the session histories of nodes through which it goes so that the destination can determine a route in consideration for its session history. Each node keeps the record about the following information.

– Active session information: This means the information about communication sessions in which the node is currently involved. It contains the start time, source node address and destination node address of the active communication, and the number of total active communication sessions.

– Past session information: This means the information about communication sessions in which the node was involved in times past. It contains the start time and session duration of the past communication session, and the number of past sessions recorded. To obtain useful session history, the past session information that is maintained during more than a given period is eliminated.

When a node is involved in a route, the active session information is created and maintained. When a node deletes all the route information due to a route failure or the completion of a communication session, the node updates its session history. The corresponding active session information is cleared, and the start time and duration of the session are registered in past session information.

Session history is quantified to use as a route selection criterion. Table 1 shows an algorithm to quantify session history in the form of pseudo-code. We name the quantified history 'history degree.' If a session is currently in active status, the node will be in relatively low movement. In addition, if a communication session remains active for more than certain duration, it is likely to be more long-lived. Thus, nodes with active sessions will have higher history degree.

2.2   Route Establishment

Like other on-demand ad hoc routing protocols, HAMR executes a route discovery process using route query / reply messages. Source-broadcast query message,

Table 1. Pseudo-code for computing session history

```
/* compute session history */
history_degree = 0;
recentSumOfDuration = 0;
recentNoOfPastSessions = 0;
if (noOfActiveSessions)
     for (each active session)
         if (sessionDuration > DURATION1)
             history_degree = 3;
         else
             history_degree = 2;
for (each past session)
     if ((CURRENT_TIME−session_start) < GIVEN_TIME) {
         recentSumOfDuration += session_duration;
         recentNoOfPastSessions++;
     }
     else
         remove past session information;
if((recentSumOfDuration/recentNoOfPastSessions) > DURATION2)
     history_degree++;
return history_degree;
```

Path Setup (PS) is propagated toward the destination carrying each node's session history information throughout a network for route discovery. Duplicate transmission of the PS packet with the same route identification is not permitted to find disjoint paths. Only nodes that are beyond a certain history threshold are allowed to forward the PS packet to adjacent nodes. This diminishes the volume of traffic generated by query messages similarly as the query localization does [2], and besides, enables the PS packets to be delivered via comparatively stable paths. A PS packet delivered along the shortest path will get the destination first. As soon as the destination receives the first PS packet, it sends toward the source a route reply message, Path Setup Acknowledgement (PSA), which indicates the path to the destination. After that, the destination will defer replying for the establishment of a redundant path until an appropriate time has elapsed or a specific number of PS packets have reached it. Then, the destination will choose another path with the highest session history out of them and reply to the source. We set the number of redundant paths to one. Thus, there can be two fully or partly disjointed multiple paths between source and destination.

The nodes with active sessions have higher possibility to be involved in a route than nodes without active sessions. In HAMR, this may cause concentration of traffic load and congestion. Hence, the number of active sessions that a node can maintain simultaneously is restricted to a certain number. If the number of active sessions gets to the limit, the node cannot participate in a route discovery process. In addition, to prevent a node under the history threshold from being continuously excluded from a route discovery process, the application is suspended for a specific period of time so that the node can be involved in a route. After that, if the node's history degree is still under the threshold, the node is again prohibited from joining in a route discovery process.

Fig. 1 illustrates how the PS packet is flooded through the entire network. Each PA packet gathers intermediate nodes' history information. Each intermediate node does not propagate duplicate PS packets. An italic number under each node indicates history degree. In this example network, the threshold of history degree is four. If a node's history degree is below the threshold, the node cannot take part in flooding PS packets. There are three nodes - N3, N7 and N15, not satisfying the threshold. They discard the received PS packets and do not forward. The numbers within a circle shows the propagation process and order of PS packets. There exist five possible routes. As soon as a PS packet reaches the destination N25 along the shortest route (1), N25 selects the path, N1-N5-N10-N17-N25 and sends a PSA packet back to the source N1. Then, N25 will wait until PS packets along all the possible routes reach to it. After that, N25 will select the route (3), N1-N6-N13-N11-N20-N19-N25 having the highest history degree per hops and generates the second PSA packet. This path has one larger hop count than the second shortest route N1-N5-N4-N16-N22-N25, but higher history degree.

During a route discovery process, two fully disjoint paths were established between source and destination. Data packets are delivered via the main route

N1-N5-N10-N17-N25. When a failure occurs on the main route, the source N1
turns the data traffic to the redundant route N1-N6-N13-N11-N20-N19-N25.

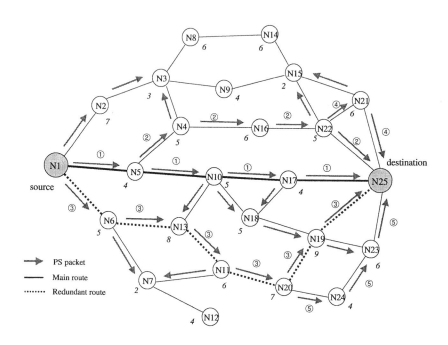

Fig. 1. Example of route establishment

2.3   Route Reconfiguation

If a node is continuously sending packets via a route, it has to make sure that
the route is well established. As soon as an upstream node detects a failure with
a route, it will progress the route reconfiguration procedure. It is assumed that
a link failure is reported from a link layer. If the remote end of a failed link is
identical with the next hop or redundant next hop, a node becomes to recognize
a route failure.

Route reconfiguration is to notify of a failure, to find an alternate route
and to forward data packets by way of the alternate route. Failure notification is
initiated when a node detects a failure on a route and does not have a redundant
path. The failure-detecting node notifies its upstream node of the route failure
using the Path Error (PE) packet. A PE packet is generated by the failure-
detecting node and is unicast to the upstream node. Any node that receives
a PE packet does the following.

- If the node is along a redundant route, it forwards the PE packet to its
  upstream node and destroys its route information.

- If the node is along a main route (shortly, main node) and the PE packet originated from a node on a redundant path, it discards the PE packet and does not forward any longer. It removes its redundant path information.
- If the node is a main node and the PE packet originated from a downstream main node, it looks for a redundant path. If the node has one, it replaces the next hop of its route table with the next hop for the redundant path. The node drops the PE packet.
- If the node is the source node and cannot find a redundant path, it initiates a route re-discovery procedure.
- Otherwise, the node removes all the route information maintained and forwards the PE packet to its upstream node.

3   HAMR Performance

This section presents simulation results for HAMR that show how well it performs. We have compared HAMR with three other routing protocols - history-aware single path routing (HASR), DSR [1, 4], and DSR with multi-path support. HASR uses the same route selection algorithm as HAMR, but establishes only one route. We have modified DSR so that it can support multi-path routing. DSR with multi-path support uses the same routing algorithm as DSR, but establishes two routes - the shortest path as a main route and the second shortest path as a redundant route.

We have used as a simulator CMU's wireless extensions for the Network Simulation 2 (ns-2) [10]. The performance metrics that we are interested in are end-to-end delay, packet delivery ratio, control traffic overhead and hop distance. The control traffic overhead indicates the ratio of routing control bytes transmitted to data bytes delivered.

3.1   Simulation Environment

Two types of simulations were executed for this analysis. Our simulation modeled a network of 50 mobile hosts placed randomly within a 500 meter x 1500 meter area. The nodes use the IEEE 802.11 radio and MAC model provided by the CMU extensions. Radio propagation range for each node was 250 meters. Constant bit rate sources were used to generate traffic data, the size of which payload is 512 bytes. Multiple runs with different seed numbers were conducted for each scenario and collected data was averaged over those runs. Each type of simulation executed for 500 seconds of simulation time. It is assumed that nodes move according to 'random waypoint' model [4] in the simulation.

One type of simulation focused on performance evaluation according to mobility speed variation. Mobility speed varies from 3 m/s to 20 m/s across the range of simulations. Fifteen sessions with randomly selected sources and destinations were simulated. Each source transmits data packets at a maximum rate of 5 packet/sec. Each node moves at a pause time of 20 seconds.

The other type of simulation measured the effect of data traffic load on the performance. We varied the number of sessions from 3 to 20. Each source transmits data packets at a maximum rate of 5 packet/sec. Mobility speed is fixed at 10 m/s across the range of simulations. Each node moves at a pause time of 10 seconds.

3.2   Simulation Results

Performance According to Mobility Speed Fig. 2 shows the end-to-end delay of data packets. Multi-path routing schemes have shorter delays than single path routing schemes because the use of a redundant path at a failure on a main route can make route reconfiguration prompt and thus, cut down queuing delay at the source node. History-aware routing schemes present better performance than non history-aware routing schemes. When mobile nodes are in low movement, they can maintain communication sessions comparatively long. Thus, the effect of history-aware routing becomes clearer at high mobility speeds than at low mobility speeds.

Fig. 3 displays the packet delivery ratio. All the routing schemes become more inefficient as the mobility speed increases. Throughout the simulations, DSR showed the poorest packet delivery ratio.

Fig. 4 presents the control traffic overhead. According to our simulation results, DSR has the largest overhead while HAMR has the smallest overhead. As the mobility speed increases, the difference becomes more obvious.

Fig. 5 shows the average hop distances of four routing schemes. As the mobility speed increases, average hop distances draw long. The simulation result indicates that DSR has the shortest hop distance since it employs only the shortest route established during a route discovery process. Two history-aware routing schemes present similar results, but HASR shows a little better result.

Performance According to the Number of Sessions This subsection describes the performance evaluation as a function of the number of communication sessions. Fig. 6 illustrates the end-to-end delay of data packets. In light traffic

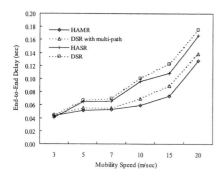

Fig. 2. End-to-end delay

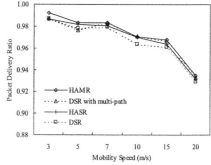

Fig. 3. Packet delivery ratio

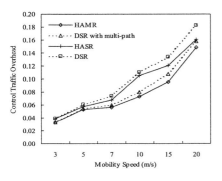

Fig. 4. Control traffic overhead

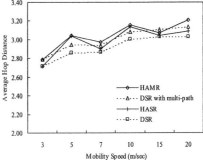

Fig. 5. Average hop distance

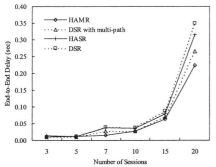

Fig. 6. End-to-end delay

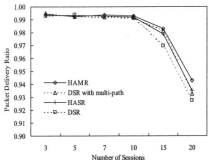

Fig. 7. Packet delivery ratio

load, the difference between four routing schemes is not apparent. When the number of session goes over fifteen, the difference becomes distinct. The simulation result indicates that HAMR outperforms other routing schemes in heavy traffic load.

Fig. 7 presents the packet delivery ratio. In light traffic load, the packet delivery ratios of all the routing schemes are fairly good. The ratios become rapidly inefficient as the traffic load passes over a certain point. In heavy traffic load, HAMR showed better performance.

Fig. 8 presents the control traffic overhead. In low traffic load, the difference between history-aware routing schemes and DSR schemes is not conspicuous. But as the number of sessions goes over ten, history-aware routing schemes display better performance. On the whole HAMR showed smaller overhead than three other routing schemes.

Fig. 9 shows the average hop distances of four routing schemes. As the number of session increases, average hop counts become large. The simulation results indicate similar results as in the previous subsection.

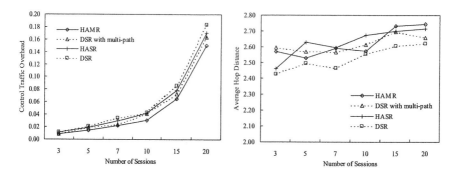

Fig. 8. Control traffic overhead   **Fig. 9.** Average hop distance

4   Conclusions

This paper introduced a new routing metric, session history. History-Aware Multi-path Routing (HAMR) employs session history as one of routing metrics and provides disjoint paths by permitting multiple replies to a route query based on session history. We conducted a performance evaluation of HAMR through a simulation using Network Simulator.

We evaluated HAMR's performance comparing with three different routing schemes - history-aware single path routing, DSR and DSR with multi-path. Two types of simulations were conducted for this analysis. One type of simulation focused on performance evaluation according to mobility speed variation. The other type of simulation measured the effect of data traffic load on the performance. History-aware routing schemes presented better performance for end-to-end delay and control traffic overhead than non history-aware routing. The effect became clearer at high mobility speeds rather than at low mobility speeds, and at heavy data traffic load rather than at light data traffic load. HAMR showed shorter end-to-end delay and lower control traffic overhead for the sake of the use of relatively stable multiple paths. On the whole, HAMR outperformed three other routing schemes.

One of weak points in HAMR is that if nodes in a network don't have enough session history information, history-aware routing may be ineffective. We have to study routing algorithms to compensate for the weakness. In addition, we plan to develop session history computing algorithm and parameter values optimal to various network environments.

References

[1] Broch J., Johnson D. B., Maltz D. A.: The Dynamic Source Routing Protocol for Mobile Ad Hoc Networks. Internet Draft, draft-ietf-manet-dsr-00.txt, (work in progress), (1998)

[2] Castaneda R., Das S. R.: Query Localization Techniques for On-demand Routing Protocols in Ad Hoc Networks. Proceedings of ACM/IEEE International Conference on Mobile Computing and Networking (MOBICOM), Seattle, WA, (1999) 186-194

[3] Internet Engineering Task Force (IETF): Mobile Ad Hoc Networks (MANET) Working Group Charter. http://www.ietf.org/html.charters/manet-charter.html.

[4] Jhonson D., Maltz D. A.: Dynamic source routing in ad hoc wireless networks. in Mobile Computing, Kluwere Academic Publishers, (1996)

[5] Leiner B., Nielson D., Tobagi F.: Issues in Packet Radio Network Design. Proceedings of the IEEE, vol. 75, no. 1. (1987)

[6] Malkin G.: RIP Version 2 - Carrying Additional Information. Internet Draft, draft-ietf-ripv2-protocol-v2-05.txt, (work in progress), (1998)

[7] Moy J.: Link-State Routing, Routing in Communications Networks. edited by M. E. Steenstrup, Prentice Hall, (1995) 135-157

[8] Perkins C., Bhagwat P.: Highly dynamic destination-sequenced vector routing (DSDV) for mobile computers. ACM SIGCOMM, (1994)

[9] Perkins C., Royer E.: Ad Hoc On-Demand Distance Vector Routing. Proceeding of IEEE WMCSA, (1999)

[10] The VINT Project: The UCB/LBNL/VINT Network Simulator-ns (version2). http://www.isi.edu/nsnam/ns.

[11] Toh C-K: Associativity-based routing for ad hoc mobile networks. Wireless Personal Communications, vol. 4, no. 2, (1997) 1-36

A Load-Balancing Approach in Ad-Hoc Networks

Sanghyun Ahn[1,*], Yujin Lim[2,**], and Jongwon Choe[3]

[1] Department of Computer Science and Statistics
University of Seoul, Seoul, Korea
ahn@venus.uos.ac.kr
[2] Computer Science Department
University of California Los Angeles, CA, USA
yujin@ieee.org
[3] Department of Computer Science
Sookmyung Women's University, Seoul, Korea
choejn@cs.sookmyung.ac.kr

Abstract. In the case of link congestion, most of the existing ad hoc routing protocols like AODV and DSR do not try to discover a new route if there is no change in the network topology. Hence, with low mobility, traffic may get concentrated on some specific nodes. Since mobile devices have low battery power and low computing capability, traffic concentration on a specific node is not a desirable phenomenon. Therefore, in this paper, we propose a new routing approach called SLAR (Simple Load-balancing Ad hoc Routing) which resolves the traffic concentration problem by letting each node check its own load situation and give up its role as a packet forwarder gracefully in the case of high traffic load. We compare the performance of SLAR with that of AODV and DSR in terms of the forwarding traffic distribution.

1 Introduction

The ad hoc network is a wireless network composed of mobile nodes with neither fixed communication infrastructure nor fixed base stations. In early days, ad hoc networks have been used for rescue operations and as the communication mechanism among soldiers in battle fields. However, in recent days, the proliferation of mobile communication devices like PDAs and laptop computers has made the ad hoc network one of the needed technologies especially in an infrastructure-less communication environment. Nodes in an ad hoc network can move freely, hence the network topology can change continuously. Also, due to the characteristics of wireless channels like the limited data transmission range, low bandwidth,

---

* Corresponding author. This work was supported by grant No. R04-2001-00054 from the Korea Science & Engineering Foundation.
** This work was supported by the Post-doctoral Fellowship Program of Korea Science & Engineering Foundation.

H.-K. Kahng (Ed.): ICOIN 2003, LNCS 2662, pp. 672–681, 2003.

high error rate, and limited battery power, routing within the ad hoc network is a hard problem to deal with.

Ad hoc routing protocols can be classified into two categories, the table-driven approach [1] and the on-demand approach. The most prominent protocols among the on-demand protocols are AODV [2] and DSR [3]. AODV maintains only one route for a destination, and tries to find a new route, by initiating the route discovery mechanism, only when the route becomes stale. On the other hand, DSR operates based on the source routing scheme, and maintains more than one route for a destination. Therefore, if the current route becomes invalid, DSR selects one of the alternate routes as a new route without triggering the route discovery mechanism.

According to the study in [4], in those two on-demand protocols, the packet delivery ratio increases and the routing overhead decreases as the node mobility decreases. One of the interesting things that we can observe in [4] is that the packet delivery delay increases as the node mobility decreases. The primary reason for this is that, since AODV and DSR do not consider the load balancing, especially with low mobility, traffic may get concentrated on some specific nodes. Traffic concentration may not only cause a high transmission delay, but also force some specific nodes to consume their battery power on forwarding packets not destined to themselves.

Therefore, in this paper, we propose a new routing approach, called Simple Load-balancing Ad hoc Routing (SLAR), which considers the load balancing aspect of the ad hoc network. In our proposed scheme, the concept of load balancing differs from that of the wired network. In the wired network, the main objective of load balancing is to reduce congestion to enhance the overall network performance. However, in the ad hoc network, the most noticeable problem of not considering load balancing is that traffic may get concentrated on some specific nodes under the low mobility situation. This is due to the fact that AODV and DSR do not discover a new route as long as the current route is valid, which is the case with low mobility, and this in turn makes some nodes on valid routes congested. Hence, to overcome this problem, in SLAR, each node determines whether it is under heavy forwarding load condition, and in that case it gives up forwarding packets and let some other nodes take over the role. In an ad hoc network, since nodes have limited CPU, memory and battery power, the message overhead for load balancing is more critical than that of the wired network, i.e., in the ad hoc network, the network-wide optimized load balancing approach of the wired network is inappropriate. Therefore we propose a new load balancing scheme based on the autonomy of each node, which may not be the network-wide optimized solution but may reduce the overhead incurred by load balancing and prevent from severe battery power consumption caused by forwarding packets. Since SLAR is designed not as an entirely new routing protocol but as an enhancement of any existing ad hoc routing protocols like AODV and DSR, in that aspect, SLAR is independent of ad hoc unicast routing protocols. We compare SLAR with AODV and DSR, and show that SLAR outperforms them in terms of traffic distribution.

The paper is organized as follows; in section 2, previously proposed ad hoc routing protocols considering load balancing are described. In section 3, we describe the operation of our newly proposed protocol SLAR. In section 4, the performance of SLAR is shown, and section 5 concludes this paper.

2   Related Work

LBAR [5] considers load balancing in the ad hoc network environment. LBAR tries to find the least loaded route by checking the network traffic condition. To do this, LBAR uses the node activity and the traffic interference as the cost, where the traffic interference means the sum of neighbors' traffic loads.

In LBAR, a source broadcasts a SETUP message and this message collects the traffic load information along its corresponding path. Also the SETUP message records the node IDs which it has passed through. The destination node waits for SETUP messages for a given time, and sends back a confirm message of the least loaded route to the source. All the routes and corresponding cost information collected by SETUP messages are kept by the destination to select an alternate route in case of link disconnection.

LBAR adopts the concept of load balancing used in the wired network, i.e., the network-wide optimized load balancing to resolve network congestion. LBAR incurs an additional overhead to collect neighbors' traffic load information periodically as well as its own load information. Also destination nodes have the burden to maintain all the route information. Lastly, in case of link disconnection, the alternate route selected by the destination node may be stale since the route information maintained at the destination may be out-of-date due to the dynamically changing ad hoc network environment.

3   Simple Load-Balancing Ad-Hoc Routing (SLAR)

In a highly mobile situation, AODV and DSR can achieve the load balancing effect since new shortest routes are found whenever the network topology changes. The worst situation in the aspect of load balancing in an ad hoc network happens when the node mobility is low, because AODV and DSR do not have the mechanism to find a new route in the face of link congestion. This may cause severe congestion, which will eventually degrade the network performance.

To overcome this problem, we propose SLAR in which each node determines whether it is suffering from traffic concentration or not, and, if so, it tries to reduce traffic load by giving up forwarding some packets. To do this, the giving-up node changes its state to GIVE_UP and sends a GIVE_UP message to one of the source nodes to notify it to discover a new route detouring the giving-up node. Within the GIVE_UP message, the source and the list of destinations whose routes pass through the giving-up node are specified, and, when the source receives the GIVE_UP message, it initiates the route discovery mechanism to the destinations specified in the GIVE_UP message. The giving-up node ignores any RREQ messages received while it is in the GIVE_UP state. By doing this, any

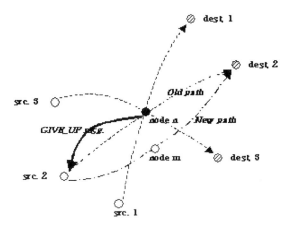

Fig. 1. SLAR operation

new routes passing through the giving-up node will be prevented from being established, so the traffic passing through the giving-up node will be reduced eventually. SLAR can be implemented as additional modules to the existing routing protocols and is independent of any ad hoc routing protocols.

Fig. 1 shows the operation of SLAR when node n faces traffic concentration. Node n is one of the intermediate nodes between node pairs (src1, dest1), (src2, dest2) and (src3, dest3). If the amount of forwarding traffic at node n reaches a specified upper threshold, node n notices that it is in traffic concentration situation and changes its state to GIVE_UP and sends a GIVE_UP message to src2 since the packet sent by src2 to dest2 is the first packet received by n after n enters into the GIVE_UP state. This GIVE_UP message notifies src2 to find a new route to dest2 detouring node n. Upon receiving the GIVE_UP message, src2 initiates the route discovery mechanism by broadcasting a RREQ message to find a new route to dest2. To prevent from the service disruption between src2 and dest2, node n just forwards data packets regardless of their sources and destinations even when it is in the GIVE_UP state (but it ignores any further RREQ messages so that no new route passing through n is established). Once a new route between src2 and dest2 is established, src2 sends data packets to dest2 along the newly established route. One thing to notice here is that node n in the GIVE_UP state does not discard the corresponding routing table entry as long as data packets from src2 to dest2 come, since the timer for the entry will not be expired due to those data packets. As a result, even though a new route is not available, n can continue forwarding data packets since the timer for the entry is still on due to those packets. On the other hand, if a new route is found, the routing table entry is discarded due to the timer expiration since there is no more data packets from src2 to dest2. Node n in the GIVE_UP state ignores all RREQ messages even if they are not from src2, so once n is in the GIVE_UP state it is not involved in any route discovery procedure. Node n can return to

```
[ When a node ⁿ receives a packet ]
     If RREQ packet
          If ⁿ is in GIVE_UP state
               ignore the packet
          else
               process the packet using the underlying routing protocol
     else if GIVE_UP packet
          If the GIVE_UP packet is destined to ⁿ
               initiate the route discovery mechanism
                                        of the underlying routing protocol
          else
               forward the packet to the specified destination
     else  /* if data or other control packet */
          process the packet using the underlying routing protocol

[ At the end of a time interval ]
     If # of forwarding packets for the time interval ≥ upper threshold
          change its state to GIVE_UP
          send a GIVE_UP packet to the src of the first newly received packet
     If # of forwarding packets for the time interval ≤ lower threshold
          change the state to Normal
```

Fig. 2. Description of SLAR operation

the normal state if its traffic load reaches below the specified lower threshold. In Fig. 2, the operation of SLAR is described in pseudo codes.

SLAR achieves load balancing by allowing each node to resolve its own congestion situation so that the limited resource of mobile nodes can be used in a fair way. Irrespective of the degree of node mobility, SLAR works well in terms of traffic distribution as we will see in section 4. Also SLAR can be easily realized only by adding SLAR-specific modules to any existing ad hoc routing protocols.

4   Performance Evaluation

To evaluate the performance of SLAR, we used GloMoSim simulator [6]. GloMoSim is a simulation package for mobile network systems written with the distributed simulation language from PARSEC [7]. The simulation environment consists of 50 mobile nodes in the range of 1000m × 1000m, and the transmission range of each node is set to 250m, and the channel capacity is set to 2Mbps. The moving direction of each mobile node is chosen randomly. And we applied the free space propagation model in which the power of the signal decreases $1/d^2$ for the distance d, and assumed the IEEE 802.11 as the medium access control protocol. AODV and DSR are used as the underlying unicast routing protocols, and the source and destination node pairs are randomly selected. Each source generates two 512-byte packets per second in constant bit rates (CBR). The total simulation time was set to 1000 seconds. The performance evaluation factors that we have considered are the number of sources, the average node mobility speed, the upper threshold (i.e., the number of forwarding packets needed for a node to transit to the GIVE_UP state), the packet delivery ratio (i.e., (the number of successfully received data packets at the destination) / (the number of data packets sent by the source)), and the end-to-end delay.

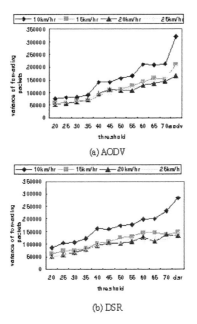

Fig. 3. Variance of the number of forwarding packets with changing node mobility

Fig. 3, 4 and 5 show the performance with changing mobility speed (with 15 sources) (the performance of the original AODV and DSR is shown at the rightmost part of the x axis in each figure). Fig. 3 shows the variance of the number of forwarding packets per node for a unit time interval (10 seconds). We used this criterion to show the degree of forwarding traffic distribution. Here, the threshold on the x axis implies the upper threshold in terms of the number of forwarding packets processed by a node for a unit time interval. As we can see in Fig. 3, SLAR outperforms AODV and DSR significantly. And as the node mobility increases, the variance decreases. This is due to the nature of the ad hoc routing protocol that with high mobility the route discovery procedure occurs more frequently, giving the natural traffic distribution effect. Also we can see that a lower threshold value yields a lower variance, i.e., better traffic distribution.

Fig. 4 shows the packet delivery ratio with changing mobility speed. SLAR works slightly better than AODV (Fig. 4. (a)) and slightly worse than DSR (Fig. 4. (b)). Since DSR maintains more than one route per (src, dest) pair, when a source receives a GIVE_UP message, it just selects one of the alternate routes without discovering a new route detouring the giving-up node, and as a result the possibility of the newly chosen route going through the congested area becomes higher. On the other hand, in AODV, the source receiving a GIVE_UP message tries to find an alternate route detouring the giving-up node, hence the congestion may be released and the packet delivery ratio can be increased.

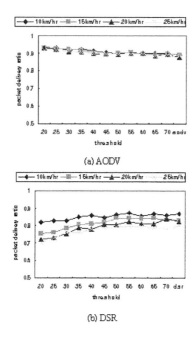

Fig. 4. Packet delivery ratio with changing node mobility

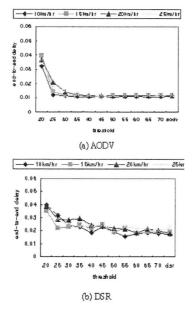

Fig. 5. End-to-end delay with changing node mobility

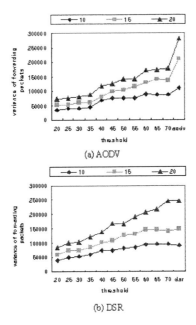

Fig. 6. Variance of the number of forwarding packets with changing the number of sources

Fig. 5 shows the end-to-end delay per packet. With a very low threshold, the route discovery procedure occurs too often (i.e., traffic increases) that the end-to-end delay increases. However, except for very low threshold cases, SLAR performs almost the same as AODV and DSR. Therefore, we can conclude that SLAR outperforms AODV and DSR since it works pretty well in terms of traffic distribution (i.e., fairness) without compensating for other performance criteria like the packet delivery ratio and the end-to-end delay.

Fig. 6, 7 and 8 show the performance comparison with varying the number of sources with the average node mobility speed 15km/hr. As Fig. 6 shows, with the increased number of sources, the degree of traffic distribution gets worse because of the increased traffic, however SLAR outperforms AODV and DSR in most of the cases. Fig. 7 and 8 show the packet delivery ratio and the end-to-end delay, respectively. Except when the threshold is 20 and the number of sources are 20, SLAR works better than AODV. On the other hand, SLAR works worse than DSR. The reason for the worse performance is due to the overhead of processing GIVE_UP messages. Again, from these figures, we can conclude that SLAR is similar to AODV and DSR in the overall packet delivery performance and at the same time gives much better traffic distribution performance so that every mobile node can get some degree of fairness, which is one of the important aspects that we should not overlook in the ad hoc network environment.

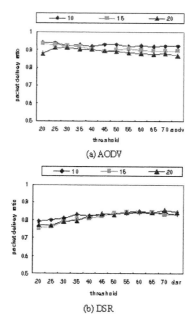

Fig. 7. Packet delivery ratio with changing the number of sources

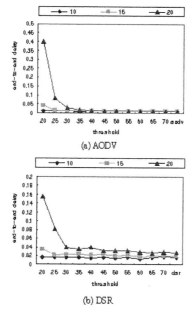

Fig. 8. End-to-end delay with changing the number of sources

5   Conclusion

Ad hoc routing protocols like AODV and DSR discover a new route only when the network topology changes, and this feature makes some specific nodes over-burdened with forwarding packets especially when the network mobility is low. This traffic concentration problem is undesirable since mobile devices have limited resources. To overcome this problem, we proposed a new routing approach, SLAR (Simple Load-balancing Ad hoc Routing), in which each node autonomously determines its traffic condition and asks a chosen source to find an alternate route detouring itself. We showed the performance of SLAR by comparing with AODV and DSR and got the results that SLAR outperforms AODV and DSR significantly in terms of traffic distribution without compensating for the overall packet delivery related performance. The most prominent achievement that SLAR provides is the fairness among mobile devices, since each mobile device in the ad hoc network has almost equal responsibility and right as a source and a destination and, most of all, as a packet forwarder. It will be unfair if a node consumes most of its power forwarding packets not destined to itself.

References

[1] C. E. Perkins and P. Bhagwat, "Highly dynamic destination-sequenced distance-vector routing (DSDV) for mobile computers", Computer and Communication, Oct. 1994, pp234-244.
[2] Charles E. Perkins, Elizabeth M. Royer, and Samir R. Das, "Ad hoc on-demand distance vector (AODV) routing", IETF Internet-draft, Nov. 2001.
[3] David B. Johnson and Davis A. Maltz, "The dynamic source routing protocol for mobile ad hoc networks", IETF Internet-draft, Oct. 1999.
[4] S. R. Das, C. E. Perkins, and E. M. Royer, "Performance comparison of two on-demand routing protocols for ad hoc networks", IEEE INFOCOM, March 2000, pp3-12.
[5] Audrey Zhou and Hossam Hassanein, "Load-balancing wireless ad hoc routing", Proc. Canadian Conference on Electrical and Computer Engineering, vol. 2, 2001, pp1157-1161.
[6] UCLA Computer Science Department Parallel Computing Laboratory and Wireless Adaptive Mobility Laboratory, "GloMoSim: A scalable simulation environment for wireless and wired network systems", http://pcl.cs.ucla.edu/projects/domains/glomosim.html
[7] R. Bagrodia, R. Meyer, M. Takai, Y. Chen, X. Zeng, J. Martin, and H. Y. Song, "PARSEC: A parallel simulation environment for complex systems", IEEE Computer, vol. 31, no. 10, pp77-85, Oct. 1998.

Minimizing Both the Number of Clusters and the Variation of Cluster Sizes for Mobile Ad Hoc Networks*

Pi-Rong Sheu and Chia-Wei Wang

Department of Electrical Engineering
National Yunlin University of Science & Technology
Touliu, Yunlin 640, Taiwan, R.O.C.
sheupr@yuntech.edu.tw

Abstract. The research of mobile ad hoc networks has attracted a lot of attentions recently. In particular, extensive research efforts have been devoted to the design of clustering strategies to divide all nodes into a clustering architecture such that the transmission overhead for the update of routing tables after topological changes can be reduced. The performance of a clustering architecture has been demonstrated to be closely related to the number of its clusters. In this paper, we will propose an efficient clustering algorithm to minimize the number of clusters and the variation of cluster sizes. Computer simulations show that both the number of clusters and the variation of cluster sizes generated by our clustering algorithm are less than those generated by other existing clustering algorithms that also aim to minimize the number of clusters.

1 Introduction

A mobile ad-hoc network (MANET) is formed by a group of mobile hosts (or called mobile nodes) without an infrastructure consisting of a set of fixed base stations. A mobile host in a MANET can generate as well as forward packets. Two mobile hosts in such a network can communicate directly with each other through a single-hop route in the shared wireless media if their positions are close enough. Otherwise, they need a multi-hop route to finish their communications. MANETs are found in applications such as short-term activities, battlefield communications, disaster relief situations, and so on. Undoubtedly, MANETs play a critical role in situations where a wired infrastructure is neither available nor easy to install.

The research of MANETs has attracted a lot of attentions recently. In particular, since host mobility causes frequent unpredictable topological changes, the task of finding and maintaining routes in MANETs is nontrivial. Therefore, extensive research efforts have been devoted to the design of clustering strategies

---

* This work was supported by the National Science Council of the Republic of China under Grant # NSC 91-2218-E-224-007

H.-K. Kahng (Ed.): ICOIN 2003, LNCS 2662, pp. 682–691, 2003.

to divide all nodes into a clustering architecture such that the transmission overhead for the update of routing tables after topological changes can be reduced [5]. In fact, the researches have demonstrated that routing on top of clustered topologies is much more scalable than flat routing [5].

In appearance, a clustering architecture is similar to a single-hop cellular architecture [4]. Fig. 1 shows a clustering architecture for a MANET. There exists a link between two nodes if the two nodes are in the transmission range of each other. The black nodes are the cluster heads of the clustering architecture. Nodes within a circle belong to the same cluster. Each node in a MANET is assigned a unique identifier (ID) that is a positive integer. We assume a cluster's identifier is the same as its cluster head's node identifier. For example, the identifier of the cluster with node 15 as its cluster head is 15. Nodes 3, 8, 13, 15, and 16 all belong to cluster 15. A cluster head in each cluster acts as a coordinator to resolve channel assignment, perform power control, maintain time division frame synchronization, and enhance the spatial reuse of bandwidth. The major characteristics of a clustering architecture are as follows. Firstly, there is only one cluster head in each cluster. Secondly, each node in a clustering architecture is either a cluster head or adjacent to one or more cluster heads. A node belonging to two or more clusters is called a gateway node (or a border node). Thirdly, any two cluster heads are not adjacent to each other. Fourthly, any two nodes in the same cluster are at most two hops away from each other.

The performance of a clustering architecture has been demonstrated to be closely related to the number of its clusters and gateway nodes [3]. This is because the overhead of broadcasting task, where packets initiated at a source is retransmitted by only cluster heads and gateway nodes, can be significantly reduced when the number of clusters and border nodes is decreased. Therefore, in the paper we will address the problem of reducing the number of clusters in a clustering architecture for MANETs.

Since the major power of a MANET is battery power, a node may become inactive because of the exhaustion of its battery power. The inaction of node may lead to communication between two nodes far away becomes broken. Similarly, when a cluster head exhausts its battery power and becomes inactive, the cluster which it belongs to will be broken up. Thus, it becomes a significant work to provide a stable clustering architecture for MANETs. Recall that a cluster head plays a role as a coordinator in its cluster. Hence, a cluster head will consume more battery power than an ordinary node. In fact, it is feasible to assume that the power consumption rate of a cluster head is proportional to its node degree. Thus, reducing the node degree of a cluster head will extend its lifetime and improve its cluster's stability. Therefore, one of the ways to provide a stable clustering architecture is to make the number of nodes in each cluster to be equal as much as possible. This is the second objective in this paper.

To sum up, in this paper, we will propose an efficient clustering algorithm whose objective is to minimize the number of clusters and the variation of cluster sizes. A small variation of cluster sizes implies that the difference between the numbers of nodes in the largest cluster and in the smallest cluster is small.

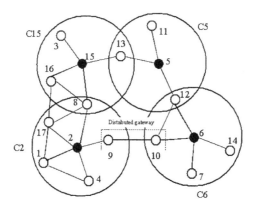

Fig. 1. A clustering architecture

Computer simulations show that both the number of clusters and the variation of cluster sizes generated by our clustering algorithm are less than those generated by other existing clustering algorithms that also aim to minimize the number of clusters.

The rest of this paper is organized as follows. In Section 2, we address backgrounds and related researches. In Section 3, our proposed clustering algorithm is presented. In Section 4, the performance of our clustering algorithm is evaluated by computer simulations. Finally, in Section 5, we make some conclusions.

2   Background and Related Researches

In this section, the background and related researches will be addressed. Generally speaking, a clustering architecture can be classified into two different kinds of types: overlapping and non-overlapping. In an overlapping clustering architecture [3] [4], a node that is not a cluster head may belong to more than one cluster and such a node is named a gateway node. Communication between any two adjacent clusters has to rely on their common gateway nodes. A node belonging to only one cluster is called an ordinary node. For example, in Fig. 1, nodes 8, 12 and 13 are gateway nodes and the other white nodes are ordinary nodes. On the other hand, in a non-overlapping clustering architecture [6], each node belongs to only one cluster and if it is not a cluster head, then it is named an ordinary node.

Now let us observe the clustering architecture shown in Fig. 1. There are no gateway nodes between cluster 2 and cluster 6. To overcome the problem, the concept of distributed gateway (DG) is proposed in [4]. A DG is a pair of ordinary nodes that belong to different clusters but there exists a link between them. For example, the pair of node 9 and node 10 may form a DG. It is shown in [4] that the DG technique can be extremely effective when connectivity is weak. Furthermore, the technique can be applied to both the overlapping and non-overlapping clustering architecture. For simplification, we will adopt the non-overlapping architecture and the DG technique in the following discussion.

In the following, some related clustering algorithms will be reviewed. Especially, we will focus on those clustering algorithms whose objective is to minimize the number of clusters. Two well-known clustering algorithms, the lowest-ID clustering algorithm and the highest-connectivity clustering algorithm, have been proposed in [4]. In the lowest-ID clustering algorithm, each node is assigned a distinct ID. Periodically, each node broadcasts its own ID to its neighbors.

- A node has ID lower than its neighbors is a cluster head.
- The lowest-ID neighbor of a node is its cluster head.
- A node which can hear more than one cluster head is a gateway node.
- Otherwise, a node is an ordinary node.

In the highest-connectivity clustering algorithm, each node is assigned a distinct ID. Periodically, each node broadcasts the list of nodes that it can hear.

- A node that has not elected its cluster head yet is an "uncovered" node; otherwise, it is a "covered" node.
- A node is elected as a cluster head if it is the most highly connected node of its "uncovered" neighbors (if there is a tie, the node with lowest ID prevails).
- A node that has already elected another node as its cluster head gives up its role as a cluster head.

To minimize the number of clusters and gateway nodes in a k-hop clustering architecture, a connectivity based k-hop clustering algorithm has proposed in [3]. In fact, when $k = 1$, the algorithm is similar to the highest-connectivity clustering algorithm above. That is, in both the algorithms, the node degree is adopted as the primary criterion in electing a cluster head while the node identifier is the secondary criterion. In [3], this idea is applied to a k-hop clustering architecture and the node degree of a node is re-defined as the number of the node's k-hop neighbors.

In order to support an efficient k-hop clustering routing, an efficient k-hop clustering (EKC) algorithm for selecting suitable cluster heads has proposed in [6]. The goal of EKC is to form less clusters and more stable clusters. To minimize the number of clusters, EKC tries to avoid as much as possible that nodes whose degrees are one are elected as clusterheads.

3   Our Proposed Clustering Algorithm

3.1   Our Custer Head Election Criteria

Basically, the problem of finding a clustering architecture with the minimum number of clusters is equivalent to the so-called *minimum independent dominating set problem* in graph theory [2]. Two of the main approaches to solve the problem are as follows. One is to try to avoid that nodes whose degrees

are too low are elected as cluster heads [1]. The other is to let nodes with high
node degrees become cluster headers as much as possible. Obviously, the highest-
connectivity clustering algorithm adopts the second approach while the efficient
k-hop clustering algorithm seems to use the first approach. As will be shown
in the following, neither of them can yield a clustering architecture with the
minimum number of clusters in many cases.

The main idea behind our clustering algorithm is to cleverly combine the
above two approaches to form a new cluster head election scheme. To be more
specific, we will use the first approach as our first criterion in electing a cluster
head. Then, if a tie occurs, then the second approach will be involved. The
computer simulations given in Section 4 have demonstrated that our idea is
efficient in reducing the number of clusters and the variation of cluster sizes.

First, let us define a node whose degree is k as a degree k node; e.g. a node
whose degree is one is called a degree-1 node, a node whose degree is two is called
a degree-2 node, and so on. Next, we will define and explain the most important packet, $CRITERION(weighted\_value, nc\_degree, id)$, used in our proposed
clustering algorithm. The packet has three parameters. We use the first parameter, $weight\_value$ to attempt to keep nodes with too low degrees from becoming
cluster heads. Thus, we might keep too many small clusters from being formed.
This will naturally decrease the number of clusters in the whole MANET. The
intention of the second parameter is to reduce the number of clusters by means
of electing nodes with higher degrees as cluster heads. Because many nodes
may have the same $weight\_value$ and nc_degree, ties may happen frequently between nodes if the election of cluster heads only depends on the two parameters.
Therefore, we need to introduce the third parameter, node identifier. When a tie
occurs, the node with the highest identifier will become the cluster head.

The parameter $weighted\_value$ can be calculated as follows. When we attempt
to keep any node with degree $\leq D_{avoid}$ from being elected as a cluster head,
where D_{avoid} is a predefined integer, each $degree$-n node whose $n \leq D_{avoid}$ will
broadcast a DEGREE_$n(id)$ packet, where id is its identifier, to all its neighboring
nodes. Then, each time when a node receives a non-duplicated DEGREE_$n(id)$
packet, the node will add $\frac{D_{avoid}}{n}$ to its $weighted\_value$, which is initially set to zero.
For example, if we try to avoid that any node with degree less than or equal to
3 will be elected as a cluster head, then each of the degree-1 nodes (degree-2
nodes, degree-3 nodes) will broadcast a $DEGREE\_1(id)(DEGREE\_2(id), DE$-
$GREE\_3(id))$ packet to its neighboring nodes. Then, each time when a node
receives a non-duplicated a $DEGREE\_1(id)$ $(DEGREE\_2(id), DEGREE\_3(id))$
packet, the node will add 3 (1.5, 1) to its $weight\_value$ It is not hard to observe
that the performance of our clustering algorithm is related to the value of D_{avoid}.
In Section 4, we will study, by means of computer simulations, the influence of
D_{avoid} on the performance of our clustering algorithm and determine the suitable
values of D_{avoid} for different network sizes.

The parameter nc_degree is defined to be the non-clustered degree of the
sending node, i.e., the number of non-clustered neighboring nodes of the sending
node. Recall that each node can know its own degree through receiving beacons

from its neighboring nodes. If each node is required to append its own cluster identifier to its beacons, then any node can calculate the value of its *nc_degree* parameter since its *nc_degree* is equal to the number of its neighboring nodes with cluster identifier being zero. Finally, the parameter *id* is the sending node's identifier.

In the process of electing a cluster head, we will first elect the node with the largest *weighted_value* as the cluster head. If there is a tie, then the second parameter will be involved in the election. The node with the largest *nc_degree* will become the cluster head. Finally, if there is still a tie, then the node with the highest identifier prevails.

3.2   Our Clustering Algorithm

Before describing our clustering algorithm in detail, we make the following assumptions that are common in designing clustering algorithms for MANETs [4]:

(a) The network topology is static during the execution of the clustering algorithm.
(b) A packet broadcasted by a node can be received correctly by all its one-hop neighbors within a finite time.
(c) Each node has a unique ID and knows its degree (the number of its one-hop neighbors). At the same time, each node knows the ID and degree of its every one-hop neighbor.

In addition to the *CRITERION(weighted_value, nc_degree, id)* packet, which have already been defined in the above, we need to define two new packets used in our clustering algorithm. The *CH(cid, nc_degree)* packet is used by a node to declare itself as a cluster head. Parameter *cid* is a node's cluster identifier and is initially zero. The *JOIN(cid, id)* packet is used by a node to inform the cluster head which it wants to join, where *cid* is the identifier of the cluster which it wants to join and *id* is its own node ID. Besides, each cluster head has a *Member_Table* to record its cluster members. Each node has an *NC_Neighbor_Table* to record its non-clustered neighbors. Each time a beacon packet with *cid* equal to zero is received, a node will update its *NC_Neighbor_Table*.

Our clustering algorithm:

• Each node broadcasts its own a *CRITERION(weighted_value, nc_degree, id)* packet to all its neighboring nodes and receives multiple *CRITERION (weighted_value, nc_degree, id)* packets from its neighbors.
• If a node discover that it has a larger *weighted_value* than all its neighbors, then it sets its *cid* to its own node ID and broadcasts a *CH(cid, nc_degree)* packet to declare itself as a cluster head. If a tie occurs, then the cluster head will be the one with the highest *nc_degree*. If there is still a tie, then the node with the highest ID will be the final cluster head.
• When a node receives multiple *CH(cid, nc_degree)* packets and it dose not belong to any cluster (i.e., its *cid* = 0), it will elect the cluster head with the lowest *nc_degree* (to reduce the variation of cluster sizes) (in case of a tie, the

cluster head with the highest *cid* prevails) and set it own *cid* to the *cid* of the elected cluster head. Then the node broadcasts a *JOIN(cid, id)* packet to join the cluster.

- When a cluster head receives a *JOIN(cid, id)* packet with *cid* equal to its own *cid*, it will record the *id* in the received *JOIN(cid, id)* packet in its *Member_Table*.

- When a non-clustered node receives a *JOIN(cid, id)* packet, it will remove the node with *id* equal to the *id* of the received *JOIN(cid, id)* packet from its *NC_Neighbor_Table*.

- Each time a non-clustered node removes a node from its *NC_Neighbor_Table*, it will check whether its *NC_Neighbor_Table* becomes empty or not. If empty, it sets its *cid* to its own node *id* and broadcasts a *CH(cid, nc_degree)* packet to declare itself as an orphan cluster.

- A node will terminate the clustering algorithm when it has joined or formed a cluster (i.e., its *cid* $\neq$ *0*).

Now, let us use the MANET shown in Fig. 2 as an example to illustrate the operation of our clustering algorithm. We set D_{avoid} to two in this example. Thus, the *weighted_value* of node 4 is three, the *weighted_value* of node 11 is two, the *weighted_value* of node 1, 2, and 7 are one, and those of the other nodes are zero. Since node 4 has the largest *weighted_value* among nodes 3, 5 and 9, it sets its *cid* to 4 and broadcasts a *CH*(4, 3) packet to declare itself as a cluster head. Similarly, node 11 sets its *cid* to 11 and broadcasts a *CH*(11, 4) packet to declare itself as a cluster head.

Because node 1 and node 2 have the same *weighted_value* and *nc_degree*, the third parameter *id* is used to solve the tie. Node 2 is the winner. Node 2 sets its *cid* to 2 and broadcasts a *CH*(2, 4) packet to declare itself as a cluster head. When receiving multiple *CH(cid, nc_degree)* packets, each non-clusterhead node has to select a proper cluster to join. Since cluster head 4 has a lower *nc_degree* than cluster head 11, node 3 sets its *cid* to 4 and broadcasts a *JOIN*(4, 3) packet to join cluster 4. Similarly, node 9 joins cluster 4 rather than cluster 2 for the same reason. On the other hand, although cluster head 11 and cluster head 2 have the same *nc_degree*, node 6 will join cluster 11 because cluster 11 has a higher *cid* than cluster 2. The final clustering architecture constructed by our clustering algorithm is as shown in Fig. 2, where three clusters are formed.

Fig. 3 shows the resulted clustering architecture when we apply the highest-connectivity clustering algorithm in [4] to the MANET in Fig. 2. As expected, this one is also the resulted clustering architecture generated by the connectivity based *k*-hop clustering algorithm in [3] when the parameter *k* is one. We can observe that both of the algorithms yield four clusters. When the MANET given in Fig. 2 executes the efficient *k*-hop clustering algorithm in [6], the produced clustering architecture is shown in Fig. 4, where there are also four clusters and two of them are orphan clusters. In this example, since our clustering algorithm avoids that two degree-2 nodes, node 8 and node 10, become cluster heads, the number of our clusters is less.

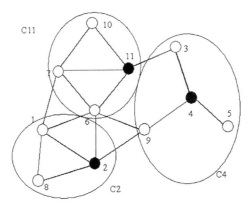

Fig. 2. A clustering architecture established by our clustering algorithm

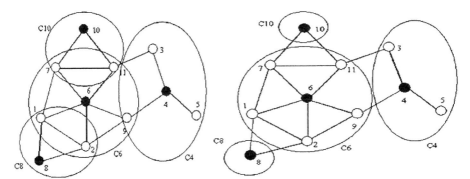

Fig. 3. A clustering architecture established by the highest-connectivity clustering algorithm or the connectivity based k-hop clustering algorithm with k=1

Fig. 4. A clustering architecture established by EKC with k=1

Note that when we set D_{avoid} to one, our clustering algorithm is similar to the efficient k-hop clustering scheme [6] if its parameter k is equal to one. Therefore, our clustering algorithm looks like a general case of the efficient k-hop clustering scheme in a one-hop clustering architecture [6].

4   Computer Simulations

In this section, we will evaluate the performance of our clustering algorithm. Since the performance of our clustering algorithm heavily deponds on the value of D_{avoid}, it is crucial to select a proper value for D_{avoid}. In [7], we have used computer simulations to determine the suitable values of D_{avoid} to be 3, 7, and 10, for three different network sizes, i.e., a 40-node network, an 80-node network, and a 120-node network. Next, we will compare the performance of our clustering algorithm with other existing clustering algorithms in terms of the number of clusters and the variation of cluster sizes. The clustering algorithms we will compare are the lowest-ID clustering algorithm [4], the highest-connectivity clustering algorithm [4], and EKC [6]. Note that the last two algorithms also focus on minimizing the number of clusters.

About the number of clusters, the simulation results are given in Fig. 5. From these curves, it can be observed that the number of clusters generated by our clustering algorithm is about 23.25% less than that generated by the lowest-ID clustering algorithm, about 8.08% less than that generated by the highest-connectivity clustering algorithm, and about 6.64% less than that generated by EKC.

Fig. 6 shows the simulation results on the variation of cluster sizes. We can observe that the clustering architecture established by our clustering algorithm has always a smaller variation of cluster sizes than those established by the other three clustering algorithms under all the three different network sizes.

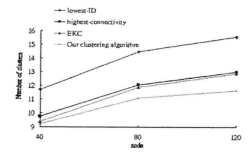

Fig. 3. Comparisons between different clustering algorithms in terms of the number of clusters

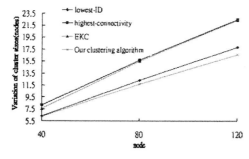

Fig. 4. Comparisons between different clustering algorithms in terms of the variation of cluster sizes

5   Conclusions

The performance of a clustering architecture has been demonstrated to be closely related to the number of its clusters. Furthermore, reducing the node degree of a cluster head will extend its lifetime and improve its cluster's stability. In this paper, we have proposed an efficient clustering algorithm to minimize the number of clusters and the variation of cluster sizes. In our clustering algorithm, we first elect the node with the largest *weighted_value* as the cluster head. If there is a tie, then the second parameter *nc_degree* will be involved in the election. The node with the largest *nc_degree* will become the cluster head. Finally, if there is still a tie, then the node with the highest identifier prevails. The computer simulations have demonstrated that our clustering algorithm is efficient in reducing the number of clusters as well as the variation of cluster sizes.

References

[1] Alimonti, P., Calamoneri, T.: Improved Approximations of Independent Dominating Set in Bounded Degree Graphs. In Proc. of 22nd WG, LNCS 1197. Springer-Verlag (1997) 2-16

[2] Clark, B. N., Colbourn C. J., Johnson, D. S.: Unit Disk Graphs. Discrete Mathematics, Vol. 86. (1990) 165-177

[3] Chen, G., Nocetti, F. G., Gonzalez, J. S., Stojmenovic, I.: Connectivity Based K-Hop Clustering in Wireless Networks. Proc. of the 35th Hawaii International Conference on System Sciences (2002)

[4] Gerla, M., Tsai, J.: Multicluster, Mobile, Multimedia Radio Network. ACM/Baltzer Journal of Wireless Networks, Vol. 1. (1995) 225-238

[5] Iwata, A., Chiang, C. C., Pei, G., Gerla, M., Chen, T. W.: Scalable Routing Strategies for Ad Hoc Wireless Networks. IEEE J. Select. Areas Commun., Vol. 17. (1999) 1369-1379

[6] Su, D. C., Hwang, S. F., Dow, C. R., Wang, Y. W.: An Efficient K-Hop Clustering Routing Scheme for Ad-Hoc Wireless Networks. Journal of the Internet Technology, Vol. 3. (2002) 139-146

[7] Wang, C. W., Sheu, P. R.: A Study on Stable Clustering for Mobile Ad Hoc Networks. Technical Report, National Yunlin University of Science & Technology (2002)

Name Service in IPv6 Mobile Ad-Hoc Network

Jaehoon Jeong, Jungsoo Park, Hyoungjun Kim, and Kishik Park

Protocol Engineering Center, ETRI
161 Gajong-Dong, Yusong-Gu, Daejon 305-350, Korea
{paul,pjs,khj,kipark}@etri.re.kr
http://www.adhoc.6ants.net/

Abstract. In this paper, we propose an architecture of name service system which can provide mobile nodes in IPv6 mobile ad-hoc network with the name-to-address resolution and service discovery. Because mobile ad-hoc network has dynamic topology, the current DNS is not appropriate to name service in mobile ad-hoc network. We suggest the design and implementation of name service system suitable for the mobile ad-hoc network, autoconfiguration technology related to name service, and service discovery based on the name service system.

1 Introduction

Mobile Ad-hoc Network (MANET) is the network where mobile nodes can communicate with one another without communication infrastructure such as base station or access point. Recently, according as the necessity of MANET increases, ad-hoc routing protocols for multi-hop MANET have been being developed by IETF Manet working group [1]. With this trend, IPv6 that has many convenient functions including stateless address autoconfiguration [2] and multicast address allocation [3] has become mature and been being deployed in the whole world. The users in MANET will be able to communicate more easily through the IPv6 zero-configuration that provides easy configuration [3, 6]. Accordingly, if we adopt IPv6 as the network protocol of MANET, we will create a number of useful services for MANET.

DNS is one of the most popular applications in the Internet. It provides the name-to-address resolution among nodes in the Internet. DNS must be a necessity of MANET but the current DNS is inappropriate to MANET that has dynamic topology because the current DNS works on the basis of dedicated and fixed name servers.

In this paper, we propose an architecture of name service system which can provide mobile nodes in IPv6 mobile ad-hoc network with the name-to-address resolution and autoconfiguration technology for easy configuration related to name service including the generation of unique domain name of mobile node and generation of zone file for name service. We also suggest service discovery performed through the name service system of this paper and DNS service resource record (SRV) [5].

H.-K. Kahng (Ed.): ICOIN 2003, LNCS 2662, pp. 692–701, 2003.

This paper is organized as follows; Sect. 2 presents related work. In Sect. 3, we explain the architecture of the suggested name service system, Ad-hoc Name Service System for IPv6 MANET (ANS), operation of ANS System, and name service through ANS. In Sect. 4, we show the result of name resolution through ANS in MANET testbed. Finally, in Sect. 5, we conclude this paper and present future work.

2 Related Work

2.1 Autoconfiguration Technology for Zero-Configuration Networking

IETF Zeroconf working group has defined the technology by which the configuration necessary for networking is performed automatically without manual administration or configuration in the environment such as small office home office (SOHO) networks, airplane networks and home networks, which is called zero-configuration or auto-configuration [6]. The main mechanisms related to the autoconfiguration technology are as follows; (a) IP interface configuration, (b) Name service (e.g., Translation between host name and IP address), (c) IP multicast address allocation, and (d) Service discovery.

2.2 Link-Local Multicast Name Resolution

Link-Local Multicast Name Resolution (LLMNR) has been devised for the resolution between domain name and IP address in the link-local scoped network [4]. The procedure of the resolution from domain name to IPv6 address in a subnet is as follows; Sender is the resolver that sends LLMNR query in link-local multicast and Responder is the name server that sends the LLMNR response to Sender in unicast. When Sender receives the response, it verifies if the response is valid. If the response is valid, Sender stores it in LLMNR cache and passes the response to the application that initiated the DNS query. Otherwise, Sender ignores the response and continues to wait for other responses.

3 Ad-hoc Name Service System for IPv6 MANET (ANS)

ANS is the name service system that provides the name resolution and service discovery in IPv6 MANET which is site-local scoped network. We assume that every network device of mobile node can be configured automatically to have site-local scoped IPv6 unicast address by ad-hoc stateless address autoconfiguration [3]. In this section, we described the architecture of ANS System, operation of ANS system and name service through ANS.

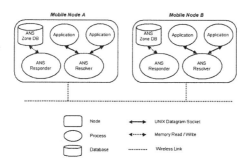

Fig. 1. ANS System for Name Service in IPv6 MANET

3.1   ANS System

ANS System consists of ANS Responder that works as name server in MANET and ANS Resolver that performs the role of DNS resolver for name-to-address translation. Fig. 1 shows the architecture of ANS System for name service in MANET. Each mobile node runs ANS Responder and Resolver. An application over mobile node that needs the name resolution can get the name service through ANS Resolver because ANS provides the applications with name resolution functions through which they can communicate with ANS Resolver for name resolution.

ANS Resolver sends DNS query in the multicast address that ANS Responder in each mobile node has joined for name service. When ANS Responder receives DNS query from ANS Resolver in other mobile nodes, after checking if it is responsible for the query, it decides to respond to the query. When it is responsible for the query, it sends the appropriate response to ANS Resolver in unicast.

3.2   Architecture of ANS System

Architecture of ANS Responder. Fig. 2 shows the architecture of ANS Responder, which is composed of Main-Thread and DUR-Thread.

Main-Thread manages ANS Zone database (DB) for name service and processes DNS queries to send the corresponding response to the querier. It initializes ANS zone file that contains DNS resource records into ANS Zone DB. When it receives a DNS query, it checks if it is responsible for the query. If it is responsible, it sends the response corresponding to the query to ANS Resolver that sent the query.

DUR-Thread performs the dynamic update request (DUR) during the verification of the uniqueness of DNS resource record. The verification is initiated by ANS Resolver on the other node that has receivied multiple responses with the same domain name and resource record type for the DNS query that it sent in multicast. The ANS Resolver sends the first response to every ANS Responder that sent a response except the Responder that sent a response first. Every

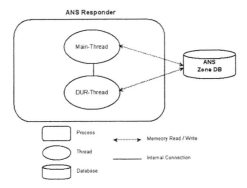

Fig. 2. Architecture of ANS Responder

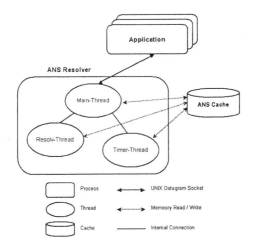

Fig. 3. Architecture of ANS Resolver

ANS Responder that receives a response managed by itself performs the verification of the uniqueness of the resource record related to the response through DUR-Thread. If DUR-Thread detects the duplication of the resource record, it invalidates the record in its ANS Zone DB.

Architecture of ANS Resolver. Fig. 3 shows the architecture of ANS Resolver, which consists of Main-Thread, Resolv-Thread and Timer-Thread.

When Main-Thread receives DNS query from application on the same node through UNIX datagram socket, it first checks if there is the valid response corresponding to the query in ANS Cache. If there is the response, Main-Thread sends the response to the application. Otherwise, it executes Resolv-Thread that will perform name resolution and asks Resolv-Thread to respond to the application through the name resolution.

When Resolv-Thread receives the request of name resolution from Main-Thread, it makes DNS query message and destination multicast address corresponding to the domain name of the query and then sends the message in the multicast address. If Resolv-Thread receives a response message from an ANS Responder, it returns the the result of the response to the application that asked the name resolution through UNIX datagram socket. Whenever a new resource record is received by Resolv-Thread, it caches the response in ANS Cache. When a record is registered in ANS Cache, ANS Cache timer is adjusted for ANS Cache management.

Whenever ANS Cache timer expires, Timer-Thread checks if there are entries that expired in ANS Cache. Timer-Thread invalidates the entries in order to make the resource records of the expired entries unused for name resolution. After the work, Timer-Thread restarts ANS Cache timer.

3.3 Operation of ANS System

Operation of ANS Responder. ANS Responder performs the name service running as daemon process. ANS Responder operates in 9 steps as follows;

1) ANS Responder starts as daemon process.
2) ANS Responder generates a unique domain name per network device. Without the intervention of network manager to manage DNS, mobile node can generate a unique domain name per network device automatically on the basis of network device identifier [8]. We describe the generation of unique domain name in detail in Sect. 3.4.
3) ANS Responder generates zone file for name service. We explain the generation of zone file in detail in Sect. 3.4.
4) ANS Responder loads resource records of zone file into ANS Zone DB.
5) ANS Responder verifies the uniqueness of the resource records related to each domain name through dynamic update request [4, 7]. Fig. 4 shows the dynamic update request during the initialization of ANS Zone DB. Node 1 verifies the uniqueness of its resource records through dynamic update request. DUR query message is sent in multicast of which address is IPv6 site-local solicited name multicast address. The multicast address is generated through the procedure of Fig. 5. In the example of Fig. 4, Node 1 can register only RR-1 (Resource Record 1) in ANS Zone DB as valid resource record because RR-2 has been already managed by Node 2 and can not be managed by Node 1.
6) ANS Responder joins multicast group corresponding to each domain name [4]. The multicast address corresponding to each multicast group is generated through the procedure of Fig. 5. Because the multicast address which is used for sending and receiving the DNS query related to the domain name can be determined with the domain name, we can reduce the number of the packets flooded for DNS query.
7) ANS Responder waits for DNS message.
8) When ANS Responder receives a DNS message, it checks if the message is a DNS query.

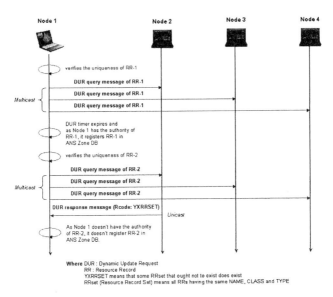

Fig. 4. Dynamic Update Request during the Initialization of ANS Zone Database

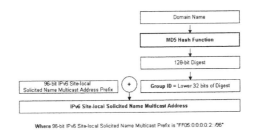

Fig. 5. Generation of IPv6 Site-local Solicited Name Multicast Address

9) If the DNS message is a query, ANS Responder processes the query and sends the response to ANS Resolver that sent the query and goes to Step 7. If the DNS message is a response of which the name belongs to this node, ANS Responder performs the dynamic update request to verify the uniqueness of the resource record related to the name [4]. If ANS Responder finds the name conflict through the dynamic update request, it invalidates the resource record corresponding to the duplicate name in its ANS Zone DB. After performing the dynamic update request, ANS Responder goes to Step 7.

Operation of ANS Resolver. ANS Resolver processes the DNS query as daemon process. ANS Resolver operates in 14 steps as follows;

1) ANS Resolver starts as daemon process.
2) ANS Resolver waits for the request of name resolution from an application on the same node.
3) When ANS Resolver receives the resolution request, it checks first if there is the result of the request in ANS Cache.
4) If there has been already the result of the request in ANS Cache, ANS Resolver passes the result to the application.
5) Unless there is the result of the request in ANS Cache, ANS Resolver creates Resolv-Thread that will perform name resolution and passes the request to Resolv-Thread.
6) Resolv-Thread receives the name resolution request from Main-Thread.
7) Resolv-Thread generates DNS query message.
8) Resolv-Thread makes the IPv6 site-local solicited name multicast address corresponding to the name like Fig. 5.
9) Resolv-Thread sends the query message in multicast and starts query timer.
10) Resolv-Thread waits for the response until query timer expires.
11) When Resolv-Thread wakes up, it checks if the response arrived.
12) If Resolv-Thread woke up because of the expiration of query timer, it checks if the number of retransmission of query message is less than or equal to the allowed maximum number of retransmission, MAX_RETRANS [4]. If the number of retransmission of query message, #Retransmission, is greater than MAX_RETRANS, Resolv-Thread returns the error to the application and exits. Otherwise, it goes to Step 9.
13) If Resolv-Thread woke up because of the receipt of response message, it checks if the response is valid. If the response is invalid, Resolv-Thread discards the response and checks #Retransmission. If #Retransmission is greater than MAX_RETRANS, Resolver returns the error to the application and exits. Otherwise, it goes to Step 9.
14) If Resolv-Thread received multiple valid response messages from other ANS Responders, it sends the first response to every ANS Responder that sent a response except the ANS Responder that has sent a response first. This triggers the dynamic update request at each ANS Responder that receives this response message. After this, Resolv-Thread stores the response in ANS Cache, passes the response to the application and exits. If Resolv-Thread received only one valid response, it stores the response in ANS Cache, passes the response to the application and exits.

3.4 Name Service in ANS

Generation of Unique Domain Name. The mechanism of name generation makes a unique domain name with user-id, device-id (network device's address extended into EUI-64 identifier) and domain like Fig. 6 [8]. user-id is the user identifier selected by user and device-id is EUI-64 identifier derived from the network device's built-in 48-bit IEEE 802 address. domain indicates the kind of network where a node is positioned, which should include "EUI-64" sub-domain which indicates that the domain name is based on EUI-64. We define

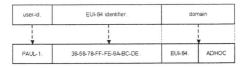

Fig. 6. Format of Unique Domain Name based on EUI-64 Identifier

```
/* Configuration File of ANS Responder (ans.conf) */
  options {
    user-id "PAUL-1";
    domain "ADHOC.";
  };
```

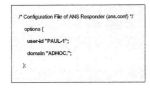

Fig. 7. Configuration File of ANS Responder

the domain for ad-hoc network as EUI-64.ADHOC. For example, when user-id is "PAUL-1", device-id is "36-56-78-FF-FE-9A-BC-DE", and domain is "EUI-64.ADHOC", a unique domain name would be "PAUL-1.36-56-78-FF-FE-9A-BC-DE.EUI-64.ADHOC". The advantage of the above mechanism guarantees that no name conflict happens without the verification procedure of the uniqueness of domain name although users in other nodes use the same user-id. It is possible only when all the nodes in MANET should use the above name generation mechanism, but even though a node configures its domain name by manual configuration or other methods, mobile nodes can detect the name conflict through the dynamic update request. user-id and domain are registered in options statement of the configuration file of ANS Responder (ans.conf) like Fig. 7.

Generation of Zone File. ANS Responder generates zone file that contains the domain name generated above and the site-local scoped IPv6 address of the network device corresponding to the name like Fig. 8. The IPv6 address of the name, "PAUL-1.36-56-78-FF-FE-9A-BC-DE.EUI-64.ADHOC", is "FEC0::3656:78FF:FE9A:BCDE".

Service Discovery through ANS. ANS can provide service discovery in use of DNS SRV resource record which has been defined for specifying the location

```
$TTL  20
$ORIGIN  ADHOC.

PAUL-1.36-56-78-FF-FE-9A-BC-DE.EUI-64  IN  AAAA   FEC0:0:0:0:3656:78FF:FE9A:BCDE
PAUL-1.36-56-78-FF-FE-9A-BC-DE.EUI-64  IN  A      169.254.1.1
PAUL-1                                 IN  CNAME  PAUL-1.36-56-78-FF-FE-9A-BC-DE.EUI-64.ADHOC.
_MULTIMEDIA-1._TCP                     IN  SRV    0  0  4500  PAUL-1.ADHOC.
```

Fig. 8. Zone File of ANS Responder

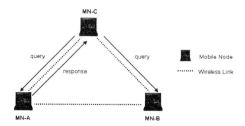

Fig. 9. Topology of MANET Testbed

of services [5]. The service discovery allows a client to get the information which is necessary to access a server. The specification for a service consists of service name, protocol of transport layer, domain which the service is located in, weight, priority, port number and domain name for IPv6 address. Fig. 8 shows an example of SRV resource record of the service named "_MULTIMEDIA-1._TCP".

Security Consideration. We assume the truested nodes generate their own domain name with the name generation mechanism of this paper. Therefore, there can not be the name conflict within the group of the trusted nodes.

In order to provide securer name service in ANS, we can use IPsec ESP with a null-transform to authenticate ANS response, which can be easily accomplished through the configuration of a group pre-shared key for the trusted nodes [4].

4   Experiment in MANET Testbed

4.1   Network Topology

Fig. 9 shows the topology of mobile network for the experiment of ANS name resolution. Three mobile nodes are connected through ad-hoc routing protocols. We implemented IPv6 AODV for unicast routing and MAODV for multicast routing [9, 10, 11, 12]. Table. 1 describes the name information of mobile nodes in the testbed.

4.2   Experiment

The DNS query issued by mobile node MN-C could be resolved as follows; "ans-sender" is a simple program that uses ANS name resolution functions in order to resolve domain name through ANS Resolver. ans-sender on MN-C could

Table 1. Name Information of Mobile Nodes

Node	User ID	EUI-64 ID	Domain	IPv6 Address
MN-A	PAUL-1	36-56-78-FF-FE-9A-BC-DE	EUI-64.ADHOC	FEC0::3656:78FF:FE9A:BCDE
MN-B	PAUL-2	02-01-02-FF-FE-FD-40-05	EUI-64.ADHOC	FEC0::0201:02FF:FEFD:4005
MN-C	PAUL-3	02-02-2D-FF-FE-1B-E8-51	EUI-64.ADHOC	FEC0::0202:2DFF:FE1B:E851

resolve the domain name of MN-A, "PAUL-1.36-56-78-FF-FE-9A-BC-DE.EUI-64.ADHOC", into the IPv6 address, "FEC0::3656:78FF:FE9A:BCDE".

5   Conclusion

In this paper, we proposed an architecture of name service system called as ANS (Ad-hoc Name Service System for IPv6 MANET) which can provide mobile nodes in IPv6 MANET with the name-to-address resolution and autoconfiguration technology for easy configuration related to name service including the generation of unique domain name of mobile node and generation of zone file for name service. We also explained service discovery performed through the name service system of this paper and DNS service resource record (SRV). With ANS, users can run name service easily in the unmanaged or unadministrated networks where there are no network manager and dedicated name server such as home network and small office home office (SOHO) as well as ad-hoc network.

In future work, we will append security service to ANS to prevent the security attacks.

References

[1] IETF Manet working group, http://www.ietf.org/html.charters/manet-charter.html

[2] S. Thompson and T. Narten, "IPv6 Stateless Address Autoconfiguration", RFC2462 , December 1998.

[3] Jaehoon Jeong and Jungsoo Park, "Autoconfiguration Technologies for IPv6 Multicast Service in Mobile Ad-hoc Networks", ICON2002, August 2002.

[4] Levon Esibov and Dave Thaler, "Linklocal Multicast Name Resolution (LLMNR)", (work in progress) draft-ietf-dnsext-mdns-13, November 2002.

[5] A. Gulbrandsen, P. Vixie and L. Esibov, "A DNS RR for specifying the location of services (DNS SRV)", RFC2782, February 2000.

[6] IETF Zeroconf working group, http://www.ietf.org/html.charters/zeroconf-charter.html

[7] P. Vixie, S. Thomson, Y. Rekhter and J. Bound, "Dynamic Updates in the Domain Name System (DNS UPDATE)", RFC2136, April 1997.

[8] Jae-Hoon Jeong, Jung-Soo Park and Hyoung-Jun Kim, "Unique DNS Name Generation", draft-jeong-name-generation-01, February 2003.

[9] Charles E. Perkins, Elizabeth M. Belding-Royer and Samir R. Das, "Ad hoc On-Demand Distance Vector (AODV) Routing", (work in progress) draft-ietf-manet-aodv-13, February 2003.

[10] Elizabeth M. Royer and Charles E. Perkins, "Multicast Ad hoc On-Demand Distance Vector (MAODV) Routing", draft-ietf-manet-maodv-00, July 2000.

[11] C. Perkins, E. Royer and S. Das, "Ad Hoc On Demand Distance Vector (AODV) Routing for IP version 6", draft-perkins-manet-aodv6-01, November 2001.

[12] Implementation of IPv6 AODV and MAODV, http://www.adhoc.6ants.net

CMDR: Conditional Minimum Drain Rate Protocol for Route Selection in Mobile Ad-Hoc Networks[*]

Dongkyun Kim[1], J.J Garcia-Luna-Aceves[2], Katia Obraczka[2],
Juan-Carlos Cano[3], and Pietro Manzoni[3]

[1] Department of Computer Engineering, Kyungpook National University, Korea
dongkyun@knu.ac.kr
[2] Computer Engineering Department, University of California at Santa Cruz, USA
{jj,katia}@cse.ucsc.edu
[3] Department of Computer Engineering, Polytechnic University of Valencia, Spain
{jucano,pmanzoni}@disca.upv.es

Abstract. Untethered nodes in mobile ad-hoc networks strongly depend on the efficient use of their batteries. In the previous work [12], we proposed a new metric, the *drain rate*, to be used to forecast the lifetime of nodes according to current traffic conditions and introduced a mechanism, called the Minimum Drain Rate (MDR) that can be used in any of the existing MANETs routing protocols as a route establishment criterion to achieve a dual goal: to extend both nodal battery life and connection duration. In this paper, we present a modified version, called Conditional MDR (CMDR), which also minimizes the total transmission power consumed per packet besides the above dual goal. Using the ns-2 simulator and the dynamic source routing (DSR) protocol, we compared CMDR against Conditional Max-Min Battery Capacity Routing (CMMBCR) in the literature.

1 Introduction

Mobile ad-hoc networks (MANET) [1] are wireless networks with no fixed infrastructure. A critical issue for MANETs is that nodes activity is power-constrained. Developing routing protocols for MANETs has been an extensive research area during the past few years [2]. However, the majority of the routing proposals have not focused on the power constraints of untethered nodes. Various works on power-aware protocols have appeared recently, see [4, 5, 6, 7, 8]. However, only a few proposals especially focused on the design of routing protocols providing efficient power utilization. The *Minimum Total Transmission Power Routing* (MTPR) scheme [9] tries to just minimize the total transmission power consumption of nodes participating in an acquired route. Since the transmission

---

[*] This work was supported by the Post-doctoral Fellowship Program of Korea Science & Engineering Foundation (KOSEF) and by the *Oficina de Ciencia y Tecnologia de la Generalitat Valenciana*, Spain, under grant CTIDIB/2002/29.

H.-K. Kahng (Ed.): ICOIN 2003, LNCS 2662, pp. 702–712, 2003.

power required is proportional to d^α, where d is the distance between two nodes and α between 2 and 4 [3], MTPR selects the routes with more hops. However, because MTPR fails to consider the remaining power of nodes, it might not succeed in extending the lifetime of each host. S. Singh et al. [10] proposed the *Min-Max Battery Cost Routing* (MMBCR) scheme, which considers the residual battery power capacity of nodes as the metric in order to extend the lifetime of nodes. MMBCR allows the nodes with high residual capacity to participate in the routing process more often than the nodes with low residual capacity. In every possible path, there exists a weakest node which has the minimum residual battery capacity. Hence, MMBCR tries to choose a path whose weakest node has the maximum remaining power among the weakest nodes in other possible routes to the same destination. However, MMBCR does not guarantee that the total transmission power is minimized over a chosen route.

Finally, a hybrid approach was devised by C.K Toh [11] that relies on the residual battery capacity of nodes. The *Conditional Max-Min Battery Capacity Routing* (CMMBCR) mechanism considers both the total transmission energy consumption of routes and the remaining power of nodes. When all nodes in some possible routes have sufficient remaining battery capacity (i.e., above a threshold γ), a route with minimum total transmission power among these routes is chosen(i.e., MTPR applied). However, if all routes have nodes with low battery capacity (i.e., below the threshold), a route including nodes with the lowest battery capacity must be avoided to extend the lifetime of these nodes with MMBCR applied. This scheme does not guarantee that the nodes with high remaining power will survive without power breakage when heavy traffic is passing through the node because CMMBCR also relies on residual battery capacity as in MMBCR. Especially, the performance totally depends on selected γ threshold value.

We consider that it is not possible to efficiently determine γ, either we choose to use a fixed value or a variable value for it. CMMBCR either needs a centralized server to keep track of the energy status of all the mobile nodes or there should be some other mechanisms to proactively exchange nodes remaining power information among all nodes. In fact, if we try to determine γ as an absolute value (e.g., y Joules), there is no easy way to decide the threshold value without considering the current network status, e.g., the network traffic. Especially, when CMMBCR uses an absolute γ threshold, CMMBCR gets the nodes with the residual battery capacity less than the selected γ value to participate in the MMBCR procedure. This causes many nodes to spend their energy due to the usage of longer route. However, if the traffic load is light, it can save more energy consumption even to allow nodes with residual battery capacity less than the selected γ value to participate in the MTPR procedure. Hence, we consider that there is the ambiguity on the selection of the γ value.

In our previous work [12], we proposed a new metric, the *drain rate*, to be used with residual battery capacity to predict the lifetime of nodes according to current traffic conditions. We described the *Minimum Drain Rate* (MDR) mechanism which incorporates the drain rate metric into the routing process.

This mechanism is basically a power-aware route selection algorithm that could be applied to any MANET routing protocol when performing route discovery.

In this paper, since MDR does not guarantee that the total transmission power is minimized over a chosen route, we propose a new version called *Conditional Minimum Drain Rate* (CMDR). With CMDR, we try to solve the non-trivial problem to simultaneously prolong the lifetime of both nodes and connections, and to minimize the total transmission power consumed per packet. We compare the performance of CMDR against CMMBCR. The paper is organized as follows. Section 2 presents the details of the CMDR protocol as well as the basic MDR protocol. Section 3 presents the simulations results against CMMBCR based on ns-2 simulator. Finally, we conclude this paper in Section 4.

2 Description of Proposed Protocol

2.1 MDR: Minimum Drain Rate Protocol

If a node is willing to accept all route requests only because it currently has enough residual battery capacity, much traffic load will be injected through that node. In this sense, the actual drain rate of power consumption of the node will tend to be high, resulting in a sharp reduction of battery power. As a consequence, it could exhaust the node's power supply very quickly, causing the node to halt soon. To mitigate this problem, we proposed the *drain rate* as the metric which measures the *energy dissipation rate* in a given node [12]. Each node n_i monitors its energy consumption caused by the transmission, reception, and overhearing activities and computes the energy drain rate, denoted by DR_i, for every T seconds sampling interval by averaging the amount of energy consumption and estimating the energy dissipation per second during the past T seconds. The actual value is calculated by utilizing the well-known exponential weighted moving average method applied to the drain rate values DR_{old} and DR_{sample} which represent the previous and the newly calculated values like $DR_i = \alpha \times DR_{old} + (1 - \alpha) \times DR_{sample}$.

The ratio $\frac{RBP_i}{DR_i}$, where RBP_i denotes the *residual battery power* at node n_i, reveals when the remaining battery of node n_i is exhausted, i.e., it indicates how long node n_i can keep up with routing operations with current traffic conditions based on the residual energy. The corresponding cost function can be defined as: $C_i = \frac{RBP_i}{DR_i}$. The maximum lifetime of a given path r_p is determined by the minimum value of C_i over the path, that is: $L_p = \min_{\forall n_i \in r_p} C_i$. Finally, the Minimum Drain Rate (MDR) mechanism is based on selecting the route r_M, contained in the set of all possible routes r_* between the source and the destination nodes, that presents the highest maximum lifetime value, that is: $r_M \doteq r_p = \max_{\forall r_i \in r_*} L_i$.

2.2 CMDR: Proposed Conditional Minimum Drain Rate Protocol

MDR still does not guarantee that the total transmission power is minimized over a chosen route, as in MMBCR. We therefore propose a modified version called *Conditional Minimum Drain Rate* (CMDR). The CMDR mechanism is based on choosing a path with minimum total transmission power among all the possible paths constituted by nodes with a lifetime higher than a given threshold, i.e., $\frac{RBP_i}{DR_i} \geq \delta$ as in the MTPR approach. In case no route verifies this condition, CMDR switches to the basic MDR mechanism.

Formally, given r_* as the set of all possible routes between a given source and a destination, and $\bar{r}_* \subset r_*$ a subset where $\forall r_i \in \bar{r}_*, L_i \geq \delta$, if $\bar{r}_* \neq \emptyset$, then the chosen route (r_M) is the one that minimizes the total transmission power with the MTPR protocol applied. Otherwise $r_M \doteq r_p = \max_{\forall r_i \in r_*} L_i$, as in the MDR mechanism. Note that, in order to overcome the ambiguity of γ threshold selection as discussed earlier, we take advantage of a threshold δ, an absolute time value, which takes into account the current traffic condition. This threshold represents how long each node can stand its current traffic with its remaining battery power (RBP) and drain rate (DR) without power breakage. Since the values assigned to δ can influence the performance of the CMDR mechanism, Section 3.2 describes how to properly assign a value to δ.

3   Performance Study

From the results of simulation comparison of MTPR, MMBCR, and MDR, the MDR protocol is the best candidate protocol for achieving the dual goal of nodal and connection lifetimes (refer to our previous work [12]). In this section, we focus on investigation of the different performances of the two conditional versions, namely CMDR and CMMBCR by using the *ns-2* simulator with the CMU wireless extension [13].

We mainly concentrate our study on estimating *nodes' expiration time net*. The *net* expresses how long a node has been active before it cannot work due to lack of battery capacity. The *net* directly affects the lifetime of an active route and possibly of a connection, we therefore also evaluate the *connection's expiration time* (*cet*). For other simulation parameters as well as energy consumption model, we applied the same environment and energy model to following simulations as in our previous work (refer to [12] for details).

Note that we used a fixed transmission range of 250 meters; currently only a few NICs can be configured to use several discrete power levels. Hence, MTPR selects the shortest path among possible routes, thus behaves exactly like the protocol using minimum-hops paths, because the shortest path minimizes the total transmission power consumed per packet.

3.1   Comparison of CMDR and CMMBCR: Using γ as the Threshold

In this section, we compare the CMDR mechanism against the CMMBCR one by using γ as the threshold value. The γ value is used as a boundary to decide when to adopt the conditional or the basic version of the MDR and the MMBCR mechanisms. In other words, instead of using δ, CMDR can be modified to

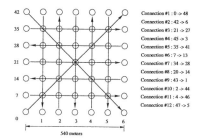

Fig. 1. The dense network scenario: 49 nodes equally distributed over a 540 m x 540 m area

choose a path with minimum total transmission power among all the possible paths constituted by nodes with a residual battery power higher than a given threshold, i.e., $RBP_i \geq \gamma$, as in the MTPR approach. In case no route verifies this condition, CMDR switches to the MDR mechanism. This threshold is expressed as a percentage of the initial battery power of a node. We used three values for γ, respectively 25 %, 50 % and 75 %.

Results with a Dense Network Scenario We first evaluate the various mechanisms in a dense network scenario. The network consists of 49 mobile nodes equally distributed over a 540 m x 540 m area (see Figure 1). We concentrate on two different situations: a completely static environment and a dynamic environment.

As expected, when there exist available routes satisfying the threshold γ, the two protocols apply MTPR to select the best route. Otherwise, MDR is able to show better performance than MMBCR in terms of expiration time of both nodes and connections, regardless of node mobility (see Figure 2 and Figure 3) because of the same reasons pointed out earlier in the performance study section of MDR. In addition, CMDR outperforms CMMBCR in terms of throughput and mean connection's expiration time (see Table I). CMDR obtained increased throughput because it allowed connections to survive longer than CMMBCR did.

Results with a Sparse Network Scenario We also evaluate the performance in a sparse network to see how network density can affect the performance with and without node mobility. The sparse network consists of 50 nodes placed in an area of 1 km x 1 km. Each node is initially placed at a randomly selected position.

Unlike the case of a dense network, a sparse network limits the number of routes available between source and destination nodes, resulting in the fact that the two approaches select similar paths whenever they find them. Therefore, CMDR and CMMBCR have similar performance results, regardless of node mo-

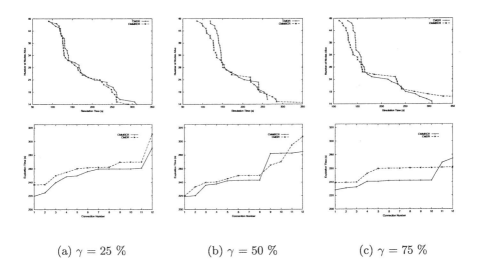

(a) $\gamma = 25$ %   (b) $\gamma = 50$ %   (c) $\gamma = 75$ %

Fig. 2. Static environment scenario, No mobility, Expiration times of nodes and connections

Table 1. Static and Dynamic environment scenarios; E2E is End-to-end including the time spent in queues at all nodes. cet is the connection expiration time. S1 and S2 represent CMMBCR and CMDR, respectively

	No Mobility						Mobility (10 m/s)					
	$\gamma = 25$ %		$\gamma = 50$ %		$\gamma = 75$ %		$\gamma = 25$ %		$\gamma = 50$ %		$\gamma = 75$ %	
	S1	S2	S1	S2	S1	S2	S1	S2	S1	S2	S1	S2
E2E Delay	0.039	0.034	0.039	0.041	0.041	0.038	0.022	0.022	0.024	0.023	0.022	0.022
Hop Count	4.73	4.71	4.77	4.74	4.77	4.75	2.19	2.17	2.20	2.18	2.24	2.19
Throughput	8955	9327	8952	9073	8808	9093	19911	20166	19609	19638	19418	19524
Mean cet	252.43	262.09	251.29	255.41	245.88	254.58	558.43	565.77	550.74	563.10	540.25	547.64

bility. However, CMDR still shows a little better performance than CMMBCR (see Figure 4, Figure 5, and Table II).

3.2   Performance of CMDR according to the Threshold δ

The results from the previous simulations show that CMDR outperforms CMMBCR even when we used the same threshold selection scheme, i.e., remaining battery power despite the ambiguity of the threshold. In this section, we investigate the performance according to absolute time values of δ.

Dense Network Scenario In a dense network environment as shown in Figure 1, regardless of node mobility, CMDR with lower δ values approaches the performance of MTPR, while CMDR with higher δ values approaches the performance of MDR (see Figure 6.a and Figure 6.b). In particular, CMDR with a higher threshold shows better performance in terms of mean expiration time of nodes, but worse performance in terms of mean expiration time of connections, because the different rate of the MDR participation alters the performance. In addition, a lower threshold derives high deviation of mean expiration time of nodes and connections than a higher threshold does. Therefore, the threshold δ can be used as a performance-protection threshold. In other words, if all nodes are equally important and should not be overused, a higher value of δ is expected. In the static network, some connections cannot progress because network partitions easily occur. However, in a dynamic network, the node mobility allows

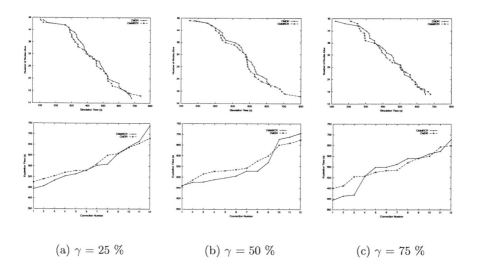

(a) $\gamma = 25\ \%$ (b) $\gamma = 50\ \%$ (c) $\gamma = 75\ \%$

Fig. 3. Dynamic environment scenario, Mobility (10 m/s), Expiration times of nodes and connections

Table 2. Sparse network scenario; E2E is End-to-end including the time spent in queues at all nodes. cet is the connection expiration time. S1 and S2 represent CMMBCR and CMDR, respectively

	No Mobility						Mobility (10 m/s)					
	$\gamma = 25\ \%$		$\gamma = 50\ \%$		$\gamma = 75\ \%$		$\gamma = 25\ \%$		$\gamma = 50\ \%$		$\gamma = 75\ \%$	
	S1	S2	S1	S2	S1	S2	S1	S2	S1	S2	S1	S2
E2E Delay	0.082	0.079	0.079	0.083	0.074	0.080	0.063	0.043	0.047	0.042	0.058	0.042
Hop Count	2.68	2.67	2.66	2.69	2.76	2.71	3.06	2.96	3.08	3.04	3.06	3.00
Throughput	11549	11631	11564	11645	11256	11532	14317	14632	14298	14466	14162	14377
Mean cet	320.54	322.66	321.31	323.56	312.77	319.40	434.50	449.40	446.69	455.36	445.43	472.13

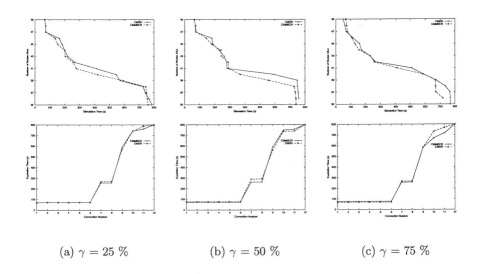

(a) $\gamma = 25$ %   (b) $\gamma = 50$ %   (c) $\gamma = 75$ %

Fig. 4. Sparce network scenario, No Mobility, Expiration time of nodes and connections

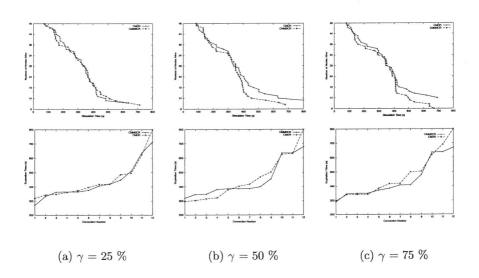

(a) $\gamma = 25$ %   (b) $\gamma = 50$ %   (c) $\gamma = 75$ %

Fig. 5. Sparce network scenario, Mobility (10 m/s), Expiration time of nodes and connections

new paths to appear and network partitions are resolved. Therefore, the lifetime of connections in the dynamic network, when compared to the static network,

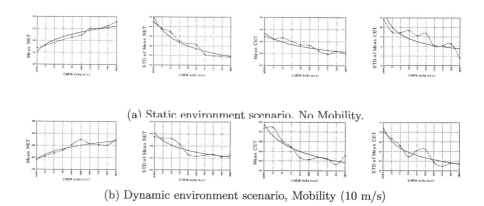

(a) Static environment scenario. No Mobility.

(b) Dynamic environment scenario, Mobility (10 m/s)

Fig. 6. Dense network scenario. CET and NET is connection and node expiration time, respectively. STD is standard deviation. The smoothed solid line shows the trend of data

significantly increases. Moreover, due to the same reason, we used δ values of different scale, specially large values, when we consider node mobility. Since we obtained very similar results with MDR when we simulated the performance with δ greater than 200 seconds and 400 seconds in the static and dynamic networks, respectively, we do not show the results for other values of δ. In addition, because the static network makes nodes participate in forwarding more frequently than the dynamic network does, the lifetime of nodes in the static network is also smaller than that in the dynamic network.

Sparse Network Scenario In a sparse static network, when node mobility is not considered, CMDR exhibits similar performance to the case of a dense network. This is because the network is prone to network partitions and there can exist some groups of nodes that can be dense (see Figure 7.a). However, when we include node mobility, we did not obtain very different results for different values of δ due to the limited number of routes available. Furthermore, different thresholds did not produce highly different values of deviation from the mean expiration time of nodes and connections. Therefore, δ does not play a crucial role in the sparse network with node movement (see Figure 7.b). However, although we obtain similar behavior, CMDR with a higher threshold still shows a little better performance in terms of mean expiration time of nodes, but a little worse performance in terms of mean expiration time of connections.

4   Conclusions

In our previous work, we introduced a mechanism, called the Minimum Drain Rate (MDR) that can be used in any of the existing MANETs routing protocols

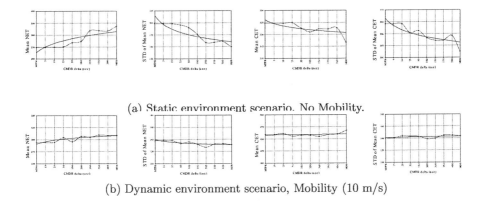

(a) Static environment scenario. No Mobility.

(b) Dynamic environment scenario, Mobility (10 m/s)

Fig. 7.   Sparse network scenario. CET and NET is connection and node expiration time, respectively. STD is standard deviation.The smoothed solid line shows the trend of data

as a route establishment criterion. In this paper, we also presented an MDR modified version, called Conditional MDR (CMDR), which also tries to minimize the total transmission power as in CMMBCR. However, we pointed out that CMMBCR has the ambiguity when selecting its performance protection threshold. With our novel absolute time threshold, our simulation study based on the ns-2 simulator showed that CMDR is better than CMMBCR in terms of performance and threshold selection.

References

[1] IETF "Manet working group charter," http://www.ietf.org/html.charters/manet-charter.html.
[2] E.Royer and C.-K. Toh, "A Review of Current Routing Protocols for Ad Hoc Mobile Wireless Networks," IEEE Personal Communications Magazine, Vol. 6, No. 2, April 1999.
[3] Theodore S. Rappaport, "Wireless Communications: Principles and Practice," Prentice Hall, July 1999.
[4] S-L. Wu, Y-C Tseng and J-P Sheu, "Intelligent Medium Access for Mobile Ad Hoc Networks with Busy Tones and Power Control," IEEE JSAC, Vol. 18, No. 9, September 2000.
[5] J. Gomez, A. T. Campbell, M. Naghshineh and C. Bisdikian, "Conserving Transmission Power in Wireless Ad Hoc Networks", IEEE ICNP 2001.
[6] S. Singh and C. S. Raghavendra, "PAMAS - Power Aware Multi-Access protocol with Signalling for Ad Hoc Networks," ACM SIGCOMM Computer Communication Review, July 1998.
[7] Peng-Jun Wan, Gruia Calinescu, Xiangyang Li and Ophir Frieder, "Minimum-Energy Broadcast Routing in Static Ad Hoc Wireless Networks", IEEE INFOCOM 2001.

[8] Laura Feeney and M. Nilsson, "Investigating the Energy Consumption of a Wireless Network Interface in an Ad Hoc Networking Environment," IEEE INFOCOM 2001.

[9] K. Scott and N. Bambos, "Routing and channel assignment for low power transmission in PCS," IEEE ICUPC 1996.

[10] S. Singh, M. Woo, and C. S. Raghavendra, "Power-aware with Routing in Mobile Ad Hoc Networks," ACM Mobicom 1998.

[11] C.-K. Toh, "Maximum Battery Life Routing to Support Ubiquitous Mobile Computing in Wireless Ad Hoc Networks," IEEE Communications Magazine, June 2001.

[12] D. K. Kim, J. J. Garcia-Luna-Aceves, K. Obraczka, J. C. Cano, and P. Manzoni, "Power-Aware Routing Based on The Energy Drain Rate for Mobile Ad Hoc Networks," IEEE ICCCN 2002.

[13] K. Fall and K. Varadhan, "ns Notes and Documents," The VINT Project. UC Berkeley, LBL, USC/ISI, and Xerox PARC, February 2000.

Medium Access Control Protocol Using State Changeable Directional Antennas in Ad-Hoc Networks

JeongMin Lee[1], YeongHwan Tscha[2], and KyoonHa Lee[1]

[1] Department of Computer Science and Engineering
Inha University at 253 Yonghyun-dong, Nam-gu, Incheon 402-751, Korea
verion@nate.com,khlee@inha.ac.kr
[2] School of Computer, Information and Communication Engineering
Sangji University San 41, Usan-Dong, Wonju, Kang won-Do, Korea
yhtscha@mail.sangji.ac.kr

Abstract. Recently environment of wireless communication is improved very fast. But in a field of Ad hoc, improvement is not so fast as characteristic of in-frastructureless. Even in the Medium Access Control protocol area of Ad hoc network, it is hard to solve traditional problems, hidden terminal and expose node problems. In this paper, we propose new Medium Access Control proto-col for solving these problems using state changeable and directional multi an-tennas.

1 Introduction

Wireless communications have been very pervasive. The numbers of mobile phones and wireless Internet users have increased significantly in recent years. For satisfying user's requirement, wireless technology is improved very fast. Consequently, many methods are proposed and change to the other one.

The conversion cost, time and effort are very high, according as changing method As ad hoc network does not use infrastructure, it has advantage to reduce cost, time and effort. But unlike cellular networks, there is a lack of centralized control and global synchronization in ad hoc networks. Hence TDMA and FDMA schemes are not suitable. In addition, many MAC (Media Access Control) protocols do not deal with host mobility. We need new MAC protocol consequently.

In ad hoc wireless networks, since multiple mobile nodes share the same media, access to the common channel must be made in a distributed fashion, through the presence of a MAC protocol. Given the fact that there are no static nodes, nodes cannot rely on a centralized coordinator. The MAC protocol must contend for access to the channel while at the same time avoiding possible collisions with neighboring nodes. The presence of mobility, hidden terminals, and exposed nodes problems must be accounted for when we come to designing MAC protocols for ad hoc wireless network.

H.-K. Kahng (Ed.): ICOIN 2003, LNCS 2662, pp. 713–722, 2003.
© Springer-Verlag Berlin Heidelberg 2003

2   Related Work

Wireless media can be shared and any nodes can transmit data at any point in time. This could result in possible contention over the common channel. If channel access is probabilistic, then the resultant attainable throughput is low. In ad hoc networks, every node can possibly move, and hence, there is no fixed network node to act as the central controller. Consequently we consider about hidden terminal and exposed node problems.

2.1   Multiple Access with Collision Avoidance (MACA)

MACA aims to create usable, ad hoc and single frequency network using omnibus antenna. MACA proposed to resolve the hidden terminal and exposed node problems.

MACA uses a three-way handshake, Request To Send (RTS)- Clear To Send (CTS)-DATA. The sender first sends an RTS frame to the receiver for reserving the channel. This blocks the sender's neighboring nodes from transmitting.

The receiver then sends CTS frame to the sender for granting transmission. This results in blocking the receivers neighboring nodes from transmitting, thereby avoiding collision. The sender can now proceed with data transmission.

In MACA collisions do occur especially during the RTS-CTS phase. There is no carrier sensing in MACA. Each mobile host basically adds a random amount of time to the minimum interval required to wait after overhearing an RTS or CTS control message. In MACA, the slot time is the duration of an RTS packet. If two or more stations transmit an RTS concurrently, resulting in collision. These stations will wait for a randomly chosen interval and try again, doubling the average interval every attempt. The station that wins the competition will receive CTS from its responder, thereby blocking other stations to allow the data communication session to proceed.

2.2   Dual Busy Tone Multiple Access (DBTMA)

Professor Fouad Tobagi from Stanford University first proposed the use of a busy tone. He proposed Busy Tone Multiple Access (BTMA) to solve the hidden terminal problem. However, BTMA relies on a wireless last-hop network architecture, where a centralized base station serves multiple mobile hosts. When the base station is receiving packets from a specific mobile host, it sends out a busy tone signal to all other nodes within its radio cell. Hence, hidden terminals sense the busy tone and refrain from transmitting.

Zygmunt Haas from Cornell applied this concept further for use in ad hoc wireless networks. In DBTMA (Dual Busy Tone Multiple Access), two out-of-band busy tones are used to notify neighboring nodes of any on-going transmission. In addition, the single shared channel is further split into data and control channels. Data packets are sent over the data channel, while control packets (such as RTS and CTS) are sent over the control channel. Specifically, one busy tone signifies transmit busy, while another signifies receive busy. These two busy tones are spatially separated in frequency to avoid interference.

The principle of operation of DBTMA is relatively simple. An ad hoc node wishing to transmit first sends out an RTS message. When the receiver node receives this message and decides that it is ready and willing to accept the data it sends out receive busy tone message followed by a CTS message. All neighboring nodes that hear receive busy tone are prohibited from transmitting. Upon receiving the CTS message, the source node sends out a transmitting busy tone message to surrounding nodes prior to data transmission. Neighboring nodes that hear transmit busy tone are prohibited from transmitting and will ignore any transmission received. Analytical and simulation work reveal superior performance compared to pure RTS-CTS MAC schemes.

3   Proposed Method

DBTMA protocol was revealed superior performance compared to pure RTS-CTS, MACA and other MAC protocols use omnibus antenna for ad hoc networks. But DBTMA protocol requires additional control channel and it increases complexity and cost. Recently use of adaptive antennas has been considered in ad hoc networks. For example [9] and [10] showed that using directional antennas can adaptable transmit data. But it cannot solve user data frame collision, as a result efficiency get worse. Figure 1 shows example of data collision problem. In this paper, we propose new protocol for solving this problem and traditional problems, hidden terminal and exposed node problems. In additional, node can communicate with other nodes using each antenna at same time.

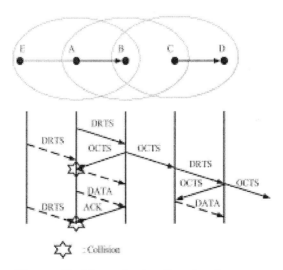

Fig. 1. Problem of exist directional antenna

3.1   Network Model

In this paper, we proposed Ad-Hoc network model on the assumption the following hypotheses.

- Every node uses single channel.
- Each node has m (where m > 1) directional antennas.
- Each node has a transceiver (transmitter + receiver) that operates on half-duplex mode.
- Each node can switch specific antenna which on-state or not and can ignore signals from off-state antenna.
- Each node can broadcast signals via on-state antennas and can send unicast signals through specific antenna.
- In case of signal comes from different antennas in the same node, node that received signal can choose the strongest signal.
- A collision is occurred when some signals arrived at same antenna or many nodes try to connect via same directional antenna. At this time, these nodes try to reconnect according to back-off algorithm (for instance [3], [5]) after delay time.

Besides this, we assume that nodes are always success in transmission except collision case and there are no out of order nodes. [3]

3.2   Basic Idea

The problem of exist protocols using directional antenna is that DATA frame has a collision with signals from other antenna received. [6] After node decides to send or to receive DATA frame with neighbor one, the node can change unusable antennas change to off state. In the other words, nodes have one transceiver, hence node must choose receiving or sending frame at the same time. We use this algorithm for protocol. At first send node changes antennas to off-state, except antenna that uses to send RTS frame. Node that receives RTS frame, changes whole antennas to off-state, except one which antenna receives RTS frame. This idea removes interference by signals from other nodes during transmission DATA frame.

3.3   Protocol

Proposed protocol is similar to other protocols that use CTS-RTS dialog. This protocol consists of three frames, Request To Send (RTS), Clear To Send (CTS) and DATA frame.

1. Request To Send: Indicate send node wish to transmit data to receive node.
2. Clear To Send: Allow DATA transmission to send node.
3. DATA: DATA frame with user information.

DA	SA	Frame Type (CONTROL.)	LEN	DATA/PAYLOAD

Fig. 2. Protocol

RTS and CTS frames are control frame and those use reservation before DATA frame transmission. All frames have Destination Address (DA) field, Source Address (SA) field and Frame Type field commonly. In DATA frame, it has additional field that contains DATA length (LEN) field and user information (DATA/PAYLOAD) field. Exist protocols broadcast CTS frame, but proposed protocol sends by unicast. Proposed protocol consists of sending node, receiving node and passive listener by function. Figure3 shows the operation of these nodes.

As send node have sending DATA frame in initial IDLE state, node sends RTS frame through antenna which have the same direction of receive node. Send node changes other antennas to off state and move to W_CTS at the same time. If collision is occurred, send node uses back-off algorithm. After waiting specific time, node tries to send RTS frame again. As the number of retrying is exceeded,

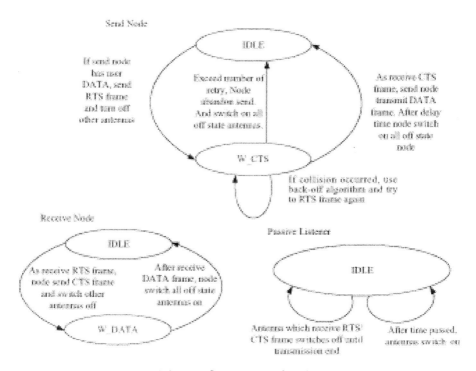

Fig. 3. Operations of nodes

send node assumes that receive node communicate with other node and return to IDLE state. When send node returns to IDLE state, node change all off state antennas to on state.

In case of receiving CTS frame, send node starts to transmit DATA frame to receive node. After sending DATA frame, send node waits during delay time for transmitting DATA frame and returns to IDLE state. After receive node turns all antennas to off state except one which receives RTS frame, receive node sends CTS frame to sends node and changes to W_DATA with waiting DATA frame. As arriving DATA frame, receive node accepts DATA frame and return to IDLE state. And all off state antennas are returned to on state.

As neighbor, passive listener perceives exchanging frames between send node and receive node. Passive listener changes an antenna to off-state as listening RTS or CTS frame between neighborhood nodes. Therefore passive listener has on state antenna and can communicate with other node by this antenna.

3.4 Hidden Terminal and Exposed Node Problems

MAC protocol should concern about two traditional problems, hidden terminal and exposed node problems.

Hidden Terminal Problem Figure 4, node C and E are hidden terminal relation with node D. As they try to send RTS frame to node D at the same time, there are two type of scenario exist.

First, RTS frames are arrived in the same antenna of D. In this case, collision is occurred and node C and E used back-off algorithm. As one of RTS frame is arrived early than another, other antennas are changed to off-state. Hence another RTS frame cannot reach to antenna. Second, RTS frame are arrived different antennas of D. In this case, strongest signal is chose. After arrived RTS frame to D correctly, any collision is not occurred between frames.

Exposed Node Problem Figure 4, node B and C is exposed node relation. Each receive node A and D are exist different cell. And node B and C send RTS frame by unicast, hence node Band C does not make interference with each other. Consequently exposed node problem dose not occur. Node F is passive listener relation with node C and D. Hence antennas that have same direction with node C and D are changed off-state. But except node C and D, node F can communicate with other nodes with other antennas.

4   Performance Evaluation

As RTS frame arrives receive node without collision, proposed protocol can guarantee that successful transmission of DATA frame. Hence possibility of successful transmission of DATA frame, P_s, such as

$$P_s = e^{-(\gamma+\tau)\lambda(\rho\pi R^2 - 1)} \tag{1}$$

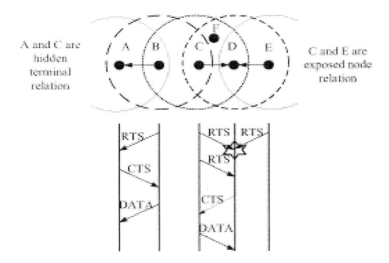

Fig. 4. Example of operation

where
γ : Time for RTS frame.
τ : Delay time.
ρ : Nodal density.
λ : Arrival rate of DATA frame (passion distribution).
R   : Radius of cell.
Busy time between continued IDLE states is represented such as,

$$B = Ts \cdot Ps + T_f \cdot (1 - P_s) \tag{2}$$

$$T_f = \gamma + 2\tau - \frac{\left(1 - e^{-\tau \lambda \rho \pi R^2}\right)}{\lambda \rho \pi R^2} \tag{3}$$

$$T_s = 2\gamma + 3\tau + \delta \tag{4}$$

where
δ : Time for transmitting DATA frame.

Ts means spending time for successful transmission of DATA frame and T_f means waiting time after collision of RTS frame. Hence average utility time U is

$$U = \delta \cdot P_s \tag{5}$$

And average idle time is

$$I = \frac{1}{(\lambda \rho \pi R^2)} \tag{6}$$

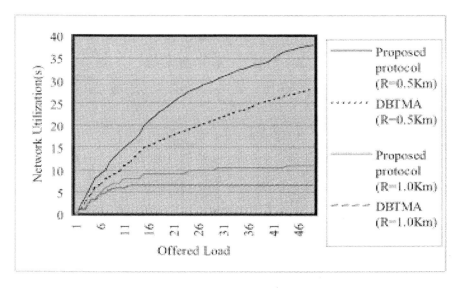

Fig. 5. Utilization of network

And capacity of channel is,

$$S_c = \frac{U}{(B+I)} \tag{7}$$

$$M = \frac{A_{cell}}{2\sqrt{3}R'^2} \tag{8}$$

We assume that cell is hexagon [3], thereby the number of hexagon cell in network, M is [1]

$$R' = \frac{R + (\frac{\sqrt{3}}{2}R)}{2} \tag{9}$$

$$\tag{10}$$

Where
Acell   : Size of cell
So, utilization of whole network that S is

$$S = M\frac{1}{\rho\pi R^2} \cdot \int_0^R 2\pi x\rho S_c dx = M \cdot S_c \tag{11}$$

We compare proposed protocol with DBTMA, which has the best performance in related works using below parameter and it shows Figure 5[3]. We assume that network has 400 nodes in 6x6Km², link-speed is 2.048Kbps, CTS/RTS frame length is 48bits, DATA frame length is 1024bit and R is 0.5km or 1km. We ignore processing time of protocol and assume that unsuccessful frame by collision is deleted in receive node.

In Figure 5, offered load means traffic of whole network. Network utilization shows total number of DATA frames which successfully transmit in each cell, and this is bigger than 1. As Figure 5 shows, proposed protocol has maximum 25% better performance than DBTMA protocol. Because DBTMA is necessary to the time required for avoiding collision amount of 2r, before transmit DATA frame. But proposed protocol needs just time amount of r, hence possibility of occurring collision is decreased. Generally increasing R, make reduction of network utilization about 1/R2 rate. Number of cell in network decrease with same rate as increase R. Radius of cell, R is most important part of efficiency, neither frame length nor link speed. Accord as link speed faster, this situation is more appeared.

5 Conclusion

In this paper, we propose new MAC protocol using directional antenna in ad hoc network. Proposed protocol divides nodes into three groups, send node, receive node and passive listener. Each node adjusts their directional antenna for solving hidden terminal and exposed node problems. Furthermore, even if during neighbor nodes transmit DATA frame, node can communicate with the other nodes. At last proposed protocol proves 25% improvement compare with DBTMA that shows best performance among previous MAC protocols. We will study more about design detail protocol that concerns moving nodes and performance evaluation with simulation, application to routing protocol.

References

[1] L. Kleinrock and F. Tobagi : Packet switching in radio channels: Part I - carrier sense multiple-access modes and their throughput-delay characteristics. IEEE Trans. Commun. vol.23, no.12 (1975) 1417–1433

[2] V. Bharghavan, A. Demers, S. Shenker, and L. Zhang : MACAW: A media access protocol for wireless LAN's. ACM SIGCOMM'94. (1994) 212–225

[3] Z. Hass and J. Deng : Dual busy tone multiple access (DBTMA) - performance evaluation. IEEE VTC'99. (1999) 16–20

[4] C. Fullmer, J. J. Garcia-Luna-Aceves : Solution to hidden terminal problems in wireless networks. ACM SIGCOMM'97. (1997) 39–49

[5] Draft Standard IEEE 802.11 : Wireless LAN medium access control (MAC) and physical layer (PHY) specifications (1997)

[6] Y. Ko, V. Shankarkumar and N. Vaidya : Medium access control protocols using directional antennas in ad hoc networks. IEEE INFOCOM'2000. (2000) 13–21

[7] Y. Ko and N. Vaidya : Location-aided routing (LAR) in mobile ad hoc networks. ACM MOBICOM'98 (1998)

[8] C. Perkins Ed. : ad hoc networking. Addison-Wesley. (2001)

[9] J. Ward and R. Compton : Improving the performance of slotted ALOHA packet radio networks with adaptive arrays. IEEE Trans. Commun. vol.41 (1993) 460–470

[10] T.-S. Yum and K.-W. Hung : Design algorithms for multihop packet radio networks with multiple directional antennas stations. IEEE Trans. Commun. vol.40 no.11 (1992) 1716–1724

Reliable Routing Algorithm
in Mobile Ad-Hoc Networks

Scott S. Guo[1] and Fangchun Yang[2]

[1] School of Information Technology and Engineering, University of Ottawa
Ottawa, Ontario, Canada, K1N 6N5
sguo@site.uottawa.ca
[2] School of Computer Science and Technology
Beijing University of Posts and Telecommunications
100876 Beijing, P.R.China
fcyang@bupt.edu.cn

Abstract. A mobile ad hoc network (MANET) [6] consists of a set of mobile hosts capable of communicating with each other without the assistance of any base stations. In this paper, we propose a novel routing algorithm, Backup Source Routing (BSR), to establish and maintain backup paths based on the concepts of similar path. The BSR algorithm selects a backup path that is piggybacked with the primary path in the header of data packets in order to achieve the most reliable routes between any pair of communicating mobile nodes. We have developed an analytical model and approximation method for this metric, and obtained an analytical expression to evaluate the performance advantage of our routing strategy. Evaluation using numerical analysis of the backup source routing in this framework demonstrates that there are definite advantages to be gained from providing backup routes.

1   Introduction

In an ad hoc network, individual mobile nodes cooperate over wireless channels in order to dynamically establish and maintain routing with each other without any fixed network interaction and centralized administration. Since nodes in the network move freely and randomly, routes often get disconnected. The central challenge in such dynamic wireless networks is that routing protocols must respond to dynamics in the network topology in order to maintain and reconstruct the routes in a timely manner as well as to establish the durable routes. Various routing strategies have been designed to address this problem of choosing the most reliable paths and minimizing the reactive cost to topological changes [1], such as ABR (Associativity Based Routing) [5], Dynamic Source Routing Protocol [3] [4], and MDSR (Multipath Dynamic Source Routing) [2].

In this paper, we propose Backup Source Routing (BSR), a backup route extension for the popular on-demand routing protocol DSR routing algorithm to establish and maintain backup paths based on the concepts of similar path. The BSR algorithm selects a backup path that is piggybacked with the primary

H.-K. Kahng (Ed.): ICOIN 2003, LNCS 2662, pp. 723–733, 2003.

path in the header of data packets in order to achieve the most reliable routes between any pair of communicating mobile nodes. The key advantage is the reduction in the frequency of route discovery flood, which is recognized as a major overhead in on-demand protocols. In order to provide the basis for the backup path selection, we define a new routing metric, the route reliability, and also develop a framework for modelling the time interval between successive route discoveries for on-demand protocols based on simple assumption on the lifetime of a single wireless link, under which an analytical expression is obtained to evaluate the performance advantage of our routing strategy. Evaluation using numerical analysis of the backup source routing in this framework demonstrates that there are definite advantages to be gained from providing backup routes. Based on the analytical framework, we design the algorithms for BSR protocol in Route Discovery phase and Route Maintenance phase.

The rest of this paper is organized as follow. In section 2, we develop a mathematical model from which the reliability metric can be defined to allow the evaluation of BSR. In section 3, we present an approximation method derived from the analytical model. Section 4 discusses in detail the algorithms of Backup Source Routing Protocol that utilizes this metric. Finally, we draw our conclusion in Section 5.

2   Model and Terminology

An ad hoc network can be modelled by a directed graph $G(V, E)$, where the vertices in set V represent the mobile nodes and the edges in set E correspond to the unidirectional communication links. Each edge (u,v) of the graph is assigned a non-negative weight $w(u,v)$ from a cost class W, which represents certain link configurations or state parameters, relevant to routing. A directed path π on graph G can be equivalently defined by a sequence of incident edges or by a sequence of adjacent vertices. Let $V(\pi)$ denote the node set of path π, and let $E(\pi)$ denote the node set of path π. A path is simple if it does not contain a cycle. In this paper, all the paths mentioned are simple paths.

Let π be a directed path. If we sort the node in increasing sequence from the source node to the target node of the directed path π, we define the rank, $r_\pi(u)$, to be the index of the node u in the sequence. Then $\pi_{uv} = (u, \ldots, v)$ is a sub directed path of π if $r_\pi(u) < r_\pi(v)$. If $\pi = (v_1, v_2, \ldots, v_n)$ and $\pi' = (v_n, u_1, u_2, \ldots, u_m), v_i \neq u_j, i = 1, 2, \ldots, n, j = 1, 2, \ldots, m$, then we say $\rho = \pi + \pi' = (v_1, v_2, \ldots, v_n, u_1, u_2, \ldots, u_m)$ is a concatenation of π and π'. The cost $C(\pi)$ of a path $\pi = (u_1, u_2, \ldots, u_n)$ is a sum of the weights of its edges, i.e., $C(\pi) = \sum_{i=1}^{n-1} w(u_i, u_{i+1})$. A primary path from a source node s to a target node t is a directed path from s to t with the minimal total weight.

2.1   Backup Route

Given a directed graph $G(V, E)$, π_1 and π_2 are both directed paths from a source node s to a target node t. The paths π_1 and π_2 are λ-link-similar if $|E(\pi_1) \cap$

$E(\pi_2)|$ equates to λ, and λ is called the link-similarity degree of π_1 and π_2, denoted as $L(\pi_1, \pi_2)$. Similarly, the paths π_1 and π_2 are μ-node-similar path if $|V(\pi_1) \cap V(\pi_2)|$ equates to μ, and μ is called the node-similarity degree of π_1 and π_2, denoted as $N(\pi_1, \pi_2)$. The paths π_1 and π_2 are disjoint paths if π_1 and π_2 are 0-link similar and 2-node similar paths with common source and target nodes. Let L_1 and L_2 be sub-paths of π_1 and π_2 respectively, and if L_1 and L_2 are disjoint paths, L_1 and L_2 are called sub-disjoint paths of π_1 and π_2. The total number of sub-disjoint paths of π_1 and π_2 is denoted as $D(\pi_1, \pi_2)$.

Definition 1. *Let π be the primary path of a directed graph from a source node s to a target node t. Then π' is called the backup path of π, if π and π' have the common source and target nodes, and satisfy the condition, $\forall u, v \in V(\pi) \cap V(\pi'), r_\pi(u) < r_\pi(v) \Leftrightarrow r_{\pi'}(u) < r_{\pi'}(v)$.*

Let L_i denote the common directed links of (π, π'), where $i = 1, 2, \ldots, \lambda$, and $\lambda = L(\pi, \pi')$, and let (P_i, P_i') denote sub-disjoint paths of (π, π'), where P_i and P_i' have the length of n_i and n_i' respectively, $i = 1, 2, \ldots, d$, and $d = D(\pi, \pi')$. The following theorem shows the topology of a backup route.

Theorem 1. *For any backup route (π, π'), $\Gamma = \{L_1, L_2, \ldots, L_l, P_1 \cup P_1', P_2 \cup P_2', \ldots, P_d \cup P_d'\}$ is a partition of $E(\pi) \cup E(\pi')$, where $\lambda = L(\pi, \pi')$ and $d = D(\pi, \pi')$.*

Proof. For any arbitrary backup route (π, π'), $\pi = (s = u_0, u_1, \ldots, u_n, t = u_{n+1})$ and $\pi' = (s = v_0, v_1, \ldots, v_m, t = v_{m+1})$, assume that $u_i, 0 < i \leq n+1$, is the first node searching from s in the sequence π that belongs to π', i.e., $u_i = v_j, 0 < j \leq m+1$. We observe that for any $0 < k < j, v_k \notin \pi$, since if $v_k \in (u_0, u_1, \ldots, u_{i-1})$, it contradicts the assumption that u_i is the first node in both π and π'; if $v_k \in (u_{i+1}, \ldots, u_{n+1})$, we can conclude both inequations $r_\pi(v_j) = r_\pi(u_i) < r_\pi(v_k)$ and $r_{\pi'}(u_k) < r_{\pi'}(v_j)$ are satisfied at the same time, which contradicts the definition of backup path. If $i=j=1$, then $(u_0, u_1) = (v_0, v_1)$ is one of the common links of π and π'; otherwise, $(u_0, \ldots, u_i)$ and $(v_0, \ldots, v_j)$ are one of the sub-disjoint paths of π and π'. Continue the searching process above from node u_i, then we can get the entire common link set and sub-disjoint path set of π and π'. Each of the components is link-disjoint, and the link union of all the components is equal to $E(\pi) \cup E(\pi')$. ☐

2.2   Lifetime of Backup Route

The lifetime for each communication link L in the network can be described using a random variable X_L. Assume that X_L has independent, identical, exponential distribution with a mean value of unit time. It is easy to derive that the lifetime for a path P consisting n serial wireless links is also an exponentially distributed random variable, denoted as X_P, with a mean value of $1/n$. Therefore, the distribution of the backup route lifetime $T_{\pi, \pi'}$ can be determined from the network topology of (π, π'), which is characterized in Theorem 1.

Theorem 2. *The cumulative distributed function of $T_{\pi,\pi'}$ is given by:*

$$F_{T_{\pi,\pi'}}(t) = 1 - e^{-\lambda t} \prod_{j=1}^{d} (e^{-n_j t} + e^{-n'_j t} - e^{-(n_j + n'_j)t}) \tag{1}$$

Proof. Consider all the exponentially distributed random variables of $X_{L_i}, X_{P_j}, X_{P'_j}, i = 1, 2, \ldots, \lambda, j = 1, 2, \ldots, d$, which depict the lifetimes of common directed links, sub-disjointed paths of (π, π'). Their cumulative distributed functions are $F_{X_{L_i}}(t) = 1 - e^{-t}$, $F_{X_{P_j}}(t) = 1 - e^{-n_j t}$ and $F_{X_{P'_j}}(t) = 1 - e^{-n'_j t}$ respectively. Therefore the lifetime of the backup route is a random variable $T_{\pi,\pi'}$ that can be expressed as follow based on the topology of (π, π') described in Theorem 1.

$$T_{\pi,\pi'} = Min(X_{L_1}, \ldots, X_{L_\lambda}, Max(X_{P_1}, X_{P'_1}), \ldots, Max(X_{P_d}, X_{P'_d})) \tag{2}$$

The cumulative distributed function of $T_{\pi,\pi'}$, is abstained as:

$$
\begin{aligned}
F_{T_{\pi,\pi'}}(t) &= P[T_{\pi,\pi'} \le t] \\
&= P[Min(X_{L_1}, \ldots, X_{L_\lambda}, Max(X_{P_1}, X_{P'_1}), \ldots, Max(X_{P_d}, X_{P'_d})) \le t] \\
&= 1 - \prod_{i=1}^{\lambda} P[X_{L_i} > t] \prod_{j=1}^{d} P[Max(X_{P_j}, X_{P'_j}) > t] \\
&= 1 - \prod_{i=1}^{\lambda}(1 - P[X_{L_i} \le t]) \prod_{j=1}^{d}(1 - P[Max(X_{P_j}, X_{P'_j}) \le t]) \\
&= 1 - \prod_{i=1}^{\lambda}(1 - F_{X_{L_i}}(t)) \prod_{j=1}^{d}(1 - F_{X_{P_j}}(t) F_{X_{P'_j}}(t)) \\
&= 1 - \prod_{i=1}^{\lambda}(e^{-t}) \prod_{j=1}^{d}(1 - (1 - e^{-n_j t})(1 - e^{-n'_j t})) \\
&= 1 - e^{-\lambda t} \prod_{j=1}^{d}(e^{-n_j t} + e^{-n'_j t} - e^{-(n_j + n'_j)t}) \qquad \square
\end{aligned}
$$

Once a path is discovered and used, the longer it lasts, the less frequently a rerouting process would have to be initiated. The strategy of using backup route is to find long-lived backup route instead of single primary path in order to reduce the frequency of initiating route discovery, which can critically affect the communicating performance especially in a stressful environment. Thus the mean lifetime of a backup route, defined as the reliability of backup route, is taken as the routing metric, which can be obtained by the equation below, derived directly from Theorem 2.

$$E[T_{\pi,\pi'}] = \int_0^\infty t \, dF_{T_{\pi,\pi'}} \tag{3}$$

2.3   Numerical Analysis

In this section we present some numerical results showing the performance benefit of backup routing using the analysis presented previously. To carry out a practical and efficient analysis, we first defined a numerical analysis profile which is a list defining parameters used in the analysis.

Table 1. Numerical Results of Lifetime of single path and backup route with different lengths of the primary path

$|\pi|$	$E[T_\pi]$	Mean $E[T_{\pi,\pi'}]$	Number of Samples	Lifetime Improvement
2	0.500	0.686	18	37%
3	0.333	0.473	48	42%
4	0.250	0.365	100	46%
5	0.200	0.300	180	50%
6	0.167	0.255	294	53%
7	0.143	0.223	448	56%
8	0.125	0.200	648	59%
9	0.111	0.180	900	61%
10	0.100	0.164	1210	64%

- The lifetime of each link has independent, identical, exponential distribution with mean of unit time.
- The primary path length is $|\pi| \in [2, 10]$.
- The backup path length is $|\pi'| \in [|\pi|, 2|\pi|]$.
- The link-similarity degree of (π, π') is $\lambda \in [0, |\pi| - 1]$.
- The lengths of sub-disjointed paths of (π, π') are $|\pi| = n_i, |\pi'| = n'_i, i = 1, 2, \ldots, D(\pi, \pi')$, and satisfy the following constraints: a) $|\pi| = \lambda + \sum_i^d n_i$, b) $|\pi'| = \lambda + \sum_i^d n'_i$, c) $n_i \leq n'_i$.

Based on the numerical analysis profile, we randomly generate total of 3846 samples of backup routes using different combinations of parameters. The lifetime of each backup route (π, π') can be calculated from Theorem 2. The mean lifetime of single path and backup route over all the samples are presented in Table 1. Note that when using the backup route routing, the performance enhancement relative to single path routing is significant especially in the case more intermediate mobile nodes involved.

3   Approximation of Routing Metric

We seek an approximation here that would provide an efficient computation of route lifetime. This is done by using the concept of cost function. Towards the end, we also validate such approximation that is to be used in our BSR algorithm.

3.1   Heuristic Cost Function

A key observation to elicit the cost function of backup route is that the backup route (π, π') would have a better performance (more reliable), if both π and π'

have shorter lengths, and π' is less link-similar to but having more disjoint sub-paths with its primary path π. Therefore, we propose the following cost function $C(\pi, \pi')$ to estimate the reliability.

Definition 2. *Let π be the primary path. The cost function of the backup route (π, π') is defined as:*

$$C(\pi, \pi') = |\pi| + L(\pi, \pi') + \frac{|\pi'|}{D(\pi, \pi')} \tag{4}$$

Given a primary path π, the routing metric is to select a backup path π' to construct the longest-lived backup route (π, π') such that their mean lifetime $E[T_{\pi,\pi'}]$ can attain its maximum value. Since the value of the cost function $C(\pi, \pi')$ has a inverse relationship with $E[T_{\pi,\pi'}]$, the Most Reliable Backup Route (MRBR) has to be chosen from the backup route with the minimal value of $C(\pi, \pi')$.

3.2   Validity of the Approximation

We made a linear regression to fit a line in a set of two-dimensional points $(E[T_{\pi,\pi'}], 1/C(\pi, \pi'))$ on the data generated from our numerical results by applying the least squares errors method. We determined the quality of the linear relation between $E[T_{\pi,\pi'}]$ and $1/C(\pi, \pi')$ by calculating the product-moment correlation coefficient also called Pearson's correlation. The regression equation can be written as:

$$E[T_{\pi,\pi'}] = \frac{\alpha}{C(\pi, \pi')} + \beta \tag{5}$$

The regression coefficients α and β in the regression equation can be found from the following two equations:

$$\alpha = \frac{\sum_{i=1}^{n} \bar{t}_i/c_i - (\sum_{i=1}^{n} 1/c_i)(\sum_{i=1}^{n} \bar{t}_i)/n}{\sum_{i=1}^{n} (1/c_i)^2 - (\sum_{i=1}^{n} 1/c_i)^2/n} \tag{6}$$

$$\beta = \frac{\sum_{i=1}^{n} \bar{t}_i}{n} - \alpha \frac{\sum_{i=1}^{n} \bar{t}_i/c_i}{n} \tag{7}$$

where $\bar{t}_i$ and c_i are the values calculated from the Equation 3 and Equation 4 respectively based on the ith generated backup route in our numerical analysis profile, and n is the total number of samples. When the values of α and β are found, the regression equation can be written in Equation 5 using these values. The regression line, shown in Fig. 1, is the line of the best fit for the data available. In other words, the error, which is the vertical distance of each of the points from the regression line, is the smallest using this line.

The validity of the approximation is indicated by the correlation coefficient, which varies from -1 to 1 but one usually takes its absolute value (ACC). To qualify for a line fitting, the ACC value should be at least equal to 0.95. An

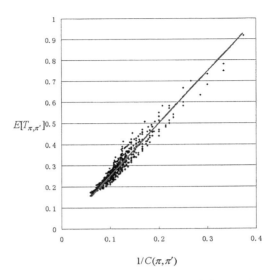

$1/C(\pi, \pi')$

Fig. 1. Verification of the validity of Approximation

ACC value of 1.0 indicates perfect linear correlation, i.e., the data points are exactly on a line. We can consider that we have a very good fit since the ACC value is 0.97 from our numerical results.

3.3   Properties of the Cost Function

Based on the heuristic cost function, we can derive some useful properties, which are the basis of the algorithms of backup source routing.

Optimum Substructure An important property of the cost function is its optimality. If (π, π^*) is the Most Reliable Backup Route (MRBR), any segment in the backup route of (π, π^*) is also be MRBR. This is explained in Theorem 3.

Theorem 3. *Let (π_{ac}, π_{ac}^*) be MRBR from node a to c and node b is a common node of π_{ac} and π_{ac}^*, such that $\pi_{ac} = \pi_{ab} + \pi_{bc}$, $\pi_{ac}^* = \pi_{ab}^* + \pi_{bc}^*$. If $D(\pi_{ab}, \pi_{ab}^*) > 0$ and $D(\pi_{bc}, \pi_{bc}^*) > 0$, then (π_{ab}, π_{ab}^*) and (π_{bc}, π_{bc}^*) are both MRBR from a to b, b to c respectively.*

Proof. Let $l_1 = L(\pi_{ab}, \pi_{ab}^*)$, $d_1 = D(\pi_{ab}, \pi_{ab}^*)$, $x_1 = |\pi_{ab}|$, $y_1 = |\pi_{ab}^*|$, and $l_2 = L(\pi_{bc}, \pi_{bc}^*)$, $d_2 = D(\pi_{bc}, \pi_{bc}^*)$, $x_2 = |\pi_{bc}|$, $y_2 = |\pi_{bc}^*|$. Then we have $L(\pi_{ac}, \pi_{ac}^*) = l_1 + l_2$, $D(\pi_{ac}, \pi_{ac}^*) = d_1 + d_2$, $|\pi_{ac}| = x_1 + x_2$ and $|\pi_{ac}^*| = y_1 + y_2$. We define $\phi_1(l_1, d_1, x_1, y_1)$ and $\phi_2(l_2, d_2, x_2, y_2)$ as $\phi_1(l_1, d_1, x_1, y_1) = C(\pi_{ab}, \pi_{ab}^*) = x_1 + l_1 + y_1/d_1$, and $\phi_2(l_2, d_2, x_2, y_2) = C(\pi_{ac}, \pi_{ac}^*) = x_1 + x_2 + l_1 + l_2 + (y_1 + y_2)/(d_1 + d_2)$. Therefore, $d\phi_2/d\phi_1$ can be obtained as:

$$\frac{d\phi_2}{d\phi_1} = \frac{\partial\phi_2}{\partial\lambda_1}\frac{\partial\lambda_1}{\partial\phi_1} + \frac{\partial\phi_2}{\partial d_1}\frac{\partial d_1}{\partial\phi_1} + \frac{\partial\phi_2}{\partial x_1}\frac{\partial x_1}{\partial\phi_1} + \frac{\partial\phi_2}{\partial y_1}\frac{\partial y_1}{\partial\phi_1}$$

$$= 1 + \frac{d_1}{d_1 + d_2} + 1 + \frac{(y_1 + y_2)d_1^2}{y_1(d^1 + d^2)^2} > 0$$

If there exists $\phi_1(l_1', d_1', x_1', y_1') < \phi_1(l_1, d_1, x_1, y_1)$, we can conclude from $d\phi_2/d\phi_1 > 0$ that $\phi_2(l_1', d_1', x_1', y_1') < \phi_2(l_1, d_1, x_1, y_1)$, which contradicts with the assumption that the backup route (π_{ac}, π_{ac}^*) are MRBR. Therefore, if (π_{ac}, π_{ac}^*) are MRBR, (π_{ab}, π_{ab}^*) are MRBR. In the same way, we can get (π_{bc}, π_{bc}^*) are also MRBR.   □

Corollary 1. *If the most reliable backup route (π, π^*) has been already equipped for a node pair (v_1, v_μ) and all the common nodes of path π and π^* are $(v_1, \ldots, v_\mu)$, where μ is the node-similarity degree of (π, π^*), then the MRBR for each (v_i, v_j) can be equipped by $(\pi_{v_i v_j}, \pi_{v_i v_j}^*)$, $i, j \in 1, 2, \ldots, \mu$.*

Convergence The linear regression illustrates the inverse relationship between the mean interval of route discoveries $E[T_{\pi, \pi'}]$ using backup route (π, π'), and the heuristic cost function $C(\pi, \pi')$ of the backup route. From the observation, when the value of $C(\pi, \pi')$ is not too large, reducing the value of $C(\pi, \pi')$ will achieve a significantly enhanced reliability of backup route. However when reaching some thresholds, more than triple of the length of the primary path, i.e., $3|\pi|$, these advantages are quite minor. Mathematically, the observation can be expressed by Equation 8 .

$$\lim_{C(\pi,\pi')>C(\pi,\pi'')>3|\pi|} \left(\frac{E[T_{\pi,\pi'}] - E[T_{\pi,\pi''}]}{C(\pi,\pi') - C(\pi,\pi'')}\right) \to 0 \tag{8}$$

In other words, traffic will still be delivered through the backup route (π, π'), even though a new better backup route (π, π'') was found, so long as $(\pi, \pi'') < (\pi, \pi')$ and (π, π'') is greater than $3|\pi|$. This property prevents the network from oscillating by converting a portion of traffic from one backup route to the other too frequently once a new backup route with less value of the cost function was found in a gratuitous mode. At the same time, it guarantees the lifetime of the backup route in use is still close to the lifetime of the MRBR. Therefore one may refine our backup route cost function in Equation 9.

$$C(\pi, \pi') = Max(3|\pi|, |\pi| + L(\pi, \pi') + \frac{|\pi'|}{D(\pi, \pi')}) \tag{9}$$

4   Backup Source Routing Algorithm

BSR establishes and utilizes a backup route to help minimizing route recovery process and control message overhead. A backup route consists of the primary path (the shortest delay path) and a backup path. Like DSR, the set of routes are discovered on demand in BSR. BSR consists of two phases: (a) Route Discovery and (b) Route Maintenance. Route Discovery is only invoked when needed, and Route Maintenance operates only when the route is used actively to send individual packets. The details are described in the following sections.

4.1   Route Discovery

BSR is an on-demand routing protocol that constructs backup route using request-reply cycles. When the source needs a route to the destination but no routes information is known, it floods the ROUTE REQUEST (RREQ) messages to entire network. Several duplicates that have traversed through different routes would reach each intermediate node, which then select the primary path and the backup path in their routing cache. Finally, the destination node constructs its backup route and sends ROUTE REPLY (RREP) messages back to the source via the chosen routes.

RREQ Propagation The main goal of BSR in the route discovery phase is to equip each intermediate node with the most reliable backup route. To achieve this goal in an on-demand routing scheme, we have to modify the forwarding mechanism used in the DSR protocol, because in the route discovery phase of DSR, a large mount of duplicate packets are dropped, and thus makes difficult establishing backup route for each intermediate node in the network. In order to avoid this problem, we introduce a different packet forwarding approach. When a source has a data packet to send but have no route information to the destination, it transmits a RREQ message. Instead of dropping every duplicate RREQs, intermediate nodes forward packets as a local broadcast by appending its own address to the source route in the RREQ only if all of the following conditions are met: (1) the node is not the target (destination) node of the RREQ message; (2) the node is not listed in the source route; (3) time-to-live is great than zero; (4) the path in the duplicate packet can produce a new backup route with lower value of cost function. The main idea is to forward duplicate packets at each node i in a way that the value of the backup route cost function over all nodes will decrease so that the resulting backup route equipped for each node will be eventually close to the optimality, which satisfies the necessary MRBR condition stated in Theorem 3.

Route Selection Method In our algorithm, each intermediate node i selects two paths, which construct a backup route from the source node. One of which is the shortest delay path, the path taken by the first RREQ that reaches a node. We use the shortest delay path to minimize the route acquisition latency required by on-demand routing protocols. When receiving this first RREQ, the intermediate node records the entire path as its primary path. From all the subsequent duplicate packets (with the same request ID) all received, each intermediate node selects the path with the minimal value of heuristic cost function defined in Equation 9, and records it as its backup path. When the target node receives a RREQ, it should delay a short period to receive more subsequent duplicate packets to construct backup route and then returns a RREP message to the source, giving a copy of the accumulated route record as approximate MRBR from the RREQ.

4.2   Route Maintenance

A link of a route can be disconnected because of mobility, congestion, packet collision or a combination of the above. It is important to recover broken routes immediately to maintain effective routing. In BSR, when a node fails to deliver the data packets to next hop using its primary path, it tries to use its backup path piggybacked in its header to deliver the data to the destination. If a node fails to deliver the data packets using both its primary path and backup path, it simply discards the data packets. At the same time, the node sends a ROUTE ERROR (RERR) message to the upstream direction of the route that is still using the reverse backup route in the packets. The idea is to reduce the probability of losing RERR messages. The RERR package contains the backup route to the source node, and the immediate upstream and downstream nodes of the broken link. When back warding these RERR messages, the intermediate nodes remove the broken link in their cache. Upon receiving these RERR messages, the source node reconstructs the backup route to the target using the updated routing information. If there is not enough information, it initiates a new routes recovery process.

5   Conclusion

We have proposed the BSR algorithm, which explores the most reliable backup routes that can be useful in case a primary path breakage. The key advantage is the reduction in the flooding frequency of route discovery, which is recognized as a major overhead in all on-demand protocols. In order to use the idea practically, we present an approximation method to measure the metric by heuristic cost function, which is proved effective using the numerical analysis. Based on the cost function and its properties, we design the algorithms for BSR protocol in Route Discovery phase and Route Maintenance phase.

References

[1] Elizabeth M. Royer and Chai-Keong Toh, A Review of Current Routing Protocols for Ad Hoc Mobile Wireless Networks, IEEE Personal Communications, April, 1999, pp.46-55.
[2] Asir Nasipuri, Robert Castaneda, Samir R. Dan, Performance of Multipath Routing for On-Demand Protocols in Mobile Ad Hoc Networks, Mobile Network and Applications, 2000.

[3] D. B. Johnson and D. A. Maltz, Dynamic Source Routing in Ad hoc Wireless Networks, in Mobile Computing, chapter 5, 1996.

[4] D. B. Johnson, Davis A. Maltz, The Dynamic Source Routing Protocol for Mobile Ad Hoc Networks, IETF Draft, `http://www.ietf.org/internet-drafts/draft-ietf-manet-dsr-03.txt`.

[5] C. K. Toh, Long-lived Ad-Hoc Routing based on the concept of Associativity, March 1999 IETF Draft, `http://www.ietf.org/internet-drafts/draft-ietf-manet-log-lived-adhoc-routing-00.txt`.

[6] The Internet Engineering Task Force. Web site at `http://www.ietf.org/`.

Part V

Network Security

An Efficient Hybrid Cryptosystem Providing Authentication for Sender'S Identity[*]

Soohyun Oh, Jin Kwak, and Dongho Won

School of Information and Communications Engineering
Sungkyunkwan University, Korea
{shoh,jkwak,dhwon}@dosan.skku.ac.kr

Abstract. To provide the confidentiality of messages transmitted over the network, the use of cryptographic system is increasing gradually and the hybrid cryptosystem is widely used. In this paper, we propose a new hybrid cryptosystem capable of providing implicit authentication for the sender's identity by means of the 1-pass key agreement protocol that offers mutual implicit key authentication, the hash function, pseudo random number generator and the symmetric cryptosystem. Also, we describe some examples such as the *Diffie-Hellman* based system and the *Nyberg-Rueppel* based system. The proposed hybrid cryptosystem is more efficient than general public key cryptosystems in the aspect of computation work and provides implicit authentication for the sender without additional increase of the communication overhead.

1 Introduction

With the rapid development of network and communication technologies, network based message transfers including e-mail are widely used. Message transfers over the network are economical in the aspect of time and cost, even though there are more vulnerable to eavesdropping, forgery and alteration by other people than document transmission in real world. For this reason, cryptosystems are more in use to protect outgoing messages against the third party.

Cryptosystems are divided into the symmetric cryptosystem and the public key cryptosystem depending on the key used. The symmetric cryptosystem is efficient but it is disadvantageous in that the key distribution process needs to be completed. On the other hand, the public key cryptosystem is advantageous in that the key distribution process is not necessary because the key used for encryption can be made public by placing it in a public directory. However, it is less efficient than the symmetric cryptosystem since it takes much time for encryption/decryption.

Recently, a hybrid cryptosystem which takes advantages of both systems is widely used. That is, the efficient symmetric cryptosystem is used for the actual message encryption and the session key of the symmetric cryptosystem is

---

[*] Research supported by the KISA(Korea Information Security Agency) under project 2001-S-092.

H.-K. Kahng (Ed.): ICOIN 2003, LNCS 2662, pp. 737–746, 2003.

encrypted with the public key cryptosystem. In addition, as the public key cryptosystem allows anyone to use an open encryption key to generate a ciphertext, it was impossible to authenticate the sender of a message, nor ensure the integrity of the contents of the received message.

The goal of this paper is to propose a new hybrid cryptosystem, which uses the 1-pass key agreement protocol and provides the implicit authentication for the sender's identity. The proposed hybrid cryptosystem is more efficient than general public key cryptosystems in the aspect of computation work and provides implicit authentication for the sender without additional increase of the communication overhead since 1-pass key agreement protocol with mutual implicit key authentication is used.

2   Related Works

Since 1976, the year when the concept of the public key cryptosystem was announced in [4], several public key cryptosystems such as RSA[16] and ElGamal[5] have been proposed. As the public key cryptosystem does not require the additional key distribution process it is appropriate to be applied to such open networks as Internet. However, it requires a large volume of computation in encryption/decryption and the security of the system has not been proven. To resolve these problems of the public key cryptosystem, a hybrid cryptosystem is widely used. In addition, a recent trend for the public key cryptosystem is to develop the provably secure public key cryptosystem. The research for the methods to transform the existing public key cryptosystem into the public key cryptosystem, which satisfies the *chosen ciphertext security(CCS)*, is actively underway. These methods are proposed in [7, 8, 12, 13, 14, 15] etc.

Most of the public key cryptosystems, which satisfy *CCS* proposed so far, are constructed by public key cryptosystem, symmetric cryptosystem, pseudo random number generator and hash function. There are hybrid cryptosystems, which use the public key cryptosystem only for encrypting the message encryption key and use the symmetric cryptosystem to encrypt the actual messages. Therefore, these methods of transformation resolve the problems of the existing public key cryptosystem in the aspect of efficiency and security. Among the public key cryptosystems satisfy the *CCS*, the encryption and decryption processes for the recently proposed *REACT(Rapid Enhanced-security Asymmetric Cryptosystem Transform)* are as follows [12].

First, the sender A selects a random number R and generates a ciphertext $C_1 = E_{pk}^{asym}(R)$ using the asymmetric cryptosystem $E^{asym}()$. The sender A computes a session key $K = G(R)$ and generates a ciphertext $C_2 = E_K^{sym}(m)$ using the symmetric cryptosystem $E^{sym}()$. Finally, to generate verification information, the sender A computes a hash value $C_3 = H(C_1, C_2, R, m)$ and transmits ciphertext $C = (C_1, C_2, C_3)$ to receiver B. To decrypt the message m, first, the receiver B decrypts C_1 using his/her secret key and obtains the random number $R = D_{sk}^{asym}(C_1)$. And, the receiver B computes a session key $K = G(R)$ and decrypts $m' = D_K^{sym}(C_2)$. Finally, the receiver B checks $C_3 = H(C_1, C_2, R, m')$

to verify whether the message has been changed or not. If it is correct, m' is accepted as a valid plaintext.

3   Proposed Hybrid Cryptosystem

3.1   Overview

The proposed cryptosystem consists of a key transmission system, which transmits a session key used for the message encryption; a pseudo random number generator, which receives a random number and generates a message encryption key; a symmetric key cryptosystem, which encrypts and decrypts an actual message; and a hash algorithm, which generates verification information to guarantee the integrity of the ciphertext.

(1) *Key Transmission System(KTS)*

A key transmission system(KTS) is composed of three algorithms $KPG()$, $SKG()$ and $SKR()$.

- $KPG(1^k)$: This algorithm is a key-pair generation algorithm, which is used for the public key cryptosystem. It is a probabilistic algorithm that on inputs a security parameter k, randomly outputs a public key-secret key pair *(pk, sk)*. (Here, $k \in N$)
- $SKG(pk_R, sk_S, R, r)$: This algorithm is a session key generation algorithm. It is a probabilistic algorithm that on inputs the receiver's public key pk_R, the sender's secret key sk_S and random number R, r, outputs an encrypted session key Σ.
- $SKR(sk_R, pk_S, \Sigma)$: This algorithm is a session key recovery algorithm, which recovers the session key used for the message encryption. It is a deterministic algorithm that on inputs the receiver's secret key sk_R,the sender's public key pk_S and encrypted information Σ, outputs an random value R.

(2) *Pseudo Random Number Generator(G)*

A pseudo random number generator G is an algorithm that on inputs a random number R, outputs a session key $K = G(R)$ used for the actual message encryption. It can be implemented by using a hash function such as $SHA\text{-}1$[18] and $MD5$[9].

(3) *Symmetric Key Cryptosystem(Π)*

A symmetric cryptosystem Π consists of two algorithms $ENC()$ and $DEC()$ which depend on the k-bit string K, the secret key.

- $ENC_K(m)$: It is a message encryption algorithm using the key K. It outputs a ciphertext c corresponding to the $m \in \{0,1\}^l$, in a deterministic way.

- $DEC_K(c)$: It is a message decryption algorithm using the key K. It outputs the plaintext m associated to ciphertext c.

It can be implemented by performing an *exclusive-OR* operation or using a block cipher such as *AES(Advanced Encryption Standard)* [6].

(4) *Hash Function(H)*

A hash function H is an algorithm that verifies whether the received message has been changed or not. It is a function that on inputs output value of the key transmission system, the output value of symmetric cryptosystem, an initial random number and a plaintext, outputs the hash value. It can be implemented by using a hash algorithm such as *SHA-1* and *MD5*.

3.2   Encryption / Decryption

The encryption and decryption processes of the proposed hybrid cryptosystem are showed in Fig. 1. Before the protocol is started, the sender and the receiver generate his/her secret key-public key pairs using the *KPG( )* algorithm and then publish their public keys in the public directory.

[Encryption]

1) The sender A selects random numbers R, r and generates $C_1 = SKG(pk_R, sk_S, R, r)$. Here, $SKG = R \oplus KAP(pk_R, sk_S, R, r)$ and $KAP()$ is a 1-pass key agreement protocol providing mutual implicit key authentication.
2) The sender A computes a session key $K = G(R)$.
3) The sender A generates a ciphertext $C_2 = ENC_K(m)$.
4) To generate verification information, the sender A computes a hash value $C_3 = H(C_1, C_2, R, m)$.

[Decryption]

1) The receiver B obtains the random number R from C_1 with sender's public key and his/her private key by using the *SKR( )* algorithm.
2) The receiver B computes $K = G(R)$ and decrypts $m' = DEC_K(C_2)$.
3) The receiver B checks $C_3 = H(C_1, C_2, R, m')$ to confirm whether the message has been changed or not. If it is correct, m' is accepted as a valid plaintext.

4   Some Examples

4.1   *Diffie-Hellman* Based System

Diffie-Hellman proposed the concept of the public key cryptosystem for the first time in [4] and proposed a key agreement protocol based on the problem of

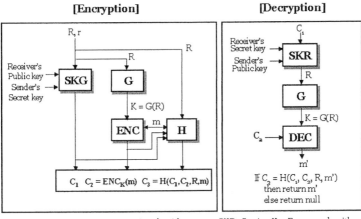

[Encryption]

[Decryption]

- SKG : Session Key Generation algorithm
- ENC : Symmetric encryption algorithm
- G : Pseudo random number generator

- SKR : Session Key Recovery algorithm
- DEC : Symmetric decryption algorithm
- H : Hash function

Fig. 1. Encryption/Decryption schemes of the proposed hybrid cryptosystem

discrete logarithm. The basic *Diffie-Hellman* protocol is a 2-pass protocol and generates different keys for every session. However, as a randomly selected number is used for the session key generation, the problem is that no authentication is provided for the participants in the protocol. To solve these problems, the static *Diffie-Hellman* system, which uses a static secret key of the user instead of a random number, was proposed. However, the disadvantage of this system is that the same session key is always set for both users.

In this paper, to provide *mutual implicit key authentication* with 1-pass protocol, a *dhHybridOneFlow* protocol of *ANSI X9.42*[1], which combines the advantage of both basic *Diffie-Hellman* protocol and static *Diffie-Hellman* protocol, is used in key transmission system of the proposed cryptosystem. The *dhHybridOneFlow* is 1-pass protocol, but it provides *mutual implicit key authentication* and *key freshness*. System parameters used in the crytosystem are as follows.

[System Parameters]

- p : a large prime with $2^{511} < p < 2^{512}$
- g : a primitive element in Z_p (i.e. $g^{p-1} \equiv 1 \ mod \ p$)
- x_i : User i's secret key
- y_i : User i's public key, $y_i \equiv g^{x_i} \ mod \ p$
- G : Pseudo random number generator
- H : Hash function
- $E_K()/D_K()$: Symmetric encryption/decryption algorithms using the key K

The encryption and decryption processes of the *Diffie-Hellman* based cryptosystem are as follows(see Fig. 2). First, to encrypt a message m, the sender A

Sender A		Receiver B
m : plaintext		
Choose $R, r \in Z_p$		
Compute		
$t_A = g^r \ mod \ p$		$R = C_1 \oplus H(y_A^{x_B} \| \ t_A^{x_B})$
$C_1 = R \oplus H(y_B^{x_A} \| \ y_B^r)$	$C = (t_A, C_1, C_2, C_3)$	$K = G(R)$
$K = G(R)$	$- - - - - - - >$	$m' = D_K(C_2)$
$C_2 = E_K(m)$		
$C_3 = H(C_1, C_2, t_A, m)$		Accept m' as a valid plaintext iff
		$C_3 = H(C_1, C_2, t_A, m')$

Fig. 2. *Diffie-Hellman* based cryptosystem

selects random numbers $R, r \in Z_{p-1}$ and computes $t_A = g^r \ mod \ p$. The sender A generates ciphertext $C_1 = R \oplus H(y_B^{x_A} \| \ y_B^r)$ and computes a session key $K = G(R)$. Finally, A generates a ciphertext $C_2 = E_K(m)$ and computes verification information $C_3 = H(C_1, C_2, t_A, m)$. And then, A transmits ciphertext $C = (t_A, C_1, C_2, C_3)$ to receiver B. To decrypt original message m, first, the receiver B computes the random number $R = C_1 \oplus H(y_A^{x_B} \| \ t_A^{x_B})$ and the session key $K = G(R)$ and obtains the plaintext $m' = D_K(C_2)$. Finally, B checks $C_3 = H(C_1, C_2, t_A, m')$ to confirm whether the message has been changed or not. S/he accepts m' as a valid plaintext if and only if it is correct.

In the *Diffie-Hellman* based system, the value of $y_B^{x_A}$ is the same in all sessions, so it is possible to compute the value in advance and save it for the participant who uses an encryption communication frequently. Even though the value of $y_B^{x_A}$ is exposed, since r is chosen randomly in each session, so the security of the ciphertext is not influenced.

To generate the initial random number R used for computing a session key, the receiver uses his own secret key and the sender's public key, allowing implicit authentication for the sender. That is, it could be guaranteed that other people than the sender can't generate a valid ciphertext.

4.2 *Nyberg-Rueppel* Based System

K. Nyberg and *R. Rueppel* proposed the message recovery-typed digital authentication for the first time in [10] and [11], which is based on discrete logarithm problem. *R. Rueppel* and *P.C. Oorschot* proposed a 1-pass key agreement protocol, which provides a mutual implicit key authentication[17] using the signature scheme in [10, 11].

While the existing *Diffie-Hellman* protocol requires 2-pass communication to provide a *mutual implicit key authentication*, this method can provide mutual implicit key authentication with 1-pass communication traffic and, also provides an *explicit entity authentication* of the sender for the receiver if a timestamp or sequence number is included in the key token. Therefore, by applying the *Nyberg-Rueppel*'s key agreement protocol to the key transmission system of the proposed

Sender A		Receiver B
m : plaintext		
Choose $R, r \in Z_{p-1}$		
Compute		
$\quad e = g^{R-r} \bmod p$		
$\quad y = r + x_A \cdot e \bmod p$		
$\quad C_1 = R \oplus H(y_B^R)$	$C = (e, y, C_1, C_2, C_3)$	$R = C_1 \oplus H((g^y \cdot y_A^{-e} \cdot e)^{x_B} \bmod p)$
$\quad\quad K = G(R)$	$- - - - - - - - - >$	$K = G(R)$
$\quad\quad C_2 = E_K(m)$		$m' = D_K(C_2)$
$C_3 = H(C_1, C_2, e, y, m)$		
		Accept m' as a valid plaintest iff
		$C_3 = H(C_1, C_2, e, y, m')$

Fig. 3. *Nyberg-Rueppel* based cyptosystem

cryptosystem can provide an authentication for sender's identity. System parameters of the cryptosystem based on the *Nyberg-Rueppel*'s key agreement protocol are the same as that of the *Diffie-Hellman* based system, and the encryption and decryption processes are as follows(see Fig. 3).

First, to encrypt a message m, the sender A selects random numbers $R, r \in Z_{p-1}$ and computes $e = g^{R-r} \bmod p$ and $y = r + x_A \cdot e \bmod p$. And, A generates a ciphertext $C_1 = R \oplus H(y_B^R)$ and computes $K = G(R)$. Finally, A generates a ciphertext $C_2 = E_K(m)$ and computes verification information by $C_3 = H(C_1, C_2, e, y, m)$. And then, A transmits ciphertext $C = (e, y, C_1, C_2, C_3)$ to receiver B. To decrypt original message m, first, the receiver B computes the random number $R = C_1 \oplus H((g^y \cdot y_A^{-e} \cdot e)^{x_B} \bmod p)$ and the session key $K = G(R)$. And, B decrypts a $m' = D_K(C_2)$. and checks $C_3 = H(C_1, C_2, e, y, m')$ to confirm whether the message has been changed or not. S/he accepts m' as a valid plaintext if and only if it is correct.

In the cryptosystem based on the *Nyberg-Rueppel*'s key agreement protocol, (e, y) plays a role of digital signature. So, if a timestamp or sequence number is included in the ciphertext to prevent a *replay attack*, it is possible to provide an *explicit entity authentication* of the sender's identity.

5   Security Analysis

(1) Can an Attacker Impersonate a Valid User and Send a Ciphertext?

Unlike a general public key cryptosystem, the proposed hybrid cryptosystem uses a secret key of the sender to generate a session key used for the message encryption. That is, it could be implicitly guaranteed that only the sender can generate a valid ciphertext. Therefore, if an attacker does not know the secret key x_A of the sender A, he cannot impersonate a ciphertext generated by user A and send a valid ciphertext.

In the *Diffie-Hellman* based system, if an attacker wants to impersonate a user A and send a ciphertext, he must compute $g^{x_A x_B} \bmod p$ from the public key y_A of the sender A and public key y_B of the receiver B. The difficulty of this computation is equivalent to the *Diffie-Hellman* problem, so it is difficult for an attacker to impersonate a valid user and send the ciphertext is as much as to solve the *Diffie-Hellman* problem.

In the *Nyberg-Rueppel* based system, if an attacker wants to impersonate a user A and send a ciphertext, he also must compute a digital signature (e, y). However, it is impossible to generate a valid value of (e, y) if the user does not know the secret key of the user A. Therefore, it is impossible for an attacker to disguise as another user and send a valid form of ciphertext.

(2) If a Secret Key of the Sender Is Compromised, Can other Users Obtain a Plaintext from a Ciphertext?

The proposed cryptosystem uses not only a public key of the receiver but also a secret key of the sender to encrypt a message encryption key. However, even though the secret key of the sender is compromised, other users cannot get a random number used for generating a session key.

In the *Diffie-Hellman* based system, if the private key of the sender x_A is compromised, anyone can get $y_B^{x_A}$. However, other users who do not know the receiver's secret key x_B or a random number r used for encryption of session key cannot recover the value of R. That is, it is difficult for other users who do not know the receiver's secret key x_B or a random number r, to get $y_B^r \equiv t_A^{x_B} \equiv g^{r \cdot x_B} \bmod p$ as much as to solve the *Diffie-Hellman* problem.

In the *Nyberg-Rueppel* based system, if a sender's secret key x_A is compromised, anyone can compute $r = y - x_A \cdot e \bmod p$ and $g^R = e \cdot g^r \bmod p$. However, one must compute $y_B^R = g^{R \cdot x_B} \bmod p$ from $g^R \bmod p$ and y_B to recover the random number R, which was used for the session key generation, and this computation is as difficult as to solve the *Diffie-Hellman* problem. As a result, even though the secret key of the sender is compromised, other users than the receiver cannot decrypt the ciphertext.

(3) If the Previous Session Key Is Compromised, Can the other Ciphertexts Be Secure?

If all ciphertexts are generated by the same session key for every sessions and only one session key is compromised, all ciphertexts become open. So, it is recommended to use different keys for every session to encrypt messages.

Because the key transmission system of the proposed cryptosystem selects different random numbers for every session, even if one session key is compromised, the security of the other session keys is not influenced. That is, even though one session key is compromised, other ciphertexts are still secure. Besides, even though the random number R used for the session key generation is compromised, other ciphertexts are still secure.

In the *Diffie-Hellman* based system, assume that a ciphertext of the previous session is $C_1' = R' \oplus H(y_B^{x_A} \| y_B^{r'} \bmod p)$ and a ciphertext of the current session is $C_1 = R \oplus H(y_B^{x_A} \| y_B^{r} \bmod p)$. In this case, if a random number R' of the previous session is exposed, anyone can easily calculate $H(y_B^{x_A} \| y_B^{r'}) = C_1' \oplus R'$. However, in case of $r \neq r'$, it is impossible to obtain the random number R of the current session from C_1. That is, if a different random number is used for every session, the security of the current ciphertext is not influenced at all even though the previous session key or the random number is exposed.

Also, as the *Nyberg-Rueppel* based system uses a random number selected differently in each session to generate a session key, even if old session keys or random number used for session key generation are compromised, the security of the ciphertexts of the other sessions will not be affected.

(4) Can an Attacker Change the Sending Ciphertext into a Valid Form of another Ciphertext?

In the original *RSA* public key cryptosystem, it is impossible for an attacker to get the entire plaintext from the ciphertext. However, due to its multiplicative property, it is possible for an attacker to change the ciphertext into a valid form of another ciphertext. This kind of cryptosystem is called a *malleable* cryptosystem[3]. To resolve these kinds of problems of *RSA*, *OAEP*[2] is proposed by *Bellare* and *Rogaway*. *OAEP* adds verification information to check the integrity of the ciphertext. The proposed cryptosystem also includes a hash value to verify the integrity of the ciphertext, so it is impossible for an attacker to change the ciphertext silently.

6   Conclusion

The usage of a cryptosystem to guarantee the confidentiality of a message on the network is on the increase and a hybrid cryptosystem which combines advantages of the symmetric cryptosystem and the public key cryptosystem is widely used.

In this paper, an efficient hybrid cryptosystem is proposed and some examples are described. Unlike the existing hybrid cryptosystem, proposed hybrid cryptosystem provides the implicit authentication for the sender's identity by using the 1-pass key agreement protocol. The proposed cryptosystem can be implemented by using the 1-pass key agreement protocol such as *dhHybridOneFlow* protocol or *Nyberg-Rueppel* key agreement protocol, a symmetric cryptosystem, a pseudo random number generator and a hash function.

The proposed cryptosystem is efficient hybrid system and also provides the implicit authentication for sender, so it is impossible for other users to impersonate the senders and send the message. In addition, even though the private key of the sender is compromised, other users than the specified receiver cannot get the plaintext from the ciphertext. Also, even though a session key or a random number of a session is exposed, the security of the ciphertext of another session is not influenced. The proposed cryptosystem includes verification information to check the integrity of the message, so it is impossible for other users to change the ciphertext. The proposed cryptosystem could be efficiently used for applications that need to have confidentiality and integrity for messages sent on the network. Besides, it is also possible to use and implement the system by using a 1-pass key agreement protocol, which provides the mutual implicit key authentication, as well as the *dhHybridOneFlow* protocol or the *Nyberg-Rueppel* key agreement protocol.

References

[1] ANSI X9.42, "Agreement of sysmmetric Key on Using Diffie- Hellman Cryptography" (2001)
[2] M. Bellare, P. Rogaway, "Optimal Asymmetric Encryption," Advances in Cryptology-Eurocrypto'94, Springer-verlag, LNCS 950, pp. 92–111 (1994)
[3] M. Bellare, A. Sahai, "Non-Malleable Encryption : Equivalence between Two Notions, And an Indistinguishability-Based Charaterization," Advances in Cryptology-Crypto'99, LNCS 1666, pp. 519–536 (1999)
[4] W. Diffie, M. E. Hellman, "New directions in cryptography," IEEE Transaction on Information Theory, IT-22, 6, pp. 644–654 (1976)
[5] T. ElGamal, "A public key cryptosystem and a signature scheme based on discrete logarithms," IEEE Trans, info, Theory, Vol 31, pp 469–472 (1985)
[6] FIPS-197, "Advanced Encryption Standard" (2001)
[7] E. Fujisaki and T. Okamoto, "How to Enhance the Security of Public-Key Encryption at Minimun cost," Second International Workshop on Practice and Theory in Public Key Cryptography, PKC'99, LNCS 1560, pp. 53–68 (1999)
[8] E. Fujisaki and T. Okamoto, "Secure Integration of Asymmetric and Symmetric Encryption Scheme," Advances in Cryptology-Crypto'99, LNCS 1666, pp. 537–554 (1999)
[9] Internet Engineering Task Force(IETF) RFC 1321, "Message Digest 5(MD5)"
[10] K. Nyberg, R. A. Rueppel, "A new signature scheme based on DSA giving message recovery," Proc. 1st ACM Conf. on Comput. Commun. Security, pp. 58–61 (1993)
[11] K. Nyberg, R. Rueppel, "Meessage recovery for signature schemes based on the discrete logarithm problem," Advances in Cryptology-Eurocrypt'94, Springer-verlag, LNCS 950, pp. 182–193 (1994)
[12] T. Okamoto and D. Pointcheval, "REACT : Rapid Enhanced-security Asymmetric Cryptosystem Transform," CT-RSA 2001, LNCS 2020, pp, 159–174 (2000)
[13] T. Okamoto and D. Pointcheval "OCAC : an Optimal Conversion for Asymmetric Crptosystems," P1363
[14] D. Pointcheval, "HD-RSA : Hybrid Dependent RSA - a New Public key Encryption Scheme," IEEE P1363
[15] D. Pointcheval, "New Public key Cryptosystem based on the Dependent-RSA Problems," Advances in Cryptology-Eurocrypt'99, LNCS 1592, pp.239-254 (1999)
[16] R. Rivest, A. Shamir and L. Adleman, "A method of obtaining digital signature and public key cryptosystem," ACM Communication 21 No. 2, pp. 120-126 (1978)
[17] R. A. Rueppel, P. C. van Oorschot, "Modern Key Agreement Techniques," Computer Communications, pp. 458-465 (1994)
[18] Secure hash standard, National Bureau of Standards FIPS Publication 180 (1993)

Network-Based Intrusion Detection
with Support Vector Machines

Dong Seong Kim[1] and Jong Sou Park[1]

Hankuk Aviation University
200-1, Hwajon-dong, Doekyang-gu, Koyang, Kyounggi-do, Korea
{dskim,jspark}@mail.hangkong.ac.kr

Abstract. This paper proposes a method of applying Support Vector Machines to network-based Intrusion Detection System (SVM IDS). Support vector machines(SVM) is a learning technique which has been successfully applied in many application areas. Intrusion detection can be considered as two-class classification problem or multi-class classification problem. We used dataset from 1999 KDD intrusion detection contest. SVM IDS was learned with triaing set and tested with test sets to evaluate the performance of SVM IDS to the novel attacks. And we also evaluate the importance of each feature to improve the overall performance of IDS. The results of experiments demonstrate that applying SVM in Intrusion Detection System can be an effective and efficient way for detecting intrusions.

1   Introduction

We propose network-based intrusion detection systems using Support Vector Machines(SVM). Intrusion is generally defined as violating confidentiality, Integrity, and Availability of computer or computer network system[1]. Intrusion detection system(IDS) detects automatically computer intrusions to protect computers and computer networks safely from malicious uses or attacks[2]. IDS should cope with a novel attacks as well as previously known attacks or misuses. Also IDS should give minimum overhead to computer system and IDS itself for processing audit data. But existing intrusion detection systems tend to have a low detection rates at novel attacks and high overhead itself to process audit data. SVM is relatively a novel classification technique proposed by Vapnik[3]. In many applications, SVM has been shown to provide higher performance than traditional learning machines, and has been introduced as powerful tools for solving classification problems such as pattern recognition and speech recognition fields [4,5,6]. Intrusion generally can be considered as a binary classification problem, distinguishing between normal and attacks. Intrusion can also be considered as multi-class classification problems. So we propose Network-based Support Vector Machine Intrusion Detection System(SVMIDS) . We used 1999 KDD dataset which was used in contest of intrusion detection areas [7,8]. We performed experiments on network-based SVM IDS with the various Kernel functions and regularization parameter C values. We also performed experiments with various

H.-K. Kahng (Ed.): ICOIN 2003, LNCS 2662, pp. 747–756, 2003.

numbers of features of dataset to evaluate the characteristics of feature values. Through analysis of features by using SVM, IDS can process network packet more effectively. The rest of this paper is organized as follows. A brief description of the Intrusion detection and the theory of SVM will be described in Section 2. The proposed structures of network-based SVM IDS will be suggested in Section 3. The various experiments of SVM IDS and their results are presented in Section 4. Some concluding remarks are given in Section 5.

2 Related Works

2.1 Intrusion Detection

Intrusion is defined as any set of actions that attempt to compromise the integrity, confidentiality or availability of a resource. Intrusion detection system(IDS) attempts to detect an intruder break into computer system or legitimate user misuse system resources in real-time. Intrusion detection techniques based on intrusion model can be divided into two major types, misuse detection and anomaly detection[9,10]. In misuse detection, intrusions are well defined attacks on known weak points of a computer system. Misuse can be detected by watching for certain action being performed on certain object, concretely by doing pattern matching on audit-trail information. Misuse detection can cope with a known attack but not novel attacks because misuse detection depends on the intrusion database for known attack patterns. In anomaly detection, intrusions are based on observations of deviations from normal system usage patterns. They are detected by building up a profile of the system being monitored, and detecting significant deviations from this profile. A model is built which contains metrics that are derived from system operation. Metrics are computed from available system parameters such as average CPU load, number of network connections per minute, number of processes per user, etc. Anomaly detection also contains some problems. Since anomaly detection techniques use significant variation from user profiles or normal profile, intruders can manipulate to modify system profiles gradually. And Intrusion detection techniques based on data-source can be divided into two main types, host-based detection and network-base detection [10]. Host-based intrusion detection use audit data such as user's CPU usage time, command log, system call of system, etc. Because host-based intrusion detection can not have any information of network events in lower layers, it usually does not detect network-based attacks. According to the operating system or the platform of each host requires a different system, so it takes higher cost to develop. And a host-based intrusion detection use resources of itself, such as CPU time, storage, etc., so it lowers the performance of the system. On the other hand, network-based intrusion detection (NIDS) use audit data such as network packet data, especially network packet headers and payloads. Most NIDS generally collects network packets through using Network Interface Card (NIC) and takes passive analysis, so NIDS do not require software to be loaded and managed on a variety of hosts. NIDS detects attacks that host-based detection systems miss, e.g., many IP-based Denial-Of-Service(DoS)

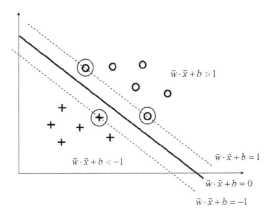

Fig. 1. Optimal seperating hyperplane is the one that seperates the data with maximal margin[3]

attacks and fragmented packet(Teardrop) attacks. We propose network-based intrusion detector which can detect both anomaly and misuse intrusions using Support Vector Classification Methods.

2.2 Support Vector Classification

Support Vector Machines (SVM) have been successfully applied to a wide range of pattern recognition problems[4,5,6]. SVM are attractive because they are based on well developed theory. A support vector machine finds an optimal separating hyper-plane between members and non-members of a given class in an high dimension feature space. Although the dimension of feature space is very large, infinite sometimes, they still show good generalization performances. This conflicts with our sense about "curse of dimension" and over-fit. Theories have been used to interpret this paradox, but more works still have to be done for these kinds of learning machines [5]

Two-Class Classification. Support vector machines are based on the structural risk minimization principle[11] from the statistical learning theory. In their basic form, SVM learn linear decision rules described by a weighted vector and a threshold . The idea of structural risk minimization is to find a hypothesis for which one can guarantee the lowest probability of error. Geometrically, we can find two parts representing the two classes with a hyper-plane with normal and distance from the origin as depicted in Figure 1.

$$D(\vec{x}) = sign\{\vec{w} \cdot \vec{x} + b\} = \begin{cases} +1, \text{ if } \vec{w} \cdot \vec{x} + b > 0 \\ -1, \qquad \text{else} \end{cases} \qquad (1)$$

Suppose such separating hyper-planes exist. We define the margin of a separating hyper-plane as the minimum distance between all input vectors and the hyper-plane. The learning strategy of support vector machines is to choose the classifier hyper-plane with maximal margin. Although this might seem a reasonable heuristic approach, it is actually well founded in statistical learning theory. Basically, bounds on the generalization error are given in terms of the classification function's complexity, which we can minimize for linear functions by maximizing the margin. In particular, these bounds do not depend on the dimension of the input space, which makes this strategy appealing for high-dimensional data. To modify this approach for linearly inseparable data, we introduce slack variables and simultaneously maximize the margin and minimize the classification error on the training set. This includes introducing a regularization parameter that controls the trade-off between these two objectives. The resulting optimization problem is a quadratic program, usually solved in the form of its Lagrangian dual :

$$Min \quad \alpha \qquad \sum_{i=1}^{m}\sum_{j=1}^{m} y_i y_j \alpha_i \alpha_j x_i \cdot x_j - \sum_{i=1}^{m} \alpha_i \qquad (2)$$

$$s.t. \qquad \sum_{i=1}^{m} y_i \alpha_i = 0, \qquad 0 \leq a_i \leq C, \quad i = l, L, M \qquad (3)$$

In our experiment to solve two-class classification problem for intrusion detection, we used the well-known software SVMlight [19] for two-class intrusion detection classification.

Multi-class Classification. Support Vector Machines were originally designed for two-class classification, commonly called as binary classification. How to effectively extend it for multi-class classification is still an on-going research issue [12]. Currently there are two types of approaches for multi-class SVM, one-against-all[13] and one-against-one approaches[14]. One is by constructing and combining several binary classifiers while the other is by directly considering all data in one optimization formulation. More detailed description for multi-class classification methods are presented in [12]. To solve multi-class attack problems in our experiments, we used libsvm[20] which use one-against-one methods.

3   Network-Based Support Vector Machines Intrusion Detection System

3.1   Structure of Network-Based SVM IDS

The overall structure and component of SVM IDS are depicted in Figure 2[18]. We describe the overall structure of SVM IDS briefly. First, training dataset and test dataset is collected. We used training set and test dataset from KDD 99 dataset in intrusion detection areas [7]. These sets are preprocessed and to be used as standard input form of SVM. In this progress, we can select the features of the training set and test set. As the result of feature selection, we can estimate

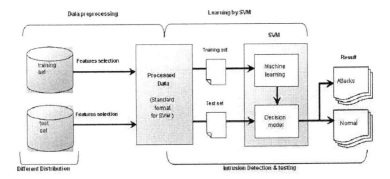

Fig. 2. Structure of Network-based SVM IDS

what features is important and we can use these characteristics of feature to improve the performance of SVM IDS. Training dataset is divided into two sets as learning dataset, validation dataset. With learning set, SVM is learned and as a result of learning, a decision model is generated. Actually decision model is a decision function, called hyper-planes in feature space, with a weight vector value and some numbers of support vectors. With validation dataset, which are different from learning set, we can evaluate how the decision model we already made perform classification task well. With test set, we can evaluate the performance of SVM decision model such about various novel attacks. The performance of a decision model depends on the detection rates, false positive rates, number of misclassifications.

3.2 Dataset and Preprocessing

Dataset. We used the datasets used in the 1999 KDD intrusion detection competition [7]. The competition task was to build a network intrusion detector, a predictive model capable of distinguishing between "bad"connections, called intrusions or attacks, and "good" normal connections. This database contains a standard set of data to be audited, which includes a wide variety of intrusions simulated in a military network environment. More detail descriptions are presented in [8]. In their tasks, attacks fall into four main categories [16] :

– DOS: denial-of-service, e.g. syn flood;

– R2L: unauthorized access from a remote machine, e.g. guessing password;

– U2R: unauthorized access to local root privileges, e.g., various "buffer over-flow" attacks;

– Probing: surveillance and other probing, e.g., port scanning.

And Stolfo et. al. defined higher-level features that help in distinguishing normal connections from attacks[15]. There are several categories of derived features, such as "same host" features, "same service" features, "content" features. And the distribution of training set is different from that of test set because test set contains additionally 14 novel attacks that do not appear in training set.

Preprocessing. Those datasets were labeled on individual connections records called attack instances whether it's normal or attacks. In dataset, derived feature values are denoted as a discrete type, e.g., tcp, ftp, SF or continuous type, e.g., 5, 0.01, etc. So we preprocessed dataset to correspond with the standard input form of SVM. In experiments, we deleted features or add some features to evaluate the characteristics of the deleted or added features.

3.3   Learning and Testing of SVM

Learning Processes. Through training process, SVM became a decision model, actually a decision function with some numbers of support vectors and weight vector values. We used two datasets, learning sets and validation set. Learning sets are used to generate a decision function of SVM. And we can evaluate the generated decision function of SVM by using the validation set. Learning set and validation sets are made from training set randomly, but learning sets are different from the validation sets. We can evaluate the performance of decision function of SVM through validating the values of detection rate and misclassification rates. If detection rates are too low to accept, new learning process must be done. This process was iterated until a high detection rates are obtained. In learning process, we used various C values, e.g., 1, 500, 1000 and kernel functions such as linear, 2-poly and Radial Basis Function (RBF). Through modification of regularization parameter C values and kernel function, we can validate which kernel function is an effective and efficient one in this application area.

Test Processes. The distributions of training sets are different from test sets. In test set, there are about 14 new attacks which are not contained in training set. So we can evaluate the model made in learning progress how the models cope with various novel attacks. In KDD'99 contest, the evaluations of detection rates and false alarm rates were carried out, and the results were open to the public [8]. So we compared our results to KDD'99 results relatively and we demonstrated the improvement of our methods.

4   Experiments of SVM IDS and Results

We now describe the processes of experiments as three phases. In first phase, we processed the training sets and test sets suitable for the standard form of SVM. In second phase, SVM IDS was under learning using the preprocessed learning sets, we validated a decision model by using the preprocessed validation sets. In third phase, we evaluated the detection rates of the decision model that had constructed in second phase by using test sets. In addition to these processes, we manipulated the number of the input features to investigate the importance of each input features. We also performed experiments on 5 classes classifications classified by DARPA projects.

4.1   Data Preprocessing

We used KDD 1999 dataset in intrusion detection of application areas. These dataset are for the purpose of evaluation to various proposed intrusion detection systems. The dataset are only classified dataset which labeled as normal or attack name, so these dataset is not suitable for standard form of SVM input. We transformed these dataset into a standard form of SVM input. The total numbers of training set is 4,898,431 instances and the total number of 10% labeled test set is 311,029 instances. We extracted learning set and validation set from training set randomly. But learning set is different from validation set. And we also extracted sub test sets from test set randomly

4.2   SVM Learning and Validation Experiments

We used SVMlight[19] for binary classification. We used linear, 2poly, and RBF functions as kernel functions. Our input dataset have 41 features and dimension is large, so we used various combination of kernel function and C values to find out the most suitable and efficient kernel to our application field. And the C values are regarded as a regularization parameter. This is the only free parameter in the SVM formulation. This parameter provides us trade-off between margin maximization and classification violation. Generally speaking, smaller parameter C makes a decision model simpler, if parameter C is infinite; all training data are classified correctly. The result of validation experiments are displayed in Table 1.

Table 1. The results of validation experiments

kernel	C	val 1.	val 2.	val 3.	val 4.
linear	0	93.56%	93.12%	93.45%	90.52%
2poly	0	49.97%	50.03%	50.02%	50.07%
	100	48.65%	49.97%	49.97%	49.97%
	1000	49.97%	50.03%	50.02	50.07%
RBF	0	50.03%	50.02%	49.97%	50.02%
	100	85.61%	85.69%	86.85%	82.59%
	1000	86.04%	85.63%	87.49%	82.35%

Table 2. The results of testing experiments

kernel	C	test 1.	test 2.	test 3.	test 4.
linear	0	90.16%	89.51%	87.71%	88.84%
RBF	1000	78.10%	77.23%	76.24%	76.51%

Table 3. The results of features deletion experiments

feature.	1	2	3	4	5	6	7	8	9	10	11	12	13	14
det.(%)	77.20	78.10	76.24	77.23	78.05	76.55	78.05	76.10	77.20	76.23	78.10	77.40	78.30	76.11
feature.	15	16	17	18	19	20	21	22	23	24	25	26	27	28
det.(%)	76.2	78.05	78.34	78.05	76.40	75.96	78.03	76.24	76.51	78.05	77.23	78.22	77.34	78.25
feature	29	30	31	32	33	34	35	36	37	38	39	40	41	
det.(%)	76.1	76.68	78.05	78.05	78.05	78.05	77.32	76.70	79.00	78.14	77.86	76.12	76.65	

Table 4. The results of five-classes classification experiments

	All	Normal	Probe	DOS	U2R	R2L
Validation	95.58%	99.2%	29.54%	94.5%	16%	32%
Testing	93.59%	99.3%	36.65%	91.6%	12%	22%

4.3 SVM Testing Experiments

We used the model which had highest detection rates in section 4.2. There are 14 new attacks in test sets, they were not contained in the training set. We displayed the results of detection rates to the test data in Table 2.

4.4 Feature Deletion Experiments

We performed features deletion experiments and the results are summarized in Table 3. Through this feature deletion, we tried to find out the importance of features in intrusion detection system. But the results of these experiments are not so promising to find out which input features are important or useless.

4.5 Five Class Classification Experiments

We performed experiments on five class classifications. The results are displayed in Table 4. Detection rates about Normal, Probe, and DOS are comparatively high, but detection rates about U2R and R2L are so low because the numbers of instances in R2L and U2R attack are so small.

4.6 Comparison of Training and Testing of Results to '99 KDD Contest

We analyzed the performance of IDS using SVM previously. But for more correct verification, comparable method to our approach is needed. But in this paper,

Table 5. The performance comparisions between SVM IDS and   KDD '99 winner

	SVM IDS	KDD 99 winner
Normal	99.3%	99.5%
Probe	36.65%	83.3%
DOS	91.6%	97.1%
U2R	12%	13.2%
R2L	22%	8.4%

we did not perform those work, so we compared our performance of SVM IDS to that of KDD 1999 contest winner relatively.

5   Conclusion and Future Works

In this paper, we proposed a network-based SVM IDS, and demonstrated it through results of 3 kinds of experiments. And we show that SVM IDS can be an effective choice of implementing IDS. In this paper, we used labeled dataset, it does not mean that our approach is not real-time intrusion detection. If network packet can be preprocessed in real-time, our approach can be use in real-time applications. And because there are a large mount of network packets to be processed in a real-time environment, we plan to implement hardware based SVM IDS chip to process faster than software. And there are some miss-classified input vectors and those degrade the performance of SVM IDS, so we need to improve the performance by applying Genetic Algorithm (GA) based feature extraction to enhance the performance of SVM. We can also apply decision tree method to get feature extraction instead of GA.

Acknowledgements

This work is supported by "Integrated Circuit Design Education Center(IDEC)", "Information Technology Research Center(ITRC)" supported by the Ministry of Information &Communication of Korea(supervised by IITA), "Internet Information Retrieval" Regional Research Center Program supported by the Korea Science and Engineering Foundation.

References

[1] J. P. Anderson., : Computer Security Threat Monitoring and Surveillance, James P Anderson Co.., Technical report, Fort Washington, Pennsylvania, April (1980)
[2] D. E. Denning. : An Intrusion Detection Model, IEEE Trans. S. E., (1987)
[3] V. Vapnik. : The Nature of Statistical Learning Theory, Springer, Berlin Heidelberg New York (1995)

[4] SVM Application List, http://www.clopinet.com/isabelle/Projects/SVM/applist.html

[5] Christopher J. C. Burges. : A Tutorial on Support Vector Machines for Pattern Recognition, Data Mining and Knowledge Discovery (1998)

[6] . G. Guo, S. Z. Li, and K. Chan. : Face Recognition by Support Vector Machines, Fourth IEEE International Conference on Automatic Face and Gesture Recognition,(2000)196-201

[7] KDD Cup 1999 Data,
http://kdd.ics.uci.edu/databases/kddcup99/kddcup99.html

[8] Results of the KDD '99 Classifier Learning Contest, http://www-cse.ucsd.edu/users/elkan/clresults.html

[9] S. Kumar.: Classification and Detection of Computer Intrusions. Ph.D. Dissertation (1995)

[10] H. Debar, M. Dacier and A.Wespi. : A revised taxonomy for intrusion-detection systems, IBM Research Technical Report. (1999)

[11] V. N. Vapnik. : Statistical Learning Theory. John Wiley & Sons (1998)

[12] Chih-Wei Hsu, Chih-Jen Lin : A comparison of Methods for Multi-class Support Vector Machines, National Taiwan University.(2001)

[13] B. Schoelkopf, C. Burges, and V. Vapnik.: Extracting support data for a given task. In U. M. Fayyad and R. Uthurusamy, editors, Proceedings, First International Conference on Knowledge Discovery&Data Mining. AAAI Press, MenloPark, CA (1995)

[14] S. Knerr, L. Personnaz, and G. Dreyfus. : Single-layer learning revisited: a stepwise procedure for building and training a neural network. In J. Fogelman, editor, Neurocomputing: Algorithms, Architectures and Applications. Springer-Verlag. (1990)

[15] Salvatore J. Stolfo, Wei Fan, Wenke Lee, Andreas Prodromidis, and Philip K. Chan. : Cost-based Modeling and Evaluation for Data Mining With Application to Fraud and Intrusion Detection: Results from the JAM Project, Technical Rep. (2000)

[16] Intrusion Detection Attacks Database, http://www.cs.fit.edu/~mmahoney/ids.html

[17] Aurobindo Sundaram : An Introduction to Intrusion Detection, ACM crossroad Issue 2.4 April (1996)

[18] J. Lee, D. S. Kim, S. Chi, J. S. Park : Using the Support Vector Machine to Detect the Host-based Intrusion, IRC 2002 international conference (2002)

[19] SVM-Light Support Vector Machine, http://svmlight.joachims.org

[20] LIBSVM, http://www.csie.ntu.edu.tw/~cjlin/libsvm/

A Conference Key Distribution Scheme in a Totally-Ordered Hierarchy

Min-Shiang Hwang[1] and Wen-Guey Tzeng[2]

[1] Graduate Institute of Networking and Communication Engineering
Chaoyang University of Technology
168 Gifeng E. Rd., Wufeng, Taichung County, Taiwan 413 , R.O.C.
mshwang@mail.cyut.edu.tw
http://www.cyut.edu.tw/~mshwang
[2] Department of Computer and Information Science, National Chiao-Tung University
1001 Ta Hsueh Road, Hsinchu 300, Taiwan, R.O.C.

Abstract. A new conference key distribution scheme in a in a totally-ordered Hierarchy is proposed in this study. A group of users with a different security class can generate a common secret key over a public channel so that a secure electronic conference can be held. The security classes form a hierarchy such that users in a higher security class can obtain a secret session key with users in the same or lower security classes. However, the opposite situation is not permitted.

1 Introduction

In 1976, Diffie and Hellman proposed a simple but effective key distribution scheme for distributing a secret common key [3]. Two users thus use the common secret key (also referred to as session key) to communicate with each other in a secure manner over insecure channels. In 1982, Ingemarsson et al. generalized the key distribution scheme [4] to the conference key distribution scheme (CKDS). In CKDS, a group (two or more) of participants generate a common secret key over the public channel to hold a secure conference.

Since 1987, a number of studies have been carried out concerning conference key distribution systems [1, 2, 5, 6, 7, 8, 9]. Koyama and Ohta proposed identity-based conference key distribution schemes with authentication in three configurations, i.e., ring network, complete graph network, and star network [6]. Unfortunately, Yacobi made an impersonation attack on the protocols for the complete graph and star networks. Although Koyama and Ohta improved the protocols to counter Yacobi's attack [7], their schemes still remain vulnerable to a conspiracy attack [9]. In 1992, Koyama and Ohta modified their schemes to counter this attack again [5]. Two other investigations proposed schemes for broadcasting secret messages in both complete graph and star network. Chiou and Chen proposed a method which is based on Chinese remainder theorem [2]. Laih et al. proposed a scheme based on cross production [8].

In this work, a new conference key distribution scheme is proposed in a hierarchy. A group of users with different security classes generate a common secret

H.-K. Kahng (Ed.): ICOIN 2003, LNCS 2662, pp. 757–761, 2003.
© Springer-Verlag Berlin Heidelberg 2003

key over a public channel so that a secure electronic conference can be held. The security classes form a hierarchy such that users in a higher security class can obtain a secret session key with users in the same or lower security classes; however, the opposite situation is not permitted. Our scheme is based on the ID-based cryptosystem.

The system has the following characteristics:

- It is a large social group system and has many users in the system.
- Each user should be associated with a security class. The security classes form a hierarchy.
- Each user can be located anywhere.
- The scheme is used in a ring network environment.

2   The Proposed Scheme

The totally-ordered hierarchy is the simplest multilevel hierarchy such that for any two distinct security classes C_1 and C_2 either $C_1 \leq C_2$ or $C_2 \leq C_1$. This hierarchy, although a special case of the partially-ordered hierarchy, has many real world applications. For example, documents are often classified into top-secret, secret, confidential, and unclassified security classes in the order *top-secret* > *secret* > *confidential* > *unclassified* and classify users into Administrators (A), Programmers (P), and Ordinary users (O) in the order $A > P > O$ in database systems.

In our scheme, we assume that m users are in the system. Each user has a security class. These classes form a totally-ordered hierarchy. Let l_i be an integer which denotes security level of the security class owned by the user i (U_i). A high l_i mean that the user has a high ranking in the hierarchy. Let n be the product of two large primes p and q; e, d be the public key and secret key such that $ed \bmod \phi(n) = 1$, here $\phi(n)$ denotes the Euler's totient function of n; c be a prime; g be an integer which is a primitive element over both $GF(p)$ and $GF(q)$; ID_i be the identification information of user i; s_i be an integer $s_i = ID_i^{d^{m-1}} \bmod n$. s_i is known only to user i, and n, e, ID_i are public and common to all the users.

The following is the procedure of our scheme with authentication in the ring topology. After m rounds, the scheme generates a common secret session key for m users.

Step 1: U_i generates a secret random number r_i and sends (x_{il}, y_i, z_i) to $U_{(i+1)}$, where

$$x_{il} = g^{e r_i^{\min\{l, l_i\}}} \bmod n, \qquad l = 1, 2, \cdots, l_{i-1},$$
$$\text{if } i - 1 < 1 \text{ then } l_{i-1} = l_{m+i-1},$$
$$y_i = s_i g^{c r_i} \bmod n,$$
$$z_i = 1.$$

Step j $(2 \leq j \leq m-1)$**:** U_i receives $(x_{(i-1)l}, y_{i-1}, z_{i-1})$. If $i = 1$ then U_1 receives (x_{ml}, y_m, z_m). U_i checks whether the following congruence holds:

$$(y_{i-1}^e / x_{(i-1)1}^c z_{i-1}^{ec})^{e^{m-j}} = \prod_{1 \leq k \leq j-1} ID_{i-k} \bmod n. \tag{1}$$

If this check succeeds, user i then sends (x_{il}, y_i, z_i) to $U_{(i+1)}$, where

$$x_{il} = x_{(i-1)l}^{er_i^{\min\{l, l_i\}}} \bmod n, \qquad l = 1, 2, \cdots, l_{i-j},$$
$$\text{if } i - j < 1 \text{ then } l_{i-j} = l_{m+i-j}$$
$$y_i = y_{i-1}^e s_i^{e^{j-1}} x_{i-1}^{cr_i} \bmod n,$$
$$z_i = x_{i-1} z_{i-1}^e \bmod n.$$

Step m: U_i receives $(x_{(i-1)l}, y_{i-1}, z_{i-1})$. If $i = 1$ then U_1 receives (x_{ml}, y_m, z_m). U_i checks whether Equation (1) for $j = m$ holds. If the check succeeds, user i then computes conference key CK:

$$CK_l = x_{(i-1)l}^{r_i^{\min\{l, l_i\}}} \bmod n, \qquad l = 1, 2, \cdots, l_i$$
$$= g^{e^{m-1} \prod_{i=1}^{m} r_i^{\min\{l_i, l\}}} \bmod n.$$

An illustrative example of our scheme for totally-ordered hierarchy in Figure 1 is provided in Table 1.

3   Security

The security of the proposed schemes is based on the difficulty of deriving secret information such as (p, q, d, s_i, r_i) from public keys, transmitted messages and other user's secret keys.

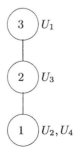

Fig. 1. An example of our scheme for totally-ordered hierarchy

Table 1. Illustrative example of our scheme for a totally-ordered hierarchies

U_1 $(l_1 = 3)$	U_2 $(l_2 = 1)$	U_3 $(l_3 = 2)$	U_4 $(l_4 = 1)$
$x_{11} = g^{er_1}$	$x_{21} = g^{er_2}$ $x_{22} = g^{er_2}$ $x_{23} = g^{er_2}$	$x_{31} = g^{er_3}$	$x_{41} = g^{er_4}$ $x_{42} = g^{er_4}$
$x_{11} = g^{e^2 r_1 r_4}$ $x_{12} = g^{e^2 r_1^2 r_4}$	$x_{21} = g^{e^2 r_1 r_2}$	$x_{31} = g^{e^2 r_2 r_3}$ $x_{32} = g^{e^2 r_2 r_3^2}$ $x_{33} = g^{e^2 r_2 r_3^2}$	$x_{41} = g^{e^2 r_3 r_4}$
$x_{11} = g^{e^3 r_1 r_3 r_4}$	$x_{21} = g^{e^3 r_1 r_2 r_4}$ $x_{22} = g^{e^3 r_1^2 r_2 r_4}$	$x_{31} = g^{e^3 r_1 r_2 r_3}$	$x_{41} = g^{e^3 r_2 r_3 r_4}$ $x_{42} = g^{e^3 r_2 r_3^2 r_4}$ $x_{43} = g^{e^3 r_2 r_3^2 r_4}$
$CK_1 = g^{e^3 r_1 r_2 r_3 r_4}$ $CK_2 = g^{e^3 r_1^2 r_2 r_3^2 r_4}$ $CK_3 = g^{e^3 r_1^3 r_2 r_3^2 r_4}$	$CK_1 = g^{e^3 r_1 r_2 r_3 r_4}$	$CK_1 = g^{e^3 r_1 r_2 r_3 r_4}$ $CK_2 = g^{e^3 r_1^2 r_2 r_3^2 r_4}$	$CK_1 = g^{e^3 r_1 r_2 r_3 r_4}$

Secrecy of (p, q, d, s_i) in our schemes is based on the difficulty of factoring a large number n. If an intruder is able of obtaining p and q from n, he/she could then compute $\phi(n)$, d from e and n. He/She also could obtain s_i by computing $ID_i^{d^{m-1}} \bmod n$, here ID_i, m, and n are a public to all users.

As generally known, computing the secret key of x from the equation $y = g^x \bmod n$ from y, g, and n is a relatively difficult task. Moreover, in our proposed scheme, it is quite difficult for an intruder to obtain the user generated random number r_i directly from the equation $x_i = g^{er_i} \bmod n$ in our schemes. Furthermore, it is difficult for an intruder to obtain s_i directly from the equation $y_i = s_i g^{cr_i} \bmod n$.

4   Conclusions

We have proposed a new conference key distribution scheme in the totally-ordered hierarchy. The main merit of our scheme is that a group of users with different security class can generate a common secret session key over a public channel so that a secure electronic conference can be held or confidential messages can be exchanged.

Acknowledgements

The authors wish to thank many anonymous referees for their suggestions to improve this paper. Part of this research was supported by the National Science Council, Taiwan, R.O.C., under contract no. NSC91-2213-E-324-003.

References

[1] M. Burmester and Y. Desmedt, "A secure and efficient conference key distribution system", *Advances in Cryptology, EUROCRYPT'94*, pp. 275–286, 1994

[2] G. H. Chiou and W. T. Chen, "Secure broadcasting using the secret lock", *IEEE Transactions on Software Engineering*, vol. 15, no. 8, pp. 929–934, 1989.

[3] W. Diffie and M. E. Hellman, "New directions in cryptography", *IEEE Transactions on Information Theory*, vol. 22, 644–654, 1976.

[4] I. Ingemarsson, D. T. Tang, and C. K. Wong, "A conference key distribution system", *IEEE Transactions on Information Theory*, vol. IT-28, no. 5, pp. 714–720, 1982.

[5] K. Koyama, "Secure conference key distribution systems for conspiracy attack", *Advances in Cryptology, EUROCRYPT'92*, pp. 449–453, 1992

[6] K. Koyama and K. Ohta, "Identity-based conference key distribution system", *Advances in Cryptology, CRYPTO'87*, pp. 194–202, 1987.

[7] K. Koyama and K. Ohta, "Security of improved identity-based conference key distribution systems", *Advances in Cryptology, EUROCRYPT'88*, pp. 11–19, 1988.

[8] C. S. Laih, L. Harn, and J. Y. Lee, "A new threshold scheme and its applications on designing the conference key distribution cryptosystem", *Information Processing Letters*, vol. 32, no. 3, pp. 95–99, 1989.

[9] A. Simbo and S. Kawamura, "Cryptanalysis of several conference key distribution schemes", *Advances in Cryptology, ASIACRYPT'91*, pp. 155–160, 1991.

Efficient Bit Serial Multiplication in $GF(2^m)$ for a Class of Finite Fields

Soonhak Kwon[1] and Heuisu Ryu[2]

[1] Institute of Basic Science, Sungkyunkwan University
Suwon 440-746, Korea
shkwon@math.skku.ac.kr
[2] Electronics and Telecommunications Research Institute
Taejon 305-350, Korea
hsryu@etri.re.kr

Abstract. We propose a design of a bit serial multiplication by using an optimal normal basis of type II in a finite field $GF(2^m)$, which has the properties of modularity, regularity and scalability. Our multiplier provides a fast and an efficient hardware architecture for various VLSI implementations such as smart cards and IC cards. We also show that the two operations xy and xy^2, where x and y are in $GF(2^m)$, can be computed simultaneously after m clock cycles in one shift register arrangement. Moreover, the related irreducible polynomial of our basis is very often primitive (approximately 80 percent of known examples). Therefore our multiplier is more suitable for a cryptographic purpose than the multipliers using an optimal normal basis of type I or an all one polynomial basis, where the related irreducible polynomials are never primitive polynomials.

Keywords: Finite field, bit serial multiplication, dual basis, optimal normal basis of type II, all one polynomial.

1 Introduction

Arithmetic of finite fields, especially finite field multiplication, found various applications in coding theory, cryptography and digital signal processing. Therefore an efficient design of a finite field multiplier is needed. Though one may design a finite field multiplier in a software arrangement, a hardware implementation has a strong advantage when one wants a high speed multiplier. Moreover, arithmetic of $GF(2^m)$ is easily realized in a circuit design using a few logical gates. A good multiplication algorithm depends on the choice of basis for a given finite field. One of the widely used finite field multipliers is the Berlekamp's bit serial multiplier [1, 2, 3, 8]. Because of its low hardware complexity, it has been used in Reed-Solomon encoders which have been utilized in various practical applications such as a deep space probe and a compact disc technology. A finite field $GF(2^m)$ with 2^m elements is regarded as a m-dimensional vector space over $GF(2)$.

H.-K. Kahng (Ed.): ICOIN 2003, LNCS 2662, pp. 762–771, 2003.

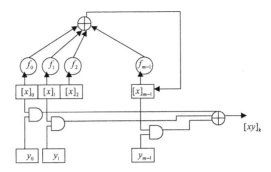

Fig. 1. Bit serial arrangement of dual basis multiplier

Definition 1. *Two bases* $\{\alpha_1, \alpha_2, \cdots, \alpha_m\}$ *and* $\{\beta_1, \beta_2, \cdots, \beta_m\}$ *of* $GF(2^m)$ *are said to be dual if the trace map,* $Tr : GF(2^m) \rightarrow GF(2)$, *with* $Tr(\alpha) = \alpha + \alpha^2 + \cdots + \alpha^{2^{m-1}}$, *satisfies* $Tr(\alpha_i \beta_j) = \delta_{ij}$ *for all* $1 \leq i, j \leq m$, *where* $\delta_{ij} = 1$ *if* $i = j$, *zero if* $i \neq j$. *A basis* $\{\alpha_1, \alpha_2, \cdots, \alpha_m\}$ *is said to be self dual if* $Tr(\alpha_i \alpha_j) = \delta_{ij}$.

Let α be an element in $GF(2^m)$ be such that $\{1, \alpha, \alpha^2, \cdots, \alpha^{m-1}\}$ is a basis of $GF(2^m)$ over $GF(2)$. Let $\{\beta_0, \beta_1, \cdots, \beta_{m-1}\}$ be the dual basis of $\{1, \alpha, \alpha^2, \cdots, \alpha^{m-1}\}$. For any $x \in GF(2^m)$, by considering both polynomial basis and its dual basis expression of x, we may express x as $x = \sum_{i=0}^{m-1} x_i \alpha^i = \sum_{i=0}^{m-1} [x]_i \beta_i$. Let $y = \sum_{i=0}^{m-1} y_i \alpha^i$ be another element in $GF(2^m)$. Then we have the dual basis expression $xy = \sum_{k=0}^{m-1} [xy]_k \beta_k$, where $[xy]_k = Tr(\alpha^k xy) = Tr(\alpha^k x \sum_{i=0}^{m-1} y_i \alpha^i) = \sum_{i=0}^{m-1} y_i Tr(\alpha^{i+k} x) = \sum_{i=0}^{m-1} y_i [\alpha^k x]_i$. Note that $[\alpha^k x]_i$ is the ith coefficient of the dual basis expression of $\alpha^k x$. On the other hand, we have $[\alpha x]_i = Tr(\alpha^i \alpha x) = [x]_{i+1}$, $0 \leq i \leq m-2$. Also letting $f_0 + f_1 X + f_2 X^2 + \cdots + f_{m-1} X^{m-1} + X^m \in GF(2)[X]$ be the irreducible polynomial of α over $GF(2)$, we get $[\alpha x]_{m-1} = Tr(\alpha^{m-1} \alpha x) = Tr(\sum_{i=0}^{m-1} f_i \alpha^i x) = \sum_{i=0}^{m-1} f_i Tr(\alpha^i x) = \sum_{i=0}^{m-1} f_i [x]_i$. Therefore, the coefficients of the dual basis expressions of $x, \alpha x, \cdots, \alpha^{m-1} x$ are recursively computed by above relations. The multiplication $[xy]_k = \sum_{i=0}^{m-1} y_i [\alpha^k x]_i$ is realized by the following shift register arrangement of Berlekamp [1].

On the other hand, there are not so many bit serial multipliers using normal basis. Normal basis multipliers are quite effective in such operations as exponentiation and inversion since a squaring operation is just a cyclic shift in normal basis expression. There are various types of bit parallel normal basis multipliers [5, 6, 7, 10]. Among them, so called optimal normal basis multipliers [4, 5, 6] require least number of gates than other types of parallel multipliers. There are two kinds of optimal normal bases [4], namely, type I and type II. Our aim in this paper is to present a design of a bit serial multiplier using a type II optimal normal basis which satisfies the following properties. First, a squaring operation is just a permutation in our basis. Therefore, exponentiation and inverse finding operations are much faster than Berlekamp's multiplier. Second, it needs no basis conversion process which is usually required in dual basis multipliers.

2   Normal Basis and Optimal Normal Basis of Type II

Let α be an element of $GF(2^m)$ of degree m and let $f(X) = f_0 + f_1 X + \cdots + f_{m-1} X^{m-1} + X^m$ be the irreducible polynomial of α over $GF(2)$. Then we have

$$0 = f(\alpha) = f_0 + f_1 \alpha + f_2 \alpha^2 + \cdots + f_{m-1} \alpha^{m-1} + \alpha^m.$$

From this, it is clear that $0 = f(\alpha)^{2^i} = f(\alpha^{2^i})$ for all $0 \le i \le m-1$ since $f_0, f_1, \cdots, f_{m-1}$ are in $GF(2)$ and the characteristic of $GF(2^m)$ is two. In other words, all the zeros of $f(X)$ are $\alpha, \alpha^2, \alpha^{2^2}, \cdots, \alpha^{2^{m-1}}$. If all the conjugates, $\alpha, \alpha^2, \alpha^{2^2}, \cdots, \alpha^{2^{m-1}}$, of α are linearly independent over $GF(2)$, then they form a basis for $GF(2^m)$ over $GF(2)$.

Definition 2. *A basis of $GF(2^m)$ over $GF(2)$ of the form $\{\alpha, \alpha^2, \cdots, \alpha^{2^{m-1}}\}$ is called a normal basis.*

It is a standard fact [4] that there is always a normal basis in $GF(2^m)$ for any $m \ge 1$. If an element x in $GF(2^m)$ is expressed with respect to a normal basis $\{\alpha, \alpha^2, \cdots, \alpha^{2^{m-1}}\}$, i.e. if $x = x_0 \alpha + x_1 \alpha^2 + \cdots + x_{m-1} \alpha^{2^{m-1}}$, then one easily notices $x^2 = x_{m-1} \alpha + x_0 \alpha^2 + x_1 \alpha^{2^2} + \cdots + x_{m-2} \alpha^{2^{m-1}}$. That is, x^2 is a right cyclic shift of x with respect the basis $\{\alpha, \alpha^2, \cdots, \alpha^{2^{m-1}}\}$. On the other hand, a normal basis expression of a product xy of two different elements x and y in $GF(2^m)$ is not so simple. This is because the expression $\alpha^{2^i} \alpha^{2^j} = \sum_{k=0}^{m-1} a_k^{ij} \alpha^{2^k}$ may be quite complicated if one does not choose a normal basis properly. Therefore, to find an efficient bit serial multiplication using a normal basis, one has to choose a normal basis so that the coefficients a_k^{ij} in the expression $\alpha^{2^i} \alpha^{2^j}$ are zero for many indices i, j and k. There are not so many normal bases satisfying this condition, but we have one example of such normal basis and it is stated in the following theorem. A detailed proof can be found in [4].

Theorem 1. *Let $GF(2^m)$ be a finite field of 2^m elements where $2m + 1 = p$ is a prime. Suppose that either $(\star)$ 2 is a primitive root $\pmod{p}$ or $(\star\star)$ -1 is a quadratic nonresidue $\pmod{p}$ and 2 generates all the quadratic residues $\pmod{p}$. Then letting $\alpha = \beta + \beta^{-1}$ where β is a primitive pth root of unity in $GF(2^{2m})$, we have $\alpha \in GF(2^m)$ and $\{\alpha, \alpha^2, \cdots, \alpha^{2^{m-1}}\}$ is a basis over $GF(2)$.*

Definition 3. *A normal basis in theorem 1 is called an optimal normal basis of type II.*

It is known [4] that we have an optimal normal basis of type II in $GF(2^m)$ when $m = 2, 3, 5, 6, 9, 11, 14, 18, 23, 26, 29, 30, 33, 35, 39, 41, 50, 51, 53, 65, 69, 74, \cdots$. The number of $m \le 2000$ for which a type II optimal normal basis exists is 324. Using the assumptions in the previous theorem, one finds easily

$$\alpha^{2^s} = (\beta + \beta^{-1})^{2^s} = \beta^{2^s} + \beta^{-2^s} = \beta^t + \beta^{-t},$$

where $0 < t < p = 2m + 1$ with $2^s \equiv t \pmod{p}$. Moreover, replacing t by $p - t$ if $m + 1 \leq t \leq 2m$, we find that $\{\alpha, \alpha^2, \cdots, \alpha^{2^{m-1}}\}$ and $\{\beta + \beta^{-1}, \beta^2 + \beta^{-2}, \cdots, \beta^m + \beta^{-m}\}$ are same sets. That is, $\alpha^{2^s}, 0 \leq s \leq m - 1$ is just a permutation of $\beta^s + \beta^{-s}, 1 \leq s \leq m$. Above observation leads to the following definition.

Definition 4. *Let β be a primitive pth ($p = 2m + 1$) root of unity in $GF(2^{2m})$. For each integer s, define α_s as*

$$\alpha_s = \beta^s + \beta^{-s}.$$

Then for each integer s and t, we easily find

$$\alpha_s \alpha_t = (\beta^s + \beta^{-s})(\beta^t + \beta^{-t}) = \alpha_{s-t} + \alpha_{s+t}.$$

In other words, a multiplication of two basis elements has a simple expression as a sum of two basis elements.

Lemma 1. *An optimal normal basis $\{\alpha^{2^s} | 0 \leq s \leq m - 1\}$ of type II in $GF(2^m)$, if it exists, is self dual.*

Proof. After a permutation of basis elements, it can be written as $\{\alpha_s | 1 \leq s \leq m\}$. Note that $Tr(\alpha_i \alpha_j) = Tr((\beta^i + \beta^{-i})(\beta^j + \beta^{-j})) = Tr(\alpha_{i-j} + \alpha_{i+j})$. If $i = j$, then we have $Tr(\alpha_{2i}) = Tr(\alpha^{2^s})$ for some $0 \leq s \leq m - 1$. Thus the trace value is $\alpha + \alpha^2 + \cdots + \alpha^{2^{m-1}} = 1$ because of the linear independence. If $i \neq j$, then there are s and t with $0 \leq s, t \leq m - 1$ such that $\alpha_{i-j} = \alpha^{2^s}$ and $\alpha_{i+j} = \alpha^{2^t}$. Therefore $Tr(\alpha_{i-j} + \alpha_{i+j}) = Tr(\alpha_{i-j}) + Tr(\alpha_{i+j}) = Tr(\alpha^{2^s}) + Tr(\alpha^{2^t}) = 1 + 1 = 0.$ □

From now on, we assume $\{\alpha^{2^s} | 0 \leq s \leq m - 1\}$ is an optimal normal basis of type II in $GF(2^m)$ and $\{\alpha_s | 1 \leq s \leq m\}$ is a basis obtained after a permutation of the basis elements of the normal basis. For a given $x = \sum_{i=1}^{m} x_i \alpha_i$ with $x_i \in GF(2)$, by using lemma 1, we have

$$x_s = \sum_{i=1}^{m} x_i Tr(\alpha_s \alpha_i) = Tr(\alpha_s \sum_{i=1}^{m} x_i \alpha_i) = Tr(\alpha_s x),$$

for all $1 \leq s \leq m$. We extend above relation by defining x_s as $x_s = Tr(\alpha_s x)$ for all integers s, where $\alpha_s = \beta^s + \beta^{-s}$ from definition 4. Then for any integer s, we have the following properties.

Lemma 2. *We have $\alpha_s = 0 = x_s$ if $2m + 1$ divides s, and*

$$\alpha_{2m+1+s} = \alpha_s = \alpha_{2m+1-s} = \alpha_{-s}, \quad x_{2m+1+s} = x_s = x_{2m+1-s} = x_{-s},$$

for all s.

Proof. We have $\alpha_s = \beta^s + \beta^{-s} = 0$ if and only if $\beta^s = \beta^{-s}$, that is, $\beta^{2s} = 1$. And this happens whenever $2m + 1 = p$ divides s since β is a primitive pth root of unity. Now we have $\alpha_{2m+1+s} = \beta^{2m+1+s} + \beta^{-(2m+1+s)} = \beta^s + \beta^{-s} = \alpha_s$ because $\beta^{2m+1} = 1$. Also $\alpha_{2m+1-s} = \beta^{2m+1-s} + \beta^{-(2m+1-s)} = \beta^{-s} + \beta^s = \alpha_s$. The result for x_s instantly follows from the result for α_s.   □

3   Multiplication Algorithm

Using lemma 1 and 2, we are ready to give the following assertion, which explains a bit serial multiplication in $GF(2^m)$ using an optimal normal basis of type II.

Theorem 2. Let $x = \sum_{i=1}^{m} x_i \alpha_i$ and $y = \sum_{i=1}^{m} y_i \alpha_i$ be elements in $GF(2^m)$. Then we have $xy = \sum_{k=1}^{m} (xy)_k \alpha_k$, where the kth coefficient $(xy)_k$ satisfies

$$(xy)_k = \sum_{i=1}^{2m+1} y_i x_{i-k}.$$

Proof. By lemma 1, $\{\alpha_k | 1 \leq k \leq m\}$ is self dual. Thus

$$(xy)_k = Tr(\alpha_k xy) = Tr(\alpha_k x \sum_{i=1}^{m} y_i \alpha_i)$$

$$= \sum_{i=1}^{m} y_i Tr(\alpha_k \alpha_i x) = \sum_{i=1}^{m} y_i Tr(\alpha_{i-k} x + \alpha_{i+k} x)$$

$$= \sum_{i=1}^{m} y_i (Tr(\alpha_{i-k} x) + Tr(\alpha_{i+k} x)) = \sum_{i=1}^{m} y_i (x_{i-k} + x_{i+k})$$

$$= \sum_{i=1}^{m} y_i x_{i-k} + \sum_{i=1}^{m} y_i x_{i+k}.$$

On the other hand, the second summation of above expression can be written as

$$\sum_{i=1}^{m} y_i x_{i+k} = \sum_{i=1}^{m} y_{m+1-i} x_{m+1-i+k}$$

$$= \sum_{i=1}^{m} y_{m+i} x_{m+i-k} = \sum_{i=m+1}^{2m} y_i x_{i-k},$$

where the first equality follows by rearranging the summands and the second equality follows from lemma 2. Therefore we get

$$(xy)_k = \sum_{i=1}^{m} y_i x_{i-k} + \sum_{i=1}^{m} y_i x_{i+k} = \sum_{i=1}^{2m} y_i x_{i-k}.$$

Since $y_{2m+1} = 0$ by lemma 2, $(xy)_k = \sum_{i=1}^{2m+1} y_i x_{i-k}$ is obvious from above result.   □

4   Bit Serial Arrangement Using an Optimal Normal Basis of Type II

We present the designs of the multiplications, xy and xy^2 in $GF(2^m)$. One of the merits of our basis is that the computation of xy^2 is nothing more complicated than the computation of xy. On the other hand, in most of the dual basis multipliers, the computation of xy^2 is more complicated than the computation of xy.

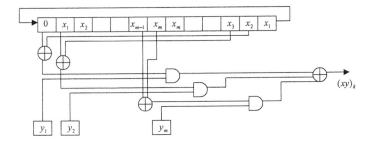

Fig. 2. Bit serial arrangement for computing xy using an optimal normal basis of type II

4.1   Computation of xy in $GF(2^m)$

Using theorem 2, we may express $(xy)_k$ as a matrix multiplication form of a row vector and a column vector,

$$(xy)_k = (x_{1-k}, x_{2-k}, \cdots, x_{2m-k}, x_{2m+1-k})(y_1, y_2, \cdots, y_{2m}, y_{2m+1})^T,$$

where $(y_1, y_2, \cdots, y_{2m}, y_{2m+1})^T$ is a transposition of the row vector $(y_1, y_2, \cdots, y_{2m}, y_{2m+1})$. Then we have

$$(xy)_{k+1} = (x_{-k}, x_{1-k}, \cdots, x_{2m-1-k}, x_{2m-k})(y_1, y_2, \cdots, y_{2m}, y_{2m+1})^T.$$

Since $x_{-k} = x_{2m+1-k}$ by lemma 2, we find that $(x_{-k}, x_{1-k}, \cdots, x_{2m-1-k}, x_{2m-k})$ is a right cyclic shift of $(x_{1-k}, x_{2-k}, \cdots, x_{2m-k}, x_{2m+1-k})$ by one position. From this observation, we may realize the multiplication algorithm in the shift register arrangement shown in Fig. 2. The shift register is initially loaded with $(x_0, x_1, \cdots, x_{2m})$ which is in fact $(0, x_1, \cdots, x_m, x_m, \cdots, x_1)$. After k clock cycles, we get $(xy)_k$, the kth coefficient of xy with respect to the basis $\{\alpha_1, \cdots, \alpha_m\}$.

Compared with Berlekamp's dual basis multiplier in Fig. 1, our design of bit serial multiplier requires $m+1$ more flip-flops (in the expression of the input x). However, since our multiplier is using a normal basis, such arithmetical operations as squaring, exponentiation and inversion can be very efficiently computed. Moreover we do not need a basis conversion process which is required in dual basis multipliers. Note that a bit parallel multiplier using an optimal normal basis of type II is discussed in [6], where it is suggested to find a bit serial version of the bit parallel construction in [6]. So our multiplier gives a modest answer for the suggestion. It should be mentioned that our multiplier can be modified to give a high throughput architecture where the input y is loaded serially and the partial sum is stored in each flip-flops at each cycle. This is also true for Berlekamp's multiplier. The strong point of our construction is that our algorithm is easily adapted to give a low complexity and low latency systolic array for multiplication, which will appear elsewhere.

4.2   Computation of xy^2 in $GF(2^m)$

In the theory of error correcting codes, the following types of product sum operations in $GF(2^m)$, $xy + z$ and $xy^2 + z$, are the most frequently used arithmetic operations. Those product sum operations are easily realized by using extra registers and XOR gates once you have the circuits which compute xy and xy^2. In previous section, we found a bit serial arrangement of the multiplication xy in $GF(2^m)$ using an optimal normal basis of type II. Since a squaring operation is a permutation in our basis, the operation xy^2 is computed by the following two steps. First, compute y^2 by permutating the coefficients of y. Second, compute xy^2 using the bit serial arrangement in Fig. 2. However, we do not have to actually permutate the coefficients of y to compute xy^2. We only need the second step by slightly adjusting the wiring in Fig. 2. The idea is derived from the following theorem.

Theorem 3. $(xy^2)_k = \sum_{i=1}^{m} y_i x_{2i-k} + \sum_{i=1}^{m} y_{m+i} x_{2i-k-1}.$

Proof. Similar to the proof of theorem 2.   □

Using above theorem, we may express $(xy^2)_k$ as a matrix multiplication form,

$$(xy^2)_k = (x_{1-k}, x_{2-k}, x_{3-k}, x_{4-k}, \cdots, x_{2m-1-k}, x_{2m-k}, x_{2m+1-k})$$
$$\times (y_m, y_1, y_{m-1}, y_2, \cdots, y_1, y_m, 0)^T,$$

where we used $y_{m+i} = y_{m+1-i}$ from lemma 2. Then we have

$$(xy^2)_{k+1} = (x_{-k}, x_{1-k}, x_{2-k}, x_{3-k}, \cdots, x_{2m-2-k}, x_{2m-1-k}, x_{2m-k})$$
$$\times (y_m, y_1, y_{m-1}, y_2, \cdots, y_1, y_m, 0)^T.$$

Since $x_{-k} = x_{2m+1-k}$ from lemma 2, we find that $(x_{-k}, x_{1-k}, \cdots, x_{2m-1-k}, x_{2m-k})$ is a right cyclic shift of $(x_{1-k}, x_{2-k}, \cdots, x_{2m-k}, x_{2m+1-k})$ by one position. Therefore, the multiplication algorithm is realized in the following shift register arrangement shown in Fig. 3, where $\lceil a \rceil$ denotes the least integer satisfying $a \le \lceil a \rceil$.

The shift register is initially loaded with $(0, x_1, \cdots, x_m, x_m, \cdots, x_1)$ which same to the case of Fig. 2. Also the input to y is same to that of Fig. 2. Again, like the case of Fig. 2, we get all the coefficients $(xy^2)_1, \cdots, (xy^2)_m$ after m clock cycles. In fact, the only difference between Fig. 2 and Fig. 3 is the wiring of the circuit. Therefore, by adding extra logical gates and wiring to the arrangement in Fig. 2, we may have a multiplier which can compute both of xy and xy^2 simultaneously in the same clock cycles. Note that, in the case of the dual basis multiplier, we do not have a similar structure for the multiplication xy^2 and the computation of xy^2 is much more time consuming than the computation of xy.

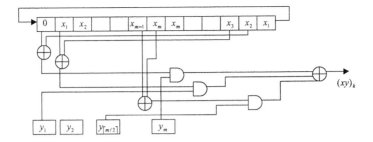

Fig. 3. Bit serial arrangement for computing xy^2 using an optimal normal basis of type II

5   Comparison with a Type I Optimal Normal Basis Multiplier

A bit serial multiplication using an optimal normal basis of type I is already well known [10, 11, 12]. Since a type I optimal normal basis has quite a simple arithmetical property than that of a type II optimal normal basis, there are abundance of papers on the type I optimal normal basis [5, 7, 13]. On the other hand, we have not so many papers on the type II optimal normal basis. When $m + 1 = p$ is a prime and 2 is a primitive root $\pmod{p}$, we have the irreducible polynomial $f(X) = 1 + X + X^2 + \cdots + X^m \in GF(2)[X]$, which is usually called the all one polynomial (AOP). Letting $\alpha \in GF(2^m)$ be any zero of the AOP $f(X)$, we have $\{\alpha, \alpha^2, \alpha^3, \cdots, \alpha^m\} = \{\alpha, \alpha^2, \alpha^{2^2}, \cdots, \alpha^{2^{m-1}}\}$. Above basis is called the optimal normal basis of type I in $GF(2^m)$. If an element x is in $GF(2^m)$ where a type I optimal normal basis (or an AOP basis) exists, one usually writes x as $x = \sum_{i=0}^{m} x_i \alpha^i$ with respect to the extended AOP basis $\{1, \alpha, \alpha^2, \cdots, \alpha^m\}$. The extended AOP basis is not really a basis because of the redundancy (linear dependence) of the basis elements. However, using the property $\alpha^{m+1} = 1$, one may have many nice bit parallel and serial multiplication structures. In fact, it is trivial to find the multiplication of $x = \sum_{i=0}^{m} x_i \alpha^i$ and $y = \sum_{i=0}^{m} y_i \alpha^i$ as

$$xy = \sum_{i=0}^{m} x_i \alpha^i \sum_{j=0}^{m} y_j \alpha^j = \sum x_i y_j \alpha^{i+j} = \sum_{i=0}^{m} \sum_{j=0}^{m} y_j x_{i-j} \alpha^i,$$

where it is defined, for all integers s and t, that $x_s = x_t$ if and only if $s \equiv t \pmod{m+1}$. From above relation, we have

$$(xy)_k = \sum_{i=0}^{m} y_i x_{k-i} = (x_k, x_{k-1}, \cdots, x_{k-m})(y_0, y_1, \cdots, y_m)^T,$$

which explains a bit serial multiplication. This is already noticed in [10] and it is mentioned again in [11]. But the explicit figure of the circuit appears only in [12].

The design uses $2m + 2$ flip-flops, which is two more than that of Berlekamp's multiplier. Therefore, the hardware complexity of a bit serial multiplier using an extended AOP basis is lower than that of our multiplier using an optimal normal basis of type II. However, the extended AOP basis multiplier gives an output after $m+1$ clock cycles, whereas our multiplier gives an output after m clock cycles. Also, since the extended AOP basis is a nonconventional basis having $m+1$ basis elements in $GF(2^m)$, one needs extra logical operations to convert the basis to an ordinary basis. But our multiplier has no such problem. Moreover, an AOP basis (or an optimal normal basis of type I) appears quite less frequently than a type II optimal normal basis. In fact, a table in [4, p. 100] shows that the number of $m \leq 2000$ for which an AOP basis exists is 118. For example, we have an AOP basis when $m = 2, 4, 10, 12, 18, 28, 36, 52, 58, 60, 66, 82, 100, 106, \cdots$. On the other hand, the same table says that the number of $m \leq 2000$ for which a type II optimal normal basis exists is 324. There is a type II optimal normal basis when $m = 2, 3, 5, 6, 9, 11, 14, 18, 23, 26, 29, 30, 33, 35, 39, 41, 50, 51, 53, 65, 69, 74, \cdots$. From this observation, one may deduce that a type II optimal normal basis is three times more likely to occur than an AOP basis. Therefore our bit serial multiplier using an optimal normal basis of type II provides an efficient multiplication architecture for many finite fields, where an AOP basis does not exist and no other comparable architecture is known yet. It should be mentioned that by slightly modifying the expression

$$(xy)_k = \sum_{i=0}^{m} y_i x_{k-i} = (x_k, x_{k-1}, \cdots, x_{k-m})(y_0, y_1, \cdots, y_m)^T,$$

it is not at all difficult to derive (replace k by $2k$ and rearrange the order of summation.)

$$(xy)_{2k} = \sum_{i=0}^{m} y_{k+i} x_{k-i} = (x_k, x_{k-1}, \cdots, x_{k-m})(y_k, y_{k+1}, \cdots, y_{k+m})^T,$$

which gives a low complexity and a high throughput bit parallel systolic multiplier using an extended AOP basis. But the result appeared quite recently in [13], where a rather complicated argument is given for above result.

6   Conclusions

In this paper, we proposed a bit serial multiplier in $GF(2^m)$ using a type II optimal normal basis. We showed that our multiplier has the properties of modularity, regularity and scalability. Therefore our architecture is suitable for VLSI implementations such as smart cards and IC cards. Since our basis is self dual, we do not have to worry about the basis conversion process which is required in Berlekamp's dual basis multiplier. Moreover, our multiplier has a simple squaring operation because we use a normal basis like other Massey-Omura type multipliers. When compared with a bit serial multiplier with an extended AOP basis multiplier, our multiplier is applicable to three times more broad class of finite fields.

Our multiplier has one more advantage over AOP basis (or type I optimal normal basis) multiplier. That is, suppose that $\{\alpha, \alpha^2, \cdots, \alpha^{2^{m-1}}\}$ is a normal basis of $GF(2^m)$. In some applications such as pseudo random number generation, it is desirable that one should choose the basis element α to be a primitive element, i.e. α is a generator of the multiplicative group $GF(2^m) - 0$, or equivalently the irreducible polynomial of α is a primitive polynomial. In the case of a type I optimal normal basis, this condition is never satisfied since one always has $\alpha^{m+1} = 1$. However if $\{\alpha, \alpha^2, \cdots, \alpha^{2^{m-1}}\}$ is a type II optimal basis, it is very likely that α is a primitive element. In fact, a table in [4, p.111] shows that among the 100 values of $m \leq 515$ for which an optimal normal basis of type II exists, the corresponding α is a primitive element for 82 values of m. For example, α is a primitive element when $m = 3, 5, 6, 9, 11, 14, 23, 26, 29, 30, 33, 35, 39, 41, 51, 53, 65, 69, 74 \cdots$. Though we do not have a complete information of the distribution of the primitive elements α at this moment, we may conclude, from known examples, that approximately 80 percent of type II optimal normal bases $\{\alpha, \alpha^2, \cdots, \alpha^{2^{m-1}}\}$ have α as a primitive element.

References

[1] E. R. Berlekamp, "Bit-serial Reed-Solomon encoders," *IEEE Trans. Inform. Theory*, **28**, pp. 869–874, 1982.
[2] M. Wang and I. F. Blake, "Bit serial multiplication in finite fields," *SIAM J. Disc. Math.*, **3**, pp. 140–148, 1990.
[3] M. Morii, M. Kasahara and D. L. Whiting, "Efficient bit-serial multiplication and the discrete-time Wiener-Hopf equation over finite fields," *IEEE Trans. Inform. Theory*, **35**, pp. 1177–1183, 1989.
[4] A. J. Menezes, *Applications of finite fields*, Kluwer Academic Publisher, 1993.
[5] Ç.K. Koç and B. Sunar, "Low complexity bit-parallell canonical and normal basis multipliers for a class of finite fields," *IEEE Trans. Computers*,**47**, pp. 353–356, 1998.
[6] B. Sunar and Ç.K. Koç, "An efficient optimal normal basis type II multiplier," *IEEE Trans. Computers*,**50**, pp. 83–87, 2001.
[7] C. Paar, P. Fleischmann and P. Roelse, "Efficient multiplier architectures for Galois fields $GF(2^{4n})$," *IEEE Trans. Computers*, vol. 47, pp. 162–170, 1998.
[8] D. R. Stinson, "On bit-serial multiplication and dual bases in $GF(2^m)$," *IEEE Trans. Inform. Theory*, **37**, pp. 1733–1736, 1991.
[9] M. A. Hasan and V. K. Bhargava, "Division and bit-serial multiplication over $GF(q^m)$," *IEE Proc. E*, **139**, pp. 230–236, 1992.
[10] T. Itoh and S. Tsujii, "Structure of parallel multipliers for a class of finite fields $GF(2^m)$," *Information and Computations*, **83**, pp. 21–40, 1989.
[11] G. Drolet, "A new representation of elements of finite fields $GF(2^m)$ yielding small complexity arithmetic circuits," *IEEE Trans. Computers*,**47**, pp. 938–946, 1998.
[12] S. T.J Fenn, M. G. Parker, M. Benaissa and D. Taylor, "Bit-serial multiplication in $GF(2^m)$ using irreducible all one polynomials," *IEE Proc. Comput. Digit. Tech.*, **144**, pp. 391–393, 1997.
[13] C. Y. Lee, E. H. Lu and J. Y. Lee, "Bit parallel systolic multipliers for $GF(2^m)$ fields defined by all one and equally spaced polynomials," *IEEE Trans. Computers*, **50**, pp. 385–393, 2001.

New Approach for Configuring Hierarchical Virtual Private Networks Using Proxy Gateways

Hayato Ishibashi[1], Kiyohiko Okayama[2], Nariyoshi Yamai[3],
Kota Abe[1], and Toshio Matsuura[1]

[1] Media Center, Osaka City University
3-3-138 Sugimoto, Sumiyoshi-ku, Osaka 558-8585, Japan
{ishibashi,k-abe,matsuura}@media.osaka-cu.ac.jp
[2] Faculty of Engineering, Okayama University
3-1-1, Tsushima-naka, Okayama 700-8530, Japan
okayama@cne.okayama-u.ac.jp
[3] Computer Center, Okayama University
3-1-1, Tsushima-naka, Okayama 700-8530, Japan
yamai@cc.okayama-u.ac.jp

Abstract. VPN is one of key technologies on the Internet that allows users to access securely to resources in a domain via unsecure networks. For hierarchically nested security domains, such as an R&D division domain in a corporate domain, In such organizations, some existing VPN schemes with multiple security gateway traversal function is applicable for a user to access to the innermost security domain from the Internet. However, most of existing schemes have some drawbacks in terms of security, efficiency and availability. In this paper, we propose a new way to remedy these shortcomings using proxy gateways. The proposed method connects two deeply embedded security domains by a series of virtual paths to create a single VPN link; and by incorporating a proxy gateway to accommodate communication between clients and the security gateway, this permits secure and highly efficient communications without modifying the client or server.

1 Introduction

As the Internet has continued to evolve, it is supporting an increasingly diverse range of information. Since this includes personal information and other data of a highly confidential nature, it is exceedingly important to safeguard the security of communications.

Authentication and encryption technologies are generally used to provide this security. One approach is to set up a Virtual Private Network (VPN) that links remote locations over the Internet. A VPN is a technology enabling users to set up a virtual link (referred to below as a VPN link) between two points that appears to be a direct connection. This approach has attracted enormous interest because VPNs incorporate authentication and encryption technologies, and therefore support secure delivery of information over the public network.

H.-K. Kahng (Ed.): ICOIN 2003, LNCS 2662, pp. 772–782, 2003.

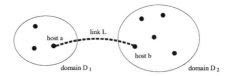

Fig. 1. Schematic of the VPN concept

There are various ways of implementing VPNs. They can be classified as those configured based on host-to-host VPN links and those based on host-to-network (or network-to-network) VPN links. The former requires that both application clients and servers incorporate special VPN software, while many implementations of the latter approach do not require VPN software in the application server. For this reason, we will focus in this work on the latter approach to implementing VPNs. In this approach to setting up VPNs, a domain having a uniform access policy (referred to hereafter as a security domain) is defined, and security gateways (SGWs) are set up at all points of contact or interfaces with the outside. The SGWs control whether or not access requests from the outside to the inside (or from the inside to the outside) will be allowed, and when access is allowed, they relay the communication. It is commonplace in larger organizations for each division and department to have its own access policy, so naturally security domains are hierarchically structured much like Internet domains. In this kind of arrangement, when a client on the Internet wants to communicate with a server in the innermost security domain, it must pass through a succession of SGWs one at a time beginning with the outermost security domain SGW.

There are a number of existing VPN configuration schemes that can be employed in these hierarchical security domain situations[1][2][3]. However, these schemes have some drawbacks in terms of security, efficiency and availability.

To solve these drawbacks, here we propose a hierarchical VPN configuration method that uses proxy gateways. The proxy gateways that perform protocol conversion are deployed between clients and SGWs, and this allows existing VPN software to be used in clients without any modification. Furthermore, by establishing a single VPN link just between the proxy gateway and SGWs adjacent to the server, the communication overheads can be minimized even if the number of security domain levels increases.

2   Hierarchical VPN Model and Current Implementations

2.1   Hierarchical VPN Model

The VPN is a technology permitting two points to be interconnected by a virtual link (VPN link) over a network in a manner that appears to users to be a direct connection. This is illustrated in Figure 1 where Host A located in Security Domain D1 and Host B Located in Security Domain D2 are interconnected by Virtual Link L, yet logically it can appear that Host A is in Security Domain D2 (or that Host B is in Security Domain D1).

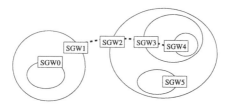

Fig. 2. Hierarchically nested security domains

Given this configuration, there are cases where only Host A logically appears to belong to Security Domain D2 (terminal connection), and cases where all hosts actually in Security Domain D1 appear as if they belong to Security Domain D2 (inter-LAN connection). Since the latter case can be considered as a special case of the former case where host a works as a router, the following discussion will assume a terminal-type connection. Moreover, since both authentication and message encryption are required in most real-world situations, we will assume a VPN implementation that includes these security-related functions.

Obviously the simplest security policy in an organization such as a university or company is a uniform policy that covers the entire organization. In this case, the entire organization constitutes a single security domain. However, in organizations of certain magnitude, this is rarely the case, there are data that can only be accessed by certain people within the organization. One can also imagine that there are certain kinds of data that cannot be accessed by other departments within the same company, yet can be accessed by certain organizations outside the company. Examples would include a university hospital connected to the university intranet that wants to share data with other hospitals yet not give access to other departments at the university, or a corporate research lab that wants to exchange data with some other external research institute in pursuing a collaborative research project. In the former example, the entire university constitutes one security domain, and the affiliated hospital also constitutes a more circumscribed security domain within the larger domain of the university.

As these examples make clear and as illustrated in Fig. 2, it is commonplace for multiple security domains to exist in a single organization, and often these security domains are hierarchically nested inside of one another. In this work, therefore, we will be mainly concerned with this kind of arrangement where security domains are hierarchically nested.

2.2   Deficiencies in the Existing VPN Implementation Methods

There are a number of existing VPN configuration schemes that can be employed in these hierarchical security domain situations: (1) SOCKS5[4] multi-stage proxy mechanism can be added as a standalone function in the SOCKS version 5[1] protocol (**Conventional Method 1**), (2) SOCKS version 5 protocol can be extended for multiple SGW traversal[2] (**Conventional Method 2**), and (3) SSL proxy servers can be deployed in each security domain as SGWs[3]

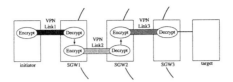

Fig. 3. VPN links implemented by conventional method 1

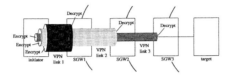

Fig. 4. VPN links implemented by conventional method 2

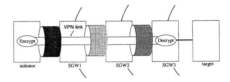

Fig. 5. VPN links implemented by conventional method 3

(**Conventional Method 3**). In the rest of this paper we shall refer to these three conventional methods as CM1, CM2, and CM3, respectively.

Figures 3, 4, and 5 show how VPN links would be established by CM1 though CM3, respectively, assuming an outside initiator (client) is seeking to access a target (host) protected by three security domain levels. In CM1 and CM2, the initiator is a socks client, SGW1 – SGW3 are socks servers, and the target corresponds to an application server. In CM3, the initiator is the application client corresponding to an SSL, SGW1 – SGW3 are SSL proxy servers, and the target is an application server corresponding to an SSL. Note that in all of these schemes, the initiator and each SGW must have a routing table so that they know which SGW to connect to next in order to reach the target. Here we will refer to the SGW that can be directly communicated with by the initiator as the starting SGW, the SGW that can be directly communicated with by the target as the ending SGW, and all SGW between the starting and final SGWs as transit SGWs.

All three conventional methods (CM1 through CM3) are similar in that the initiator traverses hierarchically nested security domain SGWs to set up a VPN link between the initiator and the final SGW, and confidential communications are sent from the initiator to gateway using the VPN link encryption capability. However, the ways in which the VPN links are established differ as follows:

CM1: VPN links are set up between adjacent SGWs hop by hop (in this case, direct communication is supported) starting with the SGW closest to the initiator based on the routing table maintained by the initiator and all the SGWs.

CM 2: VPN links are established between the initiator and the each SGW one hop at a time for all SGWs on the path from the initiator to the target.

CM 3: VPN links based on CM1 are regarded as virtual paths only with authentication capability, and a single VPN is established between the initiator and the final SGW after virtual paths are set up on all segments between the initiator and the final SGW.

In CM1 and CM2, authentication and/or encryption is performed on each VPN link as required, but these operations are carried out independently on each VPN link. This means that if encrypted communication is sent across all the VPN links, the encryption and decryption must be repeatedly performed at the initiator and each SGW along the way, which causes increasingly diminished communications efficiency as the number of security domain levels increases. One way around this problem is using SSL or some other software for encryption between the initiator and the target instead of encrypting every VPN link of CM1 or CM2. Although this would minimize the encryption overhead due to the increasing number of security domain levels, it creates a new problem in that additional VPN software not part of CM1 or CM2 must be included in both the initiator and in the target.

Alternatively, in CM3 VPN links are normally only set up between the initiator and the final SGW, so here again—just the same as when SSL or other VPN software is combined with CM1 or CM2—the encryption overhead due to increasing number of security domain levels is minimized. However, in the case of CM3 there is the problem that, since the SSL protocol would have to be extended in order for the initiator to perform authentication at each SGW, a library would have to be added to the initiator for this purpose.

3   Proposed Method Using Proxy Gateways

3.1   Preconditions

In order to solve the various problems outlined in Section 2.2 in supporting efficient communications over a VPN penetrating hierarchically nested security domains, at least the following two conditions must be satisfied:

Condition 1: Encryption overhead should be minimized.

Condition 2: The initiator should be capable of utilizing existing VPN software considering convenience and cost to the user.

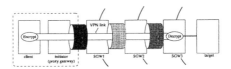

Fig. 6. VPN links implemented by the proposed method

Moreover, considering the management and operation of VPNs in the actual Internet environment, it would also make sense to consider ways to satisfy the following two additional conditions:

Condition 3: Good VPN management functions should be supported, including management of users, the ability to require or waive authentication independently for each SGW, and so on.

Condition 4: Considering situations where private addresses are used in security domains where targets are also located, it should be possible to apply the IP Network Address Translator (NAT)[5, 6].

3.2   Methods of Establishing VPN Links

Turning first to communication efficiency, Condition 1 in Section 3.1 can be readily satisfied by incorporating SSL (or comparable software) in CM1 or CM2, or by adopting CM3. The problem with these approach is, as we noted earlier, that this would require adding additional software to the initiator, which would contradict Condition 2. Especially the options of combining SSL software with CM1 or CM2 are problematic, because the target would also have to be modified to support SSL.

In this work we propose a new approach in which a proxy gateway for protocol conversion is added to the VPN link setup procedures of CM3. Figure 6 shows a reference model of our proposed method in which VPN links are established between an initiator and target separated by three security domain levels.

In the proposed method, a proxy gateway is inserted between the client and the starting SGW, and the proxy gateway receives a request to set up a VPN link using an existing VPN implementation method from the client. The proxy gateway then assumes the role of the initiator, and sets up virtual paths on each segment to the final SGW the same as CM3. After that, VPN links are established between the client and the final SGW by an existing VPN implementation method. Now we shall examine in greater detail the steps involved in setting up VPN links by our proposed method:

1. The client attempts to establish a connection to the final SGW using the protocol of the existing VPN implementation method.

2. The initiator (proxy gateway) detects the packet sent by the client, and stores it rather than sending it on. The initiator then retrieves the routing table based on the IP address of the packet, and attempts to establish a connection with the outermost security domain SGW (the starting SGW).

3. Once a connection is established, the initiator sends its own IP address and port number and the IP address and port number of the target to the SGW.

4. The SGW determines the authentication method based on the received information, and performs authentication in conjunction with the initiator.

5. If the authentication is successful (including cases where authentication is not required), the SGW checks the routing table to see if it is the final SGW. If it is not the final SGW, a connection is set up with the next SGW based on the routing table, and subsequent packets are transparently delivered.

6. Steps (3) through (5) are then repeated moving to the next-nearest SGW one at a time until the final SGW is reached.

7. The final SGW sends a message to the initiator indicating that the connection setup is complete. Once this message is received by the initiator, the packets that were received from the client and stored are delivered to the final SGW over the virtual path.

8. The client attempts to set up a VPN link to the final SGW using an existing VPN implementation protocol; if successful, the final SGW delivers the packets received from the client to the target. Encryption on the VPN links between the client and the final SGW is based on existing VPN implementation methods.

One will note in the above procedure that the information to authenticate the initiator is only known by the SGW that needs to know the information, so Condition 3 is satisfied. One will also observe that, even if NAT is used in the security domain when the target is located, direct communication is only between adjacent SGWs (including the initiator), so NAT is not needed for communication on these segments. And since packets are delivered back and forth between the client and the target transparently over the virtual path without any special processing, Condition 4 is satisfied.

We can thus conclude that the proposed method based on the above sequence of steps fully satisfies all the conditions outlined in Section 3.1.

4   Implementation and Evaluation

We stressed earlier in Section 3 that the proposed method could be implemented by existing VPN methods in the client, which is obviously very convenient and beneficial for users. To evaluate the effectiveness of the proposed method in terms of encryption overhead on communication, we constructed an implementation of the new method to evaluate its performance. This section describes the implementation in detail, and discusses how the performance evaluation was carried out.

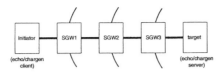

Fig. 7. The experimental setup

4.1 Implementation Method

Although the procedures of setting up VPN links in our proposed method are close to CM3, we have decided to implement our method based on CM1. Based on a CM1 implementation, we can simply treat VPN links of CM1 as virtual paths of the proposed method. Since any application can be supported by using the "runsock" command in the client, and since CM1 does not have to be incorporated in the application or set for each service, this should not present any practical problems. Consequently, we implemented an SGW and a proxy gateway by modifying the source code (using C language) of a SOCKS5 server (used by CM1) running on FreeBSD 3.2R and 4.1R. CM1-based socks servers can define different combinations of either invoking or not invoking authentication and/or encryption as various methods, then have the capability to negotiate using these method ID numbers when VPN links are set up. We defined our proposed approach as a new method, then by operating the server in accordance with the sequence of steps outlined in Section 3.2 when a method is selected through negotiation, we realized an SGW by adding the following functions:

1. Proxy gateway authentication capability.
2. Capability to notify the proxy gateway of the next hop SGW (necessary for authentication).

Then when a conventional method is selected, the proposed method SGW can also be operated as a CM1-based SGW. Furthermore, the proxy gateway can be implemented by providing the capability to negotiate with an SGW using the new method ID number indicating our proposed method in accordance with VPN setup requests from clients, and by adding the capability to perform authentication at each SGW until the final SGW is reached to the CM1-based socks server.

Regarding the client, on the other hand, the socks client source code was not completely modified. For authentication between the initiator and each SGW, and for authentication and encryption between the client and final SGW, Kerberos[7, 8] authentication and encryption capabilities were used, the same as in CM1.

4.2 Performance Evaluation

Experimental Method The experimental setup for evaluating the performance of the proposed method is illustrated in Fig. 7. Three SGWs were deployed between the initiator and the target, then actual communications were

Table 1. Trial results

Data	CM1		Proposed Method	
(bytes)	Connect time	Data trans. time	Connect time	Data trans. time
	(sec)	(sec)	(sec)	(sec)
128	1.361	8.631	1.529	5.771
512	1.401	15.210	1.566	9.850
1024	1.409	21.106	1.487	14.995

carried on between the initiator and the target. As observed earlier in Section 4.1, only implementations of CM1 have been widely disclosed to the public, so the present trials evaluated only our proposed method and CM1. Furthermore, each host can connect over 100Mbps or 10Mbps links using a campus LAN network.

For conducting the trials, echo (Port 7) service was employed and the following procedure was repeated 1,000 times: the echo client was connected to the echo server, relatively small amounts of data were sent from the client to the server, then the same data were sent back so it was received by the client from the server.

Clearly the proposed method and CM1 will respond differently to the trials depending on the amount of data sent, so we tried three different sizes of data in the trials. That is, the echo client sent 128, 512, and 1,024 bytes to the echo server. Considering that throughput could be diminished by the TCP slow start effect even when ample bandwidth is available, we had the initiator launch two clients which were executed in parallel for all trials to reduce this effect.

Actual measurements were conducted 100 times in the trials for all combinations of VPN link setup methods (the proposed method and CM1) and data services (three kinds of trials for each test), and we calculated the average times required for the initiator to initiate a connection and actually set up a VPN link (connect time) and the time after the VPN link was established until the data was finished transmitting (data send time). Authentication was performed at all SGWs, and the data was encrypted on all VPN links in the trials.

Experimental Results and Considerations The required average connect and data transmission times for the trials are shown in Table 1. The data figures in the tables represent the amounts of data sent or received in one iteration of the trials.

It is apparent in Table 1 that the data transmission times are substantially less for the proposed method than for CM1 in all cases, and the difference grows larger as more data is sent. The difference can be attributed to the encryption overhead, because in CM1 encryption is implemented separately on three different VPN links that are set up between the initiator and the final SGW, whereas in the proposed method encryption is just performed on a single VPN link that is set up between the initiator and the final SGW. Comparing the data transmission times of CM1 and the proposed method, we find that, regardless of the amount of data transmitted, CM1 always takes about 1.5 times longer than the proposed method.

Now turning to the connection times, the proposed method took between 80 to 170 milliseconds longer than CM1 to establish a connection in all trials. This can be attributed to the proxy gateway which is added in the proposed method. The differences in the connection times are much less than the differences in the data transmission times: in the types of communication procedures tested in the trials, the connection time differences should not present any practical problems.

We can conclude based on these results that the proposed method provides substantially less encryption overhead and better communication efficiency than CM1. We did not evaluate the performance of CM2 or CM3 in this work, but assuming the same authentication and encryption schemes are used, when encryption is only performed on VPN links between the initiator and final SGW in CM2 and the communication efficiency of CM3 is considered to be about the same as that of the proposed method, or when encryption is performed on all VPN links in CM2, encryption in the initiator is performed multiple times, so in principle we can assume that the performance of CM2 and CM3 would be worse than that of CM1.

5 Conclusions

In this paper we proposed a new and more efficient method of sending encrypted communications by introducing a proxy gateway that can accommodate existing VPN software for setting up VPN links across hierarchically nested security domains. By adopting this approach, special software does not have to be added to the clients. An SGW and proxy gateway were also implemented by extending SOCKSv5, and the effectiveness of the new approach was demonstrated through comparative trials.

Further study includes efficient management of the routing table that is currently maintained by both the initiator and SGWs by hands.

References

[1] Leech, M., Ganis, M., Lee, Y., Kuris, R., Koblas, D., Jones, L.: SOCKS Protocol Version 5, RFC1928 (1996)
[2] Kayashima, M., Terada, M., Fujiyama T., Ogino T.: SOCKS V5 Protocol Extension for Multiple Firewalls Traversal, Internet Draft, draft-ietf-aft-socks-multiple-traversal-00.txt (1997)
[3] Kayashima, M., Terada, M., Fujiyama T., Koizumi M., Kato, E.: "VPN Construction Method for Multiple Firewall Environment," Transactions of the Institute of Electronics, Information, & Communication Engineers, D-I, Vol. J82-D-I, No. 6, (1999) 772-778 (in Japanese)
[4] NEC: SOCKS Home Page, http://www.socks.nec.com/index.html.
[5] Egevang, K., Francis, P.: The IP Network Address Translator (NAT), RFC1631 (1994)

[6] Srisuresh, P., Holdrege, M.: The IP Network Address Translator (NAT) Terminology and Considerations, RFC2663 (1999)
[7] Kohl, J., Neuman, C.: The Kerberos Network Authentication Service (V5), RFC1510 (1993)
[8] Linn, J.: The Kerberos Version 5 GSS-API Mechanism, RFC1964 (1996)

A Key Management Scheme
Integrating Public Key Algorithms
and GATE Operation of Multi-point Control
Protocol (MPCP) for Ethernet Passive Optical
Network (EPON) Security

Kyung-hwan Ahn[1], Su-il Choi[2], Jae-doo Huh[2], and Ki-jun Han[1] *

[1] Computer Engineering, Kyungpook National University
Daegu, 702-701, Republic of Korea
khan@netopia.knu.ac.kr,kjhan@knu.ac.kr
[2] Access Network Technology, Electronic Telecommunication Research Institue
Deajeon, 305-350, Republic of Korea
{csi,jdhuh}@etri.re.kr

Abstract. An Ethernet Passive Optical Network (EPON) is a point-to-multipoint optical network and requires security services between the Optical Line Terminal (OLT) and the Optical Network Units (ONUs). In this paper, we propose an efficient key management scheme by integrating the Rivest-Shamir-Adleman (RSA) public key algorithm and the GATE operation of Multi-Point Control Protocol (MPCP). As EPON has some specific properties such as an Auto Discovery process which might be utilized for security, key management service could be much simpler.

1 Introduction

While in recent years the telecommunications backbone has experienced substantial growth, little has changed in the access network. The tremendous growth of Internet traffic has accentuated the aggravating lag of access network capacity. The "last mile" still remains the bottleneck between high-capacity local area networks (LANs) and the backbone network. The most widely deployed broadband solutions today are digital subscriber line (DSL) and cable modem (CM) networks. Although they are improvements over 56 kb/s modems, they are unable to provide enough bandwidth for emerging services such as IP telephony, video on demand (VoD), interactive gaming, or two-way videoconferencing. A new technology is required: one that is inexpensive, simple, scalable, and capable of delivering bundled voice, data, and video services to an end-user subscriber over a single network. EPONs, which represent the convergence of low-cost Ethernet equipment and low-cost fiber infrastructure, appear to be the best candidate for the next-generation access network [1].

---

* Correspondent author.

H.-K. Kahng (Ed.): ICOIN 2003, LNCS 2662, pp. 783–792, 2003.

All transmissions in an EPON are performed between an OLT and ONUs. The OLT resides in the local exchange (central office), connecting the optical access network to the metro backbone. The ONU is located at either the curb (Fiber To The Curb solution) or the end-user location (Fiber To The Home and Fiber To The Building) and provides broadband voice, data, and video services. In the downstream direction (from OLT to ONUs), an EPON is a point-to-multipoint network, and, in the upstream direction, it is a multipoint-to-point network. An EPON has a broadcasting downstream channel and serves noncooperative users. An EPON cannot be considered a peer-to-peer network in that ONUs cannot communicate directly with each other or even learn of each other's existence. Since a malicious ONU may be placed in promiscuous mode and read all downstream packets, security is strongly needed in EPON [1].

Properties desired for EPON security services may include confidentiality (protection against eavesdropping), integrity (protection against data modification), authentication (assurance that the source of the data is accurately identified to the recipient), access control (assurance that only authorized users can access network resources), and nonrepudiation (inability of the sender to disavow data the recipient receives). To provide most of these security services requires some sort of encryption. Encryption uses an encryption algorithm and a key to change some input, called plaintext, to some output, called ciphertext. Two classes of encryption algorithms exist: secret key algorithms and public key algorithms [7]. Secret key algorithms are much faster but require the communicating parties to share a secret key between them. Alternatively, public key algorithms have the unique feature of being able to provide security without revealing a secret key associated with an individual; however, they are extremely slow and require a large number of arithmetic operations.

With considering this nature, secret key algorithms are better than the public key algorithms for data transfer between OLT and ONUs in EPON. For secret key algorithms in EPON to work, the two parties must share the same key, and that key must be protected from access by others. Furthermore, frequent key changes are usually desirable to limit the amount of data compromised if an attacker learns the key. Therefore, the strength of EPON security rests with the key management technique, a term that refers to the means of delivering a key to two parties who wish to exchange data, without allowing others to see the key. In this paper, we propose an efficient key management scheme for EPON security by integrating public key algorithms and GATE operation of MPCP.

We organize the rest of the paper as follows. In section 2, we describe the concept of EPON. We then discuss security threats and security services in EPON in section 3. Section 4 proposes an efficient key management scheme in EPON. Finally, we present conclusions in section 5.

2   An EPON Network

Passive optical networking has been considered for the access network for quite some time, even well before the Internet spurred bandwidth demand. The Full

Service Access Network (FSAN) Recommendation (ITU G.983) defines a PON-based optical access network that uses asynchronous transfer mode (ATM) as its layer 2 protocol. In 1995, when the FSAN initiative started, ATM had high hopes of becoming the prevalent technology in the LAN, Metropolitan Area Network (MAN), and backbone. However, since that time, Ethernet technology has leapfrogged ATM. Ethernet has become a universally accepted standard, with over 320 million port deployments worldwide, offering staggering economies of scale. High-speed Gigabit Ethernet deployment is widely accelerating and 10 Gigabit Ethernet products are available. Ethernet, which is easy to scale and manage, is winning new ground in MANs and Wide Area Networks (WANs) [1].

Considering that 95 percent of LANs use Ethernet, ATM PON may not be the best choice to interconnect two Ethernet networks. One of ATM's shortcomings is the fact that a dropped or corrupted cell will invalidate the entire IP datagram. Also, ATM imposes a cell tax on variable-length IP packets. On the other hand, Ethernet looks like a logical choice for an IP data-optimized access network. An EPON is a PON that carries all data encapsulated in Ethernet frames. Newly adopted quality of service (QoS) techniques have made Ethernet networks capable of supporting voice, data, and video. These techniques include full-duplex support, prioritization (P802.1p), and virtual LAN (VLAN) tagging (P802.1Q). Ethernet is an inexpensive technology that is ubiquitous and inter-operable with a variety of legacy equipment [1].

The standards work for Ethernet in the local subscriber access network is being done in the IEEE P802.3ah Ethernet in the First Mile (EFM) Task Force. This group received approval to operate as a Task Force from the IEEE-SA Standards Board in September 2001. The P802.3ah EFM Task Force is bringing Ethernet to the local subscriber loop, focusing on both residential and business access networks. While at first glance this appears a simple task, in reality the requirements of local exchange carriers are vastly different from those of enterprise managers for which Ethernet was designed. In order to evolve Ethernet for local subscriber networks, P802.3ah is focused on four primary standards definitions [1].

The IEEE 802.3 standard defines two basic configurations for an Ethernet network. In one case, it can be deployed over a shared medium using carrier sense multiple access with collision detection (CSMA/CD) protocol. In another case, stations may be connected through a switch using full-duplex links. Properties of an EPON are such that it cannot be considered either shared medium or a point-to-point network; rather, it is a combination of both. Because Ethernet is broadcast by nature, in the downstream direction (from network to user) it fits perfectly with the EPON architecture. In the upstream direction (from user to network), the ONUs should share the channel capacity and resources [1].

2.1 Downstream Transmission in EPON

In the downstream direction, Ethernet frames transmitted by OLT pass through a 1:N passive splitter and reach each ONU [1]. Here, Ethernet frames have variable length packets up to 1518 bytes. Splitting ratios are typically between 4 and

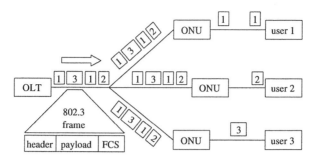

Fig. 1. Downstream transmission in EPON

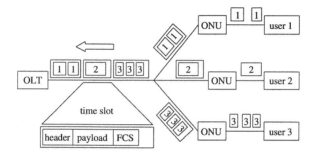

Fig. 2. Upstream transmission in EPON

64. OLT broadcasts IEEE 802.3 frames to all ONUs. ONU receives broadcast frames from OLT and filters them by the destination address as shown in Fig. 1.

2.2 Upstream Transmission in EPON

In the upstream direction, due to directional properties of a passive combiner (optical splitter), data frames from any ONU will only reach the OLT, not other ONUs. In that sense, the behavior of EPON in the upstream is similar to that of a point-to-point architecture. However, unlike in a true point-to-point network, EPON frames from different ONUs transmitted simultaneously still may collide as illustrated in Fig. 2. Thus, in the upstream direction, the ONUs need to share the trunk fiber channel capacity and resources [1].

The MPCP specifies a mechanism between an OLT and ONUs connected to a point-to-multipoint PON segment to allow efficient transmission of data in the upstream direction. Functions performed in MPCP are to provide timing reference to synchronize ONUs, to control the Auto Discovery process, and to allocate bandwidth to ONUs.

3   EPON Security Services

In this section we first describe the security threats which may be caused in EPON and then present security services and methods needed in EPON.

3.1   Security Threats in EPON

In an EPON, the process of transmitting data downstream from the OLT to multiple ONUs is fundamentally different from transmitting data upstream from multiple ONUs to the OLT. Therefore, EPON has different security threats and a different vulnerability.

As EPON medium is broadcast in downstream, security threats in EPON are as follows. First, every ONU located in an end-users' premises can eavesdrop downstream traffic unnoticed and undisturbed 24h/day using a standard Gigabit Ethernet interface. Second, the attacker can know the MAC addresses and Logical Link Identifiers (LLIDs) used by neighbors. Third, the attacker could infer the amount and type of traffic to neighbors by monitoring LLIDs and Ethernet MAC addresses. Fourth, downstream MPCP messages can reveal upstream traffic characteristics of each ONU [4].

As EPON medium is multipoint-to-point in upstream, EPON has some security threats. First, the attacker can masquerade as another ONU (using MAC address and/or LLID) and gain access to privileged data and resources in the network. Second, the attacker can steal the service by overwriting the signal with higher power. Third, the intruder can flood the network with either valid or invalid messages affecting the availability of the network resources or OAM information. Fourth, if the intruders succeed in hacking into OAM channels, they can try to change EPON system configurations or get access to TMN network. Fifth, the end user may try to disturb the EPON by sending any optical signals upstream. This could generate restarts which make it easier to hack the protection. Sixth, upstream traffic of one ONU may be detectable from other ONU access points. Seventh, the attacker could intercept upstream frames by using reflections from EPON, modify frames, and send them to OLT [4].

3.2   Security Services in EPON

Security services can be classified based on their transmission directions. In the downstream direction, security services required are confidentiality and privacy [4]. Confidentiality is the protection of transmitted data from eavesdropping or monitoring. This is typically accomplished through encrypting data messages. Privacy is not just confidentiality, though; it also includes anonymity. In order to protect from a privacy violation such as analysis of traffic volume or destination EPON encrypts an entire Ethernet frame. In the upstream direction, EPON should provide authentication, access control, message authentication, and data integrity [4]. EPON can use IEEE 802.1x in order to provide a means of authentication and access control. Message authentication and data integrity can be based on an encrypting frame check sequence (FCS). We believe that secret key

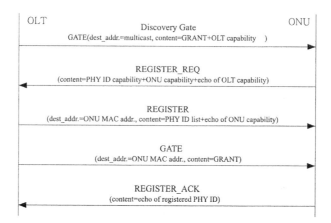

Fig. 3. Discovery GATE operation

algorithms are appropriate for data transfer between OLT and ONUs in EPON because they are much faster than public key algorithms.

4   A Key Management Method for EPON

We present an efficient key management scheme by integrating the RSA public key algorithm and the GATE operation of MPCP for EPON security in this section.

4.1   GATE Operation

MPCP is the mechanism for medium access control in EPON. MPCP specifies a control mechanism between a Master unit (OLT) and Slaves units (ONUs) connected to a point-to-multipoint segment to allow efficient transmission of data. Functions performed are the ONU registration process, bandwidth assignment to end stations, and timing reference synchronization. For these functions, MPCP carries out the Auto Discovery GATE operation and the General GATE operation.

As shown in Fig. 3 MPCP controls the ONU registration process in Auto Discovery GATE operation. First, an OLT broadcasts a Discovery GATE message to all ONUs. Second, at power-up or reset, an ONU enters the Discovery state. Once an ONU receives the Discovery GATE message from OLT, it issues REGISTER_REQ message to respond to the Discovery GATE message. As ONUs can wake-up simultaneously, MPCP must resolve contention in the REGISTER_REQ message. Third, OLT sends the REGISTER message and the GATE message to the ONU. Fourth, the ONU transmits the REGISTER_ACK message to OLT [3].

Bandwidth allocation in MPCP performed through General GATE operation is depicted in Fig. 4. In order to assign bandwidth to ONUs, OLT sends a GATE

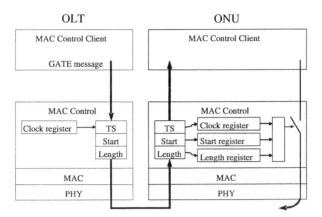

Fig. 4. General GATE operation

message containing slot start time and slot length to each ONU. The ONU transmits only during the time indicated in the GATE message. If the ONU needs more bandwidth, it generates a REPORT message to request additional bandwidth.

4.2 Key Distribution

Key distribution refers to procedures by which keys are securely provided to parties (ONUs) legitimately asking for them. The fundamental problem of key distribution is to establish keying material to be used in symmetric mechanisms whose origin, integrity, and confidentiality can be guaranteed [2]. As EPON has some specific properties which might be utilized for security, key distribution could be much simpler when considering these properties such as Auto Discovery process. During the Auto Discovery process, OLT periodically broadcasts the Discovery GATE message to all ONUs and unregistered ONUs perform the Auto Discovery process. Unless ONUs perform the Discovery process they are not able to use network resources. Therefore, it is desirable for EPON to distribute a session key between OLT and ONUs during the Auto Discovery process.

As shown in Fig. 5, OLT broadcasts the Discovery GATE message containing its public key (KU_{OLT}) and nonce (N_1) encrypted by its private key (KR_{OLT}) which is denoted by $E_{KR_{OLT}}[N_1]$. An ONU generates a session key (K_S) and responds with a REGISTER_REQ message. All fields except session key are encrypted by K_S using a secret key algorithm. The session key is encrypted using the OLT public key $(E_{KU_{OLT}}[K_S])$. After OLT knows the session key by decrypting its private key, it can decrypt the ONU's REGISTER_REQ message using the session key. The OLT sends the REGISTER message containing ONU's nonce (N_2). The GATE message and the REGISTER_ACK message encrypt using K_S and send it to OLT and ONU, respectively. During the Auto Discovery process, Both OLT and ONU perform not only the registration process but also

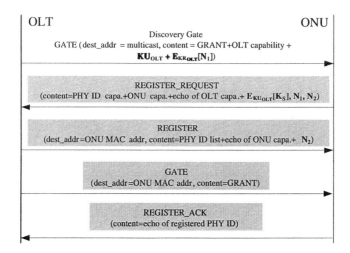

Fig. 5. Key distribution procedure

the key distribution. In order to provide both confidentiality and privacy, we encrypt all MPCP messages except Discovery GATE message using a session key.

Antti Pietiläinen [5] has proposed a key exchange method which uses a temporary MAC address of ONU during the Auto Discovery process, and encrypts the REGISTER_REQ message with the exception of the field of the temporary MAC address using the public key of OLT. However, if all MPCP messages except Discovery GATE message are encrypted, privacy service will be inherently provided, so the temporary MAC address of ONU is not needed in this case. In addition, the speed of this method is slow because it uses a public key algorithm. In our scheme, however, all MPCP messages except Discovery GATE message are encrypted using a secret key algorithm, and, as a result, EPON can work more securely and quickly. Also, our scheme is simpler since it uses the permanent MAC address of ONU during the registration process.

4.3   Key Replacement

The secure management of the key is one of the most critical elements in EPON security. The more frequently session keys are replaced, the more secure they are, because the attacker has less ciphertext to work with for any given session key. Therefore, session key should be changed periodically. Table 1 shows a typical key replacement period [6].

OLT periodically assigns bandwidth to ONU through the General GATE operation. In order to request additional bandwidth, ONU can issue a REPORT message periodically. In the General GATE operation process, the session key can be replaced between OLT and ONU such as in Fig. 6. New session key exchanges

Table 1. Key replacement period

Encryption algorithm	Period
Churning (APON)	2 seconds
DES (Data Encryption Standard)	12 hours
WEP (Wired Equivalent Privacy)	packet
AES (Advanced Encryption Standard) 128 bit	3×10^{17} years

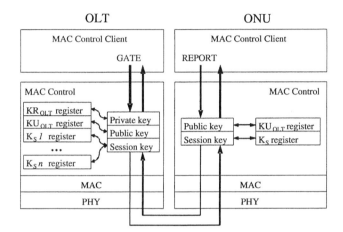

Fig. 6. Session key replacement

through General GATE and REPORT messages. Then OLT and ONU update their session key register.

4.4 Key Recovery

EPON uses the RSA public key algorithm to distribute session key and the secret key algorithm to encrypt data transmitted between OLT and ONUs. Therefore OLT and ONU distribute its public key and session key, respectively. In this process, public key and/or session key may become mismatched due to hardware malfunction or network transmission error. This is why the key recovery function is required for EPON security services.

If OLT detects its public key error by verifying FCS, it generates a new public key and transmits it to ONU by including the key in the Discovery GATE message. As soon as ONU receives OLT's public key, it compares the public key in its register with the one received. If the two keys are the same, ONU discards the new public key. Otherwise, ONU replaces the old public key with the new one.

When the session key becomes erroneous, it can be recovered by the following procedure. In General GATE operation, All ONUs are synchronized to a common time reference, and each ONU is allocated a time slot which is capable of carrying several Ethernet frames. If there are session key errors in the bandwidth allocation process, ONU is unable to decrypt the General GATE message and to carry Ethernet frames within the given time slot. As a result, OLT detects their session key error. On the other hand, ONU recognizes session key error when ONU receives the Discovery GATE message but doesn't receive the General GATE message. As it is impossible for ONU to decrypt the General GATE message it can't achieve in General GATE operation. Therefore, ONU transmits the REPORT message with a new session key to OLT and recovers the session key.

5   Conclusions

The broadcasting characteristics in the downstream and multipoint-to-point transmission characteristics in the upstream of the EPON require very strong security services. Security services can be provided by encrypting data transmitted between OLT and ONU in EPON. Secret key algorithms are appropriate for security of EPON, and they require the transport of a session key which is established between OLT and ONUs. In this paper, we propose an efficient key management scheme for EPON security services by integrating RSA public key algorithm and GATE operation of MPCP. We could provide a key management service with EPON in a much simpler way using such properties as the Auto Discovery process which might be utilized for security.

References

[1] Glen Kramer and Gerry Pesavento: Ethernet Passive Optical Network(EPON): Building a Next-Generation Optical Access Network. IEEE Commun. Mag., vol. 40. Feb. (2002) 66–73
[2] Walter Fumy and Peter Landrock: Principles of Key Management. IEEE Journal On Selected Areas In Communications, vol. 11, no. 5, Jun. (1993) 785–793
[3] B. Gaglianello, O. Haran, D. Sala, L. Khermosh and C. Ribeiro: ONU Auto Discovery. IEEE 802.3ah meeting in May (2002)
[4] Olli-Pekka Hiironen and Antti Pietilainen: Security Threats and Defense Models in EPON. IEEE 802.3ah meeting in Sept. (2002)
[5] Antti Pietiläinen: Key exchange during auto discovery. IEEE 802.3ah meeting in Sep. (2002)
[6] Jin Kim: Authentication and Privacy in EPON. IEEE 802.3ah meeting in Jul. (2002)
[7] William Stallings: Cryptography And Network Security, Prentice Hall, (1999)

New Security Paradigm
for Application Security Infrastructure

Seunghun Jin[1], Sangrae Cho[1], Daeseon Choi[1], and Jae-Cheol Ryou[2]

[1] Electronics and Telecommunications Research Institute
161 Gajeong-dong, Yuseong-gu, Daejeon, 305-350 Korea
{jinsh,sangrae,sunchoi}@etri.re.kr
[2] Department of Computer Science, Chungnam National University
220 Kung-dong, Yuseong-gu, Daejeon, 305-764 Korea
jcryou@home.cnu.ac.kr

Abstract. The recent and upcoming computing environment is characterized by distribution, integration, collaboration and ubiquity. The existing security technology alone can not successfully provide necessary security services for this environment. Therefore, it is necessary that the provision of security services reflects the characteristics of such an environment. In this paper, we analyze security requirements for existing and upcoming applications and services. We then survey deployed security services and identify the required information security services to satisfy the result of the security requirement analysis. Hence we suggest UASI (Unified Application Security Infrastructure) as a new security paradigm. UASI is a framework, which describes how a single security infrastructure can provide all the necessary security services for the ubiquitous computing environment in a seamless manner.

1 Introduction

Today a business has an increase demand to share data, integrate processes and services to offer customized and comprehensive services to its customers. And as a customer, people may want to obtain any information that they need whatever computing device, platform or application they use in anytime and anywhere. Therefore, it is common phenomenon in a computing environment that the hardware such as PCs, servers and smart devices is being connected and the software is being integrated. Such a computing environment can be characterized by distribution, integration, collaboration and ubiquity.

In a business, enterprise applications face the radical change to meet the new business requirements to integrate existing and new applications to enhance business process management. Moreover, to support the collaboration with partner organizations, Application-to-Application has been emerged as the dominant solution.

The enterprise applications operate on sharing data, disclosing internal business processes and distributing person's private profile information between

H.-K. Kahng (Ed.): ICOIN 2003, LNCS 2662, pp. 793–802, 2003.

primitive services. This will cause many serious security problems such as disclosure of important data, an invasion of a person's privacy and misuse of disclosed business process by malicious attacker. To cope with these security problems, proper security services such as authentication, authorization and audit must be provided.

When the services and information are integrated and collaborated, a set of security services are required and deployed. However, new security vulnerabilities may be introduced if the deployed security services are not integrated and managed properly. The security services including ID management, Single-Sign-On (SSO), authentication and authorization are current effort to deal with such a new security problem. However, the administration cost and security vulnerabilities increase if security services are not integrated in spite of the integration of computing environment. Security services listed above are not a new concept, of course, but what is new is how these services are integrated to provide unified security services through a single security infrastructure.

We have analyzed security requirement for existing and upcoming applications and services that requires authentication, authorization, auditing, etc. Considering present and future trend of application environment, we have derived the security requirements of information service by analyzing the Internet service environment and application integration separately. Furthermore, we have researched necessary security requirement for the integrated management framework for various information security services.

Our objective in this paper is to suggest new unified security framework called UASI as a security infrastructure to provide new security services and to satisfy the requirement of integration, unification and central management for existing security solutions. UASI is defined as a single security infrastructure that provides a group of security services providing comprehensive and compound application functions to clients in the customized form ubiquitously. UASI is under development to satisfy present and future security requirement. We also try to demonstrate why the integration of security services are only way to solve existing security problems. The suggested security framework will demonstrate how security services can be provided in centralized manner. It also shows what benefits can be acquired using UASI in the aspect of current application environment.

The rest of the paper is organized as follows. In Sect. 2, we survey a trend in current application environment and its security solutions. Section 3 describes what problems may occur when applications are integrated, and answers why the present security technology is inadequate and new security framework needs to be specified. In Sect. 4, the comprehensive security services that are required for application integration are identified and explored. UASI is introduced in Sect 5. The basic structure of UASI and its framework are proposed. Finally, Sect. 6 draws a conclusion and suggests future research directions.

2   Analysis of Current Application Environment

In this section, we have briefly surveyed the trend of existing application environment. We have also investigated the security solutions for the environment.

2.1   The Emergence of Various Applications

Recently a trend in development of enterprise application is integration. EAI (Enterprise Application Integration) is the concept that achieves the increase productivity through the automation of business process and reduction in administration costs by integrating applications' GUI, data and processes within an enterprise[5]. EAI mainly focuses on transactions and messages, which is standardized across systems. Process management automates business processes within an organization. Internal integration must be highly structured and controlled.

B2Bi (Business to Business integration) is an attempt to achieve cost reduction through the integration of business process automation and application integration between partner enterprises in purchase, sales and production[6]. The primary purpose of B2Bi is document exchanges between organizations. The document is defined with numerous data definitions and standards. Process automation tends to be point-to-point and fragmented. Unlike EAI, B2Bi uses the Internet communication which is error-prone and slow.

XML (eXtensible Markup Language) is a markup language much like HTML and was designed to describe data. XML was created so that richly structured documents could be used over the web[4]. Strictly speaking, XML is not application. However, it is basic building block for Web services and many Internet-based applications.

Web services are another example of the technical challenge to share and exchange information and services by means of using a standard-based interface to functionality exposed for use over the Internet. In other words, Web services supports application-to-application communication[7]. Web services takes a technology and vendor neutral approach to provide universal availability. For this reason, Web services can be applied to build both B2B and EAI systems.

2.2   Security Solutions for Existing Applications

EAM (Enterprise Access Management) is aimed to manage information security within an organization. However, every information security services in a organization should be reconstructed using the components that is provided by EAM. Namely EAM does not have the functional requirement to integrate existing security services. Currently Enterprise Application Security Integration (EASI) is the solution that has the concept discussed so far. EASI provides unified interface that can be added to existing applications. Unified policy based management tool is provided for a security service. However, EASI do not provide management tool and service for a user.

Unified authentication and identity management becomes critical services in information security. Microsoft provides the Passport service – one of the unified authentication and identity management services – to over 160 million users. However, they lack of providing other services such as pseudonym and anonym for privacy protection and management of third party attribute.

LibertyAlliance, which is proposed by Sun suggests a solution to federate among existing authentication and identity management services to provide unified authentication and identity management service[2]. However, LibertyAlliance is just a framework, which means that they do not provide security services and therefore there is no result of reducing the burden of service providers.

SAML (Security Assertion Mark-up Language) has been standardized to provide a mechanism to exchange authentication and authorization information between organizations. However, it is only a message delivery standard. There are still problems existing to federate between organizations for policy tuning, attribute mapping and security mechanism integration.

3   The Need of New Security Requirements

3.1   The Security Issues in Present and Future Application Environment

In the Sec. 2, we have discussed present and future application environment and the various security solutions. It is a trend in application to collaborate and integrate with other applications through B2Bi. After all, as every user and service are integrated and collaborated, information security services can be provided seamlessly by integrating and collaborating as well.

An individual user uses an uncountable number of services and they also use diverse applications. When a user tries to use a service, he has to register in the service with his private information. Furthermore, he need to be assigned with identifier to be identified by the service and he also has to remember a password for authentication. The procedure becomes inconvenient and complex as the number of services increases.

If the authentication mechanism other than password is employed, then a client should have different authentication programs for the corresponding service. It becomes very difficult to update the distributed private information.

Combining a person's identifier and personal information forms the identity. The present security service for an application manages the one's identity for individual application separately. It is the main cause why the identifier is lost and personal information is disclosed. It also leads to an invasion of a person's privacy.

With respect to a service provider, managing and authenticating personal information and collecting bills becomes great burden of administration. Of course, the number of registered users is a valuable asset to a business. However, managing personal information to provide security services can be a burden to develop applications. Moreover, it is time-consuming process for a service provider to validate individual person's personal information.

3.2   Security Requirements for the Application Environment

To solve the problems described in 3.1, additional service can be deployed to carry out authentication and identity management on behalf of each application. Unified management service manages personal identity information and it provides that information to service providers. In addition, it carries out authentication and manages person's attribute information that is provided by a third party.

Service users can provide and manage personal information in one place if authentication and identity management service is integrated. This enables the user to use every service with single authentication and therefore the user only need to remember single identifier and related password for authentication. With related to service providers, since the administration of subscribers, billing, authentication, etc can be managed by outsourcing, it reduces managing and deploying a security services and only focuses on building better services.

It is desirable for users and administrators to provide and manage every security service through the structure of single security infrastructure. For this, the provision of total security service is required by organically integrating information security services. In addition, it is most desirable that such a service is managed through single administration framework.

In application integration, there is integration between processes. When a user requests a service, the service uses other services to provide the requested service. In this case, a service requestor can be a user or other services. Integrating among processes gives rise to a new set of security service requirements. For automated process integration, SSO is a mandatory. In addition, authorization management should be integrated between processes that are integrated. A process should have ability to delegate an execution privilege to other processes based on information collected during a process execution. It is also required to manage authorization list due to process's execution path. Privilege delegation and identity impersonation is essential requirement. When a process calls other processes, impersonation is required to authenticate on behalf of original process and privilege should be delegated to access to necessary resource to carry out the requested services.

4   The Requirements of Security Services

In this section, we have identified the list of security services that should be provided for application integration security

The primary security service is the identity management. ID management is to manage and distribute identifier and personal information. The concept of authentication is to identify and verify the user's identity. ID management is the first security service to tackle.

SSO enables a user to login to an application once and then access to other applications without requesting the login procedure again. If SSO is not provided, a user has to login many application services in an environment where services are chained and collaborated.

Authentication is one of the most fundamental security requirements. As applications are integrated, there are many ways that authentication can be carried out. The new security framework should provide the wide range of authentication mechanisms including biometric identification. In addition, unified authentication scheme ensures to produce security assertions, which can be used in SSO, after a user is authenticated.

Authorization, which is called either access control or privilege management, is a complicated process compared to the authentication. In a traditional authorization model, the issue is how a privilege of a subject can be assigned efficiently. When the subject takes an action to access to an object, the policy defined previously can be employed to make an access decision. However, this simple authorization model becomes very complex as a target application is distributed, integrated and collaborated[14]. When applications are collaborated, several processes form a single service. In this case, authorization information for individual process is required to make an access decision. In the case of integration, processes form a workflow. Authorization process on workflow should be considered as well.

Consolidated billing is a new security requirement. Billing means the payment for the service provided. This service is expected to accelerate the charging for use of Internet service. The reason is that the new security framework makes it possible to provide high value-added and personalized service, which is not possible previously. A user does not need to register and pay for each service. The consolidated billing system will enable the user to pay for various services at once. The cost of managing user information, demanding and collecting a payment will be reduced.

If the authorization is the process of monitoring before a certain action is executed, audit is the process of monitoring after the action is executed. Therefore security requirements of audit are similar to that of authorization. As in the authorization, a subject and an object are distributed. The point where an action is executed and recorded as audit and the point where audit record is managed and analyzed are totally distributed. In addition, as the several services are integrated, it is necessary to collect many records that occur at various services and analyze those records completely.

5   UASI (Unified Application Security Infrastructure)

5.1   Basic Structure

UASI is the comprehensive infrastructure that provides application information security services that is required in the present and future. In Fig. 1, the basic concept of UASI is described. UASI is the collection of ASIs that supports for single trust domain. Users, applications and administrators belonging to the domain trusts ASI, which is located within the domain and relies on every application security services. UASI takes responsibility of trust relationship and integration and collaboration between domains.

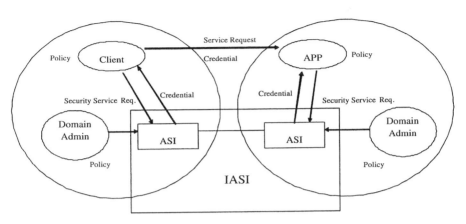

Fig. 1. The basic concept of UASI

Users, applications and domain administrators report their own policy to UASI. Then users and applications requests required information security services to UASI. UASI responses to information security services using a form of credential. A credential is data unit that contains a kind of certificate that proves their identity. The example will be SAML assertion, Kerberos ticket and X.509 Certificate.

UASI provides every required application information security service and single management and service interface. In addition, UASI satisfies the following requirements.

- Utilization of existing information security solution: An existing information security solution can be reused as many as possible,
- Extensibility for adapting new security requirements: ASI provides the extensibility to cope with a new application security environment,
- Minimization of additional cost for information security service: It is critical not to make service response time slow and administration cost for UASI should be maintained at a minimum.

5.2 UASI Framework

To satisfy UASI's own requirement, we have devised ASI framework. ASI framework consists of both existing and upcoming information security solutions. It is the framework that can support for managing and utilizing integrated service and management interface to an application.

In Fig. 2, the basic concept of ASI framework is described. The security services are attached to ASI framework. Within the security services, there is existing information security solution that has adaptors to be integrated into the ASI framework whereas service connection interface is defined for newly added security services in the future. In integrated framework, a service is called through integrated service interface and policy is enforced by using integrated

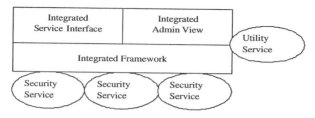

Fig. 2. ASI Framework

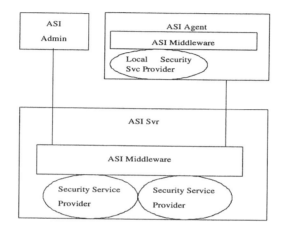

Fig. 3. The Structure of UASI

administration interface. Utility service means it is the service that support for integration between services that are not included in the integrated framework.

In a real environment, UASI is realized by connecting disperse ASIs. For example, for an organization a ASI can be a trust domain that provides members of the organization with security services. For the Internet, there can be a public ASI system provides the same services to the Internet community. UASI is constructed by connecting such a ASI system.

In Fig. 3, it shows the structure of UASI system within a single domain. ASI server is the one that provides the most of the security services. ASI server consists of ASI framework, utility service provider and information security service providers. To distribute the computational overhead, utility service provider and information security service providers can be operated in a different server. UASI framework has a middleware, which helps manage and call such a distributed service.

ASI admin is the integrated single administration view that provides every system administration task. ASI agent is used by a client system and it provides service interface to use. It should be noted that security service that can be executed in the client side can be done in the local system without going all the

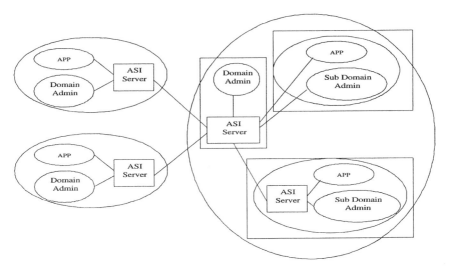

Fig. 4. UASI domain

way to the ASI server by using ASI agent. Such an attempt can substantially reduce the overhead of the server and increase the service response time. Such a security service in the client side is managed by ASI administration. The existing security solutions in a local system can be incorporated into ASI domain.

Figure 4 describes the structure of UASI domain. The domain means the individual and independent trust third party. There can be a subordinate relationship between ASI domains. In this case, the superior domain is called super-domain and any domain that is super-domain's subordinate is called sub-domain.

The domain is basically the same as the ID domain that indicates the boundary of ID system. Within the same ID system, a sub-domain is formed when either sub identifier is used or a separate authorization domain exists.

One ASI server exists for a single domain. ASI server is trusted by every component in the domain. In the case of sub-domain, it can have a separate ASI server or it can share with ASI server that is located in a super-domain. However, even sub-domain must have a domain administrator.

The integration between domains can be accomplished via ASI server. Two ASI servers execute the functions such as negotiation, policy tuning and attribute mapping for the integration.

6   Conclusion

In application security service, unified security solution means information security service and related information is managed through consolidated administration view. Moreover, single policy framework exists and it is the core part that can manage distributed policies.

In this paper, we have analyzed security requirements for existing and upcoming applications and services. Then we have surveyed deployed security services and identify the required information security services to satisfy the result of the security requirement analysis. Hence we have proposed UASI as a new security paradigm. UASI is a framework, which describes how a single security infrastructure can provide all the necessary security services for the ubiquitous computing environment in a seamless manner.

References

[1] Geiger: .Net My Services and .Net Passport User Authentication Overview. Microsoft white paper, September, (2001)
[2] Hodges, J.: Liberty Architecture Overview. Liberty Alliance Project documentation, July, (2002)
[3] Jones, R.: EAM Ain't EASY. Information Security Magazine, January (2002), SAML 1.0 Specification Set, OASIS, May (2002)
[4] Harold, E. R., Means, W. S.: XML in a Nutshell. 2nd Edition, O'Reilly Inc.
[5] Pinkston, J.: The Ins and Outs of Integration. eAI Journal, August (2001)
[6] Olsen, G.: An Overview of B2B Integration. eAI Journal, May (2000)
[7] Fremantle, P., Ferguson, D. F., Kreger, H., Weerawarana, S.: Understanding the Web Services Vision. Web Services Journal, Vol: 02 Issue 07
[8] Zhang, L., Ahn, G. J., Chu, B. T.: A Role-Based Delegation Framework for Healthcare Information systems. SACMAT'02, pp. 125-134, June (2002)
[9] Atluri, V., Chun, S. A., Mazzoleni, P.: A Chinese Wall Security Model for Decentralized Workflow Systems. CCS'01, pp. 47-58, November (2001)
[10] Powell, D.: Enterprise Security Management (ESM): Centralizing Management of Your Security Policy. SANS Institute, December (2000)
[11] Heffner, R.: Enterprise Application Security Integration. IT Trends 2002, December (2001)
[12] Lewis, J.: The Emerging Infrastructure for Identity and Access Management. Open Group In3 Conference, January (2002)
[13] Clauβ, S., Köhntopp, M.: Identity management and its support of multilateral security. Computer Networks 37 (2001) 205-219
[14] Varadharajan, V., Crall C., Pato, J.: Authorization in enterprise wide distributed tems: design and application. In Proceedings of the 14th IEEE Computer Security Application Conference, Scottsdale, Arizona, 7.11 December (1998) pp 178.189

Network Packet Filter Design and Performance

Bernd Podey, Thomas Kessler, and Hans-Dietrich Melzer

University of Rostock
Richard Wagner Str. 31,18119 Rostock, Germany
ben@comlab.uni-rostock.de

Abstract. Nearly all studies about future bandwidth-needs pointing out that today's access technologies will reach there design limit in the near future. So we have to think about new access networks, which cover the needs in bandwidth, quality of service and bring the three today's networks –cable-tv, telephone networks and Internet– together. The Quality of Service and Security needs depend on each other in those networks. The demands to security devices in such networks are extraordinary high. Already simple packet filtering decreases the throughput in routers or gateways to non acceptable rates. Linux include a powerful packet filtering software, -iptables-, that is used in small and medium networks for firewalls, and workstations. The aim of this paper is to present solvations for packet filter improvements. The paper point out two alternatives: software improvements and filter rules reorganization and will focus on reorganization if filter rules. We describe an automatic filter configuration for Linux packet filter based on connection parameters.

1 Introduction

In the history of modern telecommunication always the technical progress and rarely the users needs stands in the center of development. This cause much problems in implementing and accepting new features and most predictions about future needs were failed. Another problem is that the user needs expand as the same rate the technology develop. Five years ago ISDN-speed was sufficient for internet connections. And with the implementation of xDSL all speed problems should be over. But in analogy to road traffic where the pure existence of new, fast roads generates more traffic fast internet connections entail applications which use this fast connections. And there are already applications available which will use new and much faster access networks. These Applications are for instance Video on Demand and Broadcast TV over IP (multicast connections). With the implementation of these new features to the access network, the bandwidth demands expands at rates that none of today's access network will provide. Some studies, [2, 3, 4] , show the bandwidth development in Access Networks till year 2010, with 4 MBit/s in 2005 and 10 Mbit/s in 2010.

The access gateway is the central point of future access networks and the design will have a wide influence to the network features and –performance. The possible use of Open-Source software on that gateways will have advantages in

H.-K. Kahng (Ed.): ICOIN 2003, LNCS 2662, pp. 803–814, 2003.

security, interoperabilty and in develop of new features and services. This need research in system–, network– and firewall performance as well as the influence on QoS Parameters. The gateway have to be capable for gigabit speed connections. This paper shows standard network packet filtering (iptables) performance and some improvements for the software and the filter design.

2   Basics

2.1   Access Network

Each definition of future access networks is different, so we have to explain what the access network contains. Today's access networks begin at users home and end at the digital switching center.

Design Fundamentals *Provider Choice* – To avoid actual problems in telecommunication the user should have a free choice between the different service providers independent of the access network provider. The best way to separate the data streams is to build VPNs through the access network to the service provider. These VPNs provide the necessary security and connection control and will be built up from the user switch to the access gateway as described in Fig. 1.

Flat Network Structure – The access network will not contain any unnecessary Layer 3 (router) devices.

IPv6 – Selection of IPv6 instead of IPv4 because of the number of addresses and the build-in QoS–, autoconfig– and security Features.

Network Devices The digital switching center will migrate to the **M**ultiservice **A**ccess **G**ateway, and the users endpoint(TA) will be the **U**ser **M**ultiservice **S**witch. Figure 1 describes the new access network.

Multi–Service–Access–Gateway – The so called **M**ultiservice **A**ccess **G**ateway permit the supply of different services and provides the connection with the service-provider backbones. It gets an IP-Address in each of the local subnetworks and the local VPNs terminate in it. This Network Layout makes sure that the local networks are separated from the rest of the Internet. IPv6 supports this layout directly by local and public aggregateable addresses.

User–Multiservice–Switch – The in the users home placed **U**ser **M**ultiservice **S**witch is a modular multiport switch with an High-Speed Uplink and modules for ISDN/POTS and CaTV. The switch is QoS capable and supports multicast submissions. The switch has to build up VPNs to the Service Provider using GRE-Tunnel and IPSEC. IPSEC Tunnels are not useful, because they can't carry multicast connections, which are used for video broadcast.

Endpoint – An endpoint could be a display, computer or any other device which included a network interface card. The endpoint supports IPv6 and, where necessary, multicast connections. [5, 7, 8, 9, 6]

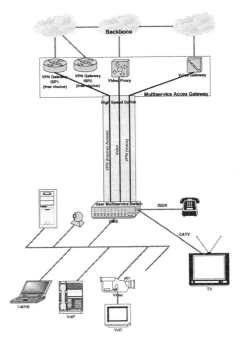

Fig. 1. Access Network – Schematic View

2.2 Quality of Service

The connection Quality of Service is defined by four parameters:

- Bandwidth – number of packets arrived per second
- Delay (time between packet send and receive)
- Jitter – Variation of the delay
- Packet-Loss – packets which don't arrive

The use of security devices in access networks will have a wide influence to these parameters. (Fig. 2)

- Security Devices – Firewalls, VPN Gateways
  - Availability of Services
    In general a firewall will not use any QoS informations for switching packets, so there is no guarantee for connection QoS-parameters. The use of QoS-parameters for routing decisions could cause security problems in case of denial of service attacks with prioritized packets.
  - QoS Parameters
    Delay: Just like switches or routers, a firewall inserts a constant delay, the switching delay.

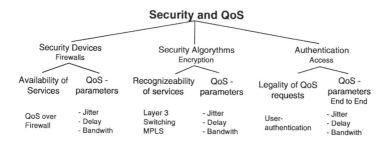

Fig. 2. Security and QoS Parameters

Jitter: If the switching process needs to much time, there could be a random delay due to the fact that the processor has reached 100% utilization.

Bandwidth: Security devices itself will not cause higher connection bandwidth, but they decrease the maximum bandwidth due to the limit of processor speed and the filter design.

– Security Algorithms

  • Recognizability of Services

  The use of end-to-end security makes it impossible to detect the correct service based on TCP-Port on the switches and routers. Devices which QoS classification based on portnumbers will fail.

  • QoS Parameters

  Delay: Encryption causes a delay because all packets has to pass the security algorithms. This delay depends on the encryption-type and the –implementation.

  Jitter: At the processing of packets there could be randomized delay at generation, transmission and reception of packets.

  Bandwidth: At the use of encryption the bandwidth needs expands at least one IP-header per packet (IPSec tunnel).

– Authentication

  • Legality of QoS Requests

  Inside a QoS network authentication is the key to assure that only the one user gets QoS who should get it.

  • QoS Parameters

  Delay: The delay during end to end authentication has influence in duration of connection setup. The

  Jitter: has no real influence to the authentication.

  Bandwidth: Authentication needs a low but guaranteed bandwidth through the network to ensure that the authentication packets arrives at the communication partner.

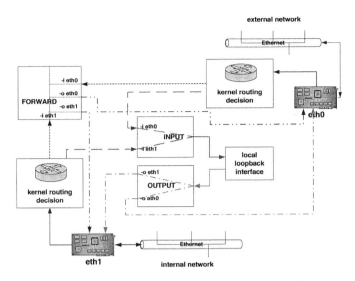

Fig. 3. Path of Packet's through the Linux Box

2.3   Security Devices and QoS

This paper concentrates on security devices, e.g. firewalls, and their impact to the Quality of Service Parameters. The example device is an open-source firewall with Linux as operating system as described in [10]. The packet filter firewall is part of the Linux kernel and will be administrated by the *iptables* command. The direct placement of the firewall to the kernel will improve the system performance and –security.

Iptables is a table based packet filter. The tables are called "chains". Every packet will be tested until it reached the end of the table or one matched rule. The basic structure of packet flow through the Linux Box is shown in fig. 3.

The packet filter itself could not base decisions on packet precedence. The only thing you can do is to mangle packet's TOS-field depending on TCP or UDP-ports and place these rules first in the chain and use the kernel-routing based on TOS/COS.

The first test parameter is the bandwidth depending on the filter-design (filter length and –structure). The filter-structure has a wide influence to the filter performance and will be the most easiest way to advance the QoS parameter before changing kernel source. Delay tests will follow up later in the research.

3   Netfilter Performance

The use of Linux as firewall and the need for high speed gateways poses the question on the firewall and network performance of this hard- and software combination. The performance (throughput) depends on the used processor-, mainboard-,

Table 1. Throughput with Address-Filter only

Number of Filter Rules	Throughput TCP [Mbit/s]	Throughput TCP %	Throughput UDP [Mbit/s]	Throughput UDP %
0	138	100	147	100
100	129	93.47	145	98.63
200	119	86.23	141	95.91
500	98	71.01	131	89.11
1000	72	52.17	96.2	65.44
2000	49	35.50	72	48.97
3000	35.8	25.94	54	36.73

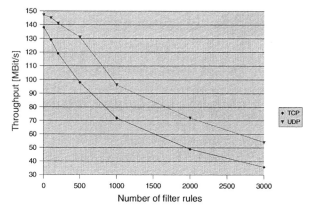

Fig. 4. Throughput TCP/UDP on address-only ruleset

ram-technology (with technology limits like 33 MHz PCI: 1056 Mbit/s max) and the used software. The aim of the performance tests in this document is to display the dependency of network speed on the number of filter-rules and not the network performance itself.

- Testhost – Pentium 200 MHz, 3Com 3c905, Software Linux Kernel 2.4.19
- Testsoftware – netperf 2.1pl13, TCP-Streamtest, UDP-Streamtest, test duration 10seconds

3.1   Filter Rules with IP Address as Argument

The first performance tests are with Filter Rules that only contains a source IP address as argument: *"iptables -I INPUT 1 -s xxx.xxx.xxx.xxx -j YYYY"*

The TCP throughput and the UDP throughput both decreasing exponential to the number of filter rules. The reduction was more than we expected. If a packet has to pass 2000 rules the throughput reduces to 35% for TCP traffic

Table 2. Throughput with Address- & Portfilter

Number of Filter Rules	Throughput TCP [Mbit/s]	Throughput TCP %	Throughput UDP [Mbit/s]	Throughput UDP %
0	138	100	133	100
100	106	76.81	132	99.25
200	85	61.59	131	98.5
500	55.8	40.43	104	78.2
1000	34.2	24.78	72.5	54.51
2000	18.6	13.48	43.8	32.93
3000	13.7	9.93	30.5	22.93

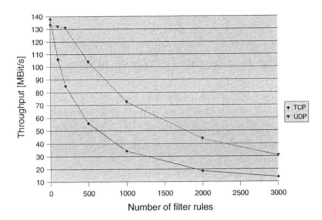

Fig. 5. Throughput TCP/UDP on UDP/TCP Port- and Protocolfilter ruleset

and 48% for UDP traffic in this setup. Tests with different firewall configuration show reductions to 22% for TCP and 30% for UDP traffic.

3.2   Filter Rules with IP Address and TCP/UDP Portnumber as Argument

Extended firewall configurations use TCP or UDP portnumbers to filter packets. This make the ruleset longer and more complicated, but it's a more realistic testing scenario: *"iptables -I INPUT 1 -p tcp(udp) -s xxx.xxx.xxx.xxx –sport vvv -d zzz.zzz.zzz.zzz –dport www -j yyy "*

With the use of IP-address and TCP/UDP port as parameters, the performance of the packet filter decreases more rapidly to the number of filter rules then with IP-address as parameter only due to the fact that the packet filter has to test more connection parameters. The rulesets in real network are a composition of IP-address based and TCP/UDP Port based rulesets. The filter length

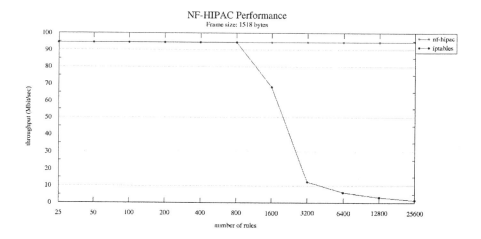

Fig. 6. Compare Netfilter to nf-hipac [11]

vary from 200 rules in small networks until a few thousand rules in enterprise networks.

4   Solvations

The previous chapter indicates the problem that long packet filters achieves poor performance in packet forwarding. This is a fact of all today's packet filters. There are in general two alternatives for problem solving, the rewriting of the filter software and the reorganization of the filter rules.

4.1   Rewrite of Filter Software

This solves the problem from root, so the packet filter in general has better performance. The software presented in [11] is the rewritten packet filter software for Linux, which based on algorithms in [1]. The software uses multi dimensional search trees for packet classification (sorting). The number of filter arguments (Source/Destination Address/Port,...) is thereby the number of dimensions. [1] sends the multidimensional problem back to the one-dimensional packet classification problem, the IP Lookup Problem (the packet classification only by destination address). They describe the built of a Fat, Inverted, Segment (FIS) tree, which is a balanced t-ary tree, for the individual dimensions. [11] present some performance tests with netfilter and nf-hipac. Fig. 6 is the testresult for a packet length of 1528 Bytes.

The results shown in fig. 6 are comparable with fig. 4 and 5. There is no measurable effect to the packet filter throughput for nf-hipac filter in this setup.

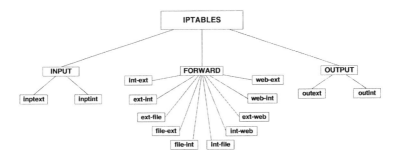

Fig. 7. Packet Filter Tree

4.2   Reorganize Filter Rules

In it's early "alpha" status nf-hipac couldn't be used in production– and server environment. The software isn't well tested and hasn't as many functions as netfilter. If you want to use netfilter in such environments without performance-losing you have to reorganize the filter rules itself. The concept presented by this paper is similar to section 4.1: minimize the executed filter rules by building a hierarchical filter setup (tree). The splitting parameter for the tree (e.g. incoming interface, address, protocol, ...) depends on the existing situation in the network. In a simple case we use the incoming and outgoing interface as parameter for splitting the tree. The maximum filter length reduces from 180 to 35 (20%) in a simple setup. Fig. 7 depicts the filter structure. Every packet has to pass a maximum of two chains, until it will be dropped or accepted.

For security and performance reasons has to be a drop rule at the end of every single chain.

Further design recommendations are:

1. Rules for delay-sensitive connections first
2. Rules for local hosts first ACCEPT then DROP port-and address-based
3. Rules for local hosts first ACCEPT then DROP address-based
4. Rules for local servers first ACCEPT then DROP
5. Rules for known external hosts first DROP then ACCEPT
6. Rules for unknown external hosts first DROP then ACCEPT

Refinement of Filter Rules The next step for improving netfilter performance is to split the tree in Fig. 7 into more dimensions depending on connection parameters like UDP/TCP-ports and IP-addresses. The requirements are:

- at least 3 addresses or ports with the same ruleset for ports or addresses
- no general address or port ranges like IP-address 0.0.0.0 or port range 0:65000
- adaptive filter design depending on network topology
- build classes of hosts with same ruleset

Table 3. Comparison of splitting setup(500 rules overal, n-rules/chain, x-chains) with standard filter (x-rules)

x chains (rules)	n rules/chain	Throughput with split [MBit/s]	Throughput standard [MBit/s]
500	1	54.783	55.8
250	2	79.31	79.025
100	5	106.472	106
50	10	121.125	121.025
25	20	127.106	127
20	25	128.001	127.5
10	50	131.865	132.6
5	100	133.766	134
2	250	135.121	137
1	500	137.022	138

The requirements are tested by the filter during setup. The design of the filter tree can change dynamically. The firewall setup is done by shell scripts, that were controlled by a single configuration file. Each script tests the defined parameters and creates a higher dimension(a new chain), if necessary. The script executes the following during setup:

1. Count all hosts and all ports for a specific ruleset
2. Test, if the counts are higher than the split parameter
3. Test for generic ports and addresses
4. Test, if more hosts or more ports are in the ruleset
5. Create a new chain and set a jump-rule for this chain into the parent chain
6. Place the rules into the new chain

The influence to the performance depends on the internal network topology/security policy and the rulesets for external networks. In this case more rules per hosts achieve better results with this concept.
A sample Calculation:
25 Hosts whith 10 rules/host= 250 rules for packets that don't match with one of the rules in normal packet filter setup
25 rules for packets that don't match with one of the rules, in the concept of this paper → 127 MBit/s instead of 79 MBit/s (61% more) throughput (Table 3)
Fig. 8 shows the results for a ruleset of 500 rules overall and the depending amounts of chains and rules/chains (e.g. with 500 rules overall and 500 chains, you will have one rule/chain). Every chain needs a jump-rule in the parent chain. The performance for packets, which doesn't match the jump-rules, depend only on the number of jump-rules (chains) and not on the number of rules per chain.(Table 3) The results come up to expectations and show the achievement potential of the new filter-design approach in this paper.

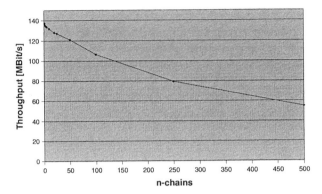

Fig. 8. Performance with tree splitting

5   Conclusion

New network services will increase the bandwidth and QoS needs in access networks. Chapter 3 gives an overview about the Packet-Filter performance of Linux
in dependence on the number of filter rules. The performance decrease rapidly
with the number of filter rules. Chapter 4 gives two alternatives for improving
packet filter performance. The one this paper concentrates in is to reorganize the
filter rules so you achieve a flat filter design and less filter rules per packet(fig. 7).
The filter length decreased to 25% with one separation only. Chapter 4.2 give
an overview about enhancement possiblities of netfilter setups and Table 3 show
the achievement potential of the new filter-design approach in this paper
Future research has to busy with optimization and automation of packet filter configuration and build the whole script-structure with install scripts. The
testing scenarios will contain more than one connection at time. Furthermore
QoS features and the network performance overall has to be tested, especially
the client dependent throughput, which attracted attention during the tests.
Furthermore research will contain delay tests for packets passing the firewall
depending on filter structure and test of QoS algorithms.

References

[1] Feldmann, A., Muthukrishnan, S.: Tradeoffs for Packet Classification. conference-
paper IEEE Infocom 2000, AT&T Labs-Research (2000)
[2] Frost and Sullivan: Market Engineering Research for the European Video on Demand Market. Frost & Sullivan (2001)
[3] Frost and Sullivan: A Strategic Review of The European Broadband Market. Frost
& Sullivan (2002)
[4] NetValue: Broadband captures private households. NetValue Germany, Essen
(2002)
[5] Williamson, B.: Developing IP Multicast Networks. Cisco Press, Indianapolis
(2000)

[6] Lee, D. C.: Enhanced IP Services For Cisco Networks. Cisco Press, Indianapolis (1999)

[7] P. Cochrane, P., Heatley, D. J. T.: Modelling Future Telecommunications Systems. first ed., Chapman & Hall (1997)

[8] Microsoft Corporation: Introduction to IP Version 6. Microsoft Corporation (2000)

[9] Atkinson, R.: Security Architecture for the Internet Protocol. Naval Research Laboratory (1995)

[10] Podey, B., Kessler, T., Melzer, H.-D.: Security Devices in High Speed Access Networks. Proceedings of the Tenth International Conference on Telecommunication Systems (2002)

[11] Bellion, M., Heinz, T.: High Performance Packet Classification for Netfilter http://www.hipac.org/ (2002)

Part VI

Network Management

Design and Implementation of Information Model for Configuration and Performance Management of MPLS-TE/VPN/QoS

Taesang Choi, Hyungseok Chung, Changhoon Kim, and Taesoo Jeong

Electronics Telecommunications Research Institute
{choits,chunghs,kimch,tsjeong}@etri.re.kr

Abstract. Multi Protocol Label Switching (MPLS) is generally considered a mature technology. Many Internet Service Providers (ISPs) and telecommunication carriers have deployed it or are considering deploying it. An easy-to-use integrated management solution is requested by these ISPs. To realize truly integrated management solution, a combined management information model is essential. In this paper, we propose an information model for integrated configuration and performance management of MPLS-Traffic Engineering (MPLS-TE)/VPN/QoS.

1 Introduction

As of today, Multi Protocol Label Switching (MPLS)[1] is considered as a mature technology. Many Internet Service Providers (ISPs) and telecommunication carriers have deployed it or are considering deploying it for various reasons: efficient usage of their valuable network and system resources and meeting customer's emerging service requirements such as provider managed IP virtual private networks (VPNs) and quality of service (QoS) guaranteed IP services for voice, video or mission critical applications.

One of the major requirements for a successful deployment is easy, efficient, scalable and reliable management of networks and services based on MPLS. This includes automated provisioning, network and service performance monitoring, fault management and billing. And service providers want a more general management solution and, thus, a common integrated information model for all these management functionalities is needed more than ever before.

In this paper, we propose an information model for integrated configuration and performance management of MPLS-TE/VPN/QoS. It is an OO-based abstract information model which means it is independent from the existing data models, encoding schemes and management protocols. Well-known data models and specification languages are the Common Information Model (CIM)[2]/the Managed Object Format (MOF)[2], the Policy Information Base (PIB)[3]/the Structure of Policy Provisioning Information (SPPI)[4] and the Management Information Base (MIB)/the Structure of Management Information (SMI)[5]. XML and ASN.1 Basic Encoding Rule (BER)[6] are commonly used for data encoding. And HTTP, Common Open Policy Service - Provisioning (COPS-PR)[7]

H.-K. Kahng (Ed.): ICOIN 2003, LNCS 2662, pp. 817–827, 2003.

and SNMP are the protocols to carry those management data. We don't want to limit our information model to any of these approaches. The main reason is that any of these data models alone can't satisfy the integrated management requirements. That is, a combination of them or even new approaches may be required. In our proposal, we defined an information model by using Unified Modeling Language (UML) [8] which is used most widely to describe OO-based information models. Our implementation approach is described in detail later in this paper.

The remainder of this paper is organized as follows. Section 2 details the design of the proposed information model. Our implementation approach is explained in the section 3. Finally, we conclude and discuss future work in the section 4.

2   Design of the Information Model

The main focus of the proposed information model is OO-based object model, protocol independency and integration of configuration and performance management. Our information model consists of four sub-models: one for MPLS-TE, MPLS-VPN, Diffserv[9]-based QoS, and Topology. Each sub-model is divided into configuration and performance parts. Topology sub-model is common to all three services. IP layer topology is a common denominator of MPLS-TE, MPLS-VPN and QoS topologies. Given the IP layer topology, MPLS-TE, MPLS-VPN or Diffserv topological information can be further added depending on an underlying network's capabilities. For example, if the underlying network supports MPLS-TE then both IP and MPLS-TE topological information is captured in the same topology information model. MPLS-TE sub-model models configuration and performance management information of MPLS-TE such as a traffic trunk, Label Switched Path (LSP) tunnel, LSP Path and traffic statistics of LSP tunnels. Similarly MPLS-VPN and QoS sub-models define conceptual management information and their relationship of respective services' configuration and performance functionalities. Further details are provided below.

Our information model can be filled in with management data via various protocols: SNMP, Telnet/CLI and Open Shortest Path First – Traffic Engineering (OSPF-TE)[10]. Its storage can also be realized in various ways: in-memory, DB, Lightweight Directory Access Protocol (LDAP)[11] or XML. Main point to emphasize here is the fact that the proposed information model is protocol and storage technologies agnostic.

Also note that we only represent an outline of the information model, that is, important classes and their relationships in this paper. Thus, the class diagram has a single row box representation rather than a three row box representation which is typical in UML. We purposely omit description of class details such as attributes, their types, methods, and their signatures. With the limited space of the paper, we try to propose ideas on the integrated information model and their relationships which play an essential role for the integrated service and network management of MPLS-TE, VPN, and QoS services.

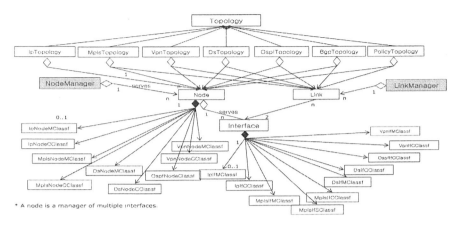

Fig. 1. Topology Configuration and Performance Information Model

2.1   Topology Information Model Common for All Services

Fig.1 shows the information model to represent various topologies such as IP, MPLS, VPN, Diffserv, OSPF, BGP and policy in an integrated way. This information model is a corner stone to glue three network-based services: MPLS-TE, VPN and QoS. A physical node, link or interface can play multiple roles depending on the capability it provides at the moment of serving. If a node is a router and provides a layer 3 MPLS-VPN service, then this node plays roles of an IP node, an MPLS node and a VPN node at the same time. If you manage each service separately, then you should have three separate topology information models. When a service provider requests an integrated management, you have to waste resources for separate information models. Also synchronization of these management information becomes complex. The proposed model not only saves resources but reduces management complexities caused by the duplication of similar information.

All topologies share Node, Interface and Link classes. Node can be an IP node, an MPLS node, an L3 MPLS-VPN node, a Diffserv Node and/or a policy Node. Similarly, Interface can be an IP interface, an MPLS interface, an L3 MPLS-VPN interface and/or a Diffserv interface. Node class contains common attributes and methods which can be applied to any node types. Role specific Node Classification class(es) can be instantiated on demand whenever they are needed. Thus Node class and its part, xxNodexClassf classes have a composition (strong aggregation, according to UML) relationship. Similar rule applies to Interface. Performance related classes which are denoted by "M" letters (stands for measurement) will be instantiated when respective configuration classes which are denoted by "C" letters are provisioned. That is, the lifetime of performance classes has meaning only when such nodes or interfaces exist. Thus the multiplicity of these node and interface classification classes is represented as 0..1.

Important point in this sub-model is that all three services (MPLS-TE, VPN and QoS) related nodes and interfaces characteristics are captured in it for both configuration and performance management. In this sense, this sub-model becomes a common ground for an integrated configuration and performance management. For the integrated service and network management, management operations on topologies of multiple layers are essential. This information model can hold these multiple layer topological information such as MPLS, IP, BGP, QoS and VPN topologies in an integrated way. Thus various management operations like topology auto-discovery, multiple topology view representation and storing layer specific performance data are possible.

2.2   Information Model for MPLS-TE

Fig.2 shows the information model for MPLS-TE configuration management. It shows required object classes and their relationships. There are three important classes: a traffic trunk, an LSP tunnel, and an LSP. The traffic trunk models the one defined by RFC2702, "The requirements for traffic engineering over MPLS" [12]. It represents an aggregate of customer traffic flows belonging to the same service class or classes. It can be mapped into zero or more LSP tunnels for load sharing purposes. Each LSP tunnel is mapped into one or more LSPs which are represented by dynamically calculated paths or explicitly specified paths. Other classes are tightly coupled with these main classes. For example, a traffic trunk (TtC) and LSP tunnel (LspTunnelC) need forwarding equivalence class (FEC) and a RSVP traffic profile (RsvpTp). FEC is a traffic classification filter and RsvpTP is a traffic profile attribute specification such as bandwidth, delay, jitter, etc.. Classes with letter "C" like TtC contain configuration and static information only. When a new class object is created, three objects (with letter "C", "M", and "S") are created at the same time and stores configuration-, performance-, and simulation-specific information separately for information consistency.

Classes with letter "M" denote that they are used for performance management. Traffic trunks, LSP tunnels and LSPs have their operational status information and statistics information such as packet per second every five minutes and bits per second every five minutes. Letter "S" stands for simulation. We identified some of the simulations which are very useful for the performance management. MPLS LSP path availability check, Node/Link failure, Traffic trunk and/or LSP tunnels attribute change and global optimization simulations are some of possible candidates. Simulations can be performed off-line with historical data acquired by off-line means. In such a situation, the simulation is typically considered to be a separate auxiliary mechanism to help performance management. In our proposed information model, we approached in a different way. These simulations are performed on-line with data collected and monitored live from the managed networks, which are stored in topology and performance sub-models. Since simulation results can modify the existing topological and performance information, separate place holders are needed. We modeled these

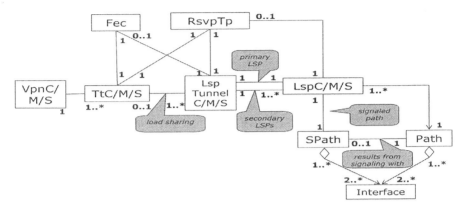

Fig. 2. MPLS-TE Configuration Information Model

"S" classes for that purpose. The path availability check function provides an efficient way of simulating a setting up process of an LSP. The CSPF (Constrained Shortest Path First) algorithm, which is resident in our integrated management system, can compute the availability of a path of an LSP without actual enforcement. The server-based CSPF can also extend its scope to add additional constraints, e.g. actual usage instead of the required bandwidth of an LSP, besides what the online CSPF allows. This feature is one of the big advantages that an offline TE management system can provide. The LSP attribute modification simulation allows network managers to evaluate the side effects of LSP and VPN attribute modification. Modification of attributes ranges from simple single value change (e.g., the affinity value) to an entire path alteration for an LSP. This simulation helps the network managers create a detour route when a particular link is congested and see the link state changes in real-time. The link and node failure simulation depends on an online protection and recovery mechanism and visualizes its effects. Features like standby secondary paths, as well as explicit or dynamically configured primary and secondary paths of LSPs are also recognized for this simulation. Depending on the situation, the simulation can just visualize the overall status of a newly optimized network status or visualize all the paths computed by the server's CSPF algorithm. The global optimization is performed by a customized algorithm based on linear programming (LP). The algorithm can find near optimal paths that satisfy a given traffic demand under some constraints, such as bandwidth, a maximum hop count, and a preferred or avoided node or link list. Our integrated management system generates a mixed integer programming formulation for a given optimization problem and solves it with a dedicated LP solver. The optimization result contains each LSP's routing paths and the traffic split ratio, in case an LSP requires multiple paths. For easy representation at network nodes, the split ratio is chosen among discrete values (0.1, 0.2 etc.). The globally optimized set of paths can then be applied to the

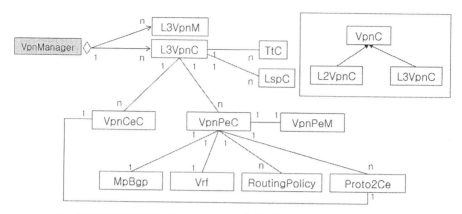

Fig. 3. MPLS-VPN Configuration and Performance Information Model

MPLS networks that permit explicit path setup. For more details, please refer to [18].

2.3 Information Model for MPLS-VPN

We modeled MPLS-based layer 2 and layer 3 VPNs. Layer3 MPLS-VPN follows IETF's RFC2547bis standard [13]. Fig.3 shows a configuration and performance management information model for layer 3 MPLS-VPN. L3 MPLS-VPN consists of PE routers and CE routers and all VPN related functionalities are configured in PE routers. Configuration information is divided into four groups: virtual routing table, multi-protocol BGP between PEs, routing protocols between PE and CE and VPN import/export routing policies in and between PEs. L3 MPLS-VPN can be a form of several topologies: hub-and-spoke, full mesh, full mesh with route reflectors, or partial mesh. It needs to know the number of PEs and CEs and their relationships. We also allow traffic engineering paths between PE routers and, thus, each L3VPN has to know the paths identifiers if any. Performance information is captured in L3VpnM and VpnPeM. L3 VPN operational status and performance traffic data are stored.

MPLS-VPNs can be managed separately from MPLS and MPLS-TE. There are a few MPLS VPN management solutions available in the market. They are either vendor specific or limited in management functionality point of view. For example one tool manages PE to CE relationships from provisioning perspective only and another tool covers provisioning and performance but limited to a single vendor. Service providers, however, want to have a more comprehensive management solution, even want to combine with MPLS-TE and QoS functionalities in vendor neutral environment. This requirement leads an integrated management solution. The first step toward such integration is defining a common information model in terms of functionality and vendor dependency. Our information model tries to achieve this goal by tightly integrating with MPLS-TE and topology information models.

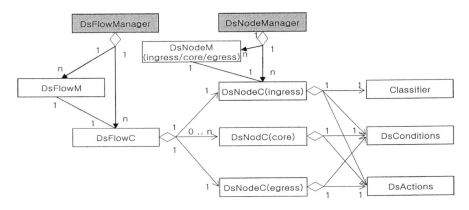

Fig. 4. MPLS-QoS Configuration and Performance Information Model

2.4   Information Model for MPLS-QoS

Diff-serv based QoS configuration and performance management information model is defined as fig.4. It consists of diff-serv flows and involved ingress, core and egress nodes. Each node has a classifier, a Diffserv condition and a Diffserv action. Classifier is meaningful in the ingress node only. Diff-serv flow specific and node specific performance information is captured in DsFlowM and DsNodeM classes. Like other sub-models, the lifecycle of these class objects are managed through corresponding manager classes.

This diff-serv based QoS can be applied for both MPLS-TE and MPLS-VPN services. Depending on the requirements of service providers, diff-serv over MPLS-TE, diff-serv aware MPLS-TE[14] or diff-serv over MPLS-VPN can be managed by using the same information model. In case of diff-serv over MPLS-TE, configuration and performance management of MPLS-TE traffic trunks and LSP tunnels can be performed via MPLS-TE information model and diff-serv DSCP field to MPLS EXP field mapping can be done via this information model. Classifier is a six-tuple generic filter and mapping between DSCP and EXP field can be represented as a classifier instance. Configuration and performance management of diff-serv aware MPLS-TE can be achieved in a similar manner except that MPLS-TE traffic trunk or LSP tunnel signaling includes Diff-serv aware MPLS-TE signaling information. Lastly, diff-serv over MPLS-VPN involves provisioning of provider-edge (PE)-to-PE MPLS-TE LSP tunnel signaling and provisioning of diff-serv over the signaled tunnel. Performance management for these flows and tunnels is also conducted no different than the other two cases.

As mentioned in this section, MPLS-TE, MPLS-VPN and MPLS-QoS are tightly coupled and managed efficiently. Such seamlessly integrated inter-service configuration and performance management is possible mainly due to the proposed information model.

3   Implementation

The proposed information model is implemented and used in the integrated configuration and performance management system called Wise<TE>[15]. We chose CORBA IDL[16] as a language to specify this information model and implemented using C++. Management information model is often realized with the management data model language, SMI in the form of the MIB. However, most data model languages like SMI, SPPI, or XML may be not sufficient to represent the proposed OO-based information model syntactically and semantically. Major decision factors include protocol independency, expandability, interoperability and API support. As shown in fig.5 below, our information model serves as the common repository of management applications which use various protocols for communication with network elements. For example, the policy server (PS) talks to network elements via COPS-PR with provisioning information retrieved from this common information repository. The resource monitoring server (RMS) and traffic measurement server (TMS) collect various topology and performance information via SNMP/CLI polling and stores in the common information repository. If the information model is designed in relation with a particular data modeling language (e.g., SNMP SMI), the choice of a protocol should be very limited. CORBA implementation allows internal implementation updates without modifying the existing interfaces and relatively simpler addition of new interfaces without affecting the existing ones. Also it is necessary to interact with other management solutions (e.g., a billing system) within a global Operation Support System (OSS) environment. CORBA is the de facto industry standard for such integration. Most service providers want to provide customer network management (CNM) interface to their customers so that customers can provision part of their services or access performance data for their subscribed services. Our proposed implementation can easily support APIs for such purposes.

Fig.5 illustrates a high level architecture of the integrated management system. The common service interface (CSI) realizes the proposed information model. The figure gives a birds-eye view of how management application components interact with the information model. As you can see, the protocol dependency is pushed to the behind of the management applications. Management applications communicate with the underlying managed network via various protocols. It is not realistic to mandate using a specific management protocol to manage today's very heterogeneous networks. Each data model and its associated language has its own role, thus we cannot simply force to converge into a single solution. For example, SNMP/MIB is the de facto IP network management protocol and data model but its usage is limited to fault and performance management functional areas. It is not well accepted as a tool for configuration management. Typically script-based solutions together with telnet/CLI as a communication protocol are used instead. Although COPS-PR/PIB was defined as a standard solution for this purpose but market didn't like that either. Discussion on XML-based standard for configuration management has just begun in the IETF. Protocols come and go but users cannot simply wait. Thus, the

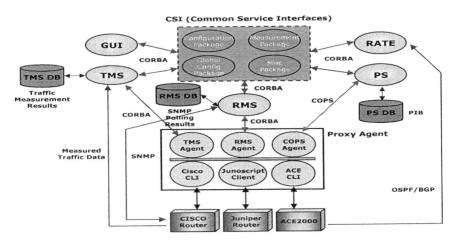

Fig. 5. Implementation Architecture of the Information Model

information model better be protocol independent. Our current implementation supports COPS-PR based provisioning and we are also adding a new provisioning mechanism based on XML/HTTP [17]. This extension is possible due to the protocol independence nature of our information model. The common repository also allows various management applications easily share and synchronize management information. It can provide public interfaces to external applications, if necessary.

Traditionally management information base in a manager-side is realized by a relational database system since it is straight forward to map from MIB to RDB. OOness of the proposed information model, however, prevents from using this traditional approach. Instead, instances of our information model reside in the main memory of the system that they are running. For reliability, we provide a persistency mechanism by storing in-memory copy into LDAP repository. In this case, static configuration information not dynamic information (e.g., operational status and performance matrices) is saved since such dynamic information becomes meaningless when the system fails. When the system downs abnormally, the copy of information just before the failure can be restored from the LDAP repository.

We have finished the prototype implementation. Fig.6 shows a snapshot of the prototype's graphical user interface. It illustrates auto-discovered MPLS topology with traffic load represented in different colors per link based on values in the legend. Also LSP tunnel creation wizard, LSP tunnel detail table and per-LSP tunnel statistics windows are shown as well. We have been testing it in our testbed. The testbed has limited size (about 9 high-end routers - 3 Cisco 7xxx, 4 Juniper Mxx, 2 Riverstone 3000 and 8000) thus we utilized it mainly for provisioning tests and used a commercial agent simulator for a scalability test. So far, we have tested it over a network with 50 nodes and found out that it showed good performance. Main concern is how the implementation can scale in terms

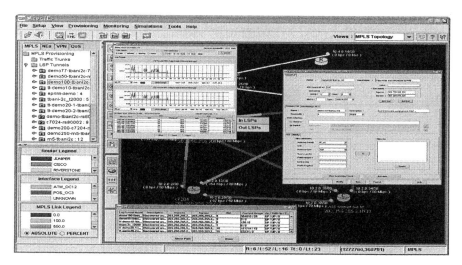

Fig. 6. Snapshot of Wise<TE> GUI

of the number of CORBA objects (e.g., topology objects, LSP tunnel objects, MPVL-VPN l3vpn objects, etc.) as the number of nodes increase. Our implementation is running in a 500MHz dual CPU PC server with intel solaris 2.8. We are planning to test our system over a network with upto several hundreds nodes. At the same time, we are preparing to test it in a real service provider network soon. More concrete test result will be available in the near future.

4   Conclusion and Future Work

MPLS-based IP network and service management is fairly new field and very few commercial solutions are available today. However, due to the complexity of the MPLS-based network and service technologies, an easy-to-use integrated management solution is strongly requested by most ISPs. To realize a truly integrated management solution, a combined management information model is essential. In this paper, we proposed an information model for integrated configuration and performance management of MPLS-TE/VPN/QoS. We described its OO-based characteristics and benefits. We also explained its implementation and associated experiences. This paper mainly focused on the configuration and performance management functional areas but we do aware that fault management is also very important area to be covered. In fact, the proposed model includes several attributes for this purpose but comprehensive extension is required. We are in the process of updating our model for this. Also performance and scalability evaluation of the implementation should follow.

References

[1] E. Rosen, A. Viswanathan, R. Callon, "Multiprotocol Label Switching Architecture ", RFC3031, IETF, Jan. 2001.

[2] Distributed Management Task Force, "Common Information Model (CIM) Specification Version 2.2", DSP 0004, June 1999.

[3] M. Fine, et al., "Framework Policy Information Base", Internet-Draft: draft-ietf-rap-frameworkpib-05.txt, IETF, July, 2001.

[4] K. McCloghrie, M. Fine, J. Seligson, K. Chan, S. Hahn, R. Sahita, A. Smith, and F. Reichmeyer, "Structure of Policy Provisioning Information (SPPI)", RFC 3159, August 2001.

[5] K. McCloghrie, D. Perkins, J. Schoenwaelder, J. Case, M. Rose, and S. Waldbusser, "Structure of Management Information Version 2 (SMIv2)", RFC 2578, STD 59, April 1999.

[6] International Organization for Standardization, "Information processing systems - Open Systems Interconnection - Specification of Abstract Syntax Notation One (ASN.1)", International Standard 8824, December 1987.

[7] K. Chan, D. Durham, S. Gai, S. Herzog, K. McCloghrie, F. Reichmeyer, J. Seligson, A. Smith, R. Yavatkar, "COPS Usage for Policy Provisioning," RFC 3084, May 2001.

[8] Object Management Group, "Unified Modeling Language (UML), Version 1.4", formal/2001-09-67, September 2001.

[9] S. Blake, D. Black, M. Carlson, E. Davies, Z. Wang, W. Weiss, "An Architecture for Differentiated Services", RFC2475, IETF, December 1998.

[10] D. Katz, D. Yeung, K. Kompella, "Traffic Engineering Extensions to OSPF", Internet Draft, draft-katz-yeung-ospf-traffic-06.txt, April 2002.

[11] Yeong, W., Howes, T., and S. Kille, "Lightweight Directory Access Protocol", RFC 1777, IETF, March 1995.

[12] D. Awduche, J. Malcolm, J. Agogbua, M. O'Dell, "Requirements for Traffic Engineering Over MPLS", RFC2702, IETF, September 1999.

[13] Rosen, Rekhter, et. al., "BGP/MPLS VPNs", Internet Draft, draft-ietf-ppvpn-rfc2547bis-02.txt, IETF, July 2002

[14] Francois Le Faucheur, et al., "Requirements for support of Diff-Serv-aware MPLS Traffic Engineering¡±, Internet-Draft: draft-ietf-tewg-diff-te-reqts-01.txt, IETF, June 2001.

[15] TS Choi, SH Yoon, HS Chung, CH Kim, JS Park, BJ Lee, TS Jeong, "Wise<TE>: Traffic Engineering Server for a Large-Scale MPLS-based IP Network", NOMS 2002, April 2002.

[16] OMG, "The Common Object Request Broker: Architecture and Specification", Revision 2.2, Feb. 1998.

[17] BJ Lee, TS Choi, TS Jeong, "X-CLI : CLI based Policy Enforcement and Monitoring Architecture using XML", Accepted for the publication of APNOMS2002 proceeding, September 2002.

[18] Y. Lee et al., "A Constrained Multipath Traffic Engineering Scheme for MPLS Networks," IEEE ICC 2002, New York, May 2002.

Definition and Visualization of Dynamic Domains in Network Management Environments

Márcio Bartz Ceccon, Lisandro Zambenedetti Granville,
Maria Janilce Bosquiroli Almeida, and Liane Margarida Rockenbach Tarouco

Federal University of Rio Grande do Sul (UFRGS) - Institute of Informatics
Av. Bento Gonçalves, 9500 - Bloco IV - Porto Alegre, RS - Brazil
{ceccon,granville,janilce,liane}@inf.ufrgs.br

Abstract. Dynamic domains are domains quickly created, used and discarded. Today, there are no facilities available to support dynamic domains in most network management systems. This paper introduces two new languages to deal with dynamic domains. The first language is used to define new domains through the selection of managed objects. The second language, on its turn, is used to visualize already created dynamic domains. Both languages are explained through examples and implementations details are presented.

1 Introduction

In network management systems, domains are an important tool used to group managed objects [1]. The most common example of domains is the network map facility found in almost any management system, where a map groups devices from the same network segment. Domains can also group functional similar devices (e.g. printers or routers). They are important because management actions can be applied to the set of managed objects of a domain, instead of repeating the same actions to each object one by one.

There are some facilities used in the definition of domains. In standard management systems, for example, a topology discovery tool can create the network map of a managed network. The network administrator can also defines his/her own domains by dragging network devices to new maps. Once defined, domains normally remain stored in a management database, for future use. However, there are some situations where certain domains should be quickly created, used and discarded. We are going to use, as an example of these situations, a scenario where a Policy-Based Network Management (PBNM) [2] system is applied to the management of a QoS-enabled network. Suppose a videoconferencing session between hosts A and B where the intermediate routers are unknown. First, the administrator creates a policy that defines the expected QoS for the videoconferencing, and stores such policy in a repository. Second, the administrator searches for the first router from A to B. The first router is added to the session's domain. The second router is also identified and added to the session's domain, and so on. When the final router from A to B is found, the opposite path, from B

H.-K. Kahng (Ed.): ICOIN 2003, LNCS 2662, pp. 828–838, 2003.

to A, is visited, since upstream and downstream paths can be different. As long as new routers are found, they are included in the session's domain. After, the stored policy is deployed in the session's domain, instead of a policy deployment applied to each session's router one by one. When the videoconferencing is over and the session is closed, the session's domain is useless, and is then discarded.

This scenario shows two interesting features of domains:

1. Sometimes, domain creation can be slow, mainly when it is done by the network administrator manually;
2. Some domains have short life-time, because they are used for a specific purpose and are then discarded.

In the first feature, the time spent in the creation of domains have to be shorter than the domain life-time, that can also be short according to the second feature. Still based in the previous PBNM example, the definition of the videoconferencing domain have to be faster then session life-time, otherwise the policy would be applied to the session routers after the videoconferencing is over. In this paper, we use the term "dynamic domains" to address domains that have to be quickly created, used and discarded.

Another important aspect of domains is their visualization. Since today GUIs are obligatory in any management system, domains have to be properly visualized. A graphical map is obviously used to show network maps. Tables, charts and matrixes can also be used to show the elements of a domain. One interesting point in domains visualization is the fact that, often, the user is limited to the customization features provided by the visualization process. For example, in network maps the administrator knows that red devices normally indicate a problem, but the administrator is unable to define the yellow color to denote DiffServ-enabled [3] devices.

In this paper we introduce two new languages used to improve the definition and visualization of dynamic domains. The first language is used in the domains definition, and the second one is used to customize the visualization process of a domain that was previously created through the definition language. We have implemented a prototype to support both languages and applied it to the management of a QoS-enabled test network. The prototype is a Web-based tool developed with PRECCX, PHP, MySQL and SNMP. We believe that the contribution of our work comes from the fact that we are introducing a solution for the commonly absent support for dynamic domains definition and visualization.

The remaining of this paper is divided as follows. Section 2 presents the management environment and the information model where the languages for dynamic domains definition and visualization, presented in Section 3, are applied. In the Section 4 we present the developed prototype, and finally the paper is finished with final remarks and future work in Section 5.

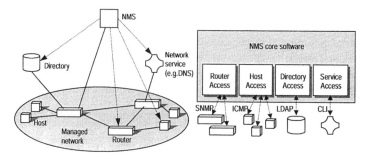

Fig. 1. Management environment

2 Management Environment and Information Model for Dynamic Domains

In a network management system, the network administrator is expected to have access to managed objects. The managed objects can be members of domains (e.g. network maps), because such domains ease the access and manipulation of the managed objects. In order to access different managed objects, it is required the use of different protocols, such as SNMP - Simple Network Management Protocol (e.g. routers interfaces access), COPS - Common Open Policy Service [4] (e.g. RSVP sessions access) or LDAP - Lightweight Directory Access Protocol [5] (e.g. network users records access). To manage a network with all these complexities and yet provide facilities for the definition of dynamic domains, we need to define three different elements: (a) the network management environment where the dynamic domains will be defined and used; (b) the information model that describes how information found in the management environment is organized; and (c) the language used to define dynamic domains that uses the management information found in the management environment.

The next subsections present the network management environment and the information model used in the definition of dynamic domains. Section 4, on its turn, will present the languages to define and visualize dynamic domains based on such management environment and information model.

2.1 Management Environment

In order to provide a language for the definition of dynamic domains, first we need to describe the environment we are dealing with. Figure 1 presents such environment. We consider the managed network as a collection of heterogeneous devices with just one central NMS, but in a distributed management environment one will find Mid-Level Managers (MLM), Bandwidth Broker (BB), Policy Decision Points (PDP) and other structures that implement a decentralized management.

Since the network resources (hosts, routers, directories, etc.) are heterogeneous, the methods to access them are different too. However, it is required

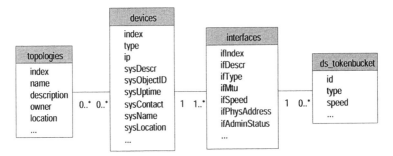

Fig. 2. Information Model

that such different methods do not increase the management complexity. Thus, we also consider that, although the ways to access the network resources are different each other, such resources are "seen" by the NMS in a standardized way. To allow that, the NMS has an intermediated layer of protocol handlers that abstracts the details of the managed objects access (figure 1 on the right). Actually, the lower layer details of the managed objects form the data model (defined, for example, through SMIv1, SMIv2SMIv2 or SPPI [6], while the upper layer common view of the managed objects forms the managed information model, which abstracts the details of the data models. On the top of the figure 1 right, we have the NMS core software responsible for the manipulation of the information model of the managed objects and domains.

2.2   Information Model

The information manipulated by the NMS core software is organized through an information model. The model is defined through classes (e.g. topologies and devices) that contains attributes (e.g. topology owner and device ip address) and relationship between classes with cardinalities. In figure 2, one can see a class to store information about the topologies of a managed network. Each topology has as collection of devices, and each device can be a member of several topologies. Thus, we also have a devices class. Inside a device we can have several network interfaces, but each interface belongs to just one device. Each interface, on its turn, can hold several DiffServ token buckets, several DiffServ droppers and several RSVP sessions. Each token bucket and dropper are associated to a single interface, while each RSVP session is part of several different interfaces from different devices.

This model, obviously, is applied to a limited set of networks. However, the set of possible managed objects could be quite greater in other environments. To manage a broader range of networks, we assume that this information model is dynamic, in the sense that it can be extended to accommodate other managed objects. For example, if a new MIB (Management Information Base) is compiled in the management system, the model is automatically extended to support the new managed objects described in the added MIB.

3   The Languages for Dynamic Domains Definition and Visualization

Here we present the dynamic domains definition language and the visualization language. Through definition expressions, administrators can define dynamic domains that are send to the visualization language in order to be properly presented.

3.1   Dynamic Domains Definition

The following features were considered in the definition of the proposed dynamic domains definition language: (a) it must consider a dynamic information model, where new classes can be added and others removed; (b) it must be easy to use; and (c) It must provide a mechanism to select objects and group them in dynamic domains (manipulation of such object through write operations is out of the scope of the language). Said that, a selection expression to define a dynamic domain is written according to the following BNF:

```
domain     ::= select expression
expression ::= term {from term}
term       ::= classdata {::classdata}
classdata  ::= class [[value]] {.attribute[value]}
```

In such BNF, `class` identifies a class from a management information model (as the one presented in subsection 3.2), `attribute` identifies an attribute from a class and `value` is used to select managed objects that math a specific value of an attribute. Since the language is not limited to a particular set of classes and attributes, the language can be used with different management information models, thus satisfying the first requirement listed previously. To better understand how the language can be used, follows some examples considering the information model from figure 2.

1) `select topologies.owner["secAdm"]`
2) `select topologies.owner["secAdm"].location["admBuilding"]`

The first expression defines a dynamic domain composed by topologies whose owner is the security administrator. The second expression will select only topologies managed by the security administrator that are located in the administrative building.

Attributes from other classes can be used when selecting object from a particular class, using the `::` element, as shown in the following examples.

3) `select topologies::devices.type["DSRouter"]`
4) `select topologies::devices["DSRouter"]::interfaces["ATM"]`

In the example number 3, a dynamic domain will be created containing topologies that have devices whose type is "DSRouter" (DiffServ router). Each class from the information model is supposed to have a default attribute. This default attribute is used when expressions like the one from example 4 is evaluated. In that case, the dynamic domain will contain topologies that have DiffServ routers with ATM interfaces, which means that `type` is the default attribute for the `devices` class and `ifType` is the default attribute for the `interfaces` class.

The first class that follows the `select` clause identifies the class that will provide the primary indexes stored in the dynamic domain that will be created. The `from` clause is used to change such first class without removing it from the original expression.

5) `select devices from topologies["admBuilding"]`

The expression number 7 will create a dynamic domain composed by all the devices from the topologies located at the administrative building (in this case, the `location` attribute is the default attribute for the `topologies` class).

3.2   Dynamic Domains Visualization

The goal of the visualization language is to allow a customized visualization of previously defined domains. The process of customizing a visualization is based on setting attributes of a visualization facility (e.g., tables, charts, graphics). In order to define a new visualization, the following steps are required.

- Create dynamic domains through the dynamic domains definition language;
- Select a **visualization facility** to present the created domains. A collection of visualization facilities is supposed to be available in the management environment, and the network administrator has to specify which one is going to be used;
- Select, among the available dynamic domains, those that will be presented through the selected visualization;
- Customize the visualization facility according to the selected dynamic domains.

The last three steps are supported through the proposed visualization language that is defined according to the following BNF.

```
visualization ::= [selection] show dmview {and dmview}
selection      ::= using visualizationfacility [customization]
dmview         ::= domain [customization]
customization ::= with attribute=value {, attribute=value}
```

In this BNF, `domain` represents a dynamic domain created with the domain definition language; `visualizationfacility` indicates the visualization facility to be used and `attribute` addresses an attribute of the selected visualization facility. To explain the visualization language through examples, first we define some domains.

```
DiffServRouters = select devices.type["DSRouter"]
IntServRouters  = select devices.type["ISRouter"]
```

In the visualization language, the `using` clause selects a visualization facility, while the `show` clause lists the dynamic domains to be presented.

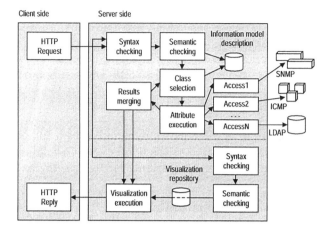

Fig. 3. Dynamic domain definition steps

(1) `using topology show DiffServRouters and IntServRouters`

In the previous example the visualization facility, named `topology`, is available to show the domains (`DiffServRouters` and `IntServRouters`). The `and` clause was used to aggregate the `DiffServRouters` domain in the topology visualization. The key feature of the visualization language is the ability to customize the visualization facilities through the `with` clause. This clause can be used to customize a visualization either for all domains to be presented or for each domain in particular.

(2) `using table with cellcolor=blue show DiffServRouters and`
                                         `IntServRouters`
(3) `using table show IntServRouters with cellcolor=yellow and`
                          `DiffServRouters with cellcolor=green`

The example number 2 shows a table whose cells are presented in blue for both `DiffServRouter` and `IntServRouters` domains. In 3, however, `IntServRouters` are presented in yellow, while `DiffServRouters` in green. Customizing a visualization facility based on dynamic domains (as in 3) overlaps a global customizations (as presented in 2).

Next section will present how both languages for the definition and visualization of dynamic domains are supported in a prototype implementation.

4   Implementation

The support for the two languages for dynamic domains was implemented as a subsystem of the QAME environment [7]. The two following subsections present how each language is supported in QAME.

Class	Attribute	ScriptFile	Function
topologies	owner	topologies.php	get_owner
topologies	location	topologies.php	get_location
...	...	...	...
devices	type	devices.php	get_type
devices	sysName	devices.php	get_sysName
devices	sysLocation	devices.php	get_sysLocation
...	...	...	...
interfaces	ifType	interfaces.php	get_ifType
...	...	...	...

Fig. 4. Complementary information with the information model

4.1 Support for Dynamic Domains Definition

A list of textboxes is used in a QAME HTML form to enter expressions to
define new dynamic domains, and a single textbox is used to enter a visualization
expression (verified in the next subsection). A PHP4 script is then executed to
evaluate the passed expressions and create the requested domains. The whole
process is done according to the steps presented in figure 3, top.

In the figure 3 example, a single HTTP request sends to the server two domain definition expressions and one visualization expressions. For each domain
definition expression the following steps are executed. First, a syntax checking of
the expression is executed. This verification is done through a C compiled program generated by PRECCX [8]. After that, the semantic checking step accesses
a description of the information model to verify if the expression is a valid one.
In our implementation, the information model description is stored in a MySQL
base. If the syntax used is correct, and the expression is consistent with the
information model stored in the MySQL base, an execution cycle starts.

A class selection and a subsequent attribute execution are responsible for
defining which access driver have to be used. This selections is done accessing
a driver information table (figure 4) in the management environment. An access
driver is a function within a PHP4 script that, using an appropriate protocol,
contacts managed objects and retrieve the requested values. For example, considering the access driver information table from figure 4, the evaluation of the expression `devices.sysLocation["LabCom"]` will issue the PHP4 script `devices.php`
and the function `get_sysLocation` will be called. Figure 5 shows a simplified code
for the `get_sysLocation` function (the actual code has some additional complexities). Notice that, in this case, the SNMP protocol was used to retrieve the
location of the managed objects, but other protocols can be used as well: it only
depends on the implementation of the used access function.

A class selection and attribute execution is evaluated to each portion of the passed expression that is separated by :: or from. For example, the expression `select devices ["DSRouter"]::interfaces["ATM"] from`
`topologies["LabCom"]` has three portions. From `devices["DSRouter"]` we will get
all the DiffServ routers. The `interfaces["ATM"]` portion will provide every ATM

```
01 list get_sysLocation (list devices, string value) {
02    list ret, string tmp, int i;
03    if (!devices) devices=AllDevices;
04    for (i=0; i<devices.number; i++);
05        tmp = snmpget (devices[j], "sysLocation");
06        if (tmp == value) ret.add (tmp);
07    }
08    return ret;
09 }
```

Fig. 5. Simplified code for the `get_sysLocation` function

interface, while `topologies["LabCom"]` returns all topologies located at "Lab-Com". Each portion is evaluated and provides an intermediate result. Since each expression may have several portions, several intermediate results will be available, but the results merging step merges the intermediate results every time a new one is available and proceeds to a new class selection and attribute execution. After the last portion is evaluated, the last intermediate result is compiled in a final result sent to a visualization facility. In our last example, the final result will provide a list of all DiffServ routers, from the topologies located at "LabCom", that have ATM interfaces. In the case of the figure 3, two dynamic domains will be passed to the visualization facility. The evaluation of the visualization expression is detailed in the next subsection.

4.2   Support for Dynamic Domains Visualization

In figure 3 bottom one can see the step to evaluate and execute a visualization. First, a syntax checking of the submitted visualization expression is triggered. This one is also executed through a C compiled code generated by PRECCX. Next, a semantic verification is done. This includes the access of a visualization repository that describes which visualization facilities are available, and lists the attributes each facility supports.

Once the visualization expression is validated, the semantic verification also picks up the visualization facility selected in the visualization expression. Each visualization facility is actually another PHP4 script issued to shown the dynamic domains created by the processes from figure 3 top. After its execution, the visualization facility sends back to the user Web browser a result of the visualization. Each facility can, obviously, use other presentation techniques and resources than standard HTML pages. For example, in our implementation, we have HTML tables generated by the table facility, but also Flash applications generated by the topology facility.

```
dev1=select devices.sysName["noc"] from topologies["LabCom"]
dev2=select devices from topologies["LabCom"]
```

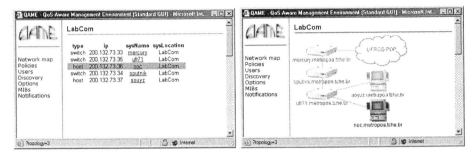

Fig. 6. Complementary information with the information model

For the previous domains, figure 6 presents on the left the results using a table facility, while right shows the same results when the topology facility is used instead.

5   Conclusions and Future Work

In this paper we presented two languages: one for dynamic domains definition and the other for dynamic domains visualization. The implementation of a subsystem for the languages in the QAME environment was also shown. We have also provided along the text some examples of the language usage.

We believe that the main goal of our work was achieved, which was the development of an initial support for dynamic domains. In the development process, we have provided further contributions. First, since the dynamic domains definition is based on a common view of the managed objects, we have implemented a common and unique management information model, instead of several different data models. We provided such information model through the use of access drivers, which are PHP4 script functions in our implementation. Besides allowing the dynamic domains definition language to work properly, the provided information model frees the network administrator of keeping in mind the specific protocols needed to access different managed devices.

We have developed the support for the dynamic domains languages in our QAME system, which aims to manage a QoS-enabled network. A common operation in this environment is the selection of devices where a network policy must be deployed. The definition and visualization languages, in the context of QAME, have easied the maintenance of the policy deployment process.

Further developed still remains. Although we are dealing with dynamic domains, storage facilities should be available (in this case, the dynamic domains would act as "standard" domains). As a future work, the storage facilities should not only store domains created through the definition language, but also should store dynamic domains definition expressions, dynamic domains visualization expressions and the final visualizations provided. Finally, although we believe that our text-based languages are easy to used, graphical wizards to help in the construction of expressions should enhance the ability to define and visualize dynamic domains.

References

[1] M. Sloman and J. Moffett, *Domain Management for Distributed Systems*, Integrated Network Management I. pp. 505-516, 1989.

[2] M. Sloman, *Policy Driven Management for Distributed Systems*, Journal of Network and Systems Management, Plenum Press. v. 2, n. 4, 1994.

[3] Y. Bernet, S. Blake, D. Grossman and A. Smith, *An Informal Management Model for Diffserv Routers*, IETF RFC 3290, May 2002.

[4] D. Durham, J. Boyle, R. Cohen, S. Herzog, R. Rajan and A. Sastry, *The COPS (Common Open Policy Service) Protocol*, IETF RFC 2748, Jan. 2000.

[5] M. Wahl, T. Howes and S. Kille, *Lightweight Directory Access Protocol (v3)*, IETF RFC 2251, Dec. 1997.

[6] K. McCloghrie, M. Fine, J. Seligson, K. Chan, S. Hahn, R. Sahita, A. Smith and F. Reichmeyer, *Structure of Policy Provisioning Information (SPPI)*, IETF RFC 3159, August 2001.

[7] L. Granville, M. Ceccon, L. Tarouco, M. Almeida *An Approach for Integrated Management of Network with Quality of Service Support Using QAME*, IFIP/IEEE DSOM, 2001.

[8] P. T. Breuer and J. P. Bowen, *The PRECC Compiler-Compiler*, In E. Davies and A. Findlay (eds). Proc. UKUUG/SUKUG Joint New Year 1993.

A Framework for Hierarchical Clustering of a Link-State Internet Routing Domain

Dohoon Kim

Kyung Hee University, Seoul 130-701, Korea
dyohaan@khu.ac.kr
http://kbiz.khu.ac.kr/dhkim

Abstract. Presented here is a Hierarchical Configuration (HC) scheme for LS (Link-State) routing protocols like OSPF and IS-IS. LS routing protocols require some unique technical constraints different from those found in other routing protocols; for example, each resulting partitioned area should be contiguous and no subnet is allowed to be separated over different areas. However, there are few literatures addressing this issue which is critical to LS network performance by scaling the routing control overhead. Our goal is to develop a HC scheme which is effective in terms of the scalability. We first devise a network model of a LS routing domain. Formulated is then a HC decision model with the protocol-specific constraints and the objective to minimize the backbone size, which is known to be crucial to stable operations of hierarchical LS routing. Lastly, we present a heuristic method and experiment results.

1   Introduction

As the number of Internet users grows, scalability issues are emerging as one of the most challenging network operation problems facing ISPs whose competitive edges are built around routing efficiency. Scalability issues exist at every layer of the Internet architecture: at IP (Internet Protocol) routing layer, for example, the explosion of flooded routing information is witnessed as network size increases. Common practice of over-engineering neither provides an economic approach at least regarding routing scalability nor accommodates ever-increasing traffic throughout the network. Considering today's competitive network service market, required are more economic and flexible measures to cope with the routing scalability. For economic management of that kind of scalability, ISPs hierarchically divide their own networks (so-called Autonomous Systems, ASs) into two tiers: local distribution areas and the backbone, thereby limiting the range of routing information exchange, and resolving the major scalability issue.

In this paper, we concentrate on the issues arising from Hierarchical Configuration (HC) of Link-State (LS) routing protocols, such as OSPF (Open Shortest Path First) and IS-IS (Intermediate System-Intermediate System). It is LS routing protocols that have been known suitable for hierarchically structured routing environment. Furthermore, many research results indicate that thanks to its ability to explicitly deal with topological information, the LS routing is gaining

H.-K. Kahng (Ed.): ICOIN 2003, LNCS 2662, pp. 839–848, 2003.

superior position over other routing protocols for QoS (Quality of Service) environment, thereby becoming prevalent not only for traditional QoS-based IP networks ([5, 6]) but also for high-speed networks using ATM (Asynchronous Transfer Mode) or MPLS (Multi-Protocol Label Switching) ([8]).

Recently, best practices for running hierarchical LS routing have been introduced (for example, [1, 14, 19, 20]). In these literatures, identified are some unique technical constraints for valid HCs as well as performance constraints for scalability. However, few literatures systematically develop such arguments upto the level of a mathematical and computerized framework for how to make an effective HC under these constraints. Our goal is to develop a method that provides valid LS HCs satisfying all the constraints while minimizing risk of unstable operation. The proposed scheme places no restriction on the structure of the network topology, so that the model and design guidelines highlighted in the following sections provide a foundation on which any ISP can manage a reliable, scalable LS routing domain (AS).

In the next section, we first briefly introduce the technical characteristics and requirements of the hierarchical LS routing. And then, discussed are the HC issues together with some lessons from the current best practices to cope with scalability challenges. Section 3 presents a conceptual network model of an LS AS, and develops a decision framework to explicitly take into account constraints arising from the protocol-specific features. A heuristic method is followed and displayed by experiment. By summarizing the presented approach and discussing the future works, we complete the paper.

2   Hierarchical Configuration of a Link-State Routing Domain

2.1   Link-State (LS) Routing and Hierarchical Configuration (HC)

In the LS routing, also known as the topology technique ([9]), each router keeps track of the explicit topology of the entire network by repeatedly flooding the information of incident LS via data frames called Link-State Advertisements (LSAs). LSAs gathered at a router collectively form a LSA DB or Topological DB (TDB) which contains the full network information such as topology, link costs, etc. Routes are then determined on-demand fashion by running Dijkstra's shortest path algorithm with the latest TDB. The complexities of synchronizing routing information (TDB synchronization) between routers and of processing information at routers increase as network grows, posing the scalability problem. In order to alleviate this problem, required is a HC which reduces routing overhead ([1, 9, 11, 16, 21]).

HC makes a routing domain logically divided, or partitioned, into a collection of subnet groups. A subnet group with member routers, each of which has an interface to a subnet in the group, forms an area. A LS routing domain is then partitioned into one specially designated backbone and multiple Local Areas (LAs), and viewed as a two-level hierarchy with the backbone at the top level.

The border of an area should be composed of only routers, each of which will be simply referred to as a gate node from now on. The constituents of the backbone are all gate nodes together with subnets, routers, and links that are not included in any LA. Furthermore, the backbone should be connected in its TDB configuration for the entire network connectivity.

Upon partitioning, all routers do not share the same TDB any more; a router has a separate TDB for each area to which it belongs. However, all routers belonging to the same area should maintain the identical TDB through the TDB synchronization. Therefore, the common TDB in an area will be referred to as the TDB of that area. For example, a gate router keeps separate TDBs not only for each LA that it belongs to but also for the backbone. Since every router has only partial information outside its own area(s), it reduces the usage of resources such as memory and CPU time; for instance, SPF calculation is performed with reduced TDB. Furthermore, HC diminishes the network bandwidth required for the TDB synchronization since the size and scope of routing information exchanged is restricted within an area.

The gain from hierarchy gives rise to some side effects on the network performance besides operational inconvenience. Recall that both intra and inter-area routing decisions are made solely based on the links in its own area, inevitably causing some degradation of quality of routes; that is, due to limitation of available routes, optimal routing may not be guaranteed. In the theoretical worst case, the quality of a route may be $O(n)$ times worse than that of the optimal route. However, in practice, the level of degradation in route quality is found not so significant as to offset the benefits from increased capacity to scalability. This, as in the previous studies ([9, 21]), allows us to consider the domain partitioning without paying attention to the routing quality.

As shown in the following sections, the overall performance of the entire AS is known to heavily depend on the backbone since the routers in the backbone play the core functions to gather, organize, and redistribute routing information across LAs. Therefore, network administrators should configure hierarchy in a way to minimize potential hazardous focal points, which can be reasonably estimated by backbone size. The following sections will introduce the decision model focusing on this point.

2.2   Hierarchical Configuration Decision of a LS Routing Domain

The capability to properly scale a LS network is determined by multiple factors: router memory requirements, CPU cycles, available bandwidth, and so on. Thus, the hierarchy does not need to be structured to provide good routes, but only must provide sound connectivity while using small amount of resources. And some simulation results indicate that among these bottleneck resources, the most critical factor for scalability is CPU cycles in a router ([1, 3, 7, 14, 15, 17, 20]). Furthermore, clear positive correlations between resources were observed in some experiments ([14, 15, 16, 20]); for example, the impacts of network size on TDB operations scale, memory size, CPU processing time, etc. are correlated in any case. Accepting these facts, we focus on router's CPU usage, which shows sharp increase according to the size of the area where the router resides.

The following is the summary of required conditions for LS HC.

Technical Conditions:

- Subnet should not be splitted over different areas ··· (1)
- LAs should be contiguous ··· (2)

These conditions are peculiar to the LS routing protocols. Here, contiguous LS area implies a continuous path can be traced from any router in an area to any other router in the same area.

Performance Conditions:

- The number of routers in a LA is less than or equal to the pre-defined bound (LA size constraint) ··· (3)
- The backbone must be contiguous ··· (4)
- All LAs must have a direct connection to the backbone ··· (5)

Even though this category of conditions does not necessarily constitute a mandatory rule, many practices recommend these conditions be satisfied since they provide a good structure to guarantee sound operations of hierarchical LS routing ([7, 14, 16, 20]). Since LS routing protocols depend on the CPU-intensive SPF algorithm, some experience has shown that 40 to 50 routers per LA should be an upper bound ([7, 14, 16]). All routers in the backbone are also generally requested to be directly connected to each other even though there is a backup measures to combine disconnected backbone parts ([11, 14, 20]). Furthermore, the last condition ensures that all LAs are attached directly to the backbone. The reasoning behind (4) and (5) is that the LS protocol expects all LAs to inject routing information into the backbone and in turn, the backbone will disseminate that routing information into other LAs.

Guidelines from the Best Practices: The goal of HC is to minimize the chance of instable operations from hierarchy while satisfying all the conditions described above. The best practices have shown that ensuring the stability of the backbone is crucial to the overall AS performance, recommending a small and simple backbone ([7, 14]). And the size of the backbone is defined in the same way as that of LA: the number of routers in the backbone. Therefore, the fewer routers are in the backbone the better overall performance do we expect.

However, the conditions (1) through (5) are likely to result in a HC with small, many LAs, which implies many gate nodes, and in turn, increases the backbone size as pointed by [11] and [21]. On the contrary, with the small backbone objective, LAs tend to be large enough to minimize the number of the backbone routers since the LS protocol requires that all the gate nodes be backbone routers. The designer should balance this trade-off between the backbone size and the size of the LA together with the number of the LAs.

Now we pose the decision problem of Link-State Hierarchical Configuration (LSHC) as determining the minimum backbone configuration on the given AS under the conditions (1) to (5). In the next section, we will see that some constraints make LSHC easy: for example, after choosing the backbone under the conditions (1) and (4), conditions (2) and (5) enforce the possible LA configurations confined in the resulting structure of components (that is, connected subgraphs) on the complementary network.

3   Model and Solution Methods

3.1   Network Model

We start this section by introducing a new network model customized to correctly describe the AS running LS routing protocol ([7, 14, 20]). Distinguishing the roles of routers and subnets, an AS is described as an undirected graph $G = (V, E)$, where $V = V_1 \bigcup V_2$ ($V_1 \bigcap V_2 = \emptyset$). V_1 and V_2 are the sets of router nodes and subnet nodes, respectively. A link in E represents a subnet-to-router or router-to-router interface. Figure 1 depicts an example of this network model together with a valid structure of HC.

Note that classic Internet models in the literature do not distinguish differences between these types of nodes, only showing the logical connections between routers ([23]). It is this distinction that allows our network model to properly address the technical requirement associated with HC: for example, condition (1) now explicitly states that all the links incident to the same subnet node should belong to the same LA or the backbone.

3.2   Hierarchical Configuration Model
for Link-State Network, LSHC

To address [LSHC] problem mathematically, we may need the following notations and definitions. Suppose that S is a non-empty subset of nodes. Let $G[S]$ be a subgraph of G induced by node subset S. The induced subgraph $G[V \backslash S]$ is denoted by simply $G - S$, the subgraph obtained from G by deleting the nodes in S together with their incident links. For our purpose, the size of a graph G, denoted by $|G|$, is defined as the number of type 1 nodes (router nodes) in G. And a node subset is called node-cut if its removal from the graph G results in multiple subgraphs, each of which is disconnected from each other. When we are given a connected subgraph G_B of G, a gate node is the node which belongs to G_B and has an incident link that does not belong to G_B. Moreover, a subgraph G_B is said to be valid if all the incident links of a type 2 node (subnet node) in G_B also belong to G_B; therefore, the boundary of a valid subgraph G_B is composed of only type 1 nodes. As to induced subgraph $G - G_B$, we define a component as a connected subgraph of $G - G_B$. Lastly, given a component C with a special node subset S, an admissible union is a family of connected subgraphs of C, each of which shares neither links nor the remaining nodes other than S with other members in the family.

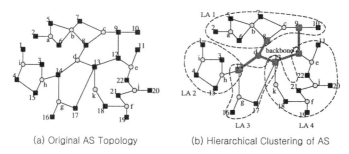

(a) Original AS Topology   (b) Hierarchical Clustering of AS

Fig. 1. Figure (a) shows a representation of a sample AS consisting of routers and subnets. Routers and subnets are distinguished and depicted in square with a numeric and circle with an alphabet, respectively. This new network model is a natural extension of typical Internet models like one proposed in [14]. This extension has clear importance regarding the implementation of a LS routing protocol since the hierarchical operations of any LS protocol explicitly demands technical conditions such as (1) and (2). Without this kind of network model, we have no means to properly take into account these technical conditions. This fact could explain the reason why there has been little literature about the optimization issues in the LS routing implementation so far. Figure (b) shows one possible example of valid HC which meets all the requirements (1) through (5). Remark that all the gate nodes should belong to the backbone. However, the vice versa is not true; that is, some backbone routers like subnet d in (b) do not have to belong to a LA. A gate node could be included more than one LA; for example router 13 belongs not only to backbone but also to LA 3 and LA 4

With the network model in 3.1 and a parameter L (upper bound on the LA size), [LSHC] can be viewed as the following combinatorial optimization problem.

[LSHC] Choose the minimal connected valid subgraph of G, $G_B = (V_B, E_B)$ whose gate node set is V_S, so that each resulting component C_i on $G_B = G - (G_B - V_S)$ satisfies the following feasibility condition.

Feasibility Conditions

1. $C_i = (V_i, E_i)$ is a component of $G - G_B$ with $V_i^S = V_i \cap V_S$ as its gate node set
2. There exists an admissible union of $B_i^1, \ldots, B_i^m$ such that $|B_i^k| \le L$ for $k = 1, \ldots, m$.

For a given component resulting from any potential backbone profile (G_B), the feasibility condition checks whether a feasible LA configuration is possible on the component or not. Validity and connectedness of the backbone configuration together with the feasibility on each resulting component in [LSHC] guarantee that the conditions for LA configuration ((1) to (3) and (5)) will be satisfied, thereby arriving at a valid HC. Furthermore, we can easily see that the following observations hold.

Proposition 1. *Suppose that a component of G, $C_i = (V_i, E_i)$ together with $V_i^S (\subset V_i)$ is given. If M, a subset of V_i^S is a node-cut, the component is an admissible union of more than two overlapping subgraphs, $B_i^1, \ldots, B_i^m$ generated by splitting C_i at M. Furthermore, if a B_i^m satisfies the size constraint (i.e., $|B_i^m| \leq L$) then it can be configured as a valid LA.*

Proposition 2. *If every gate node of a component C_i has the degree of one on C_i, the corresponding component cannot be an admissible union of more than one its subgraphs.*

Proposition 1 is based on the fact that if two LAs, A_p and A_q are possibly configured on a given component C_i resulting from a backbone configuration, the common nodes of A_p and A_q should not only constitute a node-cut of C_i but also be gate nodes. Observation 1 provides a way to configure LAs on a given component without increasing the size of the backbone. On the contrary, Observation 2 gives a condition when configuring more than one LA is impossible; if this is the case, the corresponding component could be a LA as a whole.

3.3   Solution Methods

One general and practical approach to [LSHC] starts with an initial seed backbone profile, and grows the seed until a HC satisfying all the conditions (1) to (6) is found. Even though such a sequential update of the backbone profile does not guarantee an optimal solution, this heuristic approach has strong practical implications. First, since [LSHC] is only a sub-module of an entire package for scalable and reliable HC of a LS AS ([11]), several good solutions present often more value than one optimal solution. Second, given the role of the backbone in the LS protocol, required is the network administrator's preference for HC in finalizing the backbone configuration. And in some practical situations, network administrator may need to apply the secondary criteria such as geographical considerations to evaluate or modify the hierarchical structure. Thus, incorporating speed and flexibility in the hierarchy design solution has practical importance, whereby heuristic approaches gain a strong advantage over the exact algorithms.

The proposed heuristic procedure consists of the following major modules:

- Initialization: get an initial seed of the backbone profile
- Backbone update: feed the backbone profile until all the resulting components become feasible
- Feasibility condition check: check the termination criteria of the heuristic by investigating feasibility of each component resulting from a temporary backbone profile

Even though observation 1 can be used to identify all the possible LA configurations for a given component, the counting complexity to enumerate all the

Table 1. Experiment Results

Upper bound on LA size (L)	8	9	10	11	12
Average backbone size (m_1)	17.3	17.2	13.1	11.9	11.5
(standard deviation, s_1)	(2.41)	(2.74)	(2.47)	(2.91)	(2.72)
Average number of LAs (m_2)	10.24	8.03	8.92	11.05	9.38
(standard deviation, s_2)	(1.07)	(1.33)	(1.92)	(1.41)	(1.79)
The largest size of LAs	8	9	10	11	12

node-cuts composed of gate nodes only may hinder the exact feasibility check algorithm from practical implementation. Therefore, we employ in this paper a heuristic procedure that makes the feasibility check simple. That is, after identifying every articulation gate node of a component through DFS (Depth First Search), we form an admissible union of blocks $B_i^1, \ldots, B_i^m$ and check the size condition for each B_i^k, $k = 1, \ldots, m$.

3.4 Experiment Results

For the range of L (the upper bound on LA size), we referred to [1], where L^* was mathematically calculated with the given number of routers in an AS in terms of optimizing the amount of LSA exchange within a LA. However, [1] neither distinguish the subnets and routers in their network model nor take network topology into account; these two missing parts are crucial for capturing essential difference of LS routing domain from others as we have seen in section 3.1. Therefore, the proposed L^* should be thought of only as a possible reference point.

Table 1 summarizes experimental results on the same network by changing initialization and parameter L. 20 repetitions with different initial points were conducted with L fixed at the same level in order to see the sensitivity of the heuristic to initial choice. In 20 trials, some initial points were chosen with a sophisticated manner of finding good approximation points near to the network center. Furthermore, for every 20 repetitions of the experiment, the heuristic seems to find at least one near to optimal solution to [LSHC]. In the light of the standard deviation from the results (the numbers in parentheses at the first row), the initial choice of a backbone profile does not severely affect the overall performance of the heuristic. We can also find that as L increases, average backbone size decreases, but average number of areas seems rather independent of L. Figure 2 depicts the experiment results.

4   Concluding Remarks

Critical design principles for successful hierarchical implementation of a LS routing protocol have been addressed together with a model and heuristics to formulate and solve the scalability issues. The first and most important decision

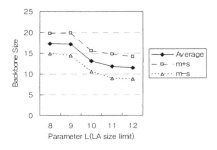
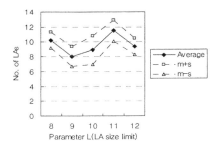

(a) Effect on the Backbone Size   (b) Effect on the Number of LAs

Fig. 2. Experimental results show that the major parameter L affects some performance indices differently. Assuming that the average behavior shows that of near optimal solution, we could conclude that the backbone size is negatively related to the upper bound on LA size. However, the number of LAs seems to be a less critical measure in assessing the network performance than the backbone size

when designing a HC on LS routing domain is to determine the boundary of the backbone, thereafter feasible LA structures could be configured. Ensuring that these activities are properly planned and executed will make all the differences in LS implementation since the network performance in terms of scalability and stability depends on HC.

The proposed framework and solution method provide more general rules and flexibility in introducing hierarchy. For more efficient and situation-specific implementation of [LSHC], we may have to extend the model and solution method. For instance, a network administrator should impose a set of routing policies based on geographical or organizational grounds on his/her network. Furthermore, for reliable access of LA, LA should maintain multiple gate nodes to prevent disconnection due to failure in primary gate node. One may also want to incorporate special restrictions on backbone profiles such as a restriction on the backbone topology with a little modification of the solution methods. Finally, some additional measures after hierarchy could be developed to improve the network performance and/or reliability. For example, [10] presents a scheme to increase backbone reliability with the route backup mechanism embedded in most LS routing protocols.

References

[1] Aho, A. V. and Lee, D.: Hierarchical networks and the LSA N-square problem in OSPF routing. Proceedings of IEEE GLOBECOM 00. (2000) 397–404
[2] Baccelli, E. and Rajan, R.: Monitoring OSPF routing. IEEE/IFIP International Symposium on Integrated Network Management Proceedings (2001) 825–838

[3] Behrens, J. and Garcia-Luna-Aceves, J. J.: Hierarchical routing using link vectors. Proceedings of IEEE INFOCOM 98 (1998) 702–710

[4] Bley, A., Grotschel, M., and Wessaly, R.: Design of broadband virtual private networks: models and heuristics for the B-WiN. **53** (2000) 1–16

[5] Chamberland S. and Sanso B.: Overall design of reliable IP networks with performance guarantees. Proceedings of IEEE CCECE 02 **3** (2002) 1571–1576

[6] Guerin, R. A., Orda, A., and Williams, D.: QoS routing mechanisms and OSPF extensions. Proceedings of IEEE GLOBECOM 97 **3** (1997) 1903–1908

[7] Halabi, S. and McPherson D.: Internet Routing Architectures. Cisco Press (2000)

[8] Iwata A. and Fujita N.: A hierarchical multilayer QoS routing system with dynamic SLA management. IEEE Journal on Selected Areas in Communications **18** (2000) 2603–2616

[9] Jaffe, J. M.: Hierarchical clustering with topology database. Computer Networks and ISDN Systems **15** (1988) 329–339

[10] Kim, D.: Virtual link configuration for the backbone augmentation in an Internet link-state routing domain. Issues in Information Systems **3** (2002) 334–340

[11] Kim, D. and Tcha, D. W.: Scalable domain partitioning in Internet OSPF routing. Telecommunication Systems **15** (2000) 113–128

[12] Lauck, A. G., Kalmanek, C. R., and Ramakrishnan, K. K.: SUBMARINE: an architecture for IP routing over large NBMA networks. IEEE INFOCOM 99 (1999) 98–106

[13] Martey, A. and Sturgess, S.: IS-IS Network Design Solutions. Cisco Press (2002)

[14] Moy, J. T.: OSPF: Anatomy of an Internet Routing Protocol. Addison Wesley (1998)

[15] Moy, J. T.: OSPF database overflow. Request For Comments 1765 (1995)

[16] Moy, J. T.: Multicast routing extension for OSPF. Communications of the ACM **37** (1994) 61–67

[17] Moy, J. T.: OSPF protocol analysis. Request For Comments 1245 (1991)

[18] Parkhurst W. R. P.: Cisco OSPF Command and Configuration Handbook. Cisco Press (2002)

[19] Shaikh, A., Goyal, M., Greenberg, A., Rajan, R., and Ramakrishnan, K. K.: An OSPF topology server: design and evaluation. IEEE Journal on Selected Areas in Communications **20** (2002) 746–755

[20] Thomas, M. T. II.: OSPF Network Design Solutions. Cisco Press (1998)

[21] Tsai, W. T., Ramamoorthy, C. V., Tsai, W. K., and Nishiguchi, O.: An adaptive hierarchical routing protocol. IEEE Transaction on Computers **38** (1989) 1059–1075

[22] White R., Retana, A., and McConnel, M.: IS-IS: Deployment in IP Networks. Addison Wesley (2002)

[23] Zegura, E. W., Calvert, K. L., and Bhattacharjee, S.: How to model an internetwork. Proceedings of IEEE INFOCOM 96 (1996) 594–602

Characteristics of Denial of Service Attacks on Internet Using AGURI

Ryo Kaizaki[1], Osamu Nakamura[2], and Jun Murai[3]

[1] Graduate School of Media and Governance, Keio University
5322 Endo, Fujisawa, Kanagawa, 252-5322, Japan
`kaizaki@sfc.wide.ad.jp`
`http://www.sfc.wide.ad.jp/~kaizaki/`
[2] Faculty of Environmental Information, Keio University
5322 Endo, Fujisawa, Kanagawa, 252-5322, Japan
`osamu@wide.ad.jp`
[3] Faculty of Environmental Information, Keio University
5322 Endo, Fujisawa, Kanagawa, 252-5322, Japan
`jun@wide.ad.jp`

Abstract. Denial of Service attacks are divided into two types, one is logic attack and the another one is flooding attack. Logic attack exploits security hole of the software such as operating system and web server bugs, then causes system crash or degrade in the performance. Logic attack can be defended by upgrading software and/or filtering particular packet sequences. In this paper, characteristics of the flooding attacks is described. For the monitoring tools, AGURI, that we have developed, is used. Using the traffic pattern aggregation method, AGURI can monitor the flooding attacks in real network traffic for a long term.

1 Introduction

Internet is the packet switching network, sharing the every resources such as the bandwidth of the links and router's processing unit. Resource management should be done by every end node. For example, congestion controls can be done only by end nodes. End nodes can also send data without congestion controls. Thus, usage of network resources depends on behavior of end nodes. However current Internet does not have any mechanisms to control ill behavior. During the network operations, it is very important to detect the flooding attacks as soon as possible. After detecting the flooding attacks, operators can take several actions: dropping the packets from attackers, limiting number of packets from attackers, and discovering the attackers.

Denial of Service attacks is divided into two types[1], one is logic attack and the another one is flooding attack. Logic attack exploits security holl of the software such as operating system and web server bugs, then causes system crash or degrade in the performance. Logic attack can be defended by upgrading software and/or filtering particular packet sequences.

H.-K. Kahng (Ed.): ICOIN 2003, LNCS 2662, pp. 849–857, 2003.

Comparing each packets of the flooding attack and the other normal communication traffics, the only difference is the number of the packets. Flooding attack creates enormous amount of packets. Therefore, to protect systems from flooding attacks, the same method for logick attacks can not be used. While operationing networks, operatiors can detect flooding attacks using traffic monitoring tools such as MRTG[2]. However those tools will not show detail information of flooding attack packets.

2   Traffic Monitoring for Flooding Attacks

There are several types of flooding attacks.

1. the large number of the bytes
2. the large number of the packets
3. packets with ill behavior protocols such as sync attack

The traffic with the large number of the bytes for the single destination degrades the performance of the end system and the routers that switching this traffic. And recent routers incur more damages by recieving the large number of packets rather than bytes.

That traffic can be monitored by using SNMP[3]. MRTG is good graphic interface for the detecting the unusual traffic. But it is not sufficient for detecting the flooding attacks. There is limitation of gathering the information using SNMP. The number of the bytes and the packets for the each interface on the routers can be collected. However the number of the byte and the packets to the single hosts can not be collected. If the bandwidth of the link was occupied in general condition, particular attacks could not be detected by using SNMP/MRTG monitoring, because total bandwidth of the link is not changed.

For detecting the flooding attacks, we should know the normal conditions of the networks. It needs for large number of the traffic data. SNMP is simple mechanisms for collecting the data from the routers and switches. It is needed for aggregation mechanisms for storing the data. NeTraMet and FlowScan which are flow based monitoring tools can monitor specific type of the traffic, such as number of bytes and packets in a long term on HTTP, FTP, IPv6 etc. However these tools require the fixed rule sets. So those tools can not detect unexpected traffic pattern.

2.1   AGURI

AGURI is an aggregation based traffic profiler targeted for long term measuring. AGURI adapts itself to spatial traffic distribution by aggregating small volume flows into its root. AGURI does not need a pre-defined rule set and is capable of detecting an unexpected increase of unknown packet patterns or flooding attacks.[4]

Figure 1 shows the concept of aggregation: small entries are aggregated into its root. There are two phases in the basic aggregation algorithm of AGURI

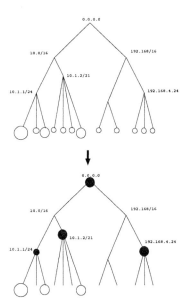

Fig. 1. Aggregation concept: small entries are aggregated into aggregates

First AGURI monitors every packets. Second, at the end, aggregates entries whose counter value is less than an aggregation threshold.

In figure 1, each circle shows enteries and its counter value is indicated by its size. Each filled dot shows sets of aggregated entries whose counter value is less than an aggregation threshold. For example, the filled dot "10.1.2/21" shows set of aggregated entries whose counter value is less than an aggregation threshold and whose IP address is included in address block "10.1.2/21".

Figure 2 shows an example of aguri's summary output. A summary consists of header part and body part.

The header part describes version, start-time of profiling, end-time of profiling and average-rate of all traffic. The header part starts with %.

The body part contains 4 profile types:

1. source ip address
2. destination ip address
3. source protocol
4. destination protocol

In the address profile, each row shows an address entry and the prefix length. The first column shows the address and the prefix length of the entries. The second column shows the culmulative byte counts. The third column shows the percentages of the entry and its subtrees.

The input for this example is a month-long packet trace taken from a transpacific link of the WIDE[5] backbone. The parameters of aguri is configured with

```
%%!AGURI-1.0
%%StartTime: Thu Mar 01 00:00:00 2001 (2001/03/01 00:00:00)
%%EndTime:   Sun Apr 01 00:00:00 2001 (2001/04/01 00:00:00)
%AvgRate: 14.91Mbps

[src address] 4992392109177 (100.00%)
0.0.0.0/0       87902964189 (1.76%/100.00%)
 0.0.0.0/1      206637364377 (4.14%/14.78%)
 0.0.0.0/2      205796877844 (4.12%/7.12%)
  60.0.0.0/6    97928228974 (1.96%/3.00%)
      62.52.0.0/16    51875058871 (1.04%/1.04%)
   64.0.0.0/8 100831910967 (2.02%/3.51%)
    64.0.0.0/9 74610984109 (1.49%/1.49%)
 128.0.0.0/2    142349668983 (2.85%/13.33%)
  128.0.0.0/3   197067746696 (3.95%/10.48%)
  128.0.0.0/5   202911635757 (4.06%/5.45%)
   133.0.0.0/8 69142535628 (1.38%/1.38%)
           150.65.136.91    54123094932 (1.08%)
  192.0.0.0/4   212653628837 (4.26%/38.41%)
  192.0.0.0/6   88855538654 (1.78%/1.78%)
   202.0.0.0/7 235853368912 (4.72%/14.70%)
    202.0.0.0/9       117196493427 (2.35%/6.77%)
           202.12.27.33     160473669718 (3.21%)
          202.30.143.128/25  60239291958 (1.21%/1.21%)
           203.178.143.127 94031811680 (1.88%)
   204.0.0.0/6 228960094456 (4.59%/17.68%)
   204.0.0.0/8 125458765333 (2.51%/7.58%)
           204.123.7.2      87103414877 (1.74%)
           204.152.184.75 165733431144 (3.32%)
   206.0.0.0/7 164036959478 (3.29%/5.51%)
    206.128.0.0/9     53526598302 (1.07%/1.07%)
   207.0.0.0/8 57628266965 (1.15%/1.15%)
  208.0.0.0/4   282590640975 (5.66%/31.72%)
   208.0.0.0/6 116047154301 (2.32%/22.20%)
   209.0.0.0/8 140888988219 (2.82%/11.78%)
           209.1.225.217    238192306019 (4.77%)
           209.1.225.218    209160635530 (4.19%)
   210.0.0.0/7 154008321340 (3.08%/3.08%)
    216.0.0.0/9           192899750315 (3.86%/3.86%)
%LRU hits: 86.82% (1021/1176)
```

Fig. 2. Example of AGURI summary output

256 nodes and 1% aggregation threshold. Among many src addresses, only 8
addresses are indentified as individual address.

Using AGURI's script, we can archive summaries with minimum disk space.
This enables long term measurements.

Thus, AGURI achieves long term traffic monitoring and detecting characteristic flows without a pre defined rule set.

We use AGURI to archive characteristic of traffic in a long term.

AGURI uses a traffic profiling technique in which records are maintained in
a prefix based tree a compact summary which is produced by erntries.

Figure3 shows tree structure of archiving summaries. In figure3, AGURI generate hourly summary "A" by aggregating minutes summaries "1"-"12".

We can see various summaries of time scale granularity.

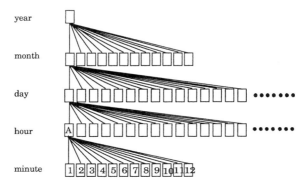

Fig. 3. Archiving structure of AGURI

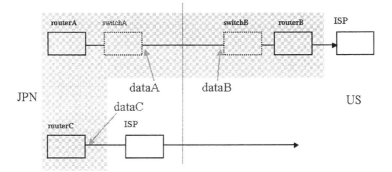

Fig. 4. Traffic monitoring points: three points on WIDE backbone. all of sampling points connected to tans-pacific link between Japan and USA

3   Basic Methodolgy

To characterise actual flooding attacks, we have to collect traffic information from internet backbone. Thus, we set AGURI on WIDE backbone to monitor flooding attacks for a long term. AGURI should detect an unexpected increase of unknown packet patterns or flooding attacks.

3.1   Information of Experimental Platform

We can collect three types of trace data to set three AGURI programs on WIDE backbone. WIDE backbone is formed with giga bit ethernet and fast ethernet.

WIDE Internet has two trans-pacific links. One its own link, the other is directly connected to an ISP which has a trans-pacific link. Figure4 shows information of traffic monitoring points.

Comparing to WIDE internal link and ISP internal links, trans-pacific links are narrow. Thus, international flooding attacks quickly fullfils those links.

Table 1. Information of sampling data

	term	data size (MB)
Data A	4 month	1340
Data B	5 month	720
Data C	18 month	5012

3.2   Information of Collected Data

Table 1 shows basical information of three data. Data A is made of 4 month long data from July to October. Data B is made of 5 month long data from June to October. Data B is made of 18 month long data from May(2001) June to October.

Each data size shows size of archiving AGURI data.

Each data consists of 2-minutes-long data.

4   Result

Using the previously described three types of data (section 3), we observed many foolding attacks.

In this section, we first define flooding attacks, and then characterize the attacks according to both type of attacks and the type of victims.

4.1   Definition of Flooding Attacks

We can not detect all flooding packet because of AGURI's aggregation concept, however we can detect flooding attacks with the large number of bytes for a long term.[6]

In this paper, we define flooding attacks as following requirements.

- Total number of all traffic increace over 20% compare with average of monthly traffic Bandwidth of trans-pacific links of WIDE backbone has margin in average, thus when remarkable flooding attacks come, there is an increace in total traffic.
- Attacks continue for over 10minutes
  In this experiment, we set 2-minutes-long as a time-series threshold. So we can not monitor flooding attack for a short term.
- A single characteristic of flow occupy over 70% in all traffic
  AGURI can profile 4types; 1)source IP address 2)destination IP address 3)source port and protocol and 4)destination port and protocol.
  We define single characteristic of flow in any profile which occupy over 70% in all traffic profile as flooding attack.

Table 2. Summary of flooding attacks on July 2002

	Data A	Data B	Data C
Number of days in case of flooding attacks	12	12	9
Number of flooding attacks	28	28	43
Maximum usage of bandwidth	97Mbps	97Mbps	14Mbps

Table 3. Breakdown of victims

	IRC servers	Router's interface	other
Destionation IP addresses(victims)	87%	12%	1%

4.2 Summary of Flooding Attacks

Table 2 shows flooding attacks we detected based on data July 2002.

Maximam usage of bandwidth are nearly equal to Maxmam width of link, Thus flooding attacks occupied almost all bandwidth of links.

All flooding attacks which we can detect spoofed their source IP address, we have trouble to trace attacker.

4.3 Destination IP Addresses

We can find outstanding characteristics of flooding attacks from profiles of destination IP addresses. We can divide victims of flooding attacks into two types, one is IRC(Internet Relay Chat) servers and the other is Router's interface. Table 3 shows breakdown of vistims.

IRC server and Router's interface have each characteristic.

- IRC servers

  Almost destination IP address of flooding attack packets are IRC servers. WIDE Internet contains three IRC servers, all of IRC servers were attacked. IRC servers keep comunication sessions each other and share same name space of comunity and user name, thus user can not use same user name, same comunity name and right of control comunity. However once servers can not connect each other, name space and comunity is independent. So anyone can gain any user name and comunity name.

  An attacker attacks IRC servers to gain user name and comunity name which they want. Thus flooding attacks toword IRC servers continue to disconnect IRC servers each other.

- Router's interface

  The other of victims are Router's interface which user can see their IP address using traceroute et. Thus once network operator sets filter against packet whose destination IP address is Router's interface, an attacker changes destination IP address to next hop of Router's interface.

```
%!AGURI-1.0
%%StartTime: Sun Oct 14 14:00:00 2001 (2001/10/14 14:00:00)
%%EndTime:   Sun Oct 14 15:00:00 2001 (2001/10/14 15:00:00)
%AvgRate: 24.30Mbps[ip:proto:dstport] 10933438650 (100.00%)
4:6:2        220337940 (2.02%)
4:6:5        220259760 (2.01%)
4:6:8        224630700 (2.05%)
4:6:11       220901820 (2.02%)
4:6:14       220496040 (2.02%)
4:6:17       219956580 (2.01%)
4:6:20       221177488 (2.02%)
4:6:23       221431646 (2.03%)
4:6:26       219968880 (2.01%)
4:6:29       219885240 (2.01%)
4:6:32       226155786 (2.07%)
4:6:35       220104840 (2.01%)
4:6:38       220877880 (2.02%)
4:6:41       220083780 (2.01%)
          :
          :
          :
4:6:101      220132020 (2.01%)
4:6:104      229349040 (2.10%)
4:6:107      220964460 (2.02%)
4:6:110      221768098 (2.03%)
4:6:119      213498789 (1.95%)
```

Fig. 5. Example of TCP flooding attacks

4.4 Destination Ports and Protocols

We can outstanding characteristic of flooding attacks in poing of destination ports and protocols. We can divide victims of flooding attacks into two types, one is ICMP and on the other hand is TCP.

Almost flooding packets are ICMP echo-reply or TCP SYN floodig, however we found outstanding characteristics of flooding attack in figure5.

In figure 5,"4:6:2" shows IPversion 4: protocol number 6(TCP): port number 2. Figure 5 shows that flooding attack packets uses port number which increase per three times.

4.5 Distributed Attack

We can see distributed attacks in figure 6 and figure 7.

Figure 6 shows data C and figure 7shows data A. Both point C and A are connected to different networks. However, both figures show same flooding attack. They show flooding attacks toward Host 1 in a same time period, and attacks stoped at the same time.

Thus, we can see distributed flooding attacks in these figures.

5   Conclusion

In this paper, characteristics of the flooding attacks are described. Attackers can exhaust network resources using any source ip address, any port number and any protocols.

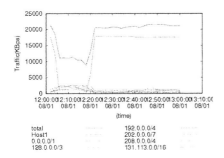

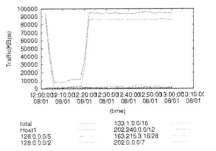

Fig. 6.   Data C: flooding attack to
Host1

Fig. 7.   Data A : flooding attack to
Host1

Thus, to operate networks against flooding attacks, we need to monitor characteristics of flooding attacks: source ip address, destination ip address, port number and protocols.

For the monitoring tools, AGURI, that we have developed, is used. Using the traffic pattern aggregation method, AGURI can monitor the flooding attacks in real network traffic for a long term.

We classify flooding attacks that we define into 4 profiles; 1)source IP address 2)destination IP address 3) source port and protocol 4) destination port and protocol.

We can found chatacteristic of flooding attacks on destination ip address and destination port and protocol.

AGURI can monitor distributed flooding attacks, port scan attacks. Also, AGURI successfully detected flooding attacks which are remarkablly one-sided to specific hosts.

References

[1] R. Needham, "Denial of Service: An Example", Communications of the ACM volume 37, November 1994
[2] MRTG: www, http://www.mrtg.org
[3] SNMP:www, http://www.ietf.org/rfc/rfc1157
[4] Kenjiro Cho, Ryo Kaizaki, Akira Kato,"AGURI: An Aggregation-Based Traffic Profiler", QofIS2001, September 2001
[5] WIDEproject:www, http://www.wide.ad.jp
[6] Ryo Kaizaki, Kenjiro Cho, Osamu Nakamura, "Detection of Denial of Service attacks using AGURI", ICT2002, June 2002

Dynamic Configuration and Management of Clustered System with JMX

Han-gyoo Kim, Ha Yoon Song, Jun Park, and Kee Cheol Lee

College of Information and Computer Engineering, Hongik University, Seoul, Korea
{hkim,song,jpark,lee}@cs.hongik.ac.kr

Abstract. To serve for high volume of requests with steady availability, servers usually incorporate clustering technology for the availability as well as load distribution, scalability, etc. In this paper, we introduce a new scheme for efficient cluster configuration and management using Java Management Extensions.

1 Introduction

Internet service environments are now rapidly changing to accommodate business and service requirements of mobile and wireless device support. Thus, to provide service for mobile environments as well as wired environment, mobile application servers need to be introduced. In order to guarantee steady availability of service, clustering technology has been widely applied. Clustering technology has the capability of distributing huge amount of requests to achieve load distribution, scalability, and availability.

A clustered system is a set of connected computers that works like a single system. It is a kind of parallel and distributed system allotting high loads among distributed computers through the high bandwidth communication links, so called interconnection network. From the view of fault tolerance, computers check each other, remove faulty units, and take over other computer's jobs in order to guarantee high availability transparently to users.

Major benefits of clustering are load balancing, scalability, and availability. Load balancing guarantees high performance of the system by distributing a large number of requests fairly to each server (computer) in order to avoid situations in which some units are overloaded while others are idle. Scalability stands for the ease of configurability in unit addition or deletion. When a new server is added to the clustered system, scalability guarantees the harmless and continuous service with updated configuration of the cluster. It also guarantees availability because the number of servers can be increased to cope with the increased requests while defected servers can be separated from the cluster without affecting the overall service. In addition, the migration capability of user requests among the servers can guarantee high availability of the clustered system [1, 2, 3]. In this paper, we introduce a new technology to configure and manage a clustered system efficiently and effectively using JMX (Java Management extensions), especially for applications based on JAVA related technologies.

H.-K. Kahng (Ed.): ICOIN 2003, LNCS 2662, pp. 858–867, 2003.

Fig. 1. Overall structure of JMX

2   Java Management Extensions

Java Management extensions (JMX) is a management framework, so called a middleware, suggested by Sun Microsystems [4, 5]. JMX basically consists of three hierarchical levels: Instrumentation level, Agent level, and Distributed Services level. The data and application programs, which require management, can be separated independently from the Manager level. JMX also has a set of APIs for management protocol. The basic structure of JMX is shown in figure 1.

2.1   Instrumentation Level

Instrumentation Level provides any Java technology based object with instant manageability. This level is aimed at the entire developer community that utilizes Java technology. This level provides management of Java technology which is standard across all industries [6].

The components of instrumentation level are MBeans (Managed Beans), Notification Model and MBean Metadata Classes. MBean is categorized by Standard MBean, Dynamic MBean, Open MBean, and Model MBean [5].

2.2   Agent Level

Agent level provides management Agents. JMX Agents are containers that provide core management services which can be dynamically extended by adding JMX resources. This level is aimed at the management solution development community, and provides management through Java technology [6].

MBean Server and Agent Services are the core parts of this level. MBean Server, a component for MBean registration, supports management interfaces for each MBean so that the management system can recognize each MBean. Agent Service is an object of management operation for MBeans registered in the server. It has Dynamic Class Loading, Monitors, Timers, and The Relation Service [5].

2.3   Distributed Services Level

Manager level provides management components that can operate as a Manager or an Agent for distribution and consolidation of management services. This level is aimed at the management solution development community, and completes the management through Java technology provided by the Agent level [6]. Manager and Agent can communicate through adapters with management protocol APIs, or via connector client to contact the connector server in the Agent. Available APIs are SNMP, IIOP protocol adaptor, and WBEM [7, 8, 9, 5].

2.4   JMX as a Solution Framework for Clustering Management

JMX is a resource management technology for any objects based on the JAVA technology. Since it is based on JAVA technologies, it can easily cooperate with other JAVA based technologies such as EJB, JINI, and JDMK(JAVA Dynamic Management Kit). JMX also supports APIs for network management such as SMNP, so that every network management feature can be available. Another benefit of JMX is that heterogeneous server architectures can be a seamless part of clustered system only if the servers incorporate JAVA technologies. The unit of management object for JMX is an application that enables the load balancer to distributes jobs over servers easily for efficient load balancing. With these features JMX can be regarded as a viable solution to configure and manage clustered systems based on JAVA technologies.

3   Architecture

In this section, we show the architecture to build up the clustered system with JMX. The manageable object for JMX is an application program based on JAVA technologies such as EJB (Enterprise JavaBeans). Each application program will be mapped on MBean, and each MBean will be registered on MBean server that resides on the same JAVA virtual machine where the application program is under execution. In addition, the Monitor of this MBean server (one of the Agent Services) manages applications with the status information of applications. Manager system can recognize and resolve the erroneous situation by the reception of events from the Monitor whenever Monitor senses exceptional situations such as the halt of MBean or the excessive processing requirement of MBean. For this purpose, Agent should be able to communicate with Manager system. The communication can be made by APIs for SNMP [10, 9]. Alternatively, the connector server on the Agent and connector client on the Manager system can build communication between Agent and Manager system. For the actual implementation of this paper, the second communication scheme was used.

Figure 2 shows the inside of the server, which is based on EJB and managed by JMX. Figure 3 shows the overall structure of the clustered system. There is one Manager in the system, which manages the whole clustered system. This simple structure can be used to configure the clustered system without

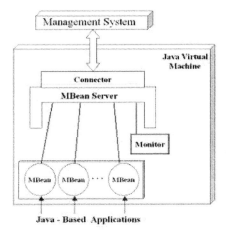

Fig. 2. Clustered server

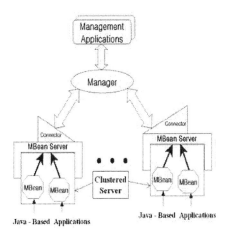

Fig. 3. Overall structure of clustered system

any degradation of the entire performance. However, the SPoF (Single Point of Failure) can occur whenever the Manager system becomes faulty, which is an example of fatal situation with very low availability. In the following sections, we will show other advanced architectures to solve this problem.

3.1 Various Architectures

In figure 4, five other architectures are suggested for dynamic clustering. Each architecture model has its own cons and pros, which are trade-offs between availability and network resources.

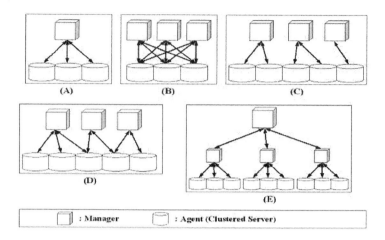

Fig. 4. Architectures for clustering

(A) is the basic architecture model discussed in the earlier section, bearing the possibility of SPoF problem. (B), (C), and (D) are alternative architectural models to solve the possible SPoF problem. Several Manager systems are employed to guarantee basic fault tolerance in case of failure of Manager systems. However, these models have the problems such as redundancy of management information and high communication overhead between servers and Managers, which may cause the resource extravagance.

(E) shows a hierarchical model with a topmost Manager system that manages each sub-Manager system. This model shows higher reliability than the other models, however, the extra hierarchy requires additional resources.

4   Managed Objects and Message Format

4.1   Managed Objects

Manager manages three sorts of objects: ServerInfo, MBeanServerInfo, and MBeanInfo. Figure 5 shows the structure of the Management object.

ServerInfo has the information of Clustered Server where the Agent resides. The information are AgentID, IP address, IsAlive, etc. MBeanServerInfo has the information of MBean Server created by the Agent. The information is AgentID, MBeanServerName, ServiceCnt, avgServiceTime, and Capacity. MBeanInfo stands for the information of each MBean registered on MBean Server. The management information is MBeanServerName, MBeanName, ServiceCnt, avgServiceTime, Capacity, and MBeanType.

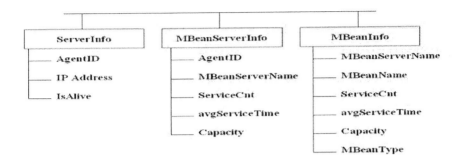

Fig. 5. Structure of management objects

4.2 Message Format

Figure 6 shows the message format for communication between Manager and Agents suggested in this paper.

The description of each fields are as follows.

- ManagerID : The identifier of Manager for communication.
- AgentID : The identifier of Agent for communication.
- MessageType : The message type which is composed of six subfields: MBean information request, MBean information transmission, event notification, MBean server status information request, MBean server status information transmission, and IsAlive.
- EventType : The information about an event thrown by MBean or MBean Server. The events are MBean Capacity Full, MBean Server Capacity Full, MBean Capacity Free, and MBean Server Capacity Free.
- MBeanName : The record of MBeanName.
- Variable Binding List : The field for request and transmission of MBean or MBean Server information.

4.3 Message Type

Six message types are defined for the clustering with JMX suggested in this paper. The descriptions of messages are as follows.

- MBean Info Request : Manager requests information of a specific MBean managed by Agent.
- MBean Info Transfer : Agent transfers the information of MBean requested by Manager.
- Event Transfer : An event from MBean or MBean Server is transferred to a Manager.
- MBean Server Status Request : Manager requests the status information of MBean Server to Agent.

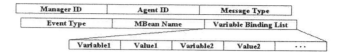

Fig. 6. Message format

- MBean Server Status Transfer : Agent transfers the status information of MBean Server requested by Manager.
- IsAlive : Manager checks the living status of clustered servers under its management.

5   Implementation and Scenario

5.1   Implementation

Actual clustered system was implemented to test the effectiveness of clustering technology described in this paper. The implemented testbed is composed of normal servers (without the Manager) and servers with the Manager. Each server has heterogeneous processors and operating systems (e.g., Windows2000 Server, Windows2000 Professional, and Linux). Manager server has connector client for communication with Agent. The detailed component diagram is shown in figure 7.

Manager server obtains management information and notifies it to the management application. Finally administrator acquires the overall status of clustered system. Each normal server executes user applications.

Each application is mapped on an MBean and registered on MBean server. For the implementation for this paper, each server has one MBean server to manage the MBeans. MBean Server registers MBeans and Agent services that manage the application programs. Among the various Agent services, Monitor plays the core role to sense the erroneous situations and to throw events for the notification of errors. Whenever each application starts a new service for related client, a new MBean will be created and registered to MBean server. Monitor checks the number of MBeans on the server or other related information specific to MBeans, and throws an event according to related situations, and thus the whole system recognizes the situation of error.

5.2   Scenario of Cluster Management

With the JMX based clustered system technologies, clustered system can cope with various situations of services. The most representative situations can be distinguished as follows.

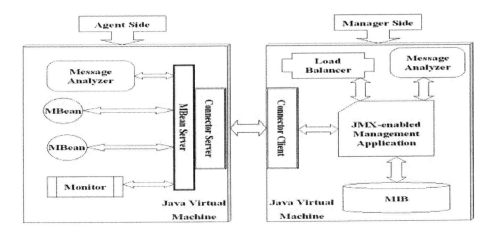

Fig. 7. Component diagram of implementation

- Addition of a new server.
- Removal of an existing server.
- Treatment of an overloaded server, in case of service overloading of a specific application by the concentration of related requests.
- Treatment of an erroneous server.

Our system can cope with each situation. In the following subsections, we show the action of our system according to each scenario.

Addition of a New Server

1. Manager checks the living status of belonging servers and their management record by periodically sending IsAlive message.
2. Whenever a start of a new server is found, Manager assigns a new ID to the Agent on the server, and sends a message requesting the information of MBeans and MBean server on the server.
3. Manager records all the information received from the server. Manager modifies the management information of the server, and notifies the addition of new server to the load balancer.
4. Load Balancer starts sending jobs to the added server.

Deletion of an Existing Server

1. Administrator sends the information of the to-be-deleted server by Management Application to the Manager.
2. Manager notifies the load balancer that the server is out of service, so that new jobs cannot be assigned to the server.

3. Manager requests information of MBeans to the server in order to check if the server is currently processing jobs.
4. The request of MBean information is repeated until there are no jobs in service on the server.
5. Manager modifies the management information of the server.
6. The server can be physically deleted.

Treatment of an Overloaded Server: Load Balancing Operation

1. Either MBean capacity or MBean Server capacity on a server exceeds the server capacity.
2. Monitor identifies the overloading situation and notifies to the Manager through an event.
3. Manager informs Load Balancer that the server cannot receive any more service request.
4. Manager requests ServiceCnt information to check if other servers can process more jobs. Each server has its own avgServiceTime, and must check if it can reduce the avgServiceTime.
5. Servers which ServiceCnt is less than their capacity can process additional requests. Manager notifies load balancer that these servers are available to serve for additional requests.

Treatment of a Server Accident

1. Monitor checks the living status of servers by sending IsAlive message periodically. Manager can check if a server is halt.
2. Manager notifies the load balancer that a server is halt. Load balancer stops the assignment of new job to the server.
3. Through the Manager application, Manager notifies the Administrator.
4. Administrator removes the server from the cluster.

6   Conclusions

In this paper we designed and implemented a new clustered system using JMX, which is a JAVA based resource management framework. Several scenarios of cluster management were introduced and the effectiveness of server systems was tested under each scenario.

It is proven that with JMX technology, configuration and management are easier than other clustered systems with traditional technology. It is possible for application-based management to be apart from server-based management. It enables the configuration of clustered system clean and effective. Our system can also dynamically cope with the various situations of cluster management for high management ability, and thus availability.

The JMX specification still has several limitations for clustered system management. For example, an Agent cannot have multiple MBean servers. We hope that the next version of JMX is improved so that more dedicated management can be available by multiple MBean servers on an Agent for better clustering environment.

Apart from the suggested clustered systems, which are based on LAN environment, WAN (Wide Area Network)-based clustered systems can be studied in the future. The ultimate integration of worldwide network will lead to a new topic of global management. One of the examples can be the technology to enable Beans based on JAVA technologies, especially EJB technology, be managed in global environment. For global management, the naming of Beans or servers will be of another issue. We expect JINI will be one of the possible solutions [11].

References

[1] Jyoti Batheja and Manish Parashar. Adaptive cluster computing using JavaSpace. In *IEEE International Conference on Cluster Computing*, 2001.

[2] K. A. Hawick and H. A. James. Dynamic cluster configuration and management using javaspaces. In *IEEE International Conference on Cluster Computing*, 2002.

[3] David Raften. Parallel sysplex cluster technology overview. *IBM*, 2002.

[4] H. Kreger. Java management extensions for application management. *IBM Systems Journal*, 40(1), 2001.

[5] Sun-Microsystems. Java management extensions instrumentation and agent specifications, v1.1. 2002.

[6] Sun-Microsystems. JMX white paper. 1999.

[7] J. A. Farrell and H. Kreger. Web services management approaches. *IBM Systems Journal*, 41(2), 2002.

[8] J. D. Case, M. Fedor, M. L. Schoffstall, and C. Davin. *Simple Network management Protocol (SNMP)*. *Internet RFC 1157*, 1990.

[9] Sun-Microsystems. Java management extensions SNMP manager APIs. 1999.

[10] William Stalling. *SNMP, SNMP v2, and CMIP*. Addison Wesley, 1993.

[11] Sun-Microsystems. Jini architecture specification v1.2. 2001.

An Optimal Algorithm for Maximal Connectivity of HFC Network

Jang Ho Lee, Changwoo Pyo, Kyun-Rak Chong, and Jun-Yong Lee

Department of Computer Engineering, Hongik University
72-1 Sangsu, Mapo, Seoul 121-791, Korea
{janghol,pyo,chong,jlee}@cs.hongik.ac.kr

Abstract. Since traditional pure coaxial cable based cable television network has been upgraded with Hybrid Fiber Coaxial(HFC) technology, the upstream channel from subscribers to a distribution center has made high speed data access available to cable TV subscribers. However, the presence of much higher levels of noise in the upstream than in the downstream, presents major challenges to the design of effective HFC data access networks. When noise from children nodes accumulated in an amplifier exceeds a certain level, the signal can no longer be separated from the noise. This paper presents an optimal noise control algorithm for HFC network management system(NMS) that minimizes the number of nodes to be cut off to keep the noise under a certain level. For a case where all nodes have equal profits, an optimal algorithm of time complexity of $O(n^2)$ is presented. The described algorithms can be used for providing privileged service for premium subscribers in HFC networks.

1 Introduction

CATV network has traditionally delivered downward broadcasting signals from distribution centers to subscribers. Recently increased utilization of upward channels has expanded broadband services [1], including the Internet [1, 4, 5] and telephone services [3]. CATV network is constructed with Hybrid Fiber and Coaxial(HFC) technology [6]. HFC network covers up to 750MHz, in which the low band below 65MHz is used for reverse channels for upward signals. The transmission speed of 10MHz is available. It is required to introduce Network Management System (NMS) to cope with expansion in large scale, diversity in required services and various technologies [4, 5].

The necessity of NMS increases when considering upward channels for bidirectional communication. Upward channel has limited bandwidth. It is also vulnerable to ingress noise. The noise characteristics of upward channel has to do with the hierarchical structure of HFC network. An Optical Node Unit (ONU) is connected to a distribution center by optical cable. The distance between them may extend to tens of Kilometers. A distribution center and ONU's form a star network. ONU transforms optical signal into radio frequency (RF) signal for downward transmission, and vice versa for upward transmission. Coaxial cables in tree structure connects an ONU and subscribers. Each node in the tree structured network represents an amplifier or a distributer.

H.-K. Kahng (Ed.): ICOIN 2003, LNCS 2662, pp. 868–875, 2003.

When a node is out of operation or a cable segment is cut, nodes below the troubled spot cannot communicate with ONU. Although the connection may be secure, upward channel is vulnerable to noise. Each amplifier in coaxial tree network combines the signals from its children. The combined signal is amplified and delivered to its parent node. Ingress noise mixed with signal is also amplified as it travels upward. If the noise exceeds a certain threshold, the signal cannot be distinguished from the noise.

This paper presents an optimal noise control algorithm required by core functions of HFC network management systems. The algorithm minimizes the number of nodes to be off to keep the noise within some level. We represented HFC network as trees, where a node denotes an amplifier. The problem is formulated as node selection problem: Given a tree T, find a subtree S of T including the root of T while minimizing the value of an objective function. We proved that this problem is the same as the 0/1 knapsack problem, which is NP-hard [2]. We also developed an algorithm for finding an optimal solution. When all nodes have equal profit, an algorithm with time complexity of $O(n^2)$ can find an optimal solution.

Section 2 defines our Node Selection Problem and shows it reduces to 0/1-knapsack problem. Section 3 presents an optimal algorithm. Section 4 shows that an algorithm with $O(n^2)$ complexity can find an optimal solution when all nodes have equal profits. Section 5 discusses several problems and further research topics.

2   Node Selection Problem

When each node i of a tree T has profit p_i, noise n_i and threshold c_i, A Node Selection Problem (NSP) is to pick a set of nodes from the tree maximizing the profit under the following conditions:

1. Let T_k be a subtree of T, whose root is the node k, and S_k, a set of nodes selected. The sum of noise of the nodes in S_k cannot exceed c_k:

$$\sum_{i \in S_k} n_i \leq c_k$$

2. If node k is not selected, then any node of T_k cannot be selected. This implies that all nodes in the path from the root of T to k should be selected if node k is to be selected.

Figure 1 shows an example. The triple appearing in the node k represents $\langle p_k, n_k, c_k \rangle$. The set S_1 of selected nodes are $\{1, 2, 3, 4, 6, 8\}$. Every node satisfies the condition 1 and 2.

Node selection problem is NP-hard. Before we prove NSP is NP-hard, let us present 0/1 knapsack problem. When an object i of n objects has profit p_i and weight w_i, 0/1 knapsack problem is to put objects maximizing profit when the

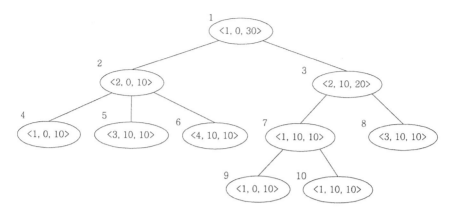

Fig. 1. An example of node selection problem

capacity of the knapsack is W.

$$\text{Maximize} \sum p_i x_i,$$

$$\text{where} \sum w_i x_i \leq W, x_i = 0 \text{ or } 1, i = 1, \ldots, n$$

Theorem 1. *NSP is NP-hard.*

Proof. We show NSP reduces to 0/1 knapsack. Suppose knapsack has capacity W. Let p_i and w_i be profit and weight of object i, $i = 1, \ldots, n$, respectively. An instance of NSP is constructed from an instance of 0/1 knapsack as follows. A tree with n leaf nodes and some internal nodes are constructed, where n leaf nodes correspond to n objects. The number of an internal node's children is determined arbitrarily. The triple $\langle p_i, n_i, c_i \rangle$ of node i is $\langle p_i, w_i, W \rangle$ if node i is a leaf. Otherwise (node i is an internal node), the triple is $\langle 0, 0, W \rangle$. Figure 2 shows a constructed tree when $n = 7$.

We present the other half of the proof: knapsack has profit P satisfying its capacity condition if and only if NSP has profit P satisfying the conditions 1 and 2. Suppose S_1 is the set of selected nodes for NSP. Then $\sum_{j \in S_1} n_j \leq W$ and $\sum_{j \in S_1} p_j$ is the maximum. Suppose L is the set of leaves of S_1. Then $\sum_{j \in S_1} n_j = \sum_{j \in L} n_j \leq W$ and $\sum_{j \in L} p_j$ is the maximum. Putting objects corresponding to the nodes in L into knapsack, $\sum_{j \in L} p_j$ is maximized and the condition $\sum_{j \in L} w_j = \sum_{j \in L} n_j \leq W$ is satisfied. Therefore NSP is NP-hard. Suppose $\langle y_1, y_2, \ldots, y_n \rangle$ is the optimal solution of 0/1 knapsack problem. Then

$$P = \sum_{i=1}^{i=n} p_i y_i, \sum_{i=1}^{i=n} w_i y_i \leq W, y_i = 0 \text{ or } 1, i = 1, \ldots, n$$

Selecting the nodes with $y_i = 1$ from the tree for NSP, maximal profit P is $\sum_{i=1}^{n} p_i y_i$. Since every internal node has noise 0 and threshold W, the sum of noise from the nodes of the overall tree T is $\sum_{i=1}^{n} n_i y_i \leq W$. Since the sum of

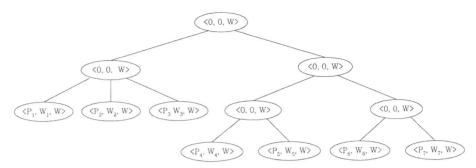

Fig. 2. A tree with 7 leaf nodes

noise of a subtree T_i is less than or equal to W, the condition 1 is stratified. The node i with $y_i = 0$ is not selected. It should be a leaf node and thus has no child, satisfying the condition 2. □

3   Optimal Algorithm

We will first introduce an operation that will be used for selecting nodes later in this section. T_i denotes a subtree whose root is the node i. The set of all sets of selected nodes from T_i is represented by $[T_i]$. The set of all tuples $\langle p, n \rangle$ is represented by D_i, where p is the sum of profits of the selected nodes from T_i and n is the sum of nose.

$$D_i = \{ \langle p, n \rangle \mid \forall S \in [T_i], p = \sum_{j \in S} p_j, n = \sum_{j \in S} n_j \le c_i \}$$

We define an operation $J(D_i, D_j, c)$.

$$J(D_i, D_j, c) = \{ \langle p, n \rangle \mid \langle p, n \rangle \in D_i \cup D_j, n \le c \}$$
$$\cup \{ \langle p + p\prime, n + n\prime \rangle \mid \langle p, n \rangle \in D_i, \langle p\prime, n\prime \rangle \in D_j, n + n\prime \le c \}$$

$J(D_i, D_j, c)$ denotes the set of all $\langle p, n \rangle$, where $n \le c$, whose nodes are selected from T_i and T_j. Given $\langle p_i, n_i \rangle$ and $\langle p_j, n_j \rangle$, $\langle p_i, n_i \rangle$ dominates $\langle p_j, n_j \rangle$ if $p_i > p_j$ and $n_i \le n_j$. Since a tuple dominated by another one generates more noise while its profit is less than the dominating one, the dominated tuple can be eliminated.

When J is applied to a set of D_i's repeatedly, $J(D_1, \ldots, D_k, c)$ is used to represent $J(\ldots J(J(D_1, D_2), c), D_3, c), \ldots, D_k, c)$. $J(D_1, \ldots, D_k, c)$ is the set of $\langle p, n \rangle$ with $n \le c$ for all nodes that can be selected from $T_1, T_2, \ldots, T_k$.

D_i can be computed by using J. Given a node i, let $i_1, i_2, \ldots, i_k$ be the children of the node i .

$$D_i = \{ \langle p_i, n_i \rangle \} \cup \{ \langle p + p_i, n + n_i \rangle \mid \langle p, n \rangle \in J(D_{i_1}, \ldots, D_{i_k}, c_i - n_i) \}$$

Since node i should be selected if any node of i's children is to be selected, the sum of noise generated from the nodes selected from $T_{i_1}, \ldots, T_{i_k}$ should not

```
void NSP (i: node) {
  if (node i is a leaf) then
      if (n_i ≤ c_i) then D_i = {⟨p_i, n_i⟩} else D_i = ∅
  else {
      let i_1, i_2, ..., i_k be the children of node i;
      for (j = 1; j <= k; j++) NSP(i_j);
      D_i = {⟨p_i, n_i⟩} ∪ {⟨p + p_i, n + n_i⟩ | ⟨p, n⟩ ∈ J(D_{i_1}, ..., D_{i_k}, c_i − n_i)}
  }
}
```

Fig. 3. NSP algorithm

exceed $c_i - n_i$. Therefore D_i can be constructed by adding $\langle p_i, n_i \rangle$ to all tuples belonging to $J(D_{i_1}, \ldots, D_{i_k}, c_i - n_i)$.

Given an NSP problem, maximal profit can be obtained from D_i where node i is the root of the tree T. After computing all D_k for all node k in T, the largest p of $\langle p, n \rangle \in D_i$ is searched for the maximal profit. NSP algorithm in Figure 3 finds maximal profit for a given tree.

For example, consider computing D_2 with $c = 10$ from the tree in Figure 1.

$$D_4 = \{\langle 1, 0 \rangle\}$$
$$D_5 = \{\langle 3, 10 \rangle\}$$
$$D_6 = \{\langle 4, 10 \rangle\}$$

Since

$$
\begin{aligned}
J(D_4, D_5, D_6, 10 - 0) &= J(J(D_4, D_5, 10), D_6, 10) \\
&= J(\{\langle 1, 0 \rangle, \langle 4, 10 \rangle\}, \{\langle 4, 10 \rangle\}, 10) \\
&= \{\langle 1, 0 \rangle, \langle 5, 10 \rangle\} \\
D_2 &= \{\langle 3, 0 \rangle, \langle 7, 10 \rangle\}
\end{aligned}
$$

The following describes how to select nodes based on the operation mentioned earlier. Suppose the maximum number of children of a node in a tree is k. Given a node i, let $i_1, \ldots, i_k$ be the children of i and $D_{i_1}, \ldots, D_{i_k}$ be the set of profit-noise pairs. When selecting nodes for a tree T_i whose root is node i, it is necessary to remember the amount of noise injected from subtrees $T_{i_1}, \ldots, T_{i_k}$. The profit-noise pair $\langle p, n \rangle$ is expanded to $\langle p, n, a_1, \ldots, a_n \rangle$, where $a_1, \ldots, a_k$ are the noise delivered from $I_{i_1}, \ldots, T_{i_k}$ respectively.

The operation $J(D_i, D_j, c)$ is also modified.

$$
\begin{aligned}
J(D_i, D_j, c) = \{ &\ \langle\ p, n, a_1, \ldots, a_k \rangle \mid \langle p, n, a_1, \ldots, a_k \rangle \in D_i \cup D_j, n \le c\} \\
\cup \{ &\ \langle\ p + p', n + n', a_1 + a_1', \ldots, a_k + a_k' \rangle \mid \\
&\ \langle\ p, n, a_1, \ldots, a_k \rangle \in D_i, \langle p', n', a_1', \ldots, a_k' \rangle \in D_j, n + n' \le c\} \\
J(D_1, \ldots, D_k, c) = J(&\ J(J(D_1, D_2, c), D_3, c), \ldots, D_k, c)
\end{aligned}
$$

Each node keeps its position among the sibling nodes. The tuple $E = \langle p, n, b_1, \ldots, b_n \rangle$ carries this information.

$$E_{i_j} = \{\langle p, n, b_1, \ldots, b_k \rangle \mid \langle p, n, a_1, \ldots, a_k \rangle \in D_{i_j}\}$$

where $b_j = n$ and $b_l = 0$ for all $l \neq j$. D_i can be defined by using E's.

$$D_i = \{\langle p_i, n_i, 0, \ldots, 0 \rangle\} \cup \{\langle p + p_i, n + n_i, a_1, \ldots, a_k \rangle \mid \langle p, n, a_1, \ldots, a_k \rangle$$
$$\in J(E_{i_1}, \ldots E_{i_k}, c_i - n_i)\}$$

For example, consider how D_2 is computed by using the new definition of J. In the following example, "$-$" means no children.

$$D_4 = \{\langle 1, 0, -, -, - \rangle\}$$
$$E_4 = \{\langle 1, 0, 0, 0, 0 \rangle\}$$
$$D_5 = \{\langle 3, 10, -, -, - \rangle\}$$
$$E_5 = \{\langle 3, 10, 0, 10, 0 \rangle\}$$
$$D_6 = \{\langle 4, 10, -, -, - \rangle\}$$
$$D_6 = \{\langle 4, 10, 0, 0, 10 \rangle\}$$
$$J(E_4, E_5, E_6, 10) = \{\langle 1, 0, 0, 0, 0 \rangle, \langle 5, 10, 0, 0, 10 \rangle\}$$
$$D_2 = \{\langle 3, 0, 0, 0, 0 \rangle, \langle 7, 10, 0, 0, 10 \rangle\}$$

Suppose $\langle p, n, a_1, \ldots, a_k \rangle \in D_i$ when node i is selected. Since the noise generated from the subtree T_{i_j} is a_j, if $n_{i_j} > a_j$ then node i_j is not selected. Since the root of T_{i_j} is not selected all nodes in subtree T_{i_j} are not selected. If $n_{i_j} \leq a_j$ then node i_j is selected. After the root of the subtree T_{i_j}, node i_j, is selected, a tuple $\langle p\prime, n\prime, a_1\prime, \ldots, a_k\prime \rangle$ such that $n\prime = a_j$ is searched from D_{i_j}. This method is recursively applied to the node i_j with the tuple $\langle p\prime, n\prime, a_1\prime, \ldots, a_k\prime \rangle$.

Suppose $D_1 = \{\langle 1, 0, 0, 0, - \rangle, \langle 6, 20, 0, 20, - \rangle, \langle 8, 10, 10, 0, - \rangle, \langle 13, 30, 10, 20, - \rangle\}$. The tuple with maximal profit and noise less than or equal to $c_1 = 30$ is $\langle 13, 30, 10, 20, - \rangle$. The noise from the first child node 2 is 10, and that from the second child, node 3, 20. Searching for the tuple with $c_2 = 10$ from D_2 of the previous example, we get $\langle 7, 10, 0, 0, 10 \rangle$. Children nodes 4, 5 and 6 transmit noise 0, 0 and 10 respectively. Since $n_4 = 0 \leq 0$, $n_5 = 10 > 0$ and $n_6 = 10 \leq 10$, nodes 4 and 6 are selected. Nodes can be selected from the subtree T_2 similarly. Figure 4 shows the node selection algorithm.

4   A Special Case of $p = 1$

This section shows a special case where the profit is 1 for all nodes. NSP algorithm has time complexity of $O(n^2)$, where n is the number of nodes.

Lemma 1. *When node i has children $i_1, \ldots, i_k$ and $|D_{i_1}| = \cdots = |D_{i_k}| = m$, the execution time of $J(D_{i_1}, \ldots, D_{i_k}, c)$ has upper bound of $O(m^2)$.*

Proof. The join operation $J(D_{i_1}, D_{i_2}, c)$ produces $m^2 + 2m$ tuples. Eliminating dominated tuples, we get $|J(D_{i_1}, D_{i_2}, c)| \leq 2m$ because every node has profit 1. $J(J(D_{i_1}, D_{i_2}, c), D_{i_3}, c)$ produces $m + 2m + m(2m) = 2m^2 + 3m$ tuples. $|J(J(D_{i_1}, D_{i_2}, c), D_{i_3}, c)| \leq 3m$ after eliminating dominated tuples. Similarly $J(D_{i_1}, \ldots, D_{i_k}, c)$ gives

$$(1 + 2 \cdots + k - 1)m^2 + (2 + 3 + \cdots + k)m = (k-1)km^2/2 + [k(k+1) - 2]m/2 = O(m^2)$$

```
void SelectNode(i:Node, c: Limit) {
  if (node i is a leaf) then
    if (n_i <= c) then selected[i] = 1 else selected[i] = 0;
  else {
    find <p,n,a_1,...,a_k> in D_i such that n <= c and p is maximum;
    if (such tuple exists) {
      let i_1,...,i_k be the children of node i;
      for (j = 1; j <= k; j++) SelectNode(i_j, a_j);
    }
    else {
      selected[i] = 0;
      for (all node j in T_i) selected[j] = 0;
    }
  }
}
```

Fig. 4. Node selection algorithm

Theorem 2. *When all nodes have profit 1, NSP algorithm has the time-complexity of $O(n^2)$, where n is the number of nodes.*

Proof. We prove when given tree is a perfect k-ary tree. Suppose the height of the tree is h, then the number of nodes $n = (k^h - 1)/(k - 1)$. Assuming the root node has level 1, we have the following equality.

$$\text{number of nodes at level } i = k^{i-1}$$
$$\text{thes size of } D \text{ of a node at level } i = k^{h-i}$$
$$\text{time required for computing } D \text{ at level } i = O(k^{2(h-i)}) \text{by lemma 1}$$
$$\text{time required for computing } D\text{'s of all nodes at level } i = O(k^{2(h-i)}) * k^{i-1}$$
$$= O(k^{2h-i})$$

Therefore, to compute D's for all nodes, it takes

$$\sum_{i=1}^{h-1} O(k^{2h-i}) = O(k^h(k^h - 1)/(k - 1)) = O(n^2)$$

$\square$

5   Conclusion

We presented an algorithm for automatic control for ingress noise at physical link layer in HFC network. Our algorithm produces optimal solutions. When all node have identical profits, the algorithm's complexity becomes $O(n^2)$. NSP algorithm allows privileged service for premium subscribers in the environment of diversified network service. This implies the necessity of heuristics. We are currently developing a heuristic algorithm giving local optimal solutions. NSP algorithm would contribute to developing automatic detection and isolation of noisy nodes for HFC network management system.

References

[1] Gary Donaldson and Doug Jones, Cable Television Broadband Network Architecture. IEEE Communications Magazine, Jun. 2001

[2] Michael R. Garey and David S. Johnson, Computers and Intractability: A Guide to the Theory of NP-Completeness. W. H. Freeman and Company, New York. 1979

[3] Andrew Paff. Hybrid Fiber/Coax in the Public Telecommunications Infrastructure. IEEE Communications Magazine, Apr. 1995

[4] Dennis Picker. Design Considerations for a Hybrid Fiber Coax High-Speed Data Access Network. In Proceedings of COMPCON. IEEE, 1996

[5] Srinivas Ramanathan and Riccardo Gusella. Toward Management Systems for Emerging Hybrid Fiber-Coax Access Networks. IEEE Network, Sep/Oct 1995

[6] Gerry White. Network Management Issues in Internetworks Based on Hybrid Fiber-Coax Cable Plants. Proceedings of the Second International Workshop on Integrated Multimedia Services to the Home. IEEE 1995

A Network Management Architecture Using XML-Based Policy Information Base*

Kwoun Sup Youn and Choong Seon Hong

School of Electronics and Information, Kyung Hee University
1 seocheon, Kihung, Yongin, Kyungki, 449-701 Korea
holiday@networking.kyunghee.ac.kr,cshong@khu.ac.kr

Abstract. XML is being used to describe components and applications in a vendor and language neutral. Therefore it already has a role in distributed system. XML is also being used as a data interchange format between components and applications in loosely coupled large-scale application. Until now, policy is described for specific applications and devices. Its use has been very limited. In current network management system, we can only invoke predefined operations and actions using policy-based Network Management. The main motivation for the recent interests in policy-based networks is to support dynamic adaptability of behavior by changing policy without recoding or stopping system. For these reasons we present the use of the XML for describing the policy and PIB(Policy Information Base) in COPS-PR. It improves flexibility and interoperability among heterogeneous network systems. It also can add new functionality into network components. In this paper, we propose a network management Architecture using XML-based policy information base.

1 Introduction

For several years, the Internet Protocol (IP) networking industry has been struggling to develop ways to manage networks. Initial attempts brought mechanisms and protocols that focused on managing and configuring individual networking devices rather than the network as a whole. While the model was worked well in early deployments of IP networks, the overhead of administrating networks was increased not only as the size of networks grew but also the capabilities of networking devices were increased. The emerging policy-based network management paradigm is a direct result of this shift [1].

The idea of policy-based management system has caught on as a way to reduce administrative costs, tighten security, and help troubleshooting by setting standards for dealing with a number of situations that range from adding new users to the network to reacting to system faults. Instead of emphasizing devices and interfaces, a policy based management system focuses on users and applications.

---

* This work was supported by grant No R05-2001-000-00976-0 from Korea Science and Engineering Foundation.

H.-K. Kahng (Ed.): ICOIN 2003, LNCS 2662, pp. 876–885, 2003.

In essence, policy-based network management system allows network mangers to express business goals as a set of rules, or policies, which are then enforced throughout the network. Also, it can automate many tasks that network mangers have had to perform manually in the past, such as configuring switches and routers to prioritize traffic from specific applications.

2   Motivation

The challenges in policy based network management system are as follows. Firstly, some policies may be user- or application-specific, such as the information to filter when bandwidth or device capabilities are limited. It is very unsuitable in large-scale network. Secondly, policies define choices in behavior in terms of the conditions under the environment that predefined operations and actions can be invoked rather than changing the functionality of the actual operations themselves.

For these reasons, we propose a network management system using XML-based policy information base. It supports dynamic adaptability of behavior by changing policy without the stop of system operation. The use of XML provides a neutral means that is independent on operating systems and applications of describing network management policies. We propose the use of the XML for describing of the policy and PIB(Policy Information Base) in COPS-PR [2,3].

3   Related Works

3.1   The COPS(COPS-PR) and PIB

We focus on configuration management based on IETF standards. The IETF has defined a policy framework consisting of management interfaces for entering policies, repositories for storing policies, policy decision points(PDPs) for evaluating policies, and policy enforcement points(PEPs) for enforcing policy decisions [2,3,10]. Based on this framework, the IETF standardized policy core information models (i.e., PCIM and PCIMe) that can be used when entering policies, when storing them in repositories, and when evaluating them at PDPs. For the transfer of policy decisions between PDP and PEP, the COPS-PR (Common Open Policy Service for Policy Provisioning) protocol was standardized. The structure of configuration information carried by COPS-PR is defined in Policy Information Base (PIBs). The language for defining PIBs has been standardized as the structure of policy provisioning information (SPPI) [11].

The basic COPS operational model shown in Figure 1(a) is described as the PEP that establishes communication with the PDP, send policy requests, receive policy decisions, and optionally send reports. At any time, the PEP can update a request; a PDP can update a decision. The PEP is responsible for deleting the request when it is no longer applicable. The PIB is a conceptual tree namespace of provisioning classes (PRCs) and provisioning instances (PRIs). There may be multiple instances of any PRC. Figure 1 (b) shows an example of PIB tree.

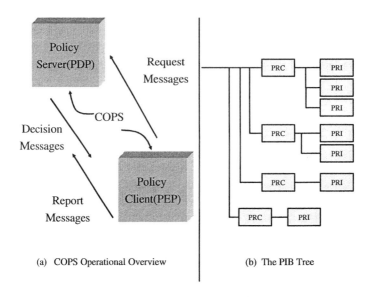

(a) COPS Operational Overview   (b) The PIB Tree

Fig. 1. COPS Operational Overview

Specific clauses in the SPPI allow the PRCs defined in generic PIBs to be reused and specialized for service-specific PIBs via well-known object oriented concepts such as inheritance. This makes the data model defined in any PIB extensible

3.2   The Use of XML in Network Management Area

XML is being used to describe components and applications in a vendor and language neutral way. Therefore it already has a key role in network management area. In this section, we present a way to represent managed data using XML. A new representation for system management data called the Common Information Model (CIM) has been proposed by DMTF [6,7]. There are fundamentally different two models for mapping CIM into XML. One is a schema mapping in which the XML schema is used to describe the CIM classes, and CIM instances are mapped to valid XML documents for that schema. The other is a metaschema mapping in which the XML schema is used to describe the CIM metaschema. Both CIM classes and instances are valid XML documents for that schema [8,9]. Similarly, there are two different models for mapping MIB into XML. One is a model-level mapping, the other is a metamodel-level mapping. Figure 2 shows the model-level mapping and a metamodel-level mapping, respectively [17].

Model-level mapping means that each MIB variable generates its own DTD fragment. The XML element names are taken directly from the corresponding MIB element names. Metamodel-level mapping means that the DTD is used to

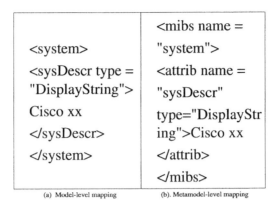

| (a) Model-level mapping | (b). Metamodel-level mapping |

Fig. 2. Two Different Models for Mapping MIB into XML

describe in a generic fashion and the notion of MIB variables. MIB element names are mapped to XML attribute or element values, rather than XML element names.

4   Proposed System

4.1   Issues and Design Objectives

As mentioned earlier, policies may be user- and application-specific, such as the information to filter when bandwidth or device capabilities are limited. Policies define choices in behavior in terms of the conditions under the environment which predefined operations and actions can be invoked rather than changing the functionality of the actual operations themselves. We use COPS-PR to communicate with policy-related information. By adding or removing PRIs, the PDP can implement the desired polices to be enforced at the device. It is important to highlight that the policies that each PIB can be implemented are predefined. For these issues, we propose the use of XML to describe policies and PIB of COPS-PR [2,3]. An XML-based policy is independent on operating system and application. It is applicable to large-scale networks. Mapping PIB to XML can dynamically add PRC. It can support dynamic adaptability of behavior by changing policy without the stop of system operation and can provide the flexibility of PIBs. In addition, we have design objectives as follows:

- A business level policy can be refined in such a way that the policy author can, at some point, actually deploy it in devices without leaving the authoring environment.
- Using the same system, a business level policy can be expressed as easily as the device level policy.

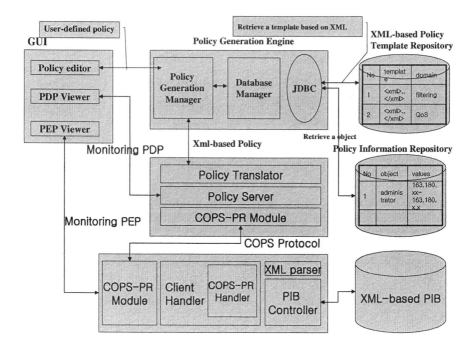

Fig. 3. An Overall Architecture of Proposed System

4.2 Proposed Architecture

In this section, we describe an overview of the architecture of our proposed system. The architecture of our system is shown as Figure 3. There are six key components in this architecture.

Graphical User Interface (GUI) The Graphical User Interface supports the creation of new policies and modifies the past polices using the XML-based generic template. It is the basic interface between the network manger and our proposed system. It also provides methods to edit or create the XML-based generic template and immediate feedback when the network manger uses an improper syntax or enters incorrect entity. It consists of following parts:

- The policy editor supports functionality to create or edit the policies.
- The PEP viewer browses XML-based PIB of PEP and retrieves value of elements in PIB.
- The PDP viewer monitors current status of policy server which receives COPS messages from PEP.

Policy Generation Engine The Policy generation engine is the most important part of our proposed system. It consists of following parts:

- Database manager that manipulates an XML based generic template according to the policy-related information.
- JDBC module that interacts with the policy information repository.
- Policy generation manager that interacts with PDP (Policy Decision Point).

It loads the XML-based policy template from template repository. Through the use of a GUI, a proper XML-based template and embedded policy-related information can be selected.

XML-Based Policy Template Repository It stores an XML-based Policy Template. The main goal of an XML-based Policy Template is to store generic policies for providing specific policy information to the Policy Generation Engine

Policy Decision Point It is one of the components in policy-based network management system. Policy Decision Point includes three components such as policy translator, policy server, and COPS-PR module. Policy translator translates higher-level abstractions of policies into device-level policy that applies to PEP (Policy Enforcement Points). The Policy server can serve as the central point for distribution policies to PEPs [2, 3, 12]. The COPS-PR module makes COPS-PR messages that are sent to PEP. Also it receives COPS-PR messages and then interprets ones.

Policy Information Repository All policy related information is stored in policy information repository and generated using policy information model. Its concept is derived from ISM (Information System Model) of POWER prototype [15]. This is a subset of the full model of the managed system. It includes sufficient information for enabling policy definition. The system related part of this model is linked into the actual managed system Through the interaction with the managed objects, it would be possible for polices to be deployed. The concept used here may easily be implemented using the Common Information Model (CIM) from the DMTF (Distributed Management Task Force) [6,7].

Policy Enforcement Point It is one of the components in policy based network management system. It controls the traffic on our networks in accordance with the business goals and polices we set at higher levels in the policy-based network management system [12]. It consists of following parts:

- COPS-PR module is similar to COPS-PR module of PDP.
- Client-handler is able to handle different client-specific requirements.
- PIB controller manipulates XML-based PIB in PEP. It provides methods to add or delete PRI in PIB. Also it supports functionality to create new PRC objects in PIB.

5   The Use of XML in Policy Template and PIB

5.1   An XML-Based Policy Template

We propose the use of XML to describe polices. It is very flexible and scalable. An XML-based policy template stores a generic policy description for providing specific policy information to Policy Generation Engine. The XML Schema of a policy template is shown as Figure 4.

The XML Schema includes five generic elements: "people", "operation", "start time", "end time", "where". These objects are defined and described in the Policy Information Repository. Individual elements represent objects of the managed domain. A specific policy is described in Figure 5. It is a form of XML instance document by XML Schema. It is described that the administrator can through the internet during working hours(08:00 12:00) and from wherever

```
<?xml version="1.0" encoding="UTF-8"?>
<xs:schema xmlns:xs="http://www.w3.org/2001/XMLSchema"
    elementFormDefault="qualified">
<xs:element name="message_template">
<xs:complexType>
<xs:sequence>
<xs:element ref="people"/>
<xs:element ref="operation"/>
<xs:element ref="start_time"/>
<xs:element ref="end_time"/>
<xs:element ref="where"/>
</xs:sequence>
</xs:complexType>
</xs:element>
<xs:element name="operation" type="xs:string"/>
<xs:element name="people" type="xs:string"/>
<xs:element name="start_time" type="xs:time"/>
<xs:element name="end_time" type="xs:time"/>
<xs:element name="where" type="xs:string"/>
</xs:schema>
```

Fig. 4. Structure of Template by Using XML Schema

```
<?xml version="1.0" encoding="utf-8"?>
<message_template
    xmlns:xsi="http://www.w3.org/2001/XMLSc
    hema-instance"
    xsi:noNamespaceSchemaLocation="templat
    e.xsd">
    <people>administrator</people>
    <operation>Internet</operation>
    <start_time>08:00</start_time>
    <end_time>17:00</end_time>
    <where>every</where>
</message_template>
```

Fig. 5. A Specific Policy by Using Template

```
-- The PIB definition by using SPPI
ipv4FilterDstAddr OBJECT-TYPE
  SYNTAX IpAddress
  MAX-ACCESS read-write
  STATUS current
  DESCRIPTION "The IP address to..."
  ::={ ipv4FilterEntry 2 }
-- Document Type Definition
<!ELEMENT ipv4FilterDstAddr #PCDATA>
<!ATTLIST ipv4FilterDstAddr SYNTAX CDATA #REQUIRED
  ACCESS CDATA #REQUIRED
  STATUS CDATA #REQUIRED>
-- XML-based PIB by using DTD (Document Type Definition)
< ipv4FilterDstAddr SYNTAX="IpAddress"
  ACCESS="read-write" STATUS="current">
</ipv4FilterDstAddr>
```

Fig. 6. Mapping SPPI-based PIB to XML-based PIB

he/she is logged in. An XML-based policy template supplies the convenience for a network manager to determine the policy description.

5.2 Mapping PIB to XML

As mentioned earlier, we generated a PIB using XML. We used a method of Model-level Mapping. It means that each PIB variable generates its own DTD fragment and the XML element names that are taken directly from the corresponding PIB element names. PIB definition using SPPI is shown in Figure 6. The IETF RFC 3159 specifies SPPI that defines numerous constructs, along with their semantics, that can be used while defining PIBs. SPPI also defines how the data definition in PIBs can be reused, thereby avoiding redundancy. We mapped SPPI-based PIB to XML-based PIB using DTD [4]. DTDs use a formal grammar to describe the structure and syntax of an XML document, including the permissible values for much of document's contents. DTD of XML-based PIB and an instance of XML document by the DTD are shown in Figure 6.

6 Execution Sequence for Proposed System

Execution Sequence for Proposed System is shown as Figure 7. We consider that a network manager creates new policy using an XML-based template as follows

1. COPS-PR module is similar to COPS-PR module of PDP.
2. It inquires a necessary XML-Based generic policy template.

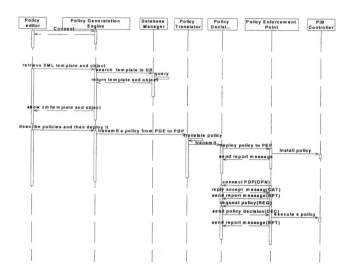

Fig. 7. Sequence Diagram of Proposed System

3. Policy Generation Engine(PGE) inquires a necessary XML-based policy template from XML-based Policy Template Repository.
4. PGE loads a required XML-based template. The manager describes a special policy with the template.
5. When a special policy is described, PGE gets necessary parameters from policy information repository.
6. Policy translator converts the described policy to a policy of device-level and then sends policy.
7. Described special policy is transferred to PEP.
8. PDP sends a device-level policy to PEP through COPS protocol.
9. PEP decides whether it is correct policy or not, after that the result is sent to PDP.
10. PDP sends the result to PGE and notifies the resulting message to manager.

7   Conclusion and Future Works

In this paper, we proposed the use of XML for describing management policies and PIB of COPS-PR. An XML-based Policy is independent of operating system and application. It is applicable to large-scale networks. PIB-to-XML mapping can dynamically add PRC (Provisioning Class). It can support dynamic adaptability of behavior by changing policy without the stop of system operation and provide the flexibility of PIBs. In future works, we plan to develop an efficient algorithm that maps high-level policy into device-level policy.

References

[1] Mark L. Stevens, Watler J. Weiss, "Policy-based Management for IP Networks", Bell Labs Technical Journal, 1999.

[2] D. Durham et al., "The COPS(Common Open Policy Service) Protocol", IETF RFC2748, Jan. 2000

[3] K. Chan et al.,"COPS Usage for Policy Provisioning,"IETF, RFC3084, Mar. 2001.

[4] W3C, "Extensible Markup Language", http://www.w3.org/XML/ .

[5] M. Fine et al.,"Framework Policy Information Base", IETF Internet-Draft, draft-ietf-rap-frameworkpib-04.txt, Nov. 2000.

[6] DMTF, "Common Information Model (CIM) Specification Version 2.2 ", June. 1999.

[7] DMTF, " CIM Schema Version 2.6 ", June 2001.

[8] DMTF, "Specification for the Representation of CIM in XML", Version 2.0, July. 1999.

[9] DMTF, "XML as a Representation for management Information", A White Paper Version 1.0, September. 1998.

[10] Heinz Johner et al, "Understanding LDAP", http://www.redbooks.ibm.com/pubs/pdfs/redbooks/.

[11] M.Wahl ," A Summary of the X.500(96) User Schema for use with LDAPv3", December. 1997

[12] A. Westerinen et al., "Terminology", IETF Internet draft, http://www.ietf.org/internet-drafts/draft-ietf-policy-terminology-02.txt, Nov.2000.

[13] K. McCloghire et al., "Structure of Policy Provisioning Information [SPPI]", IETF RFC 3159, August. 2001.

[14] R. Boutaba et al., "The Meta-Policy Information Base(M-PIB)",IETF Internet draft, Mar. 2002.

[15] M. Casassa Mont et al., "POWER Prototype: Towards Integrated Policy-Based Manage-ment", NOMS 2000, April 2000.

[16] Natarajan, R et al., "A XML based policy-driven management information ser-vice", Pro-ceedings of IM 2001, May. 2001.

[17] J. P. Martin-Flatin, "Web-based management of IP Networks and Systems", Ph.D thesis, Swiss Federal Institute of Technology, Lausanne (EPEL), Oct. 2000.

A Policy-Based QoS Management Framework for Differentiated Services Networks[*]

Si-Ho Cha[1], Jae-Oh Lee[2], Dong-Ho Lee[1], and Kuk-Hyun Cho[1]

[1] Department of Computer Science, Kwangwoon University, Korea
{sihoc,dhlee,khcho}@cs.kw.ac.kr
[2] Department of Computer Engineering
Korea University of Technology and Education, Korea
jolee@kut.ac.kr

Abstract. DiffServ is a technique to provide QoS in an efficient and scalable way. However, current DiffServ specifications have limitations in providing the complete QoS management framework. This paper proposes a policy-based QoS management framework that supports DiffServ policy beans for managing QoS of DiffServ networks. The management framework is based on IETF policy framework to which EJB technologies are adopted. High-level DiffServ QoS policies are represented as valid XML documents with an XML Schema and are translated to low-level EJB policy beans in the EJB-based policy server. The EJB-based policy server can accomplish both policy distribution and QoS monitoring by using SNMP DiffServ MIB.

1 Introduction

Today's IP networks are being transferred from best-effort service model to one that can provide different service levels for specific Quality of Service (QoS) requirements. In best-effort service model, the traffic is processed as quickly as possible, but QoS is not guaranteed. To solve this problem, the Internet Engineering Task Force (IETF) proposed two models: Integrated Services (IntServ) and Differentiated Services (DiffServ). The IntServ model is based on per-flow resource reservation and admission control through the Resource Reservation Protocol (RSVP). The main disadvantage of the IntServ is that the required per-flow state information and QoS treatment in the core IP network raise severe scalability problems [1]. The DiffServ model supports aggregate traffic classes rather than individual flows and provides different QoS to different classes of packet in IP networks. However, current DiffServ specifications have limitations in providing the complete QoS management framework. Current DiffServ RFCs and drafts are mainly for QoS provisioning and configuration only [2]. The complete QoS management framework has not yet addressed in detail. It is possible to lead to severe instabilities and even more QoS violations without the network and service management support.

---

[*] The present Research has been conducted by the Research Grant of Kwangwoon University in 2002.

H.-K. Kahng (Ed.): ICOIN 2003, LNCS 2662, pp. 886–895, 2003.
© Springer-Verlag Berlin Heidelberg 2003

In this paper, we propose a policy-based QoS management framework for DiffServ networks. Our management framework is based on the IETF policy framework [3] and Enterprise JavaBeans (EJB) technologies. The management framework uses Extensible Markup Language (XML) to describe and validate high-level QoS policies. High-level DiffServ QoS policies are represented as valid XML documents with an XML Schema and are translated to EJB beans in the EJB-based policy server. The routing topology and role information required to define QoS policies are discovered by using SNMP MIB-II. The policy distribution and QoS monitoring are processed through Simple Network Management Protocol (SNMP) DiffServ Management Information Base (MIB)[4]. Also we implement SNMP agents with IETF DiffServ MIB on the Linux-based DiffServ routers.

The rest of the paper is organized as follows. Section 2 investigates related technology. Section 3 discusses our policy-based QoS management framework for DiffServ networks. Section 4 presents the implementation of the Linux DiffServ router and the management framework described in section 3. Finally, section 5 summarizes our work and describes possible future work.

2   Related Technology

2.1   DiffServ QoS

DiffServ provides the service differentiation for aggregate IP packet streams by implementing different Per Hop Behaviors (PHBs) for different Differentiated Service Code Point (DSCP) values. The edge router marks IP packets with the DSCP and the core router forwards the packets according to a proper PHB based on the DSCP in the IP header. Traffic entering a DiffServ network through an edge router is first classified, then passed through some form of admission filter, intended to shape the traffic to meet the policy requirements associated with the classification. The shaped traffic stream is then assigned to a particular behavior aggregate by marking IP packets with the DSCP. When the stream is passed through the DiffServ network, this DSCP value triggers a selected PHB from all core routers of the network. The PHB is the DiffServ behavior expected at each node, namely, the treatment afforded to packets at each DiffServ router. The DSCP is simply the value contained within the first 6 bits of the Differentiated Services (DS) byte in the IP header. It indicates a particular PHB that should be given to the corresponding packet. Currently, four types of PHBs are specified for use within the DiffServ network: default behavior (best-effort), class selector (CLS), expedited forwarding (EF), and assured forwarding (AF).

2.2   Policy-Based Management

Policies are business rules that describe the behavior of the network in a way as independent as possible of network devices and topology. Policy-based management means to manage the behavior of a network through high-level policies. The amount of network management task can be reduced by using policies because one policy can be used for policy targets that are of various types, i.e., network nodes or interfaces, and that have been developed by a variety of vendors [5].

2.3   Network Management with EJB Technology

The EJB framework provides a platform-independent, interoperable, and server-side component-based model and low-level services, such as support for transactions, concurrency, persistence, security, multi-threaded execution [6]. EJBs can be customized at deployment time by editing their environment properties that is separated using XML. Therefore, bean integration is possible without source code changes.

The management framework based on EJB technology simplifies network management applications by handling many details of application behavior automatically, without complex programming. And by employing an EJB-based management framework, application developers on longer need to develop applications against specific operating system APIs or specific vendor's middleware APIs. So, The management framework based on EJB technology has a number of benefits as follows: encapsulation of business logic, simplified application development, extensibility, scalability, consistency, transaction management, database and directory access, container-managed persistence, and distributed object access, and so on [6]. Therefore, we adopt the EJB framework to build the policy-based QoS management framework.

3   Design

3.1   Architecture

The outline of our policy-based QoS management framework for DiffServ networks is shown in Figure 1. Our DiffServ QoS management framework conforms to the Model-View-Controller (MVC) architecture [7]. Therefore, it is highly manageable and scalable, and provides the overall strategy for the clear distribution of objects involved in managing service. The QoS management framework roughly consists of a Web server and an EJB-based policy server. To create and administer high-level QoS policies to be enforced on the DiffServ network, an administrator uses the Web browser-based interface. The topology information retrieved from the network through SNMP is stored in a Topology Database and is represented as Topology Node (TN) entity beans in the EJB container of the EJB-based policy server. A Topology Manager (TMA) session bean guides this procedure. High-level DiffServ QoS policies are represented as valid XML documents conforming to an XML Schema through a Java Servlet and are translated to low-level EJB policy beans in the the EJB-based policy server through a Policy Manager (PMA) session bean. There are three low-level policy beans for managing QoS of the DiffServ network in the EJB-policy server: Packet Classifier (PC) policy, Traffic Conditioner (TC) policy, and Queuing & Scheduling (QS) policy.

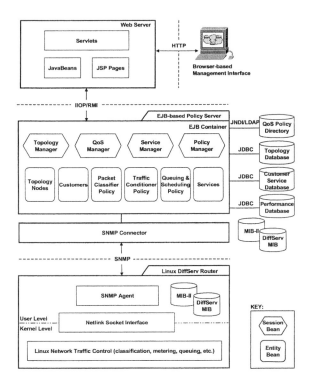

Fig. 1. The policy-based QoS management framework for DiffServ networks

The PC policy and TC policy are usually used in edge routers, while QS policy is usually used in core routers. They are stored in a QoS policy directory. A PMA session bean also distributes the low-level policy beans to DiffServ networks. A QoS Manager (QMA) session bean monitors the QoS from DiffServ networks by using the low-level policy beans. The JNDI API and the Lightweight Directory Access Protocol (LDAP) provide lookup services for the QoS policy directory. Our management framework uses SNMP to distribute and monitor low-level policies. The EJB-based policy server sets appropriately the values of DiffServ MIB according to low-level policies. It will make DiffServ routers conform to QoS policies. The QoS monitoring information is stored in a Performance Database. A Java Database Connectivity (JDBC) driver connects the EJB container to databases.

3.2   The Discovery of the Routing Topology and Role

In order to manage the QoS of DiffServ networks, the QoS management framework must obtain the routing topology and role information for all the DiffServ routers. The TMA session bean accomplishes the topology and role discovery by using SNMP MIB-II tables, an ipAddrTable and an ipRouteTable. The topology and role information is stored in a Topology database and is represented as TN entity beans in the EJB-based policy server. By using the ipRouteTable entries, the TMA bean can obtain the routing topology information. And by using the ipAddrTable, it also discovers the role information of routers. The ipAddrTable contains IP address of all network interfaces in a router and the ipRouteTable contains the IP routing table that has the next hop host and network interface for a set of destination IP addresses.

The TMA bean uses the RTRD (Routing Topology and Role Discovery) algorithm. The RTRD algorithm is given in Table 1. Δi means the list of a router's interfaces and Δn means the list of a router's neighbors. Function filter(i) and filter(n) filter interfaces and neighbors on each valid interface and route, respectively.

The direct neighbors can be learned by retrieving ipRouteNextHop fields from ipRouteTable. We can iterate every neighbor of the router, calling the same algorithm recursively on each iteration, and thus discover complete router connectivity progressively. In order to do so, the EJB-based policy server must have a local SNMP agent, and must have at least a direct connection with an edge router in the DiffServ network.

3.3   QoS Policy Management

PMA session beans and QMA session beans accomplish the QoS management by using three low-level QoS policy entity beans. Figure 2 shows the sequence diagram for a set of QoS management procedures. As demonstrated in Figure 2,

Table 1. The RTRD algorithm

Procedure RTRD (router)
  SNMPgetbulk (ipAddrTable, ipRouteTable);
  $\Delta i \leftarrow (\forall i)[\text{Interface}(i, \text{filter}(i))]$;
  $\Delta n \leftarrow (\forall n)[\text{Neighbor}(n, \text{filter}(n))]$;
  if ($\Delta i \in$ same_subnet)
    router.role $\leftarrow$ core;
  else
    router.role $\leftarrow$ edge;
  if ($\Delta n \not\ni$ router.direct_neighbor)
    return;
  else {
    next_router $\leftarrow$ router.direct_neighbor;
    while (next_router $\neq$ NULL) {
      RTRD (next_router);
      next_router $\leftarrow$ router.direct_neighbor; } }
End Procedure

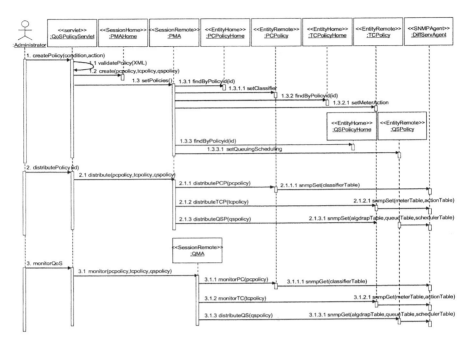

Fig. 2. UML sequence diagram for QoS management

the QoS management procedures are composed of policy creation, policy validation, policy translation, policy distribution, and QoS monitoring. The PC policy bean classifies packet flows and assigns Classifier Identifiers (CIDs) to them. A TC policy bean meters packets and performs actions on them: marker, counter, absolute dropper, and so forth. QS policy beans queue and schedule, or drop packets. By using the QoS policy entity beans, the PMA session bean provides both the policy translation and the policy distribution, and the QMA session bean provides the QoS monitoring.

Policy Validation and Translation. The most basic check of policy validation is that of validating the syntax of the policy specification. In our management framework, we defined a XML Schema file to validate high-level QoS policies described as XML documents. A Java Servlet receives the data from an administrator and creates XML policy documents and then validates them through the XML Schema. And the Java Servlet requests a PMA session bean instance to create the instances of low-level QoS policy entity beans.

The PMA session bean translates high-level QoS policies to low-level QoS policies by properly setting attributes of the three low-level QoS policy entity beans, such as a PC policy bean, a TC policy bean, and a QS policy bean. These beans represent the behavior and data of a business object for SNMP operations.

High-level policies related to edge routers may be applied to PC policy beans and TC policy beans, whereas high-level policies related to core routers may be applied to QS policy beans. The following shows a simple sample policy, an XML document for the policy, and an XML Schema for validating XML policy documents in brief.

policy1: *IF (Source-IP-Address is 192.168.1.2) THEN COS is Gold;*

A brief XML document for **policy1** has the following structure:

```
<policyList>
   <policy id = "1">
      <sourceGroup>
         <source>192.168.1.2</source>
      </sourceGroup>
      <destinationGroup>
         <destination>any</destination>
      </destinationGroup>
      <serviceGroup>
         <service>any</service>
      </serviceGroup>
      ....
      <cos>Gold</cos>
   </policy>
</policyList>
```

A brief XML Schema for policy XML documents has the following structure:

```
<?xml version="1.0"?>
<xsd:schema xmlns:xsd="http://www.w3.org/2000/10/XMLSchema">
<xsd:element name="policyList" type="policyList" />
  <xsd:complexType name="policyList">
    <xsd:sequence>
      <xsd:element name="policy" type="policy"
        minOccurs="1" maxOccurs="unbounded" />
    </xsd:sequence>
  </xsd:complexType>
  <xsd:complexType name="policy">
    <xsd:sequence>
      <xsd:element name="sourceGroup" type="sourceGroup" />
      <xsd:element name="destinationGroup" type="destinationGroup"/>
      <xsd:element name="serviceGroup" type="serviceGroup" />
      <xsd:element name="timeGroup" type="timeGroup" />
      <xsd:element name="cos" type="cosEnum" />
    </xsd:sequence>
    <xsd:attribute name="id" type="xsd:int" />
  </xsd:complexType>
  <xsd:complexType name="sourceGroup">
    <xsd:sequence>
      <xsd:element name="source" type="xsd:string"
        minOccurs="1" maxOccurs="unbounded" />
```

```
        </xsd:sequence>
    </xsd:complexType>
      ....
    <xsd:simpleType name="cosEnum">
        <xsd:restriction based="xsd:string">
            <xsd:enumeration value="Premium" />
            <xsd:enumeration value="Gold" />
            <xsd:enumeration value="Silver" />
            <xsd:enumeration value="Bronze" />
        </xsd:restriction>
    </xsd:simpleType>
</xsd:schema>
```

Policy Distribution and QoS Monitoring. In our management framework, the DiffServ QoS policies conform to the SNMP DiffServ MIB [4]. In other words, the low-level policy beans perform SNMP operations for DiffServ MIB. A QMA session bean provides the policy distribution through a PC policy bean, a TC policy bean, and a QS policy bean. A PC policy bean is usually deployed to the ingress interfaces of edge routers and applies to inbound traffic. TC policies are usually deployed to the egress interfaces of edge routers and apply to outbound traffic. A QS policy is usually deployed to the egress interfaces of core routers and applies to outbound traffic [5]. A set of those actions is accomplished by using the DiffServ MIB. The DiffServ MIB describes the configuration and management aspects of DiffServ nodes. The MIB contains the functional elements of the datapath, using various tables. The idea is that RowPointers are used to combine the various functional elements into one datapath. The DiffServ MIB tables are categorized in four architectural DiffServ elements, which are classifier, meter, action, and queue. The main advantage of using SNMP for policy distribution is that it is likely to work across all the routers in a standard manner.

Distributed policies may not behave as stated in the policy definition. The QoS resulted from a policy distribution can be different from its specification. Therefore, an administrator must monitor the QoS resulted from a policy distribution. QoS monitoring is also accomplished through using the SNMP DiffServ MIB. The QMA session bean also accesses policy definition in the PC policy bean, TC policy bean, and QS policy bean and compares the observed behavior of the network with the one defined in the policy. If degradation is verified, The QMA session bean notifies the administrator by alerting messages and updates the performance database.

4   Implementation

4.1   Linux-Based DiffServ Network

We have established a set of DiffServ routers in the Linux systems and have added an SNMP agent for the MIB-II and DiffServ MIB in each router. Support for DiffServ is already integrated into 2.4 kernels [9]. We had to reconfigure and rebuild your kernel. A SNMP agent containing MIB-II and DiffServ MIB has been implemented by using UCD-SNMP 4.2.5 [10] that provides the agent extension capability. Communication between the DiffServ agent and the Linux traffic control kernel use a netlink socket. Netlink is used to transfer information between kernel and user-space processes, over its bi-directional communications links. A SNMP agent receives management operations from the low-level policy beans in the EJB-based policy server and performs the appropriate parameter changes in the Linux traffic control kernel. As the result of these operations, policy distribution and result monitoring are possible.

4.2 DiffServ QoS Management System

Our DiffServ QoS management framework is implemented in the Windows 2000 Server platform with Java. It includes both the presentation-tier and the business-tier in the multi-tier MVC architecture [8]. The presentation-tier runs a Web server to handle administrator requests, and invokes Servlets and JSPs. We use Apache Tomcat 4.0.6 [11] for a Servlet and JSP container in our application. An EJB-based policy server within the business-tier runs an EJB server to handle EJB components. We use JBoss 2.4.10 [12] for an EJB server and use EJB 1.1 to implement EJB beans. JBoss is an Open Source, standards-compliant, application server implemented in pure Java and distributed for free. AdventNet SNMP v2c APIs [13] written in Java are used for handling EJB-based policy server's SNMP operations. The Oracle 8i Enterprise Edition 8.1.6 for Windows 2000 is used for storing performance and topology information derived from MIB tables. Figure 3 shows the user interface for our DiffServ QoS management system.

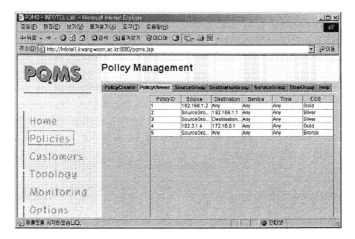

Fig. 3. The user interface for our DiffServ QoS management system

5   Conclusion

In this paper, we have proposed a policy-based QoS management framework for DiffServ networks. The management framework is designed according to multi-tier MVC architecture, IETF policy framework, and EJB technologies. In our model, high-level DiffServ QoS policies are represented as valid XML documents through Java Servlet and are translated to low-level EJB policy beans in the EJB container of the EJB-based policy server. The policy server has three low-level EJB policy beans follow as PC policy, TC policy, and QS policy. PC policies and TC policies are deployed on edge routers, and QS policies are deployed on core routers. These policies are distributed using SNMP to control and monitor QoS of DiffServ networks. To support this work, we implemented the SNMP DiffServ agent containing DiffServ MIB on the Linux-based DiffServ routers.

We are currently at the stage of implementation of the business logic on the EJB-based policy server. We plan to experiment with and demonstrate the system on laboratory testbeds using Linux-based router. Currently we are only checking the syntax validation of policy. So, we will also develop a policy conflict detection and resolution algorithm and apply that to our system.

References

[1] D. Goderis, S. Van den Bosch, Y. T' Joens, et al.: A Service-Centric IP Quality of Service Architecture for Next Generation Networks, NOMS 2002 (2002) 139-154
[2] J. Y. Kim. James W. K. Hong: Distributed QoS Monitoring and Edge-to-Edge QoS Aggregation to Manage End-to-End Traffic Flows in Differentiated Services Networks, Journal of Communications and Networks (JCN), Vol. 3, No. 4, December (2001) 324-333
[3] R. Yavatkar, D. Pendarakis, R. Guerin: A Framework for Policy-based Admission Control, RFC 2553 (2000)
[4] F. Baker, K. Chen, A. Smith: Management Information Base for the Differentiated Services Architecture, RFC 3289 (2002)
[5] Yasusi Kanada, Brian J. O'Keefe: DiffServ Policies and Their Combinations in a Policy Server, APNOMS 2001, (2001)
[6] Dinesh C. Verma: Policy-Based Networking - Architecture and Algorithms, New Riders Publishing (2001)
[7] Sun Microsystems: Telecom Network Management With Enterprise JavaBeans^{TM} Technology - A Technical White Paper, May (2001)
[8] Si-Ho Cha, Jae-Oh Lee, Kuk-Hyun Cho: Towards a Component-based Operations Support System, APNOMS 2002 (2002) 209-220
[9] Differentiated Services on Linux, h ttp://diffserv.sourceforge.net/
[10] UCD-SNMP 4.2.5, http://www.net-snmp.org/
[11] Apache: Jakarta Tomcat 4.0.6, http://jakarta.apache.org/tomcat/
[12] JBoss Org.: JBoss Application Server 2.4.10, http://www.jboss.org/
[13] AdventNet: AdventNet SNMP API v2c, http://www.adventnet.com/products/

Part VII

Network Performance

Do Not Trust All Simulation Studies
of Telecommunication Networks

Krzysztof Pawlikowski

Department of Computer Science, University of Canterbury
Christchurch, New Zealand
k.pawlikowski@cosc.canterbury.ac.nz

Abstract. Since the birth of ARPANET and the first commercial applications of computer networks, through explosion of popularity of the Internet and wireless communications, we have witnessed increasing dependence of our civilization on information services of telecommunication networks. Their efficiency and reliability have become critically important for the well-being and prosperity of societies as well as for their security. In this situation, the significance of performance evaluation studies of current and future networks cannot be underestimated. Increasing complexity of networks has resulted in their performance evaluation studies being predominantly conducted by means of stochastic discrete-event simulation. This paper is focused on the issue of credibility of the final results obtained from simulation studies of telecommunication networks. Having discussed the basic conditions of credibility, we will show that, unfortunately, one cannot trust the majority of simulation results published in technical literature. We conclude with general guidelines for resolving this credibility crisis.

1 Introduction

Since the birth of ARPANET and the first commercial applications of computer networks, through explosion of popularity of the Internet and wireless communications, we have witnessed increasing dependence of our civilization on information services of telecommunication networks. Their efficiency and reliability have become critically important for well-being and prosperity of societies as well as for their security. In the United States, the Department of Defense has listed Network Modeling and Simulation as one of the seventeen most important research areas of Information Processing; see www.darpa.mil/ipto/.

Increasing complexity of modern networks has resulted in their performance studies being predominantly conducted by means of computer simulation. A survey of over 2246 research papers on networks published in Proceedings of IEEE INFOCOM (1992-8; in total 1192 papers), IEEE Transactions on Communications (1996-8; in total 657 papers), IEEE/ACM Transactions on Networking (1996-8; in total 223 papers), and Performance Evaluation Journal (1996-8; in total 174 papers) has shown, see Figure 1, that over 51% of all publications on networks' performance reported results obtained by means of simulation, with

H.-K. Kahng (Ed.): ICOIN 2003, LNCS 2662, pp. 899–908, 2003.
© Springer-Verlag Berlin Heidelberg 2003

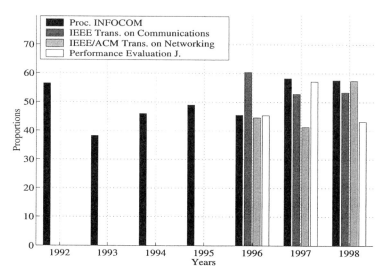

Fig. 1. Proportion of papers that reported results obtained by means of stochastic simulation; from a survey of 2246 papers published in the Proc. IEEE INFOCOM (1992-8), IEEE Trans. on Communications (1996-8), IEEE/ACM Trans. on Networking (1996-8) and Performance Evaluation J. (1996-8)

the rest of the papers relying on two other paradigms of science: theory and experimentation. Such reliance on simulation studies of telecommunication networks raises the question of credibility of the results they yield.

The main credibility issues of quantitative simulation are discussed in the next section. This is followed by a discussion of the results of a survey conducted for showing how much researchers, who use simulation as the tool of their scientific investigations, are concerned about credibility of the results they produce. The paper concludes with general guidelines for conducting fast and credible simulations of telecommunication networks.

2   Credibility of Simulation Studies of Networks

The first necessary condition of any trustworthy performance evaluation study based on simulation is to use a *valid simulation model*, with an appropriately chosen level of detail. Some experts assess that the modeling phase of a system for computer simulation consumes about 30-40% of the total effort of a typical simulation study [1]. In the case of telecommunication networks, it means a conceptually correct model of the network, based on correct assumptions about the network's internal mechanisms, their limitations, appropriate characteristics of simulated processes etc. Next, having implemented the model in software, one needs to verify this implementation, to ensure that no logical errors have been introduced. Validation and verification have been generally recognized as important stages of any credible simulation study. A good discussion of general guidelines for correct and efficient execution of these processes in simulation practice can be found, for example, in However, these are only the first steps for ensuring credibility of the final results of a simulation study, since *"succeeding in simulation requires more than the ability to build useful models ..."*, [3].

The next step is to ensure that the verified software implementation of a given valid simulation model is used in a *valid simulation experiment*. In the overwhelming majority of simulation studies of telecommunication networks, networks or their components are modeled as stochastic dynamic systems. In such a stochastic simulation-based experiment, two primary issues which have to be addressed when trying to ensure its final validity are: (i) application of appropriate source(s) of randomness, and (ii) appropriate analysis of simulation output data. In this updated and extended version of [4], having discussed the last two credibility issues in more detail, we will produce an evidence that one cannot unfortunately trust the majority of simulation studies of telecommunication networks.

It is common practice today to use algorithmic generators of (pseudo-random) uniformly distributed numbers as sources of the basic randomness for stochastic computer simulation. Such a pseudo-random number generator (PRNG) generates numbers in cycles, i.e. having generated whole cycle of numbers, it begins to repeat generation of the same sequence of numbers. Using the same pseudo-random numbers again and again during one simulation is certainly a dangerous procedure since it can introduce unknown and undesirable correlations between various simulated processes. *"Results* <of a stochastic simulation can be very> *misleading when correlations hidden in the random numbers and in the simulated system interfere constructively ..."* [5]. Thus, the practical advice is to use PRNGs that generate numbers in such long cycles that the generated numbers would not be repeated during even the longest simulation. In 2002, using a workstation equipped with a CPU operating at 2.2 GHz, I could generate 10^6 pseudo-random numbers in less than 0.14 second. It meant that PRNGs with cycle's length of about 2^{31}, that were still frequently used in 2002, would generate all numbers of the cycle in about 4.8 minutes. Assuming that the process of random number generation takes, say, 1% of the total simulation time, such a PRNG could be safely used on a 2.2 GHz CPU in a simulation lasting up to about 8 hours.

However, PRNGs with much longer cycles are required if we take seriously the fact that the primary pseudo-random numbers should pass statistical tests for being uniformly distributed. Namely, it can be shown that some tests will always reject the hypothesis about distributional uniformity of pseudo-random numbers if more than a fraction of the cycle of numbers is tested. For example, if we are concerned with two-dimensional uniformity of pseudo-random numbers, then only $O(\sqrt[3]{L})$ numbers from a linear congruential PRNG with the cycle of length L can be used. Longer sequences would not pass a test of uniformity considered in [6]. Empirical analysis of some popular PRNGs reported in [6] has specified that limit as $16\sqrt[3]{L}$. This restricts the number of pseudo-random numbers available from a PRNG with the cycle of $2^{31} - 1$ to just about 20 000, and to about 1 000 000 in the case of PRNGs with the cycle of $2^{48} - 1$. Assuming as previously that the process of random number generation takes 1% of the total simulation time, a "statistically safe" PRNG, allowing to run a simulation on a 2.2 GHz CPU for up to 8 hours, would need to have the cycle of at least 2^{81} long. Note that during any network simulation, two- or more dimensional uniformly distributed vectors have to be frequently generated ...

As the computing technology continues advancing according to Moore's law and CPUs operating with clock frequencies well over 2.2 GHz are expected to be commercially available soon[1], we need PRNGs with cycles much longer than 2^{81}, to be able to run simulation experiments over a reasonably long time intervals. Fortunately, such PRNGs have already been proposed. For example, a PRNG known as Mersenne Twister, within a class of Generalized Feedback Shift Register PRNGs, with a super astronomical cycle of $2^{19937} - 1$, and good pseudo-randomness in up to 623 dimensions (!) for up to 32-bit accuracy, has been proposed in [7]. Such a PRNG will remain "statistically safe" for any practical simulation experiment executed even on an all-optical computer, a technology that some say can be available in 10 years. And ... its portable implementation in C, on 32-bit machines, is much faster than a standard PRNG used in the ANSI C rand() function[2]; see www.math.keio.ac.jp/matumoto/emt.html for the latest news regarding the Mersenne Twister.

Thus, at this stage, there exist PRNGs that can be used as quite reliable sources of elementary randomness in stochastic simulations. We only need to use them. Unfortunately, uncontrolled distribution of various computer programs has resulted in uncontrolled proliferation of really poor PRNGs, of clearly unsatisfactory or unknown quality. Thus, the advice given by D. E. Knuth in 1969 is even more important today, in the era of Internet: *"... replace the random generators by good ones. Try to avoid being shocked at what you find ..."* [8].

2.1 Analysis of Simulation Output Data

Any stochastic computer simulation, in which random processes are simulated, has to be regarded as a (simulated) statistical experiment and, because of that, application of statistical methods of analysis of (random) simulation output data is mandatory. Otherwise, *"... computer runs yield a mass of data but this mass may turn into a mess* <if the random nature of such output data is ignored, and then> *... instead of an expensive simulation model, a toss of the coin had better be used"* [9]. John von Neumann, having noticed a similarity between computer simulators producing random output data and a roulette, coined the term *Monte Carlo simulation*.

Statistical error associated with the final result of any statistical experiment or, in other words, the degree of confidence in the accuracy of a given point estimate, is commonly measured by the corresponding interval estimate, i.e. by the

---

[1] Written in December 2002

[2] The ANSI C rand() function uses a linear congruential PRNG with modulus of 2^{31}, 1103515245 as the multiplier, and 12345 as the additive constant.

confidence interval (CI) expected to contain an unknown value with the probability known as the *confidence level*. In any correctly implemented simulation, the width of a CI will tend to shrink with the number of collected simulation output data, i.e. with the duration of simulation.

Two different scenarios for determining the duration of stochastic simulation exist. Traditionally, the length of simulation experiment was set as an input to simulation programs. In such *fixed-sample-size scenario*, where the duration of simulation is pre-determined either by the length of the total simulation time or by the number of collected output data, the magnitude of the final statistical error of results is a matter of luck. This is no longer an acceptable approach !

Modern methodology of stochastic simulation offers an attractive alternative solution, known as the *sequential scenario* of simulation or, simply, *sequential simulation*. Today, the sequential scenario is recognized as the only practical approach allowing control of the error of the final results of stochastic simulation, since "... *no procedure in which the run length is fixed before the simulation begins can be relied upon to produce a confidence interval that covers the true* < value > *with the desired probability level*" [2]. Sequential simulation follows a sequence of consecutive checkpoints at which the accuracy of estimates, conveniently measured by the *relative statistical error* (defined as the ratio of the half-width of a given CI and the point estimate), is assessed. The simulation is stopped at a checkpoint at which the relative error of estimates falls bellow an acceptable threshold.

There is no problem with running simulation sequentially if one is interested in performance of a simulated network within a well specified (simulated) time interval; for example, for studying performance of a network during 8 hours of its operation. This is the so-called *terminating* or *finite time horizon simulation*. In our example, one simply needs to repeat the simulation (of the 8 hours of network's operations) an appropriate number of times, using different, statistically independent sequences of pseudo-random numbers in different replications of the simulation. This ensures that the sample of collected output data (one data item per replication) can be regarded as representing independent and identically distributed random variables, and confidence intervals can be calculated using standard, well-known methods of statistics, based on the central limit theorem; see, for example, [2].

When one is interested in studying behavior of networks in steady-state, then the scenario is more complicated. First, since steady-state is theoretically reachable by a network only after an infinitely long period of time, the problem lies in execution of *steady-state simulation* within a finite period of time. Various methods of approaching that problem, mostly in the case of analysis of mean values and quantiles, are discussed for example in [10]. Each of them involves some approximations. Most of them (except the so-called method of regenerative cycles) require that output data collected at the beginning of simulation, during the initial warm-up period, are not used to calculate steady-state estimates. If they are included in further analysis, they can cause a significant bias of the final results. Determination of the lengths of warm-up periods can require quite elaborate statistical techniques. When this is done, one is left with a time series of (heavily) correlated data, and with the problem of estimation of confidence intervals for point estimates obtained from such data. However, although the search for robust techniques of output data analysis for steady state simulation continues ([11]), reasonably satisfactory implementations of basic procedures for calculating steady-state confidence intervals of, for example, mean values and quantiles have been already available; see, for example, [10] and [12].

There are claims that sequential steady-state simulation, and the associated with it problem of analysis of statistical errors, can be avoided by running simulation experiments sufficiently long, to make any influence of the initial states of simulation negligible. While such *brute force approach* to stochastic steady-state simulation can sometimes lead to acceptable results (the author knows researchers who execute their network simulations for a week, or longer, to get the results that, they claim, do represent steady-state behavior of simulated networks), one can still finish with very statistically inaccurate results. It should be remembered that in stochastic discrete-event simulation collecting of sufficiently large sample of output data is more important than simply running the simulation over a long period of time. For example, when analyzing rare events, the time during which the simulated network is "idle", i.e. without recording any event of interest, has no influence on the statistical accuracy of the estimates of the event. What matters is the number of events of interest recorded.

Stopping stochastic simulation too early can give misleading, or at least inconclusive, results. Figure 2 shows the final results from sequential steady-state simulation of a MAC protocol in a unidirectional bus LAN from simulation stopped when the relative error dropped below 15% (Figure 2.a) and 5% (Figure 2.b). Even more significant influence of the level of statistical error on clarity of results can be found in [4]. On the basis of this evidence, one can question the sense of drawing conclusions on the basis of results with high statistical errors, or results for which statistical errors were not measured at all !

Unfortunately, sequential stochastic simulation is still not popular among designers of simulation packages, with overwhelming majority of them advocating analysis of output data only after the simulation is finished. This makes the final statistical errors of results the matter of luck. Very few commercial packages can execute simulations sequentially. Among a few packages designed at universities and offered as freeware for non-profit research activities one should mention Akaroa2 ([14]), designed at the University of Canterbury in Christchurch, New Zealand. Recently, Akaroa2 has been linked with Network Simulator NS2, allowing sequential simulation of network models developed in NS2; see www.cosc.canterbury.ac.nz/research/RG/net_sim/simulation_group.html for more detail.

3   Credibility Crisis

It would be probably difficult to find a computer scientist or telecommunication engineer today who has not been trained how to assess and minimize errors

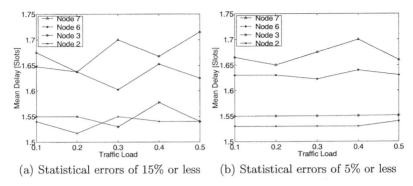

(a) Statistical errors of 15% or less   (b) Statistical errors of 5% or less

Fig. 2. Example showing influence of statistical errors on the final simulation results. The assumed confidence level=0.95. Evaluation of a MAC protocol in a unidirectional bus LAN considered in [13]

inevitably associated with statistical inference. Nevertheless, looking at further results of our survey of eight recent proceedings of INFOCOM as well as three recent volumes of IEEE Transactions on Communications, IEEE/ACM Transactions on Networking and Performance Evaluation Journal, one can note, see Figure 3, that about 77% of authors of simulation-based papers on telecommunication networks were not concerned with the random nature of the results they obtained from their stochastic simulation studies and either reported purely random results or did not care to mention that their final results were outcomes of an appropriate statistical analysis. Let us add that Figure 3 was obtained assuming that even papers simply reporting average results (say, averaged over a number of replications), without any notion of statistical error, were increasing the tally of papers "with statistically analysed results".

While one can claim that the majority of researchers investigating performance of networks by stochastic simulation simply did not care to mention that their final results were subjected to an appropriate statistical analysis, this is not an acceptable practise. Probably everybody agrees that performance evaluation studies of telecommunication networks should be regarded as a scientific activity in which one tests hypotheses on how these complex systems would work if implemented. However, if this is a scientific activity, then one should follow the *scientific method*, generally accepted methodological principle of modern science, [15]. This methodology requires that *any scientific activity should be based on controlled and repeatable experiments*.

The real problem is that the vast majority of simulation experiments reported in telecommunication network literature is not repeatable. A typical paper contains very little or no information about how simulation was conducted. Our survey has revealed that many authors did not even care to inform whether their results came from a terminating or from steady-state simulation.

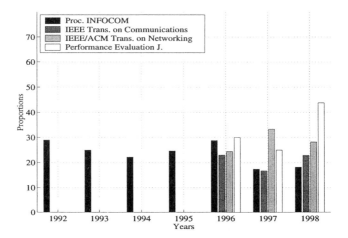

Fig. 3. Proportion of all surveyed papers based on simulation in which results were statistically analysed; from a survey of 2246 papers published in the Proc. IEEE INFOCOM (1992-8), IEEE Trans. on Communications (1996-8), IEEE/ACM Trans. on Networking (1996-8) and Performance Evaluation J. (1996-8)

While the principles of the scientific method are generally observed by researchers in such natural sciences as biology, medicine or physics, this crisis of credibility of scientific outcomes from simulation is not limited to the area of telecommunication networks. It has spanned over whole area of computer science, as well as electronic and computer engineering, despite of such early warnings like that in 1990, by B. Gaither, then the Editor-in-Chief of the ACM Performance Evaluation Review, who, being concerned about the way in which stochastic simulation was used, wrote that he was unaware of *"any other field of engineering or science <* other than computer science and engineering> *where similar liberties are taken with empirical data ..."* [16]. What can be done to change the attitude of writers (who, of course, are also reviewers) of papers reporting simulation studies of telecommunication networks ? Consequences of drawing not fully correct, or false, conclusions about a network performance are potentially huge.

3.1   A Solution ?

The credibility crisis of simulation studies of telecommunication networks could be resolved if some obvious guidelines for reporting results from simulation studies were adopted. First, the reported simulation experiments should be repeatable. This should mean that information about *the PRNG(s) used during the simulation*, and *the type of simulation*, is provided, either in a given publication or in a technical report cited in the publication. In the case of terminating simulation, its time horizon would need to be specified, of course. The next step would be to specify *the method of analysis of simulation output data*, and *the final statistical errors associated with the results*.

Negligence of proper statistical analysis of simulation output data cannot be justified by the fact that some stochastic simulation studies, in particular those aimed at evaluation of networks in their steady-state, might require sophisticated statistical techniques. On the other hand, it is true that in many cases of practical interest, appropriate statistical techniques have not been developed yet. But, if this is the case, then one should not pretend that he/she is conducting a precise quantitative study of performance of a telecommunication network. A more drastic solution of this credibility crisis in the area of computer simulation is to leave computer simulation to accredited specialists [17].

4   Final Comments

We discussed the basic issues related with credibility of simulation studies of telecommunication networks. Then, the results of a survey of recent research publications on performance evaluation of networks were used to show that the majority of results of simulation studies of telecommunication networks published in technical literature unfortunately cannot be classified as credible.

Of course, simulations of telecommunication networks are often computationally intensive and can require long runs in order to obtain results with an acceptably small statistical error. Speeding up of execution of simulation of telecommunication networks is a challenging research area which has attracted a considerable scientific interest and effort.

One direction of research activities in this area has been focused on developing methods for concurrent execution of loosely-coupled parts of large simulation models on multi-processor computers, or on multiple computers of a network. Sophisticated techniques have been proposed to solve this and related problems. In addition to efficiently managing the execution of large partitioned simulation models, this approach can also offer reasonable speedup of simulation, provided that a given simulation model is sufficiently decomposable.

In the context of stochastic simulation, there is yet another (additional) solution possible for speeding up such simulation. Namely, collecting of output data for sequential analysis can be sped up if the data are produced in parallel, by multiple simulation engines running statistically identical simulation processes. This approach to distributed stochastic simulation, known as Multiple Replications In Parallel (MRIP), has been implemented in Akaroa2 [14], a simulation controller that is offered as a freeware for teaching and non-profit research activities at universities; see www.cosc.canterbury.ac.nz/~krys.

Acknowledgements

The author would like to thank his colleagues from University of Canterbury, New Zealand: Ruth Lee and Joshua Jeong for their involvement in the cited survey of literature, as well as Don McNickle and Greg Ewing, for their contributions in research activities that led to this paper.

References

[1] Law, A. M., McComas, M. G.: Secrets of Successful Simulation Studies. Proc. 1991 Winter Simulation Conf., IEEE Press, 1991, 21-27.

[2] Law, A. M., Kelton, W. D.: Simulation Modelling and Analysis. McGraw-Hill, New York (1991).

[3] Kiviat, P. J.: Simulation, Technology and the Decision Process. ACM Trans. on Modeling and Computer Simulation, 1, no.2, 1991, 89-98.

[4] Pawlikowski, K., Jeong, H.-D. J., Lee, J.-S. R.: On Credibility of Simulation Studies of Telecommunication Networks. IEEE Comms., Jan. 2002, 132-139.

[5] Compagner, A.: Operational Conditions for Random-Number Generation. Phys. Review E., 52, 1995, 5634-5645.

[6] L'Ecuyer P.: Software for Uniform Random Number Generation: Distinguishing the Good and the Bad. Proc. 2001 Winter Simulation Conf., IEEE Press, 2001, 95-105.

[7] Matsumoto, M., Nishimura, T.: Mersenne Twister: a 623-Dimensionally Equidistributed Uniform Pseudo-Random Number Generator. ACM Trans. on Modeling and Computer Simulation, 8, no.1, 1998, 3-30.

[8] Knuth, D. E.: Art of Programming, Volume 2: Seminumerical Algorithms. Addison-Wesley, Reading (1998).

[9] Kleijnen, J. P. C.: The Role of Statistical Methodology in Simulation. In Zeigler B. P. et al. (eds.): Methodology in Systems Modelling and Simulation, North-Holland, Amsterdam (1979).

[10] Pawlikowski, K.: Steady-State Simulation of Queueing Processes: A Survey of Problems and Solutions. ACM Computing Surveys, 2, 1990, 23-170.

[11] Pawlikowski, K., McNickle, D., Ewing, E.: Coverage of Confidence Intervals in Sequential Steady-State Simulation. Simulation Practice and Theory, 6, no.2, 1998, 255-267.

[12] Raatikainen, K. E. E.: Sequential Procedure for Simultaneous Estimation of Several Percentiles. Trans. Society for Computer Simulation, 1, 1990, 21-44.

[13] Sarkar, N. I.: Probabilistic Scheduling Strategies for Unidirectional Bus Networks. M.Sc. thesis, Computer Science, University of Canterbury, Christchurch, New Zealand, 1996.

[14] Ewing, G., K. Pawlikowski and D. McNickle: Akaroa2: Exploiting Network Computing by Distributed Stochastic Simulation. Proc. 1999 European Simulation Multiconf., ISCS, 1999, 175-181.

[15] Popper, K. R.: The Logic of Scientific Discovery. Hutchison, London (1972).

[16] Gaither, B. Empty Empiricism. ACM Performance Evaluation Review. 18, 1990, no.2, 2-3.

[17] Balci, O, Ormsby, W. F.: Planning for Verification, Validation and Accreditation of Modeling and Simulation Applications. Proc. 2000 Winter Simulation Conf., IEEE Press, 2000, 829-839.

Satellite Link Layer Performance Using Two Copy SR-ARQ and Its Impact on TCP Traffic

Jing Zhu and Sumit Roy

Department of Electrical Engineering, University of Washington
Box 352500, Seattle, WA 98195, USA
{zhuj,roy}@ee.washington.edu

Abstract. The paper focuses on improving performance of land mobile satellite channels (LMSC) at higher frequencies such as K or EHF band, where shadowing is the primary impediment to reliable data transmission. Compared with short-term multipath fading, shadowing is characterized by longer time constants so that interleaving is not desirable as it introduces unacceptably large delays. To combat error bursts, an adaptive two-copy SR-ARQ scheme is proposed that uses a suitable *delay* between every retransmission. Closed-form solutions for metric of interest:*mean transmission time*, *success probability*, and *residual loss probability*are derived and validated by simulation. An optimal choice of the *delay* is determined and the performance of TCP traffic over such a link layer is evaluated by simulation and compared to normal SR-ARQ in terms end-to-end throughput.

1 Introduction

Currently, there is a great deal of interest in extending satellite communications to higher bands (K or EHF band) in order to achieve more transmission bandwidth. In [3] it was shown that the primary impediment to the land mobile satellite channel at K or EHF bands is shadowing due to blockage rather than multipath fading. In such cases, the channel can be represented by a two-state Markov process as in [1]. In bad (shadowed) states, the average SNR is too low to correctly transmit signals even with powerful forward error correction (FEC) codes while in good (unshadowed) states, the large value of Rice factor corresponding to the line-of-sight component guarantees reliable signal transmission even without much FEC protection. As is well known, interleaving is widely used with FEC to resist fading and improve the reliability of a wireless channel with burst errors. However, with increasing average length of error bursts, the interleaving depth needed may lead to unacceptable end-to-end delays. Therefore, a multiple copy (re)transmission scheme was proposed in [2], that inserts a suitable delay between copies of the transmitted packet. This paper extends the idea to a satellite channel with shadowing by modifying how the number of copies is varied with retransmission number in the interest of stability. For the two-copy

H.-K. Kahng (Ed.): ICOIN 2003, LNCS 2662, pp. 909–917, 2003.

case, simple expressions for the metrics of interest (i.e. mean transmission time, transmission success probability, and residual loss probability) are obtained.

Such adaptive transmission schemes [4] have naturally been considered since most wireless satellite channels such as LMSC are time-varying. In such methods, the coding rate, packet length and retransmission mode parameters etc. can be varied to match the transmitter for current channel conditions. Nevertheless, their performance depends critically on the efficiency and accuracy of channel state estimation (CSE) at the receiver that is fed back to the transmitter. Obviously,the long propagation delay of a satellite link implies that all variations less than one round trip time cannot be tracked. However, when average shadowing periods typically exceed a round-trip time, a long-term estimate of the average length of shadowing periods may be used to determine the optimal *delay* of our proposal.

The paper is organized as follows. In the next section, expressions for success probability for each transmission, mean transmission time, and residual loss probability are derived for delayed two-copy (DTC) SR-ARQ. Section 3 contains numerical results to quantify the improvements of the proposal. The impact of this improved LL design on end-to-end TCP throughput is assessed by comparing with normal SR-ARQ as well under the assumption of the same allowable copy number. Finally, we conclude the paper in Section 4.

2   Delayed Two-Copy (DTC) SR-ARQ

The protocol employs the basic selective repeat (SR-ARQ) strategy, except that two identical copies of a packet with a delay D is sent at each attempt (note that an attempt consists of transmissions or retransmissions of the packets). Only when *both* copies in an attempt are lost, a negative acknowledgment is produced and retransmission occurs. Compared with a normal 1-copy scheme, the equivalent code rate is thus 0.5.

The parameters used in the subsequent analysis are as follows:

X: Good-state time share parameter;
Peg: Packet error rate in good states;
Peb: Packet error rate in bad states;
m: The Mean length of bad states;
RTT: Round trip time of satellite channel;
Bw: Bandwidth of a satellite channel.

The usual alternating two-state Markov model is assumed to represent the channel state evolution in time; the duration of a 'bad' or shadowed state (i.e. the error burst length) is exponentially distributed with mean m. In the 'bad' or shadowed state, it is reasonable to assume $Peb \approx 1$, implying that no successful transmission is possible during shadowing. Further, $Peg << Peb$ and for analytical purposes, may be assumed equal to zero, i.e. all packet transmissions during 'good' or un-shadowed state are successfully received. This leads to the important simplification that the sequence of packet success and failures is a Markov process[1].

First we consider transmission of any two successive packets with a general delay (d) in between. We define an attempt to imply transmission of two copies (of the same packet) as per our scheme. We now identify two important situations: (i) the success/failure of the second copy in an attempt is related to that of the first copy in the same attempt; (ii) the success/failure of the first copy in current attempt is related to success/failure of the second copy in the previous attempt. Hence there are two relevant values of d for consideration - D and RTT.

Since the packet error rate is negligible in good states, packet losses take place in only bad states. If the previous copy is lost, the probability of correctly receiving the current one can be written as:

$$P\{S|F\} = P\{B_E, G\}P\{S_G\} = P\{B_E\}P\{G|B_E\}P\{S_G\}, \tag{1}$$

where the notations represent the following events:

S: Success on the current copy;
F: Failure on the previous copy;
B_E: The bad state during transmission of the previous copy is completed before transmission of the current copy;
G: The current copy is transmitted in a good state;
S_G: Success of the current copy in a good state.

Since the duration of bad state is exponentially distributed with the mean of m and the inter-duration between two copies is d, we have

$$P\{B_E\} = 1 - e^{-\frac{d}{m}}. \tag{2}$$

and clearly,

$$P\{S_G\} = 1 - Peg. \tag{3}$$

An exact expression for $P\{G|B_E\}$ is difficult and therefore an intuitive approximation is given next. We know that $P\{G|B_E\}$ is a function of d. Let's consider two extreme cases:

For $d \rightarrow 0$, i.e., two copies are sent next to each other with no delay. Since the end of the bad state for the previous transmission is followed by a good state, the current copy will be sent in the good state with probability 1, leading to $P\{G|B_E\} \rightarrow 1$. (Here we ignore the event that the good state ends before one LL packet duration because the good state mean duration is significantly longer than a LL packet duration).

For $d \rightarrow \infty$, the correlation between the channel states for the two copies vanishes, leading to $P\{G|B_E\} \rightarrow P\{G\}$, where $P\{G\} = X$.

---

[1] In general, the sequence of packet success and failures is a *Hidden* Markov process, and not strictly a Markov process.

We define the above two extreme cases as two mutually exclusive events: the channel state for the current copy is *completely correlated (CC)* or *completely un-correlated (CU)* to that for the previous one. The probability of *CC* is given by the correlation function of $e^{-\frac{d}{m}}$ $(= \begin{cases} 1, d = 0 \\ 0, d = \infty \end{cases})$, and the probability of *CU* is $1 - e^{-\frac{d}{m}}$. The desired result for $P\{G|B_E\}$ is given by the statistical average of the above, i.e.,

$$P\{G|B_E\} \approx 1 \cdot e^{-\frac{d}{m}} + X \cdot (1 - e^{-\frac{d}{m}}). \tag{4}$$

In conclusion, the probability of correctly receiving the current copy given that the previous is lost is expressed as

$$P_{Se}(d) = (1 - e^{-\frac{d}{m}})(e^{-\frac{d}{m}} + X(1 - e^{-\frac{d}{m}}))(1 - Peg), \; d = \{D, RTT\}. \tag{5}$$

We next consider the success probability for transmission and retransmission, denoted by P_{St} and P_{Sr} respectively. There are following events involved:

S_1: Success on the first copy;
F_1: Failure on the first copy;
S_2: Success on the second copy;
F_2: Failure on the second copy;
F_o: Failure on all previous attempts.

Obviously,

$$P\{S_2|F_1\} = P_{Se}(D), \tag{6}$$

and the state for the current copy only depends on that for the previous copy, leading to

$$P\{S_1|F_o\} = P_{Se}(RTT) \tag{7}$$

The first copy of the transmission has probability of $X(1-Peg)$ of being correctly received. Therefore,

$$\begin{aligned} P_{St} &= P\{S_1\} + P\{F_1\}P\{S_2|F_1\} \\ &= X(1 - Peg) + (1 - X(1 - Peg))P_{Se}(D). \end{aligned} \tag{8}$$

Then

$$\begin{aligned} P_{Sr} &= P\{S_1|F_o\} + P\{F_1|F_o\}P\{S_2|F_1, F_o\} \\ &= P\{S_1|F_o\} + P\{F_1|F_o\}P\{S_2|F_1\} \\ &= P_{Se}(RTT) + (1 - P_{Se}(RTT))P_{Se}(D). \end{aligned} \tag{9}$$

If the maximum number of attempts allowed is N, the packet loss probability $P_l^{(N)}$ after N attempts is

$$P_l^{(N)} = (1 - P_{Sr})^{N-1}(1 - P_{St}) \tag{10}$$

Denote by T the time from the first transmission to receipt of the acknowledgement. We have the following results of probability distribution function of T:

1)Correctly receiving the second copy:

$$\mathbf{P}[T = i(RTT + D)] = \begin{cases} (1 - X(1 - Peg))P_{Se}(D) & i = 1 \\ (1 - P_{St})(1 - P_{Sr})^{i-2}(1 - P_{Se}(RTT))P_{Se}(D) & i \geq 2 \end{cases}$$
(11)

2)Correctly receiving the first copy:

$$\mathbf{P}[T = iRTT + (i-1)D] = \begin{cases} X(1 - Peg) & i = 1 \\ (1 - P_{St})(1 - P_{Sr})^{i-2}P_{Se}(RTT) & i \geq 2 \end{cases}$$
(12)

Thus, the mean value of T is given below

$$E[T] = \frac{1}{(1 - P_l^{(N)})} \sum_{i=1}^{N} \{(iRTT + (i-1)D)\mathbf{P}[T = iRTT + (i-1)D] +$$
$$i(RTT + D)\mathbf{P}[T = i(RTT + D)]\}$$
(13)

As $N \to \infty$, $P_l^{(N)} \to 0$ and we get a closed-form solution for (13) as follows:

$$lim_{N \to \infty} E[T] = \frac{(1 - P_{St})}{P_{Sr}}[RTT + D + RTTP_{Sr} + D(1 - P_{Se}(RTT))P_{Se}(D)]$$
$$+ X(1 - Peg)RTT + (1 - X(1 - Peg))P_{Se}(D)(RTT + D)(14)$$

The analytical expressions of Eq.10 and Eq.14 are most useful as they provide performance estimate for applications with special QoS requirements. Of course, the mean error burst length must be obtained a-priori in practice using a suitable long-term channel estimator. In the following, we will assume that m is known.

3   Numerical Results

In all simulations and analysis reported here, a link layer packet duration is chosen as the unit of time. Wireless channel bandwidth (Bw) is fixed at 1Mbps.

3.1   Performance Comparison of DTC-SR-ARQ to Interleaving

In this section, we consider an error burst with fixed length m (packets) followed by a sufficiently long error-free period. The metric of interest is *additional delay*, defined as the extra delay introduced by the method employed to resist fading (DTC-SR-ARQ or Interleaving).

For DTC-SR-ARQ, the minimum delay inserted between two copies of a packet for a successful transmission is $m - 1$. The additional delay of the first copy is zero while that of the second one is m. Since either of them are equally likely to be transmitted during an error burst, the average additional delay introduced by DTC-SR-ARQ is $\frac{m}{2}$.

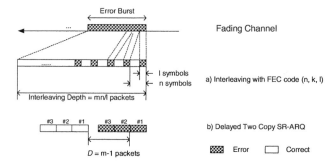

Fig. 1. Comparison of DTC-SR-ARQ to Interleaving in terms of Additional Delay

For interleaving with RS(n, k, l), the minimum interleaving length for error free transmission is $\frac{n}{l}m$ (packets), where n is the length of codeword, k is the number of information symbols in a codeword, and l is the maximum number of correctable symbols in a codeword. De-interleaving starts at receiver only after receiving all $\frac{n}{l}m$ packets, thus the additional delay is $\frac{n}{l}m$ for the first packet, and zero for the last one. The consequent average additional delay introduced by interleaving is $\frac{n}{2l}m$. Fig.2 demonstrates that the additional delay as a function of m is shorter for DTC-SR-ARQ than interleaving, with difference increasing for longer error bursts.

3.2   Delay Optimization of DTC-SR-ARQ

Fig.3 shows the success probability of the first transmission as a function of the delay D - our analytical results match simulations quite well. From Fig.3, we

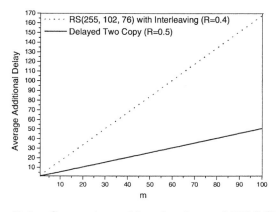

Fig. 2. Delay Comparison of Interleaving and DTC-SR-ARQ

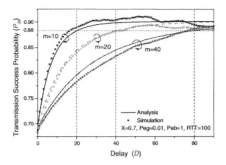

Fig. 3. Success Probability of Transmission

also see that longer error bursts need longer delay to achieve the same success probability.

Fixing the maximum number of attempts at 3, we study the residual loss probability after retransmission in Fig.4a); the residual loss probability is dramatically reduced by increasing *delay*. Fig.4b) investigates the mean transmission time T and indicates that there exists an optimal value of delay yielding the minimum mean transmission time.

The optimal delay for achieving the minimum average transmission time, using the first-order necessary conditions, is given by $\frac{dT}{dD} = 0$. However, it is tedious to explicitly solve. Figs.1-3 show that $D = 2m$ is a good pragmatic choice considering P_{Ts}, $P_l^{(N)}$, and T. Therefore, in our following simulation on TCP performance, we will use $D = 2m$.

Fig.5 studies the link layer performance of DTC-SR-ARQ with $D = 2m$ in terms of residual packet loss probability and mean transmission time. Analytical results indicate that longer average burst error length leads to higher residual packet loss probability and longer mean transmission time. We also compare these results with normal SR-ARQ for the same maximum copy number (MCN)

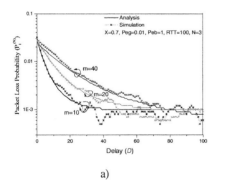

a)

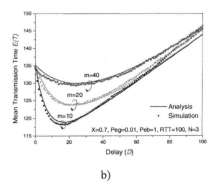

b)

Fig. 4. Packet Loss Probability a) and Mean Delay b) after N attempts ($N = 3$)

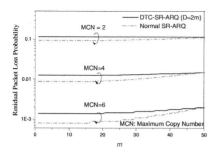
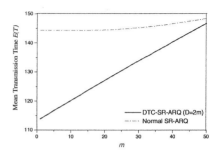

a) Residual Packet Loss Probability After N Attempts   b) Mean Transmission Time ($N= \infty$)

Fig. 5. Performance for $D = 2m$ ($RTT = 100, X = 0.7, Peg = 0.01$)

so that the maximum transmission number is MCN (say 4) for SR-ARQ and MCN/2 (say 2) for DTC-SR-ARQ. It is seen that by using DTC-SR-ARQ the mean transmission time is significantly reduced at the expense of a small increase in residual packet loss probability. Furthermore, shorter the average length of error bursts, the more the improvement in mean transmission time. In addition, by using DTC-SR-ARQ instead of normal SR-ARQ, we can reduce maximum transmission time from $MCN \times RTT$ to $MCN \times \frac{(RTT+D)}{2}$. If $m << RTT$, we have $RTT >> D$ because of $D = 2m$, leading to almost 50 % reduction in maximum transmission time.

3.3 On TCP Performance

In this section, we study the performance of TCP over two-copy delayed SR-ARQ. The delay is bounded by half the RTT and set as $D = min(2m, \frac{RTT}{2})$. Assuming a fixed maximum number of retransmission attempts(say 8), the maximum transmission time for normal is 7 and 3 RTTs respectively for regular SR-ARQ and our two-copy delayed SR-ARQ. Fig.6 shows that the TCP end-to-end throughput is improved by using our scheme, especially when the average error burst length is short. When the error burst length increases, the performance improvement using our proposal is reduced. In other words, the two-copy delayed SR-ARQ is more suitable for the fast shadow fading channel with much shorter error burst length compared to the round trip time.

4   Conclusion

In this paper, we proposed an optimized two-copy delayed SR-ARQ scheme for the shadowed satellite channel in the K or EHF bands. Shadowing leads to longer fade durations compared with multipath fading; consequently multiple copy transmission with a delay was suggested in place of interleaving to combat burst errors. Analytical results showed that success probability of each transmission is significantly improved, and mean transmission time is reduced as well.

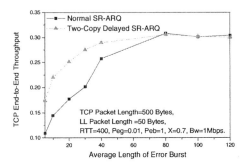

Fig. 6. TCP End-to-End Throughput Comparison (Buffer Size=20000 bytes)

Simulations performed to compare our scheme with normal SR-ARQ in terms of TCP end-to-end throughput indicate that our proposal achieves noticeable performance improvement especially for the fast shadowing channels with error burst longer than a link layer packet but shorter than one round trip time.

References

[1] E. Lutz, D. Cygen, M. Dippold, F. Dolainsky, and W. Papke, "The Land Mobile Satellite Communication Channel - Recording, Statistics, and Channel Model, IEEE Trans. on Vehicular Tech." , Vol . 40, No. 2, May 1991, pp.375-386.

[2] J. Zhu, Z. Niu, Y. Wu, "A delayed multiple copy retransmission scheme for data communication in wireless networks" ,Proc. Ninth IEEE International Conference on Networks, 2001 pp. 310 -315.

[3] J. B. Schodorf, "EHF Satellite Communications on The Move: Baseband Considerations," MIT Lincoln Lab Technical Report 1055, Feb. 2000.

[4] A. Annamalai, V. K. Bhargava, "Analysis and Optimization of Adaptive Multicopy Transmission ARQ Protocols for Time-Varying Channels," IEEE Trans. on Commu., vol. 46, no. 10, pp. 1356-1368

Survivable Network Design
Using Restricted P-Cycle

Mi-Sun Ryu and Hong-Shik Park

Information and Communications University
658-4, Hwaam-Dong, Yuseong-gu, Daejeon,305-732, Korea
{rms0,hspark}@icu.ac.kr

Abstract. We propose a new method to provide network survivability using the Restricted P-Cycle by Hamiltonian cycle (RPC), which is the improved version of p-cycle. Because p-cycle has some problem files, it is not applied to dynamic traffic or varying QoS in real time circumstance. However, RPC can significantly reduce complexity in finding proper patterns restricted by Hamiltonian cycle, which has minimal spare links to provide protection. Moreover, it can guarantee the almost same restoration performance as that of p-cycle.

1 Introduction

P-cycle provides network survivability with the new method characterized by ring-like protection switching speed because only two nodes perform any real-time actions and mesh-like capacity efficiency. P-cycle based on closed cyclic protection paths against a span failure has three to six times greater demand-carrying capacity than rings for a given transmission capacity, because p-cycle can recover not only an on-cycle span failure but also a straddling span failure [3]. Although p-cycle has good restoration performance as stated above, it has complicated complexity of solving optimal p-cycle [5]. Especially among the processes of p-cycle, the work of searching all possible ring-patterns embedded in the network is a time-consuming task. Even for the moderate node number as 10, for example, the number of all available ring patterns reaches 556014. Checking all such cycles would be nearly prohibitive. So it is impractical to apply to the network in service, that adapt to changing call demands. We propose a new procedure of searching adequate patterns using Restricted P-Cycle by Hamiltonian cycle (RPC) and compare the restoration performance and complexity with the existing method using p-cycle and find the most effective condition which gives the better network performance when adapting this procedure.

2 Restricted P-Cycle (RPC)

P-cycle has the object to find the minimal cost set of pre-configured cycle patterns, which support recovery of all possible single span failure. Despite the

H.-K. Kahng (Ed.): ICOIN 2003, LNCS 2662, pp. 918–927, 2003.

similarity to rings in use of a cycle-topology, p-cycle not only emulates ring-behavior for a certain class of failure, but also goes beyond the functionality of rings, to protect the network as a whole, as a mesh recovery scheme. Moreover, the design of a min-cost set of p-cycle to protect a given set of working flows is an NP-hard problem. Determining patterns in p-cycle cannot be calculated at once. So it is separately carries out two steps, (c) and (d), to find optimal p-cycle patterns represented in Fig.1 (A). However, this procedure also has some problems. Step (c) generates large problem files that have all available cycle pattern information embedded in the network because of the size of the set of candidate cycles to consider. Step (d) has also a difficulty to solve optimality because of having large files handled.

So we propose a new method having a simple process beyond compared to p-cycle. This procedure integrates (c) with (d) and does not need making large files for searching all possible cycle patterns embedded in the network. It has the object to search the min cost pattern shaped only as the Hamiltonian cycle. A Hamiltonian cycle, also called N-hop p-cycle, is by definition a closed cycle that traverses all the nodes in the network exactly once. In the full mesh network it can protect up to $N(N-2)$ units of working capacity, which is almost $(N-2)$ times more efficient than a corresponding ring. Namely, the Hamiltonian cycle of the working network requires the minimal spare links to provide protection against a single link failure when working and protection links are routed together [6]. Moreover, Hamiltonian cycle embedded in the network is found

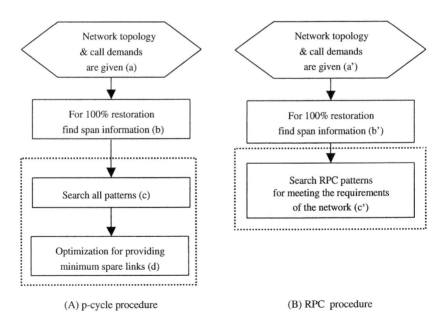

(A) p-cycle procedure   (B) RPC procedure

Fig. 1. Procedure of p-cycle (A) and RPC (B)

by a simple LP formulation. We call this procedure the Restricted P-Cycle by Hamiltonian cycle (RPC).

The step (a') and (b') in Fig.1 (A) are equal to the step (a') and (b') in Fig.1 (B). In the step (a') and (b') in Fig.1, the corresponding call demand is shortest-path routed over the topology and generates the working link quantities on each span, w_{ij}. And an optimal allocation of spare capacity, s_{ij}, is made for each span using to following IP formulation for span-restorable spare capacity [1][2].

N : The number of network nodes

L : The number of network spans

X_i : The number of working channels through span i.

$f_{i,p}$: The restoration flow though the p-th restoration route of span i

Z_j : The maximum restoration flow through span j following a single span failure

$\delta^i_{j,p}$: Value "1" if the p-th restoration route of span i uses the span j and otherwise "0"

P_i : The total number of eligible restoration routes following the failure of span i.

$$Minimize \sum_{j}^{L} Z_j$$

s.t.

$$\sum_{p=1}^{P_i} f_{i,p} \geq X_i \quad i = 1, ..., L$$

$$Z_j - \sum_{p=1}^{P_i} \delta^j_{i,p} \cdot f_{i,p} \geq 0 \quad i, j = 1, ..., L, \quad i \neq j$$

$$f_{i,p} \geq 0, \quad i = 1, 2...L, \quad p = 1, 2...P_i$$

RPC in Fig. 1(B) can integrate two complicated processes ((c) and (d)) into one simple process and do not have to generate the large problem files to get the optimized solution. The step (c') in the RPC merges the step (c) with (d) in p-cycle. So it can reduce complexity in finding min-cost Hamiltonian cycles to maximally recover a given set of working demands. In the step (c') pre-configured Hamiltonian cycle is generated heuristically, and then Hamiltonian cycles are searched for the spare links using the newly obtained minimum span contribution weights in a network's spare capacity. The span contribution weight, c_{ij}, is determined by running the k-shortest path algorithm for the failure of all network's spans and by counting the number of times that each span contributes for the restoration of other spans. The c_{ij} can be changed at any time according to dynamic call demands or QoS. However, the existing p-cycle determines adaptive patterns for static call demands and the given topology because determination of patterns is a very complicated process. So p-cycle is not suitable for dynamic call demands and varying QoS. On the other hand, in RPC the network can alter patterns dynamically by changing c_{ij} according to call demands or varying QoS. In the following IP formulation, we can get patterns,which are constrained by the Hamiltonian cycle. It uses the minimizing method (using CPLEX 8.0) to search. The reason for using the minimum contribution weight is that when a working path that is not on the cycle span fails, it has two protection paths, on the other hand, when a working path that is on cycle span fails, it has only one restoration path. Namely the minimum contribution weight can be used to search the maximum restorable Hamiltonian cycle. Finding a minimum-weight Hamiltonian cycle in the resulting network is certainly NP-complete. So, a pre-configuration design, constrained by the Hamiltonian cycle, is provided by the following IP formulation.

N : The number of network nodes.

0 : Node number 0.

c_{ij} : The contribution weight at the span that connects node i with node j .

f_{ij} : The load at the span that connects node i with node j .

x_{ij} : Connectivity at the span that connects node i with node j .

$$x_{ij} \in \{0, 1\}$$

$$Minimize \sum_i \sum_j c_{ij} x_{ij}$$

s.t.

$$\sum_i x_{ij} = 1 \ \forall j$$
$$\sum_i x_{ji} = 1 \ \forall j$$
$$\sum_i f_{ij} - \sum_j f_{ij} = 1 \ \forall j \ (except \ j \neq 0)$$
$$\sum_j f_{0j} = n$$
$$\sum_i f_{i0} = 1 \ \forall i$$
$$f_{ij} \leq (n - 1) \cdot x_{ij} \ \forall i, \ \forall j \ (except \ i \neq 0, j \neq 0)$$
$$f_{0j} = n \cdot x_{0j} \ \forall j$$
$$f_{i0} = x_{i0} \ \forall i$$

This formulation can obtain independent patterns because it is an iterative process. It does not search all available Hamiltonian cycles or all ring patterns in prior but finds proper patterns one by one using the above IP. So RPC can eliminate the complicate and difficult process used in p-cycle.

3 Comparison Complexity and Restorability

3.1 Complexity

There are two complicate steps in p-cycle. One is to find all available patterns, and the other is to determine optimal patterns using the database having available pattern information for 100restorability. The former has many possible patterns, the number of which increases geometrically as the number of nodes in the network increases. The latter is not so simple because it must be done based on the patterns searched at the prior step. RPC does not need information for all possible patterns and need not to search all available patterns. Instead, it only finds optimal patterns one by one using the previous IP formulation. Moreover, the optimization step for minimum spare links is not required. In order to compare complexity in two methods, we evaluate complexity using the number of applicable patterns. The upper limit to the number of patterns can be found as follows: assume a complete graph, which has N nodes, $_NC_2$ edges and $(N-1)$ regular degree [4].

In p-cycle case, the number of patterns to be found is:

$$\# \ of \ patterns \ (p-cycle) = \frac{1}{2} \sum_{s=3}^{n} \binom{n}{s} (s-1)!.$$

In RPC case, the number of possible Hamiltonian cycles is:

$$\# \ of \ patterns \ (HamiltonianCycle) = \frac{(n-1)!}{2}.$$

Table1 shows the number of ring patterns applicable to p-cycle and RPC. In Table1 *Ratio* is defined as *(the number of Hamiltonian cycles) / (the number of all possible ring patterns in p-cycle).*

As the number of nodes increases, the number of patterns related to two methods increases very sharply. But the number of patterns in p-cycle is about three times larger than that in RPC. Besides, the RPC need not find all possible Hamiltonian cycles. This makes RPC much simpler than p-cycle.

Table 1. The number of patterns on p-cycle and RPC

number of nodes	3	5	7	9	11	13
RPC	1	12	360	20160	1.81e+6	2.39e+8
p-cycle	1	37	1172	62814	5.48e+6	7.11e+8
Ratio(%)	100	32.43	30.72	32.09	33.10	33.61

3.2   Restorability

To evaluate the restoration performance of RPC we used two of the test networks as in [5]. One is net1 having 10 nodes and 22 spans and the other net2 having 15 nodes and 28 spans. Fig. 2 compares the restoration ratio according to spare ratio with p-cycle in the networks. The spare ratio $S = (used\ link)/(optimal\ spare\ link)$, where used link is the number of spare links which use for protection and optimal spare link is the number of optimal spare capacity allocation designed by the first IP formulation in prior. The restoration ratio $R = 1$-$(uncovered\ working\ demand)/(total\ working\ demand\ in\ all\ spans)$. In the Fig. 2, it is shown that the restoration efficiency of RPC and p-cycle are closely similar. Especially the performance of RPC is better than p-cycle when the spare ratio is lower part because Hamiltonian cycle has better performance than any other size cycles. As the number of applied spare links in the network becomes more and more within the optimal spare links per span, performance of RPC gets worse than that of p-cycle because in p-cycle case has various k ($3 \le k \le N$)-hop cycle ring patterns for optimal solution which is as many as recovery of working links within the spare links and RPC does not find more than Hamiltonian cycle. If RPC and p-cycle cannot search more than pattern, they carry out a dynamic recovery with extra spare links using KSP for protection extra working demands.

RPC has worse restorability than p-cycle. Because the patterns of RPC are found independently with one another and those of p-cycle are found dependently, p-cycle can utilize various k ($3 \le k \le N$)-hop ring patterns to obtain optimal solution. The difference between RPC with the N-hop ring shaped patterns and p-cycle with optimal solution naturally exists if there remain a few spare links in the network. P-cycle can, therefore, provide better performance but the gap is not so much as seen in Fig.2. If RPC has a few more spare links

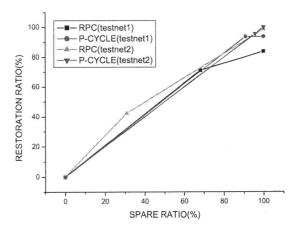

Fig. 2.   Restorability of p-cycle and RPC in the networks

in contrast to p-cycle, it can provide 100% recovery against all single span failures. The reason is that RPC applies Hamiltonian cycles, which use the minimal number of spare links to provide network protection. Though RPC has a little worse performance than p-cycle, RPC has the distinctive merit that it reduces complexity and can be applied to real-time circumstances.

For viewing the restoration performance according to the number of nodes in the networks, we used the existing networks (net1 having 10 nodes, NJ LATA having 20 nodes, NSFNET having 14 nodes, net2 having 15 nodes and net3 having 20 nodes) as larger network and for small networks we considered 2 types which have average 3.5 node degrees: one type is the centralized network where a hub switch is located at the center network and the other is the distributed network which has a backbone network and multiple access networks.

First, we can compare the RPC' restoration performance for the large networks in Fig. 3. As N grows larger, it is very different to search patterns, Hamiltonian cycle, having good performance because it is not easy to balance spare links on every span in the network. As the network size increases, the restoration ratio by only RPC pattern normally decreases. From Fig.3 we can especially see that N11(NJ LATA Net) does not find Hamiltonian cycle, so called non-Hamiltonian graph, so it executes the dynamic recovery using KSP against a single span failure. If non-Hamiltonian graph has several edge nodes with low degree, it does not embed Hamiltonian cycle and it leads to slow recovery about all working demands. After recovery by patterns shaped Hamiltonian cycle is executed and the uncovered using KSP, it is shown that restorability of all networks is very similar. But restorability supported by Hamiltonian cycle decreases as the network size increases. We can see that restorability by only Hamiltonian cycles decreases as the network size increases.

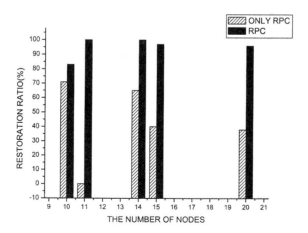

Fig. 3. RPC performance according to the network size

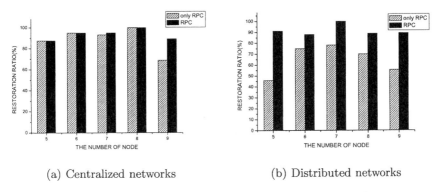

(a) Centralized networks   (b) Distributed networks

Fig. 4. RPC performance according to network topology

Second, we can compare the RPC' restoration performance for the small networks. The results for the centralized network and the distributed network are given in Fig. 4. The distributed network case has less restoration ratio than the centralized network case. They occur because spare links are not balanced well in all spans. Especially in the distributed network, it is very hard that spare links are balanced. Therefore, it is viewed as worse restorability.

As the network size is less than 8, the restorability by RPC patterns is good in the distributed networks. And in the centralized networks, as the network size is less than 9, the restorability by RPC patterns is good. The small network has normally good performance as the network size is less than around 8.

Fig. 5 shows that restorability of RPC by only RPC patterns according to the network size. We can see if N is larger, the restorability of only RPC sharply decreases and as the number of node in the network is less than 9, the restorability of only RPC is viewed better performance. *Pattern utilization* can be calculated from the number of restoration paths per link used in the pattern. *Pattern utilization = (straddling links used in patterns*2+ in-cycle links used in patterns) / (links used in pattern).* As *pattern utilization* is influenced by the node degree, we assume that the network is full mesh. *Pattern utilization* increase as the number of nodes in the network increases. From Fig. 5, we know that there is a trade-off relation between the spare ratio for 100% restoration and pattern utilization. Namely smaller networks have better performance than larger networks when RPC is applied to the network where *pattern utilization* is not so important. If we consider both performances together, we can determine the adequate N. We can divide that the existing large network into small networks using this N to improve restorability by RPC patterns and for RPC to adapt non-Hamiltonian graph network as using this simulation results.

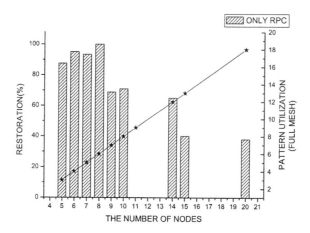

Fig. 5. Only RPC restoration ratio vs. pattern utilization according to number of network

4   Conclusion

Network survivability is a crucial requirement in the high-speed network. The existing p-cycle has the object to provide the speed like a ring and efficiency like a mesh but is not suitable for the dynamic network. However, RPC integrates search and optimization steps into a single step and applies the Hamiltonian cycle, which requires the minimal spare links to protect a single span failure. Especially RPC requires less complexity compared to p-cycle so it can handle dynamically varying traffic demands and support various QoS requirements in real time. It can also support almost same restoration performance as that of p-cycle. It is a very strong constraint that applied patterns must have the shape of the cycle. Due to this, RPC has some additional redundant spare links. Especially as the network size increases, RPC requires more redundancy. However, as the network grows larger, the pattern utilization increases greatly. So, we can find out the network size, which is good solution in terms of the restoration ratio and the required spare links. If we divide the large network using this solution network size, we can expect that the performances of the large network could be improved as the small networks are.

Acknowledgement

This work was supported in part by the KOrea Science and Engineering Foundation (KOSEF) through OIRC project.

References

[1] M. Herzberg and S. Bye : An optimal spare-capacity assignment model for survivable networks with hop limits, Proc. IEEE GLOBECOM '94 (1994) 1601–1607

[2] M. H. MacGregor and W. D. Grover: Optimized k-shortest-paths Algorithm for Facility Restoration, Software - Practice and Experience, Vol. 24, No. 9, 823–834

[3] D. Stamatelakis: Theory and Algorithms for Pre-configuration of Spare Capacity in Mesh Restorable Networks, M. Sc. Thesis, University of Alberta, Spring (1997)

[4] P. Mateti and N. Deo: On algorithms for enumerating all circuits of a graph, SIAM J. Comput., Vol. 5, No. 1, Mar. (1976) 90–99

[5] W. D. Grover and D. Stamatelakis: Cycle-Oriented Distributed Preconfiguration: Ring-like Speed with Mesh-like Capacity for Self-planning Network Restoration, Proc. IEEE ICC 1998, Atlanta, GA (1998) 537–543

[6] Hong Huan and Copeland, J.: Hamiltonian cycle protection: a novel approach to mesh WDM optical network protection, High Performance Switching and Routing, 2001 IEEE Workshop (2001) 31 –35

Taming Large Classifiers
with Rule Reference Locality

Hyogon Kim[1], Jaesung Heo[1], Lynn Choi[1], Inhye Kang[2], and Sunil Kim[3]

[1] Korea University
[2] University of Seoul
[3] Hongik University

Abstract. An important aspect of packet classification problem on which little light has been shed so far is the rule reference dynamics. In this paper, we argue that for any given classifier, there is likely a significant skew in the rule reference pattern. We term such phenomenon rule reference locality, which we believe stems from biased traffic pattern and/or the existence of "super-rules" that cover a large subset of the rule hyperspace. Based on the observation, we propose an adaptive classification approach that dynamically accommodates the skewed and possibly time-varying reference pattern. It is not a new classification method per se, but it can effectively enhance existing packet classification schemes, especially for large classifiers. As an instance, we present a new classification method called segmented RFC with dynamic rule base reconfiguration (SRFC+DR). When driven by several large real-life packet traces, it yields a several-fold speedup for 5-field 100K-rule classification as compared with another scalable method ABV. In general, we believe exploiting the rule reference locality is a key to scaling to a very large number of rules in future packet classifiers.

1 Introduction

Classifying incoming packets based on header fields is the first and the most basic step for many networking functions such as Diff-Serv traffic conditioning, firewall, virtual private network (VPN), traffic accounting and billing, load-balancing, and policy-based routing. These functions need to track flows, i.e., equivalence classes of packets, and give the same treatment to the packets in a flow. Since a flow is defined by the header values, a classifier's duty is to examine the values of packet header fields, then identify the corresponding flow. The rules that prescribe flow definitions and the associated actions reside in the rule base (a.k.a. "policy base" or "filter table") located in memory. Lying on the per-packet processing path, the classification function should minimize the accesses to memory in order to maximize packet throughput. Of particular concern is thus when the rule base becomes large and the search complexity increases. Although no such large classifiers exist as of today, rule bases of up to 100K [2] or even 1 million [1, 10] entries are believed to be of practical interest.

Recently, there has been active research on fast and/or scalable classification techniques [1, 2, 4, 5, 6, 7, 8, 9, 10]. With the exception of [9, 10], these works

H.-K. Kahng (Ed.): ICOIN 2003, LNCS 2662, pp. 928–937, 2003.
© Springer-Verlag Berlin Heidelberg 2003

focus on relatively small classifiers, e.g., with less than 20K entries. A notable property also shared by most of these works is that the dynamic aspect of rule reference behavior is not considered. For some, the lack of information on reference pattern hardly causes a problem as long as classifiers are small. For instance, the time complexity of RFC [1] algorithm is constant no matter what the rule base size is, given that table pre-computation succeeds within the given memory budget. For others, such as computational geometry formulations [5, 8, 9], it is inherently difficult to accommodate any dynamics in reference (or "point location") pattern.

As we deal with increasingly large classifiers, however, the rule reference dynamics can play a vital role in designing a scalable methodology. We firmly believe that a realistic assumption for any given classifier is that some rules are more popular than others, receiving more references. Moreover, we argue that we can even exploit it to improve the performance of large classifiers. In fact, it is the departing point and the contribution of this paper. A novel aspect of our paper is that we go as far as reconfiguring the rule base itself, to adapt the classifier to whatever rule reference pattern is. Most classifier implementations already keep a usage counter for each rule for statistics collection purpose [11]. By dynamically reordering the rule base with respect to such usage counts, we can exploit the notion of reference skew and yield good average case performance so that we can scale to a very large number of rules. One constraint in rule movement is that rules should retain the relative ordering relations or "dependency", as before the reconfiguration. However, this constraint does not seem to prohibit relatively free movements of rules required in our approach because a survey of numerous real-life rule bases [1] shows that rule intersection is very rare. We will elaborate on these issues in the rest of the paper.

Note we do not advocate a new data structure or a new search algorithm for purpose of accelerating the rule search. Rather, we propose to dynamically reflect rule reference pattern on rule base structure so that search methods scale well with the number of rules. Also, rule base optimization by eliminating redundant rules is not the focus of this paper. Our approach works for any given rule base, whether it abounds in redundancy or not. Finally, in this paper, we focus on the problem of scalable classification for the applications in which most packets get classified using non-default rules, such as in firewall.

This paper is organized as follows. In Section 2, we discuss a rule base reconfiguration method based on our approach. We also present a practical classification method where our approach is used. Section 3 shows experiment results. We demonstrate that our approach can easily reap several hundred percent speedup over existing scalable methods. The paper concludes with Section 4.

2 Dynamic Rule Base Reconfiguration

There are a few causes for a skewed reference pattern in a rule base. First, the most frequently appearing IP protocol numbers at any observing locale are almost always TCP, UDP, and ICMP, although depending on locale GRE, PIM, and others can exceed ICMP [3]. Second, probably the most direct cause for a skewed reference pattern is that within TCP and UDP, there are dominantly popular port numbers. For TCP, for example, 80 (HTTP) is an unquestionable winner. Third, the IP address fields also show non-uniform reference patterns. An evidence of non-uniform pattern at network edge has been first noted in [4]. Last but not least, there are rules with shorter address prefixes and wider port number ranges. These "super-rules" occupy a large subset of the rule base hyperspace, and have potential to capture more packets than others. Below, we design a method to exploit the skewed reference pattern.

2.1   Rule Dependency

Let the depth of a rule r be defined to be the number of rules preceding r when the rule base is sorted in the decreasing order of priority (let us assume smaller number denotes higher priority). Therefore, in this paper we will equate the depth with the priority, which the system operator can arbitrarily assign so as to meet the needs of the network. Before discussing the rule dependency, let us first define a few other terms:

Definition 1. *We say that a rule $r_1 = (< F_i^1 >, A_1)$ overlaps with a rule $r_2 = (< F_i^2 >, A_2)$ if $\forall_i F_i^1 \cap F_i^2 \neq 0$, where F_i^k and A_k is the ith element of the flow definition and the associated action of rule k, respectively.*

Intuitively, two rules overlap if there is any instance of packet header values that matches both. In our flow definition of

$$< proto, srcIPprefix, srcportrange, dstIPprefix, dstportrange >$$

all 5 fields must have non-null intersection for two rules to overlap. Usually, a mask is associated with the IP address prefix [2], where the IP address intersection is determined as follows:

min_mask ← mask1 & mask2
intersection? ((IP_address1 & min_mask) = (IP_address2 & min_mask))

where '&' is the bit-wise AND operator.

Definition 2. *The strict ordering is a constraint under which two overlapping rules r_1 and r_2 such that $priority(r_1) > priority(r_2)$ are prohibited from exchanging locations if as a consequence it becomes $priority(r_1) < priority(r_2)$.*

Definition 3. *The loose ordering is a constraint under which two overlapping rules r_1 and r_2 such that $priority(r_1) > priority(r_2)$ and $A_1 \neq A_2$ are prohibited from exchanging locations if as a consequence it becomes $priority(r_1) < priority(r_2)$.*

The loose ordering constraint can make rule relocations much easier than the strict ordering. In particular, a highly popular rule with a wildcard in one of its fields can promote past a less popular overlapping rule [3]. For instance, given

$r_1 = (< TCP, *, *, 123.45.6.0/255.255.255.0, 80 >, accept)$
$r_2 = (< TCP, 123.45.10.0/255.255.255.0, *, 123.145.6.0/255.255.255.0, 80 >$
$, accept)$
$priority(r_1) < priority(r_2)$

r_1 cannot promote beyond r_2 under the strict ordering constraint. On the other hand, we can freely change the priority relation (i.e., locations) under the loose ordering constraint. This may be important to gather only truly popular rules at the top locations of the rule base, thereby achieving a high "cache hit ratio". Since it is not very meaningful to assign actions to rules in the synthetic rule base, we enforce only the strict ordering rule in this paper. Rule base reconfiguration using the loose ordering constraint is a subject of future work. A related and important observation of [1] is that in practice the rule overlap is very rare. As we push up a rule, it implies, there are not many rules that have to be moved up piggybacked even if we enforce the strict ordering instead of the loose ordering.

Once the overlapping rules are known, we can compute the set of dependent rules $D(r)$ for a rule r as those rules satisfying: $\forall_{r' \in ID(r)} r'$ overlaps with r and $priority(r') > priority(r)$. When we promote r, each rule in $D(r)$ should be also promoted, and must retain a higher priority than r after the move. Then the complexity C of a single rule relocation in terms of the number of comparison operations is bounded as follows:

$$O(Fd) = F(d-1) \leq C \leq F \sum_{k=1}^{d-1} k = Fd(d-1)/2 = O(Fd^2)$$

where the depth of the rule is d, and F is the dimension. The worst-case occurs when all $(d-1)$ rules above the relocated rule overlaps with it, and the best case is when there is no overlap. The average number of comparisons will be determined by the probability of overlapping for a given pair of rules. It is very difficult to compute, but an important aspect is that once the rule dependency is computed, we can easily perform the subsequent relocations by reusing the already computed dependency relations. Excluding the additions of new rules, the comparison operations above need to be performed only for the first "epoch". Even if there are additions, the complexity of incrementally updating dependency relations is $O(N)$ for each addition. When we attempt a rule relocation, we retrieve its dependency relations in $O(1)$ operations (because $D(r)$ is small) and move all dependents along with it. In addition, each epoch, the entire rule base must first be sorted with respect to the reference count, and the complexity is $O(N log N)$ where N is the rule base size. An algorithm for moving up a frequently referenced rule is described below.

2.2   Reconfiguration Algorithm

The reconfiguration algorithm is done in 2 phases. In the first phase, we sort the rule base with respect to the reference counts recorded in the previous epoch. Then we attempt to move the top $TOPHITS$ rules up. If the dependency relations are not available, we find the overlapping rules as we promote the popular rule and combine them in a group and attempt to move up the group in entirety. Otherwise, we omit the dependency computation. Since the traffic pattern can change over time, we need to run the reconfiguration algorithm every so often. Fig. 1 shows the reconfiguration algorithm for the first epoch when the dependency relation is not available. *Sorted_rule_table* is an array of size N, where each element has two pieces of information: rule ID (rule) and reference count (*ref_count*). As a result of *sort*(), the array is sorted in the decreasing order of the reference count. Starting from the most popular rule, we promote *TOPHITS* rules one by one. For each promotion of a rule r located at dth position, we examine $d-1$ higher priority rules. If some higher priority rule r' has a higher reference count, r is placed just below r' and the promotion stops there. In case there is an overlapping rule, it is combined with r in the same promotion group. Each rule in the promotion group needs to be checked for overlapping with all higher priority rules. *Highest_rule*() returns the top rule in the block, while *lowest_rule*() returns the bottom rule in the block. The bottom rule is the one that originally we wanted to promote. Since overlapping rules are stacked up on the popular rule, the lowest rule is the popular rule. Note that even if there are more than one rules in a promotion group, reference count comparison is done only against the popular rule.

```
procedure reconfigure() {
/* sort the rule base by reference counts in the last epoch */
    sorted_rule_table[] = sort(RULE_BASE);
    for (i=0; i < TOPHITS; i++)
                        /* relocate top TOPHITS rules to the top */
        push_up({sorted_rule_table[i].rule});
                        /* move rule i as far up as possible */
}

procedure push_up(R) { /* R is a~set of overlapping rules */
    for (i=depth(highest_rule(R))-1; i>=0; i--) {
        r = lowest_rule(R); /* r is the rule we want to promote */
        if (RULE_BASE[depth(i)].ref_cnt > r.ref_cnt) {
            insert(R, i); /* insert R below ith rule */
            return; /* stop */
        }
        for each r in R {
            if (overlap?(r, RULE_BASE[depth(i)])) {
                /* merge */
                push_up(combine(R, RULE\_BASE[depth(i)]));
                return;
}}}}
```

Fig. 1. An example of the rule base reconfiguration algorithm

3   Performance Evaluation

In this section, we apply our approach to a well-known classification method, RFC [1]. Since RFC cannot handle the rule base size larger than 15,000, we assume that such rule bases are segmented into smaller pieces. We refer to the segmented version of RFC as SRFC, and SRFC+DR is the version augmented with the proposed dynamic rule base reconfiguration. In this section, we evaluate the general effect of dynamic rule base reconfiguration, and then compare the performance of SRFC, SRFC+DR, and ABV. Due to the lack of large, publicly available classifiers we can obtain, we synthesize them. Then we pass several large real-life packet trace through the synthesized rule bases, to measure the packet throughput with SRFC, SRFC+DR, and ABV. We first discuss how we steer the random rule base syntheses so as to maintain generality and fairness. We utilize several large real-life packet traces in rule base syntheses and in the evaluation of the three schemes.

3.1   Rule Base Synthesis

It is difficult to extract general characteristics of classifier rule bases. Rule base configuration is at the discretion of the system administrator, and it can take any structure according to the needs of a particular network. For fairness and generality, therefore, we opt to randomly generate test rule bases and repeat the performance evaluation for different rule bases a few times. This is the approach that many prior works take [1, 2, 9, 10]. We also make use of the observations made about real-life classifiers [1, 2] in order to generate rule bases as close to reality as possible. We have the following guidelines in rule generation that accommodates some of the key observations in [1, 2].

1. No rules can appear that do not match any packet in the test traffic trace.
2. In the IP address fields, wildcard can appear with a certain fixed probability. Generally, in the IP address fields, the prefix of length $0 \leq l(j) \leq 32$ can appear with a probability $P(l(j))$. The ABV paper [2] reports that in the industrial firewalls they used, most prefixes have either a length of 0 or 32, and that there are some prefixes with lengths of 21, 23, 24 and 30. In particular, the destination and source prefix fields in roughly half the rules are wildcarded [2]. We follow these guidelines when fixing $P(l(j))$.
3. The transport layer protocol field is restricted to a small set of values: TCP, UDP, ICMP, IGMP, (E)IGRP, GRE, IPINIP, and wildcard [1]. The IP protocol number is randomly chosen from a pool of these protocols with a probability associated with each.
4. The transport port number(s) for a given rule can be arbitrary. The RFC paper [1] reports roughly 10% of rules have range specifications, and in particular "> 1023" occurs in 9% of the rules. For simplicity, we assume that well-known port numbers are used as singleton (i.e., no range), and the range specifications are generated only with ephemeral port numbers with 10% probability.

Table 1. Traces used in the rule generation

Trace	Date	Duration	Total no. of Packets
T1	July 23-24, 11:01pm-4:01am	5 hrs.	159.3 million
T2	Dec. 14, 9:35am - 12:13pm	2hrs. 38 min.	211.9 million
T3	Dec. 17, 9:06am - 10:31am	1hr. 25 min.	109.6 million

5. There is one all-wildcard rule that matches all packets, and this rule is located at the bottom of the rule base. (In firewalls, for instance, this is the "all-deny" rule).

Table 1 shows the size and the duration of each packet trace used in the rule generation process. experiments.

3.2 Performance Comparison of SRFC, SRFC+DR, and ABV

The primary metric that we use in this paper is the number of memory accesses. Before discussing this per-packet processing overhead, let us briefly consider the additional memory and table computation overhead for SRFC+DR. In terms of memory usage, there is no additional overhead compared with SRFC [1], except for the temporary memory required for rule sorting and relocation. As far as the relocation is concerned, we can apply some optimization in case rule movement is limited. Namely, we can re-compute only the partitions affected by the movement. By maintaining a small free space in each partition for incoming rules, we can avoid causing cascade movement of rules and subsequent re-computation [12]. As for table computation time, the known sorting complexity is $O(N log N)$, and the reconfiguration complexity in terms of comparison operation is $O(nFd^2)$ and $O(nFd)$ without and with dependency information, respectively as we discussed in Section 2.1. Note n is the number of rules to promote, which can be configured to be small in case the reference skew is severe.

Figure 2 compares the average number of memory accesses for SRFC, SRFC+DR and ABV. For our 5-field classification, SRFC needs 13 memory accesses for each segmented table lookup. Although there is one more field (i.e. source port) compared with [1], the number of tables to lookup is the same [12]. The numbers for SRFC+DR contain the distances before the first reconfiguration as well as after. For 64- and 128-bit ABV, we used 2-level ABV hierarchy (i.e. BV-ABV-AABV). For 32-bit ABV, we used 3-level ABV hierarchy (i.e. BV-ABV-AABV-AAABV). For SRFC and SRFC+DR, $X(Y)$ denotes the number of memory accesses X and the number of rule base partitions Y. For t_1, the total number of rules is 100K, so $Y = 10$ means that each rule base partition contains 10K rules. For ABV and ABV+DR, $X(Y)$ denotes the number of memory accesses X and the memory bus width Y. We experimented with three most probable widths: $Y = 32$, 64, and 128.

To our great surprise, ABV shows the worst performance among the three experimented schemes. In particular, even SRFC in the 10K-rule segmented

Table 2. Average number of memory accesses per classification for t_1 with 100K-rule bases

Rule Set	SRFC	SRFC+DR	ABV
r0	71.617 (10)	18.225 (10)	150.015 (32)
	136.760 (20)	24.194 (20)	142.469 (64)
			115.323 (128)
r1	51.004 (10)	16.907 (10)	161.574 (32)
	101.848 (20)	21.337 (20)	143.018 (64)
			122.780 (128)
r2	71.751 (10)	18.786 (10)	202.554 (32)
	136.402 (20)	25.176 (20)	171.376 (64)
			136.968 (128)
r3	72.057 (10)	18.696 (10)	154.578 (32)
	134.974 (20)	24.889 (20)	140.767 (64)
			115.525 (128)
r4	59.718 (10)	17.625 (10)	169.281 (32)
	112.458 (20)	22.896 (20)	145.234 (64)
			121.301 (128)
Avg.	65.229 (10)	18.048 (10)	167.600 (32)
	124.488 (20)	23.698 (20)	148.573 (64)
			122.379 (128)

configuration outperforms ABV. In the 5K-rule segmentation that is much safer in terms of memory usage, SRFC records comparable results (2 better, 1 almost equal, 2 worse) with the best ABV configuration with 128-bit memory bus width. An examination of 50K-rule cases in Fig. 11-2 reveals even larger performance gap. Both in 5K- and 10K-rule segmentation SRFC performs better than ABV. This seems to demonstrate that to a certain rule base size, simply by partitioning the rule base and thereby avoiding possible memory explosion [1], RFC can outperform ABV.

Whether the segmented variant of RFC (SRFC) will retain its performance edge over ABV is not certain when the rule bases scale beyond 100K rules. Since SRFC is location-dependent (unlike RFC itself), the number of memory accesses required for a (randomly located) matching rule linearly scales with the rule base size. In contrast, ABV performance is not so much location-dependent thanks to its rule sorting algorithm [2], so the performance gap between SRFC and ABV might close with larger rule base size. In fact, the ratio of the best numbers of each scheme with growing rule base size seems to agree with our expectation.

ABV(50K, 128) / SRFC(50K, 5) = 2.37
ABV(75K, 128) / SRFC(75K, 8) = 2.01
ABV(100K, 128) / SRFC(100K, 10) = 1.88

The scalability concern for SRFC largely disappears with SRFC+DR. This is because the rule reference pattern under the rule base reconfiguration is no longer arbitrary. Rather, SRFC+DR continually condenses the most popular rules to the top partitions, so SRFC+DR is much less location-dependent than SRFC. In fact, above best performance ratio increases with the number of rules with SRFC+DR, as we can see below:

ABV(50K, 128) / SRFC+DR(50K, 5) = 5.566
ABV(75K, 128) / SRFC+DR(75K, 8) = 6.578
ABV(100K, 128) / SRFC+DR(100K, 10) = 6.780

Therefore, SRFC+DR maintains or even strengthens its lead over ABV as the rule base size increases. The performance ratio exceeds 550rule base size larger than 50K. As for SRFC, it is slower than SRFC+DR at least by a factor of 2.5. The narrowest performance gap is for 5-partition 50K-rule bases, and the gap widens as the number of rules increases. For 5-partition 100K-rule bases, the speedup is in excess of 3.5. The gap widens even more as the number of partitions doubles, due to SRFC's scalability limitation, and the speedup is over 5. SRFC memory accesses almost linearly increase with the number of partitions whereas SRFC+DR exhibits much slower growth. Finally, we observe that ABV improves with larger memory bandwidth. However, the improvement is linear rather than exponential, with exponentially growing memory bandwidth.

4   Conclusion

This paper illuminates the value of a new aspect of packet classification problem, i.e., rule reference dynamics. Circumstantial evidences point to the possibility that any classifier will exhibit (highly) skewed rule reference pattern. Based on this premise, we proceed to design a classification scheme that captures the pattern of the subject packet stream and reconstitutes its own rule base. With several large rule bases synthesized from real-life packet traces, we demonstrate that this approach indeed yields a few hundred percent speedup over other scalable schemes, such as ABV. We firmly believe that exploiting the rule reference locality is a key to scale to a very large number of rules in future packet classifiers. Finally, there are a few issues that still wait for further exploration. Event-triggered reconfiguration as opposed to periodic reconfiguration and relocation with the loose ordering constraint instead of the strict ordering constraint are expected to improve performance of our approach. We will employ the new techniques in our future work where we will attempt classification on 1 million rules and beyond. Finally, the current scheme fits better with such applications that classifies most packets with non-default ("bottom") rules.

References

[1] P. Gupta and N. McKeown, "Packet classification on multiple fields," ACM SICOMM 1999.
[2] F. Baboescu and G. Varghese, "Scalable packet classification," ACM SIGCOMM 2001.

[3] Flow analysis of passive measurement data,
    http://pma.nlanr.net/PMA/Datacube.html.
[4] M. Poletto et al., "Practical Approaches to Dealing with DDoS Attacks,"
    NANOG22, May 2001.
[5] T. V. Lakshman and D. Stiladis, "High-speed policy-based packet forwarding us-
    ing efficient multi-dimensional range matching," ACM SIGCOMM 1998, pp. 191-
    202.
[6] V. Srinivasan, S. Suri, G. Varghese, and M. Valdvogel, "Fast and scalable layer 4
    switching," ACM SIGCOMM 1998, pp. 203-214.
[7] V. Srinivasan, G. Varghese, and S. Suri, "Fast packet classification using table
    space search," ACM SIGCOMM 1999, pp. 135-146.
[8] M. M. Buddhikot, S. Suri, and M. Waldvogel, "Space decomposition techniques
    for fast layer-4 switching," Protocols for High Speed Networks, vol. 66, no. 6, pp.
    277-283, 1999.
[9] A. Feldmann and S. Muthukrishnan, "Tradeoffs for packet classification," IEEE
    Infocom 2000.
[10] T. Woo, "A modular approach to packet classification: algorithms and results,"
     IEEE Infocom 2000.
[11] P. Gupta and N. McKeown, "Packet classification using hierarchical intelligent
     cuttings," Hot Interconnects VII, 1999.
[12] H. Kim, "Exploiting reference skew in large classifiers for scalable classifier de-
     sign," techreport, 2001.
[13] K. Houle et al., "Trends in Denial of Service Attack Technology", CERT Coordi-
     nation Center, Oct. 2001. Available at
     http://www.cert.org/archive/pdf/DoS_trends.pdf.

Analytic Models of Loss Recovery
of TCP Reno with Packet Losses*

Beomjoon Kim and Jaiyong Lee

Department of Electric & Electronic Engineering, Yonsei University
134 Shinchon-dong Seodaemun-gu Seoul, Korea
{bjkim,jyl}@nasla.yonsei.ac.kr

Abstract. In this paper, we investigate the loss recovery behavior of TCP Reno over wireless links in the presence of non-congestion packet losses. We consider both random and correlated packet loss, and derive the conditions that packet loss can be recovered without retransmission timeout (RTO) by accurate modeling of loss recovery behavior of TCP Reno. Through probabilistic work with the conditions derived, we compute the fast retransmit probability for packet loss probability. According to our results, only 25% of two packet losses in a window can be recovered by two fast retransmits. In a particular case, three lost packets can be recovered by fast retransmits, but its probability is extremly low. Since more than four packet losses in a window can be recovered by fast retransmits in no cases, RTO always occurs. The continuity of correlated packet losses as well as packet loss rate can affect the fast retransmit probability. Even if overall packet loss probability is very low, successive packet losses can degrade the fast retransmit probability. We explain some of these observations in terms of the variation of the average window size with packet loss probability.

1   Introduction

Transmission Control Protocol (TCP) is widely used as the transport layer protocol in the Internet. Since it is designed on the assumption that it may be used on wireline networks where packet loss probability is negligibly low [1], TCP regards all packet losses as network congestion. The throughput of TCP, therefore, can suffer over wireless networks that are characterized by bursty and high channel error probability [1], [3].

Performance degradation of TCP over wireless links can be explained by unnecessary congestion control caused by non-congestion packet loss and frequent retransmission timeout (RTO). When a RTO takes place, especially, sender cannot only transmit data till its expiry but should start to transmit data in slow start. Hence, it takes long time for the congestion window to be restored to the size before RTO occurs. As a result, the higher frequency of RTO makes the performance of TCP worse [3], [6], [7].

---

* This work was supported by grant No.R01-2002-000-00531-0 from the interdisciplinary research program of the KOSEF.

H.-K. Kahng (Ed.): ICOIN 2003, LNCS 2662, pp. 938–947, 2003.

There have been several recent efforts to analyze TCP performance for non-congestion packet loss [3]–[11]. It has already shown that the performance of TCP Reno is dependent on the fast retransmit probability in [4], which provides the departure point of our work. In this paper, therefore, we concentrate on the loss recovery behavior of TCP Reno and describe the fast retransmit probability in terms of the number of lost packets in a window. We consider that the characteristic of packet loss is random (i.e., i.i.d.) and correlated (i.e., bursty). The difficulty to obtain the complete distribution of the congestion window makes us deploy the fact that the evolution of TCP congestion window is of a cyclical characteristic, which can be analyzed using Markov Chain as in [4], [6]. Thus, we can compute the stationary distribution of the window process numerically.

The remainder of the paper is organized as follows. In Section 2 we briefly describe the loss recovery of TCP Reno. In Section 3 and 4 we describe our model and derive the conditions for fast retransmit. We present the derived conditions probabilistically in Section 5. Section 6 contains the numerical results and their discussion. Finally, some conclusions are summarized in Section 7.

2   Loss Recovery of TCP Reno

In this section, we briefly illustrate the loss recovery operation of TCP Reno. Refer to [12]–[14] for details of TCP Reno.

There are two different ways for TCP Reno to recover a lost packet; one is by RTO and the other one is by fast retransmit and fast recovery. To trigger a fast retransmit, the sender should receive at least K duplicate acknowledgements (ACK) for a lost packet [1] (typically three duplicate ACK's).

At time t, we denote the sender's congestion window and the slow-start threshold, $ssthresh$, by $W(t)$ and $W_{th}(t)$. Suppose that a packet is lost and the sender receives the Kth duplicate ACK for the lost packet at $t = t_0$. The lost packet is retransmitted instantly by fast retransmit. After the fast retransmission, the sender sets

$$W(t_0^+) = \left\lfloor \frac{W(t_0)}{2} \right\rfloor + K, \ W_{th}(t_0^+) = \left\lfloor \frac{W(t_0)}{2} \right\rfloor. \tag{1}$$

During fast recovery, the sender increases the window by one every time another duplicate ACK arrives. If the increased window includes a new packet, the sender is allowed to transmit it. If the retransmitted packet is transmitted successfully, a normal (i.e., non-duplicate) ACK is generated. It makes the congestion window slide to the packet that is to be transmitted next and ends fast recovery. The sender continues to transmit a packet in congestion avoidance with the congestion window that is equal to W_{th} determined by (1).

When multiple packets are lost in a window, several fast retransmits and fast recoveries may be repeated. Suppose n lost packets in a window can be recovered

---

[1] We use the term "an ACK for packet B by packet A" to mean that the sender transmits packet A and the receiver delivers an ACK whose next expected packet number is B in order to inform the sender of receiving packet A.

by n fast retransmits. If t_1 is the time just before the first fast retransmit and the nth packet loss is fast retransmitted at t_2, the relation between $W(t_1)$ and $W(t_2)$ is given by

$$W(t_2) = \left\lfloor \frac{W(t_1)}{2^n} \right\rfloor. \tag{2}$$

3   Modeling

We model the loss recovery behavior of TCP in terms of 'rounds' defined in [6]. In order to obtain the stationary distribution of the window and derive the loss recovery probability, we follow the analysis of the Markov Chain in [4].

3.1   Assumptions and Definitions

The sender has infinite packets to send so that the congestion window is always fully incremented. All packets are assumed to have a same size. We do not consider the effect of *delayed acknowledgement* [14] so that the receiver delivers an ACK every time a good packet is received. As the size of an ACK packet is considerably small compared to a data packet, an ACK packet is assumed not to be lost. We denote the fast retransmit threshold by K and the maximum window size by W_{max}, which is advertised by the receiver at the connection setup time. Let l_i denote the ith packet loss in a window. If l_1 is the mth packet among the packets transmitted in the kth round, a loss window Ω is the congestion window when all normal ACK's by $(m-1)$ packets are received. The first packet that a loss window includes is always the first lost packet. For n packet losses in a loss window, we denote the number of packets that are not lost or transmitted newly in the kth round of loss recovery period by Φ_k; e.g., If Ω is equal to u packets, Φ_1 is always equal to $u - n$. For $h \geq 2$, Φ_h means the number of packets that are transmitted by the slide and the inflation of the usable window after the retransmission of the $(h-1)$th packet loss.

3.2   Derivation of Φ_n

For $\Omega = u$, Φ_1 equals to the number of packets transmitted well among u packets. Thus, we have

$$\Phi_1 = u - 1. \tag{3}$$

After the fast retransmit of l_1, Φ_2 new packets are included by the inflation of the usable window. Considering that the last duplicate ACK for l_1 makes window to be $\lfloor u/2 \rfloor + (u-2)$ and u packets are still outstanding, Φ_2 is given by

$$\Phi_2 = \lfloor u/2 \rfloor + (u-2) - u = \lfloor u/2 \rfloor - 2. \tag{4}$$

If $\Phi_2 \geq K$, l_2 can be recovered by fast retransmit. Fig. 1 shows the loss recovery behavior of TCP Reno for three packet losses in a window. Each round is explained as follows:

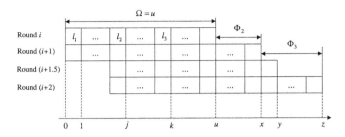

Fig. 1. Loss recovery behavior of TCP Reno for three packet losses

- Loss recovery starts at round i.
- The last duplicate ACK for l_1 is received at round $(i + 1)$.
- The first ACK for l_2 is received by retransmitted l_1 at round $(i + 1.5)$.
- The last duplicate ACK for l_2 is received at round $(i + 2)$.

When the initial loss window is denoted by $\Omega(i)$ and $R(a)$ denotes the right bound of a , each boundary value is set to have $L(\Omega(i)) = 0$, $R(\Omega(i)) = u$, $R(l_2) = j$, $R(l_3) = k$, $R(\Omega(i + 1)) = x$, $R(\Omega(i + 1.5)) = y$ and $R(\Omega(i + 2)) = z$. For every j, k that satisfies $2 \leq j \leq u - 1, j + 1 \leq k \leq u$, x, y and z are given by

$$
\begin{aligned}
x &= \lfloor u/2 \rfloor + (u - 3) \\
y &= (j - 1) + \lfloor u/2 \rfloor \\
z &= (j - 1) + \lfloor u/4 \rfloor + \Phi_2.
\end{aligned}
\tag{5}
$$

If $y \geq x$ at round $(i + 1.5)$, $y - x$ packets are transmitted. Since l_2 has not been retransmitted yet, they may generate duplicate ACK's for l_2. For two cases, Φ_3 is given by

$$
\begin{aligned}
\Phi_3 &= z - max(x, y) \\
&= \begin{cases} (j - 1) + \lfloor u/4 \rfloor - u & 2 \leq j \leq u - 2 \\ \lfloor u/4 \rfloor - 3 & j = u - 1. \end{cases}
\end{aligned}
\tag{6}
$$

Unlike Φ_1 and Φ_2, Φ_3 has relation with the position of l_2 as well as the size of Ω. When j is maximum, z has the maximum value by (5). It is because window slides most when $j = u - 1$. Therefore, the minimum value of Ω for recovery of three packet losses without RTO can be obtained by $\lfloor u/4 \rfloor - 3 = K$. For $2 \leq j \leq u - 2$, the recovery condition for l_3 is given by

$$
(j - 1) + \lfloor u/4 \rfloor - u \geq K.
\tag{7}
$$

It means that l_3 can be recovered by fast retransmit if the number of packets between l_1 and l_2 is equal to or greater than $u - \lfloor u/4 \rfloor + (K - 1)$.

Four packet losses in a window can be recovered by fast retransmit in no case. Refer to [11] for its proof.

4   Probabilistic Analysis

4.1   Packet Loss Models

For random packet loss, each packet is lost with probability p and losses are independent [3], [5]. Correlated packet loss is modeled using a first-order Markov Chain [7]–[9]. The transition probability matrix, Q_c, of the Markov Chain is given by

$$Q_c = \begin{pmatrix} p_{BB} & p_{BG} \\ p_{GB} & p_{GG} \end{pmatrix}. \tag{8}$$

The channel has two states that are the "Good" state and the "Bad" state. The packet is lost with probability 1 while in the bad state. If we set p_{GB} and p_{BG} to α and β respectively, the steady state probabilities that the channel is in the good state and the bad state, Π_G, Π_B are given by

$$\Pi_G = \frac{\beta}{\alpha + \beta}, \ \Pi_B = \frac{\alpha}{\alpha + \beta}. \tag{9}$$

Further, the average duration of the bad state is given by $1/\beta$ while the average duration of the good state is given by $1/\alpha$. The unit of the duration is the number of packets, since we assume that the state transits per every transmission time of a packet.

4.2   Markov Process

The window evolution of TCP can be analyzed by adopting a Markov Chain. The stationary distribution of Markov Chain $\{\Omega_i\}$ can be obtained numerically if the exact transition probabilities are determined. We follow the procedures in [3], [5] to get the transition probabilities except that the congestion window size after loss recovery does not always decrease to $\lfloor u/2 \rfloor$. The relation between the number of packet losses and the window size in the next cycle is defined by (2).

When RTO occurs, W_{th} of the next cycle can be inferred as follows. For $\Omega = u$, suppose two packets among packets are lost and RTO occurs. If $u \geq K + 2$, at least the first lost packet may be recovered by fast retransmit. In this case, we can sure that RTO is evoked by the second lost packet. Therefore, W_{th} in the next cycle is set to $\lfloor u/2 \rfloor$. Depending on the size of Ω and the number of lost packets, W_{th} in the next cycle may be one of $\lfloor u/2 \rfloor$, $\lfloor u/4 \rfloor$ and $\lfloor u/8 \rfloor$.

For the correlated packet loss model, it is impossible to know the state of the channel during RTO in which no packet can be transmitted. If the first packet in the next cycle is not lost, however, it is clear that the channel is in the good state. Though we do not know how many consecutive RTO's take place, the state of the channel when the next cycle starts is always good. For that reason, we consider the process $\{\Omega_i\}$ over the state space $\{2, 3, 4 \cdots, W_{max}\}$. Note that the number of successive RTO's does not affect the evolution of the congestion window process.

4.3   Fast Retransmit Probability

Let R_R be the fast retransmit probability of TCP Reno that every lost packets in a window may be recovered by fast retransmits. Then we have

$$R_R = \sum_n \sum_{u=1}^{W_{max}} R_R^{(n)}(u) \cdot \pi_n(u) \tag{10}$$

where $\pi_n(w)$ is the steady-state probability of loss window for n and $R_R^{(n)}(u)$ is given by

$$R_R^{(n)}(u) = P\{(n-1)\text{packets are lost out of}(u-1)\text{packets}\} \\ *P\{\text{no packet loss during loss recovery}\}. \tag{11}$$

For random packet loss, $R_R^{(n)}(u)$ in terms of n can be written as follows:

$$R_R^{(1)}(u) = (1-p)^u \\ R_R^{(2)}(u) = (u-1)p(1-p)^{u+\Phi_2} \\ R_R^{(3)}(u) = \binom{\lfloor u/4 \rfloor - K}{2}p^2(1-p)^{u+\Phi_2+\Phi_3}. \tag{12}$$

Finally, the total fast retransmit probability for random packet loss is given by

$$R_R(u) = (1-p)^u \left\{ 1 + (u-1)p(1-p)^{\Phi_2} + \binom{\lfloor u/4 \rfloor}{2}p^2(1-p)^{\Phi_2+\Phi_3} \right\}. \tag{13}$$

For the correlated packet loss model, the sequence of packets should be considered to calculate $R_R^{(n)}(u)$. For $n = 1$, the channel should transit to good state after a packet is lost and stay in the good state till the transmission of u packets that include the retransmitted packet are completed. For $n = 2$, let $R_R^{(2)}(u)_{su}$ be the probability when two lost packets are successive, and $R_R^{(2)}(u)_{ns}$ be the probability when two lost packets are not successive. Then, $R_R^{(2)}(u)$ is sum of $R_R^{(2)}(u)_{su}$ and $R_R^{(2)}(u)_{ns}$. For $n = 3$, l_2 and l_3 may be succesive or not for $j_{min} \leq j \leq u-2$ while they are always successive for $j = u-1$ by the given condition. Consequently, $R_R^{(n)}(u)$ for correlated packet loss is given by

$$R_R^{(1)}(u) = \beta(1-\alpha)^{\Phi_1} \\ R_R^{(2)}(u) = \beta(1-\alpha)^{u+\Phi_2-3}\{(1-\alpha)(1-\beta) + (u-2)\alpha\beta\} \\ R_R^{(3)}(u) = \begin{cases} \alpha\beta^2(1-\beta)(1-\alpha)^{\Phi_2+\Phi_3-2} & j = u-1 \\ \alpha\beta^2(1-\alpha)^{\Phi_2+\Phi_3-3}\{N_{su}(1-\alpha) + N_{ns}(1-\beta)\} & j_{min} \leq j \leq u-2 \end{cases} \tag{14}$$

where N_{su} is the number of cases that l_2 and l_3 are succesive and N_{ns} is the number of cases that they are not. The values of N_{su} and N_{ns} are determined as follows:

$$N_{su} = u - (j_{min} - 1) - 2 \\ N_{ns} = \binom{u-(j_{min}-1)}{2} - (N_{su} - 1). \tag{15}$$

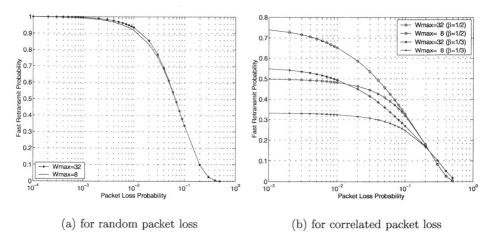

(a) for random packet loss   (b) for correlated packet loss

Fig. 2. Comparison of fast retransmit probability ($W_{max} = 8$, 32)

5   Numerical Results

In this section, we plot the fast retransmit probability defined by (10). In order to investigate the impact of correlated packet loss, we compare the fast retransmit probabilities when the characteristic of packet loss is random and correlated with $\beta = 1/2$ and $1/3$. The x-axis of each graph indicates packet loss probability, which corresponds to p for random packet loss and to Π_B for correlated packet loss. We set K to have the value of three in all cases.

Fig. 2 shows the fast retransmit probability when packet losses are random and correlated. In order to examine the effect of W_{max}, we compare the fast recovery probability for $W_{max} = 8$, 32. For random packet loss, increment of W_{max} makes no significant differences to the fast recovery probability. It is because the number of lost packets in a window for random packet loss is one in most cases. Recalling the conditions for fast retransmit of TCP Reno, two or more lost packets cannot be recovered when $W_{max} = 8$; i.e., $R_R = R_R^{(1)}$ in this case. On the other hand, two or three lost packets may be recovered when $W_{max} = 32$. Therefore, the difference in the fast recovery probability between $W_{max} = 8$ and $W_{max} = 32$ is equal to the sum of $R_R^{(2)}$ and $R_R^{(3)}$. As can be seen in fig. 3-(a), these values are too small to affect the fast recovery probability. Therefore, $R_R^{(1)}$ governs overall fast recovery probability regardless of W_{max}. A lost packet in a window can be recovered by fast retransmit if only the congestion window is equal to or larger than four. That is, it matters only whether the congestion window is larger than four or not. As packet loss probability increases, two or more packets may be lost in a window. At this time, however, window becomes so small that it is almost impossible to recover these lost packets by fast retransmits. For correlated packet loss, on the other hand, W_{max} affects

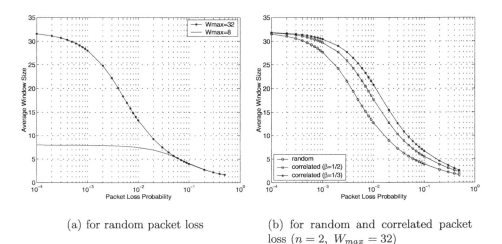

(a) for random packet loss

(b) for random and correlated packet loss ($n = 2$, $W_{max} = 32$)

Fig. 3. Comparison of average window size

the fast retransmit probability. The number of lost packets in a window tends to increase for correlated packet loss. Therefore, the weight of $R_R^{(2)}$ and $R_R^{(3)}$ on R_R grows up. In order that $R_R^{(2)}(u)$ and $R_R^{(3)}(u)$ may have a nonzero value, the conditions that are $\lfloor u/2 \rfloor - 2 \geq K$ and $\lfloor u/4 \rfloor - 3 \geq K$ should be satisfied. As we assign K to three, the corresponding values to the above inequalities are 10 and 24. Since $R_R^{(2)}(u)$ and $R_R^{(3)}(u)$ are always zero when for $W_{max} = 8$, the fast retransmit probability in this case is lower than the fast retransmit probability for $W_{max} = 32$. For a given value of W_{max}, the smaller β degrades R_R more. Note that more than three consecutive packet losses can be recovered by fast retransmit in no case.

Fig. 3-(a) shows the average window size for one packet loss when packet losses are random, $\widetilde{W}_1$, for $W_{max} = 8$, 32. Average window size for n, $\widetilde{W}_n$, is given by

$$\widetilde{W}_n = \sum_{w=1}^{W_{max}} w \cdot \pi_n(w). \tag{16}$$

For packet loss probabilities exceeding $5 \cdot 10^{-2}$, two graphs are almost overlapped. This fact makes it clear that W_{max} has little effect on fast recovery probability when packet losses are random.

Fig 4 shows $R_R^{(2)}$ and $R_R^{(3)}$ for $W_{max} = 32$. The numerical results in these figures are qualitatively similar. The reason for this similarity can be found in the average window size like $R_R^{(1)}$ for random packet loss. When packet loss probability reach 10^{-2}, the average window size values of two case meet together where the value is about 13. For packet loss probability over 10^{-2}, two graphs are almost overlapped (The figure for this case is not included). Note that the critical

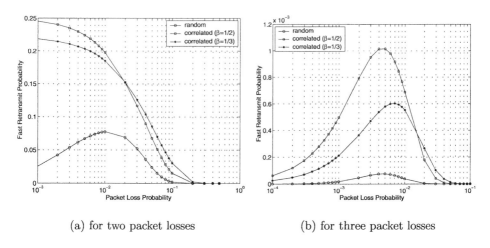

(a) for two packet losses   (b) for three packet losses

Fig. 4. Comparison of fast retransmit probability for two and three packet losses

value of the window size for recovery of two packet losses is 10. It is observed that $R_R^{(2)}$ for random packet loss increases for the packet loss probability below 10^{-2}. It is because that the frequency of the event that two packets are lost in a window is very low. For $\beta = 1/2$, $R_R^{(2)}$ is highest of the three cases except for the packet loss probability above $2 \cdot 10^{-2}$. When the packet loss probability is higher than $2 \cdot 10^{-2}$, $R_R^{(2)}$ is affected mainly by the stationary distribution of the window rather than $R_R^{(2)}(u)$. As can be seen in fig. 3-(b), the average window size shows lower values for the shorter burst of packet losses. For a given value of the packet loss probability, grouped packet losses correspond to fewer loss recovery events. Therefore, the window decreases less frequently than for scattered packet losses. As can be seen in fig. 4-(b), $R_R^{(3)}$ is conspicuously low compared to $R_R^{(2)}$ and $R_R^{(1)}$. It reflects that the condition for three lost packets to be recovered by fast retransmits is very strict; for it is related with the position of the second lost packet as well as the window size.

6   Conclusions

In this paper, we have analyzed the loss recovery mechanism of TCP Reno over a link where packet losses are not always caused by congestion. The model in our work presents the accurate loss recovery behavior of TCP Reno for the number of packet losses in a window. Based on this model, the fast retransmit feature of TCP Reno can be quantified.

We have shown that three packet losses in a window can be recovered by fast retransmits under the limited situations and four packet losses always evoke RTO. While the fast recovery algorithm inflates the congestion window in additive, every packet loss makes it decrease in multiplicative. Therefore, the loss recovery mechanism of TCP Reno does not work well when multiple packets are lost in a window. It is the main reason for the poor performance of TCP Reno over a link where multiple packet losses may exist. The maximum size of the advertised window has little effect on the fast retransmit probability, since there is no chance for a congestion window to increase close to it.

Even if the average packet loss probability is very low, the burst of the successive packet losses degrades the fast retransmit probability. It is important to alleviate the continuity of the packet losses as much as to drop the overall packet loss probability.

References

[1] H. Balakrishnan, V. N. Padmanabhan, S. Seshan, and R. H. Katz: A Comparison of Mechanisms for Improving TCP Performance over Wireless Links. IEEE/ACM Transactions on Networking, Vol. 5. (1997) 756–769

[2] K. Fall and S. Floyd: Simulation-based Comparisons of Tahoe, Reno, and SACK TCP. ACM Computer Communication Review, Vol. 26. (1996) 5–21

[3] T. V. Lakshman and Upamanyu Madhow: The Performance of TCP/IP for Networks with High Bandwidth-Delay Products and Random Loss. IEEE/ACM Transactions on Networking, Vol. 5. (1997) 336–350

[4] Anurag Kumar: Comparative Performance Analysis of Versions of TCP in a Local Network with a Lossy Link. IEEE/ACM Transactions on Networking, Vol. 6. (1998) 485–498

[5] Anurag Kumar and Jack Holtzman: Comparative Performance Analysis of Versions of TCP in a Local Network with a Mobile Radio Link.
Available: http://ece.iisc.ernet.in/ anurag/

[6] J. Padhye, V. Firoiu, D. F. Towsley, and J. F. Kurose: Modeling TCP Reno Performance: A Simple Model and Its Empirical Validation. IEEE/ACM Transactions on Networking, Vol. 8. (2000) 133–145

[7] Michele Zorzi and A. Chockalingam: Throughput Analysis of TCP on Channels with Memory. IEEE Journals on Selected Areas in Communications, Vol. 18. (2000) 1289–1300

[8] Alhussein A. Abouzeid, S. Roy, and M. Azizoglu: Stochastic Modeling of TCP over Lossy Links. in Proc. IEEE INFOCOM'2000, 1724–1733

[9] Farooq Anjum and Leandros Tassiulas: On the Behavior of Different TCP Algorithms over a Wireless Channel with Correlated Packet Losses. in Proc. ACM SIGMETRICS'99, 155–165

[10] P. P. Mishira, D. Sanghi, and S. K. Tripathi: TCP Flow Control in Lossy Networks: Analysis and Enhancements. in Proc. Computer Networks, Architecture and Applications, IFIP Transactions C-13, S. V. Raghavan, G. V. Bochman, and G. Pujolle, Eds. Amsterdam, The Netherlands: Elsevier North-Holland, (1993) 181–193

[11] B. J. Kim and J. Y. Lee: Analytic Models of Loss Recovery of TCP Reno with Packet Losses. in Proc. ICOIN'2003.

[12] V. Jacobson: Modified TCP Congestion Avoidance Algorithm. note sent to end2end-interest mailing list, (1990)

[13] W. Stevens: TCP Slow Start, Congestion Avoidance, Fast Retransmit, and Fast Recovery Algorithms. RFC2001, (1997)

[14] W. Stevens: TCP/IP Illustrated, Vol. 1 The Protocols. Addison-Wesley, (1997)

Impact of FIFO Aggregation
on Delay Performance
of a Differentiated Services Network

Yuming Jiang and Qi Yao

Institute for Communications Research
#2-34/37 TeleTech Park, Singapore Science Park II, 117674 Singapore
ymjiang@ieee.org
yaoqi@icr.a-star.edu.sg

Abstract. We investigate the impact of First-In-First-Out (FIFO) flow aggregation on delay performance of a Differentiated Services (DiffServ) network in the context of Expedited Forwarding (EF). Specifically, the focus is on investigating the impact of FIFO aggregation on edge-to-edge delay performance. Motivated by the IST - Moby Dick project, the investigation is mainly conducted through simulation. On the other hand, analytical results are also presented. We examine the conformance of simulation results with analytical results based on various findings. These findings verify analysis from different aspects.

1 Introduction

To provide end-to-end quality of service (QoS) guarantees in the Internet, the Internet Engineering Task Force (IETF) has considered a number of architectural extensions to the current best-effort service model. Among them, the Differentiated Services (DiffServ) approach [1] has attracted a lot of attention in the networking community because of its potential scalability in providing QoS guarantees. The IST - Moby Dick project [2] is a good example, which is a project of the Information Society Technologies (IST) programme of the European Commission. It makes use of DiffServ to support service guarantees in a wireless access network.

In the DiffServ architecture, packets are classified into several behavior aggregates according to their QoS requirements. Within each class, flows are aggregated. The purpose of such aggregation is to improve the efficiency of scheduling algorithms and to simplify the management of flows. As opposed to end-to-end QoS guarantees for individual flows in the Integrated Services (IntServ) architecture, the goal of the DiffServ effort is to define Per-Hop Behaviors (PHBs), such as Expedited Forwarding (EF) PHB [3] [4] [5] and Assured Forwarding (AF) PHBs [6], with which local per-hop QoS guarantees can be provided to behavior aggregates. Edge-to-edge Per-Domain Behaviors (PDBs) [7] may then be guaranteed to individual flows with the help of these defined PHBs.

H.-K. Kahng (Ed.): ICOIN 2003, LNCS 2662, pp. 948–957, 2003.

A typical implementation of a DiffServ network is to shape and police traffic at the ingress of the network, while in the core packets with a same required PHB are aggregated at a single First-In-First-Out (FIFO) buffer and serviced in the order of their arrival times at the queue. Clearly, FIFO aggregation results in a very simple implementation of the PHB. However, recent studies show that the ability of providing edge-to-edge performance guarantees in a network using FIFO aggregation is questionable [8] [9].

The purpose of this paper is to investigate the impact of flow aggregation on DiffServ networks. In particular, we investigate the impact of FIFO aggregation on edge-to-edge (EtE) delay performance in DiffServ networks. In the study, we conduct the investigation in the context of a service based on the EF PHB. The investigation is mainly carried out through simulation. On the other hand, analytical results are also presented. For simulation, a wide range of scenarios with various user traffic and interference traffic patterns, and service configurations are considered. In addition, simulation results are compared with analytical results to examine the conformance of simulation results with analytical results.

The rest of this paper is organized as follows. In the next section, we briefly review previous work. Then, in Section 3, the simulated network and scenarios are introduced. In particular, we consider a generic network topology as used in [10]. In Section 4, analytical results for edge-to-edge delay bounds are presented for the simulated network. In Section 5, a set of experiments are carried out to simulate delay performance in the generic network, and simulation results are compared with analytical results. Finally, in Section 6, the findings of this study are summarized and concluding marks are made.

2   Previous Work

Recently, several studies have been conducted to investigate the impact of flow aggregation on network edge-to-edge performance. In [10], a simulation-based case study was conducted to investigate the impact of traffic aggregation on the conformance of an EF-based service inside and across a DiffServ domain. The focus of [10] was to investigate two possible models of service contracts. One is service contracts that extend individual service agreements across domains and the other is service contracts that map individual agreements onto aggregate level service contracts. The measure of conformance in [10] is the burstiness of both individual and aggregate flows. The work in [11] extends the work in [10] by considering loss performance due to flow aggregation. Similar to [10], FIFO aggregation was assumed in [11]. However, while the work in [10] mainly relied on simulation, the work in [11] was analytical in nature. Nevertheless, both of them did not study the impact of FIFO aggregation on delay performance in DiffServ networks.

In [8] and [9], delay bounds were derived for aggregate scheduling networks, such as DiffServ networks, with FIFO flow aggregation. In [8], the considered network was strict priority based and the derivation of edge-to-edge delay bound assumed fluid traffic model. In [9], the considered network was extended to Guaranteed Rate (GR) server [12] based, which is more general than the one considered in [8]. In addition, the delay bound derived in [9] takes into account the packetization effect. Note that the derivation in both [8] and [9] does not have any assumption on network topology. Hence, the bounds derived in [8] and [9] hold in general. However, both [8] and [9] are analytical in nature, whether and how their analytical results can be used in real DiffServ networks such as IST - Moby Dick network remain untouched. The work of this paper is hence motivated to fill up the gap between analysis and real situation.

3   Network Model and Methodology

This section describes the network model and methodology we rely on to investigate the delay performance of an individual flow in presence of aggregation in a DiffServ network. They include the network topology we use in simulation, the traffic model we assume for users, and the different configurations we investigate. In all the cases, we focus on the investigation of delay performance of EF flows. Usually in an implementation of DiffServ architecture, EF packets are queued separately from packets of other traffic classes such as AF, and the EF traffic class is assigned to the highest priority level over other classes. In viewing this, for simplicity, we assume a FIFO network where all traffic belongs to the EF class and the queue for EF traffic is large enough to ensure no packet loss. We believe that it is easy to extend the obtained results to DiffServ networks where other scheduling disciplines are adopted. The reason is that most scheduling disciplines including FIFO have been proved to be Guaranteed Rate servers [12].

We first describe the network topology adopted in the investigation, which is shown in Figure 1. As depicted in the figure, the adopted network has a linear multi-hop topology, which has been widely used in previous works such as in [10] and references therein. Note that while the linear topology of Figure 1 is rather simple, it can be configured to reflect a wide range of network configurations.

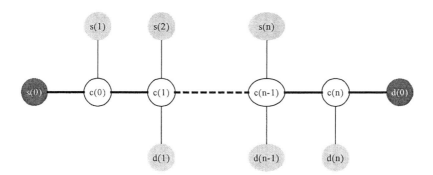

Fig. 1. Generic network topology

The linear topology shown in Figure 1 represents the path of a flow traversing a network. Although the network itself can be very complicated, the performance of the flow is only directly affected by intermediate network elements, here referred to as nodes or routers, along the path. Other network nodes affect the performance of the flow in the network indirectly. In particular, these nodes affect the flow through cross flows of the considered flow, which join and leave some nodes along the path of the flow. Hence, changing interfering traffic patterns of the cross flows is a helpful means in investigating the performance of a flow in a network with unknown topology. In this sense, for example, the access networks of the Moby Dick project [2] can also be mapped to the generic network topology.

Particularly in Figure 1, $s(0)$ and $d(0)$ respectively represent the source and destination of the considered EF flow, $c(0), c(1), \ldots, c(n)$ denote the core routers through which the considered EF flow traverses, and each pair $\{s(1), d(1)\}, \ldots,$ $\{s(n-1), d(n-1)\}$ represents the cross EF flows of the considered EF flow through the corresponding link. Throughout the rest of the paper, we refer to the traffic of the considered EF flow as tagged traffic while the traffic of the cross EF flows as cross traffic. For simplicity, we assume links along the path of the tagged EF flow have same capacity C^t, whereas all other links shown in Figure 1 for cross EF flows have capacity C^c.

We next introduce the traffic model used. As specified in the DiffServ architecture, each EF flow is assumed to be leaky bucket constrained before entering the network. Throughout this paper, we rely on this assumption. However, traffic sources of each flow can be of a variety of types, such as Poisson and Pareto. To make the arrival conforming to the assumption, a token bucket filter is enforced to each flow, i.e. either the consider flow or each of the cross flows, at its source. The token bucket filter is modeled by a two-tuple (b^i, r^i) where r^i denotes the bucket rate and b^i denotes the bucket size or burstiness of the flow after the filter at source i. The actual traffic load to the network shown in Figure 1 is controlled by the bucket rate of each token bucket filter attached to the source, and the burstiness of the traffic sent into the network is adjusted by the bucket size of the token bucket filter. For simplicity, we assume fixed packet size L for all flows, which includes all layers' headers. Each source of cross EF flows has the same leaky bucket parameters (b^c, r^c). The source of the considered EF flow has leaky bucket parameters (b^t, r^t).

Finally, the performance measures mainly include the maximum EtE delay, the average EtE delay, and the EtE delay variance or jitter. Specifically, the focus is on the maximum EtE delay since it is the key performance metric for guaranteed delay service implied by EF service.

4   Analytical Results

This section presents analytical results for the considered generic network. In particular, edge-to-edge delay bounds for the considered flow $\{s(0), d(0)\}$ through the network are derived. Assume each link j along the path of the flow has propagation delay γ_j, where $j = 0, 1, \ldots, n$ respectively denote link $\{s(0), c(0)\}$, $\{c(0), c(1)\}, \ldots, \{c(n-1), c(n)\}$. Then, we have the edge-to-edge (EtE) delay for the flow as:

$$D = \sum_{j=1}^{n} d_j + \sum_{j=0}^{n-1} \gamma_j, \tag{1}$$

where d_j is the nodal delay at router j. Note that from edge-to-edge performance point of view, the propagation delay of both $\{s(0), c(0)\}$ and $\{c(n), d(0)\}$ is sometimes not included in calculating D. Throughout this paper, we choose to include the propagation delay of link $\{s(0), c(0)\}$ in D as shown by (1).

Similar to [9], we can derive a nodal delay bound for d_j as follows:

$$d_j \leq \frac{b_j - L_j}{C^t} \cdot \frac{(P_j - C^t)^+}{P_j - r_j} + \frac{L}{C^t} \tag{2}$$

where $b_j = b_j^t + b_j^c$, $r_j = r^t + r^c$, $P_j = C^t + C^c$, $L_j = 2L$, and $(x)^+ \equiv max\{x, 0\}$. Here, b_j^t and b_j^c respectively denote the burstiness parameter of the considered EF flow and the cross EF flow before entering router/link j. From assumption made in the previous section, we have $b_j^c = b^c$ for all j. For b_j^t, however, it increases along the path of the tagged flow, although $b_0^t = b^t$ for link $\{s(0), c(0)\}$.

Note that (2) has considered the effect of link speed regulation, i.e. each link itself is a peak rate regulator. Supposing $P_j \to \infty$, we have from (2) that $d_j \leq \frac{b_j}{C^t}$.

Also note that in calculating the nodal delay bound for d_j in (2), we need to know the burstiness b_j of all incoming traffic at link j. If the network topology is known, b_j may be easily obtained. For example, for the considered network, it can be verified [13] that

$$b_j = b_j^t + b^c \leq b^t + jb^c. \tag{3}$$

Consequently, applying (3) to (2) and then (2) to (1) with some manipulation, we have derived an EtE delay bound for the considered network as

$$D \leq \sum_{j=1}^{n} \left[\frac{(b^t + jb^c) - 2L}{(C^t + C^c) - (r^t + r^c)} \cdot \frac{C^c}{C^t} \right] + \frac{nL}{C^t} + \sum_{j=0}^{n-1} \gamma_j, \tag{4}$$

where the third term represents the whole propagation delay, the second term represents the whole transmission delay, and the first term is a bound on the whole queueing delay experienced by a packet of the considered EF flow along the edge-to-edge path.

If, however, the network topology is unknown or complex, b_j is usually not easy to determine. In [9], a method is introduced to calculate b_j, based on which, a delay bound has been derived for a network of Guaranteed Rate servers with FIFO aggregation. Since the network considered here is a special case of the one considered in [9], we hence can apply the derived delay bound in [9] to the considered network.

The bound in [9] is derived with the following assumptions. First, each edge-to-edge EF flow is shaped to confirm to a leaky bucket when it arrives at the ingress edge. Second, each link j offers to the EF aggregate a guaranteed rate, i.e. C^t, for the considered generic network topology. Packets within the EF aggregate are serviced in FIFO order. Third, the amount of EF traffic on any link j does not exceed a certain ratio α of the guaranteed rate and an overall bound β is imposed on the "burstiness" of all flows *at the ingress* which are destined to a link. Specifically, it is required $\sum_{f \in F_j} r^f \leq \alpha C^t$ and $\sum_{f \in F_j} b^f \leq \beta C^t$ where F_j represents the set of EF flows constituting the EF aggregate on link j. For the considered network, F_j consists of two flows: the considered EF flow and the cross EF flow $s(j) \to d(j)$. α is hence called link utilization level. Fourth, there exists a bound on the size of any packet in the network, denoted by L. Finally, the route of any flow in the network traverses at most n nodes.

Notice that the above assumptions in [9] are also applicable here, we then have another edge-to-edge delay bound for the considered generic network. By applying specific parameter settings for Figure 1 network, we have, if the link utilization level α satisfies

$$\alpha \leq \frac{C^t + C^c}{(n-1)C^c + C^t}, \tag{5}$$

then,

$$D \leq \frac{n(u\beta + E)}{1 - (n-1)u\alpha} + \sum_{j=0}^{n-1} \gamma_j, \tag{6}$$

where $u = \frac{C^c}{C^c + (1-\alpha)C^t}$, and $E = \frac{2(1-u)L}{C^t}$.

5   Simulation Results

This section presents the simulation results. The simulations are conducted using network simulator ns-2. For every simulation, the simulated network has the topology as shown by Figure 1. The configuration of the simulated network is as follows. Each edge link $\{s(1), c(0)\}, \ldots, \{s(n-1), c(n)\}$ has capacity $10Mbps$. All other links have capacity $2Mbps$. All links are configured to have a link propagation delay of 1ms. All packets have the same size of $1000B$. All sources generate traffic according to the Pareto ON/OFF process [1] in which the shape parameter was set to 1.4, burst time to $900ms$, idle time to $100ms$ and rate to $10Mbps$. Note that, each Pareto ON/OFF source is generated with a relatively very high rate, i.e. $10Mbps$ here. The intention of doing so is to generate highly bursty traffic. Nevertheless, the actual traffic entering the network core is controlled by the leaky bucket that is attached to and performs shaping for each source. In addition all simulations are conducted in a period of 120 seconds, but only packets generated between $10s$ and $110s$ are analyzed. All flows, including both the considered EF flow and cross EF flows, start and stop at the same time.

---

[1] All simulations are repeated with exponential ON/OFF sources, but the results show little difference from the Pareto ON/OFF sources.

In the following subsections, simulation results under different configuration cases are presented. Specifically, we first investigate the impact of burstiness on EtE delay under different system load conditions. We next investigate the impact of changing cross traffic load on EtE delay. We then investigate the impact of network scale on EtE delay. In these three parts, in addition to the maximum EtE delay D^{max}, the average EtE delay $\overline{D}$, and the EtE delay variance, or jitter J, are also measured. Here, the EtE jitter is calculated as follows:

$$J = \sqrt{\sum_k (D_k - \overline{D})^2 / K}, \tag{7}$$

where D_k refers to edge-to-edge delay experienced by packet k and K is the total number of packets, all of the tagged flow, simulated between $10s$ and $110s$.

Note that, for any packet k, we always have $D_k \leq D^{max}$, which when applying to (7), we further obtain a bound on jitter as follows:

$$J \leq D^{max} - \overline{D}. \tag{8}$$

5.1   Impact of Burstiness

We first investigate the impact of burstiness on EtE delay. Simulations for this part are divided into three system load cases: low, medium, and high. For all cases, we change the bucket size for every cross traffic source and measure the maximum EtE delay as well as the average EtE delay and the EtE delay jitter experienced by the considered EF flow. Specifically, we set for the considered flow, token bucket size to $2L$ while for each cross traffic source, its bucket size changes from $1L$ to $10L$. For all cases, the number of core routers is fixed to 5. For the "low" system load case, we set token bucket rate to $0.2Mbps$ for all sources, for "medium" system load case to $0.5Mbps$ and for "high" system load case to $0.8Mbps$. The corresponding results are presented in Table 1.

Table 1 shows that changing bucket size and consequently burstiness for cross traffic has little effect on the average EtE delay no matter the system traffic load is low or medium or high. However, for the EtE maximum delay, it does increase with the burstiness of cross traffic as expected from the worst-case delay bound analysis, (4) and (6). Similarly for the EtE jitter, the EtE jitter increases with the burstiness of cross traffic. This partially explains (8), in which, with the average EtE delay keeping unchanged, the EtE jitter needs a larger D^{max} to bound when the bucket size of cross traffic increases. It can be easily verified from Table 1 that (8) holds under all cases, although the real measured EtE jitter can be smaller.

It is interesting to highlight that while the delay bound (4) can be applied to all the three load cases, the delay bound (6) is applicable only to the "low" load case. Since from (5), the delay bound (6) can be applied only when the link utilization level α satisfies $\alpha \leq \frac{12}{42}$. However, it is easy to verify that for the "medium" load case, $\alpha = 0.5$, and for the "high" load case $\alpha = 0.8$. To verify bounds (4) and (6), Table 2 is presented which compares the measured

Table 1. Impact of system traffic load

Bucket size	D^{max}			$\overline{D}$			J		
	low	medium	high	low	medium	high	low	medium	high
$1L$	25.000	29.225	29.006	25.000	27.485	27.858	0	1.993	1.462
$2L$	25.000	33.225	33.743	25.000	27.486	27.861	0	1.995	1.467
$3L$	25.000	37.225	35.993	25.000	27.487	27.869	0	1.997	1.479
$4L$	25.966	41.225	38.543	25.000	27.490	27.882	0.019	2.007	1.518
$5L$	29.966	45.225	43.280	25.002	27.494	27.901	0.097	2.018	1.600
$6L$	33.966	45.225	48.080	25.003	27.498	27.923	0.175	2.036	1.722
$7L$	37.966	46.167	52.817	25.005	27.502	27.951	0.253	2.056	1.884
$8L$	41.966	53.159	57.617	25.006	27.509	27.984	0.331	2.097	2.089
$9L$	45.966	54.167	61.649	25.008	27.515	28.021	0.409	2.137	2.331
$10L$	49.966	57.966	62.353	25.010	27.523	28.063	0.488	2.191	2.598

Table 2. Comparison of delay bounds for the low load case

Bucket size	$1L$	$2L$	$3L$	$4L$	$5L$	$6L$	$7L$	$8L$	$9L$	$10L$
D^{max}	25.000	25.000	25.000	25.966	29.966	33.966	37.966	41.966	45.966	49.966
Bound (4)	76.725	128.45	180.18	231.90	283.63	335.35	387.07	438.80	490.52	542.25
Bound (6)	196.67	255.00	313.34	371.67	430.01	488.34	546.68	605.01	663.35	721.69

maximum EtE delay with the bounds calculated from (4) and (6) for the "low" load case.

Table 2 shows that both bounds (4) and (6) can be used to bound the maximum EtE delay. In fact, for all cases presented in the subsequent subsections, it can be verified that D^{max} is bounded by (4), and by (6) if the link utilization condition (5) is satisfied. However, to save space, we omit related discussion in the next two subsections. Bound (6) is looser than bound (4). This is not surprising, since bound (6) is obtained without any knowledge of the network topology. In other words, while bound (4) is derived for the considered network, bound (6) can be used for more general cases. Table 2 also shows that theoretical maximum delays are not observed in all simulations conducted. This has three implications. One is that an improved delay bound could exist and needs further study. Another is that the worst case based on which delay bounds (4) and (6) are derived could rarely happen in real situations. Consequently, the third implication is that resource reservation based on worst case analysis could be very conservative.

5.2   Impact of Cross Traffic Load

We second investigate the impact of cross traffic load on EtE delay performance of the considered network. For this part, we set token bucket rate to $0.5Mbps$ and

Table 3. Impact of cross traffic load

Cross load	0.5M	0.6M	0.7M	0.8M	0.9M	1.0M	1.1M	1.2M	1.3M	1.4M
D^{max}	44.115	42.756	44.114	45.468	47.461	47.375	49.009	49.356	56.692	70.720
$\overline{D}$	27.963	27.497	27.597	28.012	29.262	30.683	29.983	29.100	30.967	40.199
J	1.997	2.582	2.367	2.554	2.752	3.240	2.895	2.834	3.366	9.234

token bucket size to $2L$ for the considered EF flow. For each cross traffic source, while its bucket size is fixed to $5L$, the bucket rate changes from $0.5Mbps$ to $1.4Mbps$. For all cases, the number of core routers is fixed to 5. Table 3 presents the simulated results.

From Table 3, it can be observed that although the average EtE delay and the maximum EtE delay generally increase when the cross traffic load increases, the increase is relatively slow before the cross traffic load reaches $1.3Mbps$. When the cross traffic load becomes larger than $1.3Mbps$, the increase becomes exponentially. Similar observation is made for the jitter. This is not surprising, since the capacity of each link $\{c(j), c(j+1)\}$ is $2M$ and the considered EF flow has an average rate of $0.5M$. From queueing analysis [14], the average queueing delay at such a node follows the $\frac{L}{C_j - R_j}$ law, which, when applied to the simulated network, results in what have been shown above. In addition, it can be verified that the maximum EtE delay is bounded by (4) under all cross traffic load conditions.

5.3   Impact of Hop Number

Thirdly, we simulate the impact of changing network scale on the EtE delay performance. For this part, each source is set to have the same token bucket rate as $0.8Mbps$ and bucket size as $2L$. The number of core routers or hops changes from 2 to 10. Table 4 presents the simulation results.

As can be seen from Table 4, both the average EtE delay $\overline{D}$ and the maximum EtE delay D^{max} increase when the number of hops or the network scale increases. This is what can be expected from (1). However, it is interesting that the EtE delay jitter does not follow any monotonous trend with the number of hops. Nevertheless, it can be verified that the EtE delay jitter is still bounded by (8).

Table 4. Impact of hop number

Hop NO.	2	3	4	5	6	7	8	9	10
D^{max}	17.004	20.666	28.496	33.639	40.891	48.289	53.921	63.131	67.301
$\overline{D}$	11.818	15.352	22.053	27.428	32.654	37.983	44.663	52.563	55.341
J	1.453	0.645	3.410	2.089	3.290	1.488	1.465	3.506	1.459

6   Summary

In this paper, we have investigated the impact of FIFO flow aggregation on delay performance of a DiffServ network in the context of EF service. Particularly, the focus was on investigating the impact of FIFO aggregation on edge-to-edge delay performance. The investigation was mainly conducted through simulation. On the other hand, analytical results were also presented. We examined the conformance of simulation results with analytical results based on various findings obtained. First, generally the maximum EtE delay of a flow keeps stable when the whole system EF load and/or cross load change. This, however, does not hold anymore if the system is overloaded. Second, the maximum EtE delay of a flow increases with the burstiness of EF traffic at each source. Third, the maximum EtE delay also increases with network scale. Fourth, the maximum EtE delay and the output burstiness of a flow are bounded by (4) and by (6) if a certain network utilization level is satisfied. All these have verified analysis from different aspects.

Acknowledgement

This work is supported by the IST-2000-25394 Moby Dick project [2].

References

[1] Blake, S., et al: An architecture for Differentiated Services. *IETF RFC2475*, Dec. 1998.
[2] IST Moby Dick: Mobility and Differentiated Services in a Future IP Network. http://www.ist-mobydick.org/.
[3] Jacobson, V., et al: An Expedited Forwarding PHB. *IETF RFC 2598*, June 1999.
[4] Davie, B., et al: An Expedited Forwarding PHB. *IETF RFC 3246*, March 2002.
[5] Charny, A., et al: Supplemental information for the new definition of the EF PHB. *IETF RFC 3247*, March 2002.
[6] Davie, B., et al: Assured forwarding PHB group. *IETF RFC2597*, June 1999.
[7] Nichols, K., Carpenter, B.: Definition of differentiated services per domain behaviors and rules for their specification. *IETF RFC3086*, April 2001.
[8] Charny, A., Le Boudec, J.-Y.: Delay bounds in a network with aggregate scheduling. In *Proc. First International Workshop of Quality of Future Internet Services (QOFIS'2000)*, 2000.
[9] Jiang, Y.: Delay bounds for a network of Guaranteed Rate servers with FIFO aggregation. *Computer Networks*, 40(6):683–694, Dec. 2002.
[10] Guerin, R., Pla, V.: Aggregation of conformance in Differentiated Service networks: A case study. *ACM CCR*, 31(1):21–32, Jan. 2001.
[11] Xu, Y., Guerin, R.: Individual QoS versus aggregate QoS: A loss performance study. In *Proc. IEEE INFOCOM'02*, 2002.
[12] Goyal, P., Vin, H. M.: Generalized guaranteed rate scheduling algorithms: A framework. *IEEE/ACM Trans. Networking*, 5(4):561–571, Aug. 1997.
[13] Chang, C.-S.: *Performance Guarantees in Communication Networks*. Springer-Verlag, 2000.
[14] Kleinrock, L.: *Queueing Systems*, volume 2. Wiley, New York, 1975.

A New Cell Loss Recovery Scheme for Data Services in B-ISDN

Hyotaek Lim[1] and Joo-Seok Song[2]

[1] Department of Computer Engineering, Dongseo University
Busan, 617-716, Korea
htlim@dongseo.ac.kr
[2] Department of Computer Science, Yonsei University
Seoul, 120-749, Korea
jssong@emerald.yonsei.ac.kr

Abstract. The major source of errors in high-speed networks such as Broadband ISDN(B-ISDN) is buffer overflow during congested conditions. These congestion errors are the dominant sources of errors in high-speed networks and result in the cell losses. Conventional communication protocols use error-detection and retransmission to deal with lost packets and transmission errors. However, these conventional ARQ methods are not suitable for the high-speed networks since the transmission delay due to retransmissions becomes significantly large. In this paper, a method to recover consecutive cell losses using forward error correction(FEC) in B-ISDN to reduce the problem is presented. This method recovers up to 81 consecutive cell losses.

1 Introduction

With increasing attention and demand of a B-ISDN, many researches have been conducted towards the development and standardization of a B-ISDN recently. Especially, there have been made many researches to recover cell discarding in switching nodes due to buffer overflows. Cell discarding seriously degrades transmission quality. The packet retransmission method has been used in packet networks such as LAN's to preserve the quality of the packet. However, that is not suitable for ATM networks which handle time-constraint signals such as voice and video.

Several cell loss recovery schemes based on FEC have been proposed for ATM networks. The work in [4] analyzed the reduction in packet loss probabilities due to the parity packet coding and proposes buffer management techniques at the intermediate nodes depending on the use of coding. The cell loss recovery method, which is considered to be applied to virtual paths, that uses an FEC technique in ATM networks was proposed in [1]. The method consists of cell-loss detection(CLD) and lost cell regeneration. The data cells are arranged in a matrix of $(M\text{-}1)*(N\text{-}1)$ cells. Each row of data cells is terminated b a specially designed cell, called a CLD cell, while each column is terminated by a parity cell formed based on a single-parity check code. Decoding is performed in two

H.-K. Kahng (Ed.): ICOIN 2003, LNCS 2662, pp. 958–967, 2003.
© Springer-Verlag Berlin Heidelberg 2003

steps. Lost cells are detected by row-wise operations with the aid of the CLD cell at the end of each row. Next, cells marked 'lost' are regenerated by column-wise operations with the aid of the parity cell at the end of each column. The scheme in [1] was slightly modified in [2] to extend the row size of the coding matrix and thus make the scheme more robust to burst cell loss. [3, 8] show how the use of FEC improves the performance of broadband networks by numerical analysis and simulation. [10] analyzed a hybrid automatic repeat request/forward error control(ARQ/FEC) cell-loss recovery scheme which is applied to virtual circuits(VC's) of ATM networks. In the scheme, FEC is performed based on a simple single-parity code, while a Go-Back-N ARQ is employed on top of that. [11] proposed and evaluated a novel cell-level FEC scheme at SSCS of AAL type 5 for error-free data transmission services in ATM networks. In the cell-level FEC scheme, both the length of user data (e.g., IP packet) and attached redundant data can be modified based on sender's local decision without any end-to-end parameter re-negotiation procedure. In this paper we propose a new cell loss recovery scheme that employs new cell sequence number algorithm with FEC technique. The FEC in the proposed scheme is based on the parity cell as assumed in [1]. As an alternative, we propose FEC technique to preserve the low cell loss rate. In this technique, a recipient can recover lost or erroneously received data based on parity bits which are added by the source. Recovering lost packets reduces the need for retransmissions of reliable data, and enhances the quality of real-time data that can not rely on retransmissions because of the large delays involved. The organization of the paper is as follows. In section II, the scheme to recover lost cells will be proposed and several examples of the lost-cell recovery based on message size will be shown. In section III, the new sequence number algorithms which are employed in the proposed scheme will be described. In section IV, we discuss the result of this paper.

2   A New Cell-Loss Recovery Scheme

2.1   Motivation and Coding Structure

Until now, conventional cell loss recovery methods can recover up to 16 cell losses because lost cells are identified based on 4-bit sequence numbers [1]. If the row length of the coding matrix is more than 16, the location of lost cells can't be identified. The fact that the maximum row of the coding matrix is limited by 16 means that the number of consecutive cell-loss recovery is limited to 16 cells. The consecutive cell losses are a crucial problem in ATM networks in that the cell losses in ATM networks have bursty property. For example, Fig. 1 shows that columns 1 and 2 can not be recovered because one parity cell can correct only one cell loss in the columns. However, we can recover more consecutive cell losses if we can extend the row length of the coding matrix as shown in Fig. 2. In order to extend the row length of the coding matrix, a new cell SN needs to be designed.

Also, Fig. 2 shows the cell coding matrix($M * N$). Arranging the data in a matrix and adding parity cells or error correcting codes such as Reed-Solomon(RS)

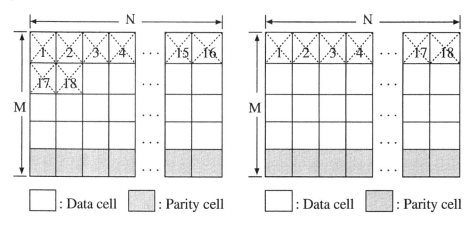

Fig. 1. Coding Matrix(row length: 16)

Fig. 2. Coding Matrix(row length: 18)

code shown in the figure are employed to recover bursts of lost cells. The i-th bit of a parity cell is the modulo 2 sum of the i-th bits of the data cells. That is, if we denote the k-th bit of the packet in the (i,j) position in the array by $c_{i,j,k}$, and consider m-bit cell size, then

$$c_{M+1,j,k} = \left(\sum_{i=1}^{M} c_{i,j,k}\right) \bmod 2, 1 \leq k \leq m, 1 \leq j \leq N.$$

Parity cells possess an OAM cell attribute so that they can be distinguished from data cells and have high priority(CLP = 0). There is no processing delay in encoding since the multiplicative and additive nature of the encoding operation can be utilized by forming a running sum of the parity packets, and adding partial sums to this running sum while information-bearing packets are being transmitted.

2.2   Cell Loss Recovery Scheme

Fig. 3 shows the proposed protocol structure which is applied FEC in IP-over-ATM environment. FEC-SAR is SAR layer which includes FEC capability. The protocol structure may use type 2 and type 3/4 as AAL. An IP packet from IP layer, a upper layer of AAL corresponds to CPCS-SDU whose size is 9,180 bytes in ATM networks. CPCS-SDU is divided into several FEC-SDU and each FEC-SDU is included in the FEC frame in which FEC code is added. The size of FEC frame is variable in order to transfer efficiently even in the small size of CPCS-SDU. FEC frame consists of user data part which contains user data and FEC code part which contains parity code or RS code. The proposed scheme in AAL is transparent to its higher layer in that any protocol modification is not required. The FEC encoding/decoding procedure is as follows.

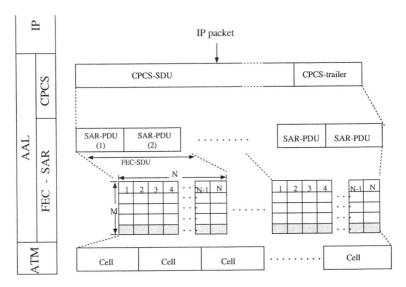

Fig. 3. Proposed Protocol Structure

Encoding Procedure The FEC encoder at the sending side first computes the number of cells in a message. Here, the message means the SAR-SDU. If the number of cells is less than or equal to 18, FEC encoder doesn't have to modify any field of a cell because all permutations of the lost cells is identified by the SN and ST fields. In this case, the tasks FEC encoder does are the generation of the parity cell, and then transferring the coding matrix to the network. If the number of cells is greater than 18, the FEC encoder uses the new SN* as shown in Table 2. The modified value of ST field is updated in FEC decoder. The following summarizes the algorithm of FEC encoder.

1. FEC encoder generates parity cells using data cells.
2. It makes the new SN* which is shown in next Section.
3. It sends coding matrix.

Decoding Procedure The FEC decoder at the receiving side buffers the cells from network. It is noted that the decoder waits until the last cell within a block comes up. The decoder finds lost cells by observing the gap of a pair of the SN and ST values in the stream of received cells, and recovers the lost cells using parity cells. The following summarizes the algorithm of FEC decoder.

1. FEC decoder buffers the cells which arrive from network.
2. It identifies the lost cell by observing SN, ST and/or LI values of the received cells depending on message size.
3. It recovers lost cells using parity cells.
4. It adjusts the modified LI value of the cell whose ST value is COM to 44.

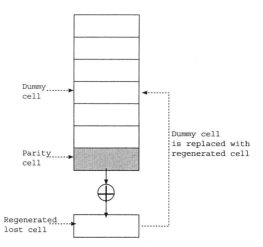

Fig. 4. Regeneration Process of Lost Cell

The FEC decoder at the receiving side uses the proposed SN* to identify lost cells. If the decoder discovers a cell is lost, the decoder inserts the dummy cell whose all bits are 0 into the location of the lost cell.

Fig. 4 shows the regeneration process of the lost cell. SN* algorithm and FEC encoding/decoding procedure is described in next Section in detail. If the number of lost cells exceeds FEC capability, some retransmission techniques such as go-back-N or selective ARQ should be used. As the size of FEC frame increases, the larger delay is needed because of the computation procedure to recover the lost cells in the received FEC frames. However, if the received FEC frames have not had any cell losses, the computation procedure is not needed any more.

3   New Sequence Number(SN*) Algorithm

3.1   Basic SN* Algorithm

Here we propose the SN* which is used in the new cell-loss recovery scheme. The SN* can be applied to AAL 3/4 sublayer without modification of any protocol. As shown in Fig. 5, SAR-PDU for AAL 3/4 is used for service classes C and D. The SN field in AAL is a 4-bit field that allows the SAR-PDUs to be numbered in a repeatedly 0-to-15. This gives the receiver a means of detecting lost or misinserted cells. The 2-bit ST field indicates whether the SAR-PDU is a beginning of message(BOM), a continuation of message(COM), an end of message(EOM), or a single segment message(SSM). The SN and ST values in a cell are used to consider a new cell SN(SN*) as shown in Table 1.

For the case when the message is less than or equal to 16 cells, all random permutations of lost cells at the receiver can be identified and recovered by the

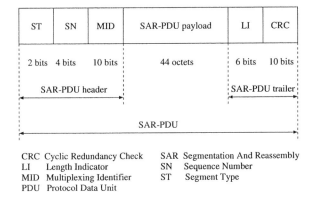

CRC	Cyclic Redundancy Check	SAR	Segmentation And Reassembly
LI	Length Indicator	SN	Sequence Number
MID	Multiplexing Identifier	ST	Segment Type
PDU	Protocol Data Unit		

Fig. 5. SAR-PDU format for AAL3/4

Table 1. Basic SN*

SN	ST	SN*
0	BOM	0
1	COM	1
0	COM	16
1	EOM	17

4-bit cell SN since the SN is incremented modulo 16. Let's consider the case when the message size consists of 18 cells. The sequence of 18-cell message at the sending side is shown below.

```
(B : BOM , C : COM , E : EOM , PC  : Parity Cell)
SN : 0  1  2  3  4  5  6  7  8  9  10  11  12  13  14  15  0  1
ST : B  C  C  C  C  C  C  C  C  C  C   C   C   C   C   C   C  E
PC : +  +  +  +  +  +  +  +  +  +  +   +   +   +   +   +   +  +
        *                                                     *
```

In this case, all positions of lost cells are not determined by 4-bit SN. For example, consider the case when the cells of which SN value is 1(*) are lost. Then, the decoder uses ST field to distinguish whether which cell(second or 18th) is lost. Thus, the row length of coding matrix can be extended to $N = 18$.

3.2 Extended SN*

In addition to the basic SN* algorithm, the row length of coding matrix can be extended to 81 using extended SN* algorithm. The 6-bit LI specifies the number of octets from the Convergence Sublayer PDU(CS-PDU) which are included in the SAR-PDU payload field(with a maximum of 44 octets). Note that when ST

Table 2. LI values

Segment Type	Permissible Value
BOM	44
COM	44
EOM	4-44, 63*
SSM	8-44

(* 63 is used in the ABORT-SAR-PDU)

field of a cell indicates COM, the LI value of the cell always indicates 44 as shown in Table 2 because the cell is at the middle of the message[2].

Also, the LI value of the cell whose ST field indicates BOM is always 44. That is, the LI field of the cell whose ST field indicates COM or BOM is meaningless because we know the value is always 44. We employ the LI field together with SN and ST fields to consider extended SN*.

The extended SN* is numbered in a repeatedly 0-to-80 cycle as shown in Table 3.

In this proposal the SN is set to zero for BOM to consider the SN*. But, it is free to set the SN to other value for BOM and then the sequence number for each of other entries in Table 3 should be changed appropriately. The only task at the sending side is to modify the LI value. The modified LI values of the PDUs whose ST value is COM are adjusted to 44 at the receiving side to preserve AAL standard.

However, a message is less than or equal to 18 cells, all random permutations of lost cells at the receiver can be identified and recovered by the SN and ST without LI as shown in Table 4 and 5. Fig. 6(a) shows a PDU-sequencing based on existing SN when sending 34 AAL-PDUs. The 4-bit SN field allows the PDUs to be numbered in a repeatedly 0-to-15 to number 34 cells. Therefore, if two cells which have same SNs are lost, we can not identify correct location of lost cells. Fig. 6(b) shows a PDU-sequencing of the proposed scheme when sending 34

Table 3. Extended SN*

SN	ST	LI	SN*	LI value modified?
0	BOM	44	0	no
1-15	COM	44	1-15	no
0	COM	44	16	no
X	COM	0-43	17-60	yes
X	COM	45-63	61-79	yes
X	EOM	X	80 (last cell no.)	no

(X : don't care)

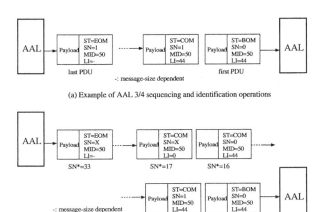

(a) Example of AAL 3/4 sequencing and identification operations

(b) Example of Proposed Sequencing

Fig. 6. AAL-PDU Sequencing

AAL-PDUs. The SN* numbered 0 through 33 is placed in the PDUs. The MID is set to 50 in this example to uniquely identify all associated PDUs. So, even if all successive PDUs are lost, the SN* can identify the lost PDUs and recover the PDUs using error correcting techniques. Thus, the SN* can identify the maximum of 81 successive lost PDUs.

3.3   Coding Example

Case 1: message size $\leq$ 18 cells When the message consists of 16 cells, the sequence of cells at the sending side is shown in Table 4.

In this case, FEC decoder can recover all of the lost cells using parity cells because SN field can identify the lost cells. In this case FEC decoder does not have to use ST field. Also, for the case when message is less than 16 cells, all random permutations of lost cells at the receiver will be identified and recovered by the 4-bit SN since the SN is incremented modulo 16. Let's consider the case when the message size consists of 17 or 18 cells. The sequence of cells at the sending side in this case is shown in Table 5.

In this case, all lost cell positions are not determined by the SN. However, all random lost cells at the receiving side can be recovered by the SN and ST fields.

Table 4. SN* Coding Example (Message Size: 16 cells)

Cell No.	0	1	2	3	4	5	6	7	8	9	10	11	12	13	14	15
SN	0	1	2	3	4	5	6	7	8	9	10	11	12	13	14	15
ST	B	C	C	C	C	C	C	C	C	C	C	C	C	C	C	E

B: BOM, C:COM, E:EOM

Table 5. SN* Coding Example (Message Size: 18 cells)

Cell No.	0	1	2	3	4	5	6	7	8	9	10	11	12	13	14	15	16	17
SN	0	1	2	3	4	5	6	7	8	9	10	11	12	13	14	15	0	1
ST	B	C	C	C	C	C	C	C	C	C	C	C	C	C	C	C	C	E
SN*	0	1	2	3	4	5	6	7	8	9	10	11	12	13	14	15	16	17

Case 2: message size > 18 cells Let's consider the case when message consists of 19 cells. In this case the cell sequence at the sending side is shown below.

```
SN : 0  1  2  3  4  5  6  7  8  9  10  11  12  13  14  15  0  1  2
ST : B  C  C  C  C  C  C  C  C  C  C   C   C   C   C   C   C  C  E
         *                                                      *
```

For this case, all random lost cells at the receiving side will not be corrected by the SN and ST fields. For example, consider the case when the two cells of which SN values is 1(*) are lost. The decoder can not determine whether which cell(second or 18th cell) is lost. In this case the FEC encoder uses LI field to discriminate the cells which have the same SN and ST values. The 34-cell message segmentation at the sending side is shown in Table 6.

Therefore the row length of coding matrix can be extended to $N = 81$ by employing the new SN*. This means that the proposed method substantially improves the recoverability of lost cells.

Table 6. SN* Coding Example (Message Size: 34 cells)

Cell No.	0	1	2	3	4	5	6	7	8	9	10	11	12	13	14	15	16
SN	0	1	2	3	4	5	6	7	8	9	10	11	12	13	14	15	0
ST	B	C	C	C	C	C	C	C	C	C	C	C	C	C	C	C	C
LI	44	44	44	44	44	44	44	44	44	44	44	44	44	44	44	44	44
SN*	0	1	2	3	4	5	6	7	8	9	10	11	12	13	14	15	16

Cell No.	17	18	19	20	21	22	23	24	25	26	27	28	29	30	31	32	33
SN	1	2	3	4	5	6	7	8	9	10	11	12	13	14	15	0	1
ST	C	C	C	C	C	C	C	C	C	C	C	C	C	C	C	C	E
LI	0	1	2	3	4	5	6	7	8	9	10	11	12	13	14	15	X
SN*	17	18	19	20	21	22	23	24	25	26	27	28	29	30	31	32	33

LI: Length Indicator, X: Don't Care

4   Concluding Remarks

In B-ISDN, the cell losses due to buffer overflow are generated in a consecutive manner, thus causing considerable quality degradation. In this paper, a scheme was presented to recover the consecutive cell losses using SN* in B-ISDN. The scheme is based on FEC transmitting parity cells, which is used by decoder at the receiving side to recover lost cells, together with data cells at the sending side and recovers up to 81 consecutive cell losses in ATM networks. The missing cells are identified by observing the gap of a pair of the SN, ST and/or LI values in the stream of received cells and are recovered by parity cells. Some coding examples were shown based on message size to recover lost cells. And finally, it was concluded that FEC provides a good solution for consecutive cell losses due to buffer overflow in high-speed networks like ATM networks.

References

[1] Hiroshi Ohta, Tokuhiro Kitami, "A cell Loss Recovery Method Using FEC in ATM Networks," *IEEE Journal on Selected Areas in Communications*, Vol. 9, No.9, Dec. 1991, pp1471-1483.

[2] Hyotaek Lim, Joo Seok Song, "Cell Loss Recovery Method in B-ISDN/ATM Networks," *IEE Electronics Letters*, Vol. 31, No.11, May. 1995, pp849-850.

[3] Ender Ayanoglu, Richard D. Gitlin, Nihat Cem Oguz, "Performance Improvement in Broodband Networks Using Forward Error Correction For Lost Packet Recovery," *Journal of High Speed Networks 2*, 1993, pp287-303.

[4] N. Shacham, "Packet Recovery in High-Speed Networks Using Coding and Buffer Management," *Proc. IEEE INFOCOM '90*, June 1990, pp124-131

[5] James R. Yee, E. J. Weldon, Jr. , "Evaluation of the Performance of Error-Correcting Codes on a Gilbert Channel," *Proc. IEEE ICC '94*, May 1994, pp655-659

[6] ITU-T, "Recommendation I.363, B-ISDN ATM Adaptation Layer(AAL) Specification" 1993.

[7] Hiroshi Ohta, Toduhiro Kitami, "Simulation Study of the Cell Discard Process and the Effect of Cell Loss Compensation in ATM Networks ," *IEICE Transaction on Communication*, Vol. E, No. 10, Oct. 1990, pp1704-1711.

[8] N. C. Oguz and E. Ayanoglu, "A Simulation Study of Two-Level Forward Error Correction for Lost Packet Recovery in B-ISDN/ATM," *Proc. IEEE ICC '93*, May 1993, pp1843-1846.

[9] Shiann-Tsong Sheu, Chih-Chiang Wu, "An Adaptive Cell Checking Controller for Wireless ATM Networks," *IEICE Transactions on Communication*, Vol. E83-B, No. 2, Feb., 2000, pp330-309.

[10] M. A. Kousa, A. K. Elhakeem, J. Yang, "Performance of ATM Networks Under Hybrid ARQ/FEC Error Control Scheme," *IEEE Transactions on Networking*, Vol. 7, No. 6, Dec., 1999, pp917-925.

[11] K. Kanai, K. Tsunoda et el., "Forward Error Correction Control on AAL 5: FEC-SSCS," *IEICE Transactions on Communications*, Vol. E-81B, No. 10, Oct., 1998, pp1821-1830.

[12] L. Zhang, D. Chow, C. H. Ng, "Cell Loss Effect on QoS for MPEG Video Transmission in ATM Networks," *Proc. IEEE ICC '99*, Jun. 1999, pp147-151.

Some Improvements of TCP Congestion Control

Yuan-Cheng Lai[1] and Rung-Shiang Cheng[2]

[1] Department of Information Management
National Taiwan University of Science and Technology
laiyc@cs.ntust.edu.tw
[2] Deptartment of Electronic Engineering, National Cheng Kung University
chengrs@locust.csie.ncku.edu.tw

Abstract. Many TCP enhanced versions, such as NewReno, SACK, FACK, and Rate-Halving, were proposed to estimate the more accurate number of the outstanding packets during fast recovery phase. In this paper, we focus on the modifications in slow-start and congestion avoidance phases. Two previous enhancements, *larger initial window* and *smooth-start*, are examined to study their effects on performance. Then we propose an algorithm, smooth congestion avoidance, in congestion avoidance phase to reduce the occurrence of packet losses. Simulation results show that all modifications can promote the TCP performance. The first two modifications particularly benefit the short-lived connections and the last modification particularly benefits the long-lived connections.

1 Introduction

TCP (transmission control protocol) is an underlying reliable layer-4 protocol that widely used by many network applications. With the maturation of the Internet and the widespread use of the TCP, the efficiency of TCP becomes an important concern. Therefore, many researches for TCP congestion control have been continuously proposed to elevate its transmission performance [1-5].

The TCP congestion control scheme maintains a proper window size to control the number of outstanding packets. The first version of TCP, standardized in RFC 793, defined the basic structure of TCP, namely, the window-based flow control scheme and a coarse-grain timeout timer [1]. The second version, TCP Tahoe, added the congestion avoidance scheme proposed by Van Jacobson [2]. The third version, TCP Reno, followed two years later extended the congestion control scheme by including fast retransmission and fast recovery schemes [3]. In summary, current TCP congestion control includes five phases: slow-start, congestion avoidance, fast retransmission, fast recovery, and timeout.

Many TCP enhanced versions, such as NewReno, SACK, FACK, and Rate-Halving, were proposed to estimate the more accurate number of the outstanding packets after the occurrence of packet losses [4,5]. Restated, their contributions are the enhancements in fast retransmission and fast recovery phases to quickly recover from the event of packet losses. However, a more important concern is how to reduce the occurrence of packet losses.

H.-K. Kahng (Ed.): ICOIN 2003, LNCS 2662, pp. 968–977, 2003.

In this paper, we focus on the enhancements in slow start and congestion avoidance phases. Two previous modifications are examined to study their effects on performance. The first modification is using a *larger initial window*. A recent proposal [6,7] suggested using TCP's initial window with a larger value, rather than one, at the beginning of a new connection and after an idle period to reduce the duration in slow-start phase, resulting in the decrease of its transmission time. The second modification is a variant of slow-start, called *smooth-start* [8,9], which provides a smooth transition from the exponential increase in the slow-start phase to the linear growth in the congestion avoidance phase. Then the paper proposes a new algorithm, *smooth congestion avoidance*, which concept is to find a proper window size and then keep the stabler window adjustment in congestion avoidance phase to reduce the occurrence of congestion.

2   Background

A TCP connection typically uses the acknowledgements (ACKs) of the packets sent previously, i.e., self-clocking, to estimate the available bandwidth. The sender believes the network is congestion-free when receiving the new ACKs and detects the congestion when receiving many duplicate-ACKs. A congestion window ($cwnd$) is adopted to limit the amount of transmitted data in one round trip time (RTT). The sender will increase or decrease the congestion window size according to the network is congested or non-congested, respectively. The control scheme of TCP can be divided into five phases, which are explained as follows.

(1) **Slow-Start:** As a connection starts or a timeout occurs, the slow-start state begins. The initial value of $cwnd$ is set to one packet in the beginning of this state. The sender increases $cwnd$ exponentially by adding one packet each time it receives an ACK. Slow-start controls the window size until cwnd achieves a preset threshold, slow-start threshold ($ssthresh$). When $cwnd$ reaches to $ssthresh$, the "congestion avoidance" state begins.

(2) **Congestion Avoidance:** Since the window size in the slow-start state expands exponentially, the packets sent at this increasing speed would quickly lead to network congestion. To avoid this, the congestion avoidance state begins when $cwnd$ exceeds $ssthresh$. In this state, $cwnd$ is added by $1/cwnd$ packet every receiving an ACK to make the window size grow linearly.

(3) **Fast Retransmission:** The duplicate ACK is cause by an out-of-order packet received in the receiver. The sender treats it as a signal of packet loss or packet delay. If three or more duplicate ACKs are received in a row, packet loss is likely. The sender performs retransmission of what appears to be the missing packet, without waiting for a coarse-grain timer to expire.

(4) **Fast Recovery:** When fast retransmission is performed, $ssthresh$ is set to half of $cwnd$ and then $cwnd$ is set to $ssthresh$ plus three packet sizes. $Cwnd$ is added by one packet every receiving a duplicate ACK. When the ACK of retransmitted packet is received, $cwnd$ is set to $ssthresh$ and the sender

reenters the congestion avoidance. Restated, *cwnd* is reset to half of the old value of *cwnd* after fast recovery.

(5) **Timeout:** For each packet sent, the sender maintains its corresponding timer, which is used to check for timeout of non-received ACK of the packet. If a timeout occurs, the sender resets the cwnd to one and restarts slow-start. The default value of clock used for the round-trip ticks is 500ms, i.e., the sender checks for a timeout every 500ms.

3 TCP Modifications

This section addresses several modifications of TCP congestion control, including a larger initial window, smooth-start and smooth congestion avoidance.

3.1 Larger Initial Window

Some researches have proposed a larger initial window, which may be two or more packets, rather than the current initial window of one packet, at the beginning of a new connection and after a long idle period [6,7]. Their results show that increasing the initial window size between two and four packets improves the perceived performance, and the packet losses appear no particular worse.

The short-lived connection delivering relatively small amount of data usually terminates before it meets the first packet loss. Thus the reduction of start-up period of a short-lived TCP connection can significantly improve its performance. The experiment shows that for many email and web page transfers that are less than 4K bytes, the larger initial window would reduce the transfer time to a few *RTT*.

3.2 Smooth-Start

During the slow-start phase, since the TCP sender has no way to perceive the available capacity, it exponentially increases window size until *cwnd* exceeds *ssthresh*. However, if *ssthresh* is overestimated, the exponential increase often causes multiple packet losses within a window of data, and consequently a timeout happens.

A modified slow-start algorithm, smooth-start, wants to slow down the window adjustment when *cwnd* approaches to *ssthresh* [8,9]. This algorithm uses a control parameter, n, to determine the separator and the gradient of increase. When the congestion window is below the separator, which is set as $ssthresh/n$ in his algorithm, the sender exponentially increases *cwnd* by adding one packet at each receiving an ACK, like the slow-start phase. Once the congestion window goes beyond this separator, $n + (i\text{-}1)$ ACKs trigger one increment of *cwnd* for the i-th subsequent RTT. In summary, the sender updates its window size as

$$cwnd = \begin{cases} cwnd + 1, & \text{if } cwnd < \frac{ssthresh}{n} \\ cwnd + \frac{1}{n+i-1}, & \text{if } cwnd \geq \frac{ssthresh}{n} \text{ and} \\ & \text{the } i \text{ - th subsequent } RTT \end{cases}$$

The number of RTTs required to increase $cwnd$ from $ssthresh/n$ to $ssthresh$ require n RTT in this smooth-start algorithm. The preliminary experiments in [9] recommend setting n to two or four to provide the better performance.

3.3   Smooth Congestion Avoidance

Because TCP needs to create losses to find the available bandwidth of the connection, the sender continuously increases its window size until a packet loss, and then halves this window size. Therefore, this process of bandwidth estimation induces a periodic oscillation of window size, reducing the performance. To reduce the periodic congestions, we propose a modified scheme, called smooth congestion avoidance, which intends to provide a more graceful rise of window size in congestion avoidance phase to obtain the better utilization of available bandwidth.

We use one parameter, separator, which is set to $3/2$ $ssthresh$. The growth of $cwnd$ is same as that in the original congestion avoidance until it reaches this separator, then increases at a slower linear rate. In such a case, $i \times cwnd$ ACKs trigger one increment of $cwnd$ for the i-th subsequent RTT. In summary, the sender updates its window size as

$$cwnd = \begin{cases} cwnd + \frac{1}{cwnd}, & \text{if } ssthresh < cwnd \leq \frac{3}{2}ssthresh \\ cwnd + \frac{1}{(\lceil \log_2 i \rceil + 1) \cdot cwnd}, & \text{if } cwnd > \frac{3}{2}ssthresh \\ & \text{and the } i\text{ - th subsequent } RTT \end{cases}$$

When a packet loss occurs, $ssthresh$ is halved. Thus the average of the new $ssthresh$ and the old one ($2 \times ssthresh$) is adopted as the separator. When $cwnd$ is less than the separator, the amount of window size is still safe because it is far away $cwnd$ at the last loss. When $cwnd$ exceeds this separator, it gradually approaches the window size at the last loss. Therefore, the sender should be more conservative in increasing $cwnd$.

Figure 1 shows the dynamics of $cwnd$ when the original TCP congestion control and three modifications are adopted. In slow-start phase, the congestion window is initially set as 1, and exponentially grows until it reaches $ssthresh$ (20 packets in this figure), and then linearly grows in the congestion avoidance phase. We can obviously observe an abrupt turn of $cwnd$ at reaching $ssthresh$. The increase of $cwnd$ in the original congestion control is not smooth.

The congestion window is initially set as four by using a larger initial window. It exponentially grows until reaching the separator ($1/2 \times ssthresh$ in this figure), then grows at a slow exponential rate during the smooth-start period. After $cwnd$ reaches $ssthresh$, the sender enters the smooth congestion avoidance. We can obviously observe the increase of $cwnd$ by using these modifications is very smooth.

4   Simulation Results and Their Implications

The network simulator ns-2, developed by Lawrence Berkeley Laboratory, is used in this investigation to run our simulations. This simulator is often used in

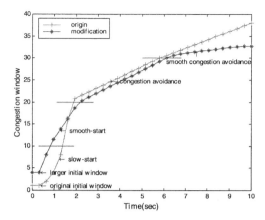

Fig. 1. The dynamics of cwnd

TCP-related studies. Figure 2 depicts the network topology where R represents a drop-tail router with a buffer which capacity is 20 packets, and S and D represent the sender and receiver, respectively. Each link is labeled with its bandwidth and one-way propagation delay. The packet size is fixed at a length of 512 bytes. N TCP senders transmit through a shared bottleneck to an equal number of receivers, and the receiver sends an ACK for each received packet.

4.1   Larger Initial Window

This simulation is constructed with 16 connections. Each one transfers 10 data files of 16k bytes, and the interval between two files is randomly distributed between 2 and 16 seconds. The average time required to transfer a 16K bytes file with various initial window sizes is shown in Figure 3(a), and the number of average lost packets per transfer is shown in Figure 3(b).

When adopting a larger window size, the connection will reduce its transmission time when the number of lost packets is less than two packets. Restated, the TCP connection receives the better performance with the larger initial window even when the number of the lost packets rises. In such a case, the sender

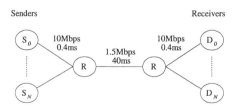

Fig. 2. Network topology

can use fast retransmission and fast recovery, rather than a timeout, to recover. Thus the sender pays a small overhead even when the number of lost packets increases.

However, the impracticably large initial window causes the large amount of lost packets. When multiple packets are dropped from a single window of data, the TCP sender usually encounters a timeout to recover for this loss event. Thus, this recovery may take a long time and nullifies the transfer time saved by using a larger initial window. That is why the transmission time jumps when the number of lost packets is larger than two packets.

4.2 Smooth-Start

Figure 4 shows the number of lost packets in using slow-start and smooth-start, where the parameter n is set to two or four. The number of lost packets increases as the slow-start threshold. When *ssthresh* becomes large, the sender will quickly increase *cwnd*, resulting in a large number of lost packets because of the limitation of capacity. The smooth-start algorithm can reduce the aggressiveness of increasing the congestion window, thus the number of lost packets obviously drops. Also when n equals four, the increase of *cwnd* becomes slower, causing a lesser number of lost packets.

4.3 Smooth Congestion Avoidance

This simulation is conducted by using 3 connections. Each connection transfers 50 seconds. Figure 5 shows the throughput of connections when they adopt original congestion avoidance and smooth congestion avoidance. Simulation results show that the throughput of a connection increases with the buffer size. When the buffer size becomes large, the probability of loss events decreases, resulting in a higher throughput.

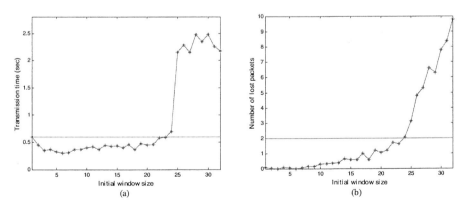

Fig. 3. The performance of connections using various initial window sizes

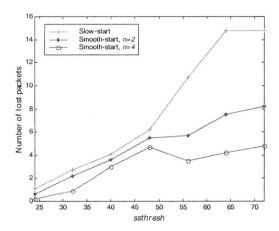

Fig. 4. The performance comparison between slow-start and smooth-start

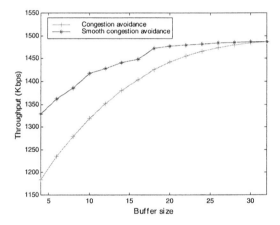

Fig. 5. The performance comparison between smooth congestion avoidance and congestion avoidance

The connection adopting smooth congestion avoidance gets the better performance than that adopting original congestion avoidance, regardless of buffer size. Since the smooth congestion avoidance algorithm increases the window size more slowly in each round when *cwnd* is beyond the separator, its window variation is more graceful, reducing the number of loss events. Also the gaps between both become small with the increase of buffer size because the number of packet loss decreases in such cases.

4.4   Combination

This simulation is conducted by adopting the original TCP congestion control, smooth-start, smooth congestion avoidance (sca in the figure), and a combined method smooth-start+sca, to short-lived connections and long-lived connections.

Figure 6 shows the mean transmission time of short-lived connections. The results show that adopting smooth congestion avoidance has less effect for short-lived connections. This is because these connections have been terminated before they enter the smooth congestion avoidance phase. Figure 7 displays the TCP state variations of two short-lived connections by adopting a combined method where the initial window equals four. The results obviously verify our claims.

As we described above, the larger initial window is accompanied with the risk of increasing the number of lost packets. Thus the smooth-start algorithm, which reduces the number of lost packets, can compensate this side effect. On the other hand, the drawback of smooth-start is that it takes a longer time to reach the congestion avoidance phase. Hence, deploying a proper larger initial window can compensate the slower growth rate of the congestion window in the smooth-start phase. Thus the method by combining smooth-start and a larger initial window has a shorter transmission time than the original TCP congestion control.

Figure 8 shows the throughput of long-lived connections by using various modifications. The results show that adopting smooth congestion avoidance has a lot of effect for long-lived connections. This is because these connections usually locate the smooth congestion avoidance phase. Figure 9 displays the TCP state variations of two long-lived connections by adopting a combined method, where the initial window equals four. The results obviously verify our claims.

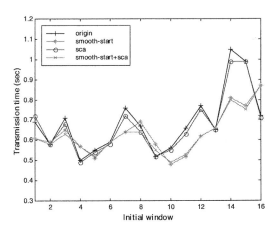

Fig. 6. Mean transmission time of short-lived connections

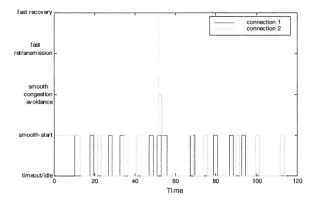

Fig. 7. State variations of short-lived connections

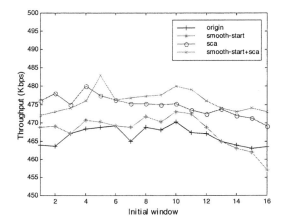

Fig. 8. Throughput of long-lived connections

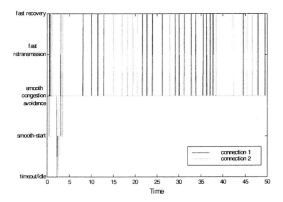

Fig. 9. State variations of long-lived connections

5   Conclusions

This paper first evaluates the effect on TCP performance for increasing the initial window size. A larger initial window can decrease the transmission time, especially for a short-lived transfer, and does not significantly increase the number of lost packets. The smooth-start algorithm can reduce the aggressiveness of increasing the congestion window, thus the number of lost packets obviously drops. Then we presented smooth congestion avoidance, by smoothly changing *cwnd* in congestion avoidance to reduce unnecessarily periodic packet losses.

The simulation results demonstrate that a method combining three modifications outperforms the original TCP congestion control. The first two modifications particularly benefit the short-lived connections and the last modification particularly benefits the long-lived connections. Furthermore, these enhancements are simply implemented and the changes are only made on the sender side.

References

[1] J. Postel, "Transmission Conrol Protocol," *RFC 793*, 1981.
[2] V. Jacobson, "Congestion Avoidance and Control," *ACM SIGCOMM '88*, pp. 273-288, 1988.
[3] W. Stevens, "TCP Slow Start, Congestion Avoidance, Fast Retransmit, and Fast Recovery Algorithms," *RFC 2001*, Jan. 1997.
[4] S. Floyd and T. Henderson,"The NewReno Modification to TCP's Fast Recovery Algorithm,"*RFC 2582*, Apr. 1999.
[5] M. Mathis, J. Mahdavi, S. Floyd, and A. Romanow,"TCP Selective Acknowledgment Options,"*RFC 2018*, Oct. 1996.
[6] M. Allman, S. Floyd and C. Partridge, "Increasing TCP's Initial Window," *RFC 3390*, Oct. 2002.
[7] M. Allman, C. Hayes, and S. Ostermann, "An Evaluation of TCP with Larger Initial Windows," *ACM Computer Communication Review*, 28(3): 41-52, July 1998.
[8] H. Wang and C. Williamson, "A New Scheme for TCP Congestion Control: Smooth-Start and Dynamic Recovery," *IEEE Proceedings on Sixth International Symposium on Modeling, Analysis and Simulation of Computer and Telecommunication Systems*, pp. 69-76, 1998.
[9] H. Wang, H. Xin, D. S. Reeves, and K. G. Shin, "A Simple Refinement of Slowstart of TCP Congestion Control," *ISCC 2000*, pp. 98-105, 2000.

Flow Allocation Algorithms for Traffic Engineering[*]

Manabu Kato[1], Hironobu Hida[2], Kenji Kawahara[2], and Yuji Oie[2]

[1] Department of Electronics and Information Engineering
Ariake National College of Technology, Omuta, 836-8585 Japan
kato@ariake-nct.ac.jp
[2] Department of Computer Science and Electronics
Kyushu Institute of Technology, Iizuka, 820-0067 Japan
{hiro,kawahara,oie}@yen.cse.kyutech.ac.jp

Abstract. The objective of traffic engineering (TE) is to optimize network resources, while satisfying traffic-oriented performance requirements. As a technology for TE, we focus on a method by which to avoid congestion on some links and balance the traffic load over the network. Given a traffic matrix representing traffic demands, we allocate their traffic flows over the network. In this paper, we propose a number of flow allocation algorithms for TE in ISP backbone networks. As an objective function, we employ the maximum link utilization. We then evaluate the effect of the proposed algorithms with respect to load balancing and compare the performance of these algorithms.

1   Introduction

In an internet service provider (ISP) network, in which the open shortest path first (OSPF) or the routing information protocol (RIP) is employed for IP routing, the weight of a link is determined by the link bandwidth or sometimes is simply set at 1 to indicate the hop count. The shortest path is selected in the latter case. However, this type of routing mechanism can cause congestion on some links.

The objective of traffic engineering (TE) is to optimize network resources such as link bandwidth, while satisfying traffic-oriented performance requirements. As a technology for TE, we focus on a method by which to avoid congestion on some links and balance the traffic load over the network.

The explicit routing mechanism of MPLS (Multi-protocol Label Switching) can be applied for TE in IP datagram networks [1, 2, 3, 4, 5]. In an MPLS-based network, each LSP (Label Switched Path) on the second layer can be explicitly established by attaching a label for the path to packets depending upon the header information of third or higher layer. On the other hand, as another approach to TE, in IP datagram networks, the metric of each link is dynamically changed according to the offered load on the link [6].

---

[*] This research was supported in part by a Grant-in-Aid for Young Scientists (B), 14750327, 2002, from the Ministry of Education, Science, Sports and Culture, Japan.

H.-K. Kahng (Ed.): ICOIN 2003, LNCS 2662, pp. 978–988, 2003.

Our major concern is to obtain the allocation of flows (a set of paths of flows) for TE in ISP backbone networks. The flow here indicates an aggregated flow of micro-flows, each of which can be a TCP connection passing from same source node to same destination node. The flow allocation is initially based upon traffic measurement for some duration, for example, one week or one month. Namely, time-dependent static traffic engineering is considered in the present paper. In contrast, a state-dependent dynamic traffic engineering is considered in [5]. If the measurement results indicate an imbalance in traffic imposed on links, flows will be reallocated on different paths for load balancing. When MPLS is applied, an allocation of flows corresponds to a set of LSPs.

In the paper, we propose some flow allocation algorithms for traffic engineering. Each of the algorithms includes the algorithms for generating an initial flow allocation and for selecting a flow to be reallocated. All of the flow allocation algorithms proposed herein are heuristic. Therefore, each of the algorithms is executed in a step-by-step manner. At each step, a flow passing through the most congested link will be selected as a candidate to be reallocated. Then, the selected flow will be reallocated on a new path which does not contain the most congested link.

As stated previously, the amount of traffic imposed on links is assumed to be given by a traffic measurement. In the present paper, the capacity of each traffic demand is determined by uniformly selecting from a range. Given a traffic matrix representing traffic demands, we attempt to calculate the paths of their traffic flows in order to minimize the maximum link utilization. Here, we assume that for a traffic demand a flow is allocated on a single path. On the other hand, in [5, 7], for a traffic demand the multiple paths of flows are considered.

As an objective function, we here employ the maximum link utilization. We then evaluate the performance of the proposed algorithms on the load balancing over the network. The maximum link utilization is also used for optimum routing in [7]. However, as described earlier, the multiple paths of flows between ingress and egress node are considered there. We herein prepare one-hundred different network topologies and one-hundred different traffic patterns and have average characteristics. Furthermore, we calculate the theoretical lower bound of the maximum link utilization and compare the performance of the algorithms.

The remainder of the paper is organized as follows. We define the problem treated herein in Section 2. Next, we propose the flow allocation algorithms in Section 3. Some numerical results are presented in Section 4. Finally, we summarize the paper in Section 5.

2   Problem Definition

In this section, we define the problem treated herein.

2.1   Network Topology

The number of edge routers (referred to as *nodes* in the following) in an ISP backbone network (referred to as *network*) is N. There are $N(N-1)/2$ links in a full-mesh network in which the links are assumed to be bidirectional. The number of n_l links are randomly selected from all of the links in the full-mesh network with parameter p that satisfies $0 \leq p \leq 1$, i.e., $n_l = \lfloor N(N-1)/2 \times p \rfloor$. Further, at least $L_{\min}$ links should be connected from a node.

We define an $N \times N$ *network topology matrix* as $D \overset{\triangle}{=} \{d_{ij}\}$ $(i, j = 1, \cdots, N)$, where $d_{ij} \in \{0, 1\}$. If there is a link from node i to node j, the element d_{ij} takes 1. Otherwise, d_{ij} takes 0. Let C_i [b/s] $(i = 1, \cdots, n_l)$ denote the link capacity (bandwidth). For simplicity, we assume that all the links have the same capacity C [b/s], i.e., $C = C_i \; \forall \; i$.

2.2   Traffic Demands

We assume that for a pair of nodes a maximum of one flow goes from one node to another. As stated previously, the flow here indicates an aggregated flow of micro flows. In this case, the traffic demands can be represented as an $N \times N$ *traffic demand matrix* $T \overset{\triangle}{=} \{t_{ij}\}$ $(i, j = 1, \cdots, N)$. The element of the matrix, t_{ij} [b/s], is represented by the capacity of the traffic demand from source i to destination j, which can be obtained below. The combination of sources and destinations is $N(N-1)$. With parameter a that satisfies $0 \leq a \leq 1$, the number of n_t source-destination pairs are randomly selected, i.e., $n_t = \lfloor N(N-1) \times a \rfloor$. After which, for each traffic demand, we determine t_{ij} by uniformly selecting from a range of $[t_{\min}, t_{\max}]$, where $t_{\min}$ and $t_{\max}$ indicate the minimum and maximum values, respectively, and, the $N(N-1) - n_t$ remaining elements are set to 0.

2.3   Objective Function

At each step of our flow allocation algorithms, a flow is selected and then real-located on a new path. Therefore, the utilization of each link, which is a ratio the total bandwidth used by the flows through the link to the link capacity, can change at the next step.

Let $u_i^{(k)}$ be the utilization of the ith $(i = 1, \cdots, n_l)$ link at the kth step. The maximum link utilization at the kth step, denoted by $u_{\max}^{(k)}$, is then defined as follows: $u_{\max}^{(k)} \overset{\triangle}{=} \max_i \{u_i^{(k)}\}$. Each flow allocation algorithm explained in section 3 will be stopped when the number of steps reaches a predetermined value, denoted by $k_{\max}$. The maximum link utilization obtained through the number of $k_{\max}$ steps, $u_{\max}$, is derived as follows:

$$u_{\max} \overset{\triangle}{=} \lim_{k \to k_{\max}} u_{\max}^{(k)}. \tag{1}$$

As mentioned in section 1, the objective of TE is to optimize network resources such as link bandwidth. For this purpose, we should avoid the situation that some links in a network are congested and other links are not utilized. From this point of view, we define the optimal state as the state in which the maximum link utilization is minimized. Therefore, we employ the maximum link utilization as an objective function. In brief, the objective of flow allocation algorithms proposed in the present paper is to

$$\text{minimize } u_{\max}. \tag{2}$$

2.4   Performance Measures

The average link utilization at the kth step, denoted by $u_{\text{ave}}^{(k)}$, is defined as $u_{\text{ave}}^{(k)} \triangleq \frac{1}{n_1} \sum_{i=1}^{n_1} u_i^{(k)}$. Further, the average link utilization when a flow allocation is finished, denoted by u_{ave}, is defined as follows:

$$u_{\text{ave}} \triangleq \lim_{k \to k_{\max}} u_{\text{ave}}^{(k)}. \tag{3}$$

Let U be a random variable of the link utilization. We define the distribution function $F^{(k)}(u)$ of the random variable U for real u that satisfies $0 \leq u \leq 1$ at the kth step as $F^{(k)}(u) \triangleq \Pr[U \leq u]$. The distribution function obtained through $k_{\max}$ steps, $F(u)$, is defined as

$$F(u) \triangleq \lim_{k \to k_{\max}} F^{(k)}(u). \tag{4}$$

The definition of the total input traffic t_{input} is $t_{\text{input}} \triangleq \sum_i \sum_j t_{ij}$. Let $h_{t_{ij}}^{(k)}$ indicate the number of hops between source i to destination j on the path for the demand t_{ij} at the kth step. The amount of the offered traffic of the flow is defined as $t_{ij} \times h_{t_{ij}}^{(k)}$. Furthermore, the total amount of the offered traffic $t_{\text{offered}}^{(k)}$ is defined as $t_{\text{offered}}^{(k)} \triangleq \sum_i \sum_j \{t_{ij} \times h_{t_{ij}}^{(k)}\}$. The total amount of the offered traffic obtained through $k_{\max}$ steps, t_{offered}, is then derived as

$$t_{\text{offered}} \triangleq \lim_{k \to k_{\max}} t_{\text{offered}}^{(k)}. \tag{5}$$

Let H be a random variable representing the number of hops between the source and the destination of a flow. The average number of hops at the kth step, denoted by $h_{\text{ave}}^{(k)}$, is defined as $h_{\text{ave}}^{(k)} \triangleq \sum_h \{h \times \Pr[H = h]\}$. The average number of hops obtained through $k_{\max}$ steps, denoted by h_{ave}, is defined as

$$h_{\text{ave}} \triangleq \lim_{k \to k_{\max}} h_{\text{ave}}^{(k)}. \tag{6}$$

We now prepare 100 different network topology matrices and 100 different traffic demand matrices. This yields 100×100 different problem instances. Solving these instances, we obtain the average characteristics of the measures defined above. In the following, we indicate the average characteristics using an overbar as $\overline{u_{\max}}$.

2.5 Theoretical Lower Bound

We derive the theoretical lower bound of the maximum link utilization to evaluate the performance of the proposed algorithms. Let $h^{\text{opt}}_{t_{ij}}$ indicate the number of hops on the shortest path between source i and destination j. For the case in which every flow is allocated along its shortest path, the optimum (lowest) total offered traffic, $t^{\text{opt}}_{\text{offered}}$, is defined as $t^{\text{opt}}_{\text{offered}} \overset{\triangle}{=} \sum_i \sum_j \{t_{ij} \times h^{\text{opt}}_{t_{ij}}\}$. Then, the theoretical lower bound of the maximum link utilization becomes as follows: $t^{\text{opt}}_{\text{offered}}/n_l$, where n_l indicates the number of links. In this case, the average link utilization is equal to the maximum link utilization. It should be noted that very few optimal solutions exist because the capacity of each traffic demand is determined by uniform selection from a range. Furthermore, when an optimal solution exists, it is very difficult to obtain.

3 Flow Allocation Algorithms

We propose three flow allocation algorithms for the purpose of obtaining a flow allocation to minimize the maximum link utilization (refer to Section 3.3). The three algorithms are referred to as *flow allocation algorithm A, B,* and *C*.

In each flow allocation algorithm, the initial flow allocation will be established according to the algorithm described in Section 3.1. After that, at each step, a flow will be selected as a candidate to be reallocated according to one of the two *flow selection algorithms*, which will be described in Section 3.2. The two flow selection algorithms are referred to as *random* and *max*. Then, the selected flow will be reallocated.

3.1 Algorithm for Generating an Initial Flow Allocation

The algorithm for generating an initial flow allocation used herein is as follows.

Algorithm for Generating an Initial Flow Allocation:
Repeat the following operations 1 to 4 until all of the flows are allocated.

1. Randomly select a source node.
2. Randomly select a traffic demand, the capacity of which is represented by c^*, among the traffic demands generated from the selected source node.
3. If the available capacity is greater than or equal to c^*, set the weight of the lth $(l = 1, \cdots, n_l)$ link to 1. Otherwise, set the weight of the lth link to ∞.
4. Calculate the shortest path for the flow using the Dijkstra algorithm, and then the flow will be allocated on the shortest path.

3.2   Flow Selection Algorithms

At each step of the flow allocation algorithms, the most congested link can be selected, and a flow passing through the selected link will be chosen. As an algorithm for selecting a flow, we use one of the two algorithms described below.

First, the flow selection algorithm *random* is described as follows:

Flow Selection Algorithm *random*:
At each step of the flow allocation algorithms, the following operations are performed:

1. Select a link in order of its utilization.
2. Randomly select a flow from among the flows passing through the selected link, as a candidate to be reallocated.

For the purpose of improving the maximum link utilization performance as soon as possible, the flow selection algorithm *max* is employed.

Flow Selection Algorithm *max*:
At each step of the flow allocation algorithms, the following operations are performed.

1. Select a link in order of its utilization.
2. Select a flow from among the flows passing through the selected link, in order of bandwidth (capacity), as a candidate to be reallocated.

3.3   Flow Allocation Algorithms

The shortest path for a traffic flow, the capacity of which is represented by c^*, is calculated using the Dijkstra algorithm. The Dijkstra algorithm uses the information on the weights of all links in the network. The weight of each link will be updated according to the network current state as follows. Given the capacity of the ith link C_i and the utilization $u_i^{(k)}$ at the kth step, the available link capacity can be represented as $C_i(1 - u_i^{(k)})$. If $C_i(1 - u_i^{(k)}) \geq c^*$, the weight of the ith link is set to 1. Otherwise, the weight of the ith link is set to ∞.

In each of the flow allocation algorithms, a flow passing through the most congested link is selected according to a flow selection algorithm. The weight of the selected link should be set to ∞. A new path that does not contain the selected link will be then selected.

We now explain the three flow allocation algorithms. The flow allocation algorithm A is the simplest of the three. This is described as follows:

Flow Allocation Algorithm A:
Input: A network topology matrix D and a traffic demand matrix T.
Output: A flow allocation (a set of paths).

[step 0]

1. Establish the initial flow allocation (refer to Section 3.1).
2. As a terminating condition, set the number of steps to be executed, k_{max}.
3. Set k as $k \leftarrow 1$.

[step k]

1. Repeat the following operations (a) to (b) until a flow is reallocated.
   (a) Select a flow, the capacity of which is represented by c^*, according to the employed flow selection algorithm, i.e., *random* or *max* (refer to section 3.2).
   (b) Reallocate the selected flow on a new path that does not contain the selected link (i.e., the weight of the selected link is set to ∞) as calculated using the Dijkstra algorithm.
2. If $k = k_{max}$, finish the algorithm. If $k < k_{max}$, update k as $k \leftarrow k+1$ and repeat the step k.

Next, we explain the flow algorithm B. In this algorithm, we introduce the threshold T for selecting links on a new path. The new path consists of links in which utilizations are less than or equal to the threshold T_h. At each step, the threshold T_h is initialized to $\min_i\{u_i^{(k)}\}$. We next determine whether the new path for the flow exists. If the new path for the flow is not found, the threshold T_h is increased by Δ, $T_h \leftarrow T_h + \Delta$. Then, as described earlier, based on the updated threshold T_h, we determine whether the new path for the flow exists. Note that the threshold T_h satisfies $T_h \leq 1.0$.

Flow Allocation Algorithm B:
Input: A network topology matrix D and a traffic demand matrix T.
Output: A flow allocation (a set of paths).

[step 0]

1. Establish the initial flow allocation (refer to Section 3.1).
2. As a terminating condition, set the number of steps to be executed, k_{max}.
3. Set the difference Δ.
4. Set k as $k \leftarrow 1$.

[step k]

1. Repeat the following operations (a) to (c) until a flow is reallocated.
   (a) Select a flow, the capacity of which is represented by c^*, according to the employed flow selection algorithm (refer to Section 3.2).
   (b) Initialize the threshold T_h as $T_h \leftarrow \min_i\{u_i^{(k)}\}$.
   (c) Repeat the following operations i to iii until either the selected flow is reallocated or the condition $T_h \leq 1.0$ is violated.

(i). Using the Dijkstra algorithm, determine whether a new path exists that does not consist of the selected link and that consists of the links of which the utilizations are less than or equal to the threshold T_h.

(ii). If a new path exists, reallocate the selected flow on the new path.

(iii). If a new path does not exist, update T_h as $T_h \leftarrow T_h + \Delta$ and then repeat the operation (c).

2. If $k = k_{max}$, finish the algorithm. If $k < k_{max}$, update k as $k \leftarrow k + 1$ and repeat the step k.

Finally, we focus on the flow allocation algorithm C. In this algorithm, an inequality (refer to (7) below) must be satisfied when a new path for the selected flow is calculated. The inequality is introduced for the purpose of constantly improving the maximum utilization performance and preventing it from increasing.

Before describing the flow allocation algorithm C, a symbol used by the algorithm is defined. Let $P_i = \{p_{i1}, \cdots, p_{ij}, \cdots\}$ be the ith $(i = 1, 2, \cdots)$ path, where p_{ij} indicates the jth link from the source on the ith path.

Flow Allocation Algorithm C:

Input: A network topology matrix D and a traffic demand matrix T.
Output: A flow allocation (a set of paths).

[step 0]

1. Establish the initial flow allocation (refer to Section 3.1).
2. As a terminating condition, set the number of steps to be executed, k_{max}.
3. Set k as $k \leftarrow 1$.

[step k]

1. Repeat the following operations (a) to (d) until a flow is reallocated. If a path satisfying a condition does not exist for all flows, finish the algorithm.

   (a) Select a flow, the capacity of which is represented by c^*, passing through the i^*th link according to the employed flow selection algorithm (refer to Section 3.2).

   (b) Using the Dijkstra algorithm, determine whether the mth new path $P_m = \{p_{m1}, p_{m2}, \cdots\}$ $(m = 1, \cdots)$ that does not consist of the selected link and that satisfies the following inequality exists.

$$\max_{p_{mj} \in P_m} \{u_{p_{mj}}^{(k)}\} < u_{i^*}^{(k)} - \frac{c^*}{C}. \tag{7}$$

   (c) If $m > 1$, select a path that minimizes the maximum link utilization among the links on the path. Namely, select a path that satisfies the following inequality:

$$\min_{m}\{ \max_{p_{mj} \in P_m} \{u_{p_{mj}}^{(k)}\}\} < u_{i^*}^{(k)} - \frac{c^*}{C}. \tag{8}$$

(d) If a new path exists, reallocate the selected flow on the new path.
2. If $k = k_{max}$, finish the algorithm. If $k < k_{max}$, update k as $k \leftarrow k+1$ and repeat the step k.

4   Numerical Results

Unless otherwise stated, associated parameters are set as follows: $N = 30$, $p = 0.2$, $L_{min} = 2$, $C = 250 \times 10^6$, $a = 0.3$, $t_{min} = 1 \times 10^6$, $t_{max} = 30 \times 10^6$, $\Delta = 10$, $k_{max} = 100$. The number of links n_l becomes 87. The number of traffic flows n_t becomes 261. In this case, the input traffic becomes $\overline{t_{input}} = 4.06 \times 10^9$.

Figure 1 shows the maximum link utilization $\overline{u_{max}}$ as a function of the sequential number of steps k. There are six combinations of the three flow allocation algorithms (denoted by 'A', 'B', and 'C') and the two flow selection algorithms ("random" and "max"). Note that the maximum link utilization of the initial flow allocation becomes $u_{max}^{(0)} = 0.970$ and the theoretical lower bound becomes 0.382. It is clear from this figure that the algorithm C outperforms the others for a relatively larger number of steps. This is because the condition under which to constantly improve the maximum link utilization is introduced in the algorithm C (refer to Eq. (7)). On the other hand, the algorithm A performs the worst. In comparison with the algorithm A, the algorithm B improves the maximum link utilization, because the threshold T_h for selecting links is introduced.

Figure 2 shows the distribution function of the link utilization $F(u)$ when $k_{max} = 100$. The utilizations for almost all the links are moderate, i.e., approximately 0.5, in the case of the algorithm C. In other words, the algorithm C can obtain a flow allocation to balance the traffic load over the network. On the other hand, in the case of the algorithm A, some links can be overloaded, whereas the other links cannot be utilized much. Therefore, the algorithm C outperforms the others.

Table 1 shows the maximum link utilization $\overline{u_{max}}$, the average link utilization $\overline{u_{ave}}$, the amount of offered traffic $\overline{t_{offered}}$, and the average number of hops $\overline{h}$. First, we focus on the maximum link utilization $\overline{u_{max}}$. Here, we compare the performance of the flow selection algorithms. This table shows that the performance of the algorithm *random* is better than that of the algorithm *max*. This is because in the algorithm *max*, since a flow passing through the congested link is selected in order of bandwidth (capacity), the maximum utilization is rapidly improved as shown in Fig. 1. However, the solution (the flow allocation) may be a local optimal. On the other hand, in the algorithm *random*, after a relatively larger number of steps, a better solution can be obtained. Next, we focus on the average link utilization $\overline{u_{ave}}$. The flow allocation algorithm C with the flow

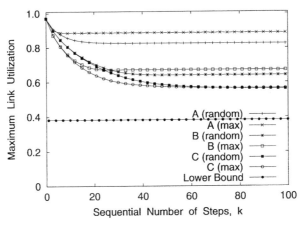

Fig. 1. The maximum link utilization $\overline{u_{\max}}$ as a function of the number of steps

Table 1. The maximum link utilization $\overline{u_{\max}}$, the average link utilization $\overline{u_{\text{ave}}}$, the total amount of offered traffic $\overline{t_{\text{offered}}}$, and the average number of hops $\overline{h}$ when $k_{\max} = 100$

Algorithm	$\overline{u_{\max}}$	$\overline{u_{\text{ave}}}$	$\overline{t_{\text{offered}}}$	$\overline{h}$
theoretical lower bound	0.382	0.382	8.31×10^9	2.06
A (*random*)	0.825	0.387	8.43×10^9	2.07
A (*max*)	0.887	0.385	8.38×10^9	2.06
B (*random*)	0.640	0.423	9.21×10^9	2.25
B (*max*)	0.671	0.418	9.10×10^9	2.16
C (*random*)	0.562	0.449	9.77×10^9	2.43
C (*max*)	0.565	0.462	10.04×10^9	2.40

selection algorithm *random* achieves a better performance than the algorithm C with *max*, while the algorithm C with *random* achieves the best maximum link utilization performance. Finally, we focus on the amount of offered traffic $\overline{t_{\text{offered}}}$. This table suggests that an increase in the amount of offered traffic cannot be ignored, because the flow may be reallocated on another path whose number of hops is longer than or equal to that of the original shortest path.

5   Conclusion

We proposed three flow allocation algorithms, A, B and C, for traffic engineering in ISP backbone networks. In each flow allocation algorithm, as a candidate to be reallocated, a flow will be selected according to one of the two flow selection algorithms, referred to as *random* and *max*.

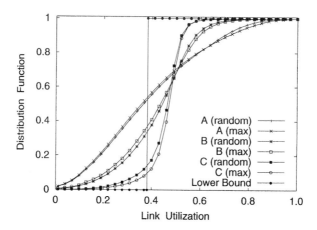

Fig. 2. Distribution function of the link utilization $F(u)$ when $k_{max} = 100$

The flow allocation algorithm C outperforms the others. In this algorithm, a condition concerned with improving the maximum utilization performance is introduced. As an algorithm for selecting a flow to be reallocated, the algorithm *random* can obtain a good solution after a relatively larger number of steps.

We conclude that the flow allocation algorithm C with the flow selection algorithm *random* is useful in the sense that this algorithm can obtain a set of flows to balance the traffic load over the network.

References

[1] Awduche, D., Chiu, A. Elwalid, A., Widjaja, I., Xiao, X.: Overview and Principles of Internet Traffic Engineering, IETF RFC 3272 (2002)
[2] Awduche, D., Malcolm, J., Agogbua, J., O'Dell, M., McManus, J.: Requirements for Traffic Engineering Over MPLS, IETF RFC 2702 (1999)
[3] Xiao, X., Hannan, A., Bailey, B.: Traffic Engineering with MPLS in the Internet, IEEE Network Magazine (March/April 2000)
[4] Li, T., Rekhter, Y.: A Provider Architecture for Differentiated Services and Traffic Engineering (PASTE), RFC 2430, IETF (1998)
[5] Elwalid, A., Jin, C., Low, S., Widjaja, I.: MATE: MPLS Adaptive Traffic Engineering, Proc. IEEE Infocom 2001 (2001) 1300 – 1309
[6] Fortz, B., Thorup, M.: Internet Traffic Engineering by Optimizing OSPF Weights, Proc. IEEE Inforcom 2000 (2000)
[7] Bertsekas, D., Gallager, R.: Data Networks, Chapter 5, Prentice-Hall (1987)

Dynamic Load Distribution in MPLS Networks*

Jeonghwa Song, Saerin Kim, and Meejeong Lee

Department of Computer Science, Ewha Womans University
{jhsong,cheer,lmj}@ewha.ac.kr

Abstract. Exploiting the efficient explicit routing capability of MPLS, we propose a dynamic multipath traffic engineering mechanism named LDM(Load Distribution over Multipath). The main goal of LDM is to enhance the network utilization as well as the network performance by adaptively splitting traffic load among multiple paths. LDM makes decisions at the flow level, and distribute traffic based on both the length and the load of a path. LDM also dynamically selects a number of good Label Switched Paths (LSPs) for traffic delivery according to the state of the entire network. We use simulation to compare the performance of LDM with the performance of several representative dynamic load distribution approaches as well as the traditional static shortest path only routing. The numerical results show that LDM outperforms the compared approaches in both the blocking ratio as well as the performance of the accepted traffic flows.

1   Introduction

Traffic engineering mainly deals with effective mapping of traffic demands onto the network topology with the purpose of enhancing the network utilization as well as the network performance perceived by the users [1]. MultiProtocol Label Switching (MPLS), in conjunction with path establishment protocols such as CR-LDP or RSVP-TE, makes it possible to set up a number of Label Switched Paths (LSPs) between a source destination pair [2, 3, 4]. MPLS with the efficient support of explicit routing enables assigning a particular traffic stream onto one of the available LSPs, providing basic mechanism for facilitating traffic engineering.

Having multiple LSPs for a destination is a typical setting which exists in an operational ISP network that implements MPLS. With multiple LSPs available for an egress node, the goal of the ingress node is to distribute the traffic across the LSPs so that the network utilization as well as the network performance perceived by users is enhanced. Load distribution over multiple paths can be based on either the historical traffic statistics collected over time or the dynamic network status information. The long-term traffic statistics based mechanisms usually utilize the global information on the incoming traffic projection as well as the network topology in order to pre-determine LSP layout and traffic assignment

---

* This Work was supported by grant No. R04-2000-000-00078-0 from the Basic Research Program of the Korea Science & Engineering Foundation

H.-K. Kahng (Ed.): ICOIN 2003, LNCS 2662, pp. 989–999, 2003.

[5, 6, 7, 8]. If the traffic variation is little, this mechanism can provide optimal traffic proportioning. It does not, though, attempt to adapt to unpredictable time-dependent traffic variations.

The dynamic network status based mechanisms utilize dynamically changing network status information in order to determine the set of paths to be used in traffic delivery and/or traffic proportioning among those paths [9, 10, 11, 12, 13]. Both [10] and [11] assume that the set of candidate paths (meaning the paths used for traffic transportation) for an ingress-egress pair is fixed, and utilize the dynamic network state information in order to equalize the congestion measure among the candidate paths. [12] supplements the work in [11] with a mechanism that dynamically selects a few good candidate paths. [9] presents a pretty thorough work that dynamically computes the candidate path set and the proportioning of traffic among the paths in the candidate set. It expands the candidate path set if congestion persists in the current set, and the paths are included in the candidate path set in the order of their length. The traffic loads are distributed over to the paths in the candidate path set in order to equalize the loading on all the paths. [13] defines a set of primary LSPs for each ingress node. If congestion occurs at one of the primary LSPs of an ingress node, the ingress node tunnels the part of traffic belonging to that LSP to the other ingress node. The overhead for setting up and maintaining LSP could be minimal with this scheme, but long detour or delay may be caused.

This paper presents a traffic engineering mechanism, named Load Distribution over Multipath (LDM) that splits traffic load to a number of available LSPs based on the dynamic network status in order to enhance the network utilization as well as the network performance perceived by the users. It is assumed that the variations in actual traffic are appreciable, and thus a pure dynamic approach is taken not requiring any a priori traffic load statistics. LDM is intended for the best-effort type traffic that does not impose any particular service requirement to the network. LDM makes the routing decision at the flow level in order to avoid packet reordering overhead. Moreover, the LSPs are set up statically based on the global network topology information in order to minimize the LSP set up overhead and latency. An LSP for an incoming traffic flow is dynamically selected based on both the current congestion level, and the length of the path in terms of the number of hops.

LDM is similar to the work in [9] in that it tries to expand the candidate path set as the congestion level of the candidate LSP set increases. There are, though, a couple of salient features that differentiate LDM from the existing works including the work in [9]:

1. As a link is more heavily utilized, it is restricted to be utilized by LSPs with less number of extra hops. The number of extra hops is defined as the difference of the hop count of an LSP from that of the smallest hop count LSP for a given source-destination pair.
2. An LSP for a particular flow is selected randomly among candidate LSPs with a probability distribution that is a function of the length as well as the utilization of the paths.

Due to the first feature, as the utilization of a certain part of a network becomes higher, less detour is allowed to take place through that area. It is based on the reasoning that the chance of degrading the performance of future traffic streams due to the waste of network resources caused by the detour is high when the detour happens through a highly utilized area. According to the second feature, the route selection for a user flow takes tradeoffs between the utilization and the length of the available paths in order to enhance the performance as well as to conserve resources.

The rest of this paper is organized in the following way. In chapter 2, we provide a complete description on the 3 main components of LDM. The simulation results that compare the performance of LDM with some of the representative dynamic traffic load distribution approaches as well as the static shortest path only routing are presented in chapter 3. Finally chapter 4 concludes our paper.

2   Load Distribution over Multipath (LDM)

LDM consists of three main components: (1) topology driven multiple LSP set up for each ingress-egress pair (2) candidate LSP set computation (3) a path selection for an incoming flow. In order to minimize the latency to start traffic forwarding, LDM at each ingress node pre-computes a number of possible paths for each egress node, and set up LSPs for those paths whenever the ingress node is initialized or there are some changes in the network topology. This process is solely dependent on the global network topology information. Among the LSPs set up by the first component, LDM selects a set of candidate LSPs to be used in the traffic distribution (note that it is called "candidate LSP set" hereinafter) based on the current network state information. Link utilization is the metric that is used to represent the dynamic network status in LDM, and it is assumed that OSPF link state information flooding can be used to exchange this information. The candidate path set could either be pre-computed when there are some significant changes in the dynamic network status or be computed on demand for a new arriving user flow request. For each of the newly arriving user flow, LDM randomly selects an LSP from the candidate LSP set according to the probability distribution determined by the path length as well as the utilization of an LSP.

2.1   Topology Driven Multiple LSP Set Up

In order to constrain the maximum amount of extra network resources as well as the number of LSPs to be set up, LDM limits the maximum number of extra hops to be δ. It is assumed that the impact of using an extra hop is relative to the scale of a network, and the scale of a network can be represented by a network diameter D. δ is determined in terms of D as the following:

$$\delta = D \times r, \quad 0 \le r \le 1.$$

Whenever an ingress node is initialized or there is a network topology change,

LDM computes all the paths that satisfy the above limitation, and set up LSPs for those paths. Due to the LSP management overhead, further limitation on the number of LSPs can be imposed by the network administrator.

2.2 Computing a Candidate LSP Set

Depending on the dynamic network status, LDM selects a subset of the LSPs for an ingress-egress pair, and distributes traffic load among those LSPs. These LSPs are referred to as the candidate LSPs. We will first define several notations in order to explain the candidate LSP set computation procedure in detail. Let L_{ij} denotes the set of all the LSPs set up between an ingress node i and an egress node j, and let the corresponding candidate LSP set by A_{ij}, then $A_{ij} \subseteq L_{ij}$. The utilization of an LSP, $u(l)$, is defined as the maximum of the utilization value of the links that constitute the LSP l, and let $h(l)$ denote the hop count of LSP l. The utilization of a candidate LSP set A_{ij} is defined as the following:

$$U(A_{ij}) = \min[u(l), \forall l \in A_{ij}].$$

Fig. 1. presents the procedure that A_{ij} is determined from L_{ij}. Initially, A_{ij} is set as follows:

$A_{ij} = \{$LSPs from i to j with the smallest hop count and with the utilization lower than η_0 $\}$

Similar to the MPLS-OMP in [9], LDM decides whether to expand the candidate LSP set based on the congestion level of candidate LSP set. If $U(A_{ij}) \geq \rho$, then LDM further expands A_{ij}. The expansion of A_{ij} continues, considering LSPs in L_{ij} in the increasing order of hop count until $U(A_{ij}) < \rho$ or there is no LSP left in L_{ij} for further consideration.

Generally, an LSP $l \in L_{ij}$ with $h(l) = (h(\text{shortest LSP}) + m)$ should satisfy the following two conditions to be eligible for A_{ij}:

1. $u(l) < \max[u(k), \forall k \in A_{ij}]$
2. $u(l) < \eta_m$, where $\eta_m < \eta_n$ for $m > n$

The first condition implies that LDM utilizes the LSPs with more extra hops if they at least have lower utilization than the LSP that has the highest utilization among the LSPs in the current A_{ij}. Moreover, since the second condition keeps the links with utilization of η_m or higher from being included in the LSPs with the path length $(h(\text{shortest LSP}) + m)$ or longer, and $\eta_m < \eta_n$ for m > n, LSPs with more extra hops are applied with more strict utilization restriction when they are considered for the candidate LSP set eligibility. The second condition actually implies a multi-level trunk reservation. That is, links with the utilization higher than η_m can only be used by the LSPs with less than m extra hops.

Note that with the candidate LSP set expansion condition i.e., $U(A_{ij}) \geq \rho$, and the first of the two eligibility conditions given above, a longer LSP is added to the candidate LSP set if the utilization of the current candidate LSP set is higher than ρ and the utilization of this longer LSP is relatively lower than the LSP with the maximum utilization among the LSPs in the current candidate LSP set. With these two conditions, longer detour using an LSP with more

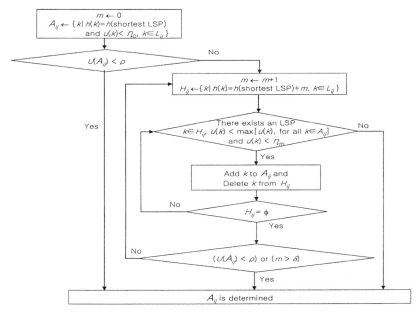

Fig. 1. Candidate LSP set computation procedure

extra hops is allowed even when that longer LSP's absolute level of congestion is already significant. The second eligibility condition enables to prevent this.

2.3   Selecting an LSP from the Candidate LSP Set to Assign a Flow

For each incoming traffic flow, LDM randomly selects an LSP from the candidate LSP set according to the probability distribution that is a function of both the length and the utilization of the LSP. Let P_l denote the probability that LSP l is selected, and let us define several additional notations to explain how to compute P_l. For a candidate LSP set A, let $A = \{l_1, l_2, \ldots l_{N_a}\}$, where N_a is the number of LSPs in A. C_0 is the constant to make the sum of the probabilities that are inversely proportionate to the hop count of an LSP. That is,

$$C_0 = \frac{1}{\displaystyle\sum_{i=1}^{N_a} \frac{1}{h(l_i)}} \quad and \quad \sum_{i=1}^{N_a} \frac{C_0}{h(l_i)} = 1.$$

Let $E = \max[u(sp), \forall sp \in \text{shortest LSPs in } A]$ and $d(l_i) = E - u(l_i)$, for $1 \leq i \leq N_a$. Then C_1 is the variable to make the sum of the probabilities that are proportionate to $d(l_i)$. That is,

$$C_1 = \sum_{i=1}^{N_a} d(l_i) \quad and \quad \sum_{i=1}^{N_a} \frac{d(l_i)}{C_1} = 1.$$

P_l is then defined as the following:

$$P_l = a_0 \frac{C_0}{h(l)} + a_1 \frac{d(l)}{C_1}, \quad where \quad a_0 + a_1 = 1 \quad and \quad 0 \le a_0, a_1 \le 1.$$

3 Simulation Experiments

The performance of LDM is studied through simulation experiments. As shown in Fig. 2., typical ISP type network model with 11 LERs and 11 intermediate LSRs is used in the simulation. Each LSR has finite output port queue with the capacity of 100 packets, and each link has bandwidth of 45Mbps in each direction. The link propagation time is assumed to be 0.1msec.

The interval of user requests for a new session follows the exponential distribution with a mean of 4 seconds, and the length of each user session also follows exponential distribution with a mean of 100 seconds. In order to simulate a realistic case, it is assumed that the 30% of the user requests occur between two hot source-destination pairs, and the rest of the requests are uniformly distributed among the rest of the edge routers. Within a session, packets are generated following a Poisson distribution. The mean of the packet inter-arrival times are varied from 5ms to 25ms in order to vary the traffic load. The size of the packet is assumed to be 1000bytes. Each simulation results are collected from the measurements of 8000 seconds of simulation time.

The values of the various LDM parameters used in the simulation are summarized in Table 1.

The network model used in the simulation (Fig. 2.) has the diameter $D=4$, and we let the maximum number of extra hops (δ) to be equal to D. To determine an appropriate value for the candidate LSP set expansion condition, ρ, we tried values ranging from 50% $\sim$ 70%, and found that around 60% is the knee point. We set $a_0 = a_1 = 0.5$, so that the LSP selection reflects the length and the utilization with the same weight. The values of η_i, for $0 \le i \le \delta$, are evenly spread in the range [70%, 90%], with the lower bound of this range being 10% higher than ρ.

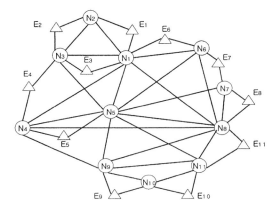

Fig. 2. Network model for the simulation

Table 1. Values of the LDM parameters used in the simulation

Parameter	Value	Parameter	Value
δ	4	η_4	70%
η_0	90%	ρ	60%
η_1	85%	a_0	0.5
η_2	80%	a_1	0.5
η_3	75%		

The performance of LDM is compared with that of the static shortest path only routing (called SP hereinafter) and some of the dynamic load distribution approaches listed in the following:

- Alternate Path routing (ALT): LSPs are utilized in the order of their hop count. If the utilization of the currently utilized path is over η_0, next hop count LSP is then utilized. For an ingress-egress pair, if the values of utilization of all the alternative LSPs are over η_0, then no further user request is allowed.
- Trunk Reservation scheme (TR): Alternative LSPs are utilized in the same order as in the ALT, except for that links with utilization over η_1 can only be used by the shortest LSP. For an ingress-egress pair, if the utilization of the shortest LSP is over η_0, and the utilization of all the alternative LSPs is over η_1, no further user request is allowed.
- Random routing (RAN): An LSP is selected randomly among all the LSPs available. When the utilization of all the LSPs for an ingress-egress pair is over η_0, no further user request is allowed.
- Lightest Load scheme (LL): Among all the LSPs for an ingress-egress pair, an LSP with the smallest utilization is selected. If the utilization is larger than η_0 for all the LSPs, no further user request can be allowed for that ingress-egress pair.

Fig. 3. ∼ 7. present the performance of the LDM and the 5 compared schemes. Call acceptance ratio, link utilization and efficiency are measured as the metrics for the efficiency of network utilization. Packet loss rate and delay are also measured as the metrics for the network performance perceived by the users. The x-axis of these graphs corresponds to the average data rate of a session. SP has the lowest call acceptance ratio as expected. Among the dynamic multi-path load distribution approaches, RAN has the lowest call acceptance ratio due to the lack of efficiency in its LSP selection. Since it utilizes multiple paths, though, it has higher call acceptance ratio than SP when the traffic load is high.

In terms of the call acceptance ratio, ALT shows slightly better performance than TR, but as presented in Fig. 4. and 5., TR shows better performance than ALT in terms of the loss and the delay for the accepted traffic. Since TR utilize LSPs in their hop count order, traffic load distribution over multiple paths occurs after the link utilization of some of the links becomes fairly high or even over η_0. This may prohibit utilizing some of the alternative LSPs with extra hops. ALT also has similar problem, but due to the trunk reservation TR tend to have fewer alternative LSPs available. Even though LDM also deploys trunk utilization, it does not suffer from this problem as can be inferred from its high call acceptance ratio. Since LDM starts distributing traffic load over multiple LSPs way before the utilization of some of the links becomes as high as in TR or ALT, and thus traffic load is more evenly distributed with LDM.

LL, which selects the lightest load LSP, has almost similar call acceptance ratio as ALT, which selects the LSP in the order of the hop count. For the performance of the accepted traffic, however, LL shows better performance than both ALT and TR as presented in Fig. 4. and 5. This shows that distributing traffic load evenly contribute to increasing the network revenue as well as to enhance the network performance for the accepted traffic. Since LL does not consider the length of the path, though, it consumes relatively larger amount of network resources as shown in the average link utilization results in Fig. 6. LDM shows the highest call acceptance ratio. In the heaviest traffic load case, it improves the call acceptance ratio about 25% compared to SP, and 5% ∼ 15% compared to the other dynamic load distribution approaches. In terms of the performance for the accepted traffic, it also shows the best performance.

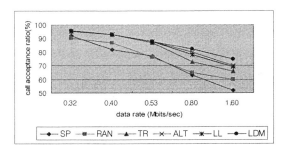

Fig. 3. Call acceptance ratio as a function of input traffic load

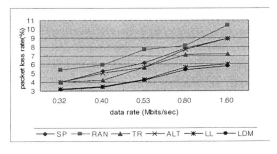

Fig. 4. Packet loss rate as a function of input traffic load

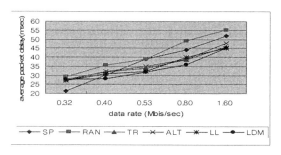

Fig. 5. Average packet delay as a function of input traffic load

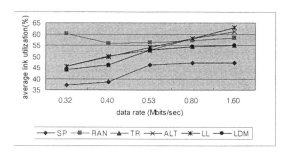

Fig. 6. Average link utilization as a function of input traffic load

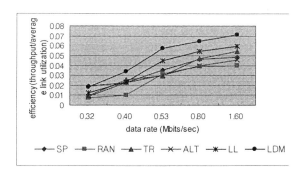

Fig. 7. Efficiency(throughput/average link utilization) as a function of input traffic load

Fig. 6. presents the average link utilization of the compared schemes. SP has the lowest average link utilization value since SP has smallest traffic load admitted into the network. As the traffic load grows larger over a certain level, TR's link utilization increases very little due to the trunk reservation restriction. As the traffic load grows, the average link utilization of LL increases most rapidly, and the link utilization of LL has the highest value when the traffic load is the heaviest among the experimented cases. Since LL tries to evenly distribute load over the entire network, the loss rate and the delay is lower than that of TR and ALT though. LDM shows a relatively rapid utilization increment as the traffic load increases in the lower traffic load range (for the packet inter-arrival time of 25ms ∼ 15ms). For further traffic load increases, though, the link utilization increment slows down in LDM. This is because with the multi-level trunk utilization, LDM restrain the longer path utilization as the congestion level of the entire network becomes more severe. LDM has lower link utilization compared to the other dynamic load distribution schemes, while balancing the traffic load comparatively well to LL as can be inferred from its high call acceptance ratio as well as the low loss ratio and the delay shown in Fig. 3.

Fig. 7. presents the efficiency of the compared schemes. Efficiency is defined as the ratio of the throughput to the average link utilization. LDM shows the highest efficiency, and LL comes to the next place. ALT, which utilizes LSPs in their hop count order, shows the lowest efficiency except for RAN.

4 Conclusions

We proposed a multipath load distribution mechanism named LDM. LDM statically sets up multiple LSPs between an ingress-egress pair, and determines the candidate LSP set dynamically. The candidate LSP set is expanded in order to include enough LSPs. The expansion is, though, restricted not to take detour through a highly utilized part of the network. Furthermore, as the level of congestion at a certain part of a network becomes higher, less detour is allowed through that area. LDM randomly selects an LSP from the candidate LSP set following a probability distribution determined by both the length and the utilization of the LSP. The simulation results show that LDM performs better than the length based LSP selection (ALT and TR) as well as the traffic load based LSP selection (LL) in terms of the network utilization as well as the network performance perceived by users.

References

[1] D. Awduche, A. Chui, A. Elwalidn, I. Widjaja, and X. Xiao, "Overview and Principles of Internet Traffic Engineering", RFC 3272, May 2000.
[2] E. Rossen, A. Viswanathan, and R. Callon, "Multiprotocol Label Switching Architecture", RFC 3031, Jan. 2001.
[3] B. Jamoussi, L. Andersson, R. Dantu, L. Wu, P. Doolan, T. Worster, N. Feldman, A. Fredette, M. Girish, E. Gray, J. Heinanen, T. Kilty, and A. Malis, "Constraint-Based LSP set up using LDP", RFC 3212, Jan. 2002.

[4] D. Awduche, L. Berger, D. Gan, T. Li, V. Srinivasan, and G. Swallow, "RSVP-TE: Extensions to RSVP for LSP Tunnels", RFC 3209, Dec. 2001.

[5] D. Mitra and K. G. Ramakrishnan, "A Case Study of Multiservice, Multipriority Traffic Engineering Design for Data Networks", Proc. GLOBECOM'99, Dec. 1999.

[6] K. W. Ross, "Multiservice Loss Models for Broadband Telecommunication Networks", Springer-Verlag, 1995.

[7] A. Sridharan, S. Bhattacharyya, C. Diot, R. Guerin, J. Jetcheva, and N. Taft, "On the Impact of Aggregation on the Performance of Traffic Aware Routing", Proc. INFOCOM, 2001.

[8] Y. Lee, Y. Seok, and Y. Choi, "A Constrained Multipath Traffic Engineering Scheme for MPLS Networks", Proc. ICC 2002, May 2002.

[9] C. Villamizar, "MPLS Optimized Multipath (MPLS-OMP)", work in progress Internet-Draft <draft-villamizar-mpls-omp-01>, Feb. 1999.

[10] Elwalid, C. Jin, S. Low, and I. Widjaja, "MATE: MPLS Adaptive Traffic Engineering", Proc. INFOCOM, 2001.

[11] S. Nelakuditi, Z-L. Zhang, and R. P. Tsang, "Adaptive Proportional Routing: A Localized QoS Routing Approach", Proc. INFOCOM'00, Mar. 2000.

[12] S. Nelakuditi and Z-L. Zhang, "On Selection of Paths for Multipath Routing", Proc. IWQoS'01, 2001.

[13] S. Patek, R. Venkateswaren, and J. Liebeherr, "Enhancing Aggregate QoS through Alternate Routing", Proc. GLOBECOM'00, 2000.

Distributed Transport Platform
for TCP-Friendly Streaming

Byunghun Song, Kwangsue Chung, and Seung Hyong Rhee

School of Electronics Engineering, Kwangwoon University, Korea
byungh@adams.gwu.ac.kr
{kchung,shrhee}@daisy.gwu.ac.kr

Abstract. Recent advances in network bandwidth and processing power of CPUs has led to the emergence of multimedia streaming frameworks, such as window media player and realvideo. These frameworks typically rely on proprietary stream establishment and control mechanisms to access multimedia context. To facilitate the development of standards-based distributed multimedia streaming applications, the OMG has defined a CORBA-based specification that stipulates the key interfaces and semantics needed to control and manage audio/video streams. But, this specification does not have a detail of implementation. Particularly, it is not able to provide QoS enabled congestion control scheme for improvement of network efficiency and bandwidth fairness over real network environments. It is a very important and difficult technical issue to provide the streaming transport platform with advanced congestion control scheme. In this paper, we propose an architecture of a distributed transport platform and deal with the design and implementation concept of our proposed architecture. Also, we present a new mechanism to improve streaming utilization by the TCP-friendly streaming scheme.

1 Introduction

Many distributed multimedia applications rely on custom and proprietary low-level stream establishment and signaling mechanisms to manage and control the presentation of multimedia content. These types of applications run the risk of becoming obsolete as new protocols and services are developed. Fortunately, there is a general trend to move from programming custom applications manually to integrating applications using reusable components based on the open distributed object computing (DOC) middleware, such as CORBA, DCOM, and Java RMI [1].

Although the DOC middleware is well-suited to handle request/response interactions among client/server applications, the stringent QoS requirements of multimedia applications have historically precluded the DOC middleware from being used as their data transfer mechanism. For instance, inefficient CORBA Internet Inter-ORB Protocol (IIOP) implementations perform excessive data-copying and memory allocation per request, which increases packet latency. Likewise, inefficient marshaling/demarshaling in the DOC middleware decreases streaming data throughput [2].

H.-K. Kahng (Ed.): ICOIN 2003, LNCS 2662, pp. 1000–1009, 2003.

If the design and performance of DOC middleware can be improved, however, the stream establishment and control components of distributed multimedia applications can benefit greatly from the portability and flexibility provided by middleware. To address these issues, the Object Management Group (OMG) has defined a specification for the control and management of Audio/Video (A/V) streams, based on the CORBA reference model [3].

The specification of CORBA A/V streaming service defines an architecture for implementing open distributed multimedia streaming applications. This architecture integrates well-defined modules, interfaces, and semantics for stream establishment and control with efficient data transfer protocols for multimedia data transmission. In addition to defining standard stream establishment and control mechanisms, the specification of CORBA A/V streaming service allows distributed multimedia applications to leverage the inherent portability and flexibility benefits provided by standard-based middleware.

But, this specification does not have a detail of implementation. Particularly, it is not able to provide QoS enabled congestion control scheme for improvement of network efficiency and bandwidth fairness over real network environment [4]. This paper focuses on a previously unexamined aspect in designing the distributed transport platform, that is, a *TCP-friendly streaming scheme in distributed transport platform*.

The rest of this paper is organized as follows: We first present the background and related work in Section 2. In Section 3, we describe the design of distributed transport platform. Section 4 illustrates the performance evaluation of the proposed platform. Our conclusions and future work are presented in Section 5.

2   Background and Related Work

This section first presents an overview of the main architectural components in the OMG's CORBA A/V streaming service. We then introduce the TCP-friendly steaming schemes for the improvement of network efficiency and bandwidth fairness.

2.1   Specification of CORBA Audio/Video Streaming Service

The specification of CORBA A/V streaming service defines *flows* as a continuous transfer of media between two multimedia devices. Each of these *flows* is terminated by a *flow data end-point*. A set of flows, such as audio flow, video flow and data flow, constitute a *stream*, which is terminated by a *stream endpoint*. A *stream end-point* can have multiple *flow data end-points*.

Figure 1 illustrates the specification of CORBA A/V streaming service. It shows multimedia stream, which is represented as a *flow* between two *flow data end-points*. One *flow data end-point* acts as a source of the data and the other *flow data end-point* acts as a *sink*. Note that the control and signaling operations pass through the GIOP/IIOP-path of the ORB, demarcated by the dashed box. In contrast, the data *stream* uses out-of-band stream, which can be implemented

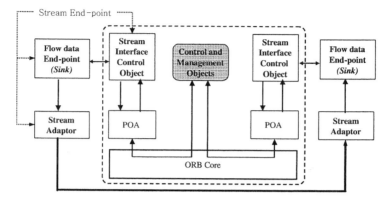

Fig. 1. Specification of CORBA A/V Streaming Service

using communication protocols (UDP, RTP, etc.) that are more suitable for multimedia streaming than IIOP. Maintaining this separation of concerns is crucial to meeting end-to-end QoS requirements [5].

Each *stream end-point* consists of three logical entities: i) a *stream interface control object* that exports an Interface Definition Language (IDL) interface, ii) a *data source (or sink)*, and iii) a *stream adaptor* that is responsible for sending and receiving frames over a network. *Control and management objects* are responsible for the establishment and control of streams. The specification of CORBA A/V streaming service defines the interfaces and interactions of the *stream interface control objects* and the *control and management objects*. Section 3 describes more details on the various components shown in Figure 1.

In short, the goals of the CORBA A/V streaming service include the following: i) Provide standardized stream establishment and control, ii) Support multiple data transfer protocols (TCP, UDP, RTP, ATM, etc.), iii) Provide the interoperability of flows, and iv) Support many types of sources (stream server) and sinks (stream client).

2.2   TCP-Friendly Streaming Schemes

The Real-time Transport Protocol (RTP) in widely used for multimedia communication in the Internet, because it offers the necessary mechanisms for collecting and exchanging information about losses and delay [6]. RTP is basically a UDP-based streaming protocol. Unfortunately, UDP does not support congestion control. For this reason, wide usage of UDP applications in the Internet might lead to the overload situation. Furthermore, the UDP causes the starvation of congestion controlled TCP traffic which reduces its transmission rate during overload situation.

Recently, there has been several proposals for TCP-friendly streaming schemes to avoid such a situation. In this paper, TCP-friendliness is the terminology used for non-TCP streaming flows. Non-TCP streaming flows are defined as TCP-friendly when "their long-term throughput does not exceed the throughput of a conformant TCP connection under the same conditions". TCP-friendly streaming can be provided on the basis of two different schemes. One is the rate-based streaming scheme and another is the window-based streaming scheme.

Particularly, many rate-based TCP-friendly streaming schemes mimic TCP's additive increase/multiplicative decrease (AIMD) behavior to achieve TCP-fairness. This approach is used to adjust the streaming rate according to simple analytical model of conformable TCP traffic. Early work in this area was presented in Jacobs's research. Afterwards, Padhye et al. present the complex analytical model for the TCP transmission rate of TCP connection with s as the segment size, p as the packet loss rate, t_{RTT} as the round trip time, t_{RTO} as the retransmission timeout. Based on there work, the average bandwidth share of TCP depends mainly on t_{RTT} and p as shown in Equation 1 [7].

$$T = \frac{s}{t_{RTT}\sqrt{\frac{2bp}{3}} + t_{RTO}\,\min\left(3\,\sqrt{\frac{3bp}{8}}\right)p(1 + 32p^2)} \tag{1}$$

3   Design of Distributed Transport Platform

The design concept of Distributed Transport Platform (DTP) is mainly based on the *DTP Overall Architecture* and the *SRTP Protocol for DTP Stream.*

3.1   Design of DTP Overall Architecture

Figure 2 illustrates the *DTP Overall Architecture*. The specification of CORBA A/V streaming service identifies two peers in stream establishment, which are known as the "A" party and the "B" party. These terms define complimentary relationships, i.e., a stream always has an A party at one end and a B party at the other [3]. Figure 2 shows a *stream controller* (*aStreamCtrl*) binding the A party together with the B party of a stream. The stream controller need not to be collocated with either end of a stream. To simplify the example, however, we assume that the controller is collocated with the A party, and is called the *aStreamCtrl.*

The principal components in the DTP architecture are as follows: i) *Stream control (StreamCtrl)* is responsible for supervising a particular stream binding, ii) *Multimedia device (MMDev)* is a representation of the hardware (microphone, loudspeakers, etc) available at a designated terminal - it is also a factory for stream end-points and virtual devices, iii) *Virtual device (VDev)* is an abstract representation of the hardware - prior to establishing an A/V stream, virtual devices at both ends exchange configuration information, iv) *Stream end-point (StreamEndPoint)* represents one end of a stream - both end-points exchange a flow specification, and v) *Flow connections, flow end-points* and *flow devices* are used to manage multiple unidirectional flows within one single stream.

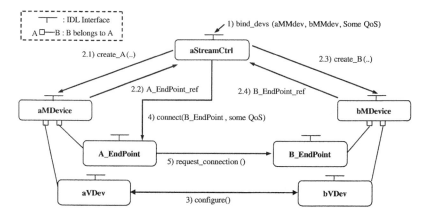

Fig. 2. DTP Overall Architecture

3.2   Design of SRTP Protocol for DTP Stream

At the heart of DTP architecture is its transport protocol design. The architecture of CORBA A/V streaming service separates its stream establishment and control protocols from its data transfer protocols, such as TCP, UDP, RTP, or ATM, thereby allowing applications to select the most suitable data transfer protocols for a particular network environment or set of application requirements.

As mentioned in the Section 2.2, TCP-friendly streaming schemes are more suitable for multimedia transmission than RTP, TCP and UDP protocols. Most of the AIMD-based TCP-friendly streaming schemes present that the sender adapts its transmission behavior based on frequent feedback messages, such as TCP. This is particularly important for the case of reliable transport where the sender needs to retransmit lost packets.

On the contrary, our proposed scheme, the Smart RTP (SRTP), was designed to use the RTP for exchanging feedback information about the round trip time and the losses at the receiver. In other words, the SRTP is TCP-friendly version of RTP protocol. This might be actually more appropriate for A/V streaming than other TCP-friendly schemes where the transmission rate is rapidly changed on the basis of very frequent feedback as follows at the end of each round:

$$X_{i+1} \leftarrow X_i + \theta \quad (X_i \leq T) \tag{2}$$

$$X_{i+1} \leftarrow \delta \cdot X_i + (1 - \delta)(X_i \cdot (1 - \sqrt{p})) \quad (X_i > T) \tag{3}$$

Where X_i is the transmission rate of round i, T is the TCP-friendly rate value calculated using Equation. 1. θ is the increase constant value, δ is a tunable parameter, which determines the amount of rate variation. ρ is the number of packet loss. RTCP messages include information about the losses and delays noticed in the network. Losses are estimated at the receiver by counting the gaps in the sequence numbers included in the RTP header of the data packets. N_{real} is number of actually received packet, N_{max} is maximum number of received packet, N_{first} is number of first received packet.

$$\rho = \frac{N_{real}}{(N_{\max} - N_{first})} \tag{4}$$

From Equation 1 it is evident that the overall TCP throughput is inversely proportional to the square root of the loss values. Hence, we propose to reduce the rate of the SRTP flows in a similar manner to TCP instead of using the exact model. Thus, after receiving a loss notification from the RTCP messages, the SRTP sender can reduce its transmission rate, whereas in AIMD the reduction is based on the decrease constant independent of number of packet loss.

Figure 3 shows the trajectory of rate allocation for two flows traversing through a single shared link. The coordinate $X = (x_1, x_2)$ represents the rate allocation for Flow 1 and Flow 2 at a given time instant. If all allocations with $x_1 + x_2 = X_{goal}$ are efficient and all allocations with $x_1 = x_2$ are fair, then the point $(X_{goal}/2, X_{goal}/2)$ is the optimal point since we want the rate allocation to be both efficient and fair.

Figure 3-(i) shows the trajectory of general AIMD congestion control, which eventually oscillates about the optimal point along the fairness line under a simplifying assumption that the rounds of x_1 and x_2 are synchronized, and whenever the sum of x_1 and x_2 exceeds the link capacity both senders receive loss feedback that is greater then zero. Figure 3-(ii), shows the trajectory of rate allocation for two flows controlled by SRTP congestion control assuming each stream has equal packet loss probability. The point X_0 represents the initial rate allocation of the flows. Both flows increase their sending rates linearly until the system reaches $X_1 = (x_1, x_2)$. This brings the system to X_2 which is closer to the fairness line.

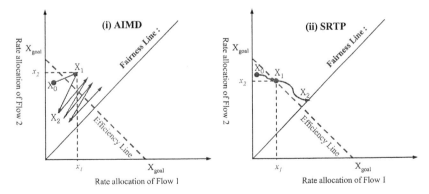

Fig. 3. Convergence Properties of AIMD and SRTP

As a corollary to the previous paragraph, SRTP competes more aggressive than AIMD. We focus on the applicability of SRTP congestion control in DTP applications. In general, DTP applications require a smooth rate variation in steady state. At the same time, as soon as the connection bandwidth decreases, the sender (A or B party) must be able to reduce its sending rate quickly. The challenge is to develop a congestion control scheme that achieves the objective of smooth rate variation while retaining the responsiveness to network dynamics. SRTP provides a very smooth rate oscillation. Therefore, the SRTP is more suitable for DTP stream.

4   Performance Evaluation

We implement our proposed distributed transport platform (DTP). This section describes the result of performance evaluations conducted on the basis of our implementation. A prototype MPEG-4 streaming system developed in Linux environments with MICO ORB [18], is used in our evaluation.

4.1   Testbed Environment

We conducted experiments using the testbed shown in Figure. 4. The $s2$ (Pentium III PC with Linux 2.4) streamed to the $s4$ (Pentium III PC with Linux 2.2) across a 1.5 Mbps link with 19 ms latency, configured using Dummynet. Background Web cross-traffic was introduced using the SURGE toolkit to emulate various network conditions. To testing the performance of the DTP service, we used a simple topology, with the bottlenecked link allowing both SRTP-based stream connections and TCP-based stream connections.

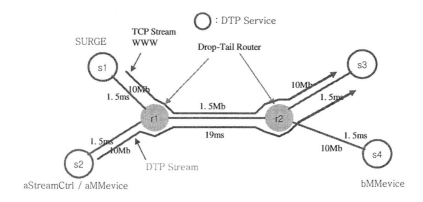

Fig. 4. Testbed Configuration

4.2   Throughput of DTP Stream

The aim of this experiment is to illustrate that DTP does not introduce appreciable overhead in transporting data. To demonstrate this, we implement a SRTP-based data streaming component and integrated it with the DTP architecture. The producer in this application establishes a stream with the consumer, using the stream establishment mechanism discussed in Section 3.1. Once the stream is established, it streams data via SRP to the consumer. Throughput evaluation is performed by three difference streams: i) *TCP stream – i.e.,* by a pair of application processes that do not use the DTP stream establishment protocol. This is the "ideal" case since there is no additional ORB-related or presentation layer overhead, ii) *ORB stream – i.e.,* the throughput obtained by a stream that used an octet stream passed through the MICO ORB. In this case, the IIOP was used for the data transfer, and iii) *DTP stream – i.e.,* this stream uses the DTP stream establishment protocol as described in Section 3.2.

We measured the throughput obtained by fixed the buffer size (2 Kbyte/s) of the sender. The results shown in Figure 5 indicate that, as expected, the DTP stream does not introduce any significant overhead in data streaming. In the case of using IIOP as the data transfer layer, the benchmark incurs additional performance overhead. This overhead arises from the dynamic memory allocation, data-copying, and marshaling/demarshaling performed by the ORB's IIOP protocol engine. Particularly, the largest disparity occurred for smaller buffer sizes, where the performance of the ORB was approximately half that of the TCP and DTP streaming implementations. Clearly, there is a fixed amount of overhead in the ORB that is amortized and minimized as the size of the data payload increases.

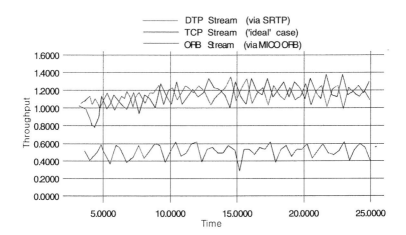

Fig. 5. Throughput Evaluation

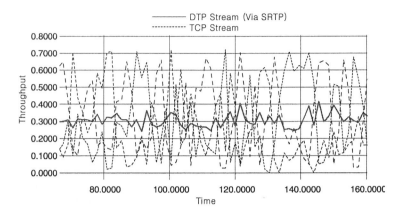

Fig. 6. One DTP and Three TCP Stream

4.3 Fairness and Smoothness of the DTP Stream

The DTP stream (via SRTP) is generally fair to TCP traffic across the wide range of network types and conditions we examined. Figure 6 shows a typical experiment with three TCP flows and one DTP stream (SRTP flow). In this case, the transmission rate of the STRP flow is slightly lower, on average, than that of the TCP flows. At the same time, the transmission rate of the SRTP flow is smooth, with a low variance. In contrast, the bandwidth used by each TCP flow varies strongly even over relatively short time periods.

Additionally, the results shown in Figure 7 indicate that, as expected, the oscillation of SRTP's windows size is slightly lower than that of the TCP flows. Based on our evaluation, the DTP provides high throughput for streaming application with no significant overhead and fairness and smoothness characteristics.

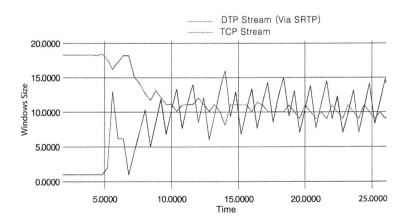

Fig. 7. Oscillation of Windows Size

5   Conclusion and Further Work

The demand for high quality multimedia streaming is growing, both over the Internet and for intranets. Distributed object computing is also maturing at a rapid rate due to middleware technologies like CORBA. The flexibility and adaptability offered by CORBA makes it very attractive for use in streaming service. This paper illustrates an approach to building standards-based, flexible, adaptive, multimedia streaming applications using CORBA.

We also emphasized the importance of TCP-Friendly scheme. It is very important for multimedia streaming to be "TCP-friendly", because a dominant portion of today's Internet traffic is TCP-based. The multimedia streaming systems are expected to react to congestion by adapting their transmission rates and provide the inter-protocol's fairness, in order to be efficiently transported over the Internet. Since our DTP architecture provides a very smooth rate oscillation, it has good-fairness characteristics.

Future work will focus on ways to guarantee an agreed QoS. This could be achieved by using such protocols as the resource reservation protocol (RSVP) or Differenciated Service (DiffServ).

Acknowledgement

This research has been conducted by the Research Grant of Kwangwoon University in 2002. It is also supported in part by Korea Science and Engineering Foundation under contract number R01-2002-000-00179-0(2002).

References

[1] J. W. Hong, J. Kim, and J. Park, "A CORBA-Based Quality of Service Management Framework for Distributed Multimedia Services and Applications," *IEEE Network*, March 1998.

[2] A. Gokhale and D. C. Schmidt, "Optimizing the Performance of the CORBA Internet Inter-ORB Protocol Over ATM," Proceeding *of SIGCOMM* 97, Aug. 1997.

[3] Object Management Group, "Control and Management of A/V Streams Specification," OMG Document Telecom, Oct. 1998.

[4] J. Padhye, V. Firoiu, D. Towsley and J. Kurpose, "Modeling TCP Throughput: A Simple Model and Its Empirical Validation," *Proceeding of SIGCOMM 98*, Aug. 1998.

[5] S. Mungee, N. Surendran, and D. Schmidt, "The Design and Performance of a CORBA Audio/Video Streaming," *Proceeding of HICSS-32*, Jan. 1999.

[6] H. Schulzrinne, "RTP: A Transport Protocol for Real-Time Application," RFC 1889, Jan. 1996.

[7] J. Widmer, R. Denda, and Martin Mauve, "A Subvey on TCP-friendly Congestion Control," *IEEE Network*, May 2001.

Directory-Based Coordinated Caching in Shared Web Proxies

Yong H. Shin[1], Hyokyung Bahn[2], and Kern Koh[1]

[1] School of Computer Science and Engineering, Seoul National University
{yhshin, kernkoh}@oslab.snu.ac.kr
[2] Department of Computer Science and Engineering, Ewha Womans University
bahn@ewha.ac.kr

Abstract. Caching at the proxy server is in use throughout the world to reduce the network congestion and access latency on the WWW. To gain the maximum benefits of caching, the sharing of the Web caches among proxy servers has recently become an important issue. The sharing scheme is generally performed by the Internet Cache Protocol (ICP), which supports the discovery of Web objects from neighboring proxy caches. However, the sharing scheme via ICP has two fatal weaknesses: 1) message multicast overhead and 2) redundant caching of same objects among collaborative proxies. In this paper, we present a new Web cache sharing scheme. Our new scheme reduces the message overhead of ICP dramatically. It also reduces the duplicated copies of the same objects in globally shared Web caches. Using trace-driven simulation, we show that the proposed scheme outperforms ICP and Cache Array Routing Protocol (CARP) in terms of various performance measures.

1 Introduction

Proxy caching is widely used to reduce the network traffic and access latency on the World Wide Web. To maximize the benefits of proxy caching, the sharing of caches among proxy servers is increasingly important. The sharing scheme is generally performed by the Internet Cache Protocol (ICP), which supports the discovery and retrieval of Web objects from neighboring proxy caches [1]. When a client requests a Web object from a proxy server that does not have the requested object in its own cache, it sends ICP queries to neighboring proxies to see whether they have the object. If one of them has the object, the proxy sends HTTP requests to it and gets the object from it. Fig. 1 shows the proxy server architectures with the sharing scheme.

The sharing scheme could significantly reduce the load of Web servers and access latency perceived by Web users. However, the sharing scheme via ICP has two fatal weaknesses. First, whenever a local cache miss occurs in each proxy, ICP multicasts queries to all the other collaborative proxies to examine whether they have the object. Thus, as the number of collaborative proxies increases, the overhead of ICP increases quadratically.

H.-K. Kahng (Ed.): ICOIN 2003, LNCS 2662, pp. 1010–1017, 2003.

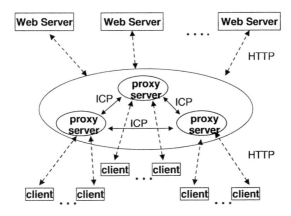

Fig. 1. Proxy Server Architecture with Sharing Scheme

Second, ICP does not coordinate the contents of each cache at all. Therefore, in the worst case, all the collaborative caches may have the same contents. Then, the cache sharing would, finally, give no benefits at all.

There have been some attempts to resolve the problems of ICP. The Cache Array Routing Protocol (CARP) partitions URL-space among the collaborative proxies, and allows each proxy to cache only the objects whose URLs are hashed to it [2]. This prevents the redundant caching of identical objects in shared Web caches. It also resolves the message multicast problem of ICP because the location of an object is decided directly by hashing. However, CARP is not appropriate for wide-area distributed proxies. For example, objects accessed frequently by local users may be cached at remote proxies, which incurs the *remote hit problem.*

In this paper, we propose a new Web cache sharing scheme called DCOORD (Directory-based Coordinated caching) that resolves the aforementioned problems of ICP and CARP. Our scheme maintains the location of a cached object in a *directory*, hence multicast is not needed. The directory is distributed such that each collaborative proxy maintains the exclusive directory information of URL-space which is hashed to it. Hence, when a requested object is not in the local cache, the proxy discovers the location of the object from the neighbor cache that the URL is hashed to.

However, DCOORD does not fix the domain of caching as is done in CARP. Our scheme allows to maintain the hot objects of local proxy users in a certain area of the local proxy cache. This area is called working set area, and resolves the remote hit problem of CARP. Without working set area, our scheme reduces the redundant caching of same objects in globally shared Web caches to store much more distinct objects. This coordination of contents among proxies could improve the global cache hit ratio. Through trace-driven simulations, we show that our scheme performs better than ICP and CARP significantly in terms of various performance measures.

2 Related Works

To avoid the multicast overhead of ICP, there have been studies that use the directory-based protocol. In the directory-based protocol, the locations of all objects in the shared caches are maintained in the *directory*, so multicasting is not needed. Baggio et al. proposed a local directory scheme to find objects in other caches and update the directories asynchronously [4]. Gadde et al. used a central server to keep track of the cache directories of all proxies, and let all proxies query the central server for cache hits in other proxies [6]. This scheme has a weakness in that the central server may be the bottleneck point. Hence, recently they have proposed a new scheme that divides the single central server into several directory servers. Fan et al. proposed a new sharing protocol named *summary cache protocol* [7]. Because the directory information is maintained in the main memory, in general, it requires too much space to maintain the full directory information as the cache size and the number of collaborating proxies increase. Hence in the summary cache protocol, each proxy keeps a summary of the URLs of cached objects of other proxies, not the exact directory information, and the summary is periodically updated. Even though the summary may incur information loss, they show that the hit ratio is almost identical with the complete directory method.

3 The Directory-Based Coordinated Caching Scheme

When a client requests a Web object to the proxy server, there can be three different scenarios as follows:

- *local hit*: If the object is in its local cache, the proxy sends it to the client directly.
- *remote hit*: If the object is not in the local cache but exists in some neighbor caches, the proxy gets the object from one of the neighbors, and sends it to the client.
- *miss*: If the object does not exist in any of the shared caches, then the proxy gets the object from the origin Web server and sends it to the client.

When we assume that the costs of a local hit and a miss are 0 and 1 respectively, a remote hit obviously incurs a cost between 0 and 1. Our scheme aims to save the total costs incurred after all of the clients' requests have been processed. We first try to reduce the number of duplicated copies of the same object in the global shared Web caches to store much more distinct objects. This effort could eventually change a miss into a hit. However, the scenario could change a local hit into a remote hit even though a remote hit is more expensive. Therefore, the scenario should be done without reducing the local hit ratios. It is known that only a small fraction of Web objects explains most of the local hits in terms of temporal locality in Web access traces, which means that local hits can be covered by a small fraction of cache spaces [3].

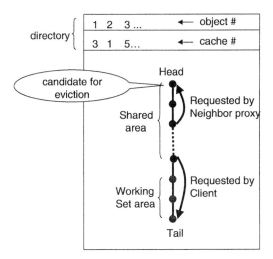

Fig. 2. The data structure of one proxy cache

Therefore, our method firstly maintains the hot working set of each proxy in a certain area of the local cache, and the remaining area is coordinated with other proxies to reduce the amount of duplicated contents.

Fig. 2 shows the data structures of a proxy cache. They consist of a directory and a list. The directory has the location information of the objects whose URLs are hashed to that proxy. We use a list structure to maintain the objects in the local cache like the LRU replacement policy. The most recently requested object by a *client* moves to the "tail" of the list and joins the working set area. Note that the working set area covers a certain fraction of objects at the tail of the list. The "head" of the list points to the least valuable object in the cache, which means that the object it points to is the candidate for eviction by the replacement policy. When an object in the shared area is requested by a *neighbor proxy*, it moves to the head of the list because it will be cached at that requesting proxy. This will eventually reduce the redundant caching of the same objects in globally shared caches. On the other hand, if an object in the working set area is referenced by a *neighbor proxy*, nothing happens in the local cache because the objects in the working set area are hot objects which have been referenced by local clients recently.

When the requested object is not in the local cache, the proxy discovers the location of the object from the neighbor proxy that the URL is hashed to. The neighbor proxy, then, updates the location of the object to the currently requested proxy because it will cache the object now and potentially preserve the object for a long time. Fig. 3 shows the algorithm that is invoked upon a request for an object at a proxy server.

```
if object i is requested by a client
{  if object i is in the local cache
       move object i to the tail of the list and send i to the client
   else
   {   k = Hash(URL of object i)
       request the location of i to neighbor proxy k
       if response from k is one of the neighbor caches
           try to fetch i from that neighbor cache
       if object i cannot be fetched from neighbors
           fetch i from the origin Web server
       send object i to the client
       add object i to the tail of the list
       while size of i is larger than free space in local cache
           remove the head of the list
       while size of the current working set is larger than
           the specified working set size
           remove the head of working set area
   }
}
else if object i is requested by one of the neighbor caches
{  send object i to that requesting neighbor cache
   if object i is in the local cache and is not in the working set
       move object i to the head of the list
}
else if location of object i is requested by neighbor cache m
{  if the location of object i is stored in the directory
       send the location to m
   else send NULL to m
   the location of object i is updated to m in the directory
}
```

Fig. 3.   The algorithm of DCOORD

4   The Results

To assess the effectiveness of DCOORD, we perform a trace-driven simulation study. We use a public Web proxy trace of NLANR (SD) [5]. The trace consists of about 3.2 million requests made over a period of 4 days. Table. 1 shows the characteristics of the trace.

In our simulation, we partition the clients in the trace into groups assuming that each group has its own proxy, and simulate the cache sharing among the proxies. This roughly corresponds to the scenario where each branch of a company or each department in a university has its own proxy cache, and the caches collaborate. Note that this scenario is generally used in the simulation of a Web cache sharing scheme [7]. We compared DCOORD with CARP and ICP. In our experiments, the "Hit Ratio" is not an appropriate performance measure,

Table 1. The characteristics of the Trace

Duration	6/28/2000 - 7/1/2000
Total Requests	3235430
Total Distinct Requests	1666801
Total Distinct Mbytes (Infinite Cache)	2680.59
The Number of Distinct Servers	129409
The Number of Groups	4

because the Hit Ratio cannot differentiate a remote hit from a local hit. As explained in Sect. 3, the benefit of a local hit is greater than that of a remote hit. Hence, we define a new performance measure, named the "Cost-Savings Ratio", to measure the benefits of remote hits and local hits properly. In this paper, we define the cost of a local hit, a remote hit, and a miss as 0, 0.2, and 1, respectively. Note that the cost of a remote hit may be different according to the closeness (i.e., tightly-coupled or loosely coupled) of each proxy. Before discussing the result itself, we point out for clarity, some of the peculiarities of the presented results and each scheme.

- In case of CARP, we hashed the URLs to equally distribute to each group of cache.
- In case of DCOORD, we specified the size of the local working set area to reflect the cost sensitivity between a local hit and a remote hit properly.
- For all of the three schemes, we used general LRU as the replacement policy in each local cache.

Fig. 4 shows the Cost-Savings Ratio of the three schemes as a function of the cache size. In the figure, the x-axis represents the cache size, which is relative size compared with the size of "Infinite Cache" in Table 1. For all cases, DCOORD performs the best, ICP the next, and CARP the worst in terms of the Cost-Savings Ratio. Specifically, DCOORD performs better than ICP and CARP by 7.2% and 10.6%, respectively.

While CARP tries to get higher hit ratios by eliminating redundant objects, it cannot exploit the locality set of each group, causing a low local hit ratio. ICP exploits the full locality set of each group, so that it could get the highest local hit ratio. However, it causes a lower remote hit ratio due to the redundant information in each cache. As shown in the figure, DCOORD shows the best performance because it tries to preserve the hot locality set of each group and at the same time minimizes redundant information.

Fig. 5 shows the total number of inter-proxy network messages for each scheme. As shown in the figure, both DCOORD and CARP consistently perform well as the number of caches increases. However, the number of ICP messages increases quadratically as the number of collaborative proxies increases. This shows that ICP is not a scalable protocol due to the message multicast overhead. In terms of message numbers, CARP is slightly better than DCOORD.

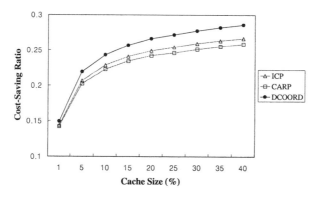

Fig. 4. Cost-Savings Ratio as the cache size increases

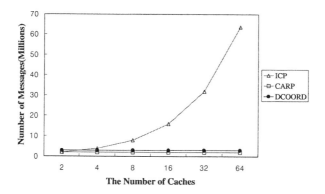

Fig. 5. The Number of Messages as the number of proxies increases

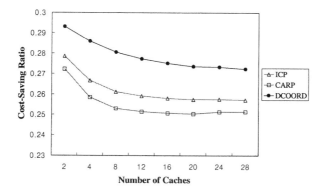

Fig. 6. Cost-Savings Ratio as the number of proxies increases

This is because DCOORD usually sends a message to the neighbor proxy that has the directory of the requested object, while CARP sends no messages at all.

Fig. 6 shows the Cost-Savings Ratio of DCOORD, ICP and CARP as the number of proxies increases. The result shows that the performance gap between DCOORD and the other schemes is consistently large, which means that DCOORD is more effective than the other schemes with a large number of collaborative proxies.

5   Conclusion

In this paper, we proposed a new Web cache sharing scheme called DCOORD. DCOORD reduces the message overhead of ICP dramatically by eliminating multicast overhead of ICP. It also reduces the duplicated copies of the same objects in globally shared Web caches, so that much more distinct objects can be cached. However, DCOORD allows the hot working set objects of local clients to be duplicated in the local caches. Through trace-driven simulations, we showed that DCOORD performs better than ICP and CARP by 7.2% and 10.6% respectively, in terms of Cost-Savings Ratio, and eliminates the message overhead of ICP between 24% to 92%.

References

[1] D. Wessels and K. Claffy: Internet Cache Protocol.
    http://ircache.nlanr.net/Cache/ICP (1997)
[2] V. Valloppillil and K. W. Ross: Cache Array Routing Protocol.
    http://ircache.nlanr.net/Cache/ICP/carp.txt
[3] M. Arlitt and C. Williamson: Web Server Workload Characterization: The Search for Invariants. Proceedings of the ACM SIGMETRICS (1996)
[4] A. Baggio and G. Pierre: Oleron: Supporting information sharing in large-scale mobile environments. Proceedings of the ERSADS Workshop (1998)
[5] National Laboratory for Applied Network Research:
    ftp://ircache.nlanr.net/Traces (2000)
[6] S. Gadde and M. Rabinovich: A Taste of Crispy Squid. Workshop on Internet Server Performance (1998)
[7] L. Fan, P. Cao, J. Almeida, and A. Z. Broder: Summary Cache: A Scalable Wide-Area Web Cache Sharing Protocol. Proceedings of ACM SIGCOMM (1998)

Characterization
of Web Reference Behavior Revisited:
Evidence for Dichotomized Cache Management

Hyokyung Bahn[1] and Sam H. Noh[2]

[1] Department of Computer Science and Engineering, Ewha Womans University
bahn@ewha.ac.kr
[2] School of Information and Computer Engineering, Hong-Ik University
samhnoh@hongik.ac.kr

Abstract. In this paper, we present the Dichotomized Cache Management (DCM) scheme for Web caches. The motivation of the DCM scheme is discovered by observing the Web reference behavior from the viewpoint of Belady's optimal replacement algorithm. The observation shows that 1) separate allocation of cache space for temporal locality and reference popularity better approximates the optimal algorithm, and 2) the contribution of temporal locality and reference popularity on the performance of caching is dependent on the cache size. With these observations, we devise the DCM scheme that provides a robust framework for on-line detection and allocation of cache space based on the marginal contribution of temporal locality and reference popularity. Trace-driven simulations with actual Web cache logs show that DCM outperforms existing schemes for various performance measures for a wide range of cache configurations.

1 Introduction

Caching mechanisms have been studied extensively to alleviate the speed gap of hierarchical storages in computer systems. More recently, due to the increase in popularity of the Internet, Web caching is becoming increasingly important. To improve the performance of caching, modeling of reference characteristics is essential. However, reference characteristics of Web environments are quite different from those of traditional caching environment because there is a human factor involved in the generation of Web references. That is, it is human that generates Web reference streams, not computer programs as in conventional paging systems. This makes management of a caching system more challenging compared to those of traditional caching systems. In traditional paging systems, *temporal locality* is dominant and hence, page reference streams can be modeled well by the LRU-stack model [1]. In Web caching environment, however, both *temporal locality* and *reference popularity* influence the re-reference likelihood of object references. Hence, an efficient cache management system must take into account both properties.

H.-K. Kahng (Ed.): ICOIN 2003, LNCS 2662, pp. 1018–1027, 2003.

Another important property that must be considered in Web caching is that the retrieval *cost* and *size* of an object to be cached are not necessarily uniform. This property makes the Web cache replacement problem much more complicated. Determining even the off-line optimal cache replacement algorithm is known to be NP-hard [2]. Hence, little effort has been made to examine how an optimal algorithm works and use it as a model for the design of a replacement algorithm. Specifically, even the influence of temporal locality and reference popularity on the re-reference likelihood of a Web object has not been explored quantitatively from the viewpoint of an optimal algorithm. This paper makes an attempt to do so.

To this end, we take the following approach. First, we simplify the problem by assuming that all objects considered by the Web cache are uniform (i.e., identical cost and size), and focus on the analysis of the re-reference likelihood of cached objects. With this assumption, we can readily observe the reference behavior of Web objects (based on a particular replacement algorithm) in contrast to the Belady's optimal algorithm (OPT) [1].

We make the following two observations from this first phase. One, the contribution of temporal locality and reference popularity on the re-reference likelihood of a Web object is dependent on the cache size, and two, partitioning of cache space for management based on temporal locality and reference popularity better approximates the optimal algorithm. Based on these observations, we devise the Dichotomized Cache Management (DCM) scheme that divides the cache space into two to separately exploit temporal locality and reference popularity based on their contribution to the reference probability.

In the second phase of our approach, we extend the DCM, which was devised under the assumption that the cache objects are uniform, to non-uniform objects. This is done via a normalization method that evaluates a cached object based on its estimated reference probability multiplied by the cost of the object per unit size. This results in a normalized assessment of the contribution to the performance measure for non-uniform objects, leading to a fair replacement algorithm.

2   Motivation

The success of any cache replacement algorithm depends heavily on how well the algorithm estimates the re-reference likelihood of cached objects. Hence, we wanted to quantify the impact of temporal locality and reference popularity on the re-reference of Web objects. To do this, we compared the hit distributions of a particular Web reference trace when using the Least Recently Used (LRU) and the Least Frequently Used (LFU)[1] replacement algorithms, and compared them with that when using Belady's off-line optimal (OPT) algorithm, assuming that all objects are uniform.

---

[1] When not explicitly mentioned otherwise, the LFU we refer to here is the perfect LFU, which does not lose its frequency count even when it is evicted from the cache.

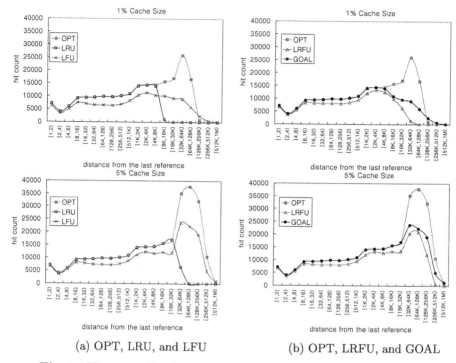

(a) OPT, LRU, and LFU   (b) OPT, LRFU, and GOAL

Fig. 1. Hit count distributions for various replacement algorithms

Figure 1(a) shows the comparison results. The x-axis in the figure represents the logical time since last reference, that is, the inter-reference gap, and the y-axis is the total hit count with the given range of inter-reference gap for the OPT, LRU, and LFU algorithms. (Note that the x-axis is in log-scale.) This particular figure is obtained with the public proxy cache logs of Digital Equipment Corporation [6] for cache sizes that are 1% and 5% of the infinite cache size. (Note that "infinite cache size" refers to the total number of distinct objects in the trace as objects are assumed to be of uniform size.)

The characteristics observable from Figure 1(a) is quite different from traditional caches such as the file system buffer cache [12], where repetitive regular reference patterns are generally observed. Observe here that for the OPT algorithm a large number of hits occur for references with large inter-reference gaps (64K-256K). Another observation, though not immediately discernable in this figure due to the log-scale in the x-axis, is that on account of temporal locality, the hit count decreases as the distance from the last reference increases. These same kind of behaviors were observable for not only this DEC trace, but for other traces as well.

Now, let us take a look at how the LRU and LFU algorithms behave. Since the LRU maintains objects based on the ordering of the distance from the last reference, the LRU follows almost exactly the hit counts of the OPT from left (smaller inter-reference gap) to right (larger inter-reference gap). This shows that temporal locality is an important factor in determining the replacement object, especially when the inter-reference gap is small. Note also that after some point in the x-axis, the number of hits drops rather quickly compared to the OPT. This implies that temporal locality alone is not sufficient to estimate the re-reference likelihood of Web objects.

Taking a look at the LFU algorithm, it follows, but with much less accuracy, the hit counts of the OPT for a wider range of inter-reference gap values. Specifically, the LFU sustains the hit counts of the OPT for large inter-reference gaps fairly well, which was not possible for the LRU. This implies that hits with large inter-reference gaps can be more effectively covered by exploiting reference popularity rather than temporal locality.

Previous research in cache replacement have pointed out that considering both the temporal locality and reference popularity at the same time can help in estimating the re-reference likelihood of objects. For example, the LRFU (Least Recently/Frequently Used) algorithm, which is one of the representative algorithms in this class, exploits temporal locality and reference popularity by a combined recency and frequency value (CRF value) [5]. This and other algorithms [3, 9] all tried to somehow unify the two properties together.

Figure 1(b) shows the hit count distributions of the LRFU. Though not directly comparable in the figure, the hit curve for the LRFU is generally located in between the curves for the LRU and LFU. Even though the LRFU generally estimates reference behavior better than the LRU and LFU at some point or another, it does not fully utilize the strong points observed for the LRU and LFU, that is, for small inter-reference gap values and large inter-reference gap values, respectively. For example, the LRU algorithm with the 1% size cache obtains more hit counts than the LRFU algorithm with the 5% size cache when the inter-reference gap values are smaller than 8K. Hence, when we have a 5% size cache, we can obtain better performance than the LRFU algorithm just by allocating the LRU 1% size and the LRFU the remaining part.

Our goal is then to come up with a scheme that will gain the most from exploiting the two properties, that is, temporal locality and reference popularity. This will result in a hit count distribution represented by GOAL in Figure 1(b), which is the hit counts taken from the higher values of the LRU and LFU algorithms of Figure 1(a). We show in Section 3 how this may be achieved.

3   The Dichotomized Cache Management (DCM) Scheme

In this section, we present the Dichotomized Cache Management (DCM) scheme. Subsection 3.1 presents DCM assuming that objects are uniform. This sets the groundwork for DCM for Web environments. The uniform object assumption is then relaxed in Subsection 3.2 via normalization. In Subsection 3.3, we show how DCM may be efficiently implemented using various data structures.

3.1 Cache Partitioning based on Re-reference Likelihood

In this subsection, we assume that all objects are of the same size and cost. As stated previously in Section 1 our goal is to manage the cache so that the performance of the management scheme reflects the best of the LRU and the LFU. This is to be done by exploiting the temporal locality and the reference popularity characteristics of the references in the cache. Note that the LRU and LFU algorithms are known to be optimal for references that show the temporal locality and reference popularity property, respectively [1].

Based on the observations made previously, we would like to employ the LRU algorithm for those objects that have small inter-reference gaps, which reflect temporal locality, while for those with large inter-reference gaps, we would like to employ the LFU algorithm.

The question then is how to divide up the cache such that objects with small inter-reference gaps retain a certain portion of the cache, while objects with large inter-reference gaps retain the remaining cache space. To achieve this we devise the Dichotomized Cache Management (DCM) scheme that divides the cache space into two based on the marginal contribution of temporal locality and reference popularity to the re-reference likelihood. Marginal contribution refers to the additional hit count that is obtained when one more unit of cache space is allocated, and is calculated as follows.

Temporal Locality: As we apply the LRU algorithm, the expected hit ratio of streams with temporal locality is $HR = \sum_{i=1}^{n} a_i$, where n is the cache size and a_i is the reference probability of position i in the LRU stack. It is possible to measure a_i by where cache hits occur in a ghost LRU stack buffer [11]. The ghost LRU stack buffer is simply a list of dataless object headers where all objects are ordered by their backward distance. Specifically, a_i is measured by the hit counts at stack position i divided by the length of reference streams.

Reference Popularity: As we apply the LFU algorithm for the streams with reference popularity, the expected hit ratio is $HR = \sum_{i=1}^{n} r_i$, where n is the cache size and r_i is the reference probability of an object i (assuming $r_i \geq r_j$, if $i \leq j$). This is reasonable because reference popularity of Web objects is known to be modeled as a Zipf-like distribution, which is an independent reference model [4]. Similarly to the temporal locality case, r_i can be measured by observing where buffer hits occur in the ghost LFU buffer.

3.2 Normalization of the Non-uniformity Factor

Now we extend the DCM algorithm for objects with arbitrary cost and size, which is a more realistic scenario for the Web caching environment. The algorithm first partitions the cache space according to the marginal contributions as explained in Subsection 3.1. (Refer to Subsection 3.3 for implementation detail.) Then, both the temporal locality cache space and reference popularity cache space are managed by the normalization method given in Definition 1.

Definition 1. Cache replacement algorithm A for non-uniform objects *normalizes* the cost and size of an object if it selects as its replacement victim the object i that has the smallest $Value(i)$, which is defined as

$$Value(i) = p(i) \cdot Weight(i)$$

where $p(i)$ denotes the reference probability of object i estimated by algorithm A and $Weight(i) = c_i/s_i$ where c_i and s_i denote the cost and size of object i, respectively.

Since the DCM uses ghost LRU and LFU buffers, $p(i)$ value can be assigned either the a_i or r_i value depending on whether it resides as part of the temporal locality cache space or the reference popularity cache space, where a_i and r_i are the probabilities of position i in the ghost LRU and LFU buffers, respectively. $Weight(i)$ can be defined differently for the performance measure of interest that reflects the primary goal of the given environment. For example, if the goal of caching is in minimizing the network traffic, the corresponding measure would be the *Byte Hit Ratio* and c_i can be defined as the retrieved size for object i when a miss to the object occurs. Similarly, if the goal is in minimizing latency perceived by Web users, the corresponding measure would be the *Delay-Savings Ratio* and c_i can be defined as the retrieval latency for object i.

3.3 Implementation of the Algorithm

To reduce the measurement overhead of ghost buffers, in our implementation of the DCM, the hit counts of the ghost buffers are observed not by individual stack positions but by groups, that is, disjoint intervals of stack positions. In evaluating each object in the cache, the DCM needs $p(i)$, that is the estimated reference probability of object i. As explained in Subsection 3.2, this value should be assigned the a_i and r_i values for the temporal locality area and reference popularity area, respectively. However, since we are grouping for the sake of efficiency, we cannot obtain all a_i and r_i values. Hence, we use an approximation method to obtain the $p(i)$ values for both areas based on previous characterization studies on Web workloads.

A number of characterization studies have shown that the reference probability of a Web object whose inter-reference time equals t is roughly proportional to $1/t$ [4, 8, 9]. Hence, we can assign the $p(i)$ of object i in the temporal locality area $p(i) = b/t_i$, where t_i is the time since last reference to object i and b is a constant. Constant b need not be considered in this case because it is a common multiple of $Value(i)$ for all objects i. As for the reference popularity area, we can use the reference count of object i divided by the length of reference streams as $p(i)$ values because long-term popularity can be modeled as a Zipf-like distribution (which assumes independent reference), and hence this value converges to the stationary probability r_i.

Another issue that should be clarified for the DCM algorithm is the overlapping problem between the temporal locality area and reference popularity area.

Even if the cache space is separately assigned, some objects may be the target of caching in both the temporal locality area and reference popularity area. In this case, only one copy is preserved in the physical cache and the remaining cache area is repeatedly allocated to either of the areas by the order of hit counts of the successive ghost buffer groups.

For caching algorithms to be practical, it is important that the time complexity of the algorithm not be excessive. When a replacement algorithm satisfies the following order preservation property, it can be implemented with a heap data structure whose time complexity for insertion and deletion operations is $O(\log_2 n)$.

Definition 2. *Order preservation property*

If $Value(a) > Value(b)$ holds at time t and neither a nor b have been referenced after t, a replacement algorithm A with *order preservation property* satisfies $Value(a) > Value(b)$ for any $t'(t' > t)$.

In the DCM algorithm, ordering in the reference popularity area satisfies the order preservation property and hence, it can be effectively implemented in $O(\log_2 n)$. However, ordering in the temporal locality area does not satisfy this property because the value of an object depends on the current time. For example, suppose there exists objects x and y such that $Weight(x) = 10$, $p(x) = 1/10$, $Weight(y) = 10000$, and $p(y) = 1/10001$, respectively, at time t. Then, $Value(x) > Value(y)$ holds at time t. However, if both x and y have not been referenced after t, then say, at time $t + 10$, $Value(y)$ becomes larger than $Value(x)$.

Hence, for the temporal locality area, we approximate the original algorithm using multiple list structures as is done by Aggarwal et al. [10]. The algorithm uses separate list structures, such that the i-th list maintains all objects whose cost per unit size value range $[2^{i-1}, 2^i)$ by the LRU order. Whenever free space is needed in the temporal locality area, the DCM compares the values of the least recently used objects in each list and replaces the object with the smallest value among them.

4   Experimental Results

In this section, we discuss the results from trace-driven simulations performed to assess the effectiveness of the DCM scheme. We used two public Web proxy cache traces from National Lab for Applied Network Research (NLANR) [7]. NLANR provides several sanitized proxy cache traces. Among them, we presented the results obtained from the SD and UC traces. (Note that results from other traces are similar to those from the SD and UC traces.) Table 4 shows the characteristics of each trace. In the experiments, we filtered out some requests such as UDP requests, "cgi_bin" requests, and requests whose size is larger than the cache size used in the simulation.

The algorithms used in the simulation are LRU, LFU, LRV [9], GD-SIZE [8], LNC-R-W3 [3], SLRU [10], and DCM. We used three performance measures,

Table 1. Characteristics of traces used in the simulations

Traces	Period	Total Requests	Unique Requests	Total Mbytes	Unique Mbytes
SD	2000.6.28 - 7.4	6913228	3695365	96153.3	56552.3
UC	2000.6.28 - 7.4	1695561	1095161	38241.9	26784.4

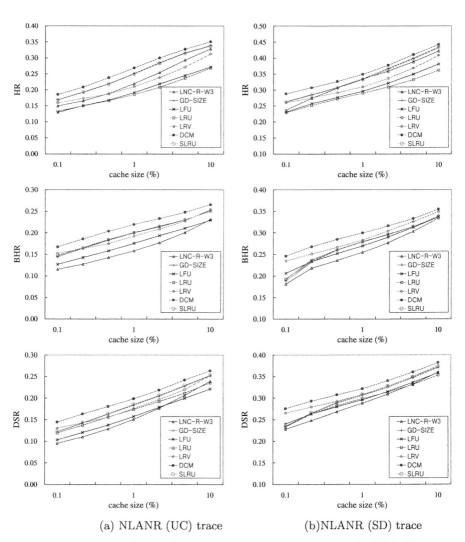

(a) NLANR (UC) trace   (b)NLANR (SD) trace

Fig. 2. Comparison of the DCM with other algorithms using the NLANR traces

that is the *Hit Ratio* (*HR*), the *Byte Hit Ratio* (*BHR*), and the *Delays-Savings Ratio* (*DSR*). Figures 2 shows the *HR*, *BHR*, and *DSR* of each algorithm as a function of the cache size. The *x*-axis, which is the cache size in logarithmic scale, is the size relative to the infinite cache size that is equal to "Total Unique Mbytes" in Table 4.

For all cases, the DCM algorithm performs the best irrespective of the cache size for all performance measures, while the other algorithms give and take amongst each other at particular cache sizes and measures. The performance gain of the DCM is as much as 47.4% (with an average of 21.3%) compared to the LRU algorithm and 30.6% (with an average of 11.9%) compared to the GD-SIZE algorithm for the traces and performance measures we considered. Specifically, the DCM algorithm offers the performance of caches more than twice its size compared with other algorithms for most cases. (Note again, that the x-axis is log-scale.)

Among the algorithms other than DCM, the GD-SIZE and SLRU algorithms consistently show good performance for all performance measures, though the performance gap between the two algorithms and DCM is somewhat wider for the BHR. The performances of GD-SIZE and SLRU are so similar that it is difficult to distinguish between them. Note that the two algorithms are similarly cost-aware and temporal locality based algorithms.

LRV shows good performance for the BHR since it predicts the re-reference likelihood of objects with the prior knowledge of workload characteristics in terms of temporal locality and reference popularity. However, the performance of LRV is inferior when the performance measure is not BHR. The reason is that the LRV algorithm cannot incorporate the cost factor when $cost$ in LRV is not proportional to the $size$ [9].

The performance of the LNC-R-W3 algorithm in our experiments is not so good as in [3]. We think that the parameters used in the LNC-R-W3 algorithm are perhaps dependent on the traces and performance measures. (Note that we use the same parameter values as in [3].)

5   Conclusion

In this paper, we presented the Dichotomized Cache Management (DCM) scheme that provides a robust framework for on-line detection and allocation of cache space based on the marginal contribution of temporal locality and reference popularity for a given proxy environment and cache size.

To accommodate objects with arbitrary cost and size, the DCM scheme normalizes the reference probability of an object by the cost of the object per unit size. This results in a normalized assessment of the contribution to the performance measure, leading to fair a replacement algorithm. Trace-driven simulations with actual proxy cache logs show that the DCM algorithm outperforms existing algorithms for various performance measures for a wide range of cache configurations. Specifically, DCM offers the performance of caches more than twice its size compared with other algorithms for most cases. We also showed how the algorithm can be effectively implemented as a proxy cache replacement module.

Acknowledgement

This Work was Supported by the Intramural Research Grant of Ewha Womans University.

References

[1] E. G. Coffman and P. J. Denning, *Operating Systems Theory*, Prentice-Hall, Englewood Cliffs, New Jersey, Ch. 6, pp. 241-283, 1973.

[2] S. Hosseini-Khayat, "On Optimal Replacement of Non-uniform Cache Objects," *IEEE Trans. Computers*, vol. 49, no. 8, pp. 769-778, 2000.

[3] J. Shim, P. Scheuermann, and R. Vingralek, "Proxy Cache Design: Algorithms, Implementation and Performance," *IEEE Trans. Knowledge and Data Eng.*, vol. 11, no. 4, pp. 549-562, 1999.

[4] L. Breslau, P. Cao, L. Fan, G. Phillips, and S. Shenker, "Web Caching and Zipf-like Distributions: Evidence and Implications," *Proc. IEEE INFOCOM*, pp. 126-134, 1999.

[5] D. Lee, J. Choi, J. Kim, S. H. Noh, S. L. Min, Y. Cho, and C. Kim, "On the Existence of a Spectrum of Policies that Subsumes the LRU and LFU Policies," *Proc. 1999 ACM SIGMETRICS Conf.*, pp.134-143, 1999.

[6] DEC Proxy Cache Traces, ftp://ftp.digital.com/pub/DEC/traces/.

[7] NLANR Proxy Cache Traces, ftp://ircache.nlanr.net/Traces.

[8] P. Cao and S. Irani, "Cost-Aware WWW Proxy Caching Algorithms," *Proc. 1st USENIX Symp. Internet Technology and Systems*, pp. 193-206, 1997.

[9] L. Rizzo and L. Vicisano, "Replacement Policies for a Proxy Cache," *IEEE/ACM Trans. Networking*, vol. 8, no. 2, pp. 158-170, 2000.

[10] C. Aggarwal, J. Wolf, and P. Yu, "Caching on the World Wide Web," *IEEE Trans. Knowledge and Data Eng.*, vol. 11, no. 1, pp.94-107, 1999.

[11] J. Choi, S. H. Noh, S. L. Min and Y. Cho, "Towards Application/File-Level Characterization of Block References: A Case for Fine-Grained Buffer Management," *Proc. 2000 ACM SIGMETRICS Conf.*, pp. 286-295, 2000.

[12] J. M. Kim, J. Choi, J. Kim, S. H. Noh, S. L. Min, Y. Cho and C. S. Kim, "A Low-Overhead, High-Performance Unified Buffer Management Scheme that Exploits Sequential and Looping References," *Proc. 4th Symp. Operating Systems Design and Implementation (OSDI 2000)*, San Diego, CA, pp. 119-134, 2000.

Author Index

Lecture Notes in Computer Science

For information about Vols. 1–2710
please contact your bookseller or Springer-Verlag

Vol. 2744: V. Mařík, D. McFarlane, P. Valckenaers (Eds.), Holonic and Multi-Agent Systems for Manufacturing. Proceedings, 2003. XI, 322 pages. 2003. (Subseries LNAI).

Vol. 2745: M. Guo, L.T. Yang (Eds.), Parallel and Distributed Processing and Applications. Proceedings, 2003. XII, 450 pages. 2003.

Vol. 2746: A. de Moor, W. Lex, B. Ganter (Eds.), Conceptual Structures for Knowledge Creation and Communication. Proceedings, 2003. XI, 405 pages. 2003. (Subseries LNAI).

Vol. 2747: B. Rovan, P. Vojtáš (Eds.), Mathematical Foundations of Computer Science 2003. Proceedings, 2003. XIII, 692 pages. 2003.

Vol. 2748: F. Dehne, J.-R. Sack, M. Smid (Eds.), Algorithms and Data Structures. Proceedings, 2003. XII, 522 pages. 2003.

Vol. 2749: J. Bigun, T. Gustavsson (Eds.), Image Analysis. Proceedings, 2003. XXII, 1174 pages. 2003.

Vol. 2750: T. Hadzilacos, Y. Manolopoulos, J.F. Roddick, Y. Theodoridis (Eds.), Advances in Spatial and Temporal Databases. Proceedings, 2003. XIII, 525 pages. 2003.

Vol. 2751: A. Lingas, B.J. Nilsson (Eds.), Fundamentals of Computation Theory. Proceedings, 2003. XII, 433 pages. 2003.

Vol. 2752: G.A. Kaminka, P.U. Lima, R. Rojas (Eds.), RoboCup 2002: Robot Soccer World Cup VI. XVI, 498 pages. 2003. (Subseries LNAI).

Vol. 2753: F. Maurer, D. Wells (Eds.), Extreme Programming and Agile Methods – XP/Agile Universe 2003. Proceedings, 2003. XI, 215 pages. 2003.

Vol. 2754: M. Schumacher, Security Engineering with Patterns. XIV, 208 pages. 2003.

Vol. 2756: N. Petkov, M.A. Westenberg (Eds.), Computer Analysis of Images and Patterns. Proceedings, 2003. XVIII, 781 pages. 2003.

Vol. 2758: D. Basin, B. Wolff (Eds.), Theorem Proving in Higher Order Logics. Proceedings, 2003. X, 367 pages. 2003.

Vol. 2759: O.H. Ibarra, Z. Dang (Eds.), Implementation and Application of Automata. Proceedings, 2003. XI, 312 pages. 2003.

Vol. 2761: R. Amadio, D. Lugiez (Eds.), CONCUR 2003 - Concurrency Theory. Proceedings, 2003. XI, 524 pages. 2003.

Vol. 2762: G. Dong, C. Tang, W. Wang (Eds.), Advances in Web-Age Information Management. Proceedings, 2003. XIII, 512 pages. 2003.

Vol. 2763: V. Malyshkin (Ed.), Parallel Computing Technologies. Proceedings, 2003. XIII, 570 pages. 2003.

Vol. 2764: S. Arora, K. Jansen, J.D.P. Rolim, A. Sahai (Eds.), Approximation, Randomization, and Combinatorial Optimization. Proceedings, 2003. IX, 409 pages. 2003.

Vol. 2765: R. Conradi, A.I. Wang (Eds.), Empirical Methods and Studies in Software Engineering. VIII, 279 pages. 2003.

Vol. 2766: S. Behnke, Hierarchical Neural Networks for Image Interpretation. XII, 224 pages. 2003.

Vol. 2769: T. Koch, I. T. Sølvberg (Eds.), Research and Advanced Technology for Digital Libraries. Proceedings, 2003. XV, 536 pages. 2003.

Vol. 2776: V. Gorodetsky, L. Popyack, V. Skormin (Eds.), Computer Network Security. Proceedings, 2003. XIV, 470 pages. 2003.

Vol. 2777: B. Schölkopf, M.K. Warmuth (Eds.), Learning Theory and Kernel Machines. Proceedings, 2003. XIV, 746 pages. 2003. (Subseries LNAI).

Vol. 2778: P.Y.K. Cheung, G.A. Constantinides, J.T. de Sousa (Eds.), Field-Programmable Logic and Applications. Proceedings, 2003. XXVI, 1179 pages. 2003.

Vol. 2779: C.D. Walter, Ç.K. Koç, C. Paar (Eds.), Cryptographic Hardware and Embedded Systems – CHES 2003. Proceedings, 2003. XIII, 441 pages. 2003.

Vol. 2781: B. Michaelis, G. Krell (Eds.), Pattern Recognition. Proceedings, 2003. XVII, 621 pages. 2003.

Vol. 2782: M. Klusch, A. Omicini, S. Ossowski, H. Laamanen (Eds.), Cooperative Information Agents VII. Proceedings, 2003. XI, 345 pages. 2003. (Subseries LNAI).

Vol. 2783: W. Zhou, P. Nicholson, B. Corbitt, J. Fong (Eds.), Advances in Web-Based Learning – ICWL 2003. Proceedings, 2003. XV, 552 pages. 2003.

Vol. 2786: F. Oquendo (Ed.), Software Process Technology. Proceedings, 2003. X, 173 pages. 2003.

Vol. 2787: J. Timmis, P. Bentley, E. Hart (Eds.), Artificial Immune Systems. Proceedings, 2003. XI, 299 pages. 2003.

Vol. 2789: L. Böszörményi, P. Schojer (Eds.), Modular Programming Languages. Proceedings, 2003. XIII, 271 pages. 2003.

Vol. 2790: H. Kosch, L. Böszörményi, H. Hellwagner (Eds.), Euro-Par 2003 Parallel Processing. Proceedings, 2003. XXXV, 1320 pages. 2003.

Vol. 2792: T. Rist, R. Aylett, D. Ballin, J. Rickel (Eds.), Intelligent Virtual Agents. Proceedings, 2003. XV, 364 pages. 2003. (Subseries LNAI).

Vol. 2794: P. Kemper, W. H. Sanders (Eds.), Computer Performance Evaluation. Proceedings, 2003. X, 309 pages. 2003.

Vol. 2795: L. Chittaro (Ed.), Human-Computer Interaction with Mobile Devices and Services. Proceedings, 2003. XV, 494 pages. 2003.

Vol. 2796: M. Cialdea Mayer, F. Pirri (Eds.), Automated Reasoning with Analytic Tableaux and Related Methods. Proceedings, 2003. X, 271 pages. 2003. (Subseries LNAI).

Vol. 2803: M. Baaz, J.A. Makowsky (Eds.), Computer Science Logic. Proceedings, 2003. XII, 589 pages. 2003.

Vol. 2805: K. Araki, S. Gnesi, D. Mandrioli (Eds.), FME 2003: Formal Methods. Proceedings, 2003. XVII, 942 pages. 2003.

Vol. 2810: M.R. Berthold, H.-J. Lenz, E. Bradley, R. Kruse, C. Borgelt (Eds.), Advances in Intelligent Data Analysis V. Proceedings, 2003. XV, 624 pages. 2003.

Vol. 2817: D. Konstantas, M. Leonard, Y. Pigneur, S. Patel (Eds.), Object-Oriented Information Systems. Proceedings, 2003. XII, 426 pages. 2003.

Vol. 2818: H. Blanken, T. Grabs, H.-J. Schek, R. Schenkel, G. Weikum (Eds.), Intelligent Search on XML Data. XVII, 319 pages. 2003.